High Performance Computing in Science
and Engineering '15

Wolfgang E. Nagel • Dietmar H. Kröner •
Michael M. Resch

Editors

High Performance Computing in Science and Engineering '15

Transactions of the High Performance
Computing Center, Stuttgart (HLRS) 2015

 Springer

Editors

Wolfgang E. Nagel
Zentrum für Informationsdienste
und Hochleistungsrechnen (ZIH)
Technische Universität Dresden
Dresden
Germany

Dietmar H. Kröner
Abteilung für Angewandte Mathematik
Universität Freiburg
Freiburg
Germany

Michael M. Resch
Höchstleistungsrechenzentrum
Stuttgart (HLRS)
Universität Stuttgart
Stuttgart
Germany

Front cover figure: Turbulent wake of a wind turbine rotor predicted by Detached Eddy Simulation. The vortex system is colored by axial velocity. Massive flow separation is present in the hub region where vortices interact with each other at different length scales. Details can be found in "Evaluation and Control of Loads on Wind Turbines under Different Operating Conditions by means of CFD", by C. Schulz, A. Fischer, P. Weihing, T. Lutz, and E. Krämer, Institute of Aerodynamics and Gas Dynamics, University of Stuttgart, Stuttgart, Germany, on page 463ff.

ISBN 978-3-319-24631-4   ISBN 978-3-319-24633-8   (eBook)
DOI 10.1007/978-3-319-24633-8

Library of Congress Control Number: 2015960436

Mathematics Subject Classification (2010): 65Cxx, 65C99, 68U20

Springer Cham Heidelberg New York Dordrecht London
© Springer International Publishing Switzerland 2016

Printed on acid-free paper

Springer International Publishing AG Switzerland is part of Springer Science+Business Media (www.springer.com)

Contents

Part I
Physics

Peter Nielaba

In this section, eight physics projects are described, which achieved important scientific results by using the HPC resources Hermit and Hornet of the HLRS. Fascinating new results are being presented in the following pages for quantum systems (elementary particle systems, ultra-cold bosonic systems, atomic and molecular collisions), soft matter systems (colloids, ionic liquids), and astrophysical systems (small scale structure of the universe).

In the last granting period, quantum mechanical properties of quarks and multi loop Feynman integrals have been investigated as well as atomic and molecular collisions and the quantum many body dynamics of trapped bosonic systems.

S. Krieg (University of Wuppertal) and the Wuppertal-Budapest collaboration in their project *HighPQCD* aim at a high precision calculation of the charmed equation of state of Quantum Chromodynamics (QCD). The principal investigators (PIs) use importance sampling methods for staggered fermions in a lattice discretized version of QCD. Following the PIs previous important investigations on the $N_f = 2+1$ flavor QCD equation of states, in the last granting period new results have been computed on the thermodynamics in the situation with a dynamical charm quark ($N_f = 2+1+1$) for $N_t = 12$, and on the neutron-proton and other mass splittings, using combined theories of quantum electrodynamics (QED) and QCD.

In the project (*NumFeyn*), A. Kurz, M. Steinhauser (both from KIT) and P. Marquard (DESY) evaluate multi-loop Feynman integrals in perturbative calculations in quantum field theories. By using Monte Carlo integration implemented in the FIESTA package (*Feynman Integral Evaluation by a Sector decomposiTion*), in the last granting period the PIs have investigated the relation between two renormalization schemes (modified minimal subtraction and the on-shell scheme) for heavy quark masses, and quantum corrections to the anomalous magnetic moment of the muon, both at four-loop accuracy.

P. Nielaba (✉)
Fachbereich Physik, Universität Konstanz, 78457 Konstanz, Germany
e-mail: peter.nielaba@unikonstanz.de

O.E. Alon, V. S. Bagnato, R. Beinke, I. Brouzos, T. Calarco, T. Caneva, L.S. Cederbaum, M.A. Kasevich, S. Klaimann, A. U. J. Lode, S. Montangero, A. Negretti, R. S. Said, K. Sakmann, O. I. Streltsova, M. Theisen, M. C. Tsatsos, S. E. Weiner, T. Wells, A. I. Streltsov from the Universities and research centers of Haifa (OEA), Sao Paulo (VSB, MCT), Heidelberg (RB, LSC, SK, MT, AIS), Ulm (IB, TC, SM, RSS), Barcelona (TC), Stanford (MAK, KS), Basel (AUJL), Hamburg (AN), Dubna (OIS), Berkeley (SEW), and Cambridge (TW) studied in their project *MCTDHB* properties of interacting ultra-cold bosonic many-body systems by their method termed *Multi-Configurational Time-Dependent Hartree method for Bosons*. MCTDHB targets at many-body effects beyond the mean-field level, in particular at a loss of coherence and fragmentation. The PIs continued their exploration of the physics of trapped interacting ultra-cold systems at the full many-body level by solving the underlying time- (in)dependent many-boson Schrödinger equation within the framework of the MCTDHB method.

In the last granting period, the PIs investigated the static properties and the many-body dynamics of vortices in two-dimensional (2D) and three-dimensional (3D) quantum objects carrying angular momentum, confined in parabolic traps and circular traps with double-well-like topologies, with particular emphasis on the loss of the coherence and the build-up of the fragmentation. In 2D harmonic traps, new many-body modes of quantized vorticity (*phantom vortices*) have been found, and connections between the resonant reaction of a Bose-Einstein condensate (BEC) stirred by a rotating laser beam and fragmentation. The properties of 2D BECs in circular traps have been explored as well, in particular the effect of time-dependent barriers on the stability against ground-state fragmentation. In the anisotropically-confined 3D systems the PIs discovered a new mechanism of vortex reconnections.

The PIs developed the MCRDHB method further, in particular the linear-response on-top of the MCTDHB method (LR-MCTDHB), providing access to the many-body excited states. By this method, the PIs classified the excited states into a mean-field-like group and a many-body-like group. In the project, MCTDHB as well has been combined with the optimal-control algorithm CRAB, in order to manipulate and control interacting quantum many-body systems, and the combined method CRAB-MCTDHB has been applied to the investigation of quantum speed limit in a bosonic Josephson junction setup. The MCTDHB package, its LR-MCTDHB extension, as well as collections of the tools for the analysis, visualization, package-making, building, teaching, and development, have been integrated in MCTDHB-Lab, a java-based environment.

In addition, the PIs propose a new analysis tool capable of simulating the outcomes of typical shots generated in the experimental detection of ultra-cold atomic systems and describe an experimental protocol capable of a direct quantitative measurement of the fragmentation in trapped bosonic systems.

B. M. McLaughlin, C. P. Ballance, M. S. Pindzola, S. Schippers and A. Müller from the Universities of Belfast (BMM), Auburn (CPB and MSP) and Giessen (SS and AM) investigated in their project *PAMOP* atomic, molecular and optical collisions on petaflop machines. The Schrödinger and Dirac equations have been solved with the R-matrix or R-matrix with pseudo-states approach, and the time dependent lattice (TDL) method has been used for charge exchange problems. Various experimentally relevant systems and phenomena have been investigated, ranging from photoionization cross sections, resonance energy positions, to Auger widths and strengths in valence- (Ca^+, W, W^+) or inner-shell- (O^+, O^{2+}, O^{3+}) systems, and charge transfer cross sections in C^{6+} collisions with H and He atoms.

The studies of the (colloidal) soft matter systems have focused on the crystal-liquid phase coexistence and their dynamical properties, and on the properties of ionic liquids in confinement.

A. Statt, F. Schmitz, P. Virnau and K. Binder from the University of Mainz in their project *colloid* developed a general method to study crystal nuclei and to obtain estimates for the free energy barrier against homogeneous nucleation. In their Monte Carlo studies, the PIs used a "softened" version of the effective Asakura-Oosawa model, with an effective potential between the colloids which is everywhere continuous, they computed the solid-liquid interface excess free energy via the ensemble switch method, and they obtained the pressure at phase coexistence from the interface velocity method. The PIs showed that the surface excess free energy can be determined accurately from Monte Carlo simulations over a wide range of nucleus volumes, and they found that the resulting nucleation barriers are independent of the size of the total system volume. In addition, the PIs results show that the nucleus shape is almost spherical, when the anisotropy of the interface tension is in the order of a few per cent, and the results were discussed in the frame of the classical nucleation theory.

J. Zhou and F. Schmid from the University of Mainz in their project *CCAC* have developed a mesoscopic colloid model based on the dissipative particle dynamics. In this model, the colloid is represented by a large spherical bead, and its surface interacts with solvent beads through a pair of dissipative and random forces, extending the tunable-slip boundary method from planar surfaces, as introduced by one of the PIs (FS) in previous works, to curved geometry. The PIs computed the diffusion constant of a single colloid in a cubic box, using the program package ESPRESSO, and found good agreement with the predictions from hydrodynamic theories.

K. Breitsprecher, N.K. Anand, J. Smiatek and C. Holm from the University of Stuttgart in their project *FFOIL* explored different models of room temperature ionic liquids (RTILs) in confined environment and bulk solution by molecular dynamics (MD) simulations with the software packages ESPRESSO and Gromacs as well as other MD-codes. In the last granting period, the PIs focused on algorithms for metal boundary conditions in various geometries, and on effects of graphite structure on the adsorbed ions in planar capacitor geometries, in particular by comparing an explicit graphene structure to an unstructured planar Lennard-Jones surface and by investigating mixtures of the ionic liquid EMIM BF4 with different concentrations of Acetonitril (ACN) in contact with carbide-derived carbon (CDC) electrodes. The PIs showed that the increased adsorption of the ionic liquid on graphite surfaces is due to the texturing influence of the honeycomb pattern.

On different length scales compared to the quantum and soft matter systems described above, the project *SSSU* has focused on the small scale structure of the universe. In this project, S. Gottlöber, C. Brook, I.T. Iliev and K.L. Dixon from the Leibniz Institute for Astrophysics at Potsdam (SG) and the Universities of Madrid (CB) and Sussex (ITI, KLD) investigated reionization and galaxy formation processes. In their project, the PIs studied the role of reionization in the early stage of cosmological evolution, on the formation and evolution of the small scale structure, by three simulations, using the CubeP^{3}M N-body code, the background cosmology based on WMAP 5-year data, the linear power spectrum of density fluctuations calculated with the code CAMB, initial conditions by the Zeldovich approximation for red shifts of 150, and radiative transfer simulations with their code C^2-Ray. In addition, the PIs use the physical model of the MaGICC project (*Making Galaxies in a Cosmological Context*) and initial conditions of the CLUES project (*Constrained Local UniversE Simulations*) to construct a model of the Local Group of galaxies, including the Milky Way, Andromeda, M33 and dwarf galaxies. As galaxy properties, rotation curves have been computed as well as stellar-to-halo mass relations and the mass of galaxy baryons as a function of their circular velocity (Baryonic Tully-Fisher relation).

Thermodynamics with $2 + 1 + 1$ Dynamical Quark Flavors

Stefan Krieg for the Wuppertal-Budapest Collaboration

Abstract We report on our calculation of the equation of state of Quantum Chromodynamics (QCD) from first principles, through simulations of Lattice QCD. We use an improved lattice action and $N_f = 2 + 1 + 1$ dynamical quark flavors and physical quark mass parameters. Now, we are in a position to present first results at $N_t = 12$.

1 Introduction

The aim of our project is to compute the charmed equation of state for Quantum Chromodynamics (for details, see [1]). We are using the lattice discretized version of Quantum Chromodynamics, called lattice QCD, which allows simulations of the theory through importance sampling methods. Our results are important input quantities for phenomenological calculations and are required to understand experiments aiming to generate a new state of matter, called Quark-Gluon-Plasma, such as the upcoming FAIR at GSI, Darmstadt.

Our simulations are performed using so-called staggered fermions. In the continuum limit, i.e. at vanishing lattice spacing a, one staggered Dirac operator implements four flavors of mass degenerate fermions. At finite lattice spacing, however, discretization effects induce an interaction between these would be flavors lifting the degeneracy. The "flavors" are, consequentially, renamed to "tastes", and the interactions are referred to as "taste-breaking" effects. Even though the tastes are not degenerate, in simulations one takes the fourth root of the staggered fermion determinant to implement a single flavor. This procedure is not proven to be correct—however, practical evidence suggests that is does not induce errors visible with present day statistics.

S. Krieg (✉)
Fachbereich C - Physik, Bergische Universität Wuppertal, 42119 Wuppertal, Germany

IAS, Jülich Supercomputing Centre, Forschungszentrum Juelich GmbH, 52425 Jülich, Germany
e-mail: krieg@uni-wuppertal.de

© Springer International Publishing Switzerland 2016
W.E. Nagel et al. (eds.), *High Performance Computing in Science and Engineering '15*, DOI 10.1007/978-3-319-24633-8_1

Fig. 1 RMS pion mass for
different staggered fermion
actions, in the continuum
limit

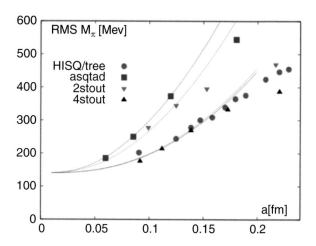

Taste-breaking is most severely felt at low pion masses and large lattice spacing, as the pion sector is distorted through the taste-breaking artifacts: there is one would-be Goldstone boson, and 15 additional heavier "pions", which results in an RMS pion mass larger than the mass of the would-be Goldstone boson. This effect is depicted in Fig. 1 for different staggered type fermion actions. As can be seen for this figure, the previously used twice stout smeared action ("2stout") has a larger RMS pion mass and thus taste-breaking effects than the HISQ/tree action. If, however, the number of smearing steps is increased to four, with slightly smaller smearing strength ("4stout"), the RMS pion mass measured agrees with that of the HISQ/tree action. In order to have an improved pion sector, we, therefore, opted to switch to this new action and to restart our production runs.

So far, the equation of state is known only in 2+1 flavor QCD. Here, the status of the field is marked by our papers on the $N_f = 2 + 1$ equation of state [2, 3] (see Fig. 2). The contribution from the sea charm quarks most likely matter at least for $T > 300$–400 MeV (for an illustration, see Fig. 3).

1.1 Reference Point: The $N_f = 2 + 1$ Equation of State

In [3] we have presented the first full calculation of the $N_f = 2 + 1$ Equation of State (EoS) of Quantum Chromodynamics (QCD) (still using our 2stout action). This result is the reference point for our calculation of the charmed EoS, and already included one continuum extrapolated result at $T = 214$ MeV for the trace anomaly using our new lattice action including a dynamical charm quark ($N_f = 2 + 1 + 1$).

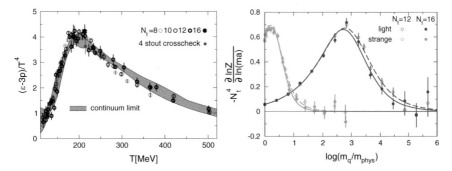

Fig. 2 *Left:* The trace anomaly as a function of the temperature. The continuum extrapolated result with total errors is given by the *shaded band*. Also shown is a cross-check point computed in the continuum limit with our new and different lattice action at $T = 214$ MeV, indicated by a *smaller filled red point*, which serves as a crosscheck on the peak's hight. *Right:* Setting the overall scale of the pressure: integration from the infinitely large mass region down to the physical point using a range of dedicated ensembles and time extents up to $N_t = 16$; the sum of the areas under the curves gives p/T^4. This result could be used for the cEoS normalization as well (see text)

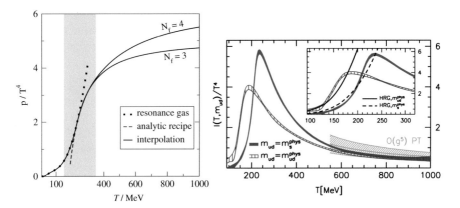

Fig. 3 *Left:* Laine and Schroeder's perturbative estimate of the effect of the charm in the QCD equation of state [4]. *Right:* Wuppertal-Budapest [2] and perturbative (up to $O(g^5)$) results for the equation of state

As visible in Fig. 2, at this temperature the charm quark is not yet relevant, since the $N_f = 2 + 1 + 1$ (continuum) data point falls right onto the (continuum) $N_f = 2 + 1$ curve. Below this temperature, we can compare the results with and without dynamical charm and can even use the $N_f = 2 + 1$ results to renormalize the $N_f = 2 + 1 + 1$ curve [5, 6].

2 Progress for the Charmed Equation of State

The $N_f = 2+1$ lattice results mentioned in the previous section agree with the HRG at low temperatures and are correct for the small to medium temperatures, and, as is shown in Fig. 3, at temperatures of about 1 GeV perturbative results become sufficiently precise. Therefore, we need to calculate the EoS with a dynamical charm only for the remaining temperatures in the region of approximately 300 MeV $< T <$ 1000 MeV.

We are using our 4stout lattice action for these calculations. The crosscheck point shown in Fig. 2 was computed using this new action. Since it perfectly agrees with the $N_f = 2 + 1$ results, even though it was computed using a dynamical charm, we can be certain that at temperatures at and below $T = 214$ MeV, we can rely on the $N_f = 2 + 1$ results.

Our preliminary results are shown in Fig. 4, all errors are statistical only. Our results span a region of temperatures from $T = 214$ MeV up to $T = 1.2$ GeV. At the low end we make contact to the $N_f = 2 + 1$ equation of state, and at large temperatures to the HTL result. Thereby, we cover the full region of temperatures, from low temperatures, where the HRG gives reliable results, to high temperatures, where we make contact with perturbation theory. The figure contains **new data points** at $N_t = 12$ generated in the last period.

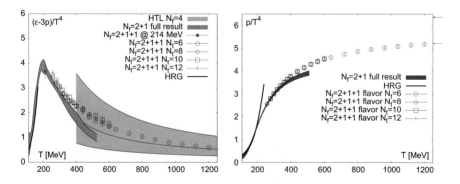

Fig. 4 *Left:* Preliminary results for the charmed EoS. For comparison, we show the HRG result, the $N_f = 2 + 1$ band, and, at high Temperatures, the HTL result [7], where the *central line* marks the HTL expectation for the EoS with the band resulting from (large) variations of the renormalization scale. *Right:* Preliminary result for the pressure, errors indicate the Stefan-Boltzmann value. All errors are statistical only

2.1 Line of Constant Physics

With the switch to a new lattice action comes the need to (re-) compute the LCP. In order to be able to reach large temperatures ($\beta > 4$), we have extended these calculations since the last report. Since we would like to span the temperature range from approximately 300 MeV $< T <$ 1000 MeV, we have to compute the LCP for a large range of couplings or lattice spacings. We split this range up in three overlapping regions (since we have to make sure that the derivative is smooth) according to the applicable simulation strategies.

At medium to coarse lattice spacings (region I) one can afford to use spectroscopy to tune the parameters. This is shown in Fig. 5. Here, we bracketed the physical point defined through M_π/f_π and $(2M_K-M_\pi)/f_\pi$ and, through interpolation, tune the light and strange quark masses to per-mill precision.

Using the parameters computed in this way, we then performed simulations on JUROPA at the SU(3) flavor-symmetrical point [8], extrapolating the results to our target couplings. There, we tuned the parameters to reproduce the extrapolated results. Since the quark masses are larger than physical, such simulations are considerably less costly than using spectroscopy as for region I, and we are thus able to compute a precise LCP down to fine lattice spacings of $a = 0.05$ fm (region II), where the HMC starts being affected by the freezing of topology (Fig. 6).

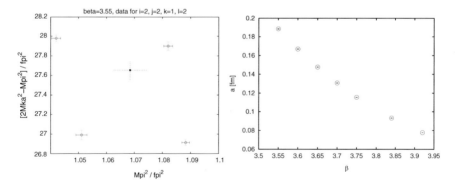

Fig. 5 Region of the LCP, for coarse to medium lattice spacing ($a > 0.08$ fm). Here, dedicated simulations bracketing the physical point archive a sub-percent accuracy for the LCP. *Left:* Bracketing of the physical point defined through M_π/f_π and $(2M_K - M_\pi)/f_\pi$. The strange quark mass is tuned (m_s/m_l is not fixed) and the ratio of the charm to strange quark mass is set at $m_c/m_s = 11.85$. *Right:* LCP computed through spectroscopy

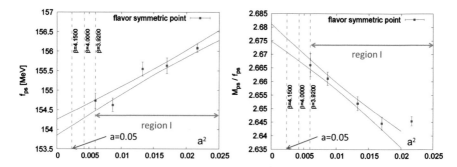

Fig. 6 Using the LCP computed from spectroscopy for coarse to medium lattice spacings (region I), dedicated simulations in the SU(3) flavor-symmetrical point [8] using these parameters are extrapolated towards the continuum. At the target coupling, the parameters are tuned until they reproduce the extrapolated value. In this way the LCP is extended to medium to small lattice spacings of $0.08 > a > 0.05$ fm (region II)

For finer lattice spacings we thus used our established step scaling procedure [3] based on the w_0 scale. To this end, we computed the observable

$$\mathcal{O} = t \frac{d}{dt} \left[t^2 E(t) \right] \Big|_{0.01 L^2}$$

at three different lattice spacings (a_0, a_1, a_2) and volumes (16^4, 20^4, 24^4) chosen to keep the physical volume fixed, extrapolated to $a_3 = 24/32 a_2$, and tuned the coupling to match the extrapolated result. Using this method, we extended the LCP to very fine lattice spacings with $a < 0.05$ fm (region III).

2.2 Additional Results

In another effort, we calculated the neutron-proton and other mass splittings from first principles [9], using simulations of the combined theories of Quantum Electro- and Quantum Chromodynamics. Here, we used Hermit for valence calculations, i.e. we analyzed configurations generated elsewhere, computing the mass difference for a number of different bare parameters. The complete result is shown in Fig. 7. Due to the long range nature of Quantum Electrodynamics (QED) these simulations face significant finite-size effects, inducing shifts in the results considerably larger than the signal. Through analytical calculations (see SOM of [9]), we were able to predict and thus subtract these effects. Another important step was the development of a new update algorithm for the QED, which reduced the autocorrelation by more than 2 orders of magnitude.

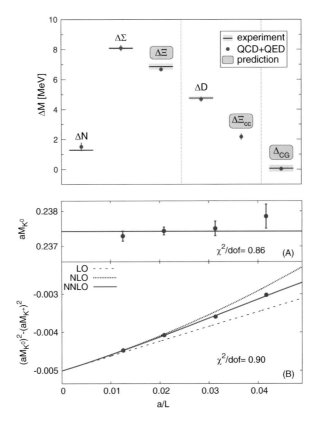

Fig. 7 *Left:* Mass splittings. The *horizontal lines* are the experimental values and the *grey shaded regions* represent the experimental error. Our results are shown by *red dots* with their uncertainties. Splittings which have either not been measured in experiment or are measured with less precision than in our calculation are indicated by a *blue shaded* region around the label. *Right:* Finite-volume behavior of kaon masses. (A) The neutral kaon mass, M_{K^0}, shows no significant finite volume dependence; L denotes the linear size of the system. (B) The mass-squared difference of the charged kaon mass, M_{K^+}, and M_{K^0} indicates that M_{K^+} is strongly dependent on volume. This finite-volume dependence is well described by an analytical results [9] (Figures taken from *Science* **347** 1452, reference [9]. Reprinted with permission from AAAS.)

3   Production Specifics and Performance

Most of our production is done using modest partition sizes, as we found these to be most efficient for our implementation.

3.1 Performance

Our code shows nice scaling properties on HERMIT and HORNET. For our scaling analysis below, we used two lattices ($N_s = 32$ and 48) and several partition sizes up to 256 nodes (HERMIT). We timed the most time consuming part of the code: the fermion matrix multiplication. The results are summarized in the following table:

No. of nodes	Gflop/node $N_s = 32$	Gflop/node $N_s = 48$
1	16.3	15.4
2	16.8	16.0
4	16.5	16.2
8	16.3	16.3
16	16.3	16.3
32	16.8	16.0
64	17.1	16.5
128	19.2	16.5
256	16.3	16.0

Test show that our scaling on HORNET is similarly good - however at a higher performance of ≈ 22 and ≈ 21 Gflop/s for the $N_s = 32$ and $N_s = 48$ lattices, respectively.

3.2 Production

Given the nice scaling properties of our code, we were able to run at the sweet spot for queue throughput, which we found to be located at a job size of 64 nodes. Larger job sizes proved to have a scheduling probability sufficiently low that benefits in the runtime due to the larger number of cores were compensated and the overall production throughput decreased. We, therefore, opted to stay at jobs sizes with 64 nodes.

4 Outlook

We believe we will be able to publish within the year. HERMIT and HORNET have proved to be essential tools enabling us to achieve this goal.

5 Publications of the Project

5.1 Peer-Reviewed

[9]   *Ab initio calculation of the neutron-proton mass difference*, Science **347** (2015) 1452–1455

[10] *Equation of state, fluctuations and other recent results from LQCD*, Proceedings of the 30th Winter Workshop on Nuclear Dynamics (WWND 2014), J.Phys.Conf.Ser. **535** (2014) 012016

5.2 Other

[11] *From quarks to hadrons and back - spectral and bulk phenomena of strongly interacting matter*, Proceedings of the XXVI IUPAP Conference on Computational Physics (CCP2014), J. Phys. Conf. Ser. **640** (2015) 012053

[12] *Recent results on the Equation of State of QCD*, Proceedings of the 32nd International Symposium on Lattice Field Theory (Lattice 2014), PoS(**LATTICE2014**) 224

References

1. Borsanyi, S., Endrodi, G., Fodor, Z., Katz, S.D., Krieg, S., et al.: The QCD equation of state and the effects of the charm. PoS **LATTICE2011**, 201 (2011). [1204.0995]
2. Borsanyi, S., Endrodi, G., Fodor, Z., Jakovac, A., Katz, S.D., et al.: The QCD equation of state with dynamical quarks. J. High Energy Phys. **1011**, 077 (2010). [1007.2580]
3. Borsanyi, S., Fodor, Z., Hoelbling, C., Katz, S.D., Krieg, S., et al.: Full result for the QCD equation of state with 2+1 flavors Phys. Lett. **B730**, 99–104 (2014). [1309.5258]
4. Laine, M., Schroder, Y.: Quark mass thresholds in QCD thermodynamics. Phys. Rev. **D73** 085009 (2006). [hep-ph/0603048]
5. Endrodi, G., Fodor, Z., Katz, S., Szabo, K.: The equation of state at high temperatures from lattice QCD. PoS **LAT2007**, 228 (2007). [0710.4197]
6. Borsanyi, S., Endrodi, G., Fodor, Z., Katz, S., Szabo, K.: Precision SU(3) lattice thermodynamics for a large temperature range. J. High Energy Phys. **1207**, 056 (2012). [1204.6184]
7. Andersen, J.O., Leganger, L.E., Strickland, M., Su, N.: Three-loop HTL QCD thermodynamics. J. High Energy Phys. **1108**, 053 (2011). [1103.2528]
8. Bietenholz, W., Bornyakov, V., Gockeler, M., Horsley, R., Lockhart, W., et al.: Flavour blindness and patterns of flavour symmetry breaking in lattice simulations of up, down and strange quarks. Phys. Rev. **D84**, 054509 (2011). [1102.5300]
9. Borsanyi, S., Durr, S., Fodor, Z., Hoelbling, C., Katz, S., et al.: Ab initio calculation of the neutron-proton mass difference. Science **347**, 1452–1455 (2015). [1406.4088]

10. Wuppertal-Budapest Collaboration, Krieg, S.: Equation of state, fluctuations and other recent results from LQCD. In: Proceedings of the 30th Winter Workshop on Nuclear Dynamics (WWND 2014) (2014); J. Phys. Conf. Ser. **535**, 012016 (2014). doi:10.1088/1742-6596/535/1/012016
11. Budapest-Marseille-Wuppertal Collaboration, Wuppertal-Budapest Collaboration, Krieg, S.: From quarks to hadrons and back - spectral and bulk phenomena of strongly interacting matter. In: Proceedings of the 26th Conference on Computational Physics (CCP 2014) (2014); J. Phys. Conf. Ser. **640**(1), 012053 (2015). doi:10.1088/1742-6596/640/1/012053
12. Borsányi, S., Fodor, Z., Hoelbling, C., Katz, S.D., Krieg, S., Ratti, C., Szabo, K.K.: Recent results on the equation of state of QCD. In: Proceedings of the 32nd International Symposium on Lattice Field Theory (Lattice 2014) (2014); PoS **LATTICE2014**, 224 (2015) [arXiv:1410.7917]

Numerical Evaluation of Multi-Loop Feynman Integrals

Alexander Kurz, Peter Marquard, and Matthias Steinhauser

Abstract The aim of this project is the evaluation of multi-loop Feynman integrals occurring in perturbative calculations within quantum field theories. The integrals under consideration enter the relation between the $\overline{\text{MS}}$ and on-shell definition of heavy quark masses and the anomalous magnetic moment of the muon. Both quantities are considered at four-loop accuracy. This report covers the period from May 2014 to April 2015.

1 Introduction

The main aim of modern particle physics is the exploration of the fundamental interaction between the elementary particles. Insight to this question is obtained by confronting high-precision calculations performed within the underlying relativistic quantum field theory with experimental data. The most powerful method to evaluate quantum corrections is perturbation theory which requires the evaluation of multi-loop integrals of the form

$$\int d^d p_1 \cdots d^d p_L \prod_i \frac{1}{k_i^2 - m_i^2} , \qquad (1)$$

A. Kurz
Deutsches Elektronen-Synchrotron DESY, Platanenallee 6, 15738 Zeuthen, Germany

Institut für Theoretische Teilchenphysik, Karlsruher Institut für Technologie, 76128 Karlsruhe, Germany
e-mail: alexander.kurz2@kit.edu

P. Marquard
Deutsches Elektronen-Synchrotron DESY, Platanenallee 6, 15738 Zeuthen, Germany
e-mail: peter.marquard@desy.de

M. Steinhauser (✉)
Institut für Theoretische Teilchenphysik, Karlsruher Institut für Technologie, 76128 Karlsruhe, Germany
e-mail: Matthias.Steinhauser@kit.edu

© Springer International Publishing Switzerland 2016   15
W.E. Nagel et al. (eds.), *High Performance Computing in Science and Engineering '15*, DOI 10.1007/978-3-319-24633-8_2

where p_i and k_i are 4-vectors. p_i are integration variables and k_i are linear combinations of p_i and possible external momenta. Note that the dimension d is given by $d = 4 - 2\epsilon$ where ϵ serves as regularization parameter which is sent to zero after the integrations are performed.

In this project a special class of integrals is considered, so-called on-shell integrals with one external momentum, q, which fulfills the relation $q^2 = m^2$. In particular we also have for the masses $m_i = m$ or $m_i = 0$. Integrals of this type have been studied in the literature up to three loops (see, e.g., [1]), a systematic study at four loops (i.e. $L = 4$) is, however, missing.

Our investigations are driven by the following physical problems which we would like to address. The first one is concerned with the definition of the heavy quark masses which appear as fundamental parameters in the underlying Lagrange density. More precisely, we want to compute the relation between the $\overline{\text{MS}}$ and on-shell definition with four-loop accuracy within Quantum Chromodynamics (QCD). For the second physical quantity we consider Quantum Electrodynamics (QED) as our fundamental theory and evaluate quantum corrections to the anomalous magnetic moment of the muon, which in the recent years has been measured with high accuracy.

Calculations within perturbation theory involve several steps which include the automatic generation of all contributing Feynman diagrams, the translation to mathematical expressions and the manipulation of the latter such that the physical quantity is expressed as a linear combination of several thousands, sometimes even millions of integrals. In a next step the so-called Laporta algorithm [2] is applied in order to reduce the number of integrals. In our case we end up with about 400 integrals, so-called master integrals, similar to the one in Eq. (1) for $L = 4$ which have to computed. In this project we apply numerical methods to compute the master integrals at the HLRS on the `Hermit` and `Hornet` computer cluster.

2 Program Package and Technical Details

The workhorse for the calculations which we are performing at the HLRS is the program package `FIESTA` [3–5], which has been developed since 2008 with the participation of the Institute for Theoretical Particle Physics (TTP) at KIT. `FIESTA` stands for Feynman Integral Evaluation by a Sector decomposiTion and applies the method of sector decomposition [6] to obtain finite expression for the coefficients of the Laurent series of Eq. (1) in $\epsilon = (4 - d)/2$. These finite expressions are multi-dimensional parameter integrals with in general large integrands of the size of a few hundred MB up to a GB.

In practice the preparation of the integrand is performed within Mathematica on the local cluster. The expressions are transferred in form of a data base to the HLRS where the time-consuming Monte-Carlo integration is performed. `FIESTA` uses a simple master slave model for the parallelization, where the integrands are distributed from the master to the slaves using MPI and each term is integrated using a single core by the slave.

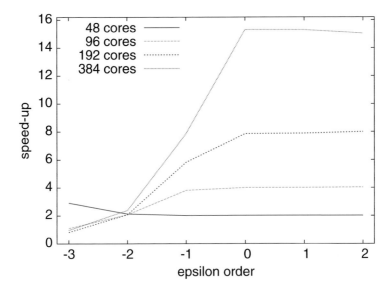

Fig. 1 Achieved speed-up for the sample integral shown in Fig. 2 needed for our calculation. The calculations have been performed on the Hornet cluster. The baseline is given by a run with 24 cores

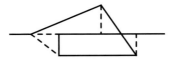

Fig. 2 Sample Feynman diagram appearing in the calculation of the $\overline{\text{MS}}$—on-shell relation at four-loop order. *Solid and dashed lines* denote massive and massless lines, respectively

In Fig. 1 we show a typical speedup behaviour which we observe for the integrals on the Hornet cluster. On the x axis we show the ϵ order of the integral corresponding to the diagram in Fig. 2 and the y axis shows the speed-up as compared to the use of 24 cores. Note that the lower orders of the expansion in ϵ (i.e. order $1/\epsilon^k$ with $k = 1, 2$ and 3) contain only a few quite simple terms and thus there is no gain from using more cores. Furthermore, their contribution to the total CPU time for the considered diagram is only marginal. For the higher ϵ terms (ϵ^k with $k = 0, 1$ and 2), however, we find a near optimal speed-up behaviour.

For our calculation we used up to 1024 cores on Hermit and up to 576 cores on Hornet. For about 20 % of the integrals we used more than 256 cores. In this way we could respect the CPU time limit of 24 h and at the same time obtain the required precision.

All integrals were calculated multiple times with different precision to ensure the convergence of the Monte-Carlo integration. We observe the expected $1/\sqrt{N}$ behaviour in the reduction of the statistical uncertainty where N is the number of sample point.

3   Heavy Quark Mass Relations to Four Loops

In the Standard Model of particle physics the masses of the quarks are free parameters and in addition due to renormalization they are also scheme dependent. Two such renormalization schemes are the $\overline{\text{MS}}$ (modified minimal subtraction) and the on-shell scheme. Within perturbation theory one can obtain relations between these two schemes.

To obtain the $\overline{\text{MS}}$-on-shell relation it is convenient to start with the relations between bare mass m_0, which is present in the original Lagrange density, and $\overline{\text{MS}}$ mass m or on-shell mass M

$$m^0 = Z_m^{\overline{\text{MS}}} m, \qquad m^0 = Z_m^{\text{OS}} M. \tag{2}$$

$Z_m^{\overline{\text{MS}}}$ and Z_m^{OS} denote the corresponding renormalization constants. $Z_m^{\overline{\text{MS}}}$ is known to four loops and can be found in [7–9]. By construction, the ratio of the two equations in (2) is finite which leads to

$$z_m(\mu) = \frac{m(\mu)}{M}, \tag{3}$$

where μ is the renormalization scale. z_m depends on $\alpha_s(\mu)$ and $\log(\mu/M)$ and has the following perturbative expansion

$$z_m(\mu) = \sum_{n \geq 0} \left(\frac{\alpha_s(\mu)}{\pi} \right)^n z_m^{(n)}(\mu), \tag{4}$$

with $z_m^{(0)} = 1$. The $\overline{\text{MS}}$-on-shell relation has been calculated at one-, two-, and three-loop order in [10–15], respectively. Fermionic four-loop corrections with two massless insertions have been computed in [16].

To obtain the complete four-loop result for $z_m(\mu)$ one has to calculate Z_m^{OS} to this order. We followed standard techniques to perform the calculation and finally obtained $z_m(\mu)$ as a linear combination of 386 four-loop integrals. The simple integrals can be computed using (semi-)analytic methods. However, for 332 integrals FIESTA has been applied as described in the previous section. We insert the numerical results in our analytic expression and add the uncertainties in quadrature. The resulting uncertainty, which is interpreted as standard deviation, is multiplied by five to obtain a conservative error estimate.

Our final result for the four-loop coefficient in the expansion (4) specified to the three heavy quark of the Standard Model, charm ("$n_l = 3$"), bottom ("$n_l = 4$") and top ("$n_l = 5$") reads [17]

$$z_m^{(4)} \Big|_{n_l=3} = -1744.8 \pm 21.5 - 703.48 \, l_{\text{OS}} - 122.97 \, l_{\text{OS}}^2$$

$$- 14.234 \, l_{\text{OS}}^3 - 0.75043 \, l_{\text{OS}}^4,$$

$$z_m^{(4)}\bigg|_{n_l=4} = -1267.0 \pm 21.5 - 500.23\, l_{\mathrm{OS}} - 83.390\, l_{\mathrm{OS}}^2$$

$$- 9.9563\, l_{\mathrm{OS}}^3 - 0.514033\, l_{\mathrm{OS}}^4\,,$$

$$z_m^{(4)}\bigg|_{n_l=5} = -859.96 \pm 21.5 - 328.94\, l_{\mathrm{OS}} - 50.856\, l_{\mathrm{OS}}^2$$

$$- 6.4922\, l_{\mathrm{OS}}^3 - 0.33203\, l_{\mathrm{OS}}^4\,, \tag{5}$$

with $l_{\mathrm{OS}} = \ln(\mu^2/M^2)$. We obtain coefficients with an accuracy of 1.2 % for $n_l = 3$, 1.7 % for $n_l = 4$ and 2.5 % for $n_l = 5$, where the given errors stem solely from the numerical integration within FIESTA. The obtained precision is at the moment sufficient for phenomenological applications. The obtained results in Eq. (5) are crucial for the precise extraction of numerical values for the heavy quark masses. Phenomenological examples are discussed in [17].

4 Anomalous Magnetic Moment of the Muon

At the classical level the leptonic anomalous magnetic moment, $a = (g-2)/2$, is predicted to be equal to zero. The deviation from zero is due to quantum corrections which have been investigated since the 1940s. Nowadays the experimental results have reached such a high precision [18, 19] that four- and even five-loop corrections [20, 21] are necessary to perform a sensible comparison.

An ongoing project of our working group is the evaluation of four-loop corrections involving closed electron loops to the anomalous magnetic moment of the muon. A sample Feynman diagram is shown in Fig. 3 for which we will present preliminary result in the following.

Fig. 3 Sample Feynman diagram contribution to $(g-2)$ of the muon at four-loop order. *Solid and wiggled lines* denote leptons and photons, respectively. In this example the *closed fermion loops* correspond to electrons whereas the external leptons are muons

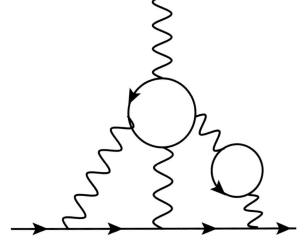

As compared to the $\overline{\text{MS}}$-on-shell relation discussed in the previous section there is the additional complication that we have two mass scales, the electron and the muon masses, m_e and m_μ. Due to the strong hierarchy, $m_e \ll m_\mu$, it is suggestive to consider an expansion in m_e/m_μ. Prescriptions to perform such an expansion at the level of integrands can be found in [22] and have been implemented by our group. This allows us to obtain the anomalous magnetic moment as a linear combination of integrals which have already been considered in Sect. 3, where the dimensionful scale is given by the muon mass, and new type of integrals, which contain m_e as mass scale. Computing also the new type of master integrals on the Hornet cluster leads to the contribution of the diagrams in Fig. 3 to the muon anomalous magnetic moment

$$(115.19 \pm 0.03) + (1.60 \pm 0.04) + \dots, \tag{6}$$

where the numbers in the first and second bracket come from the $(m_e/m_\mu)^0$ and $(m_e/m_\mu)^1$ contributions, respectively; the ellipsis indicate higher order corrections in the mass ratio which are expected to be small. Our preliminary result agrees well with the result from [23].

5   Conclusions and Outlook

It has been shown that the package FIESTA performs well on the clusters Hermit and Hornet. A good speedup behaviour is observed up to 384 cores which allowed us to obtain sufficiently precise results of all relevant integrals within the CPU time limit of 24 h. The results have been used to obtain for the first time the relation between the $\overline{\text{MS}}$ and on-shell heavy quark mass which is published in [17].

In [17] the number of colours (present in QCD) is set to the physical value, i.e. $N_c = 3$. In a next step we want to obtain results for a generic value of N_c which requires an increase of the accuracy of the master integrals. On the one hand we will further improve FIESTA. On the other hand we will try to optimize the number of cores which shall be used for a given integral.

As far as the anomalous magnetic moment of the muon is concerned we have to further study the new kind of integrals which occur in the power-corrections proportional to m_e/m_μ. In particular, we have to optimize their representation for the numerical integration on Hornet.

References

1. Lee, R.N., Smirnov, V.A.: Smirnov, Analytic epsilon expansions of master integrals corresponding to massless three-loop form factors and three-loop g-2 up to four-loop transcendentality weight. J. High Energy Phys. **1102**, 102 (2011). [arXiv:1010.1334 [hep-ph]].
2. Laporta, S.: High precision calculation of multiloop Feynman integrals by difference equations. Int. J. Mod. Phys. A **15**, 5087 (2000). [hep-ph/0102033]
3. Smirnov, A.V., Tentyukov, M.N.: Feynman integral evaluation by a sector decomposition approach (FIESTA). Comput. Phys. Commun. **180**, 735 (2009). [arXiv:0807.4129 [hep-ph]]; Preprint No. TTP08-32
4. Smirnov, A.V., Smirnov, V.A., Tentyukov, M.: FIESTA 2: parallelizeable multiloop numerical calculations. Comput. Phys. Commun. **182**, 790 (2011). [arXiv:0912.0158 [hep-ph]]; Preprint No. TTP09-39
5. Smirnov, A.V.: FIESTA 3: cluster-parallelizable multiloop numerical calculations in physical regions. Comput. Phys. Commun. **185**, 2090 (2014). [arXiv:1312.3186 [hep-ph]]
6. Heinrich, G.: Sector decomposition. Int. J. Mod. Phys. A **23**, 1457 (2008). [arXiv:0803.4177 [hep-ph]]
7. Chetyrkin, K.G.: Quark mass anomalous dimension to O (alpha-s**4). Phys. Lett. B **404**, 161 (1997). [hep-ph/9703278]
8. Vermaseren, J.A.M., Larin, S.A., van Ritbergen, T.: The four loop quark mass anomalous dimension and the invariant quark mass. Phys. Lett. B **405**, 327 (1997). [hep-ph/9703284]
9. Chetyrkin, K.G.: Four-loop renormalization of QCD: full set of renormalization constants and anomalous dimensions. Nucl. Phys. B **710**, 499 (2005). [hep-ph/0405193]
10. Tarrach, R.: The pole mass in perturbative QCD. Nucl. Phys. B **183**, 384 (1981)
11. Gray, N., Broadhurst, D.J., Grafe, W., Schilcher, K.: Three loop relation of quark (modified) Ms and pole masses. Z. Phys. C **48**, 673 (1990)
12. Chetyrkin, K.G., Steinhauser, M.: Short distance mass of a heavy quark at order α_s^3. Phys. Rev. Lett. **83**, 4001 (1999). [hep-ph/9907509]
13. Chetyrkin, K.G., Steinhauser, M.: The relation between the MS-bar and the on-shell quark mass at order alpha(s)**3. Nucl. Phys. B **573**, 617 (2000). [hep-ph/9911434]
14. Melnikov, K., van Ritbergen, T.: The three loop relation between the MS-bar and the pole quark masses. Phys. Lett. B **482**, 99 (2000). [hep-ph/9912391]
15. Marquard, P., Mihaila, L., Piclum, J.H., Steinhauser, M.: Relation between the pole and the minimally subtracted mass in dimensional regularization and dimensional reduction to three-loop order. Nucl. Phys. B **773**, 1 (2007). [hep-ph/0702185]
16. Lee, R., Marquard, P., Smirnov, A.V., Smirnov, V.A., Steinhauser, M.: Four-loop corrections with two closed fermion loops to fermion self energies and the lepton anomalous magnetic moment. J. High Energy Phys. **1303**, 162 (2013). [arXiv:1301.6481 [hep-ph]]
17. Marquard, P., Smirnov, A.V., Smirnov, V.A., Steinhauser, M.: Quark mass relations to four-loop order. Phys. Rev. Lett. **114**, 142002 (2015). [arXiv:1502.01030 [hep-ph]]
18. Bennett, G.W., et al., Muon G-2 Collaboration: Final report of the Muon E821 anomalous magnetic moment measurement at BNL. Phys. Rev. D **73**, 072003 (2006). [hep-ex/0602035]
19. Roberts, B.L.: Status of the Fermilab Muon $(g - 2)$ experiment. Chin. Phys. C **34**, 741 (2010). [arXiv:1001.2898 [hep-ex]]
20. Aoyama, T., Hayakawa, M., Kinoshita, T., Nio, M.: Tenth-order QED contribution to the electron g-2 and an improved value of the fine structure constant. Phys. Rev. Lett. **109**, 111807 (2012). [arXiv:1205.5368 [hep-ph]]
21. Aoyama, T., Hayakawa, M., Kinoshita, T., Nio, M.: Complete tenth-order QED contribution to the Muon g-2. Phys. Rev. Lett. **109**, 111808 (2012). [arXiv:1205.5370 [hep-ph]]
22. Smirnov, V.A.: Applied asymptotic expansions in momenta and masses. Springer Tracts Mod. Phys. **177**, 1 (2002)
23. Chlouber, C., Samuel, M.A.: Contribution to the eighth order anomalous magnetic moment of the Muon. Phys. Rev. D **16**, 3596 (1977)

MCTDHB Physics and Technologies: Excitations and Vorticity, Single-Shot Detection, Measurement of Fragmentation, and Optimal Control in Correlated Ultra-Cold Bosonic Many-Body Systems

Ofir E. Alon, Vanderlei S. Bagnato, Raphael Beinke, Ioannis Brouzos, Tommaso Calarco, Tommaso Caneva, Lorenz S. Cederbaum, Mark A. Kasevich, Shachar Klaiman, Axel U.J. Lode, Simone Montangero, Antonio Negretti, Ressa S. Said, Kaspar Sakmann, Oksana I. Streltsova, Marcus Theisen, Marios C. Tsatsos, Storm E. Weiner, Tomos Wells, and Alexej I. Streltsov

Abstract Here we report on further applications, developments, expansion, and proliferation of the Multi-Configurational Time-Dependent Hartree for Bosons (MCTDHB) method in the context of ultra-cold atomic systems. In this year we put our main efforts to understanding and generalizing vortices—two-dimensional (2D) and three-dimensional (3D) quantum objects carrying angular momentum—from the perspective of the many-body physics. We have studied static properties and quantum dynamics of vortices confined in simple parabolic traps and in circular traps. Particular emphasis has been put on the loss of coherence and build-up of the fragmentation. Complimentary, we continue to develop the MCTDHB method spanning several directions of the theoretical and computational physics as well as

O.E. Alon
Department of Physics, University of Haifa at Oranim, Tivon 36006, Israel

V.S. Bagnato • M.C. Tsatsos
São Carlos Institute of Physics, University of São Paulo, PO Box 369, 13560-970 São Carlos, SP, Brazil

R. Beinke • L.S. Cederbaum • S. Klaiman • M. Theisen • A.I. Streltsov (✉)
Theoretische Chemie, Physikalisch-Chemisches Institut, Universität Heidelberg, Im Neuenheimer Feld 229, 69120 Heidelberg, Germany

I. Brouzos
Center for Integrated Quantum Science and Technology, Institute for Complex Quantum Systems, Universität Ulm, 89069 Ulm, Germany

T. Calarco • S. Montangero • R.S. Said
Institute for Complex Quantum Systems, Universität Ulm, 89069 Ulm, Germany

© Springer International Publishing Switzerland 2016

W.E. Nagel et al. (eds.), *High Performance Computing in Science and Engineering '15*, DOI 10.1007/978-3-319-24633-8_3

optimal-control theory: (a) the linear-response on-top of the MCTDHB method (LR-MCTDHB) has been reformulated in a compact block form, expanded for general inter-particle interactions, and benchmarked against the exactly-solvable harmonic-interaction model; (b) a new analysis tool capable of simulating the outcomes of typical shots generated in the experimental detection of ultra-cold atomic systems has been invented, tested, and applied; (c) a novel algorithm offering a direct quantitative measurement of the possible fragmentation in bosonic systems has been proposed and applied; (d) the optimal-control Chopped RAndom Basis (CRAB) algorithm has been merged with the MCTDHB package and applied to manipulate quantum systems. Implications and further perspectives and future research plans are briefly discussed and addressed.

1 Introduction to MCTDHB

The evolution in time of quantum systems becomes nowadays one of the most perspective and interesting directions of research in modern physics. A proper understanding and description of the quantum evolution requires an access to and knowledge about the excited states of the quantum systems at hands. The evolution is described by the time-dependent many-body Schrödinger equation (TDSE) which governs the physics of the ultra-cold quantum systems [1–3]. One way is to compute the excited states one-by-one and use them to describe the dynamics. An alternative approach is to develop methods capable to predict the evolution of the entire wave-packet of the system. In the context of the physics of ultra-cold atomic systems,

T. Caneva
ICFO-Institut de Ciencies Fotoniques, Mediterranean Technology Park, 08860, Castelldefels, Barcelona, Spain

M.A. Kasevich • K. Sakmann
Department of Physics, Stanford University, Stanford, CA 94305, USA

A.U.J. Lode
Condensed Matter Theory and Quantum Computing Group, Department of Physics, University of Basel, Klingelbergstrasse 82, 4056 Basel, Switzerland

A. Negretti
Zentrum für Optische Quantentechnologien and The Hamburg Centre for Ultrafast Imaging, Universität Hamburg, Luruper Chaussee 149, 22761 Hamburg, Germany

O.I. Streltsova
Laboratory of Information Technologies, Joint Institute for Nuclear Research, Joliot-Curie 6, Dubna, Russia

S.E. Weiner
Department of Chemistry, University of California, Berkeley, CA 94720-1462, USA

T. Wells
Department of Applied Mathematics and Theoretical Physics, University of Cambridge, Cambridge, CB3 0WA, United Kingdom

wave-packet evolution becomes especially interesting because the main quantities accessible in experiments directly correspond to the density of the evolving system.

The MCTDHB method [4, 5] has been invented and designed to perform wave-packet propagation by solving the TDSE in 1D, 2D, and 3D setups:

$$\hat{H}\Psi = i\hbar\frac{\partial \Psi}{\partial t}. \tag{1}$$

$\hat{H}$ is the generic many-body Hamiltonian of N identical bosons confined in an external trap $V(\mathbf{r})$ and interacting via a general inter-particle interaction potential $W(\mathbf{r}-\mathbf{r}') \equiv W(\mathbf{R})$ of strength λ_0:

$$\hat{H}(\mathbf{r}_1,\ldots,\mathbf{r}_N) = \sum_{j=1}^{N}\left[-\frac{\hbar^2}{2m}\nabla_{\mathbf{r}_j}^2 + V(\mathbf{r}_j)\right] + \sum_{j<k}^{N}\lambda_0 W(\mathbf{r}_j-\mathbf{r}_k). \tag{2}$$

Usually, all the computations are done in dimensionless units $\hbar = 1$, $m = 1$. It is convenient to collect all one-particle terms in the single-particle Hamiltonian $\hat{h}(\mathbf{r}) = -\frac{1}{2}\nabla_{\mathbf{r}}^2 + V(\mathbf{r})$. The evolution of the many-body wave-function is monitored at each time-slice and, hence, any desirable property or quantity can in principle be computed. The MCTDHB method allows one to solve the TDSE numerically at a very high level of accuracy [4–8], utilizing efficient numerical methods, modern computational technologies [9–12], as well as the advantages of the available hardware [13–15]. The details on the numerical and algorithmical implementation of the MCTDHB method have been reported in peer-review journals, see [4–8, 16], in books [17, 18], as well as in our previous HLRS reports [19, 20].

The present contribution summarizes our findings from the recent year reported in [21–30] on some new physical phenomena predicted to exist in different ultra-cold bosonic setups, and reports on developments of new MCTDHB-based methods and technologies.

2   Many-Body Dynamics of Vortices in 2D and 3D Computed with the R-MCTDHB Package

Within this project we focus on the applications of the recently-developed recursive implementation of MCTDHB, R-MCTDHB [12], to the dynamics of 2D and 3D bosonic systems under rotation. The first part discusses *phantom vortices*, a previously unknown many-body mode of quantized vorticity in 2D systems confined in a harmonic trap with a rotating anisotropy, see [21]. The second part discusses how a 2D system *reacts resonantly* when stirred by a rotating laser beam at the right frequency, see [22]. Finally, the third part shows a previously unknown mechanism of *vortex reconnections* in systems anisotropically confined in 3D, see [23].

2.1 How Angular Momentum Hides in Phantom Vortices

Swirls in rotating liquids or gases, like tornados in the atmosphere or tiny quantum vortices in microscopic ultracold gases, are fascinating and interesting for fundamentally understanding the dynamics of complex fluid flows and turbulence [2, 31]. A standard vortex represents the coherent motion of all particles along a line: each particle spins the same way around the "eye of the hurricane".

In our paper [21] we report on how this coherent picture can change in the many-body framework (Fig. 1). The quantum fluid fragments into parts, some of which rotate while others rest. This novel form of collective motion is called "phantom vortex" due to its elusive nature: the overall gas does not resemble a swirl anymore, even though some particles do spin around a center. In this swirl, however, the eye of the hurricane is filled up with particles of another fragment at rest (Fig. 2 and videos [32]).

So, how can phantom vortices be detected? It turns out that they leave a signature in the collective behavior of the particles which can be traced through the so-called correlation functions that are experimentally accessible, see Fig. 3 and videos [32].

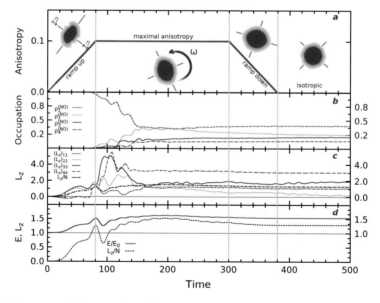

Fig. 1 Emergence of fragmentation and accumulation of angular momentum in the dynamics of a rotating 2D Bose gas. (*a*) Sketch of the anisotropy $\eta(t)$ ramping procedure with plots of $\rho(\mathbf{r})$ at representative times in each part. The potential is $V(x, y; t) = \frac{1}{2}(x^2 + y^2) + \frac{\eta(t)}{2}(x^2 - y^2)$. (*b*) The onset of fragmentation. (*c*) The orbital angular momenta reach their maximum values early in the period of maximum anisotropy ($t = 80 - 150$), but evolve to their equilibrium values before the anisotropy is removed. (*d*) Energy and angular momentum. The strong correlation between E and L_z indicates that the perturbation strictly excites angular momentum modes. All quantities shown are dimensionless. Figure reprinted from [21]

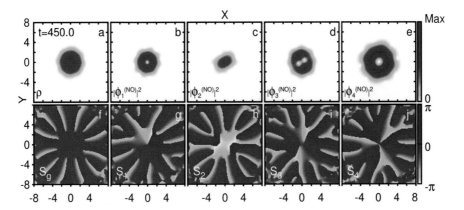

Fig. 2 Phantom vortices manifest in the condensate's fragments but are obscured in the density. (*a*) The density $\rho(\mathbf{r})$ shows a density minimum at the origin. (*b*)–(*e*) The natural orbital densities. (*f*) The phase, $S_g(\mathbf{r}|0)$, of $g^{(1)}(\mathbf{r}|0)$. (*g*)–(*j*) The phases, S_i, of the natural orbitals ϕ_i ($i = 1, 2, 3, 4$). See complementary videos [32]. All quantities shown are dimensionless. Figure reprinted from [21]

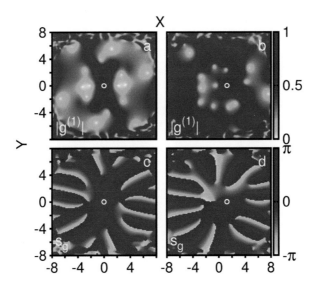

Fig. 3 Phantom vortices are exposed in the correlation function. To visualize the four-dimensional single-particle correlation function, $g^{(1)}(\mathbf{r}|\mathbf{r}')$ at $t = 219.4$, we fix a reference point at $\mathbf{r}' = (0,0)$ (*a*),(*c*) and $\mathbf{r}' = (1.25, 0)$ (*b*),(*d*). Its magnitudes (*a*),(*b*) and phases (*c*),(*d*) are plotted separately. All quantities shown are dimensionless. See complementary videos [32]. Figure reprinted from [21]

Phantom vortices are likely to play a role in the dynamics of BECs [2, 33] and also of other superfluids like Helium-4 [31] or even type-II superconductors [34]. See [21] for further details.

2.2 Resonant Stirring of a BEC Leads to Fragmentation

Resonances determine how a physical system reacts to an external perturbation and are at the heart of how many musical instruments, electronic circuits, or optical devices and lasers work. When a system is at resonance with an external drive, its reaction may be characterized by an unbounded absorption of energy from that drive—a phenomenon that can even lead to the destruction of big constructions, like bridges [35].

In the present study, [22], we investigate the resonance of an ultra-cold sample of interacting bosonic atoms stirred at different rotation frequencies by a rotating laser beam. Gases at very low temperatures are superfluids, i.e., flow with no friction [2]. Known ways in which these 2D superfluids absorb energy are center-of-mass motion, expansion, and the nucleation of vortices [34]. We find that all these excitations are present simultaneously when the stirring laser beam rotates at the resonant frequency of the system (Fig. 4). Moving the stirring frequency away from resonance results in a smaller energy increase of the system and the absence of one or more types of excitations.

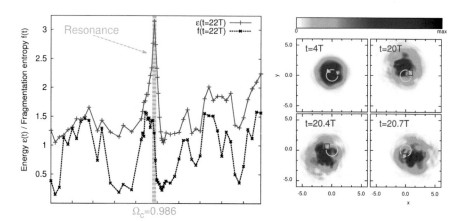

Fig. 4 Resonances in the dynamics of the 2D Bose gas with the stirring frequency Ω. The *left panel* depicts the scaled total energy of the state $\epsilon(t) = E(t)/E(0)$ (*red solid line*) and the fragmentation entropy $f(t) = \sum_{k=1}^{M} \left[\rho_k^{(NO)}(t) \log_2 \rho_k^{(NO)}(t) \right]$ (*black dashed line*) as a function of the stirring frequency Ω. The most profound resonance appears around the value $\Omega_c = 0.986$ (*gray vertical line in the left panel*), where the energy is absorbed the fastest. The *right panel* depicts screenshots of the density $\rho(\mathbf{r}; t)$ at times $t = 4T, 20T, 20.4T$, and $t = 20.7T$ after rotating with frequency $\Omega_c = 0.986$. All quantities shown are dimensionless. Figure reprinted from [22]

Interestingly, the collective behavior of the system changes: the closer the stirring frequency is to the resonance, the faster the initial superfluid splits up and *fragments* into several distinct superfluids (or order parameters) which are governed by the MCTDHB equations-of-motion [4, 5]. Like the findings on phantom vortices above, it is necessary to incorporate multiple orbitals in the description to capture the many-body physics in the dynamics. If the flow of other superfluids, like in type-II superconductors and Helium-4, is an analog of the found behavior, our findings indicate that a theory that incorporates the dynamics of multiple coupled superfluid order parameters which are interchanging particles is needed for their description on the many-body level. See [22] for further details.

2.3   *Mechanism of Vortex Reconnection*

Vortices in nature are merely spinning matter around a center or a line. They are known to appear in a variety of physical systems ranging from planet atmospheres, cups of coffee, and almost any other type of fluid. In the quantum world, they appear as topological defects and play a significant role in superfluids (fluids with no friction), superconductors [34], BECs [2, 33], and even in cosmological theories [36]. It is known that a single trapped vortex will precess around the geometrical center of its confinement while two vortices will eventually reconnect and split again.

The standard picture of the latter is two vortices mutually attracting and connecting at a certain point forming thus an X-like shape. Then, they exchange tails and split up again; they reconnect indefinitely. In a gaseous BEC in the laboratory one has the freedom to manipulate the shape of the spatial confinement (trap). The geometry of this confinement will affect the way the two vortices attract each other and their propagation and reconnection.

In our work [23], we show how the standard scenario, where two vortices meet and exchange tails from their centers in an X-reconnection, breaks down. We found that the way the vortex reconnection happens is altered, once the asymmetry of the spatial confinement is sufficiently large. The vortices, instead of following the standard X-shaped reconnection (see Figs. 5a and 6), bend until they reach the gas surface, split there and reconnect forming a shape that resembles the letter Z (see Fig. 5b and 6) before splitting up again. We hence term this novel type of vortex reconnections the "Z-reconnection". See [23] for details.

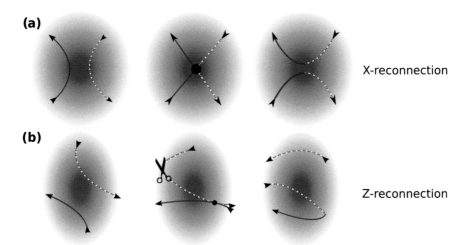

Fig. 5 Sketch of the X- and Z-reconnection for different anisotropies of the confining trap. (*a*) The conventional route to reconnection where the confining trap $V(x, y, z) = \frac{1}{2}(\omega_x x^2 + \omega_y y^2 + z^2)$ is spherically symmetric, $\epsilon = \omega_y/\omega_x = 1$. The vortices bend and move towards the geometric center of the trap where they meet and exchange tails in reconnecting. (*b*) A very similar configuration where the gas is however this time confined in an elongated trap, i.e., $\epsilon = \omega_y/\omega_x \gtrsim 1.5$. The vortices bend towards the gas surface and reconnect following a different path that resembles a 'Z' topology. Figure reprinted from [23]

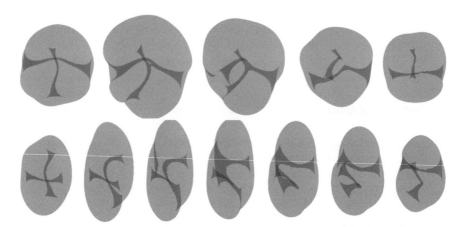

Fig. 6 Density isosurfaces of 3D vortex reconnections computed with R-MCTDHB for different trap anisotropies. The *top row* depicts isosurfaces of the density in a conventional X-reconnection event for an isotropic trap with $\epsilon = \omega_y/\omega_x = 1$. X-reconnections are seen for trap anisotropies $\epsilon \lesssim 1.5$. The *bottom row* depicts a Z-reconnection in an anisotropic trap with $\epsilon = 2$. Z-reconnections are favored by the system for $\epsilon \gtrsim 1.5$. Interestingly, the type of reconnection favored by the system depends only on the anisotropy ϵ and does not depend on the strength of the inter-particle interactions $\lambda = \lambda_0(N - 1)$. All quantities shown are dimensionless. See complementary videos [37]. Figure reprinted from [23]

2.4   Technical Details of the R-MCTDHB Studies

2.4.1   2D Computations

In 2D, we simulated $N = 100$ bosons with up to $M = 4$ orbitals. The maximal number of configurations was hence $N_{\text{conf}} = 176\,851$. The maximal number of DVR points used was $512 \times 512 = 2^{18}$, and the largest simulated domain was of extension $[-12, 12] \times [-12, 12]$. The width σ_{int} of the short-range Gaussian potential $W(\mathbf{R} = \mathbf{r}' - \mathbf{r}) = \frac{1}{2\pi\sigma_{int}^2} \exp[-(\mathbf{R})^2/(2\sigma_{int}^2)]$ was chosen as $\sigma_{int} = 0.25$ [38].

In the study on phantom vortices, the inter-particle interaction strength was $\lambda = \lambda_0(N-1) = 17.1$. The study of the resonant stirring behavior used $\lambda = \lambda_0(N-1) = 50.0$. Both studies used as an initial state the eigenstate of the interacting system in the isotropic parabolic trap. See [21, 22] for further technical details of the performed 2D computations.

2.4.2   3D Computations

In 3D, we simulated $N = 100$ bosons with up to $M = 4$ orbitals. The maximal number of configurations was $N_{\text{conf}} = 176\,851$. The maximal number of DVR points used to represent the simulated domain of extension $[-8, 8] \times [-8, 8] \times [-8, 8]$ was $128 \times 128 \times 128 = 2^{21}$.

The short-range interaction potential was taken to be a true contact interaction, $W(\mathbf{R} = \mathbf{r}' - \mathbf{r}) = \delta(\mathbf{R})$, where δ is the Dirac delta distribution (recall M is finite). The interaction strength chosen was varied from $\lambda = \lambda_0(N-1) = 0.1$ up to $\lambda = 100$ in [23]. In Fig. 6, results for $\lambda = 10$ are shown.

3   Two-Dimensional BECs in Circular Traps

In this project we investigated a 2D BEC split by a radial potential barrier. We determined the system's ground-state phase diagram as well as a time-dependent phase diagram of the splitting process. Furthermore, we go beyond the current paradigm for vortices and introduce spatially-partitioned many-body vortices, which are fragmented rather than condensed objects. Full reports have been published as a regular article entitled "Breaking the resilience of a two-dimensional Bose-Einstein condensate to fragmentation" Phys. Rev. A **90**, 043620 (2014) [24] and as a preprint entitled "Spatially-partitioned many-body vortices" [25] in arXiv:1412.4377.

We consider a repulsive BEC made of $N = 100$ bosons in the 2D circular trap shown in Fig. 7a. In dimensionless units the single-particle kinetic-energy operator in Eq. (2) reads $\hat{T}(\mathbf{r}) = -\frac{1}{2}\nabla_{\mathbf{r}}^2$. The explicit form of the one-body potential is given by $V(\mathbf{r}) = V_{trap}(\mathbf{r}) + V_{barrier}(\mathbf{r})$. Here, $V_{trap}(\mathbf{r}) = \{200e^{-(r-r_c)^4/2}, r \leq r_c = 9; 200, r > r_c\}$ is a flat trap which has the shape of "a crater" and

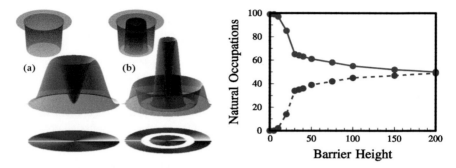

Fig. 7 Vortices in a circular trap. *Left part.* The insets (in *gray*) depict the trapping potentials. (*a*) The standard vortex with $L/N = 1$ in a 2D circular trap potential. (*b*) The many-body vortex of the first kind with $L/N = 1$ in a circular trap split by a radial barrier. Whereas the standard vortex is condensed, the many-body vortex is fragmented. Shown are the densities, the bottom parts plot the phase. *Right part.* The ground state of the BEC in the circular trap as a function of the barrier height. Shown are the occupation numbers (in percents). All bosons carry the same angular momentum, $l_1 = l_2 = 1$. See arXiv:1412.4377, [25], for more details

$V_{barrier}(\mathbf{r}) = 200e^{-2(r-R)^4}$ a ringed-shaped radial barrier of radius R. The many-boson Hamiltonian is given by Eq. (2) with the short-range repulsive interaction between the bosons modeled by a Gaussian function [38, 39] $W(\mathbf{r}-\mathbf{r}') = \frac{e^{-(\mathbf{r}-\mathbf{r}')^2/2\sigma^2}}{2\pi\sigma^2}$ with a width $\sigma = 0.25$. The interaction parameter λ_0 is taken to be positive to describe repulsive bosons. A square box of size $[-12, 12) \times [-12, 12)$ and a spatial grid of size 128×128 were found to converge the results to the accuracy reported below.

We would like to explore the ground state of a BEC in the circular trap of Fig. 7a, and ascertain when the many-body state is fragmented [40–50] or condensed [51]. This many-body property is related to the eigenvalues of the reduced one-body density matrix [52, 53]. We employ the MCTDHB method [4, 5], which most recently has been implemented and applied in higher dimensions [54, 55]. We use the R-MCTDHB [12] and the original MCTDHB [9, 10] software packages.

Figure 8 (left part) depicts the ground-state fragmentation versus the position R of the radial barrier for three different interaction strengths, $\lambda_0 = 0.002, 0.02, 0.2$. These many-body phase diagrams show that the radii R for which the ground state is fragmented are very limited, namely, that the ground state is mostly condensed within the parameter space of the problem. Increasing the interaction strength leads to two distinct effects. First, the maximal fragmentation shifts to larger values of R and, second, the width of the fragmented region also increases with the interaction. Importantly, we note that essentially 50 % fragmentation for different interaction strengths has been reached (the maxima occur for traps of different radii). Throughout this project, we have performed all computations with $M = 4$ time-adaptive (self-consistent) orbitals, and found that no more than two orbitals are macroscopically occupied, see [24] for more details.

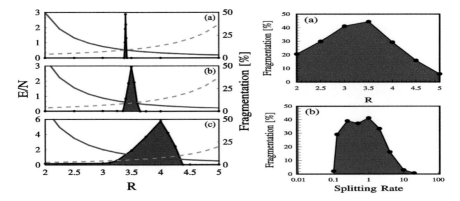

Fig. 8 Many-body phase diagrams of a BEC in a circular trap. *Left part*: Ground state. The *left panels* correspond to the interaction strengths (*a*) $\lambda_0 = 0.002$, (*b*) 0.02, and (*c*) 0.2. The BEC is mostly condensed, except for a narrow window (*magenta shaded area*) of radii R, which depends on λ_0. The static GP energies ε_{disk}^{GP} (*solid blue line*) and $\varepsilon_{annulus}^{GP}$ (*dashed orange line*) as a function of R are depicted. The maximal fragmentation on the many-body level is encountered when $\varepsilon_{disk}^{GP} = \varepsilon_{annulus}^{GP}$. *Right part*: Dynamics. Two cuts through the time-dependent phase diagram for $\lambda_0 = 0.02$. The dynamical splitting process leads to fragmentation (*a*) over the entire examined range of radii R; (*b*) over two orders of magnitude of the splitting rate. The radius of the barrier is $R = 3$. In the splitting process the system has a high affinity to fragment. See Phys. Rev. A **90**, 043620 (2014), [24], for more details

To explore the dynamical process of splitting the BEC, we prepare the system in the ground state of the trap $V_{trap}(\mathbf{r})$. In the absence of the radial barrier, the BEC is spread in the flat circular trap. One then ramps up the radial barrier such that the time-dependent one-body potential reads $V(\mathbf{r}, t) = V_{trap}(\mathbf{r}) + V_{ramp-up}(\mathbf{r}, t)$, where $V_{ramp-up}(\mathbf{r}, t) = \frac{\beta t}{200} V_{barrier}(\mathbf{r})$ and β is the splitting rate (the ramp-up process stops when the barrier reaches its maximal height, i.e., when $\beta t = 200$). This is a demanding many-body problem in 2D, because the BEC changes significantly both its shape *and* coherence, which MCTDHB can efficiently handle [9–12].

Figure 8 (right part, panel a) depicts the fragmentation at the end of the splitting process for $\lambda_0 = 0.02$ as a function of the radius R. The splitting rate is $\beta = 1$. The dynamical splitting process leads to fragmentation over the entire examined range of radii R. For most of these radii the ground state of the system (at any barrier height) is condensed. The system can thus dynamically fragment even though the ground state is condensed. Compared to the static phase diagram, see Fig. 8 (left part, panel b), the regime of dynamical fragmentation of the 2D BEC is significantly larger.

The idea to investigate many-body effects of 2D BECs in circular traps was further extended in [25]. Going beyond the standard paradigm of a vortex in BECs, which is a localized object looking much like a tiny tornado storm and well described by GP theory, we introduced the spatially-partitioned many-body vortices (MBVs). The MBVs are made of spatially-partitioned clouds, carry definite total angular momentum, and are fragmented rather than condensed quantum objects, describable only beyond GP theory. Two kinds of MBVs were introduced. MBVs of the first kind are at the global minimum of the energy for states of definite total angular momentum and comprised of bosons carrying the same angular momentum, see Fig. 7. MBVs of the second kind are fragmented excited states [56] in which macroscopic fractions of bosons carry different angular momenta, see [25] for further details. The recent experiment [57] utilizing a spatially-split circular trap, like the one in Fig. 7, encourages us to anticipated that MBVs will be further studied theoretically and searched for experimentally.

4 Excitations with LR-MCTDHB: Benchmarks and Applications

In this project the linear-response (LR) theory of the MCTDHB method (LR-MCTDHB) for computing many-body excitations of trapped BECs [Phys. Rev. A **88**, 023606 (2013); J. Chem. Phys. **140**, 034108 (2014)] was implemented for systems with general inter-particle interaction and benchmarked against the exactly-solvable harmonic interaction model (HIM). Furthermore, investigation and identification of the many-body excitations in 1D traps has been carried out. Full report has been published in an article entitled "Many-body excitation spectra of trapped bosons with general interaction by linear response" in Journal of Physics: Conference Series **594** (2015) 012039 [26]. The detailed study on the classification of excitations in BECs confined in harmonic and double-well traps has been reported in the Bachelor Thesis of Marcus Theisen, submitted to the department of Physics and Astronomy at the University of Heidelberg in (2014) [27].

4.1 Benchmark of LR-MCTDHB

The derivation of the LR theory atop the MCTDHB wave-function is rather lengthly but otherwise straightforward [58, 59]. Here we present only the final result for the resulting LR-MCTDHB theory, which takes on the form of the eigenvalue equation

$$\mathcal{L} \left(\mathbf{u}^k, \mathbf{v}^k, \mathbf{C}_u^k, \mathbf{C}_v^k \right)^T = \omega_k \left(\mathbf{u}^k, \mathbf{v}^k, \mathbf{C}_u^k, \mathbf{C}_v^k \right)^T. \tag{3}$$

The linear-response matrix $\mathcal{L}$ of the many-boson wave-function Ψ is more involved than the commonly-employed Bogoliubov–de Gennes (BdG) linear-response matrix. Physically, the response amplitudes of all modes, $\mathbf{u}^k$ and $\mathbf{v}^k$, and of all expansion coefficients, $\mathbf{C}_u^k$ and $\mathbf{C}_v^k$, combine to give the many-body excitation spectrum ω_k. In [58] we have successfully managed to explicitly construct $\mathcal{L}$ for bosons interacting by contact potential and obtained the many-body excitation spectrum.

Table 1 Excitations in the 1D HIM problem with $N = 1000$ bosons and repulsion $K = -0.0001$

	M=1	M=2	Exact analytical	n_{CM}, n_{rel}
E_{GS}	447.26949371	447.26638194	447.26638190	0, 0
ω_1	1.00000000	1.00000000	1.00000000	1, 0
ω_2	1.78907797	1.78885443	1.78885438	0, 2
ω_3	n/a	2.00004476	2.00000000	2, 0
ω_4	2.68361696	2.68328168	2.68328157	0, 3
ω_5	n/a	2.78888891	2.78885438	1, 2
ω_6	n/a	3.00007751	3.00000000	3, 0
ω_7	3.57815595	3.57771028	3.57770876	0, 4
ω_8	n/a	3.68397387	3.68328157	1, 3

Comparisons of LR-MCTDHB and the exact results for the ground, E_{GS}, and excited states, $\omega_k = E_k - E_{GS}$. The last column assigns the excitations in terms of center-of-mass and relative coordinates' quantum numbers. Some excitations are first uncovered at the $M = 2$ level of theory, i.e., they are not available (n/a) within Bogoliubov–de Gennes theory ($M = 1$). Convergence with the number or orbitals M to the exact results is clearly seen. Underlined digits indicate the difference to the exact result. All quantities are dimensionless. Table reprinted from [26].

Now we would like to report, for the first time, the implementation and application of LR-MCTDHB with general inter-particle interaction. As an illustrative system we have chosen the harmonic-interaction model (HIM) [60–63]. We recall that the HIM and a time-dependent extension of which have been used to benchmark MCTDHB [7].

In the HIM model the one-body Hamiltonian reads $\hat{h}(x) = -\frac{1}{2}\frac{\partial^2}{\partial x^2} + \frac{1}{2}\Omega^2 x^2$. The two-body interaction is $W(x - x') = K(x - x')^2$, representing thereby long-range interaction. The parameter $K < 0$ ($K > 0$) indicates repulsion (attraction) between the bosons. In what follows we set without loss of generality $\Omega = 1$. The exact excitation energies of the 1D HIM are given by [60, 62] $\omega[n_{CM}, n_{rel}] = n_{CM} + n_{rel}\delta_N$, $\delta_N = \sqrt{1 + 2NK}$, with the center-of-mass (CM) and relative coordinates' (rel) quantum numbers $n_{CM} = 1, 2, 3, \ldots$ and $n_{rel} = 2, 3, \ldots$. The results presented in Table 1 are in excellent agreement and serve to benchmark LR-MCTDHB.

4.2 Applications of LR-MCTDHB

The detailed study on the separation of excitations to mean-field-like and many-body-like can be found in the Bachelor Thesis of Marcus Theisen, submitted to the Department of Physics and Astronomy at the University of Heidelberg in 2014, see [27]. Here we briefly summarize the methodology used and the obtained physics. We consider 1D BECs made of $N = 10$ or $N = 100$ bosons each with repulsive interaction $W(x - x') = \lambda_0\delta_0(x - x')$ of strength λ_0. The BECs are confined in an harmonic $V_H(x) = \frac{\omega_H^2}{2}(x - x_0)^2$ and in a double-well $V_{DW}(x) = a\,e^{(x-x_0)^2} + V_H(x)$ traps. Here a and ω_H are constants and x_0 is a displacement of the trap origin.

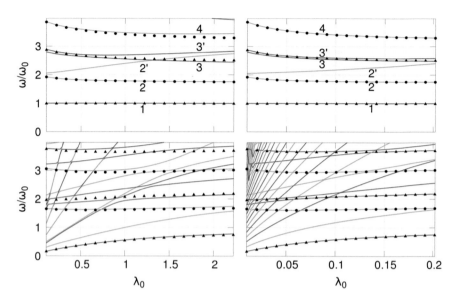

Fig. 9 LR-MCTDHB(2): harmonic (*upper panels*) and double-well (*lower panels*) traps. Excitations as a function of the interaction λ_0 are depicted for $N = 10$ bosons in *left panels* and for $N = 100$ bosons in the *right panels*. *Red and green lines* correspond to the states with *ungerade* and *gerade* symmetry, respectively. The LR-GP results are depicted as circles for *gerade* and triangles for *ungerade* states. All quantities are dimensionless. Figure reprinted from [27]

Figure 9 depicts excitations of the BECs trapped in harmonic (upper panels) and double-well potentials (lower panels) computed with the LR-MCTDHB theory for inter-boson repulsions of different strength λ_0. The left and right panels of Fig. 9 plot the LR excitation energies for the systems made of $N = 10$ and $N = 100$ bosons. In this figure it is clearly seen that there are states which are reproduced at both the mean-field (LR-GP) and many-body [LR-MCTDHB(2)] levels. However, one can also see that there are some low-energy states which appear only at the many-body level [58]. These observations allowed us to classify all the excitations according to this property—to the *mean-field-like* states and to the many-body-like states.

Here we present and discuss the excitation energies obtained by solving the many-body TDSE. We contrast the mean-field GP and many-body MCTDHB(M) dynamics. We use the following strategy to study the excitations by means of the wave-packet propagation. Initially the system is prepared in the ground state. Then we disturb it and monitor the induced dynamics. At each point of the propagation time the evolving many-body wave-function is analyzed and an information about the contributing excited states is extracted. By a proper choice of the applied perturbation one can activate the excited states of the gerade and ungerade manifolds individually.

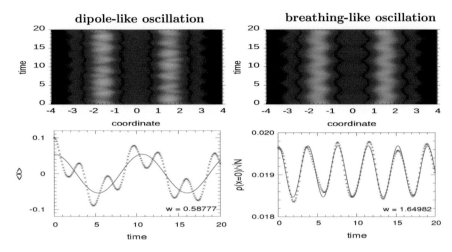

Fig. 10 Dipole-like and breathing-like excitations in the double-well trap. The *left panels* describe dipole-like oscillations. The *right panels* depict the breathing excitations. *Upper panels* show how the densities oscillate in time. *Lower panels* present the expectation value of the position operator (*left*) and the density at the origin (*right*) as a function of time. Figure reprinted from [27]

To study excitations of the *ungerade* symmetry we propose to displace suddenly the origin of the trap and to monitor the evolution of the system. The results are visualized in the left panels of Fig. 10. The upper left panel depicts the obtained density in a space-time Minkowskii-like representation. To activate excitations of the *gerade* symmetries we suddenly quench the frequency of the harmonic part of the double-well trap from $\omega_H = 1.1$ to 1.0, so the final double-well reads: $V_{DW} = 5e^{-x^2} + 0.5(1.0)^2 x^2$. In this case the center-of-mass of the system remains at the origin while the width of the many-body wave-packet evolves in time, i.e., we observe the so-called breathing modes as depicted in the upper right panels of Fig. 10. To quantify the dipole-like and breathing frequencies we compute and plot the expectation value of the position operator $\langle x(t) \rangle = \int dx \rho(x,t)x$ and the value of the density at the origin as a function of time $\Delta\rho(x=0,t)N^{-1/2}$ in the lower left and right panels. The discrete Fourier transformation applied to these evolutions allows us to extract the contributions from higher frequencies, see Table 2.

Table 2 contrasts the results of the static LR computations with dynamical wave-packet propagations. The observed excellent agreement between them for both the GP and MCTDHB approaches confirms the correctness, consistency, and equivalence of the static LR-MCTDHB approach and the dynamical methodology utilizing MCTDHB wave-packet propagation for computing excited states.

Table 2 Excitations frequencies for $N = 10$ bosons in the double-well trap

Excitation	GP		MCTDHB(2)		
no.	Static	Dynamic	Static	Dynamic	Symmetry
1	0.58049	0.57272	0.58740	0.58803	u
2	n/a	n/a	1.06608	1.06681	g
3	1.62649	1.62864	1.64982	1.65077	g
4	2.08879	2.08013	1.97639	1.97746	u
5	n/a	n/a	2.23737	2.24182	u
6	n/a	n/a	2.37206	–	g
7	n/a	n/a	2.40297	2.40331	u
8	2.94717	2.96814	2.87437	2.87328	g
9	n/a	n/a	3.06649	3.06681	g
10	3.65902	3.66367	3.57062	3.57754	u
11	n/a	n/a	3.84201	3.84708	u

The static results (LR) are taken from Fig. 9. The dynamic results are obtained as described in the caption of Fig. 10. n/a marks the many-body excitations not covered by the LR-GP (BdG) theory. u and g correspond to the *ungerade* and *gerade* symmetries of the states. $\Lambda = \lambda_0(N-1) = 10$. Table reprinted from [27].

5 Single Shot Simulations of Symmetry Breaking

In [28] an algorithm was developed that allows the simulation of single shots of an ultra-cold atom experiment where the information is usually obtained through destructive imaging. A short light pulse resonant with one of the atomic transitions of the condensate atoms is flashed at a cloud, providing thereby snapshots of the positions of all particles which are distributed according to the multidimensional probability distribution

$$P(x_1, x_2, \ldots, x_N, t) = |\Psi(x_1, x_2, \ldots, x_N, t)|^2,$$

where $\Psi(x_1, x_2, \ldots, x_N, t)$ is the many-body wave-function of the system at time t. In [28] single experimental shots are simulated by generating random deviates of the N-particle probability density $P(x_1, x_2, \ldots, x_N, t)$, obtained from solutions of the TDSE using the MCTDHB method. Drawing random deviates from $P(x_1, x_2, \ldots, x_N, t)$ is facilitated by realizing that

$$P(x_1, x_2, \ldots, x_N, t) = P(x_1)P(x_2|x_1) \cdots P(x_N|x_{N-1}, \ldots, x_1),$$

where, e.g., $P(x_2|x_1)$ is the conditional probability to detect the particle at x_2, given that there is a particle at x_1.

The position of each particle can thus be drawn from a single particle distribution instead of drawing all positions at once from a high-dimensional probability distribution. For simplicity we number these accordingly, i.e., $p_1(x) = P(x)$ is the probability density of the first particle from which x_1 is drawn, $p_2(x) = P(x|x_1)$

is that of the second particle and so on. Generally $p_1 \neq p_2 \neq \cdots \neq p_N$ (the only exception is a completely uncorrelated state such as a Gross-Pitaevskii mean-field state). The complete algorithm can be found in the preprint [28]. Apart from simulating single shots of an experiment, the algorithm in [28] also allows the evaluation of correlation functions of any order, as well as obtaining full counting statistics.

Now we consider an example that provides first principle insights into a phenomenon which is usually referred to as symmetry breaking. We provide an explanation of symmetry breaking in a special case without having to resort to an external perturbative field that breaks the symmetry. All results follow directly from the laws of quantum mechanics, so symmetry breaking emerges solely from the intrinsic fluctuations of the quantum state.

As an example we consider the ground state of a 1D attractive BEC with N bosons on a ring of length L: $V(x) = 0$, $\quad \Psi(x_1 \ldots, x_i + L, \ldots, x_N) = \Psi(x_1 \ldots, x_i, \ldots, x_N)$, for $i = 1, \ldots, N$. Due to the symmetry of the ring the exact many-body wave-function is translationally invariant and thus the single-particle density is constant. Here we use a ring of length $L = 2\pi$ and $N = 20$ bosons with a contact $\lambda_0 \delta(x - x')$ interaction, $\lambda_0(N - 1) = -4.0$. The ground state is obtained by imaginary time-propagation in a Hilbert space spanned by $M = 7$ plane waves, i.e. e^{ikx} with $k = 0, \pm1, \pm2, \pm3$.

We would like to answer the question what single experimental shots of an attractive condensate residing in the ground state on a ring would measure. The top row of Fig. 11 shows three different realizations of sampling the particle

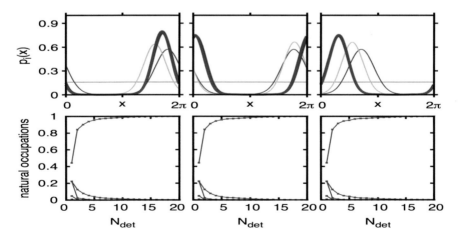

Fig. 11 Single shot realizations of the ground state of an attractive BEC. *Top row*: The probabilities $p_i(x)$ to detect the ith particle given the locations of the previously detected $i - 1$ particles for $i = 1$ (*black*) $i = 2$ (*blue*), $i = 3$ (*green*), $i = 15$ (*magenta*) and $i = 20$ (*red*). *Bottom row*: Natural occupations during the detection. Shown are the natural occupations of the system obtained by removing detected particles. After about three detected particles only one natural orbital remains significantly occupied. See [28] for more details. All quantities shown are dimensionless

positions of the ground state wave-function. Shown are the probabilities $p_n(x)$ for $n = 1, 2, 3, 15, 20$. $p_1(x)$ is always uniform as expected based on the exact many-body solution, but already the second particle has a highly peaked probability density. In different realizations the particles begin to localize in the vicinity of the first detected particle, and after a few particles have been detected the probability density of subsequent particles is essentially the same. While the ground state wave-function does have the symmetry of the Hamiltonian, each single shot localizes the particles nearby each other. So, in single shots of the experiment the symmetry is not preserved. If the average over many realizations is taken the uniform single particle density is recovered. The bottom row shows the natural occupations of the quantum state that is obtained by removing particles after their detection. It is clearly seen that the initial state has several occupied natural orbitals, but subsequent detections lead here to a build-up of coherence in the system. After about three detections only a single natural orbital is significantly occupied.

The single shot simulations require accessing all permanents that can be constructed from all particle numbers up to N particles in the space of M orbitals. No simple closed form expressions exist to estimate the complexity of the general single shot algorithm. However, in the important case where $N \gg M$ the complexity of evaluating a single shot can be shown to scale roughly as $O(N^M)$. Usually, one is interested in a large number of single shots for the purpose of obtaining meaningful statistics.

6   Protocol for Direct Measurement of Fragmentation

The mainstream of our scientific activity in the recent decade was the development of numerical methods capable of describing the physics of ultra-cold bosonic systems at the full many-body level. It turns out that key phenomena manifesting themselves beyond mean-field physics are the loss of the coherence and build-up of fragmentation. Despite the numerous theoretical predictions on fragmentation in trapped ultra-cold atomic systems, a clean, clear, and general experimental protocol capable of quantitative measurements of the fragmentation was absent.

In the preprint arXiv:1412.4049 "Interferometry with correlated matter-waves" [29] we report on a breakthrough in this direction and define such a protocol based on interferometric measurements, closing thereby a gap between predictions on fragmentation and a possibility of its quantitative experimental detection. In this project as an initial state for interferometry we use the fully condensed idealized single-hump attractive ($\lambda_0 < 0$) soliton placed in a weak harmonic trap, $V(x) = 0.1x^2$.

The first interferometric step—is to split the initially-condensed cloud into two parts by applying a laser pulse $\psi(x, t = 0)[e^{ikx} + e^{-ikx-i\chi}]$, where $\psi(x, t = 0)$ is the initial N-boson wave-function, k is the imparted momentum, and χ is the imprinted phase. If after this *sudden* imprinting of momentum and phase the split cloud were

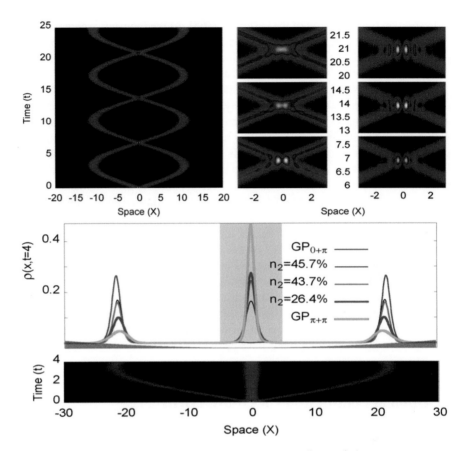

Fig. 12 Protocol to measure fragmentation. The first π-pulse $e^{ikx} + e^{-ikx-i\pi}$ splits the initially-coherent soliton confined in a weak harmonic trap, $V(x) = 0.1x^2$. The *upper-left panel* shows how the split matter-waves oscillate. The *upper-middle and upper-right panels* contrasts enlarged densities at different re-collision times computed at the many-body and mean-field (GP) levels. To measure fragmentation we apply at the moment when the two split sub-clouds re-collide a second laser-pulse π-pulse $e^{ikx} + e^{-ikx-i\pi}$. It results (*lower panel*) in three matter-waves propagating with $-2k, 0k, +2k$ momenta. The number of $0k$ atoms in the *central (gray) area*, also called the visibility υ, is proportional to the occupation of the second natural orbital n_2. Hence, one can directly access the non-condensed, i.e., fragmented fraction n_2 in the system—this is our main result. Figure reprinted from [29]

to keep the coherence and solitonic properties, χ would define the phase between the split matter-waves.

In the second interferometric step the weak harmonic trap deflects the split parts. Figure 12 confirms the solitonic-like behavior of the system—the split sub-clouds continue to oscillate by bouncing from the walls and re-colliding without significant broadening. The GP mean-field theory applied to the same initial condition mimics the overall oscillating behavior of the split system, but not the re-collision events where the many-body results reveal a blurring—degradation of the fringe visibility, indicating on fragmentation of the system [64, 65]. At the moments where the split matter-waves re-collide for the first, second, and third times as depicted in the upper-left and upper-middle panels of Fig. 12 the occupation of the second natural orbital (fragment) becomes $n_2 = 26.4\%$, 43.7%, and 45.7%, respectively.

Due to small sizes of the attractive clouds it is technically very difficult to detect the interferences directly. Fortunately, the phase between the split sub-clouds, as well as the contrast of the interference pattern, can be determined accurately within the third interferometric step—by applying a recombining (second) laser-pulse exactly at the re-collision moment. In [29] we have shown that a recombining laser pulse splits the system into three sub-clouds propagating with $0k$ and $\pm 2k$ momenta. The populations of these channels depend on the fragmentation of the system. For a perfect two-fold fragmented system in the Fock state $|n_1, n_2\rangle$, $N = n_1 + n_2$ the occupation of the second natural orbital n_2 and the population of the $0k$-channel (the visibility v) are linearly connected: $n_2^{intf}/N = 1 - \frac{3}{2}v$. So, by monitoring the populations of the momentum channels one can directly measure fragmentation. We have applied our interferometric protocol to measure fragmentation of the re-colliding solitons and obtained in [29] excellent results. Furthermore and without loss of generality, the proposed protocol to discriminate coherent and fragmented many-body systems with two-hump densities can be expanded to measure fragmentation in generic many-body systems with attractive and repulsive short- and long-range inter-particle interactions.

7 Optimal Control with CRAB-MCTDHB

Quantum dynamics is one of the frontiers of modern science attracting researchers from different fields of physics. The control of quantum dynamics is a primary goal of experimental and theoretical investigations. Inspired by a fascinating degree of control on quantum states achieved in optics we propose to integrate the available methods of optimal control [66] to the field of ultra-cold atomic systems.

In the preprint arXiv:1412.6142 "Quantum Speed Limit and Optimal Control of Many-Boson Dynamics" [30] we report on first preliminary results, where the MCTDHB method and the optimal-control CRAB algorithm have been merged together. In this work we expand our understanding of the concept of quantum speed limit (QSL)—the minimal time needed to perform a driven evolution, from a text-book example of two-level system to complex interacting many-body systems. We extend, apply, confirm, and validate this concept in a prototypical many-body system appearing in the context of ultra-cold atomic physics—a bosonic Josephson junction (BJJ) problem.

As an atomic BJJ we consider $N = 100$ bosons with contact $\lambda_0 \delta(x_i - x_j)$ interaction confined in a 1D double-well potential formed by combining two harmonic oscillators like in [6] $V_\pm(x) = \frac{1}{2}(x \pm 2)^2$, where V_+ (V_-) is the potential for $x < 0$ ($x > 0$). We investigate different values of the repulsive interaction strength ($\lambda_0 > 0$). Initially all the bosons are prepared in the ground state of the left V_+ well $|\psi_0\rangle = |L\rangle$. At time $t = 0$ the time-evolution starts in the fully symmetric double-well potential $V(x)$. This is a standard scenario for uncontrolled BJJ dynamics. In our investigation we would like to reach as a target state the ground state of the right V_- well $|\psi_T\rangle = |R\rangle$.

To control the dynamics in a time-dependent trap $V_\pm(x,t) = (x \pm 2)^2/2 + E_\pm(t)$ we propose to use the zero point energy of the left (right) well $E_\pm(t)$ as a control parameter for driving. In our calculations and optimizations we contrast the full many-body (MCTDHB) with mean-field (GP) results. To quantify the success of the driving process we use the Uhlmann fidelity $F(\rho_T, \rho) = \mathrm{Tr}\left[\sqrt{\sqrt{\rho_T}\rho\sqrt{\rho_T}}\right]$ which is based on the reduced one-body density matrix and, hence, directly available in the MCTDHB simulations.

In a linear two-level system, the QSL of a state-to-state transfer is characterized by the magnitude of the coupling J between the two involved levels with the minimum time given by half the Rabi period $T^L_{QSL} = \pi\hbar/2J$ [67]. On the Bloch sphere, the time-optimal dynamics connecting the initial and the target state occurs along a geodesic, see Fig. 13.

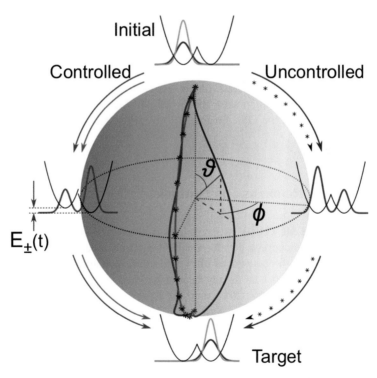

Fig. 13 BJJ dynamics. Paths for the two-mode nonlinear dynamics on a Bloch sphere: uncontrolled GPE (*red line*), driven by an analytic CCP (*green line*) and the CRAB-optimized (*blue line*). *Black asterisks* mark a geodesic. *Upper and lower sketches*: Initial and target states in the double-well (*black line*). *Left and right sketches*: controlled and uncontrolled intermediate wavefunctions and potential. Figure reprinted from [30]

In order to cancel the effect of the non-linear interaction in a two-mode treatment we propose [30] to use the following analytic compensating control pulse (CCP) $D_{CCP} = E_+(t) - E_-(t) = UN\cos^2(Jt)$. Here U, J are the standard BJJ parameters. In a more realistic experimental situation one has to adapt this CCP strategy accordingly. Namely, in cases close to the QSL, the analytic CCP does not lead to expected fidelities, and one has to apply a system-dependent optimal CCP, computed, e.g., with the CRAB algorithm [66]. As a guess pulse $G(t)$ we use the above discussed analytic CCP, $G(t) = D_{CCP}(t)$. The CRAB scheme is based on an expansion of the control correction in randomized time-dependent basis functions $D_{CRAB}(t) = G(t) + G(t) \left[\sum A_n \sin(\omega_n t) + B_n \sin(\omega_n t)\right]/\Pi(t)$, where $\Pi(t)$ assures that $D(0, T) = G(0, T)$. We have used up to 4 random frequencies and the task of the optimization is to find the coefficients A_n, B_n that maximize the fidelity.

The GPE describes the system adequately for short times and weak drivings, while for times close to the quantum speed limit and for strong interactions and fast drivings, the many-body effects (depletion-fragmentation) lead to deviations from the GPE predictions. Close to the QSL, where the mean-field description does not account for the dynamics accurately, and the simple compensating pulse strategy fails, we propose to use the optimal driving within the CRAB-MCTDHB. The calculations with the above discussed specific parameters of the potential and control pulses have been done as a proof of principles of the QSL concepts in many-body systems. We have developed a strategy and provided optimal control schemes to achieve state-to-state transfer at minimum time and validated it at all levels of complexity of the description of the problem—two-mode, nonlinear mean-field, and full many-body. These findings enable the investigation and experimental realization of different protocols at the QSL for many-body systems in different settings and experimental setups.

8   Concluding Remarks and Outlook

The present HLRS-MCTDHB progress report discusses and summarizes our achievements in applications, developments, expansion, and proliferation of the MCTDHB method in the context of ultra-cold atomic systems. The performance of the MCTDHB method has already been tested and discussed in our previous HLRS reports for the years 2012 and 2013, see [19, 20]. In this year our priority was an exploration of the vortices—2D and 3D quantum objects carrying angular momentum—from the perspective of the many-body physics. We solved the underlying time-(in)dependent many-boson Schrödinger equation within the framework of the MCTDHB method in 2D and 3D setups. We studied static properties and quantum dynamics of vortices confined in simple parabolic traps and in circular traps with double-well-like topologies. All the obtained results have a common agenda—fragmentation does play a crucial role and must be taken into account for a proper understanding and interpretation of the rich physics of vortices. In 2D harmonic setups we would like to mention the discovery of *phantom vortices*—a previously unknown many-body mode of quantized vorticity in 2D systems confined in a harmonic trap with a rotating anisotropy, and a connection between the *resonant* reaction of a 2D BEC stirred by a rotating laser beam and fragmentation. In circular 2D traps with double-well-like topologies, we went beyond the current paradigm for vortices and introduced spatially-partitioned many-body vortices which are fragmented rather than condensed objects. Unexpectedly, 2D BECs in circular traps show an outstanding resilience to ground-state fragmentation which is easily lifted by the dynamical process of splitting the BEC by a time-dependent barrier. In the anisotropically-confined 3D systems we discovered a previously unknown mechanism of *vortex reconnections.*

On the development side we would like first to mention the further progress reached in generalization, simplification, and benchmarking of the LR-MCTDHB theory which provides a direct access to the many-body excited states. The systematic application of the LR-MCTDHB method to many-body systems trapped in harmonic and double-well traps allowed us to classify all excited states into two generic groups—the mean-field-like group and the many-body-like group. An existence of the low-lying non-mean-field states in the large N-limit might impact a common belief that quantum dynamics in this limit is properly described by the mean-field theory. If this is not the case, the experimental measurements and attributions of the excitations done so far essentially exclusively by Bogoliubov–de Gennes theory might require a serious reconsideration. The second great success is the merging of the MCTDHB method with the optimal-control algorithm CRAB. The developed and implemented CRAB-MCTDHB technique opens an avenue to manipulate and control interacting quantum many-body systems. To prove the principles we apply it to study the quantum speed limit in the BJJ setup.

Since its establishment, the MCTDHB method has been targeted to discover many-body physics which is not described at the mean-field level. In all ultra-cold systems where the non-mean-field effects have been discovered, a loss of coherence along side build-up of fragmentation have also been observed. In spite of numerous theoretical predictions on appearances of the fragmentation, also presented in this report, a clean, clear, and general experimental protocol capable of quantitative measurements of the fragmentation was absent. We have attacked this problem from two complimentary sides and are happy to claim that a breakthrough in this direction has been reached. First, we proposed a new analysis tool capable of simulating the outcomes of typical shots generated in the experimental detection of ultra-cold atomic systems and, second, we described an experimental protocol for a direct quantitative measurement of the fragmentation in trapped bosonic systems.

Concluding, the MCTDHB method originally implemented, tested, and bench-marked as a high-performance software package MCTDHB [9], has opened an avenue for further developments of new theoretical approaches, numerical algorithms, and novel methodological solutions: LR-MCTDHB [10], R-MCTDHB [12], CRAB-MCTDHB [30], and ML-MCTDHB (the Multi-Layer MCTDHB [68]). We are happy to announce that the original MCTDHB package, its LR-MCTDHB extension, as well as collections of the tools for the analysis, visualization, package making, building, teaching, and development are now integrated under a common cross-platform (Linux/Win/Mac) java-based environment: MCTDHB-Lab—the laboratory to study quantum dynamics with a mouse-click interface, see for details http://QDlab.org [11].

Acknowledgements Financial support by the DFG is gratefully acknowledged. OEA acknowledges funding by the Israel Science Foundation (grant No. 600/15). SEW and TW acknowledge financial support by the DAAD-RISE program, AUJL acknowledges funding by the Swiss SNF and the NCCR Quantum Science and Technology, MCT acknowledges funding by FAPESP. AUJL, SEW, and TW acknowledge the hospitality of Vanderlei Bagnato and the CEPOF-USP. TC and SM acknowledge support from the European Commission through the grants QIBEC (No. 284584) and SIQS, and from the DFG through SFB TR21.

References

1. Dalfovo, F., Giorgini, S., Pitaevskii, L.P., Stringari, S.: Theory of Bose-Einstein condensation in trapped gases. Rev. Mod. Phys. **71**, 463 (1999); Leggett, A.J.: Bose-Einstein condensation in the alkali gases: Some fundamental concepts. Rev. Mod. Phys. **73**, 307 (2001)
2. Pethick, C., Smith, H.: Bose-Einstein Condensation in Dilute Gases, 2nd edn. Cambridge University Press, New York (2008)
3. Köhler, T., Góral, K., Julienne, P.S.: Production of cold molecules via magnetically tunable Feshbach resonances. Rev. Mod. Phys. **78**, 1311 (2006)
4. Streltsov, A.I., Alon, O.E., Cederbaum, L.S.: Role of Excited States in the Splitting of a Trapped Interacting Bose-Einstein Condensate by a Time-Dependent Barrier. Phys. Rev. Lett. **99**, 030402 (2007)
5. Alon, O.E., Streltsov, A.I., Cederbaum, L.S.: Multiconfigurational time-dependent Hartree method for bosons: Many-body dynamics of bosonic systems. Phys. Rev. A **77**, 033613 (2008)
6. Sakmann, K., Streltsov, A.I., Alon, O.E., Cederbaum, L.S.: Exact Quantum Dynamics of a Bosonic Josephson Junction. Phys. Rev. Lett. **103**, 220601 (2009)
7. Lode, A.U.J., Sakmann, K., Alon, O.E., Cederbaum, L.S., Streltsov, A.I.: Numerically exact quantum dynamics of bosons with time-dependent interactions of harmonic type. Phys. Rev. A **86**, 063606 (2012)
8. Lode, A.U.J., Streltsov, A.I., Sakmann, K., Alon, O.E., Cederbaum, L.S.: How an interacting many-body system tunnels through a potential barrier to open space. Proc. Natl. Acad. Sci. USA **109**, 13521 (2012)
9. Streltsov, A.I., Sakmann, K., Lode, A.U.J., Alon, O.E., Cederbaum, L.S.: The multiconfigurational time-dependent Hartree for bosons package, version 2.x, Heidelberg (2013). http://MCTDHB.org
10. Streltsov, A.I., Cederbaum, L.S., Alon, O.E., Sakmann, K., Lode, A.U.J., Grond, J., Streltsova, O.I., Klaiman, S.: The multiconfigurational time-dependent Hartree for bosons package, version 3.x, , Heidelberg (2006-Present). http://MCTDHB.org
11. Streltsov, A.I., Streltsova, O.I.: The multiconfigurational time-dependent Hartree for bosons laboratory, version 1.5. http://MCTDHB-lab.org; http://QDlab.org (2015)
12. Lode, A.U.J., Tsatsos, M.C.: The recursive multiconfigurational time-dependent Hartree for bosons package, version 1.0. http://ultracold.org; http://r-mctdhb.org; http://schroedinger.org (2014)
13. Cray clusters Hermit, Hornet and NEC Nehalem cluster Laki at the High Performance Computing Center Stuttgart (HLRS). https://www.hlrs.de

14. bwGRiD, member of the German D-Grid initiative, funded by the Ministry for Education and Research (Bundesministerium für Bildung und Forschung) and the Ministry for Science, Research and Arts Baden-Württemberg (Ministerium für Wissenschaft, Forschung und Kunst Baden-Württemberg). http://www.bw-grid.de

15. Hybrid computing complex K100 (Keldysh Institute of Applied Mathematics, RAS). http://www.kiam.ru

16. Streltsov, A.I., Alon, O.E., Cederbaum, L.S.: General mapping for bosonic and fermionic operators in Fock space. Phys. Rev. A **81**, 022124 (2010). http://dx.doi.org/10.1103/PhysRevA.81.022124

17. Meyer, H.-D., Gatti, F., Worth, G.A. (eds.): Multidimensional Quantum Dynamics: MCTDH Theory and Applications. Wiley-VCH, Weinheim (2009)

18. Proukakis, N.P., Gardiner, S.A., Davis, M.J., Szymanska, M.H. (eds.): Quantum Gases: Finite Temperature and Non-Equilibrium Dynamics, vol. 1. Cold Atoms Series. Imperial College Press, London (2013)

19. Lode, A.U.J., Sakmann, K., Doganov, R.A., Grond, J., Alon, O.E., Streltsov, A.I., Cederbaum, L.S.: Numerically-exact Schrödinger dynamics of closed and open many-boson systems with the MCTDHB package, HLRS report for 2012. In: Nagel, W.E., Kröner, D.H., Resch, M.M. (eds.) High Performance Computing in Science and Engineering '13: Transactions of the High Performance Computing Center, Stuttgart (HLRS) 2013. Springer, Heidelberg (2013)

20. Klaiman, S., Lode, A.U.J., Sakmann, K., Streltsova, O.I., Alon, O.E., Cederbaum, L.S., Streltsov, A.I.: Quantum Many-Body Dynamics of Trapped Bosons with the MCTDHB Package: Towards New Horizons with Novel Physics, HLRS report for 2013. In Nagel, W.E., Kröner, D.H., Resch, M.M. (eds.) High Performance Computing in Science and Engineering '14: Transactions of the High Performance Computing Center, Stuttgart (HLRS) 2014. Springer, Heidelberg (2015)

21. Weiner, S.E., Tsatsos, M.C., Cederbaum, L.S., Lode, A.U.J.: Angular momentum in interacting many-body systems hides in phantom vortices (2014). arXiv:1409.7670

22. Tsatsos, M.C., Lode, A.U.J.: Vortex nucleation through fragmentation in a stirred resonant Bose-Einstein condensate (2014). arXiv:1410.0414

23. Wells, T., Lode, A.U.J., Bagnato, V.S., Tsatsos, M.C.: Vortex reconnections in anisotropic trapped three-dimensional Bose-Einstein condensates. J. Low Temp. Phys. 0022-2291 (2015); arXiv:1410.2859 (2014)

24. Klaiman, S., Lode, A.U.J., Streltsov, A.I., Cederbaum, L.S., Alon, O.E.: Breaking the resilience of a two-dimensional Bose-Einstein condensate to fragmentation. Phys. Rev. A **90**, 043620 (2014)

25. Klaiman, S., Alon, O.E.: Spatially partitioned many-body vortices (2014). arXiv:1412.4377

26. Alon, O.E.: Many-body excitation spectra of trapped bosons with general interaction by linear response. J. Phys. Conf. Ser. **594**, 012039 (2015)

27. Theisen, M.: Excited states of Bose-Einstein condensates: linear response vs. Many-body dynamics. Bachelor Thesis, Department of Physics and Astronomy, University of Heidelberg (2014)

28. Sakmann, K., Kasevich, M.: Single shot simulations of dynamic quantum many-body systems (2015). arXiv:1501.03224

29. Streltsova, O.I., Streltsov, A.I.: Interferometry with correlated matter-waves (2014). arXiv:1412.4049

30. Brouzos, I., Streltsov, A.I., Negretti, A., Said, R.S., Caneva, T., Montangero, S., Calarco, T.: Quantum speed limit and optimal control of many-boson dynamics (2014). arXiv:1412.6142

31. Donnelly, R.J.: Quantized Vortices in Helium II. Cambridge University Press, Cambridge (1991)

32. Video references. Time evolution of the density, natural orbitals, and phase http://youtu.be/ezbdLWvSbBI; Scan of reference point through $g^{(1)}$ at $t = 115$ and $t = 450$ http://youtu.be/whRL8haF4RA and http://youtu.be/gG7dprvRWGg (2015)

33. Abo-Shaeer, J.R., Raman, C., Vogels, J.M., Ketterle, W.: Observation of Vortex Lattices in Bose-Einstein Condensates. Science **292**, 476 (2001)

34. Abrikosov, A.A.: On the Magnetic Properties of Superconductors of the Second Group. Sov. Phys. JETP **5**, 1174 (1957)
35. Tacoma Narrows Bridge (1940). http://en.wikipedia.org/wiki/Tacoma_Narrows_Bridge_ (1940)
36. Hindmarsh, M.B., Kibble, T.W.B.: Cosmic strings. Rep. Prog. Phys. **58**, 477 (1995)
37. Video references. Vortex reconnections for the isotropic trap with $\epsilon = 1$. http://youtu.be/ VLMY1eYLr0g; Vortex reconnections for the anisotropic trap with $\epsilon = 2$ and $\epsilon = 3$. http:// youtu.be/r5pY7pfMTUg and http://youtu.be/ifEFffVv93U
38. Doganov, R.A., Klaiman, S., Alon, O.E., Streltsov, A.I., Cederbaum, L.S.: Two trapped particles interacting by a finite-range two-body potential in two spatial dimensions. Phys. Rev. A **87**, 033631 (2013)
39. Christensson, J., Forssén, C., Åberg, S., Reimann, S.M.: Effective-interaction approach to the many-boson problem. Phys. Rev. A **79**, 012707 (2009)
40. Nozières, P., Saint James, D.: Particle vs. pair condensation in attractive Bose liquids. J. Phys. France **43**, 1133 (1982)
41. Nozières, P.: In: Griffin, A., Snoke, D.W., Stringari, S. (eds.) Bose-Einstein Condensation. Cambridge University Press, Cambridge, England (1996)
42. Spekkens, R.W., Sipe, J.E.: Spatial fragmentation of a Bose-Einstein condensate in a double-well potential. Phys. Rev. A **59**, 3868 (1999)
43. Streltsov, A.I., Alon, O.E., Cederbaum, L.S.: General variational many-body theory with complete self-consistency for trapped bosonic systems. Phys. Rev. A **73**, 063626 (2006)
44. Mueller, E.J., Ho, T.-L., Ueda, M., Baym, G.: Fragmentation of Bose-Einstein condensates. Phys. Rev. A **74**, 033612 (2006)
45. Bader, P., Fischer, U.R.: Fragmented Many-Body Ground States for Scalar Bosons in a Single Trap. Phys. Rev. Lett. **103**, 060402 (2009)
46. Fischer, U.R., Bader, P.: Interacting trapped bosons yield fragmented condensate states in low dimensions. Phys. Rev. A **82**, 013607 (2010)
47. Zhou, Q., Cui, X.: Fate of a Bose-Einstein Condensate in the Presence of Spin-Orbit Coupling. Phys. Rev. Lett. **110**, 140407 (2013)
48. Kawaguchi, Y.: Goldstone-mode instability leading to fragmentation in a spinor Bose-Einstein condensate. Phys. Rev. A **89**, 033627 (2014)
49. Song, S.-W., Zhang, Y.-C., Zhao, H., Wang, X., Liu, W.-M.: Fragmentation of spin-orbit-coupled spinor Bose-Einstein condensates. Phys. Rev. A **89**, 063613 (2014)
50. Kang, M.-K., Fischer, U.R.: Revealing Single-Trap Condensate Fragmentation by Measuring Density-Density Correlations after Time of Flight. Phys. Rev. Lett. **113**, 140404 (2014)
51. Penrose, O., Onsager, L.: Bose-Einstein Condensation and Liquid Helium. Phys. Rev. **104**, 576 (1956)
52. Löwdin, P.-O.: Quantum Theory of Many-Particle Systems. I. Physical Interpretations by Means of Density Matrices, Natural Spin-Orbitals, and Convergence Problems in the Method of Configurational Interaction. Phys. Rev. **97**, 1474 (1955)
53. Coleman, A.J., Yukalov, V.I.: Reduced Density Matrices: Coulson's Challenge, vol. 72. Lectures Notes in Chemistry. Springer, Berlin (2000)
54. Streltsov, A.I.: Quantum systems of ultracold bosons with customized interparticle interactions. Phys. Rev. A **88**, 041602(R) (2013)
55. Streltsova, O.I., Alon, O.E., Cederbaum, L.S., Streltsov, A.I.: Generic regimes of quantum many-body dynamics of trapped bosonic systems with strong repulsive interactions. Phys. Rev. A **89**, 061602(R) (2014)
56. Cederbaum, L.S., Streltsov, A.I.: Self-consistent fragmented excited states of trapped condensates. Phys. Rev. A **70**, 023610 (2004)
57. Corman, L., Chomaz, L., Bienaimé, T., Desbuquois, R., Weitenberg, C., Nascimbène, S., Dalibard, J., Beugnon, J.: Quench-Induced Supercurrents in an Annular Bose Gas. Phys. Rev. Lett. **113**, 135302 (2014)

58. Grond, J., Streltsov, A.I., Lode, A.U.J., Sakmann, K., Cederbaum, L.S., Alon, O.E.: Excitation spectra of many-body systems by linear response: General theory and applications to trapped condensates. Phys. Rev. A **88**, 023606 (2013)
59. Alon, O.E., Streltsov, A.I., Cederbaum, L.S.: Unified view on linear response of interacting identical and distinguishable particles from multiconfigurational time-dependent Hartree methods. J. Chem. Phys. **140**, 034108 (2014)
60. Cohen, L., Lee, C.: Exact reduced density matrices for a model problem. J. Math. Phys. **26**, 3105 (1985)
61. Gajda, M., Załuska-Kotur, M.A., Mostowski, J.: Destruction of a Bose-Einstein condensate by strong interactions. J. Phys. B **33**, 4003 (2000)
62. Yan, J.: Harmonic Interaction Model and Its Applications in Bose-Einstein Condensation. J. Stat. Phys. **113**, 623 (2003)
63. Gaida, M.: Criterion for Bose-Einstein condensation in a harmonic trap in the case with attractive interactions. Phys. Rev. A **73**, 023603 (2006)
64. Streltsov, A.I., Alon, O.E., Cederbaum, L.S.: Formation and Dynamics of Many-Boson Fragmented States in One-Dimensional Attractive Ultracold Gases. Phys. Rev. Lett. **100**, 130401 (2008)
65. Streltsov, A.I., Alon, O.E., Cederbaum, L.S.: Swift Loss of Coherence of Soliton Trains in Attractive Bose-Einstein Condensates. Phys. Rev. Lett. **106**, 240401 (2011)
66. Doria, P., Calarco, T., Montangero, S.: Optimal Control Technique for Many-Body Quantum Dynamics. Phys. Rev. Lett. **106**, 190501 (2011); Caneva, T., Calarco, T., Montangero, S.: Chopped random-basis quantum optimization. Phys. Rev. A **84**, 022326 (2011)
67. Bhattacharyya, K.: Quantum decay and the Mandelstam-Tamm-energy inequality. J. Phys. A **16**, 2993 (1983); Pfeifer. P.: How fast can a quantum state change with time? Phys. Rev. Lett. **70**, 3365 (1993); Margolus, N., Levitin, L.B.: The maximum speed of dynamical evolution. Physica D **120**, 188 (1998); Giovannetti, V., Lloyd, S., Maccone, L.: Quantum limits to dynamical evolution. Phys. Rev. A **67**, 052109 (2003); Levitin, L.B., Toffoli, T.: Fundamental Limit on the Rate of Quantum Dynamics: The Unified Bound Is Tight. Phys. Rev. Lett. **103** 160502 (2009)
68. Krönke, S., Cao, L., Vendrell, O., Schmelcher, P.: Non-equilibrium quantum dynamics of ultra-cold atomic mixtures: the multi-layer multi-configuration time-dependent Hartree method for bosons. New J. Phys. **15**, 063018 (2013); Cao, L., Krönke, S., Vendrell, O., Schmelcher, P.: The multi-layer multi-configuration time-dependent Hartree method for bosons: theory, implementation, and applications. J. Chem. Phys. **139**, 134103 (2013)

PAMOP Project: Petaflop Computations in Support of Experiments

B.M. McLaughlin, C.P. Ballance, M.S. Pindzola, S. Schippers, and A. Müller

Abstract Our computation effort is primarily concentrated in support of current and future measurements being carried out at various synchrotron radiation facilities around the globe and for charge transfer experiments. In our work we solve the Schrödinger or Dirac equation for the appropriate collision problem using the R-matrix or R-matrix with pseudo-states approach from first principles. The time dependent lattice (TDL) method is also used in our work on charge exchange. A brief summary of the methodology and ongoing developments implemented in the R-matrix suite of Breit-Pauli and Dirac-Atomic R-matrix codes (DARC) is presented. We illustrate vividly the sophistication level of large scale petaflop computations necessary to model accurately the spectra currently being measured at various synchrotron radiation facilities. The new Cray XC40 architecture installed at HLRS is playing a vital role in our computational effort.

1 Introduction

Our research efforts continue to focus on the development of computational methods to solve the Schrödinger and Dirac equations for atomic and molecular collision processes. Access to leadership-class computers such as the Cray XC40 at HLRS allows us to benchmark our theoretical solutions against dedicated collision experiments at synchrotron facilities such as the Advanced Light Source (ALS),

B.M. McLaughlin (✉)
Centre for Theoretical Atomic, Molecular and Optical Physics (CTAMOP), School of Mathematics & Physics, The David Bates Building, Queen's University, 7 College Park, Belfast BT7 1NN, UK
e-mail: b.mclaughlin@qub.ac.uk; bmclaughlin899@btinternet.com

C.P. Ballance • M.S. Pindzola
Department of Physics, 206 Allison Laboratory, Auburn University, Auburn, AL 36849, USA
e-mail: ballance@physics.auburn.edu; pindzola@physics.auburn.edu

S. Schippers • A. Müller
Institut für Atom-und Molekülphysik, Justers-Liebig-Universität Giessen, 35392 Giessen, Germany
e-mail: Stefan.E.Schippers@iamp.physik.uni-giessen.de; Alfred.Mueller@iamp.physik.uni-giessen.de

© Springer International Publishing Switzerland 2016
W.E. Nagel et al. (eds.), *High Performance Computing in Science and Engineering '15*, DOI 10.1007/978-3-319-24633-8_4

Astrid II, BESSY II, SOLEIL and PETRA III and to provide atomic and molecular data for ongoing research in laboratory and astrophysical plasma science. In order to have direct comparisons with experiment, semi-relativistic or fully relativistic computations, involving a large number of target-coupled states are required to achieve spectroscopic accuracy. These computations could not be even attempted without access to high performance computing (HPC) resources such as those available at leadership computational centers in Europe (HLRS) and the USA (NERSC, NICS and ORNL). We use the R-matrix and R-matrix with pseudo-states (RMPS) methods to solve the Schrödinger and Dirac equations for atomic and molecular collision processes.

Satellites such as *Chandra* and *XMM-Newton* are currently providing a wealth of X-ray spectra on many astronomical objects, but a serious lack of adequate atomic data, particularly in the K-shell energy range, impedes the interpretation of these spectra. With the impending launch of the Astro-H satellite in the spring of 2016, X-ray cross section data for a variety of atomic species of prominent astrophysical interest will be of paramount importance (Kallman 2015, private communication).

The motivation for our work is multi-fold; (a) Astrophysical Applications [11, 25, 28, 38, 41], (b) Fusion and plasma modelling [69, 70], (c) Fundamental interest and (d) Support of experimental measurements [6, 15] and Satellite observations. In the case of heavy atomic systems [39, 40], little atomic data exists and our work provides results for new frontiers on the application of the R-matrix; Breit-Pauli and DARC parallel suite of codes. Our highly efficient R-matrix codes are widely applicable to the support of present experiments being performed at synchrotron radiation facilities. Examples of our large scale petaflop computations for cross sections are presented below in order to illustrate the predictive nature of the methods employed as compared to experiment.

The main question asked of any method is, how do we deal with the many body problem? In our case we use first principle methods (ab initio) to solve our dynamical equations of motion. Ab initio methods provide highly accurate, reliable atomic and molecular data (using state-of-the-art techniques) for solving the Schrödinger and Dirac equation. The R-matrix non-perturbative method is used to model accurately a wide variety of atomic, molecular and optical processes such as; electron impact ionization (EII), electron impact excitation (EIE), single and double photoionization and inner-shell X-ray processes. The R-matrix method provides highly accurate cross sections and rates used as input for astrophysical modeling codes such as; CLOUDY, CHIANTI, AtomDB, XSTAR necessary for interpreting experiment/satellite observations of astrophysical objects and fusion and plasma modeling for JET and ITER.

2   Parallel R-Matrix Photoionization

The use of massively parallel architectures allows one to attempt calculations which previously could not have been addressed. This approach enables large scale relativistic calculations for trans-iron elements such as; Kr-ions, Xe-ions, Se-ions

[39, 40] and W-ions [2, 4]. It allows one to provide atomic data in the absence of experiment and takes advantage of the linear algebra libraries available on most architectures. We fill in our *"sea of ignorance"* i.e. provide data on atomic elements where none have previously existed. The present approach has the capability to cater for Hamiltonian matrices in excess of 400 K × 400 K. Examples are presented for both valence and inner-shell photoionization for systems of prime interest to astrophysics and for complex species necessary for plasma modeling in fusion tokamaks.

Further developments and refinements of the dipole codes benefit from similar modifications and developments made to the existing excitation R-matrix codes [40, 42, 44]. In this case all the eigenvectors from a pair of dipole allowed symmetries are required for bound-free dipole matrix formation. Every dipole matrix pair is carried out concurrently with weighted groups of processors assigned to an individual dipole. The method is applicable to photoionization, dielectronic recombination or radiation damped excitation and now reduces to the time taken for a single dipole formation. The method so far implemented on various parallel architectures has the capacity to cater for photoionization calculations involving ∼1500 levels. This dramatically improves (a) the residual ion structure, (b) ionization potential, (c) resonance structure and (d) can deal with in excess of 8000 close-coupled channels.

3 X-ray and Inner-Shell Processes

3.1 The Atomic Oxygen Iso-Nuclear Sequence

Theoretical studies on K-shell photoionization cross sections of neutral nitrogen and oxygen agree well with high resolution measurements made at the Advanced Light Source (ALS) radiation facility [45, 67]. Similarly cross section calculations along the atomic nitrogen iso-nuclear sequence compare favourably with high resolution measurements made at the SOLEIL synchrotron facility [1, 20, 21]. In fact the majority of the high-resolution experimental data from third generation light sources show excellent agreement with the state-of-the-art R-matrix method and with other modern theoretical approaches.

Here we concentrate on the atomic oxygen iso-nuclear sequence for the energy region in the vicinity of the K-shell. Absolute cross sections for the single and double K-shell photoionization of C-like (O^{2+}) and N-like (O^{+}) ions were measured in the 526 to 620 eV photon energy range by employing the ion-photon merged-beam technique at the SOLEIL synchrotron radiation facility [7]. High-resolution spectroscopy up to $E/\Delta E \approx 5300$ was achieved. Rich resonance structures observed in the experimental spectra (see Figs. 1 and 2) are analyzed and identified with the aid of R-matrix and MCDF methods. For these two atomic oxygen ions the strong $1s \rightarrow 2p$ and the weaker $1s \rightarrow np(n > 2)$ resonances observed are characterized [7]. A detailed comparison of the energies of the $1s \rightarrow 2p$ resonances

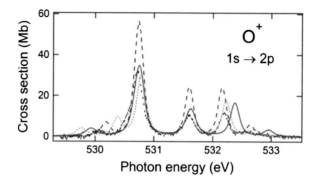

Fig. 1 Comparison in the K-shell region for the $1s \to 2p$ transitions in the O^+ ion for the single photoionization cross section measured with 100 meV band pass (*black points*) with the results of the MCDF (*blue dashed line*) and R-matrix RMPS (*red continuous line*) calculations with the previous optical potential R-matrix results of Garcia and co-workers [18] (*green dotted line*). The theoretical cross sections were reconstructed assuming relative populations of 40 % $^4S^o$, 40 % $^2D^o$ and 20 % $^2P^o$, then convolved with a 100 meV FWHM Gaussian profile. Note, the MCDF cross section calculation has been shifted by 1 eV towards higher energies [7]. The designation of these $1s \to 2p$ strong resonances (positions, widths and strengths) are presented in Table 1

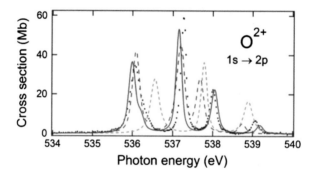

Fig. 2 Comparison in the K-shell region of the $1s \to 2p$ transitions in the O^{2+} ion for single photoionization cross section (*black points*) measured with 110 meV band pass with the results from MCDF (*blue dashed line*), R-matrix RMPS (*red continuous line*) calculations, and previous optical potential R-matrix results of Garcia and co-workers [18] (*green dotted line*). We assume relative populations of 80 % ^3P, 14 % ^1D, 4 % ^1S and 2 % $^5S^o$. The theoretical cross sections have been convolved by a 110 meV FWHM Gaussian profile [7]. The designation of these $1s \to 2p$ strong resonances (positions, widths and strengths) are given in Table 2

in the first members of the oxygen iso-nuclear sequence measured by synchrotron based experiments were made with the observations taken by the Chandra and XMM-Newton X-ray satellites [7].

The first experimental investigation on the atomic oxygen iso-nuclear sequence was performed for B-like atomic oxygen ions at the SOLEIL synchrotron radiation facility in Saint Aubin, France [43]. The results show suitable agreement for the photoionization cross sections produced by X-rays in the vicinity of the K-edge,

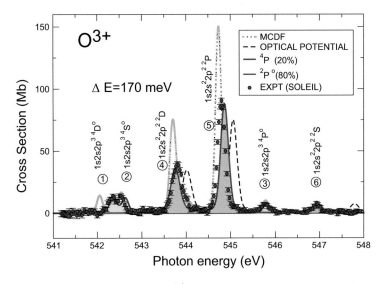

Fig. 3 Photoionization cross sections for O^{3+} ions measured with a 170 meV band pass in the region of 1s → 2p photo-excitations. *Solid points* (*magenta*), experimental cross sections. The error bars give the statistical uncertainty of the experimental data. The R-matrix (RMPS, *solid red line*, ground state, 80 % $^2P^o$, metastable, *blue line* 20 % ^4P). *Dotted line* with *solid green circles* are MCDF calculations (80 % $^2P^o$ and 20 % ^4P). The optical potential R-matrix results (*dashed black line* 80 % $^2P^o$) are from the results of Garcia and co-workers [18]. Theoretical work shown was convoluted with a Gaussian profile of 170 meV FWHM and a weighting of the ground and metastable states (see text for details) to simulate the measurements [43]

where strong $n=2$ inner-shell resonance states are observed. We note that atomic oxygen ions produced in the SOLEIL synchrotron radiation experiments are not purely in their ground state (see Figs. 1, 2, and 3). One therefore requires cross sections for the ground state and for metastable states present in the beam.

K-shell photoionization contributes to the ionization balance in a more complicated way than outer shell photoionization. In fact K-shell photoionization when followed by Auger decay couples three or more ionization stages instead of two in the usual equations of ionization equilibrium [1, 7, 21, 43].

The R-matrix with pseudo-states method (RMPS) was used to determine all the cross sections (in LS-coupling) with 390 levels of the respective atomic oxygen residual ions included in the close-coupling calculations. In the case of B-like (O^{3+}) oxygen ions, metastable states are present in the parent ion beam, theoretical PI cross-section calculations are required for both the $1s^2 2s^2 2p$ $^2P^o$ ground state and the $1s^2 2s 2p^2$ ^4P metastable states of the O^{3+} ion for a proper comparison with experiment. In order to simulate the experimental measurements cross sections have to be convoluted at the same resolution as experiment with an appropriate weighting of the initial states.

The scattering wavefunctions were generated by allowing three-electron promotions out of selected base configurations. Scattering calculations were performed with 20 continuum functions. In the case of the B-like atomic oxygen (O^{3+}) ion, $1s^2 2s^2 2p$ $^2P^o$ ground and the $1s^2 2s 2p^2$ ^4P metastable states the electron-ion collision problem was solved with a fine energy mesh of 2×10^{-7} Rydbergs ($\approx 2.72\,\mu eV$) to delineate all the resonance features in the PI cross sections.

Similarly, for C-like atomic oxygen (O^{2+}) ions, one requires cross sections for the $1s^2 2s^2 2p^2$ ^3P ground state, the $1s^2 2s^2 2p^2$ $^1D,^1S$ and $1s^2 2s 2p^3$ $^5S^o$ metastable states. Finally, for the case of N-like atomic oxygen (O^+), cross sections for ions in the initial $1s^2 2s^2 2p^3$ $^4S^o$ ground state and the $1s^2 2s^2 22p^3$ $^2D^o,^2P^o$ metastable states, need to be determined [7].

For O^{3+} ions for a direct comparison with the SOLEIL results, cross section calculations were convoluted with a Gaussian function of appropriate width and an admixture of 80 % ground and 20 % metastable states used to best simulate experiment. The peaks found in the theoretical photoionization cross section spectrum were fitted to Fano profiles for overlapping resonances as opposed to the energy derivative of the eigenphase sum method [1, 7, 21, 43]. In the case of O^{2+} ions to compare with the SOLEIL results we assume relative populations of 80 % ^3P, 14 % ^1D, 4 % ^1S and 2 % $^5S^o$ [7]. Finally for the case of O^+ ions an admixture of populations of 40 % $^4S^o$, 40 % $^2D^o$ and 20 % $^2P^o$ was used [7].

The results for all three atomic oxygen ions, O^+, O^{2+} and O^{3+}, are illustrated respectively in Figs. 1, 2 and 3 and in Tables 1 and 2.

The energies of the $1s \rightarrow 2p$ transitions for the first members of the atomic oxygen iso-nuclear series as determined from Chandra observations and synchrotron experiments are summarized in Table 3. While there is good agreement within the uncertainities in the case of O^+ ion, discrepancies observed for neutral oxygen [45] and the O^{3+} ion [43] (the Chandra observations are higher by 0.6 and 1.2 eV, respectively) are also present for the O^{2+} ion [7]. The SOLEIL measurements give results for the resonance lines lower by 0.5 eV than those deduced from the satellite spectra. We note that in the satellite observations, it is assumed that all the elements are in their gas phase. No account is taken of molecular or solid-state (grains) effects which can lead to shifts of 0.5–1.0 eV for the K resonance-line positions. We speculate this may be the difference between the synchrotron measurements and the satellite observations. Spectral resolution of the X-ray observations have a maximum resolving power of 1000 compared to 5300 in the synchrotron measurements. This is insufficient to resolve the three transitions from the ground term in O^{2+} and O^{3+} making uncertain the identification of the lines in the X-ray satellites spectra.

Table 1 N-like atomic oxygen 1s → 2p transitions, energies (eV), natural line widths Γ(meV) and oscillator strengths (f-values) for O+[1s²2s²2p³ ⁴S°,²D°,²P° → 1s2s²2p⁴ ⁴P,²D,²P,²S] lines are presented

Transition $^{2S_i+1}L_i$ → $^{2S_f+1}L_f$	Energy (eV) Merged beam[a]	Energy (eV) Satellite	Energy (eV) R-matrix	Energy (eV) MCDF	Γ (meV) Expt[a]	Γ (meV) R-matrix	f-value Expt[a]	f-value Theory
$^2P^o - {}^2D$	530.054 ± 0.028		529.936[g]	529.179[i]	86 ± 60	159[g]	0.039 ± 0.006	0.049[g]
			529.768[h]	529.930[b]		162[h]		0.048[h]
(line 1)			531.096[j]			134[j]		0.050[j]
								0.056[l]
$^2p^o - {}^2P$	530.522 ± 0.028		530.687[g]	529.738[i]		125[g]	0.036 ± 0.005	0.093[g]
			530.381[h]	530.240[b]		120[h]		0.097[h]
(line 2)			531.441[j]			103[j]		0.098[j]
								0.124[l]
$^4S^o - {}^4P$	530.720 ± 0.180[a]	530.96 ± 0.070[c]	530.764[g]	529.746[i]	158 ± 20	134[g]	0.216 ± 0.032	0.172[g]
	531.000 ± 0.500[b]	531.03 ± 0.100[d]	530.789[h]	531.640[b]		129[h]	0.192[d]	0.177[h]
(line 3)		530.97 ± 0.030[e]	533.132[j]			112[j]		0.184[i]
		530.80 ± 0.050[f]	532.093[l]					0.167[j]
		530.90 ± 0.300[k]						0.244[l]
$^2D^o - {}^2D$	531.579 ± 0.023		531.627[g]	530.602[i]	159 ± 30	163[g]	0.063 ± 0.009	0.088[g]
			531.623[h]	532.310[b]		160[h]		0.091[h]
(line 4)			532.841[j]			134[j]		0.086[j]
								0.123[l]
$^2D^o - {}^2P$	532.190 ± 0.023		532.378[g]	531.162[i]	170 ± 20	126[g]	0.044 ± 0.007	0.094[g]
			532.230[h]	532.620[b]		121[h]		0.095[h]

(continued)

Table 1 (continued)

Transition $^{2S_i+1}L_i \rightarrow {}^{2S_f+1}L_f$	Energy (eV) Merged beam[a]	Energy (eV) Satellite	Energy (eV) R-matrix	Energy (eV) MCDF	Γ (meV) Expt[a]	Γ (meV) R-matrix	f-value Expt[a]	f-value Theory
(line 5)			533.185[j]			103[j]		0.088[j]
								0.126[l]
$^2P^o - {}^2S$	532.641 ± 0.044		532.956[g]	531.641[i]	430 ± 200	154[g]	0.067 ± 0.010	0.040[g]
			532.712[h]	534.050[b]		151[h]		0.040[h]
(line 6)			533.956[j]			128[j]		0.038[j]
								0.056[l]

The SOLEIL experimental values are compared to previous measurements and to results from MCDF and R-matrix calculations, in addition to previous theoretical work

[a] SOLEIL measurements [7]
[b] Spring-8 measurements [30]
[c] Chandra observations [27]
[d] Chandra observations [78]
[e] Chandra observations [36]
[f] Chandra observations [19]
[g] R-matrix with pseudostates [7]
[h] R-matrix Optical-potential [18]
[i] R-matrix [65]
[j] R-matrix [79]
[k] XMM-Newton observations [68]
[l] MCDF [7]

Table 2 C-like atomic oxygen 1s $\rightarrow$ 2p transitions, energies (eV), natural line widths Γ (meV) and oscillator strengths (f-values) for O^{2+} [1s^{2}2s^{2}2p^2 ^{3}P, ^{1}D, ^{1}S $\rightarrow$ 1s^{2}2s^{2}2p^3 ^{3}P$^{\rm o}$, ^{3}D$^{\rm o}$, ^{3}S$^{\rm o}$, ^{1}D$^{\rm o}$, ^{1}P$^{\rm o}$, ^{1}S$^{\rm o}$] and O^{2+} [1s^{2}2s2p^4 ^{5}S$^{\rm o}$ $\rightarrow$ 1s^{2}2s2p^4 ^{5}P] lines are presented

Transition $2s_j + L_i \searrow 2s_j + L_f$	Energy (eV) Merged beam	Energy (eV) Satellite	Energy (eV) R-matrix	Energy (eV) MCDF	Γ (meV) Expt	Γ (meV) Theory	f-value Expt[a]	f-value Theory
^{3}P $-\,^3$D$^{\rm o}$	536.082 ± 0.023[a]	536.635 ± 0.116[c]	535.992[m]	536.090[l]	135 ± 8[a]	146[m]	0.098 ± 0.015	0.117[m]
			540.000[f]	531.836[k]		132[f]		0.119[g]
(line 1)			537.200[g]			128[h]		0.111[i]
			537.976[h]			143[i]		0.137[l]
			535.286[i]			207[j]		
			536.559[j]			186[k]		
^{1}S $-\,^1$P$^{\rm o}$	536.254 ± 0.029[a]		536.190[m]	536.422[l]	163 ± 26[a]	134[m]		0.291[m]
			537.947[h]	538.596[k]		112[h]		0.335[l]
(line 2)						153[k]		
^{5}S$^{\rm o}$ $-\,^5$P			536.235[m]	536.997[l]		72[m]		0.203[m]
			534.000[f]	539.315[k]		87[f]		0.232[l]
			538.243[h]			68[h]		
						186[i]		
^{1}D $-\,^1$D$^{\rm o}$			537.128[m]	537.138[l]		140[m]		0.216[m]
			548.000[f]	535.878[k]		118[f]		0.251[l]
			538.924[h]			116[h]		
						153[k]		
^{3}P $-\,^3$S$^{\rm o}$	537.269 ± 0.180[a]	537.800 ± 0.020[c]	537.156[m]	537.229[l]	102 ± 7[a]	72[m]	0.145 ± 0.022	0.097[m]
	537.408 ± 0.093[b]	536.000 ± 0.100[d]	540.000[f]	533.664[k]		58[f]		0.102[g]
(line 3)		537.950 ± 0.180[e]	538.600[g]			56[h]		0.099[i]
			536.496[h]			61[i]		0.105[l]
			537.784			106[j]		
						107[k]		

(continued)

Table 2 (continued)

Transition $^{2S_j+1}L_i \rightarrow {}^{2S_j+1}L_f$	Energy (eV) Merged beam	Energy (eV) Satellite	Energy (eV) R-matrix	Energy (eV) MCDF	Γ (meV) Expt	Γ (meV) Theory	f-value Expt[a]	f-value Theory
$^3P - {}^3P^o$	538.007 ± 0.022		538.046[m]	537.681[l]	174 ± 6[a]	139[m]	0.078 ± 0.012	0.070[m]
			541.000[f]	533.970[k]		110[f]		0.067[g]
(line 4)			540.700[g]			128[h]		0.080[i]
			540.195[h]			137[i]		0.092[l]
			537.350[i]			213[j]		
			538.888[j]			186[k]		
$^1D - {}^1P^o$	539.068 ± 0.004		539.181[m]	538.744[l]	103 ± 17[a]	134[m]	0.093 ± 0.014	0.072[m]
			545.000[f]	537.979[k]		111[f]		0.087[l]
(line 5)			541.138[g]			110[h]		
						153[k]		

The SOLEIL experimental values are compared to previous measurements and to results from MCDF and R-matrix calculations, in addition to previous theoretical work

[a]SOLEIL measurements [7]
[b]EBIT measurements [23]
[c]Chandra observations [19]
[d]XMM-Newton observations [68]
[e]Chandra observations [36]
[f]R-matrix [64]
[g]R-matrix [65]
[h]R-matrix Optical-potential [80]
[i]R-matrix [59]
[j]R-matrix [18]
[k]MCDF [9]
[l]MCDF [7]
[m]R-matrix with pseudostates [7]

Table 3 Comparison of the energy (in eV) of the 1s → 2p lines for the first members of the atomic oxygen iso-nuclear sequence in their ground state determined from the Chandra satellite observations [19, 36] and synchrotron radiation (SR) based experiments [7, 43, 45]

Element		Gatuzz and co-workers [19]	Liao and co-workers [36]	Synchrotron Experiments
O		527.548 ± 0.022	527.397 ± 0.013	526.790 ± 0.040[a]
O$^+$		530.800 ± 0.045	530.966 ± 0.025	530.720 ± 0.180[b]
O^{2+}	^{3}D	536.635 ± 0.116		536.082 ± 0.203[b]
	^{3}S	537.799 ± 0.023	537.942 ± 0.018	537.269 ± 0.180[b]
	^{3}P			538.007 ± 0.202[b]
O^{3+}	^{2}D			543.823 ± 0.213[c]
	^{2}P		546.263 ± 0.263	545.014 ± 0.181[c]
	^{2}S			547.128 ± 0.128[c]

The uncertainty in the absolute energy are included for completeness
[a] ALS, Stolte and co-workers [45]
[b] SOLEIL, Bizau and co-workers [7]
[c] SOLEIL, Bizau and co-workers, after recalibration [7, 43]

4 Valence Shell Photoionization

4.1 Tungsten (W) Atoms and Ions

Tungsten presently receives substantial scientific interest because of its importance in nuclear-fusion research. Due to its high thermal conductivity, its high melting point, and its resistance to sputtering and erosion tungsten is the favoured material for the wall regions of highest particle and heat load in a fusion reactor vessel [57]. Inevitably, tungsten atoms are released from the walls and enter the plasma. With their high atomic number, Z=74, they do not become fully stripped of electrons and therefore radiate copiously, so that the tolerable fraction of tungsten impurity in the plasma is at most 2×10^{-5} [56]. Understanding and controlling tungsten in a plasma requires detailed knowledge about its collisional and spectroscopic properties. Although not directly relevant to fusion, photoionization of tungsten atoms and ions is interesting because it can provide details about spectroscopic aspects and, as time-reversed photorecombination, provides access to the understanding of one of the most important atomic collision processes in a fusion plasma, electron-ion recombination. R-matrix theory is a tool to obtain information about electron-ion and photon-ion interactions in general. Electron-impact ionization and recombination of tungsten ions have been studied experimentally [35, 48, 66, 71, 74, 75] while there are no detailed measurements on electron-impact excitation of tungsten atoms in any charge state. Thus, the present study on photoionization of these complex systems and comparison of the experimental data with R-matrix calculations provides benchmarks and guidance for future theoretical work on electron-impact excitation.

4.2   Neutral Tungsten

Photoabsorption by neutral tungsten atoms in the gas phase has been studied experimentally by Costello et al. employing the dual-laser-plasma technique [10]. A few years later the production of W^+ and W^{2+} photo-ions from tungsten vapor was observed by Sladeczek et al. [73]. However, no experimental data have been available in the literature for tungsten ions prior to the present project. Direct photoionization of W^{q+} ions is included in the calculations by Trzhaskovskaya et al. [76] as time-reversed radiative recombination but is expected to be only a small contribution to the total photoionization cross section. Theoretical work using many-body perturbation theory (MBPT) has been carried out by Boyle et al. [8] for photoionization of neutral tungsten atoms. Theoretical treatment of photoabsorption using a relativistic Hartree-Fock (RHF) approach was reported by Sladeczek et al. [73] in conjunction with their experiments. Ballance and McLaughlin very recently performed detailed large-scale R-matrix calculations for the neutral tungsten atom [3] using the Dirac-Coulomb R-matrix method as implemented in the DARC codes [13], which the reader should consult for explicit details.

Here large-scale theoretical results are presented for photoionization of neutral atomic tungsten (W) in the photon region from threshold to 100 eV. The theoretical results were obtained from a Dirac-Coulomb R-matrix approach [15, 22, 60]. Neutral tungsten has respectively, the following ground state configurations and associated metastable levels, $5p^6 5d^4 6s^2$ ^5D_J, where $J=0$, 1, 2, 3 and 4. Our DARC calculations are compared with the limited available experimental data (which is not absolute). In the absence of very detailed electron-impact excitation (EIE) experiments for Tungsten, the current photoionization measurements theory provide a road-map for future complimentary EIE calculations.

The calculations used a model that included 645-levels arising from the eight configurations of the W^+ residual ion: namely the $4p^6 4d^{10} 4f^{14} 5s^2 5p^6 5d^4 6s$, $4p^6 4d^{10} 4f^{14} 5s^2 5p^6 5d^3 6s^2$, $4p^6 4d^{10} 4f^{14} 5s^2 5p^5 5d^4 6s^2$, $4p^6 4d^{10} 4f^{14} 5s 5p^6 5d^4 6s^2$, we opened the 4f-shell $4p^6 4d^{10} 4f^{13} 5s^2 5p^6 5d^4 6s^2$, $4p^6 4d^{10} 4f^{14} 5s^2 5p^6 5d^5$, the 4d-shell $4p^6 4d^9 4f^{14} 5s^2 5p^6 5d^4 6s^2$ and finally the 4p-shell $4p^5 4d^{10} 4f^{14} 5s^2 5p^6 5d^4 6s^2$.

For the ground and metastable initial states, of neutral W studied here, the outer region electron-ion collision problem was solved (in the resonance region below and between all thresholds) using a suitably chosen fine energy mesh of 1.2×10^{-4} Rydbergs (≈ 1.6 meV) to resolve any resonance structure in the appropriate photoionization cross sections. The jj-coupled Hamiltonian diagonal matrices were adjusted so that the theoretical term energies matched the recommended experimental values of NIST [29].

In Fig. 4, we compare the ground term statistically averaged PI cross section result with the experiments of Costello and co-workers [10] and those of Haensel and co-workers [24]. At 35 eV, the dual laser experiment exhibits the onset of 4f ionization at approximately 3–4 eV before both the theoretical values of the MBPT [8] and the present DARC PI cross section calculations, however still within the

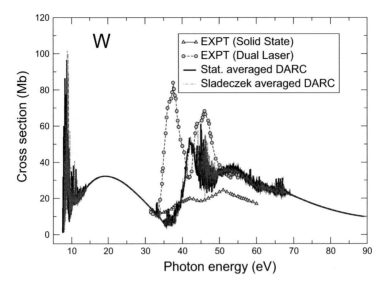

Fig. 4 Single photoionization of neutral W over the photon energy range 8–100 eV, comparing weighted averaged theoretical calculations with two different experiments. The DARC results (645-levels approximation, *solid black line*), are the statistical average of the five levels associated the ^5D term ground state. The (*dashed red-line*) is the weighted DARC results of the lowest six levels, employing the mixing coefficients reported by Sladeczek and co-workers [73]. *Solid circles* with *dashed line* are the dual-laser experimental results of Costello and co-workers [10] and the *solid triangles* are the results from the experimental work of Haensel and co-workers [24]. The statistically averaged DARC PI cross sections were Gaussian convolved at a FWHM of 250 meV

range of the other higher levels of the $6s^2$ configuration as reported in the NIST energy level table. This suggests that the metastable component of the experiment of Costello and co-workers [10] may include some of these higher levels. The Haensel experiment [24], from a solid-state target does not exhibit the strong 4f-5d, 5p-5d resonance structure to the same extent compared to all presented theories and the dual-laser experiment. The Sladeczek experiment [73] does not provide absolute values either, but in terms of a relative minimum to maximum ratio of the measurement is comparable to the Haensel and co-workers [24] experimental result. Sladeczek and co-workers [73] however report a relativistic Hartree-Fock calculation which is a mixture of the first six levels of neutral tungsten. We have employed these same six mixing coefficients with our present DARC PI cross section calculations, represented by the dashed-line in Fig. 4. Not surprisingly, as the individual PI cross sections are remarkably similar, (shifted slightly only by energy differences in the target), it provides a result in close prediction to the statistically averaged DARC result.

4.3   Singly Ionized Tungsten

For comparison with the measurements made at the ALS, state-of-the-art theoretical methods using highly correlated wavefunctions were applied that include relativistic effects [15, 22, 60]. An efficient parallel version [2, 4] of the DARC [13, 58, 77] suite of codes was applied which has been developed [16, 39, 40] to address electron and photon interactions with atomic systems providing for hundreds of levels and thousands of scattering channels. These codes are presently running on a variety of parallel high performance computing architectures world wide [42, 44]. Recently, DARC calculations on photoionization of trans-Fe elements were carried out for Se^+, Kr^+, Xe^+, and Xe^{7+} ions showing suitable agreement with high resolution ALS measurements [26, 39, 40, 49, 54].

For the Ta-like W^+ ion the present study is the first investigation on photoionization of tungsten ions and addresses singly charged W^+. Preliminary reports on our ongoing tungsten photoionization project have been presented at various conferences [51, 52, 55]. Müller and co-workers have recently made detailed measurements on the photoionization cross sections for singly ionized tungsten and compared them with large-scale DARC calculations [53] which are briefly summarised below.

First, we note that the ground level of the Ta-like W^+ ion is $5p^6 5d^4 (^5D) 6s \, ^6D_{1/2}$ with an ionization potential of (16.37 ± 0.15) eV [29]. One must assume that along with the $^6D_{1/2}$ ground level, the excited ground-configuration fine-structure levels ^6D_J with $J=3/2$, 5/2, 7/2 and 9/2 at excitation energies below 0.8 eV [29], respectively, are also populated in an ion source that produces W^+ by electron-impact ionization of neutral tungsten. Also the lowest levels of the first excited $5d^5$ and $5d^3 6s^2$ configurations have excitation energies below 2 eV and are likely populated in the ion-source plasma. The energetically lowest configurations $5p^6 5d^4 6s$, $5d^5$ and $5d^3 6s^2$ all have even parity and, hence, all the 118 excited levels within these configurations are long-lived because electric dipole transitions between any of these levels are forbidden. Any strong signal in the experimental photoionization spectrum below the threshold of (16.37 ± 0.15) eV would indicate the presence of metastable excited states in the parent ion beam, most likely within the $5p^6 5d^4 (^5D) 6s \, ^6D$, $5p^6 5d^5 \, ^6S$ and $5p^6 5d^3 6s^2 \, ^4F$ terms.

A part of the experimental and theoretical results is shown in Figs. 6 and 7. The direct and resonant photoionization processes occurring in the present energy range up to 245 eV for the interaction of a single photon with the ground-state and the lowest metastable configurations of the Ta-like tungsten ion comprise removal or excitation of either a $4f$, $5s$, $5p$, $5d$ or a $6s$ electron. For the theoretical description of W^+ photoionization suitable target wave functions have to be constructed that allow for promotions of electrons from these subshells to all contributing excited states. This is challenging for a low-charge ion such as W^+ but becomes simpler for the ions in higher charge states due to the increased effect of the Coulomb charge of the target and the slight reduction in the R-matrix box size.

Theory predicts narrow resonances at energies up to about 18 eV. Although the step width is too coarse in the experiment there is clear indication of rapid oscillations in the cross section at low photon energies. A relatively smooth energy dependence of the cross section follows at increasing photon energies. The experimental cross section goes over a very broad maximum while theory predicts a monotonically decreasing cross section. Around 35 eV the experimental result shows some structure and also narrow resonances again. The only channel calculated to provide significant amounts of resonance structure in this energy region is that for the ^{6}D term. This is further evidence of a substantial fraction of ions in their lowest-energy term present in the experiment (Fig. 5).

All calculations show the rapid increase of the cross section at photon energies above 35 eV that also characterizes the experimental result. The cross section maximum reached in each of the calculations is in close proximity of the experimental maximum. In the energy range 40 to 55 eV the details of the experimental cross section structure are not closely reproduced by the calculations or a reasonable combination of contributions from the investigated terms. These differences in the details are ascribed to the still very limited basis set of the calculations which was chosen to keep the computational effort manageable.

Beyond 55 eV the experimental cross section drops off rapidly. A similarly rapid decrease is only seen in the calculations for the ^{6}D term. The bump at about 60 eV in the experimental cross section is also seen in the theoretical data for the ^{6}S and ^{4}F terms. At energies beyond 70 eV theory overestimates the experimental single ionization cross section. Parts of the calculated ionization contributions may in fact end up in multiple-ionization channels after relaxation of the photoionized intermediate state formed by the removal of a single electron from W^+.

In energy ranges where narrow resonances could be observed, additional energy scans of the cross section were measured at 50 meV resolution. The most prominent occurrence of narrow features is in the energy range between about 30 and 36 eV. The top panel of Fig. 6 shows energy-scan results normalized to the absolute cross sections. Detailed resonance structure can be seen with the strongest peak feature occurring at about 35.5 eV just before the steep rise in the cross section due to the opening of the $4f$ subshell. The associated energy ranges of features in the theoretical cross sections are shown in the three lower panels. The energy axes of the calculated data were shifted in order to match certain features in the experimental cross section. It was felt that the experimental peak at 35.5 eV might correspond to the broad resonance structures in the ^{6}D and ^{6}S calculations occurring just below the steep rise in the cross section. Therefore, the theoretical energy scales were adjusted by -3.3 and -2.5 eV, respectively. In the ^{4}F calculations no corresponding peak could be found. The energy axis of the ^{4}F spectrum was shifted by -2.1 eV to match the steep onset of the $4f$ contribution to the cross section. Again it is the calculation for the ^{6}D initial term of W^+ ions that matches best with the experiment although the fine details of the measurements are not reproduced by theory.

Experimental and theoretical photoionization cross sections for W^+ ions are presented. The experimental cross sections were measured on an absolute scale

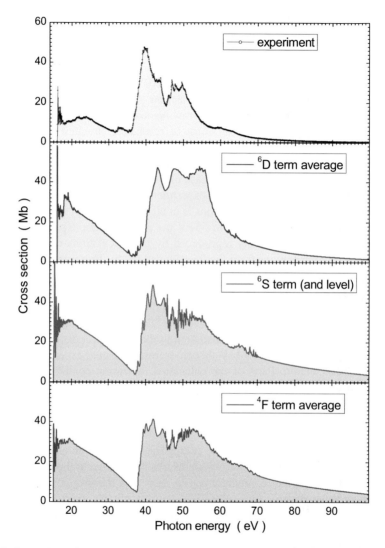

Fig. 5 Comparison of experimental and term-averaged theoretical photoionization cross sections of W^+ ions at 100 meV energy resolution. The three *lower panels* show the theoretical results obtained from 573-level DARC calculations for photoionization from the energetically lowest terms $5d^46s$ ^6D, $5d^5$ ^6S and $5d^36s^2$ ^4F, respectively

employing the photon-ion merged-beam facility at the Advanced Light Source. The comparison of the measured and calculated results is complicated by the possible presence of long-lived excited states in the parent ion beams used for the experiments. More detailed modeling of the experimental data by theory would require calculations for at least all the 119 levels in the lowest configurations of the W^+ ion which is presently beyond the availability of computer resources. There

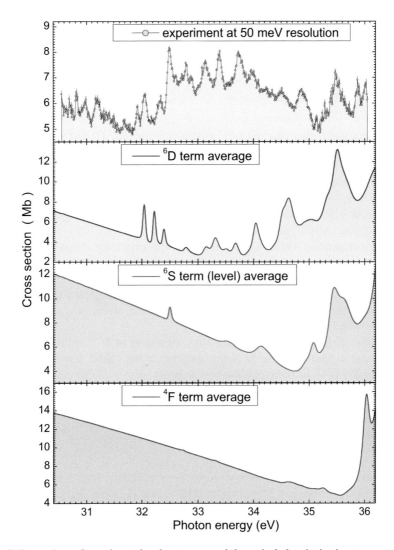

Fig. 6 Comparison of experimental and term-averaged theoretical photoionization cross sections of W^+ ions at 50 meV energy resolution in an energy range where narrow resonances occur in the cross section. The three *lower panels* show the theoretical results obtained from 573-level DARC calculations for the energetically lowest terms $5d^4 6s$ ^6D, $5d^5$ ^6S and $5d^3 6s^2$ ^4F, respectively. The theoretical spectra are shifted in energy by -3.2, -2.5 and -2.1 eV, respectively. For more details see text

are indications, though, in the measured cross section that most of the parent ions were in the ground-state ^6D term. Given the existing limitations and considering the complexity of Ta-like tungsten with its open $5d$ subshell, one can conclude that the main features of the experimental results are reasonably well reproduced by the theoretical calculations. This result for a complex singly charged ion is encouraging for applying a similar theoretical approach to other more highly charged tungsten ions.

5   Charge Exchange

The discovery of highly charged ions in the solar wind [12, 37] and their interaction with interstellar and planetary atoms, has renewed interest in accurate charge exchange data in the astrophysical community. Predicting charge exchange spectra requires knowledge of state selective ($n\ell$) charge transfer cross sections. At high collision energies the classical trajectory Monte-Carlo (CTMC) method [72] is quite accurate. At low collision energies the computational challenging atomic-orbital close-coupling (AOCC) method [17, 34] is needed.

A time-dependent lattice (TDL) method was originally developed to study excitation, ionization, and charge transfer in proton collisions with H atoms [32, 33]. The method was then extended to investigate excitation and charge transfer in proton collisions with laser excited Li atom [61, 63]. This same method was also applied to study charge transfer in proton collisions with He atoms [46] and α-particle collisions with H atoms [47].

Recently, charge exchange spectra were measured in C^{6+} collisions with He atoms using a microcalorimeter X-ray detector at the ORNL ion-atom merged-beams facility [14]. Experiments are now in progress to examine charge exchange spectra in C^{6+} collisions with H atoms.

5.1 C^{6+} -H, and He collisions

The time-dependent lattice method is used to calculate state selective charge transfer cross sections in C^{6+} collisions with H and He atoms [62]. The $C^{6+}(n\ell)$ capture cross sections for energies of 2.7, 5.2, and 8.3 keV/amu are found to be in good agreement with recent atomic orbital close-coupling calculations for H and somewhat larger than previous atomic orbital close-coupling calculations for He. Using standard radiative transition rates, Lyman β/α and Lyman γ/α line ratios are calculated using time-dependent lattice and atomic-orbital close-coupling capture cross sections for both H and He. As illustrated in Fig. 7 the theoretical line ratios for He are found to be in suitable agreement with recent experimental measurements [14]. The interested reader should consult the recent work of Pindzola and Fogle [62] for further details.

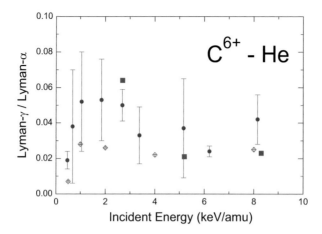

Fig. 7 Lyman-γ/Lyman-α line ratios for C^{6+} -He collisions. *Solid squares* (*red*), time-dependent lattice approach (TDL), *dashed diamonds*, the atomic-orbital close-coupling (AOCC) method [17], and *solid circles* (*blue*) experiment [14]

6 Summary and Future Work

We have vividly illustrated the power of the predictive nature of the R-matrix approach within a non-relativistic or a fully relativistic approach for photoionization cross sections, valence or inner-shell, resonance energy positions, Auger widths and strengths. Access to leadership architectures is essential to our research work such as the Cray XC40 at HLRS which provides an integral contribution to our computational effort.

In future work we intend to extend our R-matrix work to investigate several molecules such as N_2, CO and O_2 for a variety of processes, all of which are of prime interest to aeronomy and astrophysical applications. Furthermore, we draw the reader's attention to our recent work on neutral Sulfur where large-scale DARC cross section calculations were compared to photolysis experiments performed in Berlin [5]. We note that the photolysis experimental work (although contaminated with various molecular precursors) produced rich resonance structures in the experimental spectra that was reproduced well by theory. Large-scale cross sections computations have also been performed for the $2p^{-1}$ removal in Si^+ ions by photons [31]. In addition we have ongoing theoretical efforts dedicated to the modelling of cross sections for $S^+(2s^{-1}$ or $2p^{-1})$ and $SH^+(2\sigma^{-1})$ removal by photons. Finally, recent photoabsorption cross section measurements, for K-shell removal, carried out at PETRA III, on N-like neon at the K-edge, appears to be reproduced well within an RMPS approach, as is that for B-like carbon, all of which will be reported on in due course in the literature.

Acknowledgements A. Müller and S. Schippers acknowledge support by Deutsche Forschungs-gemeinschaft under project numbers M-1068/10 and Mu-1068/20 and through NATO Collaborative Linkage grant 976362. B.M. McLaughlin acknowledges support from the US National Science Foundation through a grant to ITAMP at the Harvard-Smithsonian Center for Astrophysics, under the visitor's program, the RTRA network *Triangle de le Physique* and a visiting research fellowship (VRF) from Queen's University Belfast. C.P. Ballance and M.S. Pindzola acknowledges support by NSF and NASA grants through Auburn University. This research used computational resources at the National Energy Research Scientific Computing Center in Oakland, CA, USA, and at the High Performance Computing Center Stuttgart (HLRS) of the University of Stuttgart, Stuttgart, Germany. The Oak Ridge Leadership Computing Facility at the Oak Ridge National Laboratory, provided additional computational resources, which is supported by the Office of Science of the U.S. Department of Energy under Contract No. DE-AC05-00OR22725. The Advanced Light Source is supported by the Director, Office of Science, Office of Basic Energy Sciences, of the US Department of Energy under Contract No. DE-AC02-05CH11231.

References

1. Al Shorman, M.M., Gharaibeh, M.F., Bizau, J.M., Cubaynes, D., Guilbaud, S., El Hassan, N., Miron, C., Nicolas, C., Robert, E., Sakho, I, Blancard, C., McLaughlin, B.M.: K-Shell Photoionization of Be-like and Li-like Ions of atomic nitrogen: experiment and theory. J. Phys. B At. Mol. Opt. Phys. **46**, 195701 (2013)
2. Ballance, C.P., Griffin D.C.: Relativistic radiatively damped R-matrix calculation of the electron-impact excitation of W^{46+}. J. Phys. B At. Mol. Opt. Phys. **39**, 3617 (2006)
3. Ballance, C.P., McLaughlin, B.M.: Photoionization of the valence shells of the neutral tungsten. J. Phys. B At. Mol. Opt. Phys. **48**, 085201 (2015)
4. Ballance, C.P., Loch S. D., Pindzola M.S., Griffin D.C.: Electron-impact excitation and ionization of W^{3+} for the determination of tungsten influx in a fusion plasma. J. Phys. B At. Mol. Opt. Phys. **46**, 055202 (2013)
5. Barthel, M., Flesch, R., Rühl, E., McLaughlin, B.M.: Photoionization of the $3s^2 3p^4\ ^3P$ and the $3s^2 3p^4\ ^1D,^1S$ states of sulfur: experiment and Theory. Phys. Rev A **91**, 013406 (2015)
6. Bizau, J.M., Esteva, J.-M., Cubaynes, D., Wuilleumier, F.J, Blancard, C., Compant La Fontaine, A., Couillaud, C., Lachkar, J., Marmoret, R., Rémond, C., Bruneau, J., Hitz, D., Ludwig, P., Delaunay, M.: Photoionization of highly charged ions using an ECR ion source and undulator radiation. Phys. Rev. Lett. **84**, 435 (2000)
7. Bizau, J.M., Cubaynes, D., Guilbaud, S., Al Shorman, M.M., Gharaibeh, M.F., Ababneh, I.Q., Blancard, C., McLaughlin, B.M.: K-shell photoionization of O^+ and O^{2+} ions: experiment and theory. Phys. Rev. A **92**, 023401 (2015)
8. Boyle, J., Altun, Z., Kelly H.P.: Photoionization cross-section calculation of atomic tungsten. Phys. Rev. A **47**, 4811 (1993)
9. Chen, M.H., Reed, K.J., McWilliams, D.M., Guo, D.S., Barlow, L., Lee, M., Walker, V.: K-shell Auger and radiative Transitions in the Carbon Isoelectronic Sequence $6 \leq Z \leq 54$. At. Data Nucl. Data Tables **65**, 289 (1997)
10. Costello, J.T., Kennedy, E.T., Sonntag, B.F., Cromer, C.L.: XUV Photoabsorption of laser-generated W and Pt vapours. J. Phys. B At. Mol. Opt. Phys. **24**, 5063 (1991)
11. Covington, A.M., Aguilar, A., Covington, I.R., Hinojosa, G., Shirley, C.A., Phaneuf, R.A., Álvarez, I., Cisneros, C., Dominguez-Lopez, I., Sant'Anna, M. M., Schlachter, A.S., Ballance, C.P., McLaughlin, B.M.: Valence-shell photoionization of chlorine like Ar^+ ions. Phys. Rev. A **84**, 013413 (2011)
12. Cravens, T.E.: Comet Hyakutake X-ray source: Charge transfer of solar wind heavy ions. Geophys. Res. Lett. **24**, 105 (1997)
13. DARC codes http://connorb.freeshell.org

14. Defray, X., Morgan, K., McCammon, D., Wulf, D., Andrianarijaona, V.M., Fogle, M., Seely, D.G., Draganic, I.N., Havener, C.C.: X-ray emission measurements following charge exchange between C^{6+} and He. Phys. Rev. A **88**, 052702 (2013)

15. Dyall, K.G., Grant, I.P., Johnson, C.T., Plummer, E.P.: GRASP: A general-purpose relativistic atomic structure program. Comput. Phys. Commun. **55**, 425 (1989)

16. Fivet, V., Bautista, M.A., Ballance, C.P.: Fine-structure photoionization cross sections of Fe II. J. Phys. B At. Mol. Opt. Phys. **45**, 035201 (2012)

17. Fritsch, W., Lin, C.D.: Atomic-orbital expansion study for the (quasi-)two-electron collision system O^{6+} + He and C^{6+} + He. J. Phys. B At. Mol. Phys. **19**, 2683 (1986)

18. Garcia, J., Mendoza, C., Bautista, M.A., Gorczyca, T.W., Kallman, T.R., Palmeri P.: K-shell photoabsorption of oxygen ions. Astrophys. J. Suppl. Ser. **158**, 68 (2005)

19. Gatuzz, E., Garcia, J. Mendoza, C., Kallman, T.R., Witthoeft, M., Lohfink, A., Bauitista, M.A., Palmeri, P., Quinet, P.: Photoionization modeling of oxygen K absorption in the interstellar medium: The Chandra grating spectra of XTE J1817-330. Astrophys. J. **768**, 60 (2013)

20. Gharaibeh, M.F., Bizau, J.M., Cubaynes, D., Guilbaud, S., El Hassan, N., Al Shorman, M. M., Miron, C., Nicolas, C., Robert, E., Blancard, C., McLaughlin, B.M.: K-shell photoionization of singly ionized atomic nitrogen: experiment and theory. J. Phys. B At. Mol. Opt. Phys. **44**, 175208 (2011)

21. Gharaibeh, M.F., El Hassan, N., Al Shorman, M.M., Bizau, J.M., Cubaynes, D., Guilbaud, S., Blancard, C., McLaughlin, B.M.: K-shell photoionization of B-like atomic nitrogen ions: experiment and theory. J. Phys. B At. Mol. Opt. Phys. **47**, 065201 (2014)

22. Grant, I.P.: Quantum Theory of Atoms and Molecules: Theory and Computation. Springer, New York (2007)

23. Gu, M.F., Schmidt, M., Beiersdorfer, P., Chen, H., Thorn, D.B., Träbert E, Behar, E., Kahn, S.M.: Laboratory measurement and theoretical modeling of K-shell X-ray lines from inner-shell excited and ionized ions of oxygen. Astrophys. J. **627**, 1066 (2005)

24. Haensel, R., Radler, K., Sonntag, B., Kunz, C.: Optical absorption measurements of tantalum, tungsten, rhenium and platinum in the extreme ultraviolet. Solid State Commun. **7**, 1495 (1969)

25. Hasoglu, M.F., Abdel Naby, S.A., Gorczyca, T.W., Drake J.J., McLaughlin, B.M.: K-shell photoabsorption studies of the carbon isonuclear sequence. Astrophys. J. **724**, 1296 (2010)

26. Hinojosa, G., Covington, A.M., Alna'Washi, G.A., Lu, M., Phaneuf, R.A., Sant'Anna, M.M., Cisneros, C., Álvarez, I., Aguilar, A., Kilcoyne, A.L.D., Schlachter, A.S., Ballance, C.P., McLaughlin, B.M.: Valence-shell single photoionization of Kr^+ ions: experiment and theory. Phys. Rev. A **86**, 063402 (2012)

27. Juett, A.M., Schultz, N.S., Chakrabarty D.: High-resolution x-ray spectroscopy of the interstellar medium structure at the oxygen absorption edge. Astrophys. J. **612**, 308 (2004)

28. Kallman, T.R.: Challenges of plasma modelling: current status and future plans. Space Sci. Rev. **157**, 177 (2010)

29. Kramida, A.E., Ralchenko, Y., Reader, J., NIST ASD Team (2014), NIST Atomic Spectra Database (version 5.2), National Institute of Standards and Technology, Gaithersburg, MD, USA

30. Kawatsura, K., Yamaoka, H., Oura, M., Hayaishi. T., Sekioka, T., Agui, A., Yoshigoe, A., Koike, F.: The $1s - 2p$ resonance photoionization measurement of O^+ ions in comparison with an isoelectronic species Ne^{3+}. J. Phys. B At. Mol. Opt. Phys. **35**, 4147 (2002)

31. Kennedy, E.T., Mosnier, J.-P., Van Kampen, P., Cubaynes, D., Guilbaud, S., Blancard, C., McLaughlin, B.M., Bizau, J.-M.: Photoionization cross sections of the aluminum like Si^+ ion in the region of the $2p$ threshold (94–137 eV). Phys. Rev. A **90**, 063409 (2014)

32. Kolakowska, A., Pindzola, M.S., Robicheaux, F., Schultz, D.R., Wells, J.C.: Excitation and charge transfer in proton-hydrogen collisions. Phys. Rev. A. **58**, 2872 (1998)

33. Kolakowska, A., Pindzola, M.S., Schultz, D.R.: Total electron loss, charge transfer, and ionization in proton-hydrogen collisions at 10–100 keV. Phys. Rev. A **59**, 3588 (1999)

34. Kimura, M., Olson, R.E.: Electron capture to $(n\ell)$ states in collisions of C^{4+} and C^{6+} with He. J. Phys. B At. Mol. Phys. **17**, L713 (1984)

35. Krantz, C., Spruck, K., Badnell, N.R., Becker, A., Bernhardt, D., Grieser, M., Hahn, M., Novotný, O., Repnow, R., Savin, D.W., Wolf, A., Müller, A., and Schippers S.: Absolute rate coefficients for the recombination of open f-shell tungsten ions. J. Phys. Conf. Ser. **488**, 012051 (2014)

36. Liao, J.-Y., Zhang, S.-N., and Yao, Y.: Wavelength Measurements of K Transitions of Oxygen, Neon, and Magnesium with X-ray Absorption Lines. Astrophys. J. **774** 116 (2013)

37. Lisse, C.M., Dennerl, K., Englhauser, J., Harden, M., Marshall, F.E., Mumma, M.J., Petre, R., Pye, J.P., Ricketts, M.J., Schmitt, J., Trümper J., West, R.G.: Discovery of X-ray and Extreme Ultraviolet Emission from Comet C/Hyakutake 1996 B2. Science **274**, 205 (1996)

38. McLaughlin, B.M.: Inner-shell Photoionization, Fluorescence and Auger Yields. In: Ferland, G., Savin, D.W. (eds.) Spectroscopic Challenges of Photoionized Plasma, Astronomical Society of the Pacific, ASP *Conf*. Series, vol. 247, pp. 87. San Francisco (2001)

39. McLaughlin, B.M., Ballance, C.P.: Photoionization cross section calculations for the halogen-like ions Kr^+ and Xe^+. J. Phys. B At. Mol. Opt. Phys. **45**, 085701 (2012)

40. McLaughlin, B.M., Ballance, C.P.: Photoionization cross-sections for the trans-iron element Se^+ from 18 eV to 31 eV. J. Phys. B At. Mol. Opt. Phys. **45**, 095202 (2012)

41. McLaughlin, B.M., Ballance, C.P.: Photoionization, fluorescence and inner-shell processes. In: McGraw-Hill (eds) McGraw-Hill Yearbook of Science and Technology, pp. 281. Mc Graw Hill, New York (2013)

42. McLaughlin, B.M., Ballance, C.P.:Petascale computations for large-scale atomic and molecular collisions. In: Resch, M.M. Kovalenko, Y., Fotch, E., Bez, W., Kobaysahi, H. (eds.) Sustained Simulated Performance 2014, chap. 15. Springer, New York (2014)

43. McLaughlin, B.M., Bizau, J.M., Cubaynes, D., Al Shorman, M.M., Guilbaud, S., Sakho, I., Blancard, C., Gharaibeh, M.F.: K-shell photoionization of B-like (O^{3+}) oxygen ions: experiment and theory. J. Phys. B At. Mol. Opt. Phys. **47**, 115201 (2014)

44. McLaughlin, B.M., Ballance, C.P., Pindzola, M.S., Müller, A.: PAMOP: petascale atomic, molecular and optical collisions. In: Nagel, W.E., Kröner, D.H., Resch, M.M. (eds.) High Performance Computing in Science and Engineering'14, chap. 4. Springer, New York (2014)

45. McLaughlin, B.M., Ballance, C.P., Bown, K.P., Gardenghi, D.J., Stolte, W.C.: High precision k-shell photoabsorption cross sections for atomic oxygen: experiment and theory. Astrophys, J **771**, L8 (2013) & Astrophys. J 779, L31 (2013)

46. Minami, T., Lee, T.G., Pindzola, M.S.: Coherence parameters for charge transfer in collisions of protons with helium calculated using a hybrid numerical approach. J. Phys. B At. Mol. Opt. Phys. **37**, 4025 (2004)

47. Minami, T., Lee, T.G., Pindzola, M.S., Schultz, D.R.: Total and state-selective charge transfer in He^{2+} + H collisions. J. Phys. B At. Mol. Opt. Phys. **41**, 135201 (2008)

48. Müller, A.: Fusion-related ionization and recombination data for tungsten ions in low to moderately high charge states. Atoms **3**, 120 (2015)

49. Müller, A.: Precision studies of deep-inner-shell photoabsorption by atomic ions. Phys. Scr. **90**, 054004 (2015)

50. Müller, A., Schippers, S., Phaneuf, R.A., Scully, S.W.J., Aguilar, A., Cisneros, C., Gharaibeh, M.F., Schlachter, A.S., McLaughlin, B.M.: K-shell photoionization of Be-like Boron (B^+) Ions: experiment and theory J. Phys. B At. Mol. Opt. Phys. **47**, 135201 (2014)

51. Müller, A., Schippers, S., Kilcoyne, A.L.D., Esteves, D.: Photoionization of tungstens ions with synchrotron radiation. Phys. Scr. **T144**, 014052 (2011)

52. Müller, A., Schippers, S., Kilcoyne, A.L.D., Aguilar, A., Esteves, D., Phaneuf, R.A.: Photoion-ization of singly and multiply charged tungsten ions. J. Phys. Conf. Ser. **388**, 022037 (2012)

53. Müller, A., Schippers, S., Hellhund, J., Holosto, K., Kilcoyne, A.L.D., Phaneuf, R.A., Ballance, C.P., McLaughlin, B.M.: Single-photon single ionization of W^+ ions: experiment and theory. J. Phys. B At. Mol. Opt. Phys. **48**, 235203 (2015)

54. Müller, A., Schippers, S., Esteves-Macaluso, D., Habibi, M., Aguilar, A., Kilcoyne, A.L.D., Phaneuf, R.A., Ballance, C.P., McLaughlin, B.M.: High resolution valence shell photoioniza-tion of Ag-like (Xe^{7+}) Xenon ions: experiment and theory. J. Phys. B At. Mol. Opt. Phys. **47**, 215202 (2014)

55. Müller, A., Schippers, S., Hellhund, J., Kilcoyne, A.L.D., Phaneuf, R.A., Ballance, C.P., McLaughlin, B.M.: Single and multiple photoionization of W^{q+} tungsten ions in charged states $q = 1,2,..,5$: experiment and theory J. Phys. Conf. Ser. **488**, 022032 (2014)
56. Neu, R., Dux, R., Geier, A., Gruber, O., Kallenbach, A., Krieger, K., Maier, H., Pugno, R., Rohde, V., Schweizer, S., and ASDEX Upgrade Team: Tungsten as plasma-facing material in ASDEX Upgrade. Fusion Eng. Des. **65**, 367 (2003)
57. Neu, R., Arnoux, G., Beurskens. M., Bobkov, V., Brezinsek, S., Bucalossi, J., Calabro, G., Challis, C., Coenen, J.W., de la Luna, E., de Vries, P.C., Dux, R., Frassinetti. L., Giroud, C., Groth, M., Hobirk, J., Joffrin, E., Lang, P., Lehnen, M., Lerche, E., Loarer, T., Lomas, P., Maddison, G., Maggi, C., Matthews, G., Marsen, S., Mayoral, M.L., Meigs, A., Mertens, P., Nunes, I., Philipps, V., Pütterich, T., Rimini, F., Sertoli, M., Sieglin, B., Sips, A.C.C., van Eester, D., van Rooij, G., JET-EFDA Contributors: First operation with the JET International thermonuclear reactor-like Wall. Phys. Plasmas **20**, 056111 (2013)
58. Norrington, P.H., Grant, I.P: Low-energy electron scattering by Fe XXIII and Fe VII using the Dirac R-matrix method. J. Phys. B At. Mol. Opt. Phys. **20**, 4869 (1987)
59. Olalla, E., Wilson, N.J., Bell, K.L., Martin, I., Hibbert, A.: Inner-shell photoionization of O III. Mon. Not. R. Astron. Soc. **332**, 1005 (2002)
60. Parpia, F., Froese-Fischer, C., Grant, I.P.: GRASP92: a package for large-scale relativistic atomic structure calculations. Comput. Phys. Commun. **94**, 249 (2006)
61. Pindzola, M.S.: Proton-impact excitation of laser-excited lithium atoms. Phys. Rev. A **66**, 032716 (2002)
62. Pindzola, M.S., Fogle, M.: Single charge transfer in C^{6+} collisions with H, He atoms. J. Phys. B At. Mol. Opt. Phys. **48**, 205203 (2015)
63. Pindzola, M.S., Minami, T., Schultz, D.R.: Laser-modified charge-transfer processes in proton collisions with lithium atoms. Phys. Rev A **68**, 013404 (2003)
64. Petrini, D., da Silva, E.P.: Soft X irradiation of low density plasmas: O III and O IV line emission. Rev. Mex. Astron. Astrofis. **32**, 69 (1996)
65. Pradhan, A.K., Chen, G.X., Delahaye, F., Nahar, S.N., Oelgoetz, J.: X-ray absorption via K_{α} resonance complexes in oxygen ions. Mon. Not. R. Astron. Soc. **341**, 1268 (2003)
66. Rausch, J., Becker, A., Spruck, K., Hellhund, J., Borovik Jr, A., Huber, K., Schippers S., and Müller, A.: Electron-impact single and double ionization of W^{17+}. J. Phys. B At. Mol. Opt. Phys. **44**, 165202 (2011)
67. Sant'Anna, M.M., Schlachter, A.S., Öhrwall, G., Stolte, W.C., Lindle, D.W., McLaughlin, B.M.: K-shell X-ray spectroscopy of atomic nitrogen. Phys. Rev. Lett. **107**, 033001 (2011)
68. Sala, G., Greiner, J., Bottacini, E., Haberl, F.: XMM-Newton and INTEGRAL observations of the black hole candidate XTE J1817-330. Astrophys. Space Sci. **309**, 315 (2007)
69. Saloman, E.B.: Energy levels and observed spectral lines of Xenon, Xe I through Xe LIV. J. Phys. Chem. Ref. Data **33**, 765 (2004)
70. Schippers, S., Muller, A., Esteves, D., Habibi, M., Aguilar, A., Kilcoyne, A.L.D.: Experimental absolute cross section for photoionization of Xe^{7+}. J. Phys. Conf. Ser. **194**, 022094 (2009)
71. Schippers, S., Bernhardt, D., Müller, A., Krantz, C., Grieser, M., Repnow, R., Wolf, A., Lestinsky, M., Hahn, M., Novotný, O. Savin, D.W.: Dielectronic recombination of xenon like tungsten ions. Phys. Rev. A **83**, 012711 (2011)
72. Simcic, J., Schultz, D.R., Mawhorter, R.J., Cadez, I., Greenwood, J.B., Chutjian, A., Liesse, C.M., Smith, S.J.: Measurement and calculation of absolute single-and multiple-charge-exchange cross sections for Fe^{q+} ions impacting CO and CO_2. Phys. Rev. A **81**, 062715 (2010)
73. Sladeczek, P., Feist, H., Feldt, M., Martins, M., Zimmermann, P.: Photoionization Experiments with an Atomic Beam of Tungsten in the Region of the $5p$ and $4f$ excitation. Phys. Rev. Lett. **75**, 1483 (1995)
74. Spruck, K., Badnell, N.R., Krantz, C., Novotný, O., Becker, A., Bernhardt, D., Grieser, M., Hahn, M., Repnow, R., Savin, D.W., Wolf, A., Müller, A., Schippers S.: Recombination of W^{18+} ions with electrons: absolute rate coefficients from a storage-ring experiment and from theoretical calculations. Phys. Rev. A **90**, 032715 (2014)

75. Stenke, M., Aichele, K., Harthiramani, D., Hofmann, G., Steidl, M., Völpel, R., and Salzborn E.: Electron-impact single-ionization of singly and multiply charged tungsten ions. J. Phys. B At. Mol. Opt. Phys. **28**, 2711 (1995)
76. Trzhaskovskaya, M.B., Nikulin, V.K., Clark, R.E.H.: Radiative recombination rate coefficients for highly-charged tungstens ions. At. Data Nucl. Data Tables **96**, 1 (2010)
77. Wijesundera, W.P., Parpia, F.A., Grant, I.P., Norrington, P.H.: Electron scattering by Kr XXIX (oxygen-like krypton) using the Dirace R-matrix method. J. Phys. B At. Mol. Opt. Phys. **24** 1803 (1991)
78. Yao, Y., Schultz, N.S., Gu, M.F., Nowak, M.A., and Cainizares, C.: High-resolution X-ray spectroscopy of the multiphase interstellar medium towards Cyg X-2. Astrophys. J. **696** 1418 (2009)
79. Zeng, J., Zhao,G., Yuan, J.: The $1s \rightarrow 2p$ resonance photoionization from the low-lying states of O^+ - Energies, autoionization widths, branching ratios, and oscillator strengths. Eur. Phys. J. D **28**, 163 (2004)
80. Zeng, J., Yuan, J., Lu, Q.: Photoionization of O III low-lying states: autoionization resonances energies and widths of some $1s - 2p$ excited states. J. Phys. B. At. Mol. Opt. Phys. **34**, 2823 (2001)

Monte Carlo Simulation of Crystal-Liquid Phase Coexistence

Antonia Statt, Fabian Schmitz, Peter Virnau, and Kurt Binder

Abstract When a crystal nucleus is surrounded by coexisting fluid in a finite volume in thermal equilibrium, the thermodynamic properties of the fluid (density, pressure, chemical potential) are uniquely related to the surface excess free energy of the nucleus. Using a model for weakly attractive soft colloidal particles, it is shown that this surface excess free energy can be determined accurately from Monte Carlo simulations over a wide range of nucleus volumes, and the resulting nucleation barriers are completely independent from the size of the total volume of the system. A necessary ingredient of the analysis, the pressure at phase coexistence in the thermodynamic limit, is obtained from the interface velocity method. Computing the solid-liquid interface excess free energy via the ensemble switch method, a detailed test of classical nucleation theory is possible (assuming the hypothetical spherical nucleus shape, which is not realized when the crystal nucleus is facetted). Consequences for the interpretation of experiments will be briefly discussed.

1 Introduction

Coexistence of different phases of a material is an ubiquitous phenomenon: snow crystals in the atmosphere coexist with water vapor, and the ice layer on the surface of a pond coexists with liquid water underneath, and similar phenomena are well known in the processing of oxidic or metallic materials, as well as in soft condensed matter. Thermodynamics tells us that coexistence of phases requires that their temperature and chemical potential are equal. If the volume taken by either phase is of macroscopic extent, in equilibrium also the pressure in both phases

A. Statt (✉)
Graduate School Materials Science in Mainz, Staudinger Weg 9, 55099 Mainz, Germany

Institut für Physik, Johannes Gutenberg-Universität, D-55099 Mainz, Staudinger Weg 7, Germany
e-mail: statt@uni-mainz.de

F. Schmitz • P. Virnau • K. Binder
Institut für Physik, Johannes Gutenberg-Universität, 55099 Mainz, Staudinger Weg 7, Germany

© Springer International Publishing Switzerland 2016
W.E. Nagel et al. (eds.), *High Performance Computing in Science and Engineering '15*, DOI 10.1007/978-3-319-24633-8_5

75

is the same, but this is not the case if the volume taken by the minority phase only is of nanoscopic size. For example when a fluid nanodroplet coexists with surrounding vapor, both the vapor pressure p_v and the liquid pressure p_ℓ differ from each other (and from the bulk pressure at phase coexistence p_{coex}) by terms of order $1/R$, with R being the droplet radius. This fact (Laplace pressure effect) is standard textbook knowledge [1], of course; but in the context of simulations and experiments on nanosystems these effects (namely the shift of the pressures relative to p_{coex} in both coexisting phases) can in fact be utilized as a convenient tool to learn about interfacial excess free energies in the system. While the chemical potential μ inside the nucleus and in its environment must be strictly the same in equilibrium, μ in general will also differ from its bulk coexistence value μ_{coex}, hence carrying a related valuable information as well.

These considerations turn out to be particularly useful for the case of phase coexistence between a crystalline nucleus and surrounding fluid [2]: for a crystal-fluid interface, the interfacial tension $\gamma(\mathbf{n})$ depends on the orientation vector $\mathbf{n}$ of the interface normal relative to the lattice axes. As a consequence, the nucleus shape in equilibrium in general is non-spherical, and needs to be found from $\gamma(\mathbf{n})$ by the famous Wulff construction [3]. Unfortunately, this explicit solution for the crystal nucleus shape in general is of purely formal character, since $\gamma(\mathbf{n})$ in three-dimensional systems is hardly ever known. Even in the simplistic lattice gas model on the simple cubic lattice, where the anisotropy of $\gamma(\mathbf{n})$ can be controlled by varying the temperature, $\gamma(\theta, \varphi)$ (where θ, φ are two polar angles to characterize $\mathbf{n}$) is still unknown: only $\gamma(\theta, 0)$ could recently be calculated based on a large computational effort using highly efficient codes on Graphics Processing Units (GPUs) [4]. So in a general case (in $d = 3$ dimensions) the shape of a crystalline nucleus is not explicitly known.

This fact creates a major challenge for the theory of nucleation [5, 6]. According to classical theory, in a nucleation event a free energy barrier ΔF^* needs to be overcome, which results from a competition between an (unfavorable) surface term and a (favorable) volume term. In terms of the nucleus volume V, the volume term is simply $-(p_c - p_\ell)V$, where p_c is the pressure in the crystalline nucleus, and p_ℓ the pressure in the liquid surrounding it. Hence the free energy cost to create a nucleus of volume V is

$$\Delta F = -(p_c - p_\ell)V + F_{surf}(V) = -(p_c - p_\ell)V + A_w \gamma V^{2/3}, \qquad (1)$$

where the surface term $F_{surf}(V)$ is written as an integral over the surface area A_w of a crystal nucleus of unit volume given by Wulff construction, γ then is an average surface tension

$$F_{surf}(V) = V^{2/3} \int_{A_w} \gamma(\mathbf{n}) d\mathbf{s} \equiv A_w \gamma V^{2/3} \qquad . \qquad (2)$$

Here we have assumed that the nucleus is large enough that $F_{surf}(V) \propto V^{2/3}$.

If $\gamma(\mathbf{n}) = \gamma$ is independent of $\mathbf{n}$, A_w is the surface area of a unit sphere, and we simply would have $F_{\text{surf}} = 4\pi R^2 \gamma = A_{\text{iso}} \gamma V^{2/3}$ with $A_{\text{iso}} = (36\pi)^{1/3}$.

Now the barrier ΔF^* occurs for $V = V^*$ with $\partial(\Delta F(V))/\partial V \mid_{V*} = 0$, which yields

$$V^* = \left[\frac{2A_w\gamma}{3(p_c - p_\ell)}\right]^3 \quad , \quad \Delta F^* = \frac{1}{3}A_w\gamma V^{*2/3} \quad . \tag{3}$$

Equation (3) as it is written would require the knowledge of both γ and the area A_w that follows from the Wulff construction. However, combining both equations one notes, that the product $A_w\gamma$ can be eliminated, writing

$$\Delta F^* = \frac{1}{2}\left(p_c - p_\ell\right)V^* \quad . \tag{4}$$

Equation (4) shows that the problem of obtaining A_w and γ can be completely circumvented in a simulation, when one determines both the nucleus volume V^* and the pressure difference $p_c - p_\ell$ for the surrounding liquid that is in equilibrium with the nucleus. When the volume of the surrounding liquid is macroscopic, this equilibrium is unstable (it corresponds to a saddle point in configuration space).

However, for a suitably chosen simulation box with a finite volume V_{box} this equilibrium can be shown to be stable [7, 8] in the canonical ensemble (particle number N and temperature T being fixed, as well as V_{box}). For densities $\rho = N/V_{\text{box}}$ in between the onset density of freezing (ρ_f) and of melting (ρ_m), one uses a finite-size generalization of the well-known lever rule of two-phase coexistence [1]

$$\rho V_{\text{box}} = \rho_\ell(p_\ell)(V_{\text{box}} - V^*) + \rho_c(p_c)V^* \quad . \tag{5}$$

For appropriate choices of ρ and V_{box}, the virial formula [1] is used to estimate p_ℓ (in the fluid region far outside of the crystalline nucleus) and from simulations of bulk systems (in the NpT ensemble) the equation of state for both the liquid $\{\rho_\ell(p)\}$ and the crystal $\{\rho_c(p)\}$ are known. For $V_{\text{box}} \to \infty$, we would have $p_\ell = p_c = p_{\text{coex}}$ and $\rho_\ell(p_{\text{coex}}) = \rho_f$, $\rho_c(p_c) = \rho_m$, both phases having the chemical potential μ_{coex}. For finite V_{box}, μ at coexistence, and the pressures p_ℓ, p_c are enhanced as well, but since these quantities can be estimated in the simulations, V^* can readily be extracted from Eq. (5). Note also that for this estimation of V^* a "microscopic" identification of which particles belong to the crystal (by the Steinhardt-Lechner-Dellago bond orientational order parameters [9, 10]) and which belong to the liquid is NOT needed. (There is a considerable ambiguity in this identification due to particles in the interfacial region! (Statt, 2015, Dissertation, Johannes Gutenberg Universität Mainz, unpublished; [11])

This novel concept for the estimation of crystal nucleation free energy barriers, outlined in Eqs. (1)–(5), has been implemented in large scale simulations of a soft version of the effective Asakura-Oosawa (AO) model [12, 13]. This model will be explained in Sect. 2, while in Sect. 3 we describe auxiliary methods to estimate p_{coex} (using the "interface velocity method"[14]) and a novel method for sampling the chemical potential for a fluid at melt densities [2], where standard Widom [15] particle insertion methods are inapplicable. The need for accessing a fast supercomputer such as HERMIT and HORNET at the HLRS Stuttgart will be explained. Section 4 then contains the core results of our study, simulations of phase coexistence in the two-phase coexistence region, and results on the crystal-fluid interface tension for particular interface orientations [(111), (110), (100)] which help to ascertain the role of interface tension anisotropy in this problem. Section 5 then gives a summary and outlook on further work.

2 The Model and Its Bulk Phase Behavior

Colloid-polymer mixtures are model systems that are particularly attractive both for experiment [16] and simulation [13]: Due to the large size of the colloids, one can visualize the structure of colloidal systems by confocal microscopy, and the slowness of their kinetics has allowed extensive studies of nucleation phenomena [17]. Adding polymers to a suspension containing hard sphere-like colloids creates a depletion attraction between them [16], and the strength and range of the effective attractive potential can be controlled by the fugacity and the size of the (flexible) polymers (which must be chosen such that they do not adsorb on the colloids and are under good solvent conditions in the solution).

Of course, a microscopically realistic modelling of such a system is neither feasible nor desirable. It is not feasible, since the size of the solvent particles and of the monomers of the macromolecules is less than a nm, while the size of the colloids is a factor 10^3 larger (μm). Also real colloids have some polydispersity in size, and do not act like perfect hard spheres [18]. Therefore in the theoretical modelling we do not need to use the strict AO model [12, 13], which describes the colloids as perfect hard spheres of diameter σ_c and the polymers as soft spheres of diameter σ_p (which must not overlap with the colloids, but with respect to the polymer-polymer interactions are treated like an ideal gas). Due to the singularity of the hard sphere potential this is computationally not convenient, when we wish to estimate the pressure. Thus, we choose a "soft" variant of the AO model, with an effective potential between the colloids which is everywhere continuous. Note that already in the original AO model the polymer degrees of freedom can be integrated out exactly, as long as $q = \sigma_p/\sigma_c \leq q^* = 2/\sqrt{3}-1 \approx 0.1547$ [19]. This "effective" AO model (Eff AO) has the following potential [$\beta = 1/k_B T$]

$$U_{\text{eff}}(r<\sigma_c)=\infty, \qquad (6)$$

$$\beta U_{\text{eff}}(\sigma_c<r<\sigma_c+\sigma_p)=\frac{\pi}{6}\sigma_p^3 z_p\left(1+q^{-1}\right)^3\left[1-\frac{3r/\sigma_c}{2(1+q)}+\frac{(r/\sigma_c)^3}{2(1+q)^3}\right], \qquad (7)$$

$$U_{\text{eff}}(r>\sigma_c+\sigma_p)\equiv 0, \qquad (8)$$

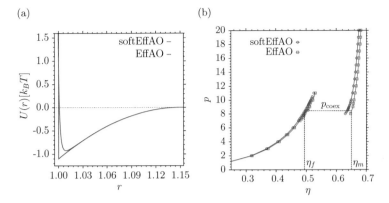

Fig. 1 (a) Potential $U(r)/k_B T$ versus distance r between two colloidal particles, for the Eff AO Model, compared to $U_{\text{eff}}(r)/k_B T$ for the soft effective AO model. Parameters chosen are $q=0.15$; $\eta_p^r=(\pi\sigma_p^3/6)z_p=0.1$. Note $\sigma_c=1$ is the unit length. (b) Normalized pressure $p\sigma_c^3/k_b T$ plotted vs. the packing fraction $\eta=(\pi\sigma_c^3/6)\rho$ of the colloids, comparing the Eff AO and soft Eff AO model. Note that there are two branches, $\eta=\eta_\ell(p)$ for the liquid phase, and $\eta=\eta_c(p)$ for the face centered cubic (fcc) crystal. Note that in the simulations the NpT ensemble is used, so p is the chosen independent thermodynamic variable, and η is obtained from Monte Carlo sampling, choosing a cubic simulation box with $N=4,000$ particles. The coexistence pressure p_{coex} is determined by the interface velocity method (see Sect. 3), and η_f, η_m are the volume fractions corresponding to ρ_f and ρ_m, respectively

z_p being the fugacity of the polymers. As announced above, we replace Eq. (6) by a soft potential

$$U_{\text{rep}}(r)=4\left(\left[\frac{b\sigma_c}{r-\varepsilon\sigma_c}\right]^{12}+\left[\frac{b\sigma_c}{r-\varepsilon\sigma_c}\right]^6-\left[\frac{b\sigma_c}{b\sigma_c+q-\varepsilon\sigma_c}\right]^{12}-\left[\frac{b\sigma_c}{b\sigma_c+q-\varepsilon\sigma_c}\right]^6\right).$$

(9)

A convenient choice of the constants b and ε is $b=0.01$, $\varepsilon=0.98857$. We denote the potential defined by Eqs. (7–9) "soft Eff AO" potential in the following. Figure 1a shows both the Eff AO and the soft Eff AO potential, and Fig. 1b compares their equation of state. As expected, both models yield very similar results.

3 Coexistence Pressure and Estimation of the Chemical Potential

Since for $p_\ell \to p_{\text{coex}}$ both the nucleus volume V^* and the associated free energy barrier diverge, a very accurate estimation of p_{coex} is an indispensable requisite for any meaningful study of nucleation. Note that near p_{coex} we may use the expansions for the chemical potential $\mu_\ell(p_\ell)$ in the liquid branch and $\mu_c(p_c)$ in the crystal branch of the equation of state

$$\mu_\ell(p_\ell)=\mu_{\text{coex}}+\frac{\pi}{6}\frac{1}{\eta_f}\Big(p_\ell-p_{\text{coex}}\Big) \quad , \tag{10}$$

$$\mu_c(p_c)=\mu_{\text{coex}}+\frac{\pi}{6}\frac{1}{\eta_m}\Big(p_c-p_{\text{coex}}\Big) \quad , \tag{11}$$

and since the chemical potentials of the crystal nucleus and surrounding fluid must be the same, we conclude that $p_\ell-p_{\text{coex}}=(\eta_f/\eta_m)(p_c-p_{\text{coex}})$ and hence

$$p_c-p_\ell=(p_c-p_{\text{coex}})\big(1-\eta_f/\eta_m\big)=(p_\ell-p_{\text{coex}})\left(\frac{\eta_m}{\eta_f}-1\right) \tag{12}$$

and hence Eq. (3) implies for p_ℓ (and p_c) near p_{coex} for the linear dimension of the nucleus

$$V^{*1/3}=\frac{2A_w\gamma}{3(p_\ell-p_{\text{coex}})\,(\eta_m/\eta_f-1)} \quad . \tag{13}$$

Thus, particular care was devoted to control both statistical and systematic errors in our determination of p_{coex}; it is this part of the present project where the by far overwhelming part of the computational resources was used.

Following Zykova-Timan et al. [14] we study slab configurations where in the initial state a crystal domain (of volume $V_c=L\times L\times L_c$) and a liquid domain (of volume $V_\ell=L\times L\times(5L-L_c)$) is present, periodic boundary conditions are applied throughout, and thus the domains are separated by two planar $L\times L$ interfaces (Fig. 2a). Since initially $L_c\approx 5L/2$, the two interfaces are far apart from each other, thus rendering them non-interacting. We simulate this configuration for a variety of pressures p in the environment of the (conjectured value) of p_{coex}. At each value of p that is tested, we use the equation of state (Fig. 1b) to choose the proper value of the lattice spacing of the fcc crystal at that pressure, such that an integer number of rows fits to the $L\times L$ interface (of (100) symmetry) without any elastic distortion to the crystal. To prepare the initial state for the NpT simulation of this slab configuration, we actually carry out a run in the NVT ensemble first, to equilibrate the liquid properly also in the crystal-liquid interfacial region. This requires about 10^{10} Monte Carlo moves (i.e. attempted single particle displacements). In the NpT run that follows, single particle moves are complemented by attempted volume changes. We have carried out both runs where all three linear dimensions L_x, L_y and L_z are allowed to fluctuate independently, and runs where $L_x=L_y=L$ were kept fixed at the equilibrium value (the first choice may seem in principle preferable, but it requires an order of magnitude more computing time). The length of these runs was 3×10^4

(a)   (b)

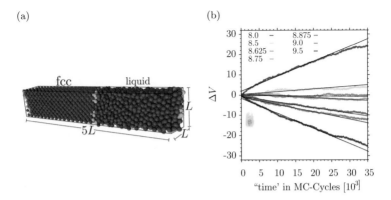

Fig. 2 (**a**) Snapshot picture of a particle configuration for the Eff AO model, using a slab geometry with linear dimensions $L_x \times L_y \times L_z$ with $L_x = L_y = L$ and $L_z = 5L$. The crystal slab (with $n=6$ lattice planes) is shown on the *left* (in *red color*), the liquid on the *right* (in *blue color*), and $N = 4000$ particles were used in total (particles in the interface, where the assignment whether they belong to the crystal or to the liquid is ambiguous, are shown in *green color*). The pressure chosen here is $p\sigma^3/k_B T = 8.0$. (**b**) Volume change ΔV as a function of Monte Carlo (MC) cycles for various pressures $p\sigma_c^3/k_B T$, as indicated in the figure, for the soft Eff AO model. Here a choice of system size with $n=8$ lattice planes was used, requiring $N = 9250$ colloids. The solid straight lines are linear fits of ΔV vs. the MC "time" Δt (100 independent runs being integrated over)

MC cycles (one cycle consisted of N attempted single particle displacement moves and one attempted volume move). Since in general the chosen value of p will differ from p_{coex}, we expect that with increasing number of MC cycles a volume change occurs (Fig. 2b): If $p < p_{\text{coex}}$, the liquid should grow on expense of the crystal, and hence $\Delta V > 0$, while for $p > p_{\text{coex}}$ the crystal should grow on expense of the liquid, and hence $\Delta V < 0$. Of course, in single runs of this simulation of crystal growth or shrinking huge statistical fluctuations occur, and thus we have found it necessary for each choice of p and L to carry out 100 runs in parallel, which are averaged over. Since for each choice of L about ten choices of p are needed, and several choices of L (corresponding to $n=5,6,7,8,9$ and ten lattice planes of the fcc crystal) needed to be studied to check for finite size effects, this study would have been completely impossible without the access to the supercomputers at the HLRS. Figure 3 shows that non-trivial finite size effects do in fact occur, and only through the proper extrapolation versus $1/n^2$ do we obtain the final estimate $p_{\text{coex}}\sigma_c^3/k_b T = 8.45 \pm 0.04$ for the soft Eff AO model. Note that a comprehensive analysis of finite size effects on interfacial tensions [20, 21] has shown that finite size effects of order $1/n^2$ should be expected, so this choice of extrapolation is based on general theoretical results.

(a)   (b)

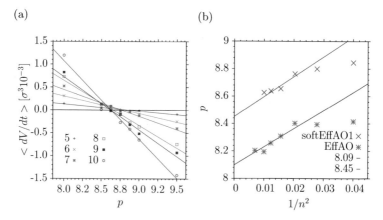

Fig. 3 (a) Velocities of volume changes $\Delta V/\Delta t$, as extracted for data as shown in Fig. 2b, plotted vs. p (in units of $k_B T/\sigma_c^3$), for different choices of n, as indicated. Note that these straight lines do not show a unique intersection point at $\Delta V/\Delta t=0$, indicating that systematic finite size effect indeed are present. (b) Pressure p where $\Delta V/\Delta t=0$ for a particular choice of n plotted vs. $1/n^2$ for the soft Eff AO model (upper set of data) and the Eff AO model (lower set of data). The extrapolation versus $1/n^2$ (i.e., inversely in the interface area) yields $p_{coex}=8.09\pm0.06$ for the Eff AO model and $p_{coex}=8.45\pm0.06$ for the soft Eff AO model

Next we turn to the estimation of the chemical potential μ of the liquid near its coexistence value μ_{coex}. Based on Eqs. (10)–(13), one might think that the knowledge of p_ℓ, the pressure in the liquid surrounding the crystal, is already sufficient for the desired analysis (and p_ℓ is directly found in simulations of nucleus-liquid coexistence from the virial expression, see Sect. 4). However, in the general case one cannot rely on the hope that all data are close enough to p_{coex} so that the linear expansions Eqs. (10), (11) are already valid. Developing a method to directly evaluate $\mu_\ell(p_\ell)$ for pressures near p_{coex} and beyond will allow us to test whether Eq. (10) is sufficiently accurate.

One might think one could obtain p_c from carrying out pressure measurement using the virial expression inside of the crystal, and hence use Eq. (3), (4) directly. However, such an approach is not feasible for nuclei of nanoscopic size: the pressure tensor is known to not be constant and also anisotropic in the fluid-crystal interfacial region [22]. Knowing the chemical potential of the liquid coexisting with the crystal allows us to exploit the relation

$$\mu_\ell(p_\ell)=\mu_c(p_c)=\mu_{coex}+\frac{\pi}{6}\int_{p_{coex}}^{p_c} dp/\eta_c(p) \quad . \tag{14}$$

Since the function $\eta_c(p)$ is known (Fig. 1b), we can find the function $\mu_c(p_c)$ when μ_{coex} has been determined, and using the value of p_c that satisfies Eq. (14) allows us to choose $\eta_c(p_c)$ in the generalized lever rule, Eq. (5).

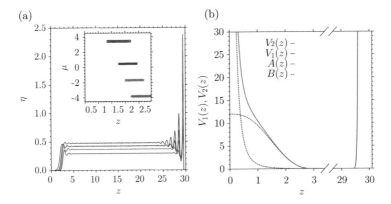

Fig. 4 (**a**) Illustration of the method to compute the chemical potential of a dense colloid fluid, using a $L \times L \times L_z$ geometry with two walls at $z=0$ and at $z=L_z$, and periodic boundary conditions in x and y direction, for the case of $L=7$, $L_z=30$, and four choices of particle number ($N=$ 750, 950, 1100 and 1250 respectively, from *bottom* to *top*). Part (**b**) shows the potentials that act on the walls. The bulk packing fraction $\eta_\ell(\mu)$ is extracted from an average of the profile for $10 \leq z \leq 20$ in this example

Unfortunately, the standard Widom [15] particle insertion method is inapplicable (acceptance rates for attempted insertions of virtual particles are many orders of magnitude too low) for η close to η_f. However, extending a proposal due to Powles et al. [23] we can exploit the fact that the chemical potential $\mu(z)=\mu=$constant, independent of a coordinate z in which the density $\rho(z)$ varies due to external potentials, in thermal equilibrium. Figure 4 shows a typical implementation of our method: we choose a $L \times L \times L_z$ geometry with $L_z=30$ and apply a soft, slowly decaying repulsive potential at the left wall (while at the right wall the standard repulsive part of the Lennard-Jones potential, truncated in the minimum and shifted there to zero, acts.) Fig. 4b shows these potentials (see (Statt, 2015, Dissertation, Johannes Gutenberg Universität Mainz, unpublished; [11]) for details). One can see that near the left wall the density smoothly decays to zero, and there is always a regime of z (roughly of width $\sigma_c=1$, in our case) where the density is in the right range that particle insertion works and accurate estimation of μ is possible. It is essential, of course, that L_z is large enough so that in the center of the system the bulk packing fraction $\eta_\ell(\mu)$, to which the "measured" chemical potential belongs, can be estimated (this condition was missed in [23]). Although the "measurement" of μ utilizes only a fraction of order $1/L_z$ of the total simulated volume, we have found that the desired accuracy is easily reached (Fig. 5). From Fig. 5 we see that Eq. (10) is indeed very accurate, in the parameter regime of interest. Each value for N in Fig. 5a and b was obtained by simulations using 32 CPUs in parallel for 24 h.

(a) (b)

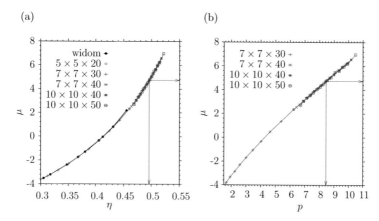

Fig. 5 (**a**) Chemical potential μ (in units of $k_B T$) plotted vs. packing fraction η, for different choices of L and L_z, as indicated, to show that finite size effects are negligible. For $\eta < 0.4$ runs with the Widom particle insertion method for a bulk $L \times L \times L$ system have been included for comparison, to demonstrate the consistency of our new method. *Arrows* on abscissa and ordinate indicate η_f and μ_{coex}, respectively. (**b**) Chemical potential μ plotted vs. pressure p, using the same data as in part (**a**). *Arrows* at the abscissa and ordinate indicate p_{coex} and μ_{coex}, respectively

4 Estimation of Crystal Nucleation Barriers

The basic idea of our new methodology is summarized in Fig. 6: We first identify the region of densities in the two-phase coexistence region, where stable compact nuclei surrounded by fluid in stable thermodynamic equilibrium can be found. We characterize the fluid, "measuring" both its density ρ_ℓ and its pressure p_ℓ (inferring then also its chemical potential from the data of Fig. 5). Using then Eq. (11) and/or Eq. (14), we find p_c and from Fig. 1 we find $\eta_c(p_c)$ Using $\eta_c(p_c)$ and $\eta_\ell(p_\ell)$ in Eq. (5), the associated nucleus volume V^* is unambiguously identified from Eq. (5). Varying then V_{box} at constant particle number N, we obtain the desired relation between nucleus volume V^* and chemical potential difference $\mu - \mu_{coex}$ (or pressure difference $p_c - p_\ell$, respectively).

Of course, this analysis makes tacitly two important assumptions: (i) when we repeat the analysis for different choices of N, we obtain results compatible with each other (clearly too small choices of N need to be avoided, so that finite size effects can be neglected). (ii) The nucleus volume V^* must be large enough, so that corrections to the $V^{2/3}$ relation of $F_{surf}(V)$ in Eq. (2) can be neglected. Figure 7a presents our final results for three choices of N ($N=$ 6000, 8000, and 10000) and yield compelling evidence that both assumptions are very well fulfilled: In a plot of ΔF^* vs. $V^{*2/3}$ the data for the three choices of N nicely fall on the same straight line through the origin. The slope of this straight line hence is our simulation estimate for γA_w.

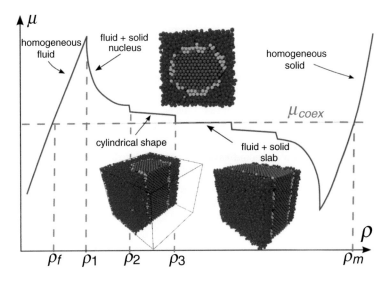

Fig. 6 Schematic plot of the chemical potential μ vs. density ρ for a system undergoing a liquid solid transition. In the thermodynamic limit, $\mu = \mu_{coex}$ over the entire two-phase coexistence region, $\rho_f \leq \rho \leq \rho_m$. In a finite simulation box, there is an overshoot of μ over μ_{coex} (and similarly p over p_{coex}) until $\rho = \rho_1$, where the droplet evaporation-condensation transition [8] takes place (note that this transition for finite L is slightly rounded, and $\rho_1 - \rho_f \propto L^{-4}$ in $d = 3$ dimensions). The region where μ decreases with increasing μ, $\rho_1 < \rho < \rho_2$, is the region that is studied here, since there a crystal nucleus coexists with surrounding fluid in thermal equilibrium. In practice, this interval is quite narrow (here data were taken only for $0.525 \leq \eta \leq 0.530$). At ρ_2 the nucleus make a transition from compact to cylindrical shape (stabilized by the periodic boundary condition), followed by the onset of slab configuration for $\rho > \rho_3$. These different states are illustrated by snapshot configurations for the Eff AO model

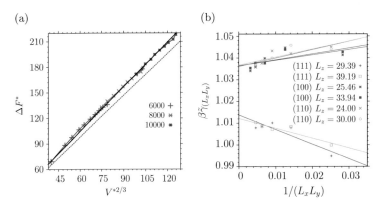

Fig. 7 (**a**) ΔF^* computed from Eq. (4), using the estimates for p_ℓ, p_c, and V^*, versus $V^{*2/3}$, using the data for $N = 6000$, 8000, and 10000, as indicated. The *upper straight line* is a fit of γA_w in the relation Eq. (3), the *lower straight line* uses $\gamma = \gamma_{111} = 1.013$ and $A_{iso} = (3\pi)^{1/3}$. (**b**) Interfacial tensions γ_{111}, γ_{110} and γ_{100} plotted vs. interfacial area. Data were obtained from the ensemble switch method

Of course, it is interesting to ask how far the standard theory (assuming a spherical droplet shape) would be off: Applying the ensemble switch method (Schmitz, 2014, Dissertation, Johannes Gutenberg Universität Mainz, unpublished; [20, 21, 24, 25]) we have obtained the fluid-crystal interface tension for planar (111), (100) and (110) interfaces (Fig. 7b). One finds the smallest value for the closed packed (111) surface, namely $\gamma_{111} = 1.013(3)$, yielding $A_{\mathrm{iso}}\gamma_{111}/3 \approx 1.633$. It turns out that the resulting straight line falls only about 7 % underneath the actual data, indicating that the spherical approximation for the nucleus shapes is very good, when the anisotropy of γ_{111}, γ_{100}, γ_{110} amounts only a few percent. Such a result could have been expected from simulations of the Ising lattice gas model for temperatures above the roughening transition temperature, where the anisotropy of the interfacial tensions also is only a few present [4], and similar the enhancement of ΔF^* over the spherical estimate is of order of 10 % as well [26].

5   Conclusions and Outlook

We have developed a general method to study crystal nuclei, obtaining the relation between their volume V^* and the supersaturation ($\mu - \mu_{\mathrm{coex}}$ or $p_c - p_\ell$, respectively) of the surrounding fluid, and thus obtain estimates for the barrier ΔF^* against homogeneous nucleation. The method has been demonstrated for the soft Eff AO model of colloid-polymer mixtures; but we believe that all aspects of the method can be transferred to other model systems as well. A further step will be the use of the same model where the fluid density is rather small, and a larger anisotropy of interfacial tensions is expected; also the use of a Lennard-Jones model is attractive, where the temperature of the system acts as an additional control parameter. Ultimately, it is hoped that the technique can be carried over to the problem of ice nucleation from water, a basic problem of the atmospheric sciences.

We have found that the nucleus shape is almost spherical, when the anisotropy of the interfacial tensions amounts to a few % only, and we believe that this is a general feature, not only specific for our particular model. Similarly, we feel that our confirmation of the concepts of classical nucleation theory (which has been particularly controversial for colloidal systems! [17]) is of general validity, but there we must make an important caveat: our study concerns the regime of rather large barriers ($\Delta F^*/k_B T \geq 60$); many other simulations (and some experiments) address the regime of much smaller barriers, $10 \leq \Delta F^*/k_B T \leq 20$, and then we can neither disregard corrections to Eqs. (1)–(4), nor can the kinetics of the nucleation processes be left outside of consideration. Nevertheless, the explicit knowledge what classical nucleation theory would predict is always an anchor point of any study of nucleation.

Acknowledgements We would like to thank the DFG for funding in the framework of the priority program on heterogeneous nucleation (SPP 1296, grant N° VI 237/4-3) and we thank the HLRS Stuttgart for generous grants of computer time at the HERMIT and HORNET supercomputers.

References

1. Landau, L.D.; Lifshitz, E.M.: Statistical Physics. Pergamon Press, Oxford (1958)
2. Statt, A., Virnau, P., Binder, K.: Phys. Rev. Lett. **114**, 026101 (2015)
3. Wulff, C.: Z. Krist. Mineral. **34**, 449 (1901)
4. Block, B., Kim, S., Virnau, P., Binder, K.: Phys. Rev. E **90**, 062106 (2014)
5. Kashchiev, D.: Nucleation: Basic Theory with Applications. Butterworth-Heinemann, Oxford, 2000
6. Kelton, K.F., Greer, A.L.: Nucleation (Pergamon, Oxford, 2009)
7. Binder, K., Kalos, M.H.: J. Stat. Phys. **22**, 363 (1980)
8. Binder, K.: Physica A **319**, 99 (2003)
9. Steinhardt, P.J., Nelson, D.R., Ronchetti, M.: Phys. Rev. B **28**, 783 (1983)
10. Lechner, W., Dellago, C.: J. Chem. Phys. **129**, 114707 (2008)
11. Statt, A., Virnau, P., Binder, K.: Mol. Phys. **113** (17–18) (2015)
12. Asakura, S., Oosawa, F.: J. Chem. Phys. **22**, 1255 (1954)
13. Binder, K., Virnau, P., Statt, A.: J. Chem. Phys. **141**, 140901 (2014)
14. Zykova-Timan, T., Horbach, J., Binder, K.: J. Chem. Phys. **133**, 014705 (2010)
15. Widom, B.: J. Chem. Phys. **39**, 2808 (1963)
16. Lekkerkerker, H.N.W., Tuinier, R.: Colloids and the Depletion Interaction. Springer, Berlin, 2011
17. Palberg, T.; J. Phys.: Condens. Matter **26**, 333101 (2014)
18. Royall, C.P., Poon, W.C.K., Weeks, E.R.: Soft Matter **9**, 17 (2013)
19. Dijkstra, M., van Roij, R., Evans, R.: Phys. Rev. E **59**, 5744 (1999)
20. Schmitz, F., Virnau, P., Binder, K.: Phys. Rev. Lett. **112**, 125701 (2014)
21. Schmitz, F., Virnau, P., Binder, K.: Phys. Rev. E **90**, 012128 (2014)
22. Rowlinson, J.S., Widom, B.: Molecular Theor of Capillarity. Clarendon Press, Oxford, 1982
23. Powles, J.G., Holtz, B., Evans, W.A.B.: J. Chem. Phys. **101**, 7804 (1994)
24. Schmitz, F., Virnau, P.: J. Chem. Phys. **142**, 144108 (2015)
25. Schmitz, F., Statt, A., Virnau, P., Binder, K.: High Performance Computing in Science and Engineering '14. In: Nagel, W.E. et al. (eds.) Springer, Berlin (2015)
26. Schmitz, F., Virnau, P., Binder, K.: Phys. Rev. E **87**, 705330 (2013)

A New Colloid Model
for Dissipative-Particle-Dynamics Simulations

Jiajia Zhou and Friederike Schmid

Abstract We propose a new model to simulate spherical colloids. This is a meso-scopic method based on the dissipative particle dynamics. The colloid is represented by a large spherical bead, and its surface interacts with the solvent beads through a pair of dissipative and stochastic forces. This new model extends the tunable-slip boundary condition [Eur. Phys. J. E **26**, 115 (2008)] from planar surfaces to curved geometry, thus allows one to study colloids with slippery surfaces. Simulation results show good agreement with the prediction of hydrodynamic theories, indicating the hydrodynamic interactions are properly accounted in our new model.

1 Introduction

Colloidal dispersions have many scientific interests and practical applications in various fields such as soft matter physics, physical chemistry, cell biology, medicine development, and micro- or nanofluidics [1, 2]. In an aqueous solution, colloidal particles in general have a size one order of magnitude larger than the solvent molecules. This separation of length scales presents a significant challenge in simulating the dynamics of colloidal particles: On one hand, explicit consideration of the solvent with molecular details imposes a huge burden on the computational time, while most interesting phenomena may not concern solvents. On the other hand, the solvents cannot be totally ignored because they mediate the hydrodynamic interactions between larger colloidal particles. This dilemma is somewhat alleviated by resorting to coarse-grained simulations, in which the solvent degree of freedom are greatly reduced while hydrodynamic interactions are still preserved by enforcing the conservation of several physical quantities, such as the particle number and the momentum. Coarse-graining approaches allow one to study larger system and reach longer time scale. Notable examples are dissipative particle dynamics (DPD) [3–6],

J. Zhou (✉) • F. Schmid
Institut für Physik, Johannes Gutenberg-Universität Mainz, Staudingerweg 9, D55099 Mainz, Germany
e-mail: zhou@uni-mainz.de; friederike.schmid@uni-mainz.de

© Springer International Publishing Switzerland 2016
W.E. Nagel et al. (eds.), *High Performance Computing in Science and Engineering '15*, DOI 10.1007/978-3-319-24633-8_6

Lattice Boltzmann (LB) [7–9], Multi-Particle Collision Dynamics (MPCD) [10, 11], and Direct Numerical Simulation (DNS) [12–15].

Once one settles on a mesoscopic model for the solvent, the next task is to build a colloid model which couples to the solvent. From a simulation point of view, one would like a colloid model which is conceptually simple and easy to implement, but also represents the correct physics. There are several approaches:

- *Discretization of Large Colloid*
  Based on the initial DPD formulation, Hoogerbrugge and Koelman [3, 4] constructed the colloid using the same solvent beads. The relative motion of these beads is "frozen" so the integrity of the large colloid is maintained. Similar approach has been implemented to study many interacting colloids in microfluidic devices [16]. This "frozen particle" model is relatively simple because the beads in the colloid interact with the solvent through the same DPD interactions, and no extra parameters are required. A conceptually similar approach has been proposed in Lattice Boltzmann simulations: the so-called raspberry model presents the colloid as a collection of the surface beads. The surface beads couple to the LB fluids by a viscous force which depends on the relative velocity of the beads to the local fluids [17]. The positions of the surface beads are maintained either by a spring force [18, 19] or by fixing the bead position with respect to the colloid center [20, 21]. One drawback of the raspberry model is that the interacting beads occurs only on the surface and the fluid is allowed to penetrate inside the hollow sphere. It was later demonstrated to cause a discrepancy between the translational and rotational diffusion [22]. This can be remedied by adding the internal coupling points [23, 24]. Raspberry model was also implemented in DPD simulations [25–28], where we used a repulsive interaction to prevent solvent penetration.
- *Using Single Large Bead*
  Español proposed the fluid particle model (FPM) [29] which treats the colloid as one single object instead of combination of small particles. To model the large colloids, two additional non-central shear components are incorporated into the dissipative forces. Similar model has been proposed by Pan et al. [30].
- *Boundary Condition*
  Colloids can also be implemented as the boundary condition, which is common in LB and MPCD simulations. Ladd had constructed the colloid as an extended hollow sphere where bounce-back collision rules are applied on the colloid surfaces [31–33]. In MPCD, the coupling between the immersed colloids and the solvent can be implemented by either bounce-back rules or thermal wall boundary condition [34–37].

Most of the colloid models are designed to realize the no-slip boundary condition. But for certain colloids with hydrophobic surfaces, slippage can occur. We have proposed tunable-slip boundary condition for flat surfaces in DPD [38–45], which allows one to set the local slip length in the simulation. In this work, we present a colloid model based on the tunable-slip boundary condition. The remainder of this article is organized as follows: In Sect. 2, we introduce the colloid model and describe relevant parameters in simulation. We present the simulation results of diffusion constant in Sect. 3. Finally, we conclude in Sect. 4 with a brief summary.

2 Marble Model

To facilitate the discussion, we first present a short introduction to the traditional DPD. We then describe our new colloid model based on the same notation.

Dissipative Particle Dynamics is an well-established simulation method for mesoscale fluid. For two solvent beads i and j, their relative displacement is denoted by $\mathbf{r}_{ij} = \mathbf{r}_i - \mathbf{r}_j$ and their relative velocity $\mathbf{v}_{ij} = \mathbf{v}_i - \mathbf{v}_j$ (see Fig. 1). The force exerted by bead j on i is given by a pair of dissipative and random forces,

$$\mathbf{F}_{ij}^{\mathrm{DPD}} = \mathbf{F}_{ij}^{\mathrm{D}} + \mathbf{F}_{ij}^{\mathrm{R}}. \tag{1}$$

The dissipative force $\mathbf{F}_{ij}^{\mathrm{D}}$ is proportional to the relative velocity between two beads,

$$\mathbf{F}_{ij}^{\mathrm{D}} = -\gamma^{\mathrm{DPD}}\,\omega^{\mathrm{D}}(r_{ij})(\mathbf{v}_{ij} \cdot \hat{\mathbf{r}}_{ij})\hat{\mathbf{r}}_{ij}, \tag{2}$$

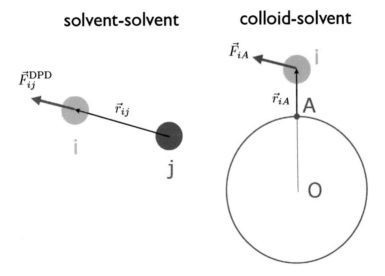

Fig. 1 Sketches to demonstrate the solvent-solvent interaction (*left*) and colloid-solvent interaction (*right*)

with a friction coefficient γ^{DPD} and a weight function ω^D,

$$
\omega^D(r) =
\begin{cases}
\left(1 - \dfrac{r}{r_c}\right)^2 & \text{if } r \le r_c^{\mathrm{DPD}}, \\
0 & \text{if } r > r_c^{\mathrm{DPD}}.
\end{cases}
\tag{3}
$$

The cutoff radius r_c^{DPD} characterizes the finite range of the interaction.

The random component $\mathbf{F}_{ij}^R$ has the form

$$
\mathbf{F}_{ij}^R = \sqrt{2k_B T \gamma^{\mathrm{DPD}} \, \omega^D(r_{ij})} \, \xi \hat{\mathbf{r}}_{ij},
\tag{4}
$$

where ξ is a random number with zero mean $\langle \xi(t) \rangle = 0$ and variance $\langle \xi(t)\xi(t') \rangle = \delta(t-t')$. The dissipative and random forces are related by the fluctuation-dissipation theorem, so to maintain the proper temperature in the simulation. The forces between two beads are the same in magnitude but opposite in direction, $\mathbf{F}_{ij}^{\mathrm{DPD}} = -\mathbf{F}_{ji}^{\mathrm{DPD}}$, hence the momentum is conserved. The conservation of momentum is essential for DPD to obtain the correct long-time hydrodynamic behavior.

In the following, physical quantities will be reported in a unit system of σ (length), m (mass), ε (energy), and a derived time unit $\tau = \sigma\sqrt{m/\varepsilon}$. We use a solvent density $\rho = 3.0\sigma^{-3}$. The friction coefficient for the solvent is $\gamma^{\mathrm{DPD}} = 5.0\,m/\tau$ and the cutoff $r_c^{\mathrm{DPD}} = 1.0\,\sigma$. The shear viscosity is measured as $\eta_s = 1.23 \pm 0.01\,m/(\sigma\tau)$.

The colloidal particle is represented by a large spherical bead of radius $R = 3.0\,\sigma$. To prevent the solvent penetration, a modified Lennard-Jones interaction is applied between the colloid and the solvent,

$$
V(r) =
\begin{cases}
4\varepsilon \left[\left(\dfrac{\sigma}{r - r_0}\right)^{12} - \left(\dfrac{\sigma}{r - r_0}\right)^6 + \dfrac{1}{4} \right] & \text{if } r - r_0 \le r_c^{\mathrm{LJ}}, \\
0 & \text{if } r - r_0 > r_c^{\mathrm{LJ}},
\end{cases}
\tag{5}
$$

where $r_0 = 2.0\,\sigma$ and $r_c^{\mathrm{LJ}} = 1.0\,\sigma$.

We image that at each time step, the solvent bead interacts with the colloid through the surface point A, which is located at the intersection of the colloid surface and the line connecting the solvent bead and the colloid center (see Fig. 1). The velocity of the surface point is given by

$$
\mathbf{v}_A = \mathbf{v}_O + \boldsymbol{\omega} \times \mathbf{r}_{OA},
\tag{6}
$$

where $\mathbf{v}_O$ is center-of-mass speed and $\boldsymbol{\omega}$ is the angular velocity of the colloid.

Similar to the interaction between two solvent beads, the forces between the solvent bead i and the surface point A consist of a dissipative and random components.

$$\mathbf{F}_{iA} = \mathbf{F}_{iA}^D + \mathbf{F}_{iA}^R. \tag{7}$$

There are at least three choices in setting the interaction between the colloid and the solvent:

1. *Traditional DPD Interaction.* One can use the traditional DPD interactions, just to imagine there is a solvent bead sitting at the surface point A,

$$\mathbf{F}_{iA}^D = -\gamma \omega^D(r_{iA})[(\mathbf{v}_i - \mathbf{v}_A) \cdot \hat{\mathbf{r}}_{iA}]\,\hat{\mathbf{r}}_{iA} \tag{8}$$

$$\mathbf{F}_{iA}^R = \sqrt{2k_B T \gamma \omega^D(r_{iA})}\,\xi \hat{\mathbf{r}}_{iA} \tag{9}$$

where γ characterizes the coupling strength and ω^D is the same as in Eq. (3).

2. *Transverse Interaction.* The second possibility is to use the transverse interactions, which project the traditional DPD interactions on the plane perpendicular to the vector $\mathbf{r}_{iA}$ [46],

$$\mathbf{F}_{iA}^D = -\gamma \omega^D(r_{iA})\,\mathscr{P}(\mathbf{v}_i - \mathbf{v}_A) \tag{10}$$

$$\mathbf{F}_{iA}^R = \sqrt{2k_B T \gamma \omega^D(r_{iA})}\,\mathscr{P}(\hat{\boldsymbol{\xi}}) \tag{11}$$

where $\mathscr{P} = \mathscr{I} - \hat{\mathbf{r}}_{iA} \otimes \hat{\mathbf{r}}_{iA}$ is the projection operator and $\hat{\boldsymbol{\xi}}$ is a vector whose three component are random numbers.

3. *Tunable-Slip Interaction.* In the tunable-slip approach, the dissipative component is directly proportional to the relative velocity $\mathbf{v}_i - \mathbf{v}_A$. We implement this approach in our simulations, to be consistent with our previous works on the flat surfaces.

$$\mathbf{F}_{iA}^D = -\gamma \omega^D(r_{iA})(\mathbf{v}_i - \mathbf{v}_A) \tag{12}$$

$$\mathbf{F}_{iA}^R = \sqrt{2k_B T \gamma \omega^D(r_{iA})}\,\hat{\boldsymbol{\xi}} \tag{13}$$

The total force exerted on the colloid is a summation over all solvent beads which are less than 1.0σ away from the colloid surface,

$$\mathbf{F}^C = \sum_i \mathbf{F}_{Ai} + \mathbf{F}^{LJ}. \tag{14}$$

Similarly, the total torque exerted on the colloid is

$$\mathbf{T}^C = \sum_{i=1} \mathbf{F}_{Ai} \times (\mathbf{r}_A - \mathbf{r}_O). \tag{15}$$

The total force and torque are then used to update the position and velocity of the colloid in one time step using the Velocity-Verlet algorithm. All simulations were carried out using the open source package ESPResSo [47].

3 Results and Discussions

To test our new colloid model, we simulate a single colloid in a cubic box of size L. We measure the diffusion constant of the colloid using three different methods and compare with the hydrodynamic theory.

The first method computes the diffusion constant through the autocorrelation functions. We obtain two correlation functions from the simulations: the translational and rotational velocity autocorrelation functions

$$C_v(t) = \frac{\langle \mathbf{v}(0) \cdot \mathbf{v}(t) \rangle}{\langle \mathbf{v}^2 \rangle}, \tag{16}$$

$$C_\omega(t) = \frac{\langle \boldsymbol{\omega}(0) \cdot \boldsymbol{\omega}(t) \rangle}{\langle \boldsymbol{\omega}^2 \rangle}, \tag{17}$$

where $\mathbf{v}(t)$ and $\boldsymbol{\omega}(t)$ are the translational velocity and rotational velocity of the colloid at time t, respectively. Using the Green-Kubo relation, we obtain the diffusion constant of the colloid by integrating the translational velocity autocorrelation function

$$D = \frac{1}{3} \int_0^\infty dt \, \langle \mathbf{v}(0) \cdot \mathbf{v}(t) \rangle. \tag{18}$$

Figure 2 shows the simulation results for $\gamma = 5.0 \, m/\tau$ in log-log plots. The autocorrelation functions show different behavior at short and long time scales. At short times, both autocorrelation functions decay exponentially. At long times, hydrodynamic interactions lead to a slow relaxation of algebraic decay, which is called long-time tail [48]. Mode-coupling theory predicts the coefficient of algebraic decay at long times to be $-\frac{3}{2}$ for the translational velocity and $-\frac{5}{2}$ for the rotational velocity [49]. These predictions are plotted as green lines in Fig. 2. The simulation results are consistent with the theoretical prediction for $t > 10 \, \tau$, but both autocorrelation functions exhibit large fluctuation. This is due to the poor statistics for long-time values of the correlation function. One can improve the results by running very long simulations.

The second method to measure the diffusion constant is by calculation of the mean squared displacement. From Einstein relation

$$\lim_{t \to \infty} \langle (\mathbf{r}(t) - \mathbf{r}(0))^2 \rangle = 6Dt, \tag{19}$$

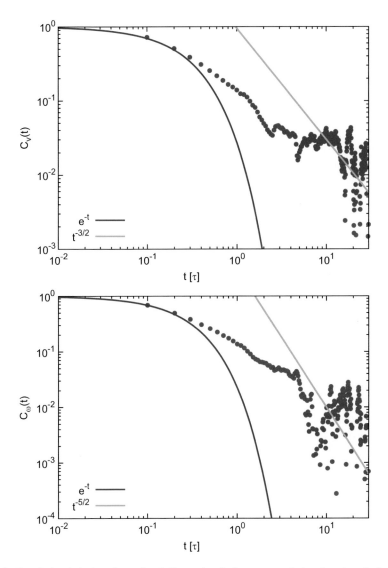

Fig. 2 Translational (*top*) and rotational (*bottom*) velocity autocorrelation functions in log-log plots. The simulation is performed for the following parameters: colloid radius $R = 3.0\sigma$, temperature $k_B T = 1.0\,\varepsilon$, and colloid-solvent parameter $\gamma = 5.0\,m/\tau$

Fig. 3 The mean squared
displacement of a spherical
colloid with radius $R = 3.0\,\sigma$
in a cubic simulation box of
sizes $L = 20\,\sigma$. The *line*
shows a linear fit to data
$t > 25\tau$

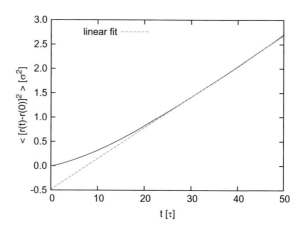

where $\mathbf{r}$ is the position of the colloid center. Figure 3 shows a typical mean
squared displacement as a function of the time. At short times, the colloid exhibits
a ballistic motion where the mean squared displacement increases as t^2. At later
times, the ballistic motion is replaced by a diffusive motion where the mean squared
displacement is proportional to t. A linear fit of the mean squared displacement at
$t > 25\,\tau$ gives the diffusion constant.

The diffusion constant can also be obtained by a simulation experiment. We apply
a constant force to the colloid. To prevent accelerating the whole system, we also
apply a small force to each fluid bead, and the total force on the fluid is opposite
to the force on the colloid, and with the same magnitude. At late times, the colloid
reaches a constant velocity $\mathbf{v}_f$ with respect to the fluid. In this stationary state, the
external driving force is balanced by the viscous friction

$$\mathbf{F}_{\mathrm{ext}} = -\gamma_C\, \mathbf{v}_f, \qquad (20)$$

where γ_C is the friction coefficient. The fluctuation-dissipation theorem relates the
friction coefficient to the diffusion constant by

$$D = \frac{k_B T}{\gamma_C}. \qquad (21)$$

Figure 4 collects all simulation results obtained by three different methods
[velocity autocorrelation function (vacf), mean squared displacement (msd), and
force measurement (force)]. The diffusion constants are measured for different
values of colloid-solvent friction coefficient γ. Error bars are calculated by three
independent runs with different initial configurations. The force measurement shows
relatively small error bars in comparison to the other two approaches.

Upon increasing γ, the surface property changes from slippery to no-slip, and
the diffusion constant decreases accordingly. No-slip boundary condition can be
realized by using a friction coefficient $\gamma > 5m/\tau$. By adjusting the γ value, one

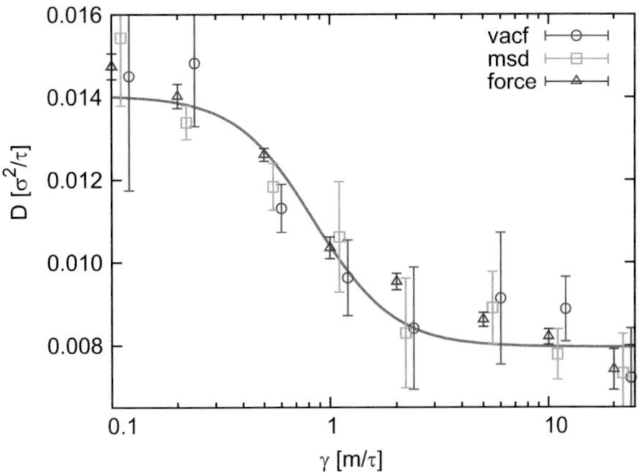

Fig. 4 The diffusion constant D for a spherical colloid of radius $R = 3.0\,\sigma$ as a function of the surface-solvent friction coefficient γ. The simulation box has a size of $L = 20\,\sigma$. The *solid curve* is a guide to the eyes. The data from autocorrelation function and mean squared displacement are shifted slightly in x-axis for a better view

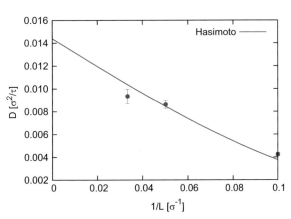

Fig. 5 The diffusion constant D for a spherical colloid of radius $R = 3.0\,\sigma$ as a function of $1/L$, the reciprocal of the box size. The *curve* is the prediction from Eq. (22)

can set the surface boundary of the colloidal particle from no-slip to full-slip. This freedom provides possibilities to investigate the effect of hydrodynamic slip on the dynamics of colloidal particles [50, 51].

We use periodic boundary condition in the simulations. For small simulation box, the colloid can interact with its periodic images, resulting a box-size dependent diffusion constant. In general, the diffusion constant increases with increasing box size. Hasimoto presented an analytic expression for the diffusion constant in terms of a series expansion of $1/L$, the reciprocal of the box size [52],

$$D = \frac{k_B T}{6\pi \eta_s}\left(\frac{1}{R} - \frac{2.837}{L} + \frac{4.19R^2}{L^3} + \cdots\right). \qquad (22)$$

In Fig. 5, we plot the diffusion constant as a function of $1/L$, based on the force measurement. The simulation results show relatively well agreement to the Hasimoto formula, which is shown as the solid curve in Fig. 5.

4 Summary

We have developed a mesoscopic colloid model based on the dissipative particle dynamics. The colloid is represented by a large spherical bead, and its surface interacts with solvent beads through a pair of dissipative and random forces. We test this new colloid model by measuring the diffusion constant of a single colloid in a cubic box. The simulation results show good agreement with the predictions from hydrodynamic theories. Our model can be viewed as an extension of the tunable-slip boundary condition [38] to curved geometries, and allows one to investigate the dynamics of colloidal particles with slippery surfaces.

Acknowledgements We thank the HLRS Stuttgart for a generous grant of computer time on HERMIT. This work is funded by the Deutsche Forschungsgemeinschaft (DFG) through SFB TR6 (subproject B9) and SFB 1066 (subproject Q1).

References

1. Russel, W.B., Saville, D.A., Schowalter, W.: Colloidal Dispersions. Cambridge University Press, Cambridge (1989)
2. Dhont, J.: An Introduction to Dynamics of Colloids. Elsevier, Amsterdam (1996)
3. Hoogerbrugge, P.J., Koelman, J.M.V.A.: Europhys. Lett. **19**, 155 (1992)
4. Koelman, J.M.V.A., Hoogerbrugge, P.J.: Europhys. Lett. **21**, 363 (1993)
5. Español, P., Warren, P.B.: Europhys. Lett. **30**, 191 (1995)
6. Groot, R.D., Warren, P.B.: J. Chem. Phys. **107**, 4423 (1997)
7. Succi, S.: The Lattice Boltzmann Equation. Clarendon Press, Oxford (2001)
8. Raabe, D.: Model. Simul. Mater. Sci. Eng. **12**, R13 (2004)
9. Dünweg, B., Ladd, A.J.C.: Adv. Polym. Sci. **221**, 89 (2009)
10. Malevanets, A., Kapral, R.: J. Chem. Phys. **110**, 8605 (1999)
11. Gompper, G., Ihle, T., Kroll, D.M., Winkler, R.G.: Adv. Polym. Sci. **221**, 1 (2009)
12. Tanaka, H., Araki, T.: Phys. Rev. Lett. **85**, 1338 (2000)
13. Nakayama, Y., Yamamoto, R.: Phys. Rev. E **71**, 036707 (2005)
14. Kim, K., Nakayama, Y., Yamamoto, R.: Phys. Rev. Lett. **96**, 208302 (2006)
15. Nakayama, Y., Kim, K., Yamamoto, R.: Eur. Phys. J. E **26**, 361 (2008)
16. Steiner, T., Cupelli, C., Zengerle, R., Santer, M.: Microfluid. Nanofluid. **7**, 307 (2009)
17. Ahlrichs, P., Dünweg, B.: J. Chem. Phys. **111**, 8225 (1999)
18. Lobaskin, V., Dünweg, B.: New J. Phys. **6**, 54 (2004)
19. Lobaskin, V., Dünweg, B., Medebach, M., Palberg, T., Holm, C.: Phys. Rev. Lett. **98**, 176105 (2007)
20. Chatterji, A., Horbach, J.: J. Chem. Phys. **122**, 184903 (2005)
21. Chatterji, A., Horbach, J.: J. Chem. Phys. **126**, 064907 (2007)
22. Ollila, S.T.T., Smith, C.J., Ala-Nissila, T., Denniston, C.: Multiscale Model. Simul. **11**, 213 (2013)
23. Fischer, L.P., Peter, T., Holm, C., de Graaf, J.: Arxiv:1503.02671 [cond-mat.soft] (2015)
24. de Graaf, J., Peter, T., Fischer, L.P., Holm, C.: Arxiv:1503.02681 [cond-mat.soft] (2015)
25. Zhou, J., Schmid, F.: J. Phys.: Condens. Matter **24**, 464112 (2012)
26. Zhou, J., Schmid, F.: Eur. Phys. J. E **36**, 33 (2013)
27. Zhou, J., Schmitz, R., Dünweg, B., Schmid, F.: J. Chem. Phys. **139**, 024901 (2013)
28. Zhou, J., Schmid, F.: Eur. Phys. J. Spec. Top. **222**, 2911 (2013)

29. Español, P.: Phys. Rev. E **57**, 2930 (1998)
30. Pan, W., Pivkin, I.V., Karniadakis, G.E.: Europhys. Lett. **84**(1), 10012 (2008)
31. Ladd, A.: Phys. Rev. Lett. **70**, 1339 (1993)
32. Ladd, A.: J. Fluid Mech. **271**, 285 (1994)
33. Ladd, A.: J. Fluid Mech. **271**, 311 (1994)
34. Lee, S.H., Kapral, R.: J. Chem. Phys. **121**, 11163 (2004)
35. Padding, J.T., Wysocki, A., Löwen, H., Louis, A.A.: J. Phys.: Condens. Matter **17**, S3393 (2005)
36. Padding, J.T., Louis, A.A.: Phys. Rev. E **74**(3), 031402 (2006)
37. Whitmer, J.K., Luijten, E.: J. Phys.: Condens. Matter **22**, 104106 (2010)
38. Smiatek, J., Allen, M., Schmid, F.: Eur. Phys. J. E **26**, 115 (2008)
39. Smiatek, J., Sega, M., Holm, C., Schiller, U.D., Schmid, F.: J. Chem. Phys. **130**, 244702 (2009)
40. Smiatek, J., Schmid, F.: J. Phys. Chem. B **114**, 6266 (2010)
41. Smiatek, J., Schmid, F.: Comput. Phys. Commun. **182**, 1941 (2011)
42. Zhou, J., Belyaev, A.V., Schmid, F., Vinogradova, O.I.: J. Chem. Phys. **136**, 194706 (2012)
43. Asmolov, E.S., Zhou, J., Schmid, F., Vinogradova, O.I.: Phys. Rev. E **88**, 023004 (2013)
44. Zhou, J., Asmolov, E.S., Schmid, F., Vinogradova, O.I.: J. Chem. Phys. **139**, 174708 (2013)
45. Nizkaya, T.V., Asmolov, E.S., Zhou, J., Schmid, F., Vinogradova, O.I.: Phys. Rev. E **91**, 033020 (2015)
46. Junghans, C., Praprotnik, M., Kremer, K.: Soft Matter **4**, 156 (2008)
47. Limbach, H., Arnold, A., Mann, B., Holm, C.: Comput. Phys. Commun. **174**, 704 (2006)
48. Alder, B.J., Wainwright, T.E.: Phys. Rev. A **1**, 18 (1970)
49. Hansen, J.P., McDonald, I.R.: Theory of Simple Liquids, 3rd edn. Academic, London (2006)
50. Swan, J.W., Khair, A.S.: J. Fluid Mech. **606**, 115 (2008)
51. Khair, A.S., Squires, T.M.: Phys. Fluids **21**, 042001 (2009)
52. Hasimoto, H.: J. Fluid Mech. **5**, 317 (1959)

Force Field Optimization for Ionic Liquids: FFOIL

Annual Report 2015

Konrad Breitsprecher, Narayanan Krishnamoorthy Anand, Jens Smiatek, and Christian Holm

Abstract In the reporting period July 2014 to June 2015, we performed molecular dynamics (MD) simulations with the software packages *ESPResSo* and *Gromacs* as well as custom MD-code to investigate different models of room temperature ionic liquids (RTILs) in confined environment and bulk solution. The application in mind is an IL-based electric double-layer capacitor, a non-faradaic energy storage device with advantages in power density and cycle stability over electrochemical cells. In this field, we have developed and applied force fields for ion-ion and ion-electrode interactions on different levels of detail. In 2014, we focused on algorithms for metal boundary conditions in various geometries, needed for realistic modeling of the charging process in a capacitor device. Further, we studied the effects of graphite structure on the adsorbed ions in planar capacitor setups. A possible one-dimensional force field for the interaction between ion and electrode was tested and compared to a fully modeled graphite surface with explicit carbon atoms. This allowed us to show the increased adsorption of the ionic liquid on graphite surfaces due to the texturing influence of the honeycomb pattern. More recently, we performed MD-simulations with electrolyte mixtures and accurate models for carbide-derived carbon electrode systems. In this ongoing work, we study the effect of solvent concentration in ionic liquids on the capacitance of the porous electrode material.

1 Introduction

IL based capacitors belong to the class of energy storage devices known as *double layer capacitors* or *supercapacitors*. They consist of a liquid electrolyte confined between two electrodes of various geometry and material. Ionic liquids are known to

K. Breitsprecher (✉) • N.K. Anand • J. Smiatek • C. Holm
Institute for Computational Physics, University of Stuttgart, Stuttgart, Germany
e-mail: konrad@icp.uni-stuttgart.de

© Springer International Publishing Switzerland 2016
W.E. Nagel et al. (eds.), *High Performance Computing in Science and Engineering '15*, DOI 10.1007/978-3-319-24633-8_7

be good candidates for the dielectric material, as they have advantageous properties for the realization of such a capacitor device [1]: From a molecular perspective, their size and asymmetry in structure and charge distribution prevents the formation of regular patters, reducing the melting point in applicable ranges below 100 °C or even room temperature. In addition, the coulombic interactions of the molten salt in the liquid phase hinders quick evaporation and leads to a very low vapor pressure [2], which is an important safety issue for capacitor applications. Compared to aqueous or organic electrolytes, the applied potential can be larger before unwanted faradaic processes take place, which leads to more energy being saved [3]. The variety of synthesized ILs allows to pick a certain IL to reach desired properties.

The other, even more important part of supercapacitors in application is the electrode. There, materials with a huge surface area are required in order to achieve high energy densities. Examples for promising electrode materials are graphene sheets, carbon nanotubes or activated carbon, reaching surface areas of more than $1000\,\text{m}^2/\text{g}$ [4]. The outstanding property of supercapacitors however is their excellent power density. The (theoretical) absence of chemical reactions and the energy storage mechanisms based on ion separation makes them superior over batteries concerning charging times and cycle stability. Depending on the electrode material, the charging process of the capacitor will be dramatically different. In case of planar electrodes, charging goes along with the rearrangement of layers of ions in front of the electrode interface [5, 6]. This can happen orders of magnitudes faster than the mechanisms taking place in a porous network such as activated carbon, where the width of the pores and the ion sizes can come close [7]. In this context, mixtures of ionic liquids and organic solvents may lead to improved properties in porous electrodes [8].

1.1 Report Structure

In this report, we give an overview of the projects related to computing time on the HLRS clusters *Hermit* and later *Hornet* in the reporting period from July 2014 to June 2015. Note that extracts of the report on the finished *Project A: Graphite structure and planar capacitor simulation* and the shown data in the Appendix are part of a publication under submission. For the work in progress in *Project B: CDC electrodes with IL mixtures*, we provide a brief summary of the methodology and details about computing resources. Finally we give an outlook with details about the projects planning to use computing time on *Hornet* in the upcoming period.

2 Project A: Graphite Structure and Planar Capacitor Simulation

2.1 Simulation Methods

2.1.1 Short-Range Electrode Interaction

In this project, we compared an explicit graphene structure to a computationally more efficient unstructured planar Lennard-Jones representation of the surface. The graphene consists of three layers of carbon atoms fixed in a hexagonal lattice structure with a C–C bond length of $1.42\,\text{Å}$ and plane distances of $3.35\,\text{Å}$ (see Appendix 1, Fig. 1). For the carbon short-range interactions, we used the common Lennard-Jones parameters $\sigma_C = 3.37\,\text{Å}$ and $\epsilon_C = 0.23\,\text{kJ/mol}$ [9] with Lorentz-Berthelot mixing rules. To obtain the planar representation, we averaged the short-range interaction of the explicit electrode model in the x-y-plane for several distances from the electrode and fitted the data with a generic Lennard-Jones function. As expected, a 9-3-Lennard-Jones function or the Steele potential [10, 11] common for liquid-surface interactions couldn't be fitted to the data since the potential is a superposition of three graphene layers. We found that

$$U(z) = 4\epsilon_P \left(\left(\frac{\sigma_P}{z} \right)^{9.32} - \left(\frac{\sigma_P}{z} \right)^{4.66} \right)$$

with the parameters $\sigma_P = 3.58\,\text{Å}$ and $\epsilon_P = 24.7\,\text{kJ/mol}$ resulted in a precise representation of the data.

2.1.2 Electrode Polarization Methods

In addition to the investigation on structural properties, we compared different approaches to model the dielectric interfaces. In front of the conducting boundaries, the effect of surface charge induction introduces an additional local attraction of the charged species towards the electrodes [12]. In general, this can be modeled in two different levels of detail: The computationally much more efficient way is to distribute a constant charge on the electrode surface. This approach however can not be used to accurately simulate the charging process, as the final surface charge is a result of an applied potential between the electrodes rather than an input parameter. One method we applied to model the correct dynamical behaviour of the electrodes is the Electrostatic Layer Correction with Image Charges (ELCIC). It accounts for 2D-electrostatics and charge induction by evaluating the image charges in every time step [13–15]. Another approach to obtain the same effect is the Induced Charge

Computation (ICC) method, which iteratively determines the induced charges given by a discretised, closed surface of point charges [16, 17]. In contrast to ELCIC, ICC is used with the 3D-periodic electrostatic solver p3m [18, 19].

2.1.3 Simulation Details

The IL selected was a coarse grained model of BMIM PF6 presented by Roy and Maroncelli [20, 21]. For production runs, we performed NVT simulations of 12 to 18 ns with a Langevin thermostat at 400 K, a fixed number of 320 ion pairs and a timestep of 2 fs. For the systems with planar Lennard-Jones electrodes, we choose a quadratic surface area of 30 Å^2 and adjusted the system length to obtain a molar volume of 2.247 m^3/mol in the bulk [20]. The graphene systems had a slightly adjusted cross sectional area to account for the periodicity of the graphene pattern. All simulations were performed with the software package *ESPResSo 3.2* [22].

2.2 HPC Setup

For the computation of bonded and short-range interactions, ESPResSo provides a domain decomposition scheme for parallelization. The small system size and particle number in the capacitor cells limits the amount of cores, because the size of a computing domain cannot be smaller that the interaction range of the short-range potentials. However, due to the large number of individual simulation setups and the need for several applied electric potentials to investigate the capacitance behaviour, the computing resource strategy relied on a spread of several low-core (4–8) runs rather than large, highly parallel jobs.

2.3 Results

In the evaluation of the voltage dependence of the ion configuration in the interfacial electrode regions, the influence of the electrode structure was determined by analysing ion density profiles (Figs. 2 and 3 in Appendix 1), orientation profiles (Fig. 4 in Appendix 1, the differential capacitance (Fig. 5 in Appendix 1), the inter-layer ion exchange process (Fig. 6 in Appendix 1) and the in-plane radial distribution of the adsorbed ions (Fig. 7 in Appendix 1). An extended discussion of the appended data will be found in a future article, currently under submission.

The two different approaches to account for the electrode charge confirmed that it is valid to use the constant charge ensemble to analyse equilibrium properties of supercapacitors. Investigations concerning dynamical properties, like charging behaviour or applied alternating voltage however require the electrode surface charge to be adaptive. Further, we demonstrate that due to the underlying ionic models, anions and cations adapt to the structural pattern of the graphene electrodes in different ways. Both show increased adsorption on graphite compared to the unstructured wall, which results in larger mass- and charge density in the interfacial ion layer. The lateral behaviour of the ion layers, as well as the in-plane structure of the ionic liquid are affected by the choice of the electrode model. It is clear that the ions will adapt to the local structure of the surface and that the oscillating interfacial layers known to be present in ILs will further carry this influence. The comparison of the in-plane radial distributions of the first ionic layer revealed the details of this influence: The direct comparison between the anions and cations showed that a more anisotropic ion model disturbs the adaption to the electrode structure. The in-plane order of the simple spherical anions is much more affected by the honeycomb structure than the anisotropic, three-bead cations.

2.3.1 Method Benchmarks

Among the electrode models under investigation, the particle free planar Lennard-Jones electrodes in combination with a homogeneous electric field yields the best benchmarking results and was used as the reference system for the comparison in Table 1.

The computationally most efficient way to simulate the 2D periodic system at constant potential is provided by the ELCIC-method, because it benefits from the direct calculation of the homogeneously distributed induced charge. Introducing particle based electrodes comes at a sizable computational cost, the same applies for the iterative ICC method to simulate at constant potential.

Table 1 Relative runtimes of the electrode methods under investigation

LJ + ELC	LJ + ELCIC	GR + ELC
100 %	83 %	49 %
GR + DISCRETE	GR + ELCIC	GR + ICC
43 %	38 %	21 %

LJ 1D-Lennard-Jones electrodes, *GR* explicit carbon particle electrodes, *ELC* 2D-electrostatics without charge induction, *ELCIC/ICC* electrostatics with charge induction, *DISCRETE* constant point charges on carbon atoms

3 Project B: Carbide-Derived Carbon Electrodes with Mixtures of EMIM BF4 and Acetonitril

3.1 Simulation Setup

3.1.1 Motivation

Although planar capacitor layouts are helpful to study a clean electric double layer without electrode geometry perturbations, high energy densities for storage devices can only be achieved with high surface area electrode material. In this projects, we investigate mixtures of the ionic liquid EMIM BF4 with different concentrations of Acetonitril (ACN) in contact with carbide-derived carbon (CDC) electrodes [23]. System snapshots can be found in Appendix 2, Figs. 8 and 9. Preceding work on similar systems gave valuable insight from a simulation point of view on the storage mechanisms taking place in this setup [7, 24–26]. From experiments it was found that similar values of pore size and ion diameter can lead to a peak in capacitance [27]. Our choice of ionic liquid and CDC-800 [23] network aims towards this configuration. With our study, we try to eludicate the role of the organic solvent on capacitance, pore composition and charging dynamics.

3.1.2 Models and Methods

Again, we use a coarse grained ion model with an imidazole-based cation (EMIM) and a spherical anion (BF4) [28]. The Acetonitril is a linear three-bead model [8], intramolecular constraints are calculated with the RATTLE/SHAKE algorithm [29]. Each of the two CDC-800 electrodes consists of 3821 fixed carbon particles. The configuration was obtained with quenched molecular dynamics and reproduces realistic CDC geometries with tunable pore size [23]. The underlying simulation software is a custom FORTRAN code specialized on molecular dynamics simulations of capacitors. The outstanding feature is the charge induction method for the constant potential ensemble, described by Siepmann and Sprink [30] and later Reed et al. [31]. The method, according to the current ion configuration, iteratively determines the induced surface charge with a minimization routine and maintains the applied potential between the electrodes. To account for smeared out charges on the curved graphene network, electrostatic interaction between gaussian charges on the carbons and point charges on the moving particles is possible. Similar to the limitations of the ICC method, the expensive iterative calculation significantly reduces the performance of the simulation code.

3.2   HPC Resource Usage

For core load balancing, the simulation code uses pairwise particle distribution rather than domain decomposition, so the parallelization is not limited by a minimal domain size. Our average systems contains about 3000 charged moving particle compounds (depending on the solvent concentration) and 7642 inducible fixed carbon atoms. The parameter range under investigation includes four different mass percentages of ACN (0–40 %), two temperatures (298 and 340 K) and three voltages (0, 1 and 2 V). The simulation needs careful equilibration and preceding NPT bulk simulation to obtain the correct density of the mixture at each temperature and solvent concentration. Due to the nature of the porous electrodes, the system requires long trajectories to cover the complete charging process (similar setups report charging times of about 18 ns [25]). Once the equilibrium at a certain applied voltage is reached, low frequency surface charge fluctuations again require long simulations runs, especially for differential capacitance calculations.

4   FFOIL Outlook

In the upcoming project period, we will carry on the investigation on charging dynamics and capacitance measurements in the CDC-Capacitor for the mentioned parameters in temperature, solvent concentration and applied voltage.

As a new feature in the software package *ESPResSo*, we developed a procedure for convenient integration of electrode surfaces in molecular dynamics simulations. Using mesh files (.stl) as input for the electrode geometry, the feature can take care of carbon particle placement and the setup of the ICC method to maintain zero potential drop between the electrodes. Also, it detects the surface, sets the boundary conditions according to the given applied potential and solves Laplace equation (vacuum case) in the simulation volume. The generated volume data contains the external potential, its gradient is the applied electric field acting on the particles during integration. We superimpose the generated field from the vacuum case with the charge induction provided by the ICC method (which reacts on the ion charges and maintains a zero potential drop) and end up with the target potential. This setup therefore allows to visualize electric field lines of the applied potential in the capacitor with the help of further tools in cooperation with the *Visualization Research Center in Stuttgart* (VISUS). We will test and apply this method for a number of interesting electrode shapes such as slit-pore geometries or rough, 'gaussian landscape' electrodes.

A further study in our group involves *polyionic liquids* (PILs), which are formed from monomeric units of ionic liquids. These liquids have been recently recognized as innovative polyelectrolytes. Dense incorporation of ionic liquid moieties into the macro-molecular architecture provides additional physical properties like enhanced mechanical stability, conductivity, and durability over the IL species and consequently expands their application as polymeric ion conductor, smart dispersants and stabilizers, powerful absorbents for solvent purification and as complex solvents. One of the interesting aspects is that PILs show good solubilities in various solvents of different polarity and hydrophobicity, depending on the cationic and anionic pair of ionic liquid. Combining the adaptable solubility in organic solvents and the superior processability afforded by the polymer nature, PILs are able to form transparent films of different thickness, making them model systems to investigate the bulk nature of polyelectrolytes. This project deals with studying the solution properties of vinyl based PILs under aprotic, protic and apolar solvents using atomistic simulations with clap based force fields.

In another project, we intend to develop a polarizable atomistic force field for ionic liquids. Nowadays, polarization effects in all-atom simulations have become of specific importance. Our aim is to establish a so-called 'Drude polarization' model for simple ionic liquids like 1-Methyl-3-Methylimidazolium [MMIM] chloride [Cl]. The main parameter for the Drude polarization model, which is the polarization value of the considered atoms, shall be obtained by quantum mechanical simulations. Hereby, it is planned to study the short time dynamics for a set of roughly 4–5 ion pairs in pre-equilibrated configurations. The fluctuation of the atomic polarization due to environment and rearrangement effects will be fully evaluated with regard to ab-initio Molecular Dynamics simulations. Herewith, quantum mechanical effects are taken into account via electronic density functional theory. It is planned to use the software package CP2K for the simulation of the system as well as the analysis of the data. For the development and the refinement of the polarizable atomistic force field, it is furthermore necessary to run a large number of classical calibration simulations to match experimental results and to refine the polarizable force field. These simulations will be performed with the software packages GROMACS or NAMD and will include 500 ion pairs to obtain statistically relevant data.

Appendix 1: Project A

See Figs. 1, 2, 3, 4, 5, 6, and 7.

Fig. 1 Snapshot of the planar capacitor system with BMIM (*blue*) PF4 (*red*) and the graphite electrodes (*gray*)

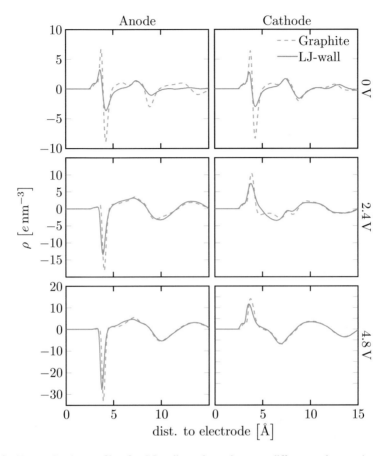

Fig. 2 Charge density profiles for LJ-walls and graphene at different voltages. A stronger influence of the electrode structure at low voltages is observed

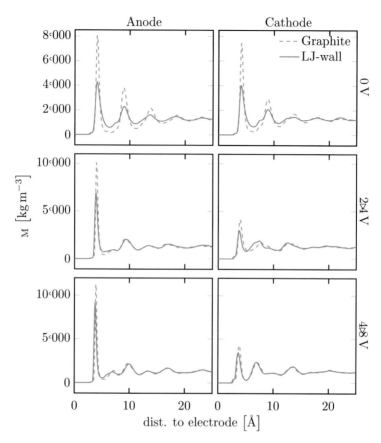

Fig. 3 Mass density profiles in front of the electrodes for LJ-walls and graphene electrodes at different voltages. Increased ion adsorption on the structured electrode appears at all voltages

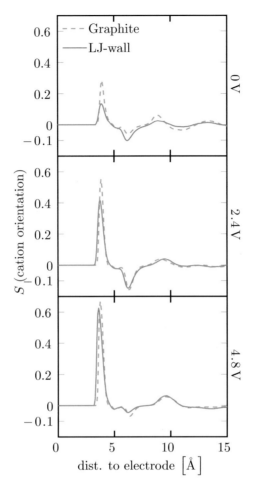

Fig. 4 Averaged lateral cation orientation in front of the cathode. Comparison between LJ-walls and graphene electrodes at different voltages. (Second Legendre polynomial $S = \frac{3}{2}\cos^2\alpha - \frac{1}{2}$; α: angle between normal vector of the wall and normal vector of the cation)

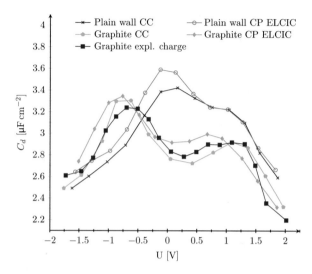

Fig. 5 The differential capacitance C_d measured for five different electrode models. Planar electrodes with constant charge (Plain wall CC) and constant potential (Plain wall CP ELCIC), graphene electrode model with constant charge (homogeneous electric field: Graphite CC; explicit charges on the carbon atoms of the innermost graphene sheet: Graphite expl. charge) and graphite electrodes with constant potential (Graphite CP ELCIC)

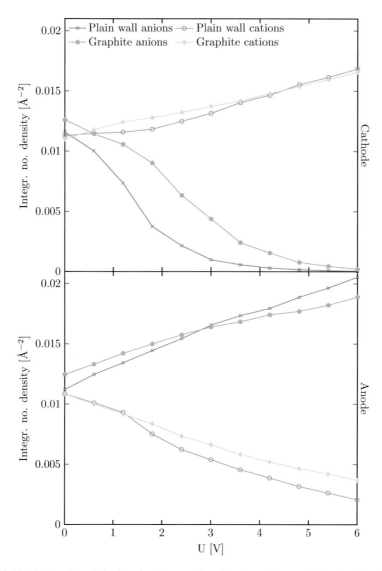

Fig. 6 First layer voltage behaviour for planar walls and graphene. In case of structured electrodes, the migration of co-ions is shifted to higher potentials. This effect is more pronounced for the spherical anion

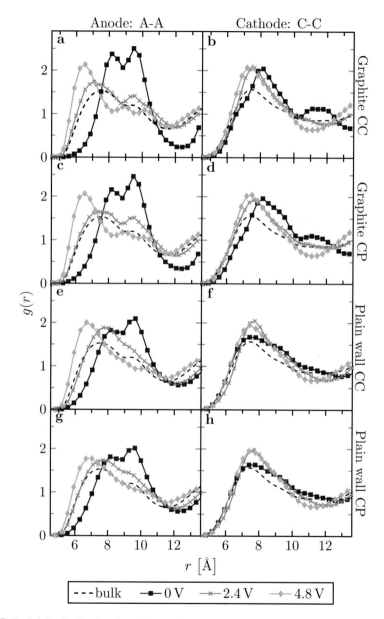

Fig. 7 Radial distribution functions (RDF) of the adsorbed ions in the first layer for three different voltages and different electrode models: Graphite electrodes and planar LJ-approximation at constant charge (CC) and constant potential (CP). *Left column*: PF6-PF6 RDF at the anode; *Right column*: BMIM-BMIM RDF at the cathode

Appendix 2: Project B

See Figs. 8 and 9.

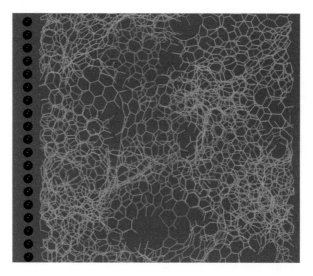

Fig. 8 Detailed view of the carbide-derived carbon network without ions. The average pore size of 8 Å and the particle diameters are close, which requires long trajectories to reach the final molecular composition in the pores. To obtain feasible particle numbers and simulation time, the system only represents the very surface of the electrode-electrolyte interface. In application, the CDC network can extend to macroscopic length scales which is out of scope for detailed atomistic simulations

Fig. 9 Snapshot of the carbide-derived carbon system with EMIM (*blue*), BF4 (*purple*), Acetonitril (*green*) and the CDC network (*cyan*)

References

1. Ohno, H.: Electrochemical Aspects of Ionic Liquids. Wiley, New Jersey (2005)
2. Bier, M., Dietrich, S.: Mol. Phys. **108**, 211 (2010)
3. Galiński, M., Lewandowski, A., Stepniak, I.: Electrochim. Acta **51**(26), 5567 (2006). doi:10.1016/j.electacta.2006.03.016. http://www.sciencedirect.com/science/article/pii/S0013468606002362
4. Simon, P., Gogotsi, Y.: Nat. Mater. **7**(11), 845 (2008)
5. Breitsprecher, K., Košovan, P., Holm, C.: J. Phys. Condens. Matter **26**, 284108 (2014). doi:10.1088/0953-8984/26/28/284108
6. Breitsprecher, K., Košovan, P., Holm, C.: J. Phys. Condens. Matter **26**, 284114 (2014). doi:10.1088/0953-8984/26/28/284114
7. Merlet, C., Rotenberg, B., Madden, P.A., Taberna, P.L., Simon, P., Gogotsi, Y., Salanne, M.: Nat. Mater. **11**(4), 306 (2012). doi:10.1038/NMAT3260
8. Merlet, C., Salanne, M., Rotenberg, B., Madden, P.A.: Electrochim. Acta **101**, 262 (2013)
9. Cole, M., Klein, J.: Surf. Sci. **124**(2), 547–554 (1983)
10. Steele, W.A.: Surf. Sci. **36**(1), 317 (1973). http://www.sciencedirect.com/science/article/B6TVX-46SXMDM-10V/2/7af1864fa233d6ca3b05170ca5882861
11. Steele, W.A.: J. Phys. Chem. **82**(7), 817 (1978). doi:10.1021/j100496a011
12. Arnold, A., Breitsprecher, K., Fahrenberger, F., Kesselheim, S., Lenz, O., Holm, C.: Entropy **15**(11), 4569 (2013). doi:10.3390/e15114569. http://www.mdpi.com/1099-4300/15/11/4569
13. Arnold, A., de Joannis, J., Holm, C.: J. Chem. Phys. **117**, 2496 (2002)
14. de Joannis, J., Arnold, A., Holm, C.: J. Chem. Phys. **117**, 2503 (2002)
15. Tyagi, S., Arnold, A., Holm, C.: J. Chem. Phys. **129**(20), 204102 (2008). http://www.link.aip.org/link/?JCP/129/204102/1
16. Tyagi, C., Süzen, M., Sega, M., Barbosa, M., Kantorovich, S., Holm, C.: J. Chem. Phys. **132**, 1154112 (2010)
17. Kesselheim, S., Sega, M., Holm, C.: Comput. Phys. Commun. **182**(1), 33 (2011)
18. Deserno, M., Holm, C.: J. Chem. Phys. **109**, 7678 (1998)
19. Arnold, A., Fahrenberger, F., Holm, C., Lenz, O., Bolten, M., Dachsel, H., Halver, R., Kabadshow, I., Gähler, F., Heber, F., Iseringhausen, J., Hofmann, M., Pippig, M., Potts, D., Sutmann, G.: Phys. Rev. E **88**, 063308 (2013). doi:10.1103/PhysRevE.88.063308
20. Roy, D., Patel, N., Conte, S., Maroncelli, M.: J. Phys. Chem. B **114**(25), 8410 (2010). doi:10.1021/jp1004709. http://www.pubs.acs.org/doi/abs/10.1021/jp1004709. PMID: 20536202
21. Roy, D., Maroncelli, M.: J. Phys. Chem. B **114**(39), 12629 (2010). doi:10.1021/jp108179n. http://www.pubs.acs.org/doi/abs/10.1021/jp108179n
22. Arnold, A., Lenz, O., Kesselheim, S., Weeber, R., Fahrenberger, F., Röhm, D., Košovan, P., Holm, C.: In: Griebel, M., Schweitzer, M.A. (eds.) Meshfree Methods for Partial Differential Equations VI. Lecture Notes in Computational Science and Engineering, vol. 89, pp. 1–23. Springer, Berlin (2013). doi:10.1007/978-3-642-32979-1_1. http://www.springer.com/mathematics/computational+science+%26+engineering/book/978-3-642-32978-4
23. Palmer, J., Llobet, A., Yeon, S.H., Fischer, J., Shi, Y., Gogotsi, Y., Gubbins, K., Carbon **48**(4), 1116 (2010). http://www.sciencedirect.com/science/article/pii/S0008622309007684
24. Merlet, C., Péan, C., Rotenberg, B., Madden, P.A., Daffos, B., Taberna, P.L., Simon, P., Salanne, M.: Nat. Commun. **4** (2013)
25. Péan, C., Merlet, C., Rotenberg, B., Madden, P.A., Taberna, P.L., Daffos, B., Salanne, M., Simon, P.: ACS Nano **8**(2), 1576 (2014)
26. Péan, C., Daffos, B., Merlet, C., Rotenberg, B., Taberna, P.L., Simon, P., Salanne, M.: J. Electrochem. Soc. **162**(5), A5091 (2015)
27. Largeot, C., Portet, C., Chmiola, J., Taberna, P.L., Gogotsi, Y., Simon, P.: J. Am. Chem. Soc. **130**(9), 2730 (2008). doi:10.1021/ja7106178

28. Merlet, C., Salanne, M., Rotenberg, B.: J. Phys. Chem. C **116**(14), 7687 (2012). doi:10.1021/jp3008877. http://www.pubs.acs.org/doi/abs/10.1021/jp3008877
29. Andersen, H.C.: J. Comput. Phys. **51**, 24 (1983)
30. Siepmann, J.I., Sprik, M.: J. Chem. Phys. **102**(1), 511 (1995). http://www.scitation.aip.org/content/aip/journal/jcp/102/1/10.1063/1.469429
31. Reed, S.K., Lanning, O.J., Madden, P.A.: J. Chem. Phys. **126**(8), 084704 (2007). http://www.scitation.aip.org/content/aip/journal/jcp/126/8/10.1063/1.2464084

The Small Scale Structure of the Universe

Stefan Gottlöber, Chris Brook, Ilian T. Iliev, and Keri L. Dixon

Abstract We describe a series of reionization and galaxy formation simulations performed at HLRS within the "Small Scale Structure of the Universe" (SSSU) project.

1 Introduction

In 2013, we submitted a project proposal to the JSC Juelich asking for computational time to continue our project "The Small Scale Structure of the Universe". We have been informed from Juelich that our project had received high ratings from the referees but, as the requested amount of computing time would be not available in Juelich, JSC suggested to move the whole project to HLRS where we got the requested time at Hermit. Unfortunately, the procedure of moving our project to a new computer and a different environment took substantially longer time than expected. In particular, we encountered some problems in the running and continuation of simulations started in Juelich. Because we could not run the planned simulations timely, these runs were done elsewhere. These complications resulted in some changes to the project plan.

First of all we concentrated ourselves on the early stage of cosmological evolution, the period of reionization. The role of reionization in regulating the formation of small scale structure has been recognized already more than two decades ago [9], however only recently new observational data and improved numerical methods

S. Gottlöber (✉)
Leibniz Institute for Astrophysics Potsdam, An der Sternwarte 16, 14482 Potsdam, Germany
e-mail: sgottloeber@aip.de

C. Brook
Departamento de Física Teórica, Módulo C-15, Facultad de Ciencias, Universidad Autónoma de Madrid, 28049 Madrid, Spain
e-mail: cbabrook@gmail.com

I.T. Iliev • K.L. Dixon
Department of Physics & Astronomy, Astronomy Centre, Pevensey II Building, University of Sussex, Falmer, Brighton BN1 9QH, UK
e-mail: I.T.Iliev@sussex.ac.uk; K.Dixon@sussex.ac.uk

© Springer International Publishing Switzerland 2016
W.E. Nagel et al. (eds.), *High Performance Computing in Science and Engineering '15*, DOI 10.1007/978-3-319-24633-8_8

allowed to study the influence of reionisation on the formation and evolution of dwarfs in detail. We performed a series of radiative transfer simulations to study the reionisation history.

In a second project we were running a set of simulations within the MaGICC (http://www2.mpia-hd.mpg.de/~stinson/magicc/, "Making Galaxies in a Cosmological Context") galaxy formation project. To this end we used the physical model of the MaGICC project and the initial conditions of the CLUES project (http://www. clues-project.org/, "Constrained Local UniversE Simulations") to create an accurate model of the Local Group of galaxies, including the Milky Way, Andromeda, M33 and the myriad of dwarf galaxies.

This report is organized as follows: In the first section we discuss the effect of low-mass ionizing source suppression on the 21-cm signal from the epoch of reionization. In the second session we discuss the CLUES simulations performed with the MaGICC galaxy formation model.

2   The Reionization Period

The first billion years of cosmic evolution remain the only period in the history of the universe still largely unconstrained by direct observations. While we now have fairly detailed data on the Cosmic Microwave Background originating from the last scattering surface at redshift $z \sim 1100$ and a wealth of multi-wavelength observations at later times, $z < 6$, the intermediate period remains largely uncharted. A number of ongoing observational programs aim to provide observations of this epoch in e.g. high-z Ly-α [29], CMB secondary anisotropies [56] and redshifted 21-cm [16, 32, 42]. Improved observational constraints could provide a wealth of information about the nature of the first stars and galaxies, their properties, abundances and clustering, the timing and duration of the reionization transition and the complex physics driving in this process.

During photoionization the excess photon energy above the Lyman limit heats the gas to temperatures above 10^4 K. The temperature reached generally depends on the local level of the ionizing flux and its spectrum [47]. Typical values are $T_{IGM} = 10{,}000$–$20{,}000$ K, but it could be as high as $\sim 40{,}000$ K for hot (Pop. III) black-body spectrum. However, the hydrogen line cooling is highly efficient for $T > 10^4$ K, particularly at high redshifts, where the gas is denser on average, which would typically bring its temperature down to $T_{IGM} \sim 10^4$ K, and possibly somewhat below that due to the adiabatic cooling from the expansion of the Universe.

This increase of the IGM temperature caused by its photo-heating results in a corresponding increase of the local Jeans mass. The Jeans mass is based on the linear theory of cosmological perturbations and for 10^4 K gas it is roughly $M_J \sim 10^9 M_\odot$ (with some redshift dependence) [19, 23, 46]. The actual galaxy mass under which the gas infall, and thus star formation, is suppressed differs somewhat from this instantaneous Jeans mass since the mass scale on which baryons succeed in collapsing out of the IGM along with the dark matter must be determined, even in linear theory, by integrating the differential equation for perturbation growth over time for the evolving IGM [11, 12, 46]. In reality, determining the minimum mass necessary for a halo collapsing inside an ionized and heated region to acquire its fair share of baryons which subsequently cool further to form stars is even more complicated. It depends on the detailed, non-linear, gas dynamics of the process and on radiative cooling. There is no single mass above which a collapsing halo retains all its gas, and below which the gas does not collapse with the dark matter. Instead, simulations show that the cooled gas fraction in halos decreases gradually with decreasing halo mass [8, 9, 38, 53].

The typical halo sizes at which this transition occurs as derived by these different studies also vary. Thoul and Weinberg [53] found that photoionization suppresses star formation in halos with circular velocities below $\sim$30 km s^{-1}, and decreases the cooled gas mass fraction in larger halos, with circular velocities up to $\sim$50 km s^{-1}. Navarro and Steinmetz [38] found that the cooled gas fraction is affected by photoionization even in larger galaxies, with circular velocities up to $\sim$100–200 km s^{-1}. On the other hand, Dijkstra et al. [8] recently showed, using the same method as [53], that at high redshifts the suppression is not as effective, and somewhat smaller galaxies can still retain some cooled gas. For simplicity, we assume that star formation is suppressed in halos with masses below $10^9 M_\odot$ and not suppressed in larger halos, in rough agreement with the linear Jeans mass estimate for 10^4 K gas and the above dynamical studies.

2.1   The Simulations

We start by performing very high resolution N-body simulations of the formation of high-redshift structures. We use the CubeP^{3}M N-body code [17]. For these simulations we used two computational volumes, 500 h^{-1} Mpc = 714 Mpc and 244 h1^{-1} Mpc = 349 Mpc, both chosen so as to be representative for the large-scale reionization patchiness [26]. The corresponding particle numbers, at 6912^3 and 4000^3 are chosen to ensure reliable halo identification down to $10^9 M_\odot$ (with 25 and 40 particles, respectively for the two volumes). As discussed above, $M_{halo} \sim 10^9 M_\odot$ is roughly the Jeans mass for gas at temperature 10^4 K, typical for the post-reionization IGM. The unresolved halos can be added using a sub-grid model, as discussed in detail in [1]. This model provides the mean local halo abundance based on the cell density and here is used to include halos with masses $10^8 M_\odot < M_{halo} < 10^9 M_\odot$ in both simulations.

The background cosmology is based on WMAP 5-year data combined with constraints from baryonic acoustic oscillations and high-redshift supernovae ($\Omega_M = 0.27, \Omega_\Lambda = 0.73, h = 0.7, \Omega_b = 0.044, \sigma_8 = 0.8, n = 0.96$). The linear power spectrum of density fluctuations was calculated with the code CAMB [31]. Initial conditions were generated using the Zel'dovich approximation at sufficiently high redshifts ($z_i = 150$) to ensure against numerical artefacts [6].

The radiative transfer simulations are performed with our code C^2-Ray (Conservative Causal Ray-Tracing) [35]. The method is explicitly photon-conserving in both space and time or individual sources and approximately (to a good approximation) photon-conserving for multiple sources, which ensures correct tracking of ionization fronts without loss of accuracy, independent of the spatial and time resolution, with corresponding great gains in efficiency. The code has been tested in detail against a number of exact analytical solutions [35], as well as in direct comparison with a number of other independent radiative transfer methods on a standardized set of benchmark problems [20, 24]. The ionizing radiation is ray-traced from every source to every grid cell using the short characteristics method, whereby the neutral column density between the source and a given cell is given by interpolation of the column densities of the previous cells which lie closer to the source, in addition to the neutral column density through the cell itself. The contribution of each source to the local photoionization rate of a given cell is first calculated independently, after which all contributions are added together and a non-equilibrium chemistry solver is used to calculate the resulting ionization state. Ordinarily, multiple sources contribute to the local photoionization rate of each cell. Changes in the rate modify the neutral fraction and thus the neutral column density, which in turn changes the photoionization rates themselves (since either more or less radiation reaches the cell). An iteration procedure is thus called for in order to converge to the correct, self-consistent solution.

The N-body simulations discussed above provide us with the spatial distribution of cosmological structures and their evolution in time. We then use this information as input to a full 3D radiative transfer simulations of the reionization history, as follows. We saved series of time-slices, both particle lists and halo catalogues from redshift 50 down to 6, uniformly spaced in time, every $\Delta t = 11.53$ Myr, a total of 82 slices. Simulating the transfer of ionizing radiation with the same grid resolution as the underlying N-body (fine grid of $13,824^3$ and 8000^3, respectively for the two volumes) is still not feasible on current computers. We therefore use a SPH-style smoothing scheme using nearest neighbors to transform the data to lower resolution, with 300^3 or 600^3 cells for the $500\,h^{-1}$ Mpc and $250\,h^{-1}$ Mpc boxes and 500^3 cells for the $244\,h^{-1}$ Mpc box, for the radiative transfer simulations. We combine sources which fall into the same coarse cell, which reduces slightly the number of sources to be considered compared to the total number of halos.

We characterize our source efficiencies through a factor g_γ. We assign to each an ionizing photon production rate per unit time, $\dot{N}_\gamma$, proportional to the total mass in haloes within that cell, M as introduced in [21, 22, 25]:

$$\dot{N}_\gamma = \frac{g_\gamma M \Omega_b}{\Omega_0 m_p} \left(\frac{\Delta t}{10\,\text{Myr}} \right), \tag{1}$$

where m_p is the proton mass and $g_\gamma = f_{\text{esc}} f_\star N_\star \left(\frac{10\,\text{Myr}}{\Delta t} \right)$ is an ionizing photon production efficiency parameter which includes the efficiency of converting gas into stars, $f_\star$, the ionizing photon escape fraction from the halo into the IGM,

f_{esc} and the number of ionizing photons produced per stellar atom, $N_\star$, Δt is the time between two snapshots from the N-body simulation. All simulations include an approximate treatment of Lyman-Limit absorber systems. During the early evolution the photon mean free path is set by the neutral patches and Lyman-Limit Systems are unimportant, while at late times they set a mean free path of several tens of Mpc [48]. In the current simulations we roughly model this by imposing a hard limit on the distance an ionizing photon can travel, set at 40 comoving Mpc.

We have performed series of radiative transfer simulations with varying underlying assumptions about the source efficiencies and the suppression conditions imposed on the low-mass sources. The ones performed here include two models, as follows:

- *Partially suppressed low-mass halos (pS):*
  For this model, the low-mass halos ($10^8 M_\odot < M < 10^9 M_\odot$) contribute to reionization at all times. In neutral regions, we assign them a higher efficiency of $g_\gamma = 7.1$. In ionized regions, these small galaxies are suppressed resulting in diminished efficiency, with $g_\gamma = 1.7$. This situation arises if star formation remains ongoing, but at a lower rate. Physically, this situation could arise if the fresh gas supply is cut off or diminished by the photo-heating of the surrounding gas, but a gas reservoir remains available for star formation in the galaxy itself. In this case the high-mass sources ($M > 10^9 M_\odot$) are assigned efficiency $g_\gamma = 1.7$.
- *Mass-dependent suppression of low-mass halos (gS):*
  Instead of a sharp decrease in ionizing efficiency as in the previous case, we also consider the gradual suppression of sources in ionized regions. As before, high-mass sources are assigned $g_\gamma = 1.7$ everywhere, and low-mass ones have that same efficiency when in neutral regions. In ionized patches, the low-mass sources are suppressed in a linearly mass-dependent manner, with $10^8 M_\odot$ fully suppressed and $10^9 M_\odot$ not suppressed at all.

2.2 Specific Simulations Ran and Sample Results

The simulations ran here form part of a larger PRACE Tier-0 project (Project PRACE4LOFAR, 22+19M core-hours awarded in fifth and ninth call), which is mostly based on Curie (CEA, France), with some work done on clusters in Sweden and elsewhere under a separately-granted project. At HLRS we have ran 3 simulations. Two were ran on Hermit and were based on the 244 h^{-1} Mpc volume, with 250^3 grid size, and used the 'pS' and 'gs' suppression models, respectively. The third simulation (currently incomplete, since the allocation ran out) is based on the 500 h^{-1} Mpc volume, with 600^3 grid and is using the 'pS' suppression model. The Hermit simulations were ran on 16,384 cores (4096 MPI processes × 4 OpenMP threads). The Hornet simulation was started on 6144 cores, increasing to 12,288 and 18,432 cores as more sources formed, creating more computational work. This used 8 OpenMP threads due to higher memory footprint of the larger grid, and 768, 1536 and 2304 MPI processes, respectively. The three runs used approximately 360,000, 400,000 and 450,000 core-hours, respectively.

Once the runs were set up properly and got going there were no significant issues encountered, apart from the fairly long queues (requiring up to a week or more to get through). Both Hermit and Hornet were fairly easy to use, although the wiki help was confusing at times and probably could be organized better and more transparently. The code ran well and fairly efficiently and fast, particularly on Hornet.

On the other hand, the workspace mechanism used at HLRS is by far the most opaque and user-unfriendly we have ever encountered for the many years we have used supercomputers. Due to this we have lost our codes and setup several times, resulting in a significant waste of time and efforts, and much frustration. If such an extreme disk management policy is really required (all other centres we have used do not need it, so this is not obvious), at the very least it should be set up so that it gives some kind of automatic warning before workspaces will be deleted, and maybe a short grace period, so as to ensure no data or codes are lost. Such a (fairly minor) change should make the systems much more user-friendly without really impacting on performance or disk management significantly.

Some preliminary results from our simulations are shown in Fig. 1. This shows the evolution of the differential brightness temperature of the redshifted 21-cm line

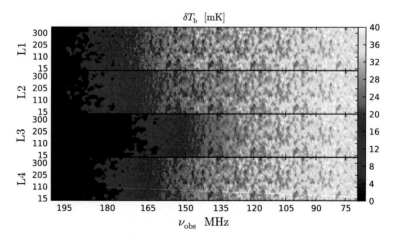

Fig. 1 Position-frequency slices from our 244 h^{-1} Mpc volume simulations. These slices illustrate the large-scale geometry of reionization and the significant local variations in reionization history for several of our simulations as seen at redshifted 21-cm line (shown is the differential brightness temperature in mK) smoothed with a 5 arcmin gaussian beam and 0.4 MHz (boxcar) bandwidth filter. The spatial scale is given in comoving Mpc. The models partly ran here are L3 (pS) and L4 (gs)

from neutral hydrogen as seen by a LOFAR-like experiment (assuming no noise and foregrounds). The full analysis of the results is currently being performed and will be published in series of papers, the first of which is currently being finalized (Dixon et al., in preparation). These simulations will form the core of a library of models for analyzing and interpreting the results from the LOFAR Epoch of Reionization Key Science Project.

3   MaGICC CLUES to Galaxy Formation and Cosmology

The "Making Galaxies in a Cosmological Context" (MaGICC) project [5, 51] formed a suite of isolated, hydro-dynamically simulated galaxies that match observed galaxy scaling relations over a large mass range, from dwarf galaxies to Milky Way analogues.

The CLUES-project (Constrained Local Universe Simulations, [13]) provides constrained simulations of the local universe designed to be used as a numerical laboratory of the current paradigm of cold dark matter cosmology. The Local Group and its environment is the most well observed region of the universe. Only in this unique environment can we study structure formation on scales as small as that of very low mass dwarf galaxies.

In this project we are combining the physical model of the MaGICC project with the initial conditions of CLUES to simulate an accurate model of the Local Group of galaxies, including the Milky Way, Andromeda, M33 and the myriad dwarf galaxies.

The simulations will be used for unprecedented analysis of the complex dark matter and gas-dynamical processes which govern the formation of galaxies. The predictions of these experiments can be easily compared with the detailed observations of our galactic neighborhood. By simulating a local environment, more constraints can be placed on our model than were available with an isolated suite of galaxies, such as the requirement to match luminosity function. Outstanding issues such as the missing satellite, cusp/core and "too big to fail" problem will be probed in search of a self-consistent solution.

Moving the detailed modeling of the galaxy formation physics that is used in isolated galaxies in the MaGICC project, to the larger volume of the Local Group provided by the CLUES initial conditions, requires significant computational expenditure, making massively parallel supercomputing facilities essential.

In this introduction to the MaGICC CLUES program, we use simulated galaxies from the well studied MaGICC program to compare with isolated galaxies from the new MagiCC CLUES simulations.

3.1   Simulations and the MagICC Model

The MagICC simulations have previously been shown to match a wide range of
scaling relations including the Tully-Fisher, luminosity-size, mass-metallicity, and
HI to stellar-mass relations at z = 0 [5]. The simulations also match the evolution
of the stellar mass-halo mass relation [27, 51], as derived by abundance matching
[37] and several relations at high redshift [40]. The simulations also expel sufficient
metals to match local observations [44, 54] of OVI in the circum-galactic medium
[5, 50].

Therefore, our first goal is to ensure that the isolated MagICC CLUES simulations retain similar properties as the MagICC simulations. In Table 1 we list
properties of 12 galaxies from the MaGICC project [4, 51], which were zoomed-
in regions of a total cosmological volume of side 68 Mpc. Both sets of galaxies
use a ΛCDM cosmology with WMAP3 parameters, i.e. $H_0 = 73 \,\mathrm{km\,s^{-1}\,Mpc^{-1}}$,
$\Omega_m = 0.24$, $\Omega_\Lambda = 0.76$, $\Omega_{baryon} = 0.04$ and $\sigma_8 = 0.76$.

The second set are from a single CLUES simulation. Again the zoom-in
technique is used, this time together with observational data imposed as constraints
on the initial conditions, in order to simulate a cosmological volume with structures
similar to those most representative in our local universe. Several dark matter-
only realizations are run until a Local Group analogue is found. Then this Local
Group region is re-simulated with baryons and at a higher resolution. The CLUES
simulation used also follows a WMAP3 cosmology.

The 12 MaGICC disk galaxies listed in Table 1 are separated into two sub-sets
labelled as Milky-Way (MW) and irregular (Irr) type galaxies, although all are disc
galaxies with stellar masses ranging from 1×10^8–$5\times10^{10} M_\odot$. From the CLUES
simulation we have selected the halos that satisfy the following conditions, in order
to compare isolated galaxies with the MagICC suit: (1) Not a sub-halo, (2) M_{halo}
$>4\times10^{10} M_\odot$. These integrate a sample of 10 well resolved isolated galaxies. Since
this is a Local Group simulation, the three most massive galaxies are loose analogues
of the Milky Way, M31 and M33, and the rest are isolated dwarf galaxies.

Halos in both simulations have been identified using Amiga's Halo Finder (AHF;
[28]). Halo masses are defined as the mass inside a sphere containing $\Delta_{vir} \simeq 350$
times the cosmic background matter density at redshift z = 0.

Table 1 Properties of isolated simulated galaxies ordered by halo mass

Name	M_{halo} ($M_\odot$)	M_{star} ($M_\odot$)	M_{HII} ($M_\odot$)	h (kpc)	μ_0 (mag as^{-1})	V_{max} (km s^{-1})	V_{flat} (km s^{-1})
g15784_MW	1.49×10^{12}	5.67×10^{10}	1.96×10^{10}	4.64	20.45	221.34	221.34
g21647_MW	8.24×10^{11}	2.51×10^{10}	5.62×10^{9}	1.40	18.44	189.61	164.13
g1536_MW	7.10×10^{11}	2.36×10^{10}	6.78×10^{9}	3.76	21.40	174.61	174.61
g5664_MW	5.39×10^{11}	2.74×10^{10}	4.19×10^{9}	2.30	20.30	196.66	151.75
g7124_MW	4.47×10^{11}	6.30×10^{9}	3.49×10^{9}	2.58	21.31	120.00	120.00
g15807_Irr	2.82×10^{11}	1.46×10^{10}	4.68×10^{9}	2.27	19.95	140.65	140.65
g15784_Irr	1.70×10^{11}	4.26×10^{9}	2.70×10^{9}	2.07	20.90	106.96	106.96
g22437_Irr	1.10×10^{11}	7.44×10^{8}	1.08×10^{9}	1.63	21.83	75.26	75.26
g21647_Irr	9.65×10^{10}	1.98×10^{8}	3.68×10^{8}	1.64	23.43	60.58	60.58
g1536_Irr	8.04×10^{10}	4.46×10^{8}	4.39×10^{8}	1.94	23.12	66.85	66.85
g5664_Irr	5.87×10^{10}	2.36×10^{8}	2.56×10^{8}	1.72	23.22	59.08	59.08
g7124_Irr	5.23×10^{10}	1.32×10^{8}	2.30×10^{8}	1.23	22.79	52.72	52.72
M-CLUES1	7.23×10^{11}	1.45×10^{10}	2.41×10^{9}	1.38	20.09	167.74	127.17
M-CLUES2	5.31×10^{11}	1.11×10^{10}	3.92×10^{8}	1.99	21.59	123.59	123.59
M-CLUES3	2.67×10^{11}	5.08×10^{9}	1.46×10^{9}	3.37	22.88	119.77	119.77
M-CLUES4	1.87×10^{11}	4.18×10^{9}	4.66×10^{7}	1.49	21.12	101.17	101.17
M-CLUES5	1.51×10^{11}	4.54×10^{9}	1.50×10^{9}	2.21	22.77	116.47	116.47
M-CLUES6	1.29×10^{11}	2.08×10^{9}	1.43×10^{9}	1.52	22.21	101.36	101.36
M-CLUES7	1.18×10^{11}	2.82×10^{9}	8.94×10^{8}	1.76	21.67	88.45	71.34
M-CLUES8	1.21×10^{11}	1.57×10^{9}	4.31×10^{8}	0.94	20.74	85.19	85.19
M-CLUES9	8.04×10^{10}	1.10×10^{9}	8.17×10^{7}	2.01	24.36	70.87	70.87
M-CLUES10	6.44×10^{10}	3.78×10^{8}	5.17×10^{7}	0.92	22.71	53.32	53.32

Disk scale lengths h and central surface brightnesses μ_0 are derived from exponential fits to the surface brightness profile in the V band

3.2   Simulated Galaxy Properties

3.2.1   Rotation Curves

Rotation curves of observed galaxies provide significantly more information regarding the angular momentum of galaxies than is contained within the Baryonic Tully-Fisher relation, allowing more stringent constraints on galaxy formation models; constraints that have not previously been applied to simulated galaxies. High resolution observations of HI velocities, combined with studies of the gas and stellar mass distributions, provide detailed information on how the different mass components are radially distributed in galaxies with a wide range of rotational velocities V_r (e.g. [3, 7, 10, 30, 41, 45]).

Figures 2 and 3 show in different symbols the gaseous, stellar, baryonic and total rotation curves of the MaGICC and M-CLUES simulated galaxies, respectively. These are measured at radii ranging from 0.7 kpc to 10×h where h is the disc scale length. The rotational velocities are calculated using the mass within spherical radial shells in the expression $V_{circ}(r) = \sqrt{(GM(r)/r)}$, where G is the gravitational constant.

The simulated galaxies reach a flat value of the velocity which persists to large radii, and they lack the strong peak at small radii that not so long ago was ubiquitous in simulations due to overcooling. A couple of galaxies, g5665_MW and MC1, have significant bulges, which is reflected in the heightened inner region of their rotation curves. Differences between the two sets of simulations are not evident,

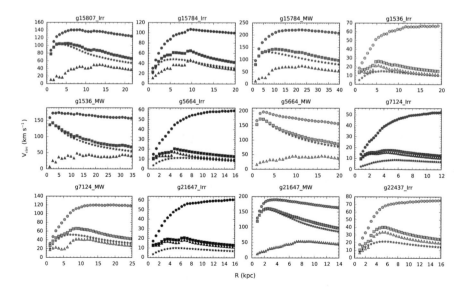

Fig. 2 The rotation curves of the 12 MaGICC disk galaxies. Different symbols represent the rotation values due to different mass components (*triangles*: cold gas; *stars*: stars; *squares*: all baryons; *circles*: total). Simulations reproduce the variety of observed rotation curves. Furthermore, like in observations, the features present in the baryonic curves are reflected in the total one

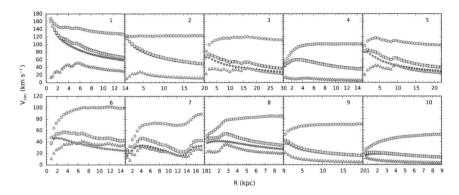

Fig. 3 The rotation curves of the ten galaxies selected from the CLUES simulation. They reach lower circular velocities than the MaGICC sample

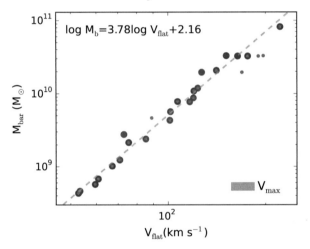

Fig. 4 *Left panel*: The baryonic Tully-Fisher relation. Total baryonic mass M_b (stars + cold gas) plotted against circular velocity V_{flat}. *Blue* points are from the MagiCC suite while the *red* points are the CLUES simulations, with the line showing linear fit, with slope = 3.76. The small *green* points show V_{max} rather than V_{flat}, which results in a slightly flatter relation, slope = 3.43, and slightly larger scatter (see text for details). *Right panel*: The stellar-to-halo mass relation. Also shown are the empirical stellar-halo mass relations of [15] (*green line*) and [36]. The MagICC galaxies are in *blue* while the CLUES simulations are in *red*

with galaxies of similar masses reaching similar maximum, flat velocities (see for example g7124_MW & MC3, g15807_Irr & MC1 or g5664_Irr & MC10).

3.2.2 Stellar Mass-Halo Mass

In the right panel of Fig. 4 the stellar-to-halo mass relation is plotted, along with the empirical relation [15, 36]. The MagICC simulations actually were tuned to match the stellar mass halo mass relation at one galaxy mass, and shown to then match the relation over a range of masses ([5, 40] and also the Nihao simulations [55] which use very similar implementation of physics as the MaGICC runs). The CLUES simulation were also calibrated, from experience, to match the relation. So, although in some sense it is not surprising that the simulations match the relation to which they were tuned, they actually match the relation over a far wider mass range than the one on which the parameter search was performed.

3.2.3 The Baryonic Tully-Fisher Relation

In the left panel of Fig. 4 we plot the Baryonic Tully-Fisher relation, that is the mass of baryons of each simulated galaxies as a function of their circular velocity. There has been significant progress over the past years in our ability to simulate these processes of disc formation within a cosmological context. Without an efficient feedback scheme, angular momentum is lost to dynamical friction during the mergers of overly dense sub-structures (e.g. [33, 39, 43]). Progress was made by implementing increasingly effective recipes for feedback from supernovae [49, 52] and the inclusion of other forms of feedback from massive stars [18, 51]. The benchmark for assessing this progress has primarily been the ability to match the Tully-Fisher relation (e.g. [14]), with recent simulations matching the relation, and in particular the Baryonic Tully Fisher relation (BTFR), for galaxies over a range of masses [2, 5]. The latest simulations shown in Fig. 4 show perhaps the best reproduction of the BTFR for simulated galaxies to date.

The maximum circular velocity found in each simulated galaxy, V_{max}, is a good approximation of the flat velocity , V_{flat}, in most cases. In the cases mentioned above where a couple of MW type galaxies have significant bulges, we show different values of V_{max} and V_{flat} in Table 1. In the left panel of Fig. 4 we plot the BTFR using V_{flat}, with the MaGICC and M-CLUES sets of simulations shown as blue and red dots, respectively. Here the baryonic mass is defined as the sum of the mass coming from stars and cold gas particles, where the latter are estimated as a multiple of the atomic HI gas mass $M_g = \eta M_{HI}$, with $\eta = 4/3$ (following e.g. [34]). In the case of MC7, the galaxy is about to undergo a merger, and we use the maximum velocity from the inner 10 kpc as V_{flat} which is the central galaxy, and use the baryonic mass from within this same radius. The scatter is very small, with the galaxy that is furthest from the fit being MC7, the one which has a very close companion galaxy with which it is dynamically interacting.

If we simply use V_{max} in each case, the relation is slightly flatter, and can be seen as small green dots in Fig. 4, with a slightly larger scatter than in the case of V_{flat}. These relations are consistent with the observational fits found in the literature (see [34] for a summary), as is the trend for a flatter relation with greater scatter when using V_{max} rather than V_{flat}.

4 Summary

After a significant delay in our project due to problems in the running and continuation of simulations started in Juelich we defined two new projects within our SSSU project at HLRS, namely radiative transfer simulations to study the reionisation history and simulating the formation of the Local Group using the physical model of the MaGICC project. Besides some problems with automatically deleted workspaces and the resulting delay in the project the simulations went very well at Hermit and Hornet. The preliminary results presented in the two sections of this report are not yet published. We plan to submit the corresponding papers by the end of the year.

Acknowledgements This work was supported by the Science and Technology Facilities Council [grant number ST/L000652/1].

References

1. Ahn, K., Iliev, I.T., Shapiro, P.R., Srisawat, C.B.: Nonlinear bias of cosmological halo formation in the early universe. Mon. Not. R. Astron. Soc. **450**, 1486 (2015)
2. Aumer, M., White, S.D.M., Naab, T., Scannapieco, C.: Towards a more realistic population of bright spiral galaxies in cosmological simulations. Mon. Not. R. Astron. Soc. **434**, 3142 (2013)
3. Begeman, K.G., Broeils, A.H., Sanders, R.H.: Extended rotation curves of spiral galaxies - dark haloes and modified dynamics. Mon. Not. R. Astron. Soc. **249**, 523–537 (1991)
4. Brook, C.B., Stinson, G., Gibson, B.K., Roškar, R., Wadsley, J., Quinn, T.: Hierarchical formation of bulgeless galaxies - II. Redistribution of angular momentum via galactic fountains. Mon. Not. R. Astron. Soc. **419**, 771–779 (2012). doi:10.1111/j.1365-2966.2011.19740.x
5. Brook, C.B., Stinson, G., Gibson, B.K., Wadsley, J., Quinn, T.: MaGICC discs: matching observed galaxy relationships over a wide stellar mass range. Mon. Not. R. Astron. Soc. **424**, 1275–1283 (2012). doi:10.1111/j.1365-2966.2012.21306.x
6. Crocce, M., Pueblas, S., Scoccimarro, R.: Transients from initial conditions in cosmological simulations. Mon. Not. R. Astron. Soc. **373**, 369–381 (2006). doi:10.1111/j.1365-2966.2006.11040.x
7. de Blok, W.J.G., McGaugh, S.S., Bosma, A., Rubin, V.C.: Mass density profiles of low surface brightness galaxies. Astrophys. J. Lett. **552**, L23–L26 (2001). doi:10.1086/320262
8. Dijkstra, M., Haiman, Z., Rees, M.J., Weinberg, D.H.: Photoionization feedback in low-mass galaxies at high redshift. Astrophys. J. **601**, 666–675 (2004). doi:10.1086/380603
9. Efstathiou, G.: Suppressing the formation of dwarf galaxies via photoionization. Mon. Not. R. Astron. Soc. **256**, 43P–47P (1992)
10. Gentile, G., Salucci, P., Klein, U., Vergani, D., Kalberla, P.: The cored distribution of dark matter in spiral galaxies. Mon. Not. R. Astron. Soc. **351**, 903–922 (2004). doi:10.1111/j.1365-2966.2004.07836.x
11. Gnedin, N.Y.: Effect of reionization on structure formation in the universe. Astrophys. J. **542**, 535–541 (2000). doi:10.1086/317042
12. Gnedin, N.Y., Hui, L.: Probing the Universe with the Lyalpha forest - I. Hydrodynamics of the low-density intergalactic medium. Mon. Not. R. Astron. Soc. **296**, 44–55 (1998)
13. Gottlöber, S., Hoffman, Y., Yepes, G.: Constrained local universe simulations (CLUES). In: Wagner, S., Steinmetz, M., Bode, A., Müller, M.M. (eds.) High Performance Computing in Science and Engineering, p. 309. Springer (2010)

14. Governato, F., Mayer, L., Wadsley, J., Gardner, J.P., Willman, B., Hayashi, E., Quinn, T., Stadel, J., Lake, G.: The formation of a realistic disk galaxy in Λ-dominated cosmologies. Astrophys. J. **607**, 688–696 (2004). doi:10.1086/383516

15. Guo, Q., White, S., Li, C., Boylan-Kolchin, M.: How do galaxies populate dark matter haloes? Mon. Not. R. Astron. Soc. **404**, 1111–1120 (2010). doi:10.1111/j.1365-2966.2010.16341.x

16. Harker, G., Zaroubi, S., Bernardi, G., Brentjens, M.A., de Bruyn, A.G., Ciardi, B., Jelić, V., Koopmans, L.V.E., Labropoulos, P., Mellema, G., Offringa, A., Pandey, V.N., Pawlik, A.H., Schaye, J., Thomas, R.M., Yatawatta, S.: Power spectrum extraction for redshifted 21-cm epoch of reionization experiments: the LOFAR case. Mon. Not. R. Astron. Soc. **405**, 2492–2504 (2010). doi:10.1111/j.1365-2966.2010.16628.x

17. Harnois-Déraps, J., Pen, U.L., Iliev, I.T., Merz, H., Emberson, J.D., Desjacques, V.: High-performance P^3M N-body code: CUBEP^{3}M. Mon. Not. R. Astron. Soc. **436**, 540–559 (2013). doi:10.1093/mnras/stt1591

18. Hopkins, P.F., Kereš, D., Oñorbe, J., Faucher-Giguère, C.A., Quataert, E., Murray, N., Bullock, J.S.: Galaxies on FIRE (feedback in realistic environments): stellar feedback explains cosmologically inefficient star formation. Mon. Not. R. Astron. Soc. **445**, 581–603 (2014). doi:10.1093/mnras/stu1738

19. Iliev, I.T., Shapiro, P.R., Ferrara, A., Martel, H.: On the direct detectability of the cosmic dark ages: 21 centimeter emission from minihalos. Astrophys. J. Lett. **572**, L123–L126 (2002)

20. Iliev, I.T., et al.: Cosmological radiative transfer codes comparison project - I. The static density field tests. Mon. Not. R. Astron. Soc. **371**, 1057–1086 (2006). doi:10.1111/j.1365-2966.2006.10775.x

21. Iliev, I.T., Mellema, G., Pen, U.L., Merz, H., Shapiro, P.R., Alvarez, M.A.: Simulating cosmic reionization at large scales - I. The geometry of reionization. Mon. Not. R. Astron. Soc. **369**, 1625–1638 (2006). doi:10.1111/j.1365-2966.2006.10502.x

22. Iliev, I.T., Mellema, G., Shapiro, P.R., Pen, U.L.: Self-regulated reionization. Mon. Not. R. Astron. Soc. **376**, 534–548 (2007). doi:10.1111/j.1365-2966.2007.11482.x

23. Iliev, I.T., Mellema, G., Pen, U., Bond, J.R., Shapiro, P.R.: Current models of the observable consequences of cosmic reionization and their detectability. Mon. Not. R. Astron. Soc. **384**, 863–874 (2008). doi:10.1111/j.1365-2966.2007.12629.x

24. Iliev, I.T., et al.: Cosmological radiative transfer comparison project - II. The radiation-hydrodynamic tests. Mon. Not. R. Astron. Soc. **400**, 1283–1316 (2009). doi:10.1111/j.1365-2966.2009.15558.x

25. Iliev, I.T., Mellema, G., Shapiro, P.R., Pen, U.L., Mao, Y., Koda, J., Ahn, K.: Can 21-cm observations discriminate between high-mass and low-mass galaxies as reionization sources? Mon. Not. R. Astron. Soc. **423**, 2222–2253 (2012). doi:10.1111/j.1365-2966.2012.21032.x

26. Iliev, I.T., Mellema, G., Ahn, K., Shapiro, P.R., Mao, Y., Pen, U.L.: Simulating cosmic reionization: how large a volume is large enough? Mon. Not. R. Astron. Soc. **439**, 725–743 (2014). doi:10.1093/mnras/stt2497

27. Kannan, R., Stinson, G.S., Macciò, A.V., Brook, C., Weinmann, S.M., Wadsley, J., Couchman, H.M.P.: The MaGICC volume: reproducing statistical properties of high redshift galaxies. ArXiv.1302.2618 (2013)

28. Knollmann, S.R., Knebe, A.: AHF: Amiga's Halo Finder. Astrophys. J. Suppl. **182**, 608–624 (2009). doi:10.1088/0067-0049/182/2/608

29. Krug, H.B., Veilleux, S., Tilvi, V., Malhotra, S., Rhoads, J., Hibon, P., Swaters, R., Probst, R., Dey, A., Dickinson, M., Jannuzi, B.T.: Searching for z ~ 7.7 Lyα emitters in the COSMOS field with NEWFIRM. Astrophys. J. **745**, 122 (2012). doi:10.1088/0004-637X/745/2/122

30. Kuzio de Naray, R., McGaugh, S.S., de Blok, W.J.G., Bosma, A.: High-resolution optical velocity fields of 11 low surface brightness galaxies. Astrophys. J. Suppl. **165**, 461–479 (2006). doi:10.1086/505345

31. Lewis, A., Challinor, A., Lasenby, A.: Efficient computation of CMB anisotropies in closed FRW models. Astrophys. J. **538**, 473–476 (2000)

32. Lonsdale, C.J., et al.: The Murchison widefield array: design overview. IEEE Proc. **97**, 1497–1506 (2009). doi:10.1109/JPROC.2009.2017564

33. Maller, A.H., Dekel, A.: Towards a resolution of the galactic spin crisis: mergers, feedback and spin segregation. Mon. Not. R. Astron. Soc. **335**, 487–498 (2002). doi:10.1046/j.1365-8711. 2002.05646.x

34. McGaugh, S.S.: The Baryonic Tully-Fisher relation of gas-rich galaxies as a test of ΛCDM and MOND. Astron. J. **143**, 40 (2012). doi:10.1088/0004-6256/143/2/40

35. Mellema, G., Iliev, I.T., Alvarez, M.A., Shapiro, P.R.: C^2-ray: a new method for photon-conserving transport of ionizing radiation. New Astron. **11**, 374–395 (2006). doi:10.1016/j.newast.2005.09.004

36. Moster, B.P., Somerville, R.S., Maulbetsch, C., van den Bosch, F.C., Macciò, A.V., Naab, T., Oser, L.: Constraints on the relationship between stellar mass and halo mass at low and high redshift. Astrophys. J. **710**, 903–923 (2010). doi:10.1088/0004-637X/710/2/903

37. Moster, B.P., Naab, T., White, S.D.M.: Galactic star formation and accretion histories from matching galaxies to dark matter haloes. Mon. Not.R. Astron. Soc. **428**, 3121–3138 (2013). doi:10.1093/mnras/sts261

38. Navarro, J.F., Steinmetz, M.: The effects of a photoionizing ultraviolet background on the formation of disk galaxies. Astrophys. J. **478**, 13–+ (1997). doi:10.1086/303763

39. Navarro, J.F., Steinmetz, M.: Dark halo and disk galaxy scaling laws in hierarchical universes. Astrophys. J. **538**, 477–488 (2000)

40. Obreja, A., Brook, C.B., Stinson, G., Domínguez-Tenreiro, R., Gibson, B.K., Silva, L., Granato, G.L.: The main sequence and the fundamental metallicity relation in MaGICC Galaxies: evolution and scatter. Mon. Not. R. Astron. Soc. **442**, 1794–1804 (2014). doi: 10.1093/mnras/stu891

41. Oh, S.-H., Hunter, D.A., Brinks, E., Elmegreen, B.G., Schruba, A., Walter, F., Rupen, M.P., Young, L.M., Simpson, C.E., Johnson, M.C., Herrmann, K.A., Ficut-Vicas, D., Cigan, P., Heesen, V., Ashley, T., Zhang, H.-X.: High-resolution Mass Models of Dwarf Galaxies from LITTLE THINGS. Astron. J. **149**, 180 (2015)

42. Parsons, A.R., et al.: The precision array for probing the epoch of re-ionization: eight station results. Astron. J. **139**, 1468–1480 (2010). doi:10.1088/0004-6256/139/4/1468

43. Piontek, F., Steinmetz, M.: The modelling of feedback processes in cosmological simulations of disc galaxy formation. Mon. Not. R. Astron. Soc. **410**, 2625–2642 (2011). doi:10.1111/j. 1365-2966.2010.17637.x

44. Prochaska, J.X., Weiner, B., Chen, H.W., Mulchaey, J., Cooksey, K.: Probing the intergalactic medium/galaxy connection. V. On the origin of Lyα and O VI absorption at z < 0.2. Astrophys. J. **740**, 91 (2011). doi:10.1088/0004-637X/740/2/91

45. Sanders, R.H., Verheijen, M.A.W.: Rotation curves of Ursa major galaxies in the context of modified newtonian dynamics. Astrophys. J. **503**, 97–108 (1998). doi:10.1086/305986

46. Shapiro, P.R., Giroux, M.L., Babul, A.: Reionization in a cold dark matter universe: the feedback of galaxy formation on the intergalactic medium. Astrophys. J. **427**, 25–50 (1994). doi:10.1086/174120

47. Shapiro, P.R., Iliev, I.T., Raga, A.C.: Photoevaporation of cosmological minihaloes during reionization. Mon. Not. R. Astron. Soc. **348**, 753–782 (2004)

48. Songaila, A., Cowie, L.L.: Approaching reionization: the evolution of the Ly α forest from z = 4 to z = 6. Astron. J. **123**, 2183–2196 (2002). doi:10.1086/340079

49. Stinson, G., Seth, A., Katz, N., Wadsley, J., Governato, F., Quinn, T.: Star formation and feedback in smoothed particle hydrodynamic simulations - I. Isolated galaxies. Mon. Not. R. Astron. Soc. **373**, 1074–1090 (2006). doi:10.1111/j.1365-2966.2006.11097.x

50. Stinson, G.S., Brook, C., Prochaska, J.X., Hennawi, J., Shen, S., Wadsley, J., Pontzen, A., Couchman, H.M.P., Quinn, T., Macciò, A.V., Gibson, B.K.: MAGICC haloes: confronting simulations with observations of the circumgalactic medium at z = 0. Mon. Not. R. Astron. Soc. **425**, 1270–1277 (2012). doi:10.1111/j.1365-2966.2012.21522.x

51. Stinson, G.S., Brook, C., Macciò, A.V., Wadsley, J., Quinn, T.R., Couchman, H.M.P.: Making galaxies in a cosmological context: the need for early stellar feedback. Mon. Not. R. Astron. Soc. **428**, 129–140 (2013). doi:10.1093/mnras/sts028

52. Thacker, R.J., Couchman, H.M.P.: Star formation, supernova feedback, and the angular momentum problem in numerical cold dark matter cosmogony: halfway there? Astrophys. J. Lett. **555**, L17–L20 (2001). doi:10.1086/321739
53. Thoul, A.A., Weinberg, D.H.: Hydrodynamic simulations of galaxy formation. II. Photoionization and the formation of low-mass galaxies. Astrophys. J. **465**, 608–+ (1996). doi:10.1086/177446
54. Tumlinson, J., Thom, C., Werk, J.K., Prochaska, J.X., Tripp, T.M., Weinberg, D.H., Peeples, M.S., OMeara, J.M., Oppenheimer, B.D., Meiring, J.D., Katz, N.S., Davé, R., Ford, A.B., Sembach, K.R.: The large, oxygen-rich halos of star-forming galaxies are a major reservoir of galactic metals. Science **334**, 948 (2011). doi:10.1126/science.1209840
55. Wang, L., Dutton, A.A., Stinson, G.S., Macciò, A.V., Penzo, C., Kang, X., Keller, B.W., Wadsley, J.: NIHAO project - I. Reproducing the inefficiency of galaxy formation across cosmic time with a large sample of cosmological hydrodynamical simulations. Mon. Not. R. Astron. Soc. **454**, 83 (2015)
56. Zahn, O., et al.: Cosmic microwave background constraints on the duration and timing of reionization from the south pole telescope. Astrophys. J. **756**, 65 (2012). doi:10.1088/0004-637X/756/1/65

Part II
Molecules, Interfaces, and Solids

Holger Fehske and Christoph van Wüllen

The computational treatments in the fields of solid state physics, chemistry, and material science were directed towards the structural, transport and vibrational properties of fascinating systems with high application potential, such as colloidal semiconductor nano-clusters or quantum dots, organic semiconductors chains, or graphene flakes. The contributions selected in this volume focus on charge carrier transport, scattering, relaxation processes and interface formation. Most calculations are based on density functional theory (DFT) and reveal that research in this area extraordinarily benefits from the supercomputing facilities of the High Performance Computing Center Stuttgart.

An outstanding example is the collaborative work by M. Walz, A. Bagrets, F. Evers and I. Kondov from the Institute of Nanotechnology and the Steinbuch Centre of Computing at the Karlsruhe Institute of Technology, which addresses electron transport in large hydrogenated graphene flakes. In spite of a manifold of investigations during the last decade the striking transport properties of graphene samples are far from being fully understood. This particularly concerns the effects of disorder, induced, e.g., by adsorbates in the process of functionalisation. Calculating the local current density related to dc-conductivity measurements the authors present an impressive ab initio treatment of transport that includes the influence of quantum interferences and mesoscopic fluctuations. Employing the all-electron DFT code FHI- AIMS in combination with an AITRANSS module, Walz et al. perform optimised (shared/distributed-memory parallelised) simulations of graphene flakes with 2500 carbon atoms on the Cray XE6/XC40 supercomputer.

H. Fehske (✉)
Institut für Physik, Lehrstuhl Komplexe Quantensysteme, Ernst-Moritz-Arndt-Universität Greifswald, Felix-Hausdorff-Str. 6, 17489 Greifswald, Germany
e-mail: fehske@physik.uni-greifswald.de

C. van Wüllen
Fachbereich Chemie, Technische Universität Kaiserslautern, Erwin-Schrödinger-Str. 52, 67663 Kaiserslautern, Germany
e-mail: christoph.vanwullen@chemie.uni-kl.de

The current flow shows complicated nontrivial patterns that originate from the inner structure of the scattering states on the mesoscopic sample. Interestingly, in a wide range of impurity concentrations, the distribution function of the current density follows a log-normal distribution which might be indicative of current rings or even Anderson localisation effects.

The importance of impurities or structural imperfections on quantum transport is also the object of investigation, when W. G. Schmidt's group from the University Paderborn solved the scattering problem for poly(3-hexylthiophene-2,5-diyl) (P3HT) molecular structures obtained within DFT. Polythiophenes are organic semiconductors frequently used for light-emitting diodes, transistors, and solar cells. Utilizing the QUANTUM-ESPRESSO-package (infinite) molecular chains are modelled by a supercell approach, where (ultrasoft) Kleinman-Bylander pseudo-potentials approximate the ion-electron interaction and the electron-electron many-body interaction is taken into account by the generalised gradient approximation using the Perdew-Burke-Ernzerhof functional. Then, making use of a modified PWCOND program for the calculation of the set of wave-functions, the quantum conductance is obtained from the transmission coefficients. Even though the approach neglects bias voltage and dissipative scattering effects, the qualitative trends obtained with respect to the influence of, e.g., structural deformations are trustworthy. So intra-chain transport is highly hindered by large torsion angles while chain bending only weakly affects the conductance. It is fascinating that structural relaxation processes favour geometries having good electron transport properties. Analysing the influence of isomer defects the authors arrive at the conclusion that the P3HT transport properties are determined by the thiophene rings rather than the hexyl side chains.

Another large-scale DFT-based ab initio study deals with strains and their consequences on the vibrational properties of core-shell (Si-Ge, InAS-InP, and CdSe-CdS) semiconductor nano-clusters (NCs), and with the mechanism of inter-band carrier relaxation in colloidal NCs. To this end, Bester and Han from the Institute of Physical Chemistry of the University Hamburg calculated (1) the vibrational eigenmodes and electron-phonon coupling matrix elements of the NCs and (2) the time evolution of a population after a pulse laser excitation, using the CPMD (DFT) code package developed at the MPI Institute in Stuttgart and at IBM, which offers an excellent scaling on the Cray XE6 Hermit cluster at HLRS on account of a hybrid MPI–OpenMP programming scheme. The authors find—amongst other things—that (a) the shell determines the atom positions of the whole core-shell complex, (b) the frequency shift observed for the NCs (compared to the bulk frequencies) result from the mode blue-shift according to strain and the red-shift created by the under-coordination of the near surface atoms (which evidently work against each other), and (c) the low-energy shoulder on the Raman spectra originate from interface vibrations with small surface character. With regard to carrier relaxation Bester and Han report a fast picosecond electronic relaxation process (from the excited P-like to ground S-like state) in InAs, CdSe and Si NCs due to a strong (weak) coupling between the electronic transitions and acoustic-type vibrons (passivant vibrations). The results indicate that a window of intermediate-sized NCs exists for InAs and CdSe, where the phonon bottleneck might be observed.

The contribution of Arrigone, Kotomi and Maier investigates oxygen vacancies in BARO$_3$ crystals. Such defects have a major influence on the properties of perovskite materials, especially on their ion conductivity. The simulation uses the hybrid exchange-correlation functional PBE0 (including 25 % of Hartree-Fock exchange) and the CRYSTAL14 code. Large supercells (with up to 135 atoms) are required to keep the defects sufficiently far apart. The changes in the local electronic and geometric structure in the vicinity of the defects were calculated, as well as the free energy of defect formation. Here, the defect-induced changes in the phonon spectrum, which has not been taken into account in previous investigations, play an important role. The calculation of the phonons (vibrations) needs huge computational resources.

Chemical reactions at semiconductor surfaces is the topic of the contribution of Rosenow, Stegmüller, Pecher, and Tonner. As a model reaction for the functionalization of silicon with organic layers, the adsorption of cyclooctyne on Si(001) was simulated. It readily reacts with the surface by forming two covalent carbon-silicon bonds, resulting favourably in an *on-top* geometry. Furthermore, the interface region between pure silicon and GaP was investigated. The interface formation energy strongly depends on which crystal plane of silicon the GaP film grows. How to get optimum parallel performance from the VASP program used in these simulation was also investigated.

Polyelectrolytes are polymers which in solution evolve charges along the polymer backbone. Using polycations and polyanions in alternating *"layer by layer"* order, layered films form by electrostatic self-assembly. This process was simulated in the contribution of Sánchez, Smiatek, Qiao, Sega and Holm using the molecular dynamics package Gromacs. The coverage fraction and roughness of the films that form were in agreement with experimental measurements, as well as how these values change with the concentration of electrolytes (salts) in the surrounding water.

Finally, the contribution by Marx, Wollenhaupt and Zoloff investigates chemical reactions that are induced by mechanical forces. (Covalent) mechanochemistry is a relatively new research field, mainly because techniques to investigate this experimentally (such as atomic force microscopy) are now at hand. External forces alter the potential energy surface of the system, so new reaction mechanisms may occur that lead to chemical selectivities different from normal thermally activated reactions. For a ring-opening of a substituted cyclopropane, it was found that although the electronic structure is not altered much, the deformation of the potential energy surface (by the mechanical work term) leads to reaction pathways that are highly distorted.

To sum up, the projects in the rapidly growing field of solid state physics and computational material science have in common, besides a high scientific quality, the strong need for computers with high performance to achieve their results. That is why the leading-edge supercomputers at the HLRS are a prerequisite for such ambitious research.

Ab Initio Transport Calculations for Functionalized Graphene Flakes on a Supercomputer

Michael Walz, Alexei Bagrets, Ferdinand Evers, and Ivan Kondov

Abstract We present ab initio transport studies of large graphene flakes focusing on the local current density $\mathbf{j}(\mathbf{r})$ as it arises from a dc-transport measurement. Such ab initio transport calculations for sufficiently large flakes can be successfully tackled only using well scaling ab initio packages capable for transport calculations in thin film geometries. We employ the FHI-AIMS/AITRANSS packages to study the effect of disorder on the local current density in graphene flakes, in particular, the effect of chemical functionalization on mesoscopic fluctuations of the current density. Simulating graphene flakes with several thousands of atoms, we clearly see the qualitative effects of quantum interference and mesoscopic fluctuations in such systems. We also discuss the parallelization and optimization techniques, which we implemented into the transport module AITRANSS to allow efficient ab initio transport calculation on Cray XE6 and XC40 supercomputers.

1 Introduction

Graphene is an atomically thin layer of carbon atoms that self-organize in a honeycomb lattice. Since several years, such layers are routinely manufactured in laboratories worldwide. Electron transport in graphene and other graphene properties are investigated intensively and form one of the hottest fields in condensed matter and materials sciences. The driving force results from prospective applications in information and nanotechnology, e.g., as field effect transistors or

M. Walz
Institute of Nanotechnology and Institut für Theorie der Kondensierten Materie, Karlsruhe
Institute of Technology, Karlsruhe, Germany
e-mail: michael.walz@kit.edu

A. Bagrets • F. Evers
Institute of Nanotechnology, Karlsruhe Institute of Technology, Karlsruhe, Germany

I. Kondov (✉)
Steinbuch Centre for Computing, Karlsruhe Institute of Technology, Karlsruhe, Germany
e-mail: ivan.kondov@kit.edu

© Springer International Publishing Switzerland 2016
W.E. Nagel et al. (eds.), *High Performance Computing in Science
and Engineering '15*, DOI 10.1007/978-3-319-24633-8_9

spin quantum bits. Also, chemical functionalization triggers proposals for additional technologies like hydrogen storage or sensor applications.

In most applications, the electron transport properties of graphene-based films are crucial. Despite intensive research, the transport properties are far from being fully understood, especially concerning the influence of disorder, as it is induced, e.g., by functionalization with adsorbates. In computational work, disorder effects have been treated so far mostly on the level of tight-binding calculations that incorporate aspects of π-electron physics; genuine ab initio treatments of transport in functionalized molecular films were not yet available. Partially, this is because electronic structure calculations based on the density functional theory (DFT) for large, sufficiently representable flakes are computationally highly demanding. They can be successfully treated only by the most advanced ab initio packages. An additional difficulty is that such packages are often not capable for transport calculations in film geometries.

Recently, the FHI-AIMS code [1] has been complemented with a versatile transport module AITRANSS [2–5], which implements the non-equilibrium Green's function formalism (NEGF) and uses DFT results as input. It enables computing ab initio transport properties of single molecules but also of more extended structures such as molecular flakes, films or nanotubes.

In this work we study the effect of disorder (e.g. hydrogen adsorbates) and chemical functionalization on the conductance of graphene flakes. In contrast to the presently available tight-binding treatments, our DFT approach includes the effects of charging and screening of impurities, and lattice distortion, i.e., strain. As a consequence, we can study the cross-talk between different impurities which is not possible with present tight-binding implementations.

The sizes of computationally tractable graphene flakes reach 2500 carbon atoms. This is sufficient to see the qualitative effects of quantum interference and mesoscopic fluctuations in such systems. Reaching such flake sizes was only possible by achieving an excellent DFT performance on the HLRS systems when using the FHI-AIMS package. In addition, we improved the parallel scaling of the transport module AITRANSS.

The structure of the paper is as follows: in Sect. 2.1, we describe our transport method as implemented in the code, and we comment on the employed parallelization techniques (Sect. 2.2). In Sect. 3.1, the key findings are summarized. For reference, we provide the numerical parameters used in a typical calculation in Sect. 3.2. Next, the achieved excellent performance on the HLRS computing systems is shown, first for the DFT simulation (Sect. 3.3), second for the transport part (Sect. 3.4). In Sect. 3.5, we present two optimizations, necessary for achieving the excellent performance.

2 Methods

Our transport calculations are performed in a two-step procedure. First, we perform a DFT calculation using the all-electron DFT code FHI-AIMS including relaxation effects. Second, we perform a transport calculation using AITRANSS with the Kohn–Sham orbitals from the previous DFT step as input. We summarize the transport calculation in the following and point out where parallelization is necessary. The implementation description of FHI-AIMS and AITRANSS is outlined in [1] and [5], respectively.

2.1 Transport Calculations

We extract the Kohn–Sham (KS) Green's function $\mathbf{G}_0^{\mathrm{KS}}(E) = \left(E\mathbb{1} - \mathbf{H}^{\mathrm{KS}} + \mathrm{i}0\right)^{-1}$ for a finite disordered graphene flake from the DFT calculations. To model the infinite extension of the system in current flow direction, we compute the self-energies $\boldsymbol{\Sigma}_{\mathrm{L/R}}$ using absorbing boundary conditions [5, 6]. The self-energies $\boldsymbol{\Sigma}_{\mathrm{L/R}}$ represent the influences of the leads. The resulting Green's function

$$\mathbf{G}(E)^{-1} = \mathbf{G}_0^{\mathrm{KS}}(E)^{-1} - \boldsymbol{\Sigma}_{\mathrm{L}}(E) - \boldsymbol{\Sigma}_{\mathrm{R}}(E) \tag{1}$$

describes the propagation of KS particles in the device in the presence of leads and is used to calculate the transmission $\mathscr{T}(E) = \mathrm{Tr}\{\boldsymbol{\Gamma}_{\mathrm{L}}\,\mathbf{G}\,\boldsymbol{\Gamma}_{\mathrm{R}}\,\mathbf{G}^{\dagger}\}$. Here, $\boldsymbol{\Gamma}_{\mathrm{L/R}}$ denote the anti-Hermitian parts of the self-energies, i.e., $\boldsymbol{\Gamma}_{\mathrm{L/R}} = \mathrm{i}(\boldsymbol{\Sigma}_{\mathrm{L/R}} - \boldsymbol{\Sigma}_{\mathrm{L/R}}^{\dagger})$. They account for the level broadenings in the finite graphene flake due to the coupling to the leads.

Using the retarded Green's function, we calculate the non-equilibrium Keldysh Green's function $\mathbf{G}^{<} = \mathrm{i}\mathbf{G}\big[f_{\mathrm{L}}\boldsymbol{\Gamma}_{\mathrm{L}} + f_{\mathrm{R}}\boldsymbol{\Gamma}_{\mathrm{R}}\big]\mathbf{G}^{\dagger}$. Assuming that the scattering states originating from the left/right lead are occupied/unoccupied, the Keldysh Green's function $\mathbf{G}^{<}(E)$ simplifies to

$$\mathbf{G}^{<}(E) = \mathrm{i}\mathbf{G}(E)\boldsymbol{\Gamma}_{\mathrm{L}}(E)\mathbf{G}^{\dagger}(E). \tag{2}$$

Here, we assumed zero temperature and an energy inside the voltage window, so that $f_{\mathrm{L}} = 1$ and $f_{\mathrm{R}} = 0$ for the occupation in the leads.

We use orthonormal basis functions $\tilde{\varphi}_i(\mathbf{r})$ [constructed via Löwdin-orthogonal-ization from the DFT basis functions $\varphi_i(\mathbf{r})$] to transform the Keldysh Green's function into real space:

$$\mathbf{G}^<(\mathbf{r}, \mathbf{r}', E) = \sum_{ij} \tilde{\varphi}_i(\mathbf{r}) \mathbf{G}_{ij}^<(E) \tilde{\varphi}_j^*(\mathbf{r}') . \tag{3}$$

The current density (per spin and energy) is then expressed as

$$\mathbf{j}(\mathbf{r}, E) = \frac{1}{2\pi} \frac{\hbar}{2m} \lim_{\mathbf{r}' \to \mathbf{r}} (\nabla_{\mathbf{r}'} - \nabla_{\mathbf{r}}) \mathbf{G}^<(\mathbf{r}, \mathbf{r}', E) . \tag{4}$$

2.2 Implementation: Parallelization Efforts

Because the computational demand of the DFT and transport steps scale very similar with flake size (see Sect. 3.5), none of the steps represents a single bottleneck for the whole simulation. Thus, we also optimized and parallelized the transport module AITRANSS to enable study of large flakes.

The AITRANSS code uses two types of parallelization:

shared-memory parallelization, in which several threads within one process have access to the same data, i.e., the same energy point E. We use threaded LAPACK as implemented in the Intel MKL for matrix operations and OpenMP for parallelization of loops over real space, etc.

distributed-memory parallelization, in which each process works with separate data sets for possibly different energy points E. The Message Passing Interface (MPI) is used for work balancing, e.g., for distributing different energy points to different processes, cutting and distributing the real-space grid.

A workflow diagram of the parallelized modules is shown in Fig. 1.

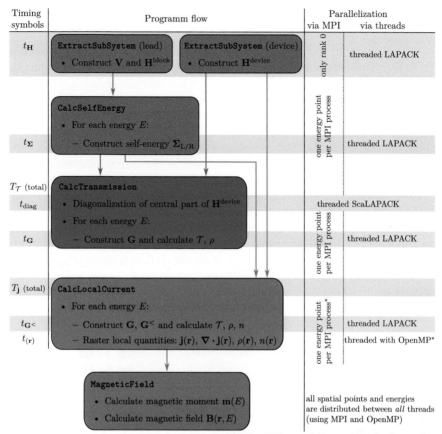

Timing symbols	Programm flow	Parallelization via MPI	via threads

$t_{\mathbf{H}}$ — ExtractSubSystem (lead) • Construct $\mathbf{V}$ and $\mathbf{H}^{\text{block}}$ | ExtractSubSystem (device) • Construct $\mathbf{H}^{\text{device}}$ — only rank 0 — threaded LAPACK

CalcSelfEnergy
• For each energy E:
$t_{\mathbf{\Sigma}}$ — Construct self-energy $\mathbf{\Sigma}_{\text{L/R}}$ — one energy point per MPI process — threaded LAPACK

$T_{\mathcal{T}}$ (total) CalcTransmission
t_{diag} • Diagonalization of central part of $\mathbf{H}^{\text{device}}$ — threaded ScaLAPACK
• For each energy E:
$t_{\mathbf{G}}$ — Construct $\mathbf{G}$ and calculate $\mathcal{T}$, ρ — one energy point per MPI process — threaded LAPACK

$T_{\mathbf{j}}$ (total) CalcLocalCurrent
• For each energy E:
$t_{\mathbf{G}^<}$ — Construct $\mathbf{G}$, $\mathbf{G}^<$ and calculate $\mathcal{T}$, ρ, n — one energy point per MPI process* — threaded LAPACK
$t_{(\mathbf{r})}$ — Raster local quantities: $\mathbf{j}(\mathbf{r})$, $\boldsymbol{\nabla}\cdot\mathbf{j}(\mathbf{r})$, $\rho(\mathbf{r})$, $n(\mathbf{r})$ — threaded with OpenMP*

MagneticField
• Calculate magnetic moment $\mathbf{m}(E)$
• Calculate magnetic field $\mathbf{B}(\mathbf{r}, E)$ — all spatial points and energies are distributed between *all* threads (using MPI and OpenMP)

* If number of MPI processes is less or equal to number of energy points.

Fig. 1 Workflow diagram of the transport module AITRANSS. The two most common module sequences are depicted in *blue* (transmission calculation) and *red* (current density and magnetic field calculation). Both sequences include the reconstruction of the Kohn–Sham Hamiltonian and the calculation of the self-energies (*purple*). *Left*: timing symbols used in the following for performance analysis, cf. Figs. 4 and 5. *Right*: overview of the parallelization techniques used in each module

3   Results

The results of this work are twofold. First, we summarize the physics results for the investigated graphene flakes in Sect. 3.1. Second, we present our development of computational techniques: (a) we demonstrate that our applied codes scale sufficiently when using a few thousand cores and (b) we discuss the improvements in our transport module that were necessary to achieve good scalability.

3.1 Overview of the Key Findings

As a preparation for this work, we calculated the local current density response $\mathbf{j}(\mathbf{r}, E)$ of pristine armchair graphene nanoribbons (AGNRs) with varying width [7]. We observe pronounced current patterns, which we call "streamlines", with three-fold periodicity in the ribbon width. They arise as a consequence of quantum confinement in the direction transverse to the current flow. Neighboring streamlines are separated by stripes of almost vanishing flow. This effect can be explained in a tight-binding toy model. The response of the current density to adatoms is very sensitive to the placement: adatoms placed within the current filaments lead to strong backscattering; while in other regions, adatoms have almost no impact.

Then, we switched to larger graphene flakes calculating the local current density $\mathbf{j}(\mathbf{r}, E)$ in the presence of hydrogen adsorbates [8], an example with 5 % hydrogen is shown in Fig. 2. We discovered pronounced local current patterns, ring currents (current vortices), that go along with orbital magnetism. Importantly, the magnitude of the ring currents can exceed the average transport current by orders of magnitude. The associated magnetic fields exhibit drastic fluctuations with large field gradients reaching up to $1\,\mathrm{T\,nm^{-1}\,V^{-1}}$. These observations are relevant for spin relaxation in systems with very weak spin–orbit interaction, e.g., organic semiconductors. In

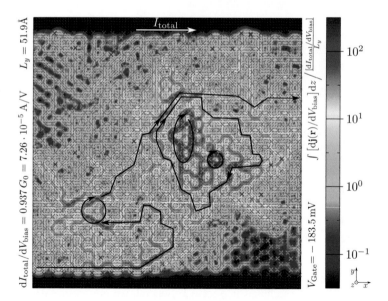

Fig. 2 Local current density response (integrated over the out-of-plane direction) in an AGNR41 (24×41) with 5 % hydrogen adsorbates. The current density exhibits very strong mesoscopic fluctuations which exceed the average current by over two orders of magnitude in the logarithmic color scale. The current density is plotted relative to average current density. Plot shows current amplitude (*color*), current direction (*arrows*), carbon atoms (*grey crosses*) and hydrogen atoms (*red crosses*). Some arbitrary current paths (*black lines*) are drawn into the plot for illustration

such systems, spin relaxation induced by bias-driven orbital magnetism competes with the hyperfine interaction. Both appear to be of similar strength. As a result of our calculation, we propose an NMR-type experiment combined with a dc-transport setup to observe the spatial fluctuations of the induced magnetic fields. We studied several impurity concentrations and different graphene flake sizes. The described physics seems to be independent and, therefore, should also be present in larger mesoscopic samples which are more common in experiments.

We also studied the statistical distribution of the current density in the graphene flakes [9]. The distribution function of the current density follows a log-normal distribution in a wide range. Its typical value is larger than the average current. Therefore, there are always significant contributions to the current density which do not contribute to the conductance, i.e., they form current rings. This work is still on-going, but so far, these features seem to be remarkably stable, in a wide range of impurity concentrations (5–30 %) and system sizes (up to 2500 carbon atoms), and even survive an averaging over several scattering states, e.g., when a finite bias voltage is applied to the system.

3.2   Typical Numerical Parameters

In Table 1, we list the numerical parameters of a typical calculation performed for hydrogenated graphene flakes. First, the graphene flake is structurally relaxed using FHI-AIMS until the remaining forces decrease below 10^{-2} eV/Å. This is, by far, the most expensive part of the calculation. Then a final DFT run for the relaxed structure

Table 1 An overview of a typical calculation performed on Cray XC40 (Hornet) for a graphene flake (with 1312 carbon atoms) whose central 24×41 carbon atoms have been functionalized with hydrogen (compare with Fig. 2)

| | FHI-AIMS | | AITRANSS | |
	Relaxation	DFT	Transmission[a]	Current density[b]
Numbers of processes N_{MPI}	2688	96	60	3
Cores per process p	1	1	4	8
Total number of cores P	2688	96	240	24
Number of nodes n	112	4	10	1
Wall time T_{wall}	36.4 h[c]	1.85 h	0.80 h	7.37 h
Core hours T_{cores}	97 800 h	178 h	192 h	177 h
Memory usage per process $\mathcal{M}$	0.55 GB	0.55 GB	14.8 GB	21.5 GB
Memory usage per node $\mathcal{M}_{node}$	13.2 GB	13.2 GB	88.7 GB	64.4 GB

Note that some information is redundant, i.e., $P = p N_{MPI} = 24 n$, $T_{cores} = T_{wall}P$, $\mathcal{M}_{node} = \mathcal{M} N_{MPI}/n$
[a]Transmission $\mathcal{T}(E)$ has been calculation at 2423 energies values
[b]The current density $\mathbf{j}(\mathbf{r}, E)$ has been calculation at 6 energies values
[c]The relaxation calculation was broken down into several jobs, each with a wall time below 24 h

is performed and the output written to disk. This is used by AITRANSS to perform a wide scan over the transmission function (the self-energies Σ are pre-calculated since they only depend on the system size, not on the impurity configuration). Eventually, a few interesting energy points are taken from the transmission function and the current density is calculated at those energy points.

3.3 Achieved DFT Performance on Cray XE6 (Hermit) and Cray XC40 (Hornet)

Within a test project on Hermit for which a budget of 40,000 core hours has been granted, we carried out code porting, tuning and performance measurements for DFT calculations of graphene to ensure that the FHI-AIMS code scales as necessary for the completion of the project goals and that the envisaged computing system Hermit was appropriate for the planned productive simulations.

The scalability of the FHI-AIMS code was demonstrated running the DFT module for pure planar graphene flakes, in which dangling bonds at the boundaries were saturated with hydrogen atoms with 170 (10×17), 345 (15×23), 735 (21×35) and 1500 (30×50) carbon atoms, each represented by a tier 1 basis set, i.e. 14 basis functions per C atom and 5 per H atom. The total number of basis functions N is 2560, 5095, 10,675 and 21,550, respectively.

Figure 3 (upper plot) presents the scaling of the code where the data from each set (color) are plotted as reduced speedup T_n/T_P. T_P is the wall time of calculation on P processor cores and T_n is the wall time on the minimum number of cores n necessary for the job to finish within wall time limit (24 h). Thus, n is 8, 16, 32 or 256, depending on the graphene flake size, i.e., each color in Fig. 3 is for a different n. The normalization with n is also necessary to allow easy comparison of the speedups for the four graphene flake sizes. Then, the data for each set have been fitted to Amdahl's law (the solid lines) showing that the serial fraction of the work α is always very small, around 1 ‰(= 0.001) and furthermore decreases with increasing the graphene flake size.

Figure 3 (lower plot) shows the scaling for graphene using the same data as in Fig. 3 (upper plot) now represented by the speedup $S(P) = \tilde{T}_1/T_P$ achieved on P cores, where $\tilde{T}_1$ is a hypothetical time for which the same calculation would take when a single core were used. Note that for a fixed number of cores used for computations, e.g., set of data points for $P = 256$, the speedup is improved when the size of the problem increases approaching the ideal speedup (Gustafson's law).

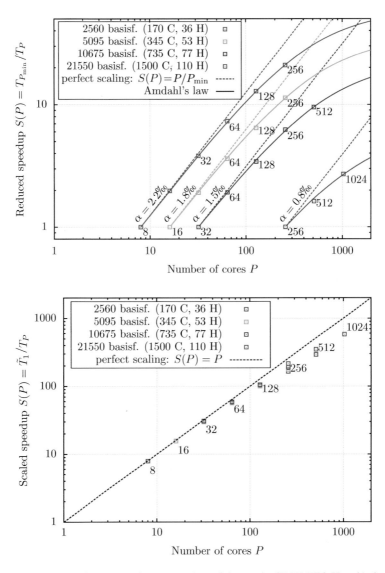

Fig. 3 Strong scaling of FHI-AIMS for pure graphene flakes on the CRAY XE6 (Hermit) cluster at HLRS (*upper plot*). Scaling of FHI-AIMS for pure graphene flakes of different size on the CRAY XE6 (Hermit) cluster at HLRS (weak scaling, *lower plot*)

3.4 Parallelization of the Transport Module AITRANSS and Achieved Performance

In Fig. 4, we present detailed measurements of the performance of our transport code for realistic system sizes. In the tests, we distinguish between calculation of the transmission and of local observables. Transmission calculations ($\mathscr{T}(E)$)

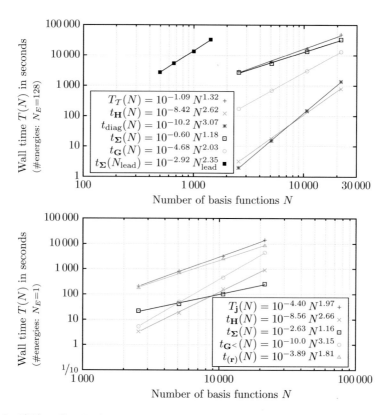

Fig. 4 $\mathcal{O}(N^n)$-scaling: Performance measurements with varying system size for a transmission (*upper plot*) and local observable (*lower plot*) calculation for a fixed number of CPU cores ($P{=}32$). Symbols: number of basis functions N, number of energy points N_E, number of CPU cores P

also include the density of states $\rho(E)$. Local observables are the local current density $\mathbf{j}(\mathbf{r}, E)$ but also its divergence $\nabla \cdot \mathbf{j}(\mathbf{r}, E)$, the non-equilibrium density $n(\mathbf{r}, E)$ and the local density of states $\rho(\mathbf{r}, E)$.

In the upper panel, the wall time for calculating $N_E = 128$ transmission and density of states of hydrogen-saturated AGNRs (the same as in Sect. 3.3) is plotted depending on the number of basis functions N. The total time $T_{\mathscr{T}}(N)$ is divided into four groups ($t_\mathbf{H}$, t_{diag}, $t_{\mathbf{\Sigma}}$, $t_\mathbf{G}$, cf. Fig. 1). Because the calculation (via 200 iterations in the decimation technique [5]) of the self-energy representing the leads depends directly on the number of basis functions N_{lead} of the leads (and only indirectly on the basis functions N of the device region), it is plotted separately. The main effort for a transmission calculation is reconstructing the self-energy of the leads; therefore it makes sense to save them on the hard disk if several different impurity configurations are processed which all use the same leads. Then, the main effort is spent for reconstructing the Green's function ($t_\mathbf{G}$). The diagonalization (t_{diag}, see Sect. 3.5.3) does not significantly contribute to the overall effort for the considered system sizes since it is performed only once and not for every energy point.

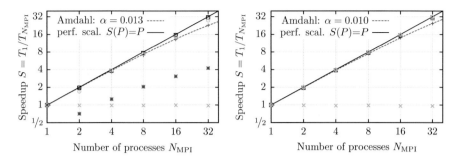

Fig. 5 MPI-parallelization: speedup for a fixed system size (735 (21×35) carbon atoms) and for a fixed number of CPU cores per MPI process ($p=8$). *Left*: speedup for transmission calculations. *Right*: speedup for local observable calculation. Symbols are the same as in Fig. 4. The p is the number of CPU cores per MPI process, and N_{MPI} is the number of MPI processes ($P = p \, N_{MPI}$)

In the lower panel of Fig. 4, the same quantities are plotted for a single ($N_E = 1$) current density calculation. The number of grid points used is proportional to the system size (one grid point every 0.2 Å, 31 in z-direction). We first note, that the main effort is dominated by the discretization of the local quantities ($t_{(\mathbf{r})}$). We were able to optimize local observable calculation to scale below N^2 employing the locality of the basis functions.[1]

In Fig. 5, we discuss the parallelization efficiency of the transmission calculation and current density calculation for a fixed system size ($21 \times 35 = 735$ carbon atoms). The speedup S for many MPI processes compared to a single process is shown and compared to Amdahl's law: $T_{N_{MPI}} = T_1 [\alpha + (1 - \alpha)/N_{MPI}]$, $\alpha \approx 1\%$. We see a good scalability for the total wall time, the self-energy construction and the Green's function construction ($T_{\mathscr{G}}$, t_{Σ}, $t_{\mathbf{G}}$). The reconstruction of the KS Hamiltonian does not speedup since only the first MPI process is involved (cf. Fig. 1). For the diagonalization, we even observe that using two processes (using ScaLAPACK) is slower than using only one process (using LAPACK). Therefore, our code now only uses ScaLAPACK starting with 4 MPI processes.

3.5 Optimization of the Transport Module AITRANSS

3.5.1 Overview: Code Optimization

In the following sections, the two most important optimizations used throughout our code development are presented. Their performance impact is summarized in Table 2. Please note, that both optimizations have larger impact for larger systems.

---

[1]Naively the evaluation of the current or its divergence scales as N^3 with N since the number of spatial grid points $\mathbf{r}$ scales linearly, for constant grid spacing, and the summation of i, j [see Eq. (3)] gives additional N^2. Please, refer to Sect. 3.5.2 for optimization details.

Table 2 Increase in wall time when specific optimizations *are removed* from the code

Optimization	Changed quantity	398 atoms (345 carbon atoms)	812 atoms (735 carbon atoms)
SpaceBlocks: $\mathbf{j}$	Local quantities: $t_{(\mathbf{r})}$	$\times$ 18.3; 557s $\rightarrow$10177s	$\times$ 41.1; 1845s $\rightarrow$75811s
MatrixInverse: $\mathscr{T}$	Constructing $\mathbf{G}$: $t_{\mathbf{G}}$	$\times$ 7.0; 203s $\rightarrow$1414s	$\times$ 10.6; 1342s $\rightarrow$14264s

To get a feeling about the dependence on the system size, two different sizes are shown. The computational parameters and the systems are identical to the ones used in Fig. 4. Symbols are used to distinguish transmission ($\mathscr{T}$) and current density ($\mathbf{j}$) calculations. The self-energy was always read from hard disk. To measure the wall time without optimization "SpaceBlocks" for 810 atoms, we doubled the grid spacing (taking only every eighth grid point) and used linear extrapolation to approximate $t_{(\mathbf{r})}$ for the full grid

The optimization "SpaceBlocks" is vital because without it calculations for systems with more than 1000 atoms become unfeasible. Also the optimization "MatrixInverse" is quite handy because it allows quick transmission scans before turning to more expensive current density calculations.

3.5.2 SpaceBlocks: Dividing Space into Blocks

Here, we discuss how to evaluate the formulas for space-depending local quantities such as the current density

$$\mathbf{j}(\mathbf{r}, E) \propto \sum_{ij} \varphi_i(\mathbf{r}) \overset{\text{as}}{\mathbf{G}_{ij}^<} (\nabla \varphi_j(\mathbf{r})), \qquad \overset{\text{as}}{\mathbf{G}_{ij}^<} = \frac{1}{2} (\check{\mathbf{G}}_{ij}^< - \check{\mathbf{G}}_{ji}^<). \tag{5}$$

In principle, the double sum runs over *all* basis functions of the underlying DFT simulation. FHI-AIMS uses numerically tabulated atom-centered orbitals (NAOs), i.e. localized basis functions. When restricting the spatial point $\mathbf{r}$ to a small region, most basis functions are vanishing inside this region. (These basis functions are, of course, nonzero elsewhere.) This locality in the basis set can be exploited in the following way.

First, we define $r_{\max}$ as the maximal radial extent of all basis functions (i.e. all basis functions are zero at points which are further away from the central atom than $r_{\max}$). Second, the 3D space is divided into little cubes with edge length $r_{\max}/n$ with n being an integer (see Fig. 6 for an example). When performing the sum of Eq. (5) for any spatial point $\mathbf{r}$ inside the blue shaded area, the only basis functions taken into account are centered around atoms in the green (and blue) shaded area. All other basis functions do not contribute to this area.

Hence, we divide the space into cubes of length $r_{\max}/n$ and distribute them to separate MPI processes. The integer n is chosen such that every MPI process works on at least five blocks to alleviate load imbalance due to different block sizes. Then, for each inner (blue) block, we restrict the Green's function $\mathbf{G}^<$ to the basis functions localized at atoms in the extended (green) block.

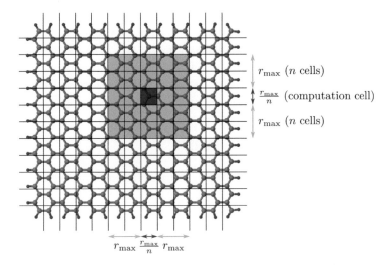

Fig. 6 Dividing space into 156 (13×12×1) non-overlapping blocks, exemplary for a graphene flake with 398 atoms. (2D model with $n = 2$, $r_{max} = 5.05\,Å$)

3.5.3 MatrixInverse: Calculating the Green's Function Inverse

As the self-energy can be read from hard disk, the most expensive part in a transmission calculation is the matrix inversion in calculating the retarded Green's function $\mathbf{G}$, cf. Eq. (1). According to Fig. 4 (upper panel), $\mathbf{G}$ can be constructed in $\mathcal{O}(N^2)$. Without this optimization the matrix inversion in calculating the Green's function would scale as $\mathcal{O}(N^3)$ and would therefore dominate for large systems.

Partitioning of the Green's function: The Green's function inverse, see Eq. (1), can be calculated by transforming the Hamiltonian so that it is diagonal in the regions where the self-energies Σ are zero. We partition the indices in the Green's function inverse such that the self-energy contribution of the leads only appears in subblock $\mathbf{D}$, i.e.,

$$\mathbf{G}^{-1} = E\mathbb{1} - \mathbf{H} - \Sigma_L(E) - \Sigma_R(E)$$

$$= \begin{pmatrix} E\mathbb{1}_{AA} - \mathbf{H}_{AA} & -\mathbf{H}_{AD} \\ -\mathbf{H}_{DA} & E\mathbb{1}_{DD} - \mathbf{H}_{DD} - \Sigma_L(E) - \Sigma_R(E) \end{pmatrix}^{-1} =: \begin{pmatrix} \mathbf{A} & \mathbf{B} \\ \mathbf{C} & \mathbf{D} \end{pmatrix}^{-1},$$

$$(6)$$

with the subscripts AA, AD, DA, DD denoting the restriction to the respective matrix subspace.

As advantage of this partitioning, the only non-trivial energy dependence appears in subblock $\mathbf{D} = E\mathbb{1}_{DD} - \mathbf{H}_{DD} - \Sigma_L(E) - \Sigma_R(E)$. The block $\mathbf{A}$ can be diagonalized for all energies in a single eigenvalue problem: the eigenvalues are given by the diagonal matrix $\tilde{\mathbf{A}} = E\mathbb{1} - \tilde{\mathbf{H}}_{AA}$ where $\tilde{\mathbf{H}}_{AA}$ denotes the diagonal matrix with the eigenvalues of $\mathbf{H}_{AA}$. The transformation matrix $\mathbf{V}$ ($\tilde{\mathbf{H}}_{AA}=\mathbf{V}^{-1}\mathbf{H}_{AA}\mathbf{V}$) is constructed by filling its columns with the (right) eigenvectors of $\mathbf{H}_{AA}$. The off-diagonal blocks stay energy independent, i.e., $\tilde{\mathbf{B}} = -\mathbf{V}^{-1}\mathbf{H}_{AD}$.

General matrix: For the matrix inversion, we first tend to a general matrix which we divide into four blocks

$$\begin{pmatrix} \mathbf{A} & \mathbf{B} \\ \mathbf{C} & \mathbf{D} \end{pmatrix}, \tag{7}$$

so that the submatrices $\mathbf{A}$ and $\mathbf{D}$ are square matrices. The inverse is given by

$$\begin{pmatrix} \mathbf{A} & \mathbf{B} \\ \mathbf{C} & \mathbf{D} \end{pmatrix}^{-1} = \begin{pmatrix} \mathbf{A}^{-1}(\mathbb{1} + \mathbf{B}\mathbf{E}^{-1}\mathbf{C}\mathbf{A}^{-1}) & -\mathbf{A}^{-1}\mathbf{B}\mathbf{E}^{-1} \\ -\mathbf{E}^{-1}\mathbf{C}\mathbf{A}^{-1} & \mathbf{E}^{-1} \end{pmatrix} \text{ with } \mathbf{E} := \mathbf{D} - \mathbf{C}\mathbf{A}^{-1}\mathbf{B} \tag{8}$$

as is easily checked by direct matrix multiplication.

Transforming $\mathbf{A}$ into diagonal form $\tilde{\mathbf{A}}$, i.e., $\mathbf{A} = \mathbf{V}\tilde{\mathbf{A}}\mathbf{V}^{-1}$, makes the calculation of the inverse $\mathbf{A}^{-1}$ trivial and we get:

$$\begin{pmatrix} \mathbf{A} & \mathbf{B} \\ \mathbf{C} & \mathbf{D} \end{pmatrix}^{-1} = \begin{pmatrix} \mathbf{V}\tilde{\mathbf{A}}^{-1}(\mathbb{1} + \tilde{\mathbf{B}}\mathbf{E}^{-1}\tilde{\mathbf{C}}\tilde{\mathbf{A}}^{-1})\mathbf{V}^{-1} & -\mathbf{V}\tilde{\mathbf{A}}^{-1}\tilde{\mathbf{B}}\mathbf{E}^{-1} \\ -\mathbf{E}^{-1}\tilde{\mathbf{C}}\tilde{\mathbf{A}}^{-1}\mathbf{V}^{-1} & \mathbf{E}^{-1} \end{pmatrix} \text{ with } \mathbf{E} := \mathbf{D}-\tilde{\mathbf{C}}\tilde{\mathbf{A}}^{-1}\tilde{\mathbf{B}} \tag{9}$$

using the abbreviations $\tilde{\mathbf{C}} := \mathbf{C}\mathbf{V}$ and $\tilde{\mathbf{B}} := \mathbf{V}^{-1}\mathbf{B}$.

Exploiting symmetries of G: In general, the Hamiltonian $\mathbf{H}$ is Hermitian and the self-energies $\mathbf{\Sigma}$ are non-Hermitian. In most cases, we can restrict ourselves to a real symmetric Hamiltonian and complex symmetric self-energies. In that case, the Green's function G is also (complex) symmetric, $\tilde{\mathbf{B}}$ and $\tilde{\mathbf{C}}$ are related by transposition, the eigenvalue problem simplifies to a real symmetric one[2] which makes the transformation matrix $\mathbf{V}$ orthogonal, i.e., $\mathbf{V}^{-1} = \mathbf{V}^T$.

Basis change for non-local quantities: If we are only interested in non-local quantities such as the transmission or the density of states, we can go one step further. Such quantities do not dependent on the spatial basis and we can transform the Green's function so that the Hamiltonian is diagonal in the subblock $\mathbf{A}$:

$$G \rightarrow \mathbf{S}^{-1}G\mathbf{S}, \qquad \mathbf{S} = \begin{pmatrix} \mathbf{V} & 0 \\ 0 & \mathbb{1}_{DD} \end{pmatrix}. \tag{10}$$

---

[2]For real symmetric eigenvalue problems, efficient implementations such as ScaLAPACK [10] or ELPA [11] exist that are parallelized over many computing nodes.

In practice, we indirectly perform this transformation by omitting the respective factors of $\mathbf{V}$ in Eq. (9). All in all, the inverse is given by:

$$\mathbf{G} = \begin{pmatrix} \tilde{\mathbf{A}}^{-1}(\mathbb{1} + \tilde{\mathbf{B}}\mathbf{E}^{-1}\tilde{\mathbf{B}}^T\tilde{\mathbf{A}}^{-1}) & -\tilde{\mathbf{A}}^{-1}\tilde{\mathbf{B}}\mathbf{E}^{-1} \\ \left[-\tilde{\mathbf{A}}^{-1}\tilde{\mathbf{B}}\mathbf{E}^{-1}\right]^T & \mathbf{E}^{-1} \end{pmatrix} \quad \text{with } \mathbf{E} := \mathbf{D} - \tilde{\mathbf{B}}^T\tilde{\mathbf{A}}^{-1}\tilde{\mathbf{B}} \quad (11)$$

using the abbreviation $\tilde{\mathbf{B}} := \mathbf{V}^T\mathbf{B}$.

Optimization traits: In Eq. (11), no matrix operations for matrices of size of $\mathbf{H}_{AA}$ appear (except for the initial eigenvalue problem): the inverse $\tilde{\mathbf{A}}^{-1}$ is trivial since $\tilde{\mathbf{A}}$ is diagonal. Therefore, this optimization is extremely useful for large systems where the contact regions to the leads are only a small part of the overall system, i.e., $N_A \gg N_D$ with $N_{A/D}$ denoting the size of the square matrices $\mathbf{A}, \mathbf{D}$, respectively.

For a short complexity analysis, we assume that multiplication and eigenvalue problem of $N{\times}N$-matrices have computational complexity $\mathcal{O}(N^3)$. Then, without above optimization, the direct matrix inversion used to calculate the Green's function has complexity $\mathcal{O}((N_A + N_D)^3) \overset{N_A \gg N_D}{\to} \mathcal{O}(N_A^3)$.

In the above optimization, the complexity of the preparation process containing the eigenvalue problem and the calculation of $\tilde{\mathbf{B}}$ is $\mathcal{O}(N_A^3 + N_A^2 N_D) \overset{N_A \gg N_D}{\to} \mathcal{O}(N_A^3)$. All the following inversions using Eq. (11) only are of complexity $\mathcal{O}(N_D^3 + N_A N_D^2 + N_A) \overset{N_A \gg N_D}{\to} \mathcal{O}(N_A N_D^2)$. The summands stand for inversion of $\mathbf{E}$, products of $N_A{\times}N_D$-matrices with $N_D{\times}N_D$-matrices like $\tilde{\mathbf{B}}\mathbf{E}^{-1}$ and inversion of $\tilde{\mathbf{A}}$, respectively.

Strictly speaking, the optimization still scales cubically in N_A due to the initial eigenvalue problem. Nevertheless, for energy sweeps over the density of states or the transmission, the complexity of each inversion step dominates and this effort could be reduced to complexity $\mathcal{O}(N_A N_D^2)$ for large systems,[3] cf. Fig. 4 (lower panel).

As stated above, the optimization only applies for non-local quantities. For local quantities such as current densities, the transformation matrix $\mathbf{V}$ cannot be omitted from Eq. (9) and we are back to cubic complexity.

4   Conclusions

In this work, we calculated the local current density in large hydrogenated graphene flakes. The current flow shows complicated patterns, a fact that is ignored in most studies focusing only on the total conductance. These patterns show large local fluctuations. The idea behind the nontrivial local current patterns is very general:

---

[3] For the AGNRs used for Fig. 4, the central part scales linearly, $N_A \in \mathcal{O}(N)$, but the contact regions scale with the square root, $N_D \in \mathcal{O}(\sqrt{N})$ because they only grow transverse to the current direction but not in current direction. This gives the observed overall complexity $\mathcal{O}(N^2)$.

scattering states of mesoscopic samples have an inner, nontrivial structure. The latter is seen in the electric current density but it is easily generalized for other observables, e.g. for heat. A mesoscopic device (like a hydrogenated graphene ribbon) which is connected to reservoirs with different temperatures will show fluctuations in the local temperatures as a result of the nontrivial structure of the scattering states.

Along the way, we parallelized and optimized our transport module AITRANSS to benefit from a supercomputer, thus to enable ab initio transport calculations for large graphene flakes. We showed that our techniques feature good scalability and discussed necessary optimizations. Future work will benefit from the fact that ab initio current density calculations for disordered systems are now available for large 2D film materials and for medium sized 3D materials. Essentially, the approach is only limited by the available computing power.

Acknowledgements We gratefully acknowledge HLRS Stuttgart for granting us computing time on the Hermit and Hornet systems.

References

1. Blum, V., Gehrke, R., Hanke, F., Havu, P., Havu, V., Ren, X., Reuter, K., Scheffler, M.: Comput. Phys. Commun. **180**(11), 2175 (2009). doi:10.1016/j.cpc.2009.06.022. http://www.sciencedirect.com/science/article/pii/S0010465509002033
2. Bagrets, A.: J. Chem. Theory Comput. **9**(6), 2801 (2013). doi:10.1021/ct4000263. http://pubs.acs.org/doi/abs/10.1021/ct4000263
3. Wilhelm, J., Walz, M., Stendel, M., Bagrets, A., Evers, F.: Phys. Chem. Chem. Phys. **15**, 6684 (2013). doi:10.1039/C3CP44286A. http://dx.doi.org/10.1039/C3CP44286A
4. Arnold, A., Weigend, F., Evers, F.: J. Chem. Phys. **126**(17), 174101 (2007). doi:http://dx.doi.org/10.1063/1.2716664. http://scitation.aip.org/content/aip/journal/jcp/126/17/10.1063/1.2716664
5. Walz, M., Bagrets, A., Evers, F.: Local current density calculations for molecular films from ab initio. J. Chem. Theory Comput. (accepted, 2015).
6. Evers, F., Arnold, A.: In: Röthig, C., Vojta, M. (eds.) CFN Lectures on Functional Nanostructures - Volume 2: Nanoelectronics. Springer, Berlin/Heidelberg (2011)
7. Wilhelm, J., Walz, M., Evers, F.: Phys. Rev. B **89**, 195406 (2014). doi:10.1103/PhysRevB.89.195406. http://link.aps.org/doi/10.1103/PhysRevB.89.195406
8. Walz, M., Wilhelm, J., Evers, F.: Phys. Rev. Lett. **113**, 136602 (2014). doi:10.1103/PhysRevLett.113.136602. http://link.aps.org/doi/10.1103/PhysRevLett.113.136602
9. Walz, M., Evers, F.: (in preparation)
10. Blackford, L.S., Choi, J., Cleary, A., D'Azeuedo, E., Demmel, J., Dhillon, I., Hammarling, S., Henry, G., Petitet, A., Stanley, K., Walker, D., Whaley, R.C.: ScaLAPACK User's Guide. Society for Industrial and Applied Mathematics, Philadelphia (1997)
11. Marek, A., Blum, V., Johanni, R., Havu, V., Lang, B., Auckenthaler, T., Heinecke, A., Bungartz, H.J., Lederer, H.: J. Phys. Condens. Matter **26**(21), 213201 (2014). http://stacks.iop.org/0953-8984/26/i=21/a=213201

Solving the Scattering Problem for the P3HT On-Chain Charge Transport

A. Lücke, U. Gerstmann, S. Sanna, M. Landmann, A. Riefer, M. Rohrmüller,
N.J. Vollmers, M. Witte, E. Rauls, R. Hölscher, C. Braun, S. Neufeld,
K. Holtgrewe, and W.G. Schmidt

Abstract The effect of oxygen impurities and structural imperfections on the coherent on-chain quantum conductance of poly(3-hexylthiophene) is calculated from *first principles* by solving the scattering problem for molecular structures obtained within density functional theory. It is found that the conductance drops substantially for polymer kinks with curvature radii smaller than 17 Å and rotations in excess of about 60°. Oxidation of thiophene group carbon atoms drastically reduces the conductance, whereas the oxidation of the molecular sulfur barely changes the coherent transport properties. Also isomer defects in the coupling along the chain direction are of minor importance for the intrachain transmission.

1 Introduction

Organic semiconductors are increasingly used for a wide range of devices ranging from organic light-emitting diodes and transistors to organic solar cells. In particular their low cost fabrication processes and the possibility to fine-tune desired functions by chemical modification of their building blocks make them interesting for numerous applications. Poly(3-hexylthiophene-2,5-diyl) also known as P3HT is one of the most important examples for organic semiconductors [1]. It is a p conducting organic polymer that is frequently used as active layer in organic solar cells as well as for organic field-effect transistors. Previous studies on the electronic properties concluded that the molecular electron and hole wave functions are delocalized over several thiophene rings and become spatially confined due to ring torsions and chain bendings [2–4] as well as fluctuations of the electrostatic potential caused by adjacent polymer chains [5]. Furthermore it was reported that defects like kinks or torsions do not result in a significant localization of the excited states[6]. There are various publications that address the P3HT transport properties: Improved transport properties for higher molecular weight (MW) chains are reported

A. Lücke • U. Gerstmann • S. Sanna, M. Landmann • A. Riefer • M. Rohrmüller • N.J. Vollmers •
M. Witte • E. Rauls • R. Hölscher • C. Braun • S. Neufeld • K. Holtgrewe • W.G. Schmidt (✉)
Lehrstuhl für Theoretische Physik, Universität Paderborn, 33095 Paderborn, Germany
e-mail: W.G.Schmidt@upb.de

© Springer International Publishing Switzerland 2016
W.E. Nagel et al. (eds.), *High Performance Computing in Science
and Engineering '15*, DOI 10.1007/978-3-319-24633-8_10

in [7–10] and rationalized by band transport in high MW films composed of long polymer chains, while hopping processes dominate in low MW polymer films. However, it was noted that using processing conditions which allow the chains to equilibrate, increases the mobility even in low MW P3HT [11]. The electron transport in high MW polycrystalline P3HT films was found to be highly anisotropic, with larger mobilities along the chains than perpendicular, i.e., parallel to the stacking direction [12]. These findings are in accordance with transfer integral calculations [13, 14] that find the intrachain direction to be the dominating charge-transport route within P3HT ordered domains [15].

Previous theoretical studies like transfer integral calculations have contributed much to the qualitative understanding of the P3HT electron transport properties. Quantitative data on the modification of the P3HT quantum conductance upon structural or chemical modification are not available yet, however. This concerns, for example, the oxygen-related degradation of the transport properties. The present study is a first step in that direction. The P3HT potential energy surface depends critically on the side chains and cannot be simply derived from smaller systems such as bithiophene [16], which may significantly influence the calculated transport properties due to the sensitivity of the transfer integrals [13, 17, 18]. Also isomeric dispersion may be important [19]. Therefore, much attention is paid here to the realistic modelling of the molecular geometry and the accurate solution of the scattering problem [20] based on the electron structure obtained within density functional theory. On the other hand, our study focusses on single polymer chains, is restricted to the ballistic limit and neglects temperature effects [21]. These are clearly highly idealized conditions. Still, information on coherent on-chain transport is essential for understanding transport characteristics in real samples, in particular since transport in bulk P3HT is largely influenced by tie molecules spanning adjacent nanocrystalline lamellae [12].

2 Methodology

The present calculations are based on density functional theory (DFT) as imple-mented in the QUANTUM-ESPRESSO-package [22]. Infinite molecular chains are modeled by a supercell approach with periodic boundary conditions. Ultrasoft pseudopotentials [23] in Kleinman-Bylander form [24] are used to model the ion-electron interaction. The electronic many-body interactions are described in generalized gradient approximation (GGA) using the PBE functional [25, 26]. Dispersion interaction is taken into account by the so-called DFT-D approach, i.e., by a semi-empirical London-type correction term [27, 28] using parameters suggested by Grimme [29]. Plane waves up to an energy cutoff of 544 eV are used as basis set for the expansion of the electron wave functions.

The ballistic conductance is obtained from the solution of the scattering problem as outlined by Choi et al. [30] and Smogunov et al. [31]. For the actual calculations we use the PWCOND program [32] that has been modified in our group for better

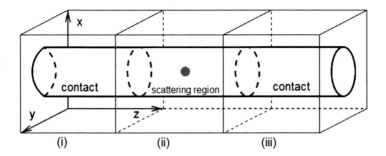

Fig. 1 Schematic setup of the scattering region that contains the investigated defect and is sandwiched between two ideal, semi-infinite contacts

performance on parallel architectures. Below we first outline the methodology and then our modifications to the code that were made for better scalability.

In order to determine the on-chain transmission of charge carriers with energy E through a defect we divide the system along the z direction into three regions as shown schematically in Fig. 1. The left and right contact (i) and (iii), respectively, are assumed to be ideal without any defects and the scattering region (ii) contains the defect. As the influence of the defect on the electronic structure may extend along the polymer chain, parts of the intact molecule should be included into the scattering region. The calculation then proceeds as follows:

1. The total electron potential for regions (i), (ii), and (iii) is obtained self-consistently within DFT for each region separately using periodic boundary conditions. Thereby infinite polymer chains serve as contacts. The scattering region around the defect must be sufficiently large to ensure a smooth match of the potentials.
2. Based on the total potential for each region the electron wave functions are determined without specifying any boundary conditions for the $z = 0$ and $z = L_z$ plane (where L_z is the length of the actual region). These boundary conditions are introduced in the next steps when we calculate the propagating modes in the contacts and the transmission through the whole system. Because of the open boundaries we can find a set of wavefunctions that are parametrized in the form $\Psi = \sum_n a_n \psi_n$. This is done separately for each of the three regions using as many free parameters a_n as the number of open boundary conditions.
3. The propagating ($\Im(k_z) = 0$) and evanescent ($\Im(k_z) \neq 0$) modes (in z-direction) in the contacts are determined exploiting Bloch's theorem for the periodically repeated potential of the contacts, i.e.,

$$\Psi_{k_z}(L_z) = e^{ik_z L_z} \Psi_{k_z}(0). \tag{1}$$

This represents a generalized eigenvalue problem for the eigenvalues $e^{ik_z L_z}$, where the eigenvectors are given by the free parameters a_{n,k_z} in $\Psi_{k_z} = \sum_n a_{n,k_z} \psi_n$ (cf. step 2). It is solved with the LAPACK routine ZGGEV.

4. The results of the previous steps enter the ansatz below that describes the electron wave function in the three regions as

$$
\Psi = \begin{cases}
\Psi_{k_z} + \displaystyle\sum_{\Im(k_z') \leq 0} r_{k_z,k_z'} \Psi_{k_z'} & z \text{ in I} \\[2ex]
\displaystyle\sum_n a_n \Psi_n & z \text{ in II} \\[2ex]
\displaystyle\sum_{\Im(k_z') \geq 0} t_{k_z,k_z'} \Psi_{k_z'} & z \text{ in III,}
\end{cases}
\tag{2}
$$

and which is solved for the coefficients $r_{k_z,k_z'}$, a_n and $t_{k_z,k_z'}$ that describe k_z modes that propagate to the right hand side with energy E in the left contact ($\Im(k_z) = 0$, $\Re(k_z) > 0$). Thereby the LAPACK routine ZGESV is utilized.

5. Finally, the quantum conductance $G(E)$ is calculated from the coefficients $t_{k_z,k_z'}$ weighted with the probability current I_{k_z} of the corresponding k_z mode

$$
G(E) = \frac{2e^2}{h} \sum_{k_z} \frac{1}{I_{k_z}} \sum_{k_z'} I_{k_z'} |t_{k_z,k_z'}|^2.
\tag{3}
$$

In order to obtain the conductance at further charge carrier energies the calculations above are repeated for different values of E starting at step 2. Eventually, the quantum conductance as function of the charge carrier energy $G(E)$ is obtained.

In the original PWCOND code the calculation of the set of wavefunctions in step 2 is parallelized over the number of FFT grid points in z direction. Thereby each MPI process calculates the wavefunction in its own subregion. The wavefunctions for the full region are afterwards determined by matching the boundary conditions between the subregions of the single processes. The corresponding data collection is done in a tree-like manner with a depth proportional to the logarithm of the number of subregions, cf. Fig. 2. This procedure is applied to each of the three regions (i), (ii), and (iii) in Fig. 1. Afterwards steps 3 and 4 of the algorithm above are performed in serial calculations. This is acceptable for systems, which mainly extend in z direction, but are small along x and y: Since the matrices in step 3 and 4 scale with the number of grid points perpendicular to the transport direction, the computational effort for steps 3 and 4 is far smaller than for step 2 in such cases. The P3HT polymer studied here, however, does not have a large aspect ratio due to its long hexyl side chains. Therefore its transport characteristics are not efficiently obtained using the existing parallelization scheme of PWCOND.

In order to make this system tractable, we therefore parallelized steps 3 and 4 of the algorithm above. Thereby it is exploited that the transmission for charge carriers with energy E can be calculated independently from the ones with energy E'. Therefore almost linear scaling is achieved by implementing parallelization over the charge carrier energies E which allows for determining the transmission $G(E)$ for the complete energy window. The speed-up depends obviously on the energy

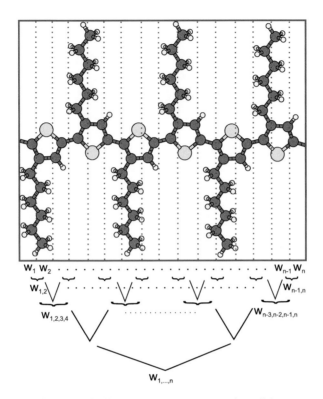

Fig. 2 Illustration of the load distribution (see text) onto several parallel processes and the tree-like access to single subregions exemplarily shown for P3HT contact calculations

region considered, but typically is of the order of two or three orders of magnitude in wall-clock time.

Additionally, we improve the efficiency of the calculations by modifying the numerical implementation of the boundary conditions between the scatter region (ii) and the contacts (i) and (iii) in PWCOND: We perform self-consistent electronic structure calculations for an extended scattering region and reduce its size afterwards for solving the scattering problem. This modification of the PWCOND module is not only useful to cut computational effort, it is also indispensable for systems where the scattering region is not periodic. It thus allows to overcome the limitations of the periodic boundary conditions for systems that have no corresponding translational symmetry. For a detailed description of these modifications and the corresponding test calculations we refer the reader to [33].

Altogether, we modified the PWCOND code in such a way that it may be used for the accurate solution of the scattering problem for large systems containing several hundred atoms. Thus the quantum conductance of real polymers containing structural and chemical defects (see, e.g., Fig. 3 for an example considered here) is accessible to parameter-free electronic structure calculations. The conductance of

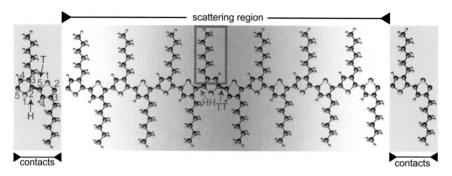

Fig. 3 Schematic setup of the transport calculations, exemplarily shown for the case of isomer defects: the scatter region contains the investigated defect consisting of a HH-TT linkage between the thiophene rings which is caused by a swapped position of the hexyl side chain (*red rectangle*). The contacts are made of ideal rr-HT-P3HT. In the *left* contact modeled by two monomers the atomic numbering is indicated. *H* denotes the second and *T* the fifth position

systems such as studied here that are modelled with about 400 atoms can presently be obtained within about 10 h wall-clock time using 1024 cores on the HLRS Cray XC40. We are presently working on further improving the scaling behaviour of the code by using `ScaLAPACK` routines for the matrix operations.

From a physics point of view, however, it has to be said the present approach still neglects bias voltage and dissipative scattering effects.While this should not affect qualitative trends with respect to the influence of, e.g., structural deformations on the conductance which we investigate here, it restricts meaningful quantitative predictions to conditions of small bias and low temperature. The present calculations model the contacts by ideal P3HT polymers, cf. Fig. 3, which have strongly dispersive and well separated valence bands. Hence p conductance suppressing bias voltage effects can be expected to be small.

3 Results and Discussion

P3HT polymer chains are characterized by large carrier mobilities ($\sim 10^3\,\mathrm{cm^2/Vs}$) [34] which may re reduced, however, due to conformational modifications likely to occur in actual blends [35, 36] as well as oxygen impurities due to, e.g., exposure to ambient conditions [37].

In the following the effect of a twist of the P3HT chains—as shown in Fig. 4—on its quantum conductance is studied. The scattering region considered here consists of 12 monomers. The atoms of the two outer monomers on each side are kept fixed during the relaxation. Since the present calculations employ periodic boundary conditions, two additional monomers are added on each side to decouple the torsion from its periodic images. The total region contains about 400 atoms within a cell that is about 63 Å long. After the electron structure of this system has been

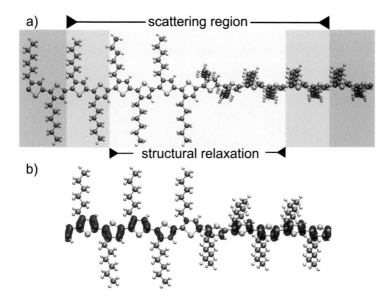

Fig. 4 (**a**) Schematic setup used to describe the electron transport across a chain torsion. The atoms of the outer four monomers of the scattering region are kept frozen during structural relaxation. Additional four monomers outside the scattering region decouple the torsion from its periodic images. (**b**) Charge density isosurfaces indicate the HOMO of a structure with a 90° torsion between two thiophene rings

determined self-consistently, the potential of the 12 innermost monomers is used for the transport calculations, exploiting the approach described in [33], which allows for combining calculations of differently dimensioned systems for self-consistent field and transport calculations, respectively.

At first we consider a rotation that affects a single thiophene-thiophene bond only. The p conductivity of P3HT is governed by the states close to the valence-band maximum (VBM). They correspond to overlapping p_z orbitals of the thiophene rings that are perpendicular to the thiophene plane. Since the overlap of these p_z orbitals is directly affected by the torsion, one expects a strong decrease of the quantum conductance with increasing torsion angle, as indeed shown earlier by transfer integral calculations [13, 17, 18].

In fact, also the quantum conductance through single twisted thiophene-thiophene bonds calculated here is reduced with increasing twist angle, see blue curve in Fig. 5. There is a pronounced conductance minimum for square angles, where no delocalized π-bonds can be formed anymore (cf. Fig. 4b). In addition to the vanishing overlap of the carbon p_z-like orbitals directly at the defect, one observes in Fig. 4b a fragmentation of the highest occupied molecular orbital (HOMO) in the adjacency of the defect. Interestingly, the calculated conductance shows for twist angles up to around 60° only a moderate conductance drop to about 80%.

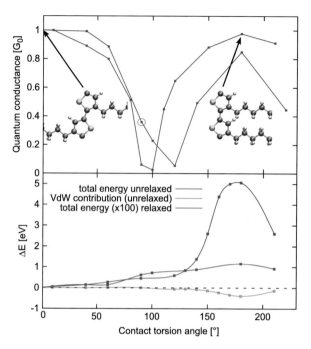

Fig. 5 *Top*: calculated conductance at the VBM through single twisted thiophene-thiophene bonds (*blue*) and configurations, where the total torsion angle around the polymer axis is realized by a sequence of twists obtained from structural relaxation (*red*). The respective geometries for zero and 180° rotation are shown as *insets*. The encircled data point corresponds to the configuration shown in Fig. 6. *Bottom*: total-energy increase due to the bond twisting for the case that only a single bond is twisted (*blue*) and for configurations where the torsion is realized by a sequence of twists obtained from structural relaxation (*red*). The dispersion energy contributions to the former case are shown in *green*

The calculated conductance vs. twist angle is not symmetric with respect to its minimum, since the corresponding torsion does not represent a symmetry transformation of the polymer, as illustrated by the 0° and 180° configurations shown in Fig. 5. Furthermore, the angles 140° and 220° are not equivalent as the polymer is slightly bent (cf. Fig. 2 in [38]).

The total energy as a function of the twist angle is shown by a blue line in the lower part of Fig. 5. For single bond twisting there is only a minor increase in energies for small torsion angles, suggesting that thermal activation may lead to corresponding geometry changes: The energy difference between the ideal polymer and the structure twisted by 10° (20°) amounts to 5 (37 meV). These torsion angles may thus be expected at room temperature if the chain motion is not strongly constrained otherwise. This is in accord with molecular dynamics simulations [39]. There is a steep increase of the total energy for torsion angles in excess of 150°. This is related to the steric repulsion of the hexyl side chains, which is only to a small extent compensated by attractive van der Waals forces (green curve in Fig. 5), and efficiently prevents substantial rotations around single bonds.

The finding that only twists larger than about 60° have a measurable impact on the polymer conductance suggests that the overall conductance decrease will be reduced if not only a single bond is twisted, but if the total torsion angle around the polymer axis is realized by a sequence of twists. Moreover, such a sequence of rotations may be energetically more favorable than a single twisted bond. For this reason we also consider polymer geometries that result from the structural relaxations where the total torsion angle is fixed by the boundary conditions, but may be distributed among several thiophene-thiophene bonds. It is found that upon relaxation the hexyl side chains increase their mutual distance which leads to energetically far more favorable structures, see red curve (instead of the blue curve) in the lower part of Fig. 5. Also the quantum conductivity calculated for these relaxed structures differs from the corresponding curve for single bond twists, see upper part of Fig. 5. Not only the conductance drop is reduced and its onset occurs now for larger rotation angles, also the minimum of the transmission curve has shifted from nearly perpendicular to about 120°. This is a consequence of the nonlinear relation between conductance drop and bond twist. In case of a total torsion of 90° the maximum twist angle between two adjacent thiophene rings is 41° (cf. Fig. 6a). Hence, there is still some wave function delocalization along

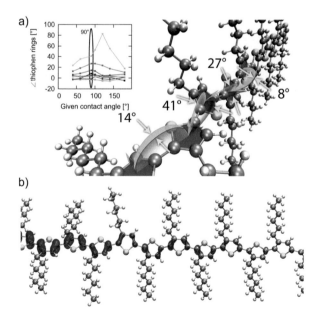

Fig. 6 (a) Atomic structure obtained upon structural relaxation for an overall torsion angle of 90° induced by the boundary conditions. The resulting twist angles for single bonds between thiophene rings are indicated. The data point belonging to the calculated quantum conductance for this configuration is highlighted in Fig. 5. The *inset* shows the twist angle distributions calculated here for various torsions. (**b**) Charge density isosurfaces indicate the HOMO for the structure above

the polymer direction. Only for total torsions as large as 120° where maximum twist angles of about 83° occur, the conductance is essentially quenched due to the vanishing overlap between the carbon p orbitals and a strongly fragmented HOMO.

We find the product of the transmission coefficients calculated for single twists to be a very rough approximation to the transmission through the complete torsional defect: The product of the transmission coefficients corresponding to the overall chain torsion of 90° realized by the structure shown in Fig. 6a for example is about twice as large as the transmission coefficient obtained from the solution of the complete scattering problem of this structure, cf. highlighted data point in Fig. 5. This may partially be related to the stronger fragmentation of the HOMO induced by an extended distortion pattern, cf. Figs. 4b and 6b. In case of single confined twists the reduced overlap between neighbouring thiophen orbitals limits the charge transfer, whereas the HOMO is interrupted for several monomers in case of the extended defect. Obviously, the transfer integral approach has to be applied with care to complex deformation patterns.

The present total-energy findings obtained for various torsion patterns indicate that the minimum energy configuration for a given torsion angle is realized by a sequence of single twists of similar amplitude, i.e., a helix configuration. However, the energy gain upon helix formation is small, of the order of a few meV, and the formation of such regular structures will in general be prevented by the surrounding P3HT matrix.

Polymer bending is a common defect for P3HT [40] that occurs along with torsions. In order to model this defect we analyze several folded P3HT chains. We characterize these foldings by their minimum curvature radius along the polymer chain. The starting configuration is structurally relaxed, where the atoms of the inner ten monomers (250 atoms) are free to move and the outer four monomers are kept frozen. This leads to polymer geometries such as shown in Fig. 7. The structures resulting for small curvatures are rather regular with straight and ordered hexyl side chains. With increasing curvature additional disorder including torsional defects is observed, see Fig. 7.

The inset in Fig. 7 shows the quantum conductance calculated for the relaxed geometries in dependence of their minimum curvature radii. Remarkably, the conductance is nearly unaffected by the structural deformations that have radii larger than about 17 Å. This can be rationalized in terms of the overlap between p orbitals of neighboring thiophene rings: While thiophene ring torsions may completely quench this overlap, chain bending leads to an overlap decrease/increase on opposite polymer sides. Additionally, we observe very efficient local relaxation mechanisms that smooth the polymer structure and prevent sharp kinks. The calculated conductance is quenched, however, for small curvature radii that also lead to bond twists caused by the structural relaxation in response to hexyl side chain related strain. For curvature radii smaller than 12 Å a similar effect on the calculated conductance is found as for twist angles larger than about 60°. The critical influence of the hexyl side chains onto the planarity of the backbone has already been pointed out in [36]. Similarly as observed for torsion patterns, a HOMO fragmentation is observed upon bending, cf. Fig. 7.

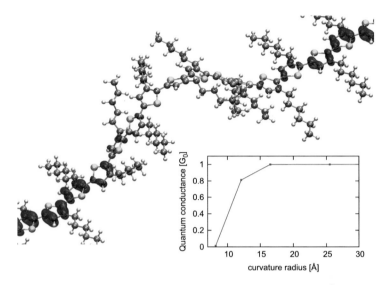

Fig. 7 Relaxed molecular structure with a minimum curvature radius of 8 Å. Charge density isosurfaces indicate the HOMO. The calculated quantum conductance for structures with different curvature radii are shown as *inset*

Crystalline regioregular (rr)-HT-P3HT is of particular interest for organic semi-conductor devices due to its large carrier mobility of up to $0.1 \, cm^2/(Vs)$ [41]. Here the identifier HT indicates the head-tail coupling between two thiophene rings where H denotes the second and T the fifth position on the ring, with numeration starting at the sulfur atom as shown in Fig. 3. In order to assess the influence of isomer defects on the (rr)-HT-P3HT transport properties, the intrachain scattering along a defective polymer is calculated, where HH and TT coupling instead of HT occurs, according to a replacement of a hexyl side chain, see Fig. 3. Ideal rr-HT-P3HT segments serve as contacts to the isomer defect. Additional HT coupled P3HT segments in the vicinity of the defect are included in the scattering region, in order to allow for a realistic modeling of the geometrical and electronic environment of the defect embedded in the polymer.

The calculated quantum conductance of the isomer defect (red) is compared with corresponding calculations for the ideal P3HT chain (blue) in Fig. 8 rhs. The data for the ideal P3HT chain can be obtained by simply counting the number of bands at a given energy (cf. blue band structure in Fig. 8 lhs). As the holes will almost exclusively move at the Fermi edge, we find only one conducting band which results in a maximum quantum conductance of $1 \, G_0$.

Close to the Fermi edge a small conductance reduction is caused by the defect. Stronger defect induced drops occur, however, for energies about 2.3 eV below the VBM, cf. Fig. 8. They will barely affect the actual electron transport that is governed by charge carrier close to the VBM. The physical origin for the sharp drop around -2.3 eV is a strong wave function localization at the corresponding energy, cf. [33].

Fig. 8 *Left*: band structure of rr-HT-P3HT along the chain direction. *Right*: quantum conductance for rr-HT-P3HT (*blue*) in comparison to the calculations for the single isomer defect shown in Fig. 3 (*red*)

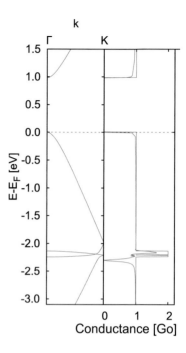

The all in all minor transport changes induced by the isomer defect confirm that the P3HT transport properties are determined by the thiophene rings rather than the hexyl side chains. The present results indicate that the conductance of (rr)-HT-P3HT is surprisingly robust with respect to the occurrence of single and repeated isomer defects. However, such disorder will affect the bulk morphology and might therefore have some influence on the performance of the polymer in electronics applications [42].

Oxygen is used as p-dopant in organic semiconductors where it forms a reversible charge transfer complex [43–45], but may also be present unintentionally due to the exposure to ambient atmosphere and aging. Chemical reactions of O_2 and P3HT may result in a covalent bonding between oxygen and carbon as well as sulfur [46]. The P3HT degradation upon exposure to oxygen [47, 48] decreases the charge carrier mobility [45]. Thereby differences are observed between the effect of oxygen and ozone [49]. Experimentally, the formation of S-O and C-O bond containing species upon P3HT exposure to air was detected [50]. In the present work we focus on the energetically most favored S and C structures identified in recent calculations [37], cf. Fig. 9. It can be seen that both oxygen defects slightly repel the nearest hexyl side chain and that the C defect slightly distorts the thiophene ring.

In order to calculate the influence of these two O defects on the P3HT intrachain transport we proceed similarly as above, i.e., we use ideal rr-HT-P3HT contacts which sandwich a scatter region of 12 monomers including the defect. The scatter region has a length of 47 Å. The calculated quantum conductance for the two defects is shown in Fig. 10 along with the results for the ideal P3HT chain. Obviously, the

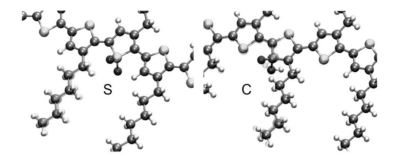

Fig. 9 Calculated relaxed geometries of two oxygen impurity configurations S and C in P3HT. The *shaded atoms* indicate the oxidation induced structure changes

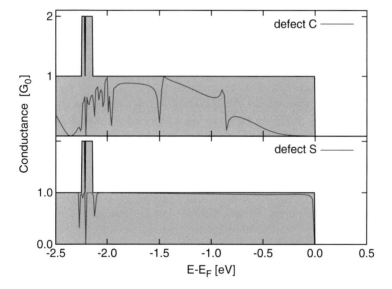

Fig. 10 Calculated quantum conductance through defect S and C, cf. Fig. 9. The *shaded region* is the maximum possible quantum conductance for rr-HT-P3HT

two defects differ strongly in their influence on the polymer conductance. The S defect barely influences the transport even for energies far below the VBM. Only conduction channels at around $-2.2\,eV$ are broken. This is in marked contrast to the C defect, which nearly quenches the hole transport up to nearly 1 eV below the VBM.

In order to understand the different transport properties of the two defects we analyze their influence on the electronic density of states (DOS), cf. [33]. It is found that, apart from minor energy shifts, the DOS of S nearly coincides with the respective calculations for the ideal polymer. Only for energies below $-2\,eV$ additional, oxygen-localized p states appear. Stronger deviations occur in case of C, where the complete valence band DOS is modified and additional oxygen p states show up at the VBM. This different behaviour is expected: The P3HT valence band is formed by p orbitals of the conjugated chain system which are little influenced by the adsorption of oxygen to the sulfur. However, oxidation of a thiophene group carbon directly affects the conjugated electron system and thus has a major influence on the transport in the valence band.

4 Conclusions

A scattering approach based on the self-consistent DFT electronic structure is employed to study the on-chain quantum conductance of P3HT. The influence of defects on the intrachain transport is highly specific. Large torsion angles hinder the formation of a delocalized π electron system and, thus, the transport along the polymer direction while the chain bending has a considerably smaller effect on the conductance. Structural relaxation tends to recover geometries with favorable electron transport properties. The calculations performed for complex distortion patterns show that their conductivity cannot simply be broken down into conductivity drops related the disturbed interaction of directly neighboring molecular segments. P3HT oxidation is found to affect the intrachain transmission strongly, when thiophene group C atoms are attacked, while sulfur oxidation barely modifies the transmission. Also the conductance drop due to isomer defects in the chain is negligible.

Acknowledgements The calculations were done using grants of computer time from the Höchstleistungs-Rechenzentrum Stuttgart (HLRS) and the Paderborn Center for Parallel Computing (PC^2). The Deutsche Forschungsgemeinschaft (FOR1700, FOR1405, SF-TRR142, SCHM 1361/21) is acknowledged for financial support.

References

1. Dang, M.T., Hirsch, L., Wantz, G.: Adv. Mater. **23**(31), 3597 (2011). doi:10.1002/adma.201100792
2. Vukmirovic, N., Wang, L.W.: J. Phys. Chem. B **113**(2), 409 (2009). doi:10.1021/jp808360y
3. McMahon, D.P., Troisi, A.: ChemPhysChem **11**(10), 2067 (2010). doi:10.1002/cphc.201000182
4. Bjorgaard, J.A., Köse, M.E.: J. Phys. Chem. A **117**(18), 3869 (2013). doi:10.1021/jp401521j
5. Vukmirović, N., Wang, L.W.: J. Chem. Phys. **134**(9), (2011). doi:10.1063/1.3560956. http://scitation.aip.org/content/aip/journal/jcp/134/9/10.1063/1.3560956
6. Beenken, W.J.D., Pullerits, T.: J. Phys. Chem. B **108**(20), 6164 (2004). doi:10.1021/jp0373321
7. Kline, R.J., McGehee, M.D.: J. Macromol. Sci. Polym. Rev. **46**(1), 27 (2006). doi:10.1080/15321790500471194
8. Coropceanu, V., Cornil, J., da Silva Filho, D.A, Olivier, Y., Silbey, R., Brédas, J.L.: Chem. Rev. **107**(4), 926 (2007). doi:10.1021/cr050140x
9. Nelson, J., Kwiatkowski, J.J., Kirkpatrick, J., Frost, J.M.: Acc. Chem. Res. **42**(11), 1768 (2009). doi:10.1021/ar900119f

10. Pingel, P., Zen, A., Abellón, R.D., Grozema, F.C., Siebbeles, L.D., Neher, D.: Adv. Funct. Mater. **20**(14), 2286 (2010). doi:10.1002/adfm.200902273
11. Kline, R.J.., McGehee, M.D., Kadnikova, E.N., Liu, J., Fréchet, J.M.J., Toney, M.F.: Macromolecules **38**(8), 3312 (2005). doi:10.1021/ma047415f
12. Crossland, E.J.W., Tremel, K., Fischer, F., Rahimi, K., Reiter, G., Steiner, U., Ludwigs, S.: Adv. Mater. **24**(6), 839 (2012). doi:10.1002/adma.201104284
13. Lan, Y.K., Huang, C.I.: J. Phys. Chem. B **112**(47), 14857 (2008). doi:10.1021/jp806967x
14. Lan, Y.K., Huang, C.I.: J. Phys. Chem. B **113**(44), 14555 (2009). doi:10.1021/jp904841j
15. Noriega, R., Rivnay, J., Vandewal, K., Koch, F.P.V., Stingelin, N., Smith, P., Toney, M.F., Salleo, A.: Nat. Mater. **12**(11), 1038 (2013). doi:10.1038/nmat3722
16. Darling, S.B., Sternberg, M.: J. Phys. Chem. B **113**(18), 6215 (2009). doi:10.1021/jp808045j
17. Grozema, F.C., van Duijnen, P.Th., Berlin, Y.A., Ratner, M.A., Siebbeles, L.D.A.: J. Phys. Chem. B **106**(32), 7791 (2002). doi:10.1021/jp021114v
18. Darling, S.B.: J. Phys. Chem. B **112**(30), 8891 (2008). doi:10.1021/jp8017919
19. Jackson, N.E., Savoie, B.M., Kohlstedt, K.L., Marks, T.J., Chen, L.X., Ratner, M.A.: Macromolecules **47**(3), 987 (2014). doi:10.1021/ma4023923
20. Datta, S.: Electronic Transport in Mesoscopic Systems. Cambridge Studies in Semiconductor Physics and Microelectronic Engineering, vol. 3, 8th edn. Cambridge University Press, Cambridge (2009)
21. Alessandro, T.: Adv. Polym. Sci. **223**, 259 (2010)
22. Giannozzi, P., Baroni, S., Bonini, N., Calandra, M., Car, R., Cavazzoni, C., Ceresoli, D., Chiarotti, G.L., Cococcioni, M., Dabo, I., Dal Corso, A., Gironcoli, S.D., Fabris, S., Fratesi, G., Gebauer, R., Gerstmann, U., Gougoussis, C., Kokalj, A., Lazzeri, M., Martin-Samos, L., Marzari, N., Mauri, F., Mazzarello, R., Paolini, S., Pasquarello, A., Paulatto, L., Sbraccia, C.M., Scandolo, S., Sclauzero, G., Seitsonen, A.P., Smogunov, A., Umari, P., Wentzcovitch, R.M.: J. Phys. Condens. Mat. **21**(39), 395502 (2009). doi:10.1088/0953-8984/21/39/395502
23. Vanderbilt, D., Phys. Rev. B **41**(11), 7892 (1990). doi:10.1103/PhysRevB.41.7892
24. Kleinman, L., Bylander, D.: Phys. Rev. Lett. **48**(20), 1425 (1982). doi:10.1103/PhysRevLett.48.1425
25. Perdew, J., Burke, K., Ernzerhof, M.: Phys. Rev. Lett. **77**(18), 3865 (1996). doi:10.1103/PhysRevLett.77.3865
26. Perdew, J., Burke, K., Ernzerhof, M.: Phys. Rev. Lett. **78**(7), 1396 (1997). doi:10.1103/PhysRevLett.78.1396
27. London, F.: Z. Phys. Chem. Abt. B **11**, 222 (1930)
28. Ortmann, F., Schmidt, W.G., Bechstedt, F.: Phys. Rev. Lett. **95**(18), 186101 (2005). doi:10.1103/PhysRevLett.95.186101
29. Grimme, S.: J. Comput. Chem. **27**(15), 1787 (2006). doi:10.1002/jcc.20495
30. Choi, H.J., Ihm, J., Phys. Rev. B **59**(3), 2267 (1999). doi:10.1103/PhysRevB.59.2267
31. Smogunov, A., Dal Corso, A., Tosatti, E.: Phys. Rev. B **70**(4), 045417 (2004). doi:10.1103/PhysRevB.70.045417
32. Sclauzero, G., Dal Corso, A., Smogunov, A.: Phys. Rev. B **85**(16), 165411 (2012). doi:10.1103/PhysRevB.85.165411
33. Lücke, A., Schmidt, W.G., Rauls, E., Ortmann, F., Gerstmann, U.: J. Phys. Chem. B **119**, 6481 (2015)
34. Grozema, F.C., Siebbeles, L.D.A.: J. Phys. Chem. Lett. **2**(23), 2951 (2011). doi:10.1021/jz201229a
35. Salaneck, W.R., Inganäs, O., Thémans, B., Nilsson, J.O., Sjögren, B., Österholm, J.E., Brédas, J.L., Svensson, S.: J. Chem. Phys. **89**(8), 4613 (1988). doi:10.1063/1.454802
36. Dkhissi, A., Ouhib, F., Chaalane, A., Hiorns, R.C., Dagron-Lartigau, C., Iratcabal, P., Desbrieres, J., Pouchan, C.: Phys. Chem. Chem. Phys. **14**(16), 5613 (2012). doi:10.1039/c2cp40170c
37. Volonakis, G., Tsetseris, L., Logothetidis, S.: Phys. Chem. Chem. Phys. **16**(46), 25557 (2014). doi:10.1039/c4cp03203a
38. Colle, R., Grosso, G., Ronzani, A., Zicovich-Wilson, C.M.: Phys. Status Solidi B **248**(6), 1360 (2011). doi:10.1002/pssb.201046429

39. Alexiadis, O., Mavrantzas, V.G.: Macromolecules **46**(6), 2450 (2013). doi:10.1021/ma302211g
40. Chang, J.F., Clark, J., Zhao, N., Sirringhaus, H., Breiby, D., Andreasen, J., Nielsen, M., Giles, M., Heeney, M., McCulloch, I.: Phys. Rev. B **74**(11) (2006). doi:10.1103/PhysRevB.74.115318
41. Sirringhaus, H., Brown, P.J., Friend, R.H., Nielsen, M.M., Bechgaard, K., Langeveld-Voss, B.M.W., Spiering, A.J.H., Janssen, R.A.J., Meijer, E.W., Herwig, P., Leeuw, D.M.D.: Nature **401**(6754), 685 (1999). doi:10.1038/44359
42. McMahon, D.P., Cheung, D.L., Goris, L., Dacuña, J., Salleo, A., Troisi, A.: J. Phys. Chem. C **115**(39), 19386 (2011). doi:10.1021/jp207026s
43. Lüer, L., Egelhaaf, H.J., Oelkrug, D., Cerullo, G., Lanzani, G., Huisman, B.H., Leeuw, D.D.: Curr. Trends Cryst. Org. Semicond. Growth Model. Fundam. Prop. **5**(1–3), 83 (2004). doi:10.1016/j.orgel.2003.12.005. http://www.sciencedirect.com/science/article/pii/S1566119904000059
44. Liao, H.H., Yang, C.M., Liu, C.C., Horng, S.F., Meng, H.F., Shy, J.T.: J. Appl. Phys. **103**(10), 104506 (2008). doi:10.1063/1.2917419
45. Schafferhans, J., Baumann, A., Wagenpfahl, A., Deibel, C., Dyakonov, V.: Org. Electron. **11**(10), 1693 (2010). doi:10.1016/j.orgel.2010.07.016. http://www.sciencedirect.com/science/article/pii/S1566119910002466
46. Seemann, A., Sauermann, T., Lungenschmied, C., Armbruster, O., Bauer, S., Egelhaaf, H.J., Hauch, J.: Org. Photovolt. Dye Sensitized Sol. Cells **85**(6), 1238 (2011). doi:10.1016/j.solener.2010.09.007. http://www.sciencedirect.com/science/article/pii/S0038092X10002999
47. Sirringhaus, H.: Adv. Mater. **21**(38–39), 3859 (2009). doi:10.1002/adma.200901136
48. Grossiord, N., Kroon, J.M., Andriessen, R., Blom, P.W.M.: Org. Electron. **13**(3), 432 (2012). doi:10.1016/j.orgel.2011.11.027. http://www.sciencedirect.com/science/article/pii/S1566119911004046
49. Chabinyc, M.L., Street, R.A., Northrup, J.E.: Appl. Phys. Lett. **90**(12), 123508 (2007). doi:10.1063/1.2715445
50. Norrman, K., Madsen, M.V., Gevorgyan, S.A., Krebs, F.C.: J. Am. Chem. Soc. **132**(47), 16883 (2010). doi:10.1021/ja106299g

Ab-Initio Calculations of the Vibrational Properties and Dynamical Processes in Semiconductor Nanostructures

Gabriel Bester and Peng Han

Abstract The computational facility made available to us in the year 2014 at the High Performance Computing Center Stuttgart (HLRS), enabled us to calculate the lattice strains, the vibrational properties, and carrier relaxation processes in colloidal semiconductor nanoclusters based on first-principles density functional theory (DFT). In this reporting period, we made two important contributions to the field by pushing the boundary in terms of computational power and in term of new algorithmic developments. We studied the phonon-assisted carrier relaxation processes in colloidal semiconductor nanoclusters using large-scale ab initio calculations and found two relaxation pathways for fast carrier relaxation. We also studied the strains and their consequences on the vibrational properties in colloidal semiconductor core-shell nanoclusters. We found a heavily compressive strain on both the cores and the shells of InAs-InP and CdSe-CdS core-shell nanoclusters, which contributes a large blue-shift of the vibrational frequencies. These results lead to a different interpretation of the frequency shifts of recent Raman experiments.

1 Introduction

Colloidal semiconductor nanoclusters (NCs), or quantum dots (QDs), are attracting significant attention due to their applications in the fields of optoelectronics, spintronics, photovoltaic, biolabeling [1–7], and to their potential to emerge as the key components of the next generation displays [8]. Therefore, the optical properties of colloidal semiconductor NCs and the fact that they can be engineered to be either excellent emitters or excellent absorbers are necessarily required. One important milestone was the synthesis of core-shell nanostructures, which contain at least two semiconductors arrangement in an onion-like geometry [9–11]. In such onion-like structures, the core material is surrounded by a different shell material in order to

G. Bester (✉) • P. Han
Max-Planck-Institut für Festkörperforschung, Stuttgart, Germany

Institut für Physikalische Chemie, Universität Hamburg, Hamburg, Germany
e-mail: gabriel.bester@uni-hamburg.de; peng.han@chemie.uni-hamburg.de

© Springer International Publishing Switzerland 2016
W.E. Nagel et al. (eds.), *High Performance Computing in Science and Engineering '15*, DOI 10.1007/978-3-319-24633-8_11

reduce the influence of the possibly imperfect surface onto the core. Indeed, very recent advances suggest that coating NCs with a semiconducting shell may solve the long-standing problem of their discontinuous optical emission ("blinking"), making them good emitter candidates.

Although colloidal semiconductor core-shell NCs have been studied experimentally in detail, theoretical understanding of the strains, vibrational properties, electronic structure and the resulting optical properties of such nanostructures is still unclear. The theoretical modeling of these type of structures, performed at the level of the effective mass approximation [12], the $k.p$ method [13], the tight-binding approach [14, 15] or the empirical pseudopotential method [16, 17] were done assuming perfect, unstrained, and unrelaxed atomic positions. Indeed, colloidal quantum dots in their fluid environment were often considered as unstrained, in direct contrast to their self-assembled (Stransky Krastanov) homologues that are known to only exist because of the presence of strain. Only very recently, ab-initio large-scale calculations showed that the effect of structural relaxation in colloidal QDs is important [18, 19] and a lack of such relaxation leads, for instance, to the appearance of unphysical imaginary vibrational frequencies and red shifts of vibrational modes [19].

To reveal the strains and their consequences on the vibrational properties, we perform large-scale ab initio density functional theory (DFT) calculations to study the structural and vibrational properties of core-shell Si(core)-Ge(shell), the inverted Ge-Si, InAs-InP, and CdSe-CdS NCs with radii ranging from 13.5 to 15.6 Å. We find that (1) The shell dictates the atom positions of the entire core-shell NCs; (2) Both the core and the shell are compressed in the InAs-InP and in the CdSe-CdS NCs; (3) The bond-length distribution in the NCs goes from homogeneous (small scattering) throughout the NCs for our group IV NCs to inhomogeneous only at the interface for our group III-Vs, to inhomogeneous in the entire shell region in our II-VIs NCs; and (4) The frequency shifts we obtain, compared to the bulk frequencies, for our NCs can all be traced back to two fundamental effects: One being the shift of the modes according to strain (given by the Grüneisen parameters), the second being the red-shift created by the undercoordination of the near-surface atoms. Both effects tend to work against each other since the NCs are typically compressively strained and the strain effect leads to a blue-shift (with a positive Grüneisen parameter).

The second topic we could address within this reporting period was the mechanism of intraband carrier relaxation in NCs. The processes responsible for this carrier relaxation has been under debate since the early days of NC growth. Initially, an inefficient intraband relaxation process was expected in strong confined NCs where the separation between energy levels are much larger than the phonon (or rather vibron) energy. The hot-electron was thought to be unable to transfer its energy to the lattice, which was referred to as the "phonon bottleneck" [20, 21]. However, such a phonon bottleneck was only observed in a few experiments [22, 23] and fast relaxation processes, which occur on sub-picosecond to picosecond (ps) time scales, have been reported in many instances [24–27]. Several, often contradictory, mechanisms have been proposed in order to explain how hot-electrons can efficiently release their energy via nonradiative pathways in strongly confined NCs. Some of the most popular ideas include Auger cooling [28–30], polaron relaxation [31–33], surface trapping [34], nonadiabatic multiphonon relaxation [35], electron (hole)-ligand vibration coupling [36], and surface and atomic fluctuation induced shifts of electronic levels [37]. However, in NCs without surface trapping and without hole (which eliminates the possibility for Auger cooling), the electron-phonon (e-ph) interaction must play the key role in the fast relaxation process even with large energy detuning.

We report here on the dynamical processes of the electronic transition from the excited (*P*-like) state to the ground (*S*-like) state (in the absence of a hole) via e-ph interaction performing large scale ab initio DFT calculations. We obtain a fast (picosecond) carrier relaxation process in InAs, CdSe and Si NCs for most sizes, in agreement with experiment. This relaxation is shown to occur by two different pathways: (1) a strong coupling between the electronic transition and acoustic-type vibrons allow the electronic relaxation, although the energy detuning between the electronic level spacing and the vibrational frequency is extremely large; and (2) a very weak, but nearly resonant, coupling between the electron transition and the passivant vibrations.

2 Computational Methods

2.1 Vibrational Eigenmodes for Large Nanoclusters

The electronic eigenstates and the self-consistent local potentials are obtained by solving the Kohn-Sham equation self-consistently within DFT. We calculate the bulk phonon states using density functional perturbation theory [38]. The vibrational properties are obtained from the eigenvalue equation [39]:

$$\sum_J \frac{1}{\sqrt{M_I M_J}} \frac{\partial^2 V(\mathbf{R})}{\partial \mathbf{R}_I \partial \mathbf{R}_J} \mathbf{U}_J = \omega^2 \mathbf{U}_I. \tag{1}$$

where M are the atomic masses (and we will vary the mass of the passivant), V the potential, $\mathbf{U}$ the eigenvectors and ω the frequencies. We use ab initio DFT implemented in the CPMD code [40] to optimize the geometry and to calculate the vibrational eigenmodes. The supercell is simple cubic with 3 Å space between the outmost atoms and the boundary. To analyze the eigenmodes in terms of bulk and surface contributions, we calculate the projection coefficients

$$\alpha_{c,s,p}^v = \frac{\sum_I^{(N_c, N_s, N_p)} |\mathbf{X}^v(I)|^2}{\sum_{I=1}^N |\mathbf{X}^v(I)|^2}, \tag{2}$$

where, N_c, N_s, N_p and N are the core, surface, passivant, and total number of atoms, $\mathbf{X}^v(I)$ represents the three components that belong to atom I from the $3N$-component eigenvector. We define the surface atoms as the atoms belonging to the outermost seven layers of the cluster (around 3 Å thick). To compare our results with Raman spectroscopy measurements, we calculate the Raman intensities using a phenomenological model proposed by Richter et al.[41]. Based on this model, the Raman intensity of nanostructures is proportional to the projection coefficient of the vibrational modes of the nanostructure onto bulk modes with a relaxation of the wave-vector selection rule[19, 41],

$$I(\omega) \propto \sum_{n,v,\mathbf{q}} \frac{|C_{n,\mathbf{q}}^{v}|^2}{(\omega - \omega^v)^2 + (\Gamma_0/2)^2},$$ (3)

where $C_{n,\mathbf{q}}^{v}$ is the projection coefficient of the NC mode v on the bulk mode n with wave vector $\mathbf{q}$ and Γ_0 is the natural Lorentzian linewidth (an empirical parameter in our work). The coefficient $C_{n,\mathbf{q}}^{v}$ are summed up to $\Delta q = 1/(2R)$.

2.2 Calculation of Electron-Phonon Coupling Matrix Elements

Within the Born-Oppenheimer approximation, the many-body Hamiltonian of the NC is decomposed into an electronic part, an ionic part and the coupling of the electron system with the lattice vibrations. The first two parts deal with the motion of electrons and ions separately while the third part describes the *electron-phonon* interaction. The e-ph interaction Hamiltonian can be expressed as a Taylor series expansion of the electronic potential [40]:

$$\Delta V^v(\mathbf{r}, \mathbf{R}) = \sum_I \frac{\partial V}{\partial \mathbf{R}_I} \cdot \mathbf{u}_I^v,$$ (4)

where V is the electronic potential and $\mathbf{R}_I$ denotes the nuclear position of atom I. The displacement vector $\mathbf{u}_I^v$, which belongs to atom I and the vibrational eigenmode v, can be written in terms of the normal coordinates Q^v [42]:

$$\mathbf{u}_I^v = \frac{1}{\sqrt{M_I}} Q^v \mathbf{X}_I^v,$$ (5)

with the occupation number representation

$$Q^v = \sqrt{\frac{\hbar}{2\omega^v}} (a_v^\dagger + a_v),$$ (6)

where M_I is the mass of atom I, $\hbar$ is the reduced Planck constant, ω^v is the frequency of the vibrational mode v, $\mathbf{X}_I^v$ are the three components of the vibrational eigenmode belonging to atom I, $a_v^\dagger$ and a_v denote the creation and the annihilation operators of the v-mode phonon. According to Eqs. (4) and (5), the e-ph coupling matrix elements for the transition from the initial state $|\psi_n, 0\rangle$ to the finial state $|\psi_m, 1^v\rangle$ with emission of a v-mode phonon has the form:

$$g^v(m, n) = \sum_I \sqrt{\frac{\hbar}{2M_I \omega^v}} \times \langle \psi_m, 1^v | \frac{\partial V}{\partial \mathbf{R}_I} \cdot \mathbf{X}_I^v (a_v^\dagger + a_v) | \psi_n, 0 \rangle, \qquad (7)$$

where the polaron state $|\psi_m, i^v\rangle$ is composed of an electronic state $|\psi_m\rangle$ and a vibrational state $|i^v\rangle$.

Based on the frozen-phonon approach for the e-ph coupling, the change of the potential caused by a phonon distortion in Eq. (7) is replaced by [43–45]:

$$\sum_I \frac{\partial V}{\partial \mathbf{R}_I} \cdot \mathbf{X}_I^v \approx \frac{V_{scf}^v(\mathbf{r}) - V_{scf}^0(\mathbf{r})}{u^v}, \qquad (8)$$

where $u^v = \sqrt{\sum_I \frac{1}{M_I} \mathbf{X}_I^{v\,2}}$ describes the frozen-phonon displacement caused by the lattice vibration, $V_{scf}^v(\mathbf{r})$ and $V_{scf}^0(\mathbf{r})$ are the self-consistent potentials with and without the phonon distortion [43–45].

In DFT, the self-consistent potential $V_{scf}(\mathbf{r})$ has the form [46]:

$$V_{scf}(\mathbf{r}) = V_{ion}(\mathbf{r}) + e^2 \int \frac{\rho(\mathbf{r}')}{|\mathbf{r} - \mathbf{r}'|} d\mathbf{r}' + \frac{\delta E_{xc}[\rho(\mathbf{r})]}{\delta \rho(\mathbf{r})}, \qquad (9)$$

where the ionic potential $V_{ion}(\mathbf{r})$ is typically decomposed into a local part $V_{ion}^{loc}(\mathbf{r})$ and a non-local part $V_{ion}^{NL}(\mathbf{r}, \mathbf{r}')$, the Hartree potential $e^2 \int \frac{\rho(\mathbf{r}')}{|\mathbf{r}-\mathbf{r}'|} d\mathbf{r}'$, and the exchange-correlation potential $\frac{\delta E_{xc}[\rho(\mathbf{r})]}{\delta \rho(\mathbf{r})}$ are obtained self-consistently with $\rho(\mathbf{r}) = \sum_n^{occ} |\psi_n(\mathbf{r})|^2$. The local part of the self-consistent potential $V_{scf}^{loc}(\mathbf{r})$, which includes the local ionic potential $V_{ion}(\mathbf{r})$, the Hartree and the exchange-correlation potentials, is computed self-consistently using standard DFT. The non-local potential is calculated using the real space representation of the Kleinman-Bylander form:

$$\langle \mathbf{r} | V_{ion}^{NL} | \mathbf{r}' \rangle = \sum_{I,l,m} \frac{\langle \mathbf{r} | \delta V_l^I \phi_{lm}^I \rangle \langle \phi_{lm}^I \delta V_l^I | \mathbf{r}' \rangle}{\langle \phi_{lm}^I | \delta V_l^I | \phi_{lm}^I \rangle}, \qquad (10)$$

where l and m label the angular and magnetic moments, $\phi_{lm}^I(\mathbf{r} - \mathbf{R}_I)$ are the pseudo-wavefunctions centered on the atom position $\mathbf{R}_I$, and the potential $\delta V_l^I(|\mathbf{r} - \mathbf{R}_I|) = V_l^I(|\mathbf{r}-\mathbf{R}_I|) - V_{loc}^I(|\mathbf{r}-\mathbf{R}_I|)$. In contrast to the local potential, the Kleinman-Bylander non-local potential in Eq. (10) is a projector. The contribution of the non-local potential to Eq. (8) takes the form,

$$\sum_I \frac{\hbar}{2M_I \omega^v} \sum_{\mathbf{r},\mathbf{r}'} [\langle \psi_m | \mathbf{r} \rangle \langle \mathbf{r} | V_{ion}^{NL,v} | \mathbf{r}' \rangle \langle \mathbf{r}' | \psi_n \rangle - \langle \psi_m | \mathbf{r} \rangle \langle \mathbf{r} | V_{ion}^{NL,0} | \mathbf{r}' \rangle \langle \mathbf{r}' | \psi_n \rangle], \qquad (11)$$

where $V_{ion}^{NL,v}$ and $V_{ion}^{NL,0}$ represent the non-local potential with and without phonon distortion.

Finally, using Eqs. (7)–(11) with the phonon distorted and un-distorted atom positions $\{\mathbf{R}_I + \mathbf{X}_I^v/\sqrt{M_I}\}$ and $\{\mathbf{R}_I\}$, the e-ph interaction matrix elements of the NCs can be calculated at the ab initio level without additional parameters [47].

2.3 Time Dependent Evolution of the Population

To model the time evolution of a population after a pulsed laser excitation with frequency ω_L and the dipole coupling Ω, we consider an n-level system consisting of non-interacting e-ph states: $|0\rangle \equiv |S, 0\rangle$, $|1\rangle \equiv |P, 0\rangle$, $|2\rangle \equiv |S, v_1\rangle$, $|3\rangle \equiv |S, v_2\rangle$, $\cdots$, $|n-1\rangle \equiv |S, v_{n-2}\rangle$. The electronic part is limited to S and P states and the vibronic part to the modes v_i.

The corresponding Hamiltonian can be written as [48–50],

$$H(t) = \sum_{j=0}^{n-1} \hbar v_j |j\rangle\langle j| + \sum_{j=2}^{n-1} g_{1,j}(|1\rangle\langle j| + |j\rangle\langle 1|)$$

$$+ \frac{1}{2}\hbar\Omega(e^{i\omega_L t}|0\rangle\langle 1| + e^{-i\omega_L t}|1\rangle\langle 0|), \qquad (12)$$

where $\hbar v_j$ denotes the eigenvalues of the jth states of the n-level system and $g_{1,j}$ are the e-ph coupling matrix elements between the electronically excited state $|1\rangle$ and the states $|j\rangle$, that we calculate ab initio as described above. In our calculation, we assume an initial full occupation of the excited state $|1\rangle$, which corresponds to the situation where an electron has been photo-excited from the ground state $|0\rangle$ (i.e. an intraband transition). By doing so, we force the system in a state that is formally not an eigenstate of the system but merely a non-interacting basis state. Only the solution of the coupled systems is a true eigenstate that can be fully occupied. As our results will later show, we are exclusively in the weak coupling regime, so that our basis states (the uncoupled e-ph states) represent a very good approximation to the true eigenstate. This allows us to analyze the results by projecting the coupled states onto the uncoupled basis. This situation is not given in self-assembled QDs, where the system is in the strong coupling regime [32], where a full polaron description should be used, instead of the forced occupation calculations performed initially [31].

By utilizing the unitary transformation [50]

$$\hat{U} = \exp\left[\frac{i\omega_L t}{2}\left(\sum_{j=1}^{n-1} |j\rangle\langle j| - |0\rangle\langle 0|\right)\right], \qquad (13)$$

and using the *Baker-Hausdorff lemma* [51], the time independent version of the Hamiltonian can be expressed as

$$
H = \begin{pmatrix}
-\delta_1/2 & \hbar\Omega & 0 & 0 & \cdots & 0 \\
\hbar\Omega & \delta_1/2 & g_{1,2} & g_{1,3} & \cdots & g_{1,n-1} \\
0 & g_{2,1} & \delta_2/2 & 0 & \cdots & 0 \\
0 & g_{3,1} & 0 & \delta_3/2 & \cdots & 0 \\
& & \cdots & & & \cdots \\
0 & g_{n-1,1} & 0 & 0 & \cdots & \delta_{n-1}/2
\end{pmatrix}
$$

with $\delta_1 = \hbar(\omega_1 - \omega_0 - \omega_L)$ and $\delta_j = \delta_1 + \hbar(\omega_j - \omega_1)$.

Replacing the time-independent observable density matrix ρ by the time-dependent expression $\rho(t) = \exp(iHt/\hbar)\rho\exp(-iHt/\hbar)$, we obtain the Liouville–von Neumann equation in the form:

$$
\frac{\partial\rho(t)}{\partial t} = -\frac{i}{\hbar}[H, \rho(t)] + L(\rho(t)), \tag{14}
$$

with the Lindblad decay term

$$
L(\rho) = \frac{1}{2}\sum_{j=2}^{n-1} \Gamma_0^j (2|0\rangle\langle j|\rho|j\rangle\langle 0| - \rho|j\rangle\langle j| - |j\rangle\langle j|\rho), \tag{15}
$$

where Γ_0^j denotes the decay rate from the state $|j\rangle$ ($|S, v_{j-1}\rangle$ in our specific case) to the ground state $|0\rangle$ ($|S, 0\rangle$ in our case) and involves the decay of the vibronic part.

2.4 Computational Details

The NCs we studied are constructed by cutting a sphere, centered on a cation with T_d point group symmetry, from the zinc blende bulk structure and removing the surface atoms having only one nearest-neighbor bond. The surface dangling bonds are terminated by pseudohydrogen atoms H* with a fractional charge of 1/2, 3/4, 5/4, and 3/2 for group VI, V, III, and II atoms, respectively. The calculations are performed using the LDA, Trouiller-Martin norm-conserving pseudopotentials with an energy cutoff of 30 Ry for III-Vs and 40 Ry for II-VIs.

During the geometry relaxation the forces are minimized to less than 3×10^{-6} a.u. under constrained symmetry for the NCs. In the calculation, the charge density $\rho_R(\mathbf{r})$ are obtained by solving the Kohn-Sham equation self-consistently and the values of $\frac{\partial\rho_R(\mathbf{r})}{\partial\mathbf{R}_J}$ are calculated using a finite difference approach. In principle we need $3N$ atomic displacements to obtain all the elements of the dynamical matrix

(N being the number of atoms). In practice we calculate a significantly lower number of displacements $(3N/24)$ and use the symmetry elements of the point group to deduce the missing elements. This is a key point to be able to treat these large structures.

All the calculations are performed with the CPMD code package developed at the Max Planck Institute in Stuttgart and at IBM [40]. The CPMD code is a high performance parallelized plane wave/pseudopotential code implementation of DFT. It offers excellent scaling among the DFT codes, using a hybrid scheme of MPI and OpenMP. In this mixed parallelization, MPI is used for parallelization across nodes while OpenMP is used for on-node parallelization across cores. In this project, all the calculations were carried out on the Cray XE6 Hermit cluster at HLRS with 2.3 GHz and 32 GB memory per nodes, and Cray Gemini interconnects. The typical job sizes for the computations reported here are between 64 to 128 tasks with runtimes around 24 h. The total memory requirements are between 4 and 192 GB, depending on system type and size. We usually run using the maximum available memory on each note, i.e., 1.5 GB per task. The details of the scaling behavior and the performance per CPU for the CPMD code have been given elsewhere [52].

3 Results

3.1 Carrier Relaxation in Colloidal Nanoclusters

In Fig. 1a, we present the vibrational density of states (VDOS) of an $In_{141}As_{140}H^*_{172}$ NC where the contributions from core, surface, and passivant atoms are shown as black, red, and green lines, respectively. The energy splitting between the S and P states will be referred to as E_{SP} and amounts to about 280 meV in this structure, which corresponds to 2255 cm^{-1}. This frequency is close to the vibrational frequency of the passivant stretching modes, but is much higher than the frequency of any bulk-like modes. In Fig. 1b–e we plot the time-dependent occupation probability of the $|P, 0\rangle$ state $(|\rho_{11}(t)|^2)$ which represents the solution of Eq. (14), where we have used a simplified three-level system for illustration purposes. We have selected three different types of vibrational modes: in Fig. 1b and c, a mode with acoustic character and a frequency of 56 cm^{-1}; in Fig. 1d a vibrational mode with optical character and a frequency of 255 cm^{-1} and in Fig. 1e, a passivant stretching mode with a frequency of 2199 cm^{-1}. In our calculations, we assume an initial full occupation of the excited state $|P, 0\rangle$, which corresponds to the situation where the additional electron has been either injected into that orbital or photo-excited via an intraband transition from the states $|S, 0\rangle$. From Figs. 1b–e, we see that the probability to find the electron in the excited state decays with an exponential envelope function modulated by an ultrafast Rabi-type oscillation. The most surprising result is that the hot-electron $|P, 0\rangle$ decay to the ground

Fig. 1 (**a**) Vibrational density of states of an $In_{141}As_{140}H^{*}_{172}$ ($R = 12.3$ Å) NC. The modes are colored according to their dominant localization: core-modes (*black*), surface-modes (*red*) and passivant-modes (*dashed green*), and broadened by $0.8\,cm^{-1}$. (**b**)–(**e**) Time-dependent occupation probability of the $|P, 0\rangle$ state calculated explicitly for the interaction with three different types of modes: (**b**), (**c**) an acoustic mode [labeled (A)] for two different time spans, (**d**) an optic mode [labeled (O)] and (**e**) a passivant stretching mode [labeled (P)]

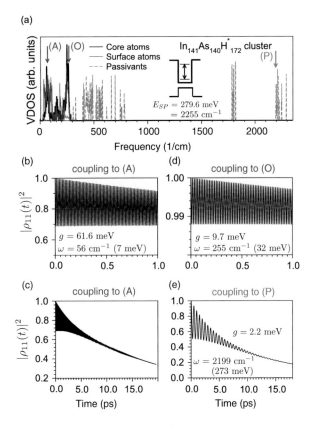

state effectively, although the vibron energy is orders of magnitude smaller than the electronic energy gap (7 meV vibron energy compared to 279.6 meV electronic energy gap in Fig. 1b); in a naïve picture —violating energy conservation.

To give a clear physical picture of this process we start in the top panels of Fig. 2 with the two-level system encountered in optics with a detuned laser, which is usually used to explain Rabi oscillations. The two electronic states $|0\rangle$ and $|1\rangle$ are coupled through the light-field, which is detuned to the red and has an energy of ω_L (Fig. 2a). The lowest three electron-photon basis states are given in (Fig. 2b) and do not account for the light-matter interaction Ω; these are the "bare states" of the system. A decay of the state $|0, \omega_L\rangle$ would correspond to a photon decay and is impossible. The state $|1, 0\rangle$ could decay to $|0, 0\rangle$ and would induce a damping of the Rabi oscillations, but we neglect it at the moment. If the light-matter interaction is switched on, the bare states turn into dressed states with modified energies as indicated in (Fig. 2c). If the system is in the strong coupling regime, the dressed states are called polaritons. The Hamiltonian describing the two level dressed states is

$$\begin{pmatrix} 0 & \Omega \\ \Omega^* & \Delta \end{pmatrix} \longrightarrow \Omega_R^2 = \Omega^2 + \Delta^2 \quad ; \quad |c_2^{max}|^2 = \frac{\Omega^2}{\Omega_R^2} \quad .$$

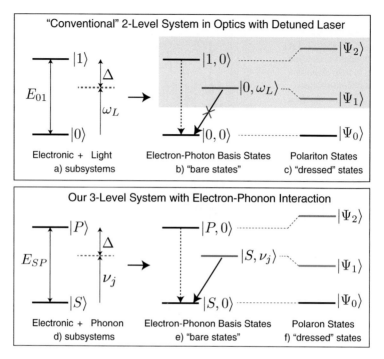

Fig. 2 Comparison between a two-level system, as usually encountered in optics, with a red detuned laser (*upper panel*) and our situation (*lower panel*)

The probability for the state $|\Psi_1\rangle$ to be in state $|1, 0\rangle$, which is not a stationary state of the system, oscillates in the time domain with the Rabi frequency Ω_R. For the initial condition where $|\Psi_1\rangle$ was in state $|0, \omega_L\rangle$ at t = 0, the maximum of the oscillating occupation probability to be in state $|1, 0\rangle$ is $|c_2^{\max}|^2$. In the weak coupling regime (Ω is small compared to Ω_R), the value of $|c_2^{\max}|^2$ is small, which means that the state $|\Psi_1\rangle$ is rather well described by the bare state $|0, \omega_L\rangle$. Our situation is qualitatively very similar (lower panel of Fig. 2). The electronic states $|S\rangle$ and $|P\rangle$ are now coupled through vibrations, which have an energy ν_j. The vibron decay between $|S, \nu_j\rangle$ and $|S, 0\rangle$ is now possible and we treat it using a Lindblad term. The direct decay $|P, 0\rangle$ to $|S, 0\rangle$ is possible trough an optical intraband transition, but is very slow and neglected here. In the strong coupling regime the dressed states are called polaron states. Our corresponding Hamiltonian is given by:

$$\begin{pmatrix} 0 & g \\ g & \Delta \end{pmatrix} \longrightarrow \Omega_R^2 = g^2 + \Delta^2 \quad ; \quad |c_2^{\max}|^2 = \frac{g^2}{\Omega_R^2} \quad , \tag{16}$$

with Rabi-like oscillation frequency Ω_R and the maximum occupation probability $|c_2^{\max}|^2$.

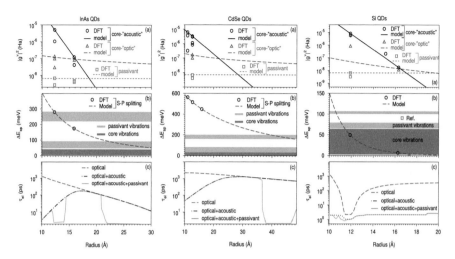

Fig. 3 (**a**) E-ph coupling strength, (**b**) energy level spacing between *P*- and *S*-like states and (**c**) electron lifetime as a function of cluster size for InAs (*left*), CdSe (*middle*) and Si (*right*) NCs. In (**a**)–(**c**), the values calculated from DFT and estimated from empirical models are plotted as *symbols* and *lines*, respectively. The energy ranges for the cluster vibrational modes and passivant modes are filled by *red* and *green colors* in (**a**) and (**b**).

With the help of the computational facility made available to us by the HLRS, we could extend our studies to InAs, CdSe and Si NCs with radius up to 50 Å. From Fig. 3, we see that the e-ph coupling is very strong for core acoustic type modes in small NCs. This leads to the unexpected situation where the electronic gap (*S-P* splitting), although very large compared to the vibron energy (i.e. very large energetic detuning), can be overcome within ps. This lack of phonon bottleneck for small cluster sizes is a consequence of quantum mechanics with its probabilistic description. The system undergoes ultrafast Rabi-like oscillations accompanied by the emission and absorption of vibrons. When the system occupies, to a certain probability, the state $|S, \nu_j\rangle$, the vibrational part decays, with a certain rate, into few, lower energy, vibrons. The e-ph coupling to passivant vibrations is weak, but that the energetic position of these modes leads to nearly resonant processes, in some ranges of diameters, where despite the weak coupling, the carrier relaxation is very fast (sub ps). Our results show that a window of intermediate sizes exists for InAs and CdSe where phonon bottleneck could be observed, while for large and small NC the electron can relax efficiently via coupling to acoustic modes or resonant coupling to passivant vibrations, even in the absence of any kind of defects. In the case of Si we obtain picosecond carrier relaxation for all sizes, dominated by coupling to core acoustic vibrations, which makes the observation of a phonon-bottleneck impossible [53].

3.2 Strains and Their Consequences on Vibrational Properties in Colloidal Core-Shell NCs

To describe the structural properties of the NCs, we plot the nearest neighbor distances as a function of their distance from the cluster center. The results for CdSe-CdS NCs are given in Fig. 4. From this figure, we see a clear progression towards a more scattered data: the core-shell structure (black squares) shows a bond-length distribution that significantly differs from the corresponding NC made of only shell material (CdS, red circles). The core CdSe material is still compressed by as much as 2.0 % but has larger bond-length than the core of the corresponding pure CdS NC. In the area of the interface, the bond-length variation is large, going from nearly CdSe bulk bond-length to 1.9 % below the CdS bulk bond length. The comparison of Fig. 4a, b shows as noticeable difference, especially in the surface/shell region, despite the fact that the NCs have only a small difference in radius. This highlights the fact, that the atomistic description leads to a shell-by-shell construction of the structure with increasing radius. A small radius increase can lead to geometrically and chemically rather different structures. The pure CdS NCs have a ratio of Cd to S atoms of 225/240 in the smaller structure and 321/312 in the larger structure. So, Cd poor in the first case and Cd rich in the second. This should be kept in mind when comparing structures with different radii.

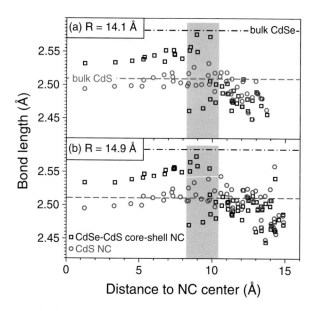

Fig. 4 Bond length distribution as a function of the distance to the NC center for (**a**) $Cd_{79}Se_{68}$-$Cd_{146}S_{172}H_{228}^{*}$ (*black square*), $Cd_{225}S_{240}H_{228}^{*}$ (*red circle*), and (**b**) $Cd_{79}Se_{68}$-$Cd_{242}S_{244}H_{300}^{*}$ (*black square*), $Cd_{321}S_{312}H_{300}^{*}$ (*red circle*). The LDA bond lengths of bulk CdS and CdSe are given as *dashed lines* and *dotted dashed lines*, respectively

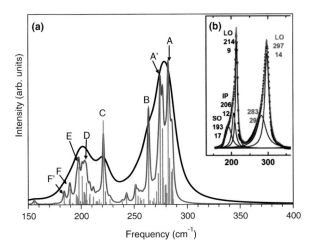

Fig. 5 (**a**) Calculated Raman spectrum [see Eq. (3)] for CdSe-CdS core-shell NC with $R = 14.9$ Å. *Green lines* represent the raw data of the intensities, *red* and *black lines* represent the Raman intensities with a broadening of 1.5 and 8.0 cm^{-1}, respectively. (**b**) Measured Raman spectrum of CdSe-CdS core-shell NC with R around 30.0 Å taken from [54]

Next, we calculated the Raman intensities of CdSe-CdS core-shell $Cd_{79}Se_{68}$-$Cd_{242}Se_{244}H^*_{300}$ NC and compared them with experimental results taken from [54] in Fig. 5. Two comments are due up front: (1) The size of our NC with a radius of 14.9 Å is significantly smaller than the experimental size with R around 30.0 Å. (2) **Peak C** (221.4 cm^{-1}) is a surface mode of CdS. Without considering peak C, we observe for the larger broadening (black line in Fig. 5a) a two peak structure, similar to the experimental result. Each of these peaks is composed of several peaks, as was also deduced from the non-Lorentzian lineshapes in the experiment [54]. We now analyze the peaks subsequently.

Peaks A (282.5 cm^{-1}) and **A′** (274.6 cm^{-1}) are vibrational modes with optical character of CdS. With a broadening of 8.0 cm^{-1}, these two peaks merge into one peak with a frequency of 279 cm^{-1} (Fig. 5), which corresponds to the peak labeled as CdS LO in the experimental work [54]. Our combined peak is red-shifted compared to the experimental results. From the two effects mainly responsible for frequency shifts, strain and undercoordination, only the latter effect is relevant since the CdS shell is already mainly unstrained in our NC (see Fig. 4). The undercoordination effect becomes weaker with increasing shell size and we expect a blue shift of peaks A and A′ if we go towards the experimental situation with a much thicker shell, until they reach the bulk value of around 300 cm^{-1} (Fig. 4a); in good agreement with the experiment. A dependence of the blue shift on the shell thickness was also reported recently [55]. The experimental peak "LO 297" in [54] is therefore a bulk-like, unstrained, CdS peak without large confinement effect.

Peak B (264.3 cm^{-1}) is a vibrational mode localized at the interface and in the CdS near-surface shell, with a small but non-vanishing surface component. It corresponds well to the "low-energy should (LES)" described in the experimental work [54] and which was observed in NCs with different sizes, shapes, surface environments and materials using Raman and photoluminescence measurements. This LES is often described as a "surface optical" phonon mode which is in rather good agreement with our identification.

Peak D (203.5 cm^{-1}) represents a vibrational mode with optical character of the CdSe core combined with CdS near-interface contributions. This mode has a blue shift (8.5 cm^{-1}) compared to the bulk LO mode in CdSe. The experimental blue shift is somewhat larger (214 cm^{-1}), which can be traced back to the fact that a structure with a larger shell will experience an even larger compression of the core and hence a larger blue shift.

Peak E (197.7 cm^{-1}) are modes with contributions from all regions of the NC: core, interface and shell. They correspond well to the "additional intermediate band or IP mode" seen in the experimental work.

Peak F (190.0 cm^{-1}) and F′ (184.0 cm^{-1}). These modes are CdS shell modes combined with optical surface breathing-type modes. They may correspond to the "SO mode" labeled in the experimental paper.

As a final remark, the qualitative picture of our frequency shifts can be understood based on two, often competing, effects of compressive strain (blue-shift of optical modes) and undercoordination (red-shift). We show that the often discussed low-energy shoulder on the Raman spectra originate from interface vibrations with small surface character, in agreement with most of the experimental interpretations and in disagreement with earlier theoretical models. The comparison to experiment—which is a necessary condition to benchmark our results and access the reliability of our prediction—requires to treat large structures and solve massive eigenvalue problems, which was thankfully possible at the HLRS.

References

1. Pijpers, J.J.H., et al.: J. Phys. Chem. C **111**, 4146 (2007)
2. Klimove, V.I., Ivanov, S.A., Nanda, J., Achermann, M., Bezel, I., McGurie, J.A., Piryatinski, A.: Nature **447**, 441 (2007)
3. Kraus, R.M., Lagoudakis, P.G., Rogach, A.L., Talapin, D.V., Weller, H., Lupton, J.M., Feldmann, J.: Phys. Rev. Lett. **98**, 017401 (2007)
4. Gaponik, N., Hickey, S.G., Dorfs, D., Rogach, A.L., Eychmüller, A.: Small **6**, 1364 (2010)
5. Talapin, D.V., Lee, J.-S., Kovalenko, M.V., Shevchenko, E.V.: Chem. Rev. **110**, 389 (2010)
6. Sambur, J.B., Novet, T., Parkinson, B.A.: Science **330**, 63 (2010)
7. Mocatta, D., Cohen, G., Schattner, J., Millo, O., Rabani, E., Banin, U.: Science **332**, 77 (2011)
8. Bourzac, K.: Nature **493**, 283 (2013)
9. Yang, C.S., Kauzlarich, S.M., Wang, Y.C.: Chem. Mater. **11**, 3666 (1999)
10. Cao, Y.W., Banin, U.: J. Am. Chem. Soc. **122**, 9692 (2000)

11. Baranov, A.V., Rakovich, Yu.P., Donegan, J.F., Perova, T.S., Moore, R.A., Talapin, D.V., Rogach, A.L., Masumoto, Y., Nabiev, I.: Phys. Rev. B **68**, 165306 (2003)
12. Jing, L., Kershaw, S.V., Kipp, T., Kalytchuk, S., Ding, K., Zeng, J., Jiao, M., Sun, X., Mews, A., Rogach, A.L., Gao, M.: J. Am. Chem. Soc. **137**, 2073 (2015)
13. Pistol, M.E., Pryor, C.E.: J. Phys. Chem. C **115**, 10931 (2011)
14. Niquet, Y.-M., Delerue, C.: Phys. Rev. B **84**, 075478 (2011)
15. Neupane, M.R., Lake, R.K., Rahman, R.: J. Appl. Phys. **110**, 074306 (2011)
16. Schrier, J., Wang, L.-W.: Phys. Rev. B **73**, 245332 (2006)
17. Luo, Y., Wang, L.-W.: ACS Nano **4**, 91 (2010)
18. Khoo, K.H., Arantes, J.T., Chelikowsky, J.R., Dalpian, G.M.: Phys. Rev. B **84**, 075311 (2011)
19. Han, P., Bester, G.: Phys. Rev. B **85**, 041306(R) (2012)
20. Bockelmann, U., Bastard, G.: Phys. Rev. B **42**, 8947 (1990)
21. Benisty, H., Sotomayor-Torrès, C.M., Weisbuch, C.: Phys. Rev. B **44**, 10945 (1991)
22. Urayama, J., Norris, T.B., Singh, J., Bhattacharya, P.: Phys. Rev. Lett. **86**, 4930 (2001)
23. Xu, S., Mikhailovsky, A.A., Hollingsworth, J.A., Klimov, V.I.: Phys. Rev. B **65**, 045319 (2002)
24. Woggon, U., Giessen, H., Gindele, F., Wind, O., Fluegel, B., Peyghambarian, N.: Phys. Rev. B **54**, 17681 (1996)
25. Klimov, V.I., McBranch, D.W.: Phys. Rev. Lett. **80**, 4028 (1998)
26. Harbold, J.M., Du, H., Krauss, T.D., Cho, K.S., Murray, C.B., Wise, F.W.: Phys. Rev. B **72**, 195312 (2005)
27. Kraus, R.M., Lagoudakis, P.G., Muller, J., Rogach, A.L., Lupton, J.M., Feldmann, J, Talapin, D.V., Weller, H.: J. Phys. Chem. B **109**, 18214 (2005)
28. Klimov, V.I., Mikhailovsky, A.A., McBranch, D.W., LeatSherdale, C.A., Bawendi, M.G.: Phys. Rev. B **61**, 13349 (2000)
29. Wang, L.-W., Califano, M., Zunger, A., Franceschetti, A.: Phys. Rev. Lett. **91**, 056404 (2003)
30. Pijpers, J.J.H., Milder, M.T.W., Delerue, C., Bonn, M.: J. Phys. Chem. C **114**, 6318 (2010)
31. Li, X.-Q., Nakayama, H., Arakawa, Y.: Phys. Rev. B **59**, 5069 (1999)
32. Grange, T., Ferreira, R., Bastard, G.: Phys. Rev. B **76**, 241304 (2007)
33. Stauber, T., Vasilevskiy, M.I.: Phys. Rev. B **79**, 113301 (2009)
34. Guyot-Sionnest, P., Wehrenberg, B., Yu, D.: J. Chem. Phys. **123**, 074709 (2005)
35. Schaller, R.D., Pietryga, J.M., Goupalov, S.V., Petruska, M.A., Ivanov, S.A., Klimov, V.I.: Phys. Rev. Lett. **95**, 196401 (2005)
36. Cooney, R.R., Sewall, S.L., Anderson, K.E.H., Dias, E.A., Kambhampati, P.: Phys. Rev. Lett. **98**, 177403 (2007)
37. Kilina, S.V., Kilin, D.S., Prezhdo, O.V.: ACS Nano **3**, 93 (2009)
38. Baroni, S., de Gironcoli, S., Dal Corso, A., Giannozzi, P.: Rev. Mod. Phys. **73**, 515 (2001)
39. Yu, P.Y., Cardona, M.: Fundamentals of Semiconductors. Springer, Berlin (2010)
40. The CPMD consortium page, coordinated by M. Parrinello and W. Andreoni, Copyright IBM Corp 1990–2008, Copyright MPI für Festkörperforschung Stuttgart 1997–2001. http://www.cpmd.org
41. Richter, H., Wang, Z.P., Ley, L.: Solid State Commun. **39**, 625 (1981)
42. Madelung, O.: Introduction to Solid-State Theory. Springer, Berlin (1996)
43. Dacorogna, M.M., Cohen, M.L., Lam, P.K.: Phys. Rev. Lett. **55**, 837 (1985)
44. Lam, P.K., Dacorogna, M.M., Cohen, M.L.: Phys. Rev. B **34**, 5065 (1986)
45. Chang, K.J., Cohen, M.L.: Phys. Rev. B **34**, 4552 (1986)
46. Martin, R.M.: Electronic Structure Basic Theory and Practical Methods. Cambridge University Press, Cambridge (2004)
47. Han, P., Bester, G.: Phys. Rev. B **85**, 235422 (2012)
48. Rau, A.R.P., Zhao, W.: Phys. Rev. A **68**, 052102 (2003)
49. Villas-Bôas, J.M., Ulloa, S.E., Govorov, A.O.: Phys. Rev. B **75**, 155334 (2007)
50. Borges, H.S., Sanz, L., Villas-Bôas, J.M., Alcalde, A.M.: Phys. Rev. B **81**, 075322 (2010)
51. Sakurai, J.J.: Modern Quantum Mechanics. Addison-Wesley, Reading MA (1994)
52. Bester, G., Han, P.: In: Nagel, W., Kröner, D., Resch, M. (eds.) High Performance Computing in Science and Engineering'11. Springer, Heidelberg (2012)

53. Han, P., Bester, G.: Phys. Rev. B **91**, 085305 (2015)
54. Tschirner, N., Lange, H., Schliwa, A., Biermann, A., Thomsen, C., Lambert, K., Gomes, R., Hens, Z.: Chem. Mater. **24**, 311 (2012)
55. Cirillo, M., Aubert, T., Gomes, R., Van Deun, R., Emplit, P., Biermann, A., Lange, H., Thomsen, C., Brainis, E., Hens, Z.: Chem. Mater. **26**, 1154 (2014)

Large-Scale Modeling of Defects in Advanced Oxides: Oxygen Vacancies in BaZrO₃ Crystals

Marco Arrigoni, Eugene A. Kotomin, and Joachim Maier

Abstract Quantum mechanical simulations have proved to be an accurate tool in the description and characterization of point defects which can substantially alter the physical and chemical properties of oxides and their applications, e.g. in fuel cells and permeation membranes. Accurate simulations should take into account both the defect energetics in the real material and the thermodynamic effects at finite temperatures. We studied and compared here the structural, electronic and thermodynamic properties of the neutral ($v_O^\times$) and the positively doubly charged ($v_O^{\bullet\bullet}$) oxygen vacancies in bulk BaZrO₃; particular emphasis was given in the evaluation of the contribution of lattice vibrations on the defect thermodynamic properties. The large-scale computer calculations were performed within the linear combination of atomic orbitals (LCAO) approach and the hybrid of Hartree-Fock method and density functional theory (HF-DFT). It is shown that phonons contribute significantly to the formation energy of the charged oxygen vacancy at high temperatures ($\sim 1\,\mathrm{eV}$ at $1000\,\mathrm{K}$), due to the large lattice distortion brought by this defect and thus their neglect would lead to a considerable error.

1 Introduction

Perovskite oxides comprise a broad family of technologically important materials, which display a wide range of functional properties, such as ferroelectricity, magnetism, their combination (multiferroics), piezoelectricity, high-temperature superconductivity, mixed ionic-electronic conductivity and electro-optic effects [1–3]. This family of compounds can be described by the ABO₃ formula, where A and B are two cations occupying dodecahedral interstices and oxygen octahedra

M. Arrigoni (✉) • J. Maier
Max Planck Institute for Solid State Research, Heisenbergstraße 1, 70569 Stuttgart, Germany
e-mail: m.arrigoni@fkf.mpg.de; s.weiglein@fkf.mpg.de

E.A. Kotomin
Max Planck Institute for Solid State Research, Heisenbergstraße 1, 70569 Stuttgart, Germany

Institute for Solid State Physics, University of Latvia, Kengaraga 8, 1063 Riga, Latvia
e-mail: e.kotomin@fkf.mpg.de

© Springer International Publishing Switzerland 2016   187
W.E. Nagel et al. (eds.), *High Performance Computing in Science
and Engineering '15*, DOI 10.1007/978-3-319-24633-8_12

respectively. Oxygen vacancies are common point defects in these materials and have been shown to influence a variety of properties, such as mechanical and optical [4, 5] as well as ionic conductivity [6].

BaZrO₃ is a cubic perovskite that has attracted considerable attention in recent years, it has been used as dielectric material for wireless communication devices, hybrid perovskite-polymer-magnetic nanocomposites and as a substrate for thin films. Additionally, Y-doped $BaZrO_3$ is a promising candidate material for protonic ceramic fuel cells operating in the range of 400–700 °C, due to its high solid state proton conductivity. Working at intermediate temperatures improves the devices durability and compatibility between components and is thus more favorable than operating at the usual temperature range of 800–1000 °C of ceramic solid oxides fuel cells. Other electrochemical applications of Y-doped $BaZrO_3$ entails sensors and hydrogen pumps fabrication. The proton conductivity arises from incorporation of protons in the doped material by a dissociative absorption of water molecules upon exposure to humid atmospheres. Thus, fully charged oxygen vacancies play here a crucial role. Although charged vacancies dominate acceptor doped oxides, vacancies with two trapped electrons (the color F centers) are more common in undoped oxides under reducing conditions where the electron chemical potential is higher. These two charge states of the oxygen vacancies represent the limiting cases in partly ionic and covalent perovskites, where electrons are neither fully localized in the oxygen vacancy by electrostatic fields, as in more ionic compounds like MgO, nor are completely localized on the dangling bonds of the nearest cations, as in more covalent materials such as silicates.

Density functional approaches have in the recent decade been proved to be an important tool in defect analyses of functional oxides, and are now routinely applied in investigations of the electronic properties and defects thermodynamics. The majority of such investigations, however, consider the thermodynamic properties of defects only at 0 K, while most physical and chemical processes occur at moderate or high temperatures.

An important contribution to the thermodynamic properties at finite temperatures is given by the vibrational partition function, which can be evaluated by calculating the material's normal modes of lattice vibrations. There are still few reports on $BaZrO_3$. Sundell et al. [7] addressed defect entropies in $BaZrO_3$ from a computational perspective through the simplified Einstein model and considered only the vibrational contributions due to changes in the vibrational properties of the eight oxygen atoms nearest to the charged oxygen vacancy, thereby neglecting long range and volume relaxations effects.

Calculations of vibrational frequencies may also reveal structural instabilities through low frequency or imaginary phonon modes, allowing for identification of eventual soft modes responsible for phase transitions. According to experimental X-ray and neutron powder diffraction analyses, $BaZrO_3$ does not undergo any phase transition down to 2 K [8]. First principles studies of $BaZrO_3$ with the LDA [8] and GGA [9] exchange-correlation functionals have, however, indicated structural instabilities at the R- and/or M-points of the Brillouin zone that would lead to an antiferrodistortive transition at low temperatures. This disagreement with experiments has been suggested to stem from neglect of zero-point motions [8] and/or anharmonic effects [10], which could suppress the phase transition. On the other hand, the employment of the hybrid PBE0 functional has shown a significantly better agreement with the experimental frequencies at the Γ-point and the absence of imaginary modes at any of the high symmetry reciprocal space points [11]. These contradictory results illustrate the importance of suitable choice of computational approach when evaluating the vibrational properties of, especially, perovskite structured oxides.

In this study, we employed the linear combination of atomic orbitals (LCAO) approach with the hybrid HF-DFT PBE0 functional [12] for investigating of the structural, electronic and vibrational properties of oxygen vacancy defects in the two different charge states (0 or +2) in bulk BaZrO$_3$. Particular emphasis is given to the vibrational contributions on the defect thermodynamic properties.

2 Methods

The fundamental quantity determining the thermodynamic stability of a defect is its free energy of formation, which, for an isolated oxygen vacancies in charge state q, takes the form:

$$\Delta_f G_{(v_O,q)}(T) = G_{(v_O,q)}(T) - G_{perfect}(T) + \mu_O(T,p_i) + q\mu_e, \tag{1}$$

where, $G_{(v_O,q)}(T)$ and $G_{perfect}(T)$ are the Gibbs free energy of the defective and non-defective system respectively, while $\mu_O(T,p_i)$ and μ_e are the chemical potentials of oxygen and electrons. $\mu_O(T,p_i)$ is taken as half that of the O$_2$ molecule Gibbs energy:

$$\mu_O(T,p_i) = \frac{1}{2}G_{O_2}(T). \tag{2}$$

For electrons, μ_e may take values from the top of the valence band to the bottom of the conduction band. Equation (1) was firstly suggested in this form by Zhang and Northrup [13]. In the present study we generalized it including vibrational contributions.

As the main focus of this work is to determine the contribution from phonons to the Gibbs free energy of defect formation, it is convenient to separate $\Delta_f G_{(v_O,q)}(T)$ into electronic, vibrational and gas phase contributions:

$$\Delta_f G_{(v_O,q)}(T) =$$
$$\Delta_f E^{el}_{(v_O,q)}(T) + \Delta_f H^{vib}_{(v_O,q)}(T) - T\Delta_f S^{vib}_{(v_O,q)}(T) + \mu_O(T,p_i) + q\mu_e, \tag{3}$$

where Δ quantities are the differences between the defective and non-defective systems given property (H, S, ecc.). $\Delta_f H^{vib}$ and $\Delta_f S^{vib}$ are determined from the calculated harmonic vibrational frequencies ($v_s(\mathbf{q}), s = 1, 2, \ldots, 3N$, where N is the number of atoms in the unit cell and $\mathbf{q}$ is a reciprocal lattice vector in the first Brillouin zone) according to:

$$G^{vib}(T) = \sum_{s,\mathbf{q}} \left(n_{s,\mathbf{q}}(T) + \frac{1}{2} \right) h\nu_s(\mathbf{q})$$

$$-T \sum_{s,\mathbf{q}} \left(\frac{n_{s,\mathbf{q}}(T)}{T} \right) h\nu_s(\mathbf{q}) - k_B ln \left(1 - e^{-\frac{h\nu_s(\mathbf{q})}{k_B T}} \right) \right) + pV,$$

(4)

where the first sum represent the vibrational internal energy U^{vib} (zero point energy included) and the second one the vibrational entropy, S^{vib}. $n_{s,\mathbf{q}}(T)$ is the Bose-Einstein distribution $\left(e^{\frac{h\nu_s(\mathbf{q})}{k_B T}} - 1 \right)^{-1}$ and pV is the product between the system pressure and volume.

2.1 Computational Details

The calculations were performed employing localized basis sets (LCAO) and the hybrid HF-DFT Perdew-Burke-Ernzerhof exchange-correlation functional (PBE0) as implemented in the CRYSTAL14 code [14, 15].

The employed atomic basis sets (BSs) consisted of Gaussian-type functions and pseudopotentials. The basis set included quasi-relativistic pseudopotentials taken from the pseudopotential library of Stuttgart-Cologne group (http://www.theochem.uni-stuttgart.de/pseudopotentials/clickpse.html) for 46 core electrons of Ba and 28 core electrons of Zr. In order to avoid spurious interactions between the diffuse functions and the core functions of neighboring atoms, the basis set diffuse exponents smaller than 0.1 bohr^{-2} were removed as well as f-electron virtual functions. The all-electron BS 8-411G was used for O atoms. The tolerance factors of 7,7,7,7 and 14 for the Coulomb and exchange integrals were used. For all the studied defects we considered only the closed shell electronic configuration and setting the SCF calculations threshold value for energy convergence to 10^{-7} eV.

The neutral oxygen vacancy was modeled by removing a single O atom, keeping the BS of the vacant oxygen atom by means of a spurious 'ghosts' atom with no atomic mass. The latter approach allows for localization of electrons within the vacancy itself. The charged oxygen vacancy was modeled by removing two electrons from the overall supercell through charge compensation by a homogeneous jellium background charge. The energetics of these removed electrons is taken into account in the expression of the defect formation energy through the electron chemical potential (Eqs. (1) and (3)); while they do not given any contribution to $G^{vib}(T)$ (Eq. (4)), which entails only the vibrational degrees of freedom of the ions. Calculations were performed with $2\times2\times2$ (40 atoms) and $3\times3\times3$ (135 atoms) supercell expansions of the 5-atoms primitive cubic cell of $BaZrO_3$, corresponding to defect concentrations of 12.5 and 3.7 % respectively. Electronic integration over the Brillouin zone was performed using a $8\times8\times8$ Monkhorst-Pack [16] k-mesh for the five atoms unit cell. For supercell expansions, the k-mesh density was reduced accordingly.

Vibrational frequencies were calculated within the harmonic approximation by numerical evaluation of the dynamical hessian matrix elements through the first derivative of the atomic energy gradients, displacing each atom along the Cartesian coordinates by 0.001 Å.

2.2 Computational Resources

Supercell calculations deal with a relatively large amount of atoms. In this study we employed 2×2×2 and 3×3×3 supercells, formed by 40 and 135 atoms respectively. The presence of defects reduce the crystal symmetry, drastically increasing the computational costs. Approximately, each geometry optimization calculation of defective BaZrO$_3$ required 20 nodes and 24 h using the smallest supercell and up to 32 nodes and 96 h with the largest one for each run. Phonons calculations are much more demanding and required at least 20 nodes and up to two weeks per single run. On average 500.000 core-hours per month were approximately used. All calculations were performed on HLRS CRAY XE6 supercomputer at Stuttgart University.

3 Results and Discussion

3.1 Defective-Free Barium Zirconate

Table 1 compares various bulk properties of BaZrO$_3$ calculated in this work with the PBE0 functional with selected literature reports. The lattice parameter is in very good agreement with the experimental value of 4.192 Å[17]. The pure

Table 1 Lattice parameter (a), various bonding lengths (d), Mulliken atomic charges in LCAO calculations (q) and band gap ΔE_g

	LCAO				PW	
	PBE0[a]	PBE [9]	PBE0 [11]	B3LYP [19]	PW91 [9]	Expt.
a, Å	4.195	4.242	4.198	4.234	4.207	4.192 [17]
d(Zr-O), Å	2.098					
d(Ba-O), Å	2.966					
d(O-O), Å	2.966					
q(Ba), e	1.66		1.87	1.81		
q(Zr), e	2.35		2.21	2.15		
q(O), e	−1.33		−1.36	−1.32		
ΔE_g, eV	5.36		5.4	4.79		5.33 [18]

[a] Current study

The values are compared with those found in the literature from different computational studies and experiments

DFT functionals predict larger lattice parameter than the hybrid PBE0 functional, except for the study [9] where the PW91 was applied. The PBE0 functional yields an indirect band gap (M→ Γ) of 5.36 eV, being in very good agreement with the experimental value of 5.33 eV [18]; whereas the hybrid B3LYP functional noticeably underestimates it [19]. Correct evaluation of the optical band gap is crucial when evaluating the occupied donor state of the neutral oxygen vacancy. As is common, pure DFT functionals heavily underestimate this gap.

Mulliken atomic charges calculations reveal non-integer atomic charges for all elements, reflecting the partial ionic character of $BaZrO_3$. Moreover, the calculations show an overlap population charge of 68 me for the Zr-O bond, and almost zero for Ba-O, indicative of larger covalent character of the former bond.

Table 2 reports the vibrational frequencies at the Γ-point calculated in this study, including the splitting between transverse and longitudinal optical modes (T.O.-L.O. splitting), and their experimental values [20]. Due to the cubic symmetry of $BaZrO_3$ (space group O_h^1 number 221), one obtains the following set of optical phonon modes at the Γ-point: $4t_{1u} + t_{2u}$. The obtained results are in good agreement with the experimental values. The $Zr-O_3$ torsion mode is both IR and Raman inactive (silent mode of t_{2u} symmetry) and cannot be detected experimentally. In particular, due to the cubic symmetry, there are no Raman allowed vibrational modes in $BaZrO_3$ and thus the L.O. frequencies are not detectable from experiments. Although the T.O.-L.O. splitting is very pronounced, we found that its contribution to the thermodynamic potentials is negligible (taking it into account, G^{vib} increases at maximum 0.1 eV in the range 0–1600 K), and was thus not accounted for in the defective calculations. To further investigate the vibrational properties and structural stability of $BaZrO_3$, we calculated the phonon dispersion relation along the Γ → X → M → Γ → R → M → Γ path in the BZ. The curves shown in Fig. 1 were obtained by interpolating the discrete points obtained with a 2×2×2 supercell. The frequencies obtained by us are in agreement with those reported by Evarestov [11] employing the same code and functional. No imaginary modes where observed, therefore we predict a stable cubic structure for $BaZrO_3$.

Table 2 Calculated and experimental vibrational translational optic frequencies at the Γ-point for defective-free $BaZrO_3$. We have also reported the split of the longitudinal optics modes

Mode, cm^{-1}	Calculated		Expt.[21]
	T.O.	L.O.	T.O.
Ba-ZrO_3 stretch	125	145	115
O-Zr-O bend	219	406	210
$Zr-O_3$ torsion	220	–	Inactive
Zr-O stretch	511	718	505

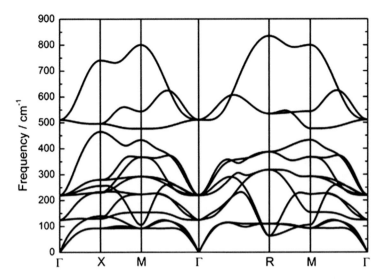

Fig. 1 Vibrational band structure for BaZrO₃ calculated with CRYSTAL in this study. The q-points sample was formed by the Γ-, M-, X- and R-points. Continuous curves were obtained using a Fourier interpolation technique

3.2 Oxygen Vacancies: Atomic and Electronic Properties

Table 3 shows structural relaxation induced by the neutral and charged oxygen vacancies in BaZrO₃ using 2×2×2 and 3×3×3 supercell expansions. The neutral vacancy leads to a slight contraction of the cell for both defect concentrations. We suggest this effect to be related to the inclusion of the 'ghost' basis functions in the vacancy that, in order to decrease the energy of the defect-induced level (the position of defect-induced level on the band structure, Fig. 2), yields to the c-axis shrinks to maximize the overlap between the 'ghost' orbitals and the p, d orbitals of the nearest Zr atoms. We in addition performed a test calculation leaving vacuum (i.e no 'ghost' basis functions) in the neutral oxygen vacancy and noticed that the c-axis, and in particular the distances between Zr atoms in the Zr-v_O-Zr complex, expands; while the other two axes shrink. A comparison of our results for the two cases, i.e. with ghost functions and leaving vacuum in the vacancy position, revealed lower formation energy of oxygen vacancy for the former .

Our study shows the new band induced by $v_O^\times$ defect is fully occupied (see Fig. 2), with its maximum occurring at the M-point of the first BZ and situated below the conduction band. The band is located deep in the band gap, with a minimum distance to the conduction band minimum of 1.31 eV (1.58 eV in the 3×3×3 supercell), in agreement with previous reports of deep F-center induced defect levels in zirconate perovskites modelled with a 'ghost' BS [22]. The F-center partly reduces the two nearest Zr atoms, as the two electrons in the defective level are mostly localized on their d-states (Fig. 2).

Table 3 Relative lattice constant ($\Delta a/a_0$, $\Delta b/b_0$, $\Delta c/c_0$) relaxations, formation volume ($\Delta_f V$), relative volume relaxation ($\Delta V/V_0$), minimum distance between the conduction band minimum and defective level ($\Delta \epsilon$) and defective band width ($\delta \epsilon$) of the oxygen vacancy in the charge state 0 ($v_O^\times$) and +2 ($v_O^{\bullet\bullet}$), calculated using 2×2×2 (8 f.u.) and 3×3×3 supercell expansion (27 f.u.)

Expansion	2×2×2		3×3×3	
Defect	$v_O^\times$	$v_O^{\bullet\bullet}$	$v_O^\times$	$v_O^{\bullet\bullet}$
$\Delta a/a_0$ (%)	−0.29	−13.4	0.08	−12.7
$\Delta b/b_0$ (%)	−0.29	−13.4	0.08	−12.7
$\Delta c/c_0$ (%)	−0.88	−10.6	−0.66	−10.7
$\Delta_f V$ (Å^3)	−1.07	−27.1	−0.37	−26.6
$\Delta V/V_0$ (%)	−1.45	−37.3	−0.50	−36.3
$\Delta \epsilon$ (eV)	1.31	0.20	1.58	0.20
$\delta \epsilon$ (eV)	0.78	0.74	0.15	0.13

The relative relaxations are given as percentile relative expansion per mole fraction of oxygen vacancies. The formation volume is defined as the difference between the defective and defective-free supercell volumes. In accordance with local site symmetry of oxygens (point group symmetry D_{4h}) in cubic perovskite BaZrO3, the symmetry of defective supercells lowers to space group D_{4h}^1

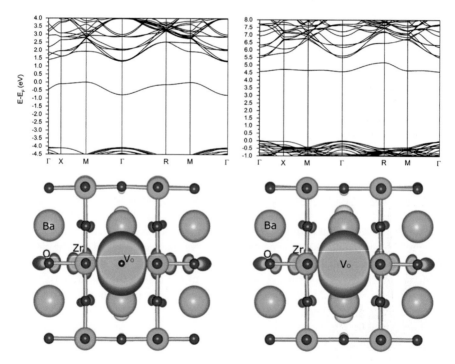

Fig. 2 *Top*: Band structure of the 2×2×2 BaZrO$_3$ supercell with a $v_O^\times$ (*left*) and $v_O^{\bullet\bullet}$ (*right*) oxygen vacancy. *Bottom*: electronic charge density projected on the defective level obtained modelling the vacancy using the ghost BS (*left*) or leaving vacuum (*right*). The isosurface threshold level is set to 0.002 e/a_0^3; points with an electron charge density value greater than the threshold lies inside the isosurface (the larger the value the warmer the color)

For the charged vacancy, Mulliken charge analyses indicated a small charge of merely 0.2 e associated with the vacancy ghost BS; hence the site is effectively positively charged and repels the nearest Zr and Ba cations while it attracts the nearest oxide anions. This is clearly shown in Table 3. In particular, $v_O^{\bullet\bullet}$ induces a large outward displacement of the two nearest Zr ions, resulting in an anisotropic relaxation that yields a tetragonal defective cell. The inward relaxation of the eight nearest O ions is significantly larger for $v_O^{\bullet\bullet}$ than $v_O^{\times}$, resulting in a general contraction in all directions, and thus a large negative formation volume. Finally, the anisotropic relaxations induced by the oxygen vacancies reduce the symmetry of defective supercells to space group D_{4h}^1 (123). For the charged vacancy, the defective induced band is unoccupied and lies much closer to the conduction band bottom with respect to the neutral defect (only 0.20 eV for both supercell sizes).

3.3   Oxygen Vacancies: Thermodynamic Properties

Figure 3 shows the vibrational density of states (DOS) for $v_O^{\times}$ and $v_O^{\bullet\bullet}$ calculated with the 2×2×2 and 3×3×3 supercells. Both defects induce a distinguishable vibrational peak at around 660 cm^{-1} in the 2×2×2 supercell, and 676 and 630 cm^{-1} in 3 × 3 × 3 supercell for $v_O^{\times}$ and $v_O^{\bullet\bullet}$, respectively, originating from Zr-O stretch relative to the two Zr atoms closest to the vacancy. This mode belongs to the A$_1$ irreducible representation of the space group and is thus both IR and Raman active. Its IR intensity, evaluated through the Berry phase approach, values 446 km/mol (2×2×2 supercell) and 464 km/mol (3×3×3 supercell) for $v_O^{\times}$, and around 704 km/mol (2×2×2 supercell) and 1394 km/mol (3×3×3 supercell) for $v_O^{\bullet\bullet}$, and should therefore be detectable in experiments. These results agree with those of a previous study on SrTiO$_3$ [23] that shown a vibrational mode at around 630 cm^{-1} appearing

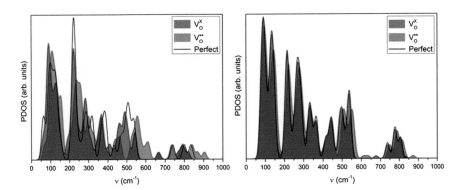

Fig. 3 Phonon density of states (PDOS) calculated for the 2×2×2 (*left*) and 3×3×3 (*right*) BaZrO3 supercells. In each plot the PDOS is reported for the perfect system (*unfilled curve*) and the defective systems containing the neutral oxygen vacancy $v_O^{\times}$ (*blue*) and the double charged oxygen vacancy $v_O^{\bullet\bullet}$ (*red*)

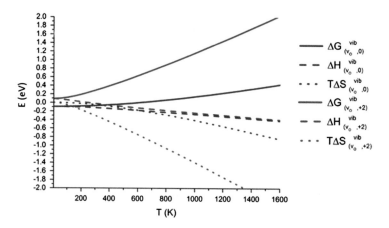

Fig. 4 Phonon contribution to the formation enthalpy, entropy and Gibbs free energy (cfr. Eq. (3)) for neutral and charged oxygen vacancies, calculated with a 3×3×3 supercell

in presence of neutral oxygen vacancies and corresponding to the relative motion of Ti and O atoms near the vacancy. In addition, the large negative formation volume of $v_O^{\bullet\bullet}$ leads to stiffening of the bonds and thus a general blue-shift of all vibrational frequencies.

Figure 4 shows the corresponding phonon contribution to the formation enthalpy, entropy and Gibbs free energy of $v_O^{\times}$ and $v_O^{\bullet\bullet}$ calculated using both 2×2×2 and 3×3×3 supercells. At T=0 K the only contribution to ΔG^{vib} is the difference in zero-point energy between the defective and perfect supercells. For $v_O^{\times}$, $\Delta H^{vib} < 0$ at 0 K with both supercells, as expected due to the removal of three vibrational degrees of freedom upon vacancy formation. For $v_O^{\bullet\bullet}$, however, $\Delta H^{vib} > 0$ at 0 K, which stems from the blue-shift of the phonon spectrum due to the considerable negative formation volume. Further, ΔH^{vib} decreases with increasing temperature for both defects, reflecting a negative contribution from phonons to the formation enthalpy. The larger blue shift of the vibrational frequencies in the 2×2×2 than in the 3×3×3 supercell, Fig. 4, slightly increases the enthalpy of the defective system. The shift is even more pronounced for $v_O^{\bullet\bullet}$ and therefore $\Delta H^{vib}_{(v_O,+2)}$ (0 K) is around 0.08 eV larger in the smaller supercell than in the larger; while for $v_O^{\times}$ the difference in $\Delta H^{vib}_{(v_O,0)}(0\,K)$ is around 0.03 eV only.

The calculated $\Delta_f S^{vib}$ is negative for both $v_O^{\times}$ and $v_O^{\bullet\bullet}$. $\Delta_f S^{vib}$ is not affected by the supercell size and its value is modest within the considered temperature range, as expected from the negligible effect of $v_O^{\times}$ on the PDOS. On the other hand, $\Delta_f S^{vib}$ is significantly larger (more negative) with the 3×3×3 than the 2×2×2 supercell, due to the larger number of vibrational modes and the wider local structural relaxation allowed by the former supercell.

The calculated $\Delta_f S^{vib}$ is considerably more negative for $v_O^{\bullet\bullet}$ than $v_O^{\times}$, as expected from both the greater local relaxations, and the more negative formation volume of the former. Therefore, the vibrational contribution to the Gibbs free energy

($\Delta_f G^{vib}$) of formation is particularly relevant for $v_O^{\bullet\bullet}$ and at high temperature. This is illustrated in Fig. 4, which shows $\Delta_f G^{vib}$ of $v_O^{\bullet\bullet}$ and $v_O^{\times}$ calculated with a 3×3×3 supercell at standard pressure.

4 Conclusions

In this contribution we performed first principles calculations to determine the electronic and thermodynamic properties of oxygen vacancies in BaZrO$_3$ with emphasis on the contribution from phonons. We employ LCAO calculations using the CRYSTAL14 code and the hybrid PBE0 functional. For defective-free BaZrO$_3$, we predicted a cell parameter, optical band gap and vibrational frequencies at the Γ-point in very good agreement with the experiments.

We find that $v_O^{\times}$ creates a deep level in the band gap (around 1.5 eV from the conduction band bottom), while $v_O^{\bullet\bullet}$ a shallower one at 0.2 eV from the conduction band bottom. Both vacancy charge states induce a new IR and Raman active lattice vibration at around 650 cm^{-1} that, according to this study, could be experimentally detectable. These results agree with recent observations in defective SrTiO$_3$ [23] and could therefore be a general feature of oxygen vacancies in perovskite-structured oxides. The charged oxygen vacancy, $v_O^{\bullet\bullet}$, induces significantly larger structural distortions than the neutral one, $v_O^{\times}$, which causes a noticeable blue shift of the vibrational spectrum. Hence, while the contribution from phonons the Gibbs free energy of formation of $v_O^{\times}$ is small, phonons increase that of $v_O^{\bullet\bullet}$ by circa 1 eV at 1000 K (under standard pressures).

Acknowledgements Authors greatly appreciated help and support from the High Performance Computer Center in Stuttgart (HLRS, project DEFTD 12939).

References

1. Walsch, A., Sokol, A., Catlow, C.R.A.: Energy storage: rechargeable lithium batteries. In: Computational Approaches to Energy Materials. Wiley, New York (2013)
2. Kuklja, M.M., Kotomin, E.A., Merkle, R., Mastrikov, Yu.A., Maier, J.: Combined theoretical and experimental analysis of processes determining cathode performance in solid oxide fuel cells. Phys. Chem. Chem. Phys. **15**, 5443–5471 (2013)
3. Donnerberg, H.J.: Atomic Simulations of Electro-Optical and Magneto-Optical Materials. Springer Tracts in Modern Physics. Springer, Berlin (1999)
4. Scott, J.F., Dawber, M.: Oxygen-vacancy ordering as a fatigue mechanism in perovskite ferroelectrics. App. Phys. Lett. **76**, 3801–3803 (2000)
5. Hwang, H.Y.: Perovskites: oxygen vacancies shine blue. Nat. Mater. **4**, 803–804 (2005)
6. Merkle, R., Maier, J.: How is oxygen incorporated into oxides? A comprehensive kinetic study of a simple solid-state reaction with SrTiO$_3$ as a model material. Angew. Chem. Int. Ed. **47**, 3874–3894 (2008)

7. Sundell, P.G., Björketun, M.E., Wahnström, G.: Thermodynamics of doping and vacancy formation in BaZrO$_3$ perovskite oxide from density functional calculations. Phys. Rev. B **73**, 104112 (2006)
8. Akbarzadeh, A.R., Kornev, I., Malibert, C., Bellaiche, L., Kiat, J.M.: Combined theoretical and experimental study of the low-temperature properties of BaZrO$_3$. Phys. Rev. B **72**, 205104 (2005)
9. Bilić, A., Gale, J.D.: Gale: Ground state structure of BaZrO$_3$: a comparative first-principles study. Phys. Rev. B **79**, 174107 (2009)
10. Magyari-Köpe, B., Vitos, L., Grimvall, G., Johansson, B., Kollár, J.: Low-temperature crystal structure of CaSiO$_3$ perovskite: an *ab initio* total energy study. Phys. Rev. B **65**, 193107 (2002)
11. Evarestov, R.A.: Hybrid density functional theory LCAO calculations on phonons in Ba(Ti,Zr,Hf)$_3$. Phys. Rev. B **83**, 014105 (2011)
12. Perdew, J.P., Ernzerhof, M., Burke, K.: Generalized gradient approximation made simple. J. Chem. Phys. **105**, 9982–9985 (1996)
13. Zhang, S.B., Northrup, J.E.: Chemical potential dependence of defect formation energies in GaAs: application to Ga self-diffusion. Phys. Rev. Lett. **67**, 2339–2342 (1991)
14. Dovesi, R., Saunders, V.R., Roetti, R.O.C., Zicovich-Wilson, C.M., Pascale, F., Civalleri, B., Doll, K., Harrison, N.M., Bush, I.J., D'Arco, P., Llunell, M., Causà, M., Noël, Y.: CRYSTAL14 User's Manual University of Torino, Torino (2014)
15. Dovesi, R., Orlando, R., Erba, A., Zicovich-Wilson, C.M., Civalleri, B., Casassa, S., Maschio, L., Ferrabone, M., De La Pierre, M., D'Arco, P., Noël, Y., Causà, M., Rérat, M., Kirtman, B.: CRYSTAL14: a program for the ab initio investigation of crystalline solids. Int. J. Quant. Chem. **114**, 1287–1317 (2014)
16. Monkhorst, H.J., Pack, J.D: Special points for Brillouin-zone integrations. Phys. Rev. B **13**, 5188–5192 (1976)
17. Pies, W., Weiss, A.: *Landolt–Börnstein*. Numerical data and functional relationships in science and technology, Group III. In: Crystal and Solid State Physics, Vol. 7. Crystal Structure Data of Inorganic Compounds. Parts a and g. A. Acta Cryst. A. International Union of Crystallography (1975)
18. Robertson, J.: Band offsets of wide-band-gap oxides and implications for future electronic devices. J. Vac. Sci. Technol. B **18**, 1785–1791 (2000)
19. Eglitis, R.I.: Ab initio calculations of the atomic and electronic structure of BaZrO$_3$ (111) surfaces. Solid State Ionics **230**, 43–47 (2013)
20. Parida, S., Rout, S.K., Cavalcante, L.S., Sinha, E., Li, M.S., Subramanian, V., Gupta, N., Gupta, V.R., Varela, J.A., Longo, E.: Structural refinement, optical and microwave dielectric properties of BaZrO$_3$. Ceram. Int. **38**, 2129–2138 (2012)
21. Perry, C.H., McCarthy, D.J., Rupprecht, G.: Dielectric dispersion of some perovskite zirconates. Phys. Rev. **138**, A1537–A1538 (1965)
22. Zhukovskii, Y.F., Kotomin, E.A., Piskunov, S., Ellis, D.E.: A comparative ab initio study of bulk and surface oxygen vacancies in PbTiO$_3$, PbZrO$_3$ and SrTiO$_3$ perovskites. Solid State Commun. **149**, 1359–1362 (2009)
23. Evarestov, R., Blokhin, E., Gryaznov, D., Kotomin, E.A., Merkle, R., Maier, J.: Jahn-Teller effect in the phonon properties of defective SrTiO$_3$ from first principles. Phys. Rev. B, **85**, 174303 (2012)

Interfacial Properties and Growth Dynamics of Semiconductor Interfaces

Phil Rosenow, Andreas Stegmüller, Lisa Pecher, and Ralf Tonner

Abstract We present computational results on dynamics and properties of semi-conductor materials and interfaces. The adsorption of cyclooctyne on silicon can be shown to proceed barrierless into an *on-top* structure. Comparing different interfaces of the GaP/Si system, a preference for mixed interfaces (i.e. not purely Si/Ga or Si/P) can be found and understood in terms of the electrostatic potential across the interface and chemical bonding specifics. In further work, the electronic structure of mixed III/V semiconductors will be studied in the way described here for GaAs and used for the prediction of optical properties.

1 Introduction

The structure of interfaces and the dynamics of growth processes determine each other and the properties of the resulting material. Due to the importance of semiconductor devices for modern everyday technology, understanding the growth, structure and properties of semiconductor/semiconductor and semiconductor/organic interfaces will be decisive for the design of improved devices. Since these interface systems are generally complex, computational studies require a large amount of resources, especially when going beyond simple model systems toward descriptions of real materials.

Here, we present some examples of our recent work in this area. In Sect. 2, we show some results from a study shedding light on the adsorption of cyclooctyne on Si(001), which is a promising candidate for the functionalization of silicon with organic layers. Sections 3–5 evolve around III/V-semiconductors grown on silicon, a prospective material class for the construction of laser devices on silicon substrates: in Sect. 3, the composition of precursor molecules for gas phase epitaxy is described, which will be used as starting points for adsorption studies. In Sect. 4, the stability of various Si/GaP interfaces is described; Since this is often the first layer in a III/V

P. Rosenow • A. Stegmüller • L. Pecher • R. Tonner (✉)
Philipps-Universität Marburg, Fachbereich Chemie, Hans-Meerwein-Straße, 35032 Marburg, Germany
e-mail: rosenow4@staff.uni-marburg.de; stegmuea@Staff.Uni-Marburg.de; pecher@staff.uni-marburg.de; tonner@chemie.uni-marburg.de

© Springer International Publishing Switzerland 2016
W.E. Nagel et al. (eds.), *High Performance Computing in Science and Engineering '15*, DOI 10.1007/978-3-319-24633-8_13

laser structure, the interface has a strong influence on the properties of the final device. The electronic properties of the actual laser material will also be studied computationally, as is outlined in Sect. 5. Finally, we will present some data on the scaling behaviour of our main DFT code VASP on the Hornet cluster in Sect. 6.

2 Surface Reactivity of Cyclooctyne on Si(001) with DFT: Predicting Growth Processes for Semiconductor/Organic Interfaces

2.1 Introduction

Due to the strong binding character of their addition to the silicon(001) surface, [1, 2] cyclooctyne (Fig. 1) and its derivates promise to be ideal building blocks for the construction of semiconductor/organic interfaces. In order to predict the growth process leading to these interfaces, one has to understand the interaction between single molecules and the surface first. We gather this understanding by conducting density functional theory (DFT) calculations in a periodic environment and determine adsorption pathways of the molecule to the surface using the Nudged Elastic Band (NEB) method [3]. Afterwards, *ab-initio* molecular dynamics (AIMD) simulations are conducted to validate the previous results for finite temperatures. The experience gained here will later be transferred to the GaP/Si material system.

2.2 Computational Details

All calculations have been carried out with periodic DFT [4, 5] as implemented in the *Vienna Ab-initio Simulation Package* (VASP) [6, 7]. The chosen level of approximation was the Generalized Gradient Approximation (GGA) with the functional by Perdew, Becke and Enzerhof (PBE) [8, 9] and addition of the DFT-D3 (BJ) correction [10, 11] to account for dispersion interactions. The electronic structure was calculated in the projector-augmented wave (PAW) formalism [12, 13] using a plane wave cutoff of 400 eV. Due to the periodic boundary conditions, the silicon surface had to be treated in a slab approach (see Fig. 2) where a thickness of six layers of silicon was chosen with the bottom two layers being frozen and saturated by hydrogen atoms at 1.48 Å distance in the direction of the next bulk

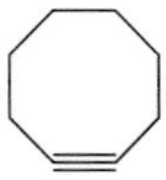

Fig. 1 Structural formula of cyclooctyne

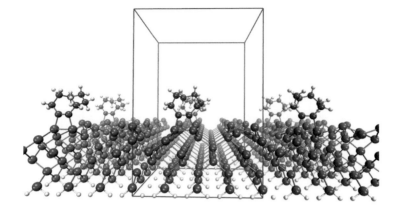

Fig. 2 Cyclooctyne bound to the Si(001) surface in the *on-top* configuration as an example for the slab approach in periodic boundary conditions, the *red box* highlighting the boundaries. Dimer formation due to surface reconstruction of the silicon atoms is also apparent in this figure

layer atoms to mimic the transition into the silicon bulk. The *on-top* reaction was calculated in a $c(4 \times 2)$ reconstructed cell with 48 silicon atoms in total, while for the *end-bridge* reaction and AIMD simulations the cell size was doubled along the short axis, leading to a (4×4) super cell with 96 silicon atoms. Reciprocal space was sampled using a Gamma-centered Monkhorst-Pack scheme, [14] where a $\Gamma(241)$ grid was used for the (4×2) cell while a $\Gamma(221)$ grid was used for the (4×4) cell.

Structural optimizations were done using the Conjugate gradient algorithm [15] while the NEB calculations utilized the limited-memory BFGS algorithm [16]. *Ab-initio* molecular dynamics simulations were done on the Born-Oppenheimer potential surface [17] with a timestep of $\Delta t = 1$ fs using a Verlet integrator [18, 19]. To account for finite temperatures, the Nosé-Hoover thermostat [20–22] was used within a *NVT* ensemble.

2.3 Optimizing Parallel Performance in VASP

Due to the high computational demand of NEB calculations and AIMD simulations, parallelization settings had to be optimized to values that enable the lowest possible cost of calculation time. The VASP code has several options for this, with the two parameters having the largest effect on computation time being KPAR and NPAR. KPAR enables the calculation at different k points at once while NPAR subdivides those groups to calculate different electronic states in the Self-Consistent Field (SCF) process in parallel. If, for example, one uses 64 cores with KPAR set to 2 and NPAR set to 4, the 64 cores are being split up into two groups of 32 cores each, who are working on two different k points simultaneously, while each of the two groups is again divided by 4, so in each subgroup, eight cores work on an individual

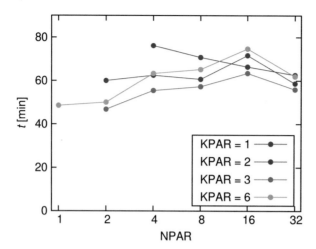

Fig. 3 Computation time of a single-point benchmark calculation in VASP as a function of the parallelization parameters NPAR and KPAR. Run on 192 cores on Cray XE6 (Hermit)

electronic state. For NEB calculations, several structures on the reaction path are being optimized at once, so a NEB calculation with 16 structures and equivalent performance to the example above would use $64 \times 16 = 1024$ cores.

To test the performance as a function of those two parameters, a benchmark single-point calculation in a (4×2) cell was set up on Cray XE6 (Hermit) using 192 cores, since most NEB calculations were done using 2048 or 4096 cores on 16 structures which corresponds to 128 or 256 cores working on a single structure. The results are shown in Fig. 3.

It can be seen that the best performance is achieved with NPAR = 2, KPAR = 3 and NPAR = 1, KPAR = 6, which both correspond to groups of 32 cores, exactly the amount of cores on one compute node. This tells us that minimizing inter-node communication is the most important factor in code performance optimization.

Subsequently, the chosen settings for the AIMD simulations, which were run on 1024 cores, were KPAR = 4 and NPAR = 8, so the cores were put in groups of 32 as well (the KPAR value arises from the larger cell size and corresponding smaller k mesh).

These setting should be well transferable to the Hornet machine with adjusting the values to the 24 core/node setting (see Sect. 6).

2.4 Nudged Elastic Band Results

There are two dominant ways in which the cyclooctyne molecule can bond to the Si(001) surface: In the *on-top* adsorption mode, a [2+2] cycloaddition takes place with two silicon atoms of a surface dimer, leading to a 4-ring structure of covalent bonds (see Fig. 2). In contrast to that, the *end-bridge* adsorption mode (Fig. 4) sees

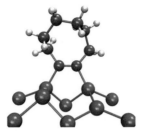

Fig. 4 The *end-bridge* adsorption mode of cyclooctyne bonding to the Si(001) surface, where the molecule reacts with two silicon dimers

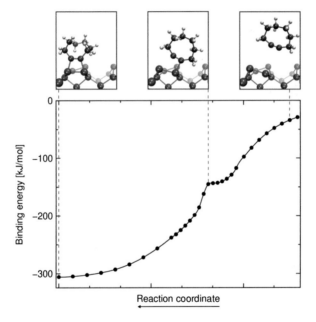

Fig. 5 Energy profile of the *on-top* adsorption path with structures shown at selected points. The binding energy is calculated with respect to the isolated and relaxed molecule and surface, respectively

the molecule bonding to one atom each of two neighbouring silicon dimers, leading to a 5-ring structure, and a change of 90° in the molecular orientation at the surface compared to the *on-top* mode. Over the course of both reactions, the C-C triple bond is being reduced to a double bond.

To find out which of the two adsorption modes is dominant during the growth process, NEB calculations were performed with the initial structure being the molecule at a position where the distance between the triple bond atoms and the closest surface atom was at least 4.5 Å and the final structures being the *on-top* and *end-bridge* modes. The resulting energy profiles are given in Figs. 5 and 6.

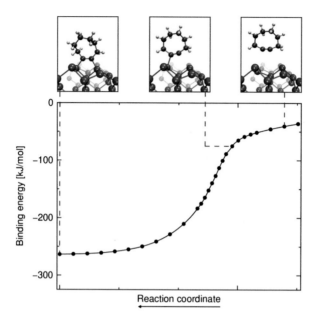

Fig. 6 Energy profile of the *end-bridge* adsorption path with structures shown at selected points. The binding energy is calculated with respect to the isolated and relaxed molecule and surface, respectively

The *on-top* adsorption (Fig. 5) shows a flat plateaux-like shape in the region where the first bond formation takes place (see central top insert), but a local minimum does not emerge and the whole adsorption process does not feature any barrier. A symmetric approach of the triple bond atoms to the lower silicon atom of the dimer reflects the proposed reaction mechanism of ethylene with the Si(001) surface, [23] but that reaction features a local minimum while this one does not. The *end-bridge* adsorption (Fig. 6) also does not feature any local minimum, while in contrast to the *on-top* adsorption, the lower silicon atom is approached in an asymmetric fashion and a plateaux region does not emerge.

Due to both processes being barrierless, the adsorption of the molecule to the surface can not be kinetically controlled. However, adsorption to the *on-top* structure follows a much steeper pathway, making it more likely. Comparison of the final adsorption energies (*On-top*: −306 kJ/mol, *end-bridge*: −263 kJ/mol) should favor the *on-top* mode from a thermodynamical point of view, i.e. if an equilibrium can be reached. This would require a thermally accessible pathway between the two structures, requiring at least one bond to be broken. From the NEB calculations discussed above, where the singly bonded state is about 130 kJ/mol above the minima, it can be estimated that an equilibrium will not be reached at room temperature. For further analysis, *ab-initio* molecular dynamics simulations were carried out.

2.5   Ab-initio Molecular Dynamics Results

Before putting molecule and slab in one simulation, each had to be thermally equilibrated first. For the molecule, 20 simulations at 300 K with 10 ps length were conducted, with the $t=0$ velocities initialized randomly according to a Maxwell-Boltzmann distribution at the given temperature. The slab was initialised in the same way at 300 K, but only one simulation was conducted, where after 10 ps of simulation time, the positions and velocities of the atoms were extracted every 3 ps to yield the 20 starting structures and velocities for further proceedings.

For the actual molecule-surface simulation, the positions and velocities were combined into 20 pairs where in each case the molecule was placed at ≈ 6 Å distance to the surface in z direction and displaced in the x and y directions by adding a random decimal fraction of the cell vector for each using a true random number generator [24]. Thermal translational motion was added to the molecule by again adding random decimal fractions for each velocity component while scaling the length of the vector to the mean Maxwell-Boltzmann velocity of cyclooctyne at 300 K being $\langle v \rangle = 108.18 \frac{m}{s}$ and ensuring that the z component was pointing in the direction of the surface.

Twenty individual simulations were performed and a simulation was considered finished when the molecule was in a clearly bound state and the two C-Si bond lengths had not changed in 500 fs aside from their vibrational motion. Out of these 20 simulations, 17 ended up in the *on-top* state and only 3 in the *end-bridge* state. While 20 simulations do not represent a statistical distribution, the results imply that the *on-top* mode might indeed be the favoured adsorption mode and the predominant reaction in the growth process leading to this semiconductor/organic interface. This is in agreement with the assumption, that the steeper path favours adsorption into the *on-top* mode.

3   Gas Phase Decomposition of MOVPE Precursors: Prelude to Adsorption and Growth Studies

The gas phase decomposition of common MOVPE-precursors has been studied using density functional methods [25–27]. By combining thermodynamic, kinetic and chemical bonding considerations, decomposition pathways and products could be identified. Descriptors based on partial charges were proposed to guide rational design of further precursors. This will be complemented by studies of adsorption and growth on a large scale, employing NEB and MD techniques as described above.

4   GaP/Si-Interfaces in Different Crystal Orientations

Interfaces play an important role in determining the properties of optoelectronic
materials. Understanding specific structures, their origin and structure-property
relations is thus crucial for material design. In the case of GaP epitaxially grown on
Si, intermixing has been observed, leading to the formation of pyramidically shaped
patterns [28]. Here, we present energetic considerations evolving around strain and
a study of the electrostatic potential across interface regions.

4.1   Strain in Growth Direction

Strain cannot be avoided in the construction of heterostructures and strongly
influences the atomic structure and electronic properties of the resulting materials.
To study the effects of strain on the lattice matched system GaP/Si, periodic cells
with interfaces in (001), (111) and (112) lattice planes have been constructed and
are shown in Fig. 7. The thickness of the resulting films varied between 16 and 24
atomic layers, depending on the crystal plane defining the interface.

To enable a meaningful comparison of formation energies for the interfaces, the
following definition for the interface formation energy ΔE_{if} has been used:

$$\Delta E_{if} = \left(N \cdot E_{GaPSi} - \frac{N}{2} \left(E_{Si}^{ref} + E_{GaP}^{ref} \right) \right) \cdot A_{(001)}^{-1}. \tag{1}$$

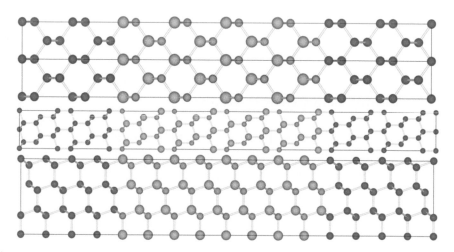

Fig. 7 Supercells used for interface study. From *top to bottom*: (001), (111) and (112) interface.
Blue: Si, *pink*: Ga, *orange*: P

Here, E_{GaPSi}, E_{Si}^{ref} and E_{GaP}^{ref} are the total energy of the supercell containing the interface, and a structurally identical reference cell with pure Si and GaP, respectively. The factor $N/2$ with $N = 2$ accounts for the presence of two interfaces in the model cell and the normalization to the interface area for a (001) interface $A_{(001)}$ enables comparison of models with different numbers of atoms.

The interface formation energy as a function of z-axis elongation, where z is the axis perpendicular to the interface plane, is shown in Fig. 8. It is evident that the interface formation energy depends strongly on the crystal plane. The (112) interface is the most stable one, with a formation energy of 0.95 eV per normalization area ($A_{(001)} = 29.5\,\text{Å}^2$) for no z-axis elongation (corresponding to the Si lattice constant). For the (111) interface, the formation energy is slightly higher (1.1 eV), while the (001) interface is the least stable with 1.9 eV. Furthermore, the non-elongated structures represent the actual structure well, as can be seen by the very low energy gain due to elongation in z-direction. It should be noted, that the optimum elongation depends on the interface orientation and is 0.20 % for (111), 0.25 % for (112) and 0.65 % for the (001) orientation. However, the order of interface stabilities remains for elongated cells.

The observations above pose an explanation for the observed formation on intermixing layers at Si/GaP (001) boundaries. Since the (111) and (112) orientations are more stable, their formation is preferred. An important aspect of interface

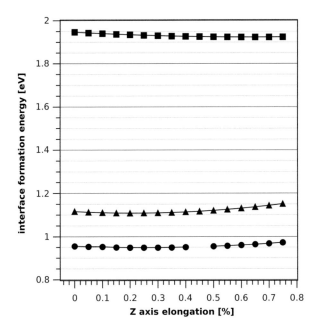

Fig. 8 Interface formation energy according to Eq. (1) for elongated supercells with interfaces in the crystal planes (001) (*squares*), (111) (*triangles*) and (112) (*circles*). Zero elongation corresponds to the experimental lattice constant of Si

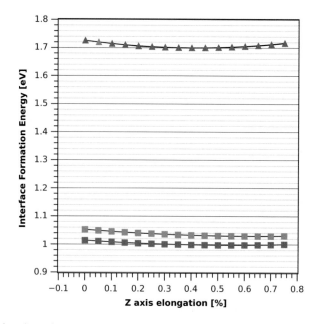

Fig. 9 Interface formation energy according to Eq. (1) for elongated supercells with intermixed interfaces in the crystal planes (001) (*squares*) and (111) (*triangles*). The intermixed atomic layers contain either exclusively Si, Ga (*green*) or Si, P (*orange*) in the intermixing layer. Data point in the (111)-line left out to show that lines coincide

stability is charge neutrality. Ga and P carry different (partial) charges, destabilizing interfaces that lead to charge accumulation. In the case of the (001) and (111) orientation, interfaces are entirely composed of either Ga or P on the GaP side. In the model cells with two interfaces here, each termination occurs once. For the (112) orientation, two mixed terminations dominated by either component is formed. If an intermixed layer is introduced into the (001) and (111) terminations, two identical interfaces in the cell result, either containing Si and Ga or Si and P. As shown in Fig. 9, the introduction of intermixed layers strongly stabilizes the interface formation for the (001) orientation, where intermixing of Si and Ga is slightly more favourable than Si and P. In the case of the (111) interface, intermixing leads to destabilization. The origin of this is the bilayer structure of this interface, which is charge neutral within itself. This internal charge neutrality is broken by intermixing leading to the observed destabilization. The (112) oriented interface is already mixed, making the introduction of one intermixed atomic layer obsolete.

In conclusion, the occurrence of intermixing in the GaP/Si-interface can be explained by the energetic stability of various orientations, which leads to a divergence from a perfectly smooth boundary.

4.2 Electrostatic Potentials Across the GaP/Si-Interface

The performance of a semiconductor device is strongly determined by the electronic structure of its constituents and the interfaces between those. By plotting the plane-averaged local potential in growth direction for the supercells described above, the electrostatic properties and internal dipole moments of the films and interfaces can be studied.

In Fig. 10 the averaged local potential in stacking direction is shown for the (001), (111) and (112) oriented interfaces. In the (001) and (111) interface cells, the local potential oscillates strongly due to the layers consisting of exclusively Ga or P, while in the intrinsically mixed (112) interface cell the oscillations are weaker. A dipole over the film can be recognized for all three cells and can be attributed to either side being dominated (or fully comprised) of one species. A corresponding effect can be seen in the Si to compensate the polarization of the GaP layer. The film dipole is less pronounced for the (112) orientation.

5 Density Functional Computations on the Electronic Structure of Dilute GaAs: Method Validation

The design of optoelectronic devices requires understanding the optical properties of the employed materials. These can be computed by microscopic theories, which build on an accurate description of the electronic bandstructure [29]. The common approach of using interpolated experimental data in the framework of $\mathbf{k} \cdot \mathbf{p}$-theory can be replaced by *ab initio* bandstructure calculations, sufficient accuracy provided.

The systems of interest are multinary III/V-semiconductors, in general dilute GaAs. Thus, simple GaAs is a reasonable choice for method calibration. By comparing a GGA, a hybrid and a meta-GGA functional, namely PBE, HSE06 and TB09 (MBJLDA), respectively, the latter has been chosen for further work, as it yields accurate results for a reasonable computational cost (in agreement with [30]). This allows the use of large supercells for the dilute compounds, enabling the study of the effect structural changes, e.g. the distribution of impurities, have on the electronic structure. From the unfolded supercell bandstructure, as is shown in Fig. 11 for a simple GaAs supercell, the input parameters for the $\mathbf{k} \cdot \mathbf{p}$-theory can be extracted. A computational approach allows not only to predict the properties of new materials, thus guiding the development process, but also to study the effect of local structural motifs on the electronic structure.

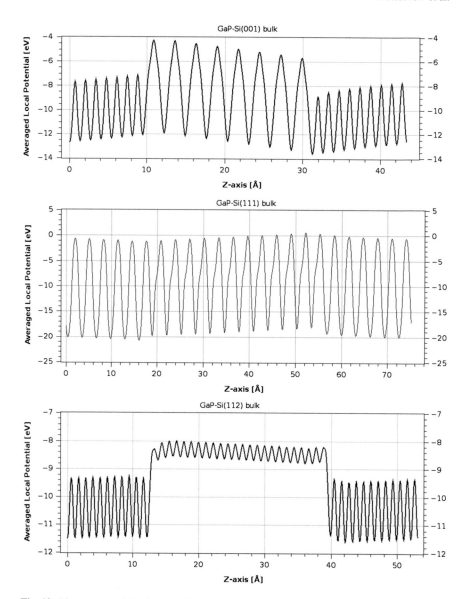

Fig. 10 Plane-averaged local potential along *z*-direction (perpendicular to interface plane) for (001) (*top*), (111) (*middle*) and (112) (*bottom*) interfacial plane

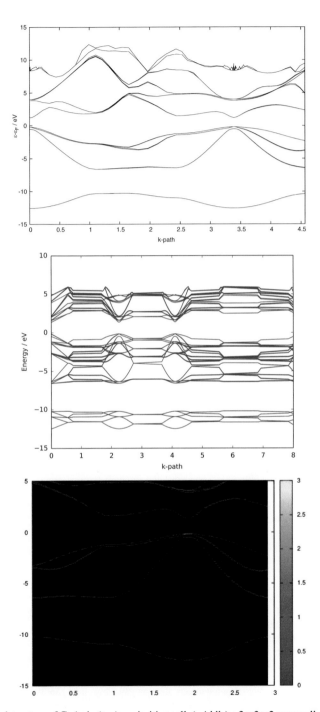

Fig. 11 Bandstructure of GaAs in (*top*) a primitive cell, (*middle*) a $2 \times 2 \times 2$ supercell and (*bottom*) the unfolded bandstructure of the supercell

Table 1 Scaling data on Hornet for a Si slab containing 1280 atoms

No. of cores	Optimized		Non-optimized	
	NCORE	t_{eff}/Ms	NCORE	t_{eff}/Ms
528	24	14.95	—	—
2064	48	11.84	24	18.74
4104	57	16.77	24	25.25

The effective time t_{eff} is the elapsed time times the number of cores

6 Scaling Tests

Some scaling tests have been performed on the Hornet cluster in order to control optimal settings for the setting NCORE. This variable is complementary to NPAR, which was described earlier (see Sect. 2). The product of both yields the number of cores per group defined by KPAR (which was set to one in this test). Two settings have been tested. The first one, henceforth dubbed non-optimized, was obtained by setting NCORE to the number of cores per node, i.e. 24. In the second one (optimized), NCORE was set close to the generally recommended value of the square root of the number of nodes. While only very few data points have already been obtained, the effect of setting NCORE to a proper value is quite impressive, leading to a speed-up of roughly one third (see Table 1). It should also be noted, that increasing the number of cores with non-optimized setting does not lead to a speed-up, but has a negative effect. For the optimized settings, it can be seen that increasing the number of cores with optimized settings leads to a speed up in effective time when going from 528 to 2064 cores. Going to 4104 cores actually deteriorates that. This can probably be attributed to the set value of NCORE being close to the square root of the number of cores but not ideal for the actual workload distribution.

Acknowledgements The authors acknowledge the research training group (Graduiertenkolleg, DFG) 1782 "Functionalization of Semiconductors", the collaborative research centre (Sonderforschungsbereich, DFG) 1083 "Structure and Dynamics of Internal Interfaces" and the Beilstein Institut, Frankfurt am Main, for support.

References

1. Mette, G., Dürr, M., Bartholomäus, R., Koert, U., Höfer, U.: Real-space adsorption studies of cyclooctyne on Si(001). Chem. Phys. Lett. **556**, 70 (2013)
2. Schober, C.: Theoretische Untersuchungen der Adsorption von Ethin und Cyclooctin auf der Si(001)-Oberfläche. Master's thesis, Philipps-Universität Marburg (2012)
3. Jónsson, H., Mills, G., Jacobsen, K.W.: In: Berne, B.J., Ciccotti, G., Coker, D.F.: Classical and Quantum Dynamics in Condensed Phase Simulations. World Scientific, Singapore (1998)
4. Hohenberg, P., Kohn, W.: Inhomogeneous electron gas. Phys. Rev. **136**(3B), B864 (1964)
5. Kohn, W., Sham, L.J.: Self-consistent equations including exchange and correlation effects. Phys. Rev. **140**(4A), A1133 (1965)

6. Kresse, G., Furthmüller, J.: Efficiency of ab-initio total energy calculations for metals and semiconductors using a plane-wave basis set. Comp. Mat. Sci. **6**, 15 (1996)
7. Kresse, G., Furthmüller, J.: Efficient iterative schemes for ab initio total-energy calculations using a plane-wave basis set. Phys. Rev. B **54**, 11169 (1996)
8. Perdew, J.P., Burke, K., Ernzerhof, M.: Generalized gradient approximation made simple. Phys. Rev. Lett. **77**, 3865 (1996)
9. Perdew, J.P., Burke, K., Ernzerhof, M.: Erratum: Generalized gradient approximation made simple. Phys. Rev. Lett. **78**, 1396 (1997)
10. Grimme, S.: Density functional theory with London dispersion corrections. WIREs Comput. Mol. Sci. **1**(2), 211 (2011)
11. Grimme, S., Ehrlich, S., Goerigk, L.: Effect of the damping function in dispersion corrected density functional theory. J. Comput. Chem. **32**, 1456 (2011)
12. Blöchl, P.E.: Projector augmented-wave method. Phys. Rev. B **50**, 17953 (1994)
13. Kresse, G., Joubert, D.: From ultrasoft pseudopotentials to the projector augmented-wave method. Phys. Rev. B **59**, 1758 (1999)
14. Monkhorst, H.J., Pack, J.D.: Special points for brillouin-zone integrations. Phys. Rev. B **13**(12), 5188 (1976)
15. Hestenes, M.R., Stiefel, E.: Methods of conjugate gradient for solving linear systems. J. Res. Natl. Bur. Stanf. **49**(6), 409 (1952)
16. Nocedal, E.: Updating quasi-newton matrices with limited storage. Math. Comput. **35**, 773 (1980)
17. Born, M., Oppenheimer, R.: Zur Quantentheorie der Molekeln. Ann. Phys. **84**, 457 (1927)
18. Verlet, R.: Computer "experiments" on classical fluids. i. thermodynamical properties of lennard-jones molecules. Phys. Rev. **159**(1), 98 (1967)
19. Verlet, L.: Computer "experiments" on classical fluids. ii. equilibrium correlation functions. Phys. Rev. **165**(1), 201 (1967)
20. Nose, S.: A unified formulation of the constant temperature molecular dynamics methods. J. Chem. Phys. **81**(1), 511 (1984)
21. Hoover, W.G.: Canonical dynamics: equilibrium phase-space distributions. Phys. Rev. A **31**(3), 1695 (1985)
22. D.J. Evans, B.L. Holian, The nose-hoover thermostat. J. Chem. Phys. **83**(8), 4069 (1985)
23. S.F. Bent, Organic functionalization of group IV semiconductor surfaces: principles, examples, applications, and prospects. Surf. Sci. **500**, 879 (2002)
24. Random.org true random number service. URL http://www.random.org/decimal-fractions. Accessed 21.04.2015
25. Stegmüller, A., Rosenow, P., Tonner, R.: A quantum chemical study on gas phase decomposition pathways of triethylgallane (TEG, $Ga(C2H5)3$) and tert-butylphosphine (TBP, $PH2(t-C4H9)$) under MOVPE conditions. Phys. Chem. Chem. Phys. **16**(32), 17018 (2014)
26. Stegmüller, A., Tonner, R.: The beta-hydrogen elimination mechanism in absence of acceptor orbitals in $EH_2(t-C_4H_9)$ (E = N-Bi). Inorg. Chem. **54**(13), 6363 (2015)
27. Stegmüller, A., Tonner, R.: A quantum-chemical descriptor for CVD precursor design: predicting decomposition rates of TBP and TBAs isomers and derivatives. Chem. Vap. Deposition, **21**, 161–165 (2015). doi:10.1002/cvde.201504332
28. Beyer, A., Oelerich, J.O., Jandieri, K., Werner, K., Stolz, W., Baranovskii, S.D.: Tonner, R., Volz, K.: Pyramidal Structure Formation at the Interface between III/V Semiconductors and Silicon (2015, submitted)
29. Koukourakis, N., Buckers, C., Funke, D.A., Gerhardt, N.C., Liebich, S., Chatterjee, S., Lange, C., Zimprich, M., Volz, K., Stolz, W., Kunert, B., Koch, S.W., Hofmann, M.R.: High room-temperature optical gain in ga(nasp)/si heterostructures. Appl. Phys. Lett. **100**, 092107 (2012)
30. Kim, Y.S., Marsman, M., Kresse, G., Tran, F., Blaha, P.: Towards efficient band structure and effective mass calculations for III-V direct band-gap semiconductors. Phys. Rev. B **82**(20), 205212 (2010)

Atomistic Simulation of Oligoelectrolyte Multilayers Growth

Pedro A. Sánchez, Jens Smiatek, Baofu Qiao, Marcello Sega, and Christian Holm

Abstract We simulate at the atomistic scale the layer-by-layer growth of a four layers thin film of poly(diallyl dimethyl ammonium chloride)/poly(styrene sulfonate sodium salt) oligomers adsorbed on a silica substrate. The simulation is intended to provide atomistic details on the structure of a swollen multilayer in solutions with different concentrations of added salt ions. The scale of the simulated system has been chosen to produce, at a minimum computing cost, a reasonable estimation of some selected mesoscopic structural parameters that are measurable with current experimental techniques.

1 Introduction

Polyelectrolytes (PEs) are polymers formed by acid or basic groups which, in adequate solvents, can dissociate and lead to a solution of charged polymer backbones—either polyanions or polycations—and their respective neutralizing counter-ions. These charged polymers can form complexes with oppositely charged substances, including other PEs. By means of this electrostatically driven assembly mechanism, it is possible to create stable layered thin films by the alternating deposition of polyanions and polycations on a charged substrate. Such thin film growth method is called layer-by-layer (LbL) deposition [14, 16], and the resulting materials are known as Polyelectrolyte Multilayers (PEMs).

P.A. Sánchez (✉) • M. Sega
University of Vienna, Sensengasse 8/9, 1090 Wien, Austria
e-mail: pedro.sanchez@univie.ac.at; marcello.sega@univie.ac.at

J. Smiatek • C. Holm
Institute for Computational Physics, Universität Stuttgart, Allmandring 3, 70569 Stuttgart, Germany
e-mail: smiatek@icp.uni-stuttgart.de; holm@icp.uni-stuttgart.de

B. Qiao
Chemical Sciences and Engineering Division, Argonne National Laboratory, Argonne, IL 60439, USA
e-mail: qiaob@anl.gov

© Springer International Publishing Switzerland 2016
W.E. Nagel et al. (eds.), *High Performance Computing in Science and Engineering '15*, DOI 10.1007/978-3-319-24633-8_14

P.A. Sánchez et al.

The experimental procedure for the LbL deposition of PEMs is rather simple: first, a substrate with a charged surface is exposed to a solution of oppositely charged PEs, allowing the adsorption of some PEs on the surface; this is usually done by simply dipping the substrate wafer into the polyelectrolyte solution. After the dipping, the supernatant PEs and counter-ions are disposed by rinsing the wafer with pure solvent. In order to obtain the deposition of a new layer, such dipping and rinsing steps are then repeated for another solution of PEs that are oppositely charged with respect to the previous layer. This alternating charge deposition method can be systematically applied to grow films of hundreds of layers. Different variations and enhancements of this simple methodology have been proposed in recent years [10, 11, 17, 36, 68].

PEMs can be made from either synthetic PEs or biomolecules like DNA, lipids or proteins. These thin film materials combine a simple synthesis procedure with a great potential for many different applications [1, 15, 67]. For instance, PEMs can be used as biocompatible protective coatings [75], as matrix materials for active components in solar cells or biosensor applications [76], as membranes in filtration and catalysis systems [6, 13, 42], or as essential elements of non-linear optical materials [2, 33], nanocapsules for drug delivery applications [35] or tissue engineering systems [24]. As a consequence of this wide range of potential applications, the accurate understanding and control of the structure and dynamic properties of PEMs materials is a relevant topic which has attracted large research efforts during the last two decades [7, 70, 71, 78].

In spite of their relatively easy preparation procedure, PEMs show a complex behavior which has turned out to be rather difficult to characterize accurately. This complexity arises from the interplay between different factors that have a significant impact on the interactions that govern the system. The long-range electrostatic interactions between the oppositely charged components of PEMs is one of the most obvious mechanisms to consider, but its effective strength is strongly affected by factors like the concentration, ionic strength and nature of the counter-ions and added salt ions present in the dipping solution, the charge density of either the adsorbing surface and the deposited PEs, or the electrostatic properties of the solvent—typically water, which has a significant polarizability. Among the non electrostatic effects one can consider, for instance, the entropy and the intrinsic stiffness of the deposited PEs.

The amount of existing studies on the formation and properties of PEMs is large. Numerous experimental and theoretical studies have been devoted to the understanding of the different growth regimes that can be obtained in the LbL deposition process [31, 38, 39, 44, 45, 56, 69]. The dependence of the formation and stability of the multilayers on the charge density of the PEs has been also largely studied [21, 22, 25, 51, 65, 69]. Other aspects that have been analyzed include the effects of the ionic strength and the nature of the ions in the dipping solution [18, 23, 25, 66], the internal distributions of ions and solvent [20, 63, 74], the short-range interactions [78], or the hydration properties [71]. A more detailed review of the research performed so far on PEMs can be found in reference [9]. Thanks to these efforts, the understanding of the growth mechanisms and properties of PEMs has experienced a great progress in recent years. Nevertheless, it is still far from being fully satisfactory due to different limitations in the existing experimental and theoretical approaches. Currently available technologies are in general unable to provide direct measurements of the internal structure and properties of PEMs at length scales smaller than few tens of nanometers [5, 64]. Most analytical models rely on strong approximations and some assumptions that have not yet been proven to be universal—like the frequently assumed equilibrium nature of the multilayer structure [37]. Such models have serious difficulties to deal with the high structural correlations observed for oppositely charged PEs, which lead to a significant intermixing and complexation between distinct layers. Theoretical studies have also taken advantage of computer simulations. In particular, most computational studies have been based to date on coarse-grained models obtained from top-down modeling approaches that include very crude approximations [8, 61]. Such approaches have been useful for the study of different generic properties of PEMs at a relatively low computing cost [46, 47, 54, 55], allowing in many cases a direct comparison of their results with experimental measurements. However, top-down models are not expected to connect accurately the mesoscopic length and time scales that frequently characterize most experimental measurements with the microscopic properties of the films. In general, an agreement with experimental data is not enough to ensure the physical representativity and accuracy of such top-down crude models, and often many different approximations should be tested in order to determine the true mechanisms that govern the system. On the other hand, atomistic simulations may provide direct insights on the microscopic properties of PEMs at the cost of large computing resources. With present computing power, this cost is large enough to keep mesoscale properties out of reach for this approach, so in most cases it is not possible a direct comparison of atomistic simulation data with many experimental results of mesoscale properties that are relevant for technological applications. This inconvenient has made atomistic simulations extremely scarce to date in studies on PEMs. Nevertheless, atomistic simulations can be used not only for the study of microscopic properties, but also as a meaningful basis for the development of representative and accurate mesoscale models when bottom-up coarse-graining strategies—like the Inverse Monte Carlo [40, 41], Force Matching [32], Boltzmann Inversion [77] and Iterative Boltzmann Inversion [62] methods— are applied. Another key point is that most experimental studies have dealt so far with high degrees of polymerization for the PEs—typically in the range of hundreds to thousands of monomers. Nevertheless, PEMs made from oligomers may have specific properties potentially interesting for different applications. For instance, PEMs made with low molecular weight materials are expected to show faster dynamics—including a faster degradation, which can be an advantage in some cases—as well as a higher sensitivity to external stimuli. The lower number of atoms present in oligomer chains makes their atomistic simulation more feasible by reducing the computational cost required to reach experimentally relevant time and length scales. In summary, atomistic simulations of oligoelectrolyte multilayers may provide valuable insights on the microscopic properties of these materials and pave the way for the development of accurate coarse-grained models based on bottom-up modeling approaches.

2 Goals and Overview of the System Under Study

The main goal of our study is to determine the atomistic details of the structure of a swollen multilayer thin film made out of polyelectrolyte oligomers in presence of different added salt concentrations, thus, establishing its structural properties with a resolution that is still unreachable to available experimental techniques. In addition, as pointed in the previous Section, such microscopic information can be also useful for the development of accurate mesoscopic models that may reach lengths and time scales currently unfeasible for atomistic simulations. For both purposes, it is desirable to validate the results obtained from the atomistic simulations by measuring mesoscopic properties that could be directly compared to experimental observables. This imposes a lower bound to the scales required for the simulated system and, consequently, to the computing power footprint of our simulations.

Regarding the modeling of the multilayer growth process, in recent works we introduced an atomistic simulation protocol intended to mimic the experimental LbL deposition of charged polymers on a substrate [58–60]. Simulations of the deposition of a monolayer [58] and a bilayer [59, 60] of very short oligoelectrolytes in a relatively small substrate were successfully performed. By taking advantage of the same simulation protocol, here we need to extend the scale of such preliminary simulation efforts to reach system dimensions comparable to the resolution of current experimental measurements.

The main characteristics of the system under study are the following. First, the polyelectrolyte pair poly(diallyl dimethyl ammonium chloride) and poly(styrene sulfonate sodium salt) (PDADMAC/PSS) has been chosen. These are strong polyelectrolytes frequently used for the synthesis of multilayer thin films. The charge density of both macromolecules—one absolute unit charge per monomer, positive for PDADMAC and negative for PSS—is strong enough to be independent from the pH of the solution. This simplifies the simulation of the effects of different ionic concentrations on the film structure. The degree of polymerization taken in both cases is 30 monomers per chain. The selected dipping solutions consist of a concentration $0.1M$ of oligoelectrolyte chains, together with their respective counterions (30 Cl^- ions per chain for PDADMAC and the same amount of Na^+ for PSS), in a water solution with ionic strength $0.1M$ of added monovalent salt (NaCl). The chosen charged substrate is a flat silica surface, which has a negative surface charge at intermediate pH values [3, 72]. Therefore, the first deposited layer corresponds to PDADMAC chains. The film obtained under these conditions is then exposed to different concentrations of added salt ions and its structural properties are determined.

In the next Section, the details of the system setup and the simulation protocol employed in this work are discussed in more detail.

3   Simulation Details

The atomistic simulation of the assembly of a PDADMAC/PSS oligoelectrolyte multilayer has been performed by means of classical molecular dynamics (MD) with the package Gromacs 4.5.3 [29] in the HLRS' Hermit Cray XE6 supercomputer. The atomistic interactions were modeled with the OPLS-AA force field [34]. The parametrization of this force field for our system is detailed in [57]. In order to reproduce accurately the dielectric constant of the solvent, the SPC/E water model was chosen [4, 27, 28]. The SETTLE algorithm [50] was applied for the imposition of geometry constraints to the water molecules. The silica substrate has been modeled as a four layers sheet silicate structure. This substrate was placed at $z = 0$ in a simulation box with periodic boundaries in the x and y directions. The dimensions of this simulation box in such directions were chosen to be comparable to the nominal radius of the scan tip apex of currently available atomic force microscopes. In addition, they were required to allow the crystalline structure of the substrate to fit to the lateral periodic boundaries, as well as to have an aspect ratio as low as possible. This resulted in a horizontal deposition area of $13.1650 \times 12.1608 \, \text{nm}^2$. In order to make the substrate surface hydrophilic, their uppermost oxygen atoms in every silicate tetrahedron of the interfacial layer were replaced with polar hydroxyl groups (–OH). Since it is known that a fixed orientation of such hydroxyl groups produces simulation artifacts [59], they were allowed to freely rotate with respect to the axis normal to the surface [30]. Finally, a surface charge was introduced by skipping one in every four of the hydroxyl group allocations. This left exposed the corresponding SiO_2^- groups, leading to an average surface charge of $\sigma = -3 \cdot 10^{-20} \, C/\text{nm}^2$, in good agreement with experimental measurements [3, 72]. This simple approach to set the distribution of surface charges is based on the assumption that their local microscopic details have no significant impact on the mesoscopic properties of the system. Such assumption is supported by a recent computational study on the mesoscopic dynamics of electro-osmotic flows with charged walls [73].

The simulation protocol focuses on reproducing the adsorption process of every layer, following the iterative experimental preparation steps as close as possible. Within the simulation protocol, such steps are mimicked by different manual preparation procedures and subsequent MD equilibration runs. The first step is the preparation of an equilibrated dipping solution of oligoelectrolyte chains of the corresponding type (PDADMAC for the deposition of the first layer, PSS for the second, and so on). In this case, the dipping solution consisted of 20 oligomer chains, their 600 corresponding counterions, 192 Na^+ and Cl^- added salt ions, and 99286 water molecules. This solution was initially prepared with the package Packmol 1.1.1 [43] in a simulation box with the same lateral dimensions of the silica substrate, $13.1650 \times 12.1608 \, \text{nm}^2$. It was equilibrated in a NPT simulation of 50 ns, with a pressure of 1 bar in the z direction applied with a semi-isotropic Parrinello-Rahman barostat [52, 53]. The final height of the equilibrated system was of approximately 19.9 nm, matching the concentrations indicated in the previous section. Once it is prepared, the dipping solution is placed in the simulation box on top of the adsorbing surface, i.e., the silica substrate, eventually covered by the previous deposited layers. A cycle of four simulation steps is then applied to this system to achieve the deposition of a new layer: first, a initial relaxation of the system is performed by means of energy minimization with the steepest descent algorithm; this is followed by three MD simulation steps in the NVT ensemble with the Noose-Hoover thermostat at a reference temperature of $T = 298\,K$ and a characteristic time $\tau = 0.5\,ps$. For these MD steps, a neighbor searching cut-off of 1.3 nm was used. Van der Waals interactions were represented by Lennard-Jones potentials shifted to smoothly decay to zero at distances larger than 0.9 nm. Electrostatic interactions were calculated with the particle mesh Ewald method (PME) [12, 19], using a Fourier grid spacing of 0.125 nm and a direct space cut-off of 1.3 nm. This latter cut-off is slightly larger than the usual value of 1.2 nm which, for grid spacings around ~ 0.1 nm, is considered to provide a good accuracy with the PME method. In order to take into account the slab geometry of the system, the Yeh-Berkowitz correction to the electrostatic force and potential [79] was applied, and a region of twice the height of the initial vertical dimensions of the system was kept empty in the upper part of the simulation box to avoid the artifacts associated to this method. The first MD step was simulated for 100 ps, using a time step of 1 fs. In the second one, that ran also for 100 ps with a time step of 2 fs, the covalent bonds of the hydrogen atoms were constrained with the LINCS algorithm [26]. Finally, the last MD step lasted for 300 ns with a time step of 2.5 fs and all the covalent bonds constrained with the LINCS method. During this last step, the number of oligoelectrolyte chains adsorbed into the substrate and film structure was monitored to ensure that a constant value—i.e., a saturation of the adsorption process—was reached. Once this adsorption simulation cycle is completed, the rinsing of the film is mimicked by explicitly removing back the dipping solution from the simulation box. This includes removing all the remaining non adsorbed chains and the surrounding ions and water molecules, leaving only the substrate, the adsorbed chains, their corresponding counterions and the water molecules below the uppermost position of the atoms belonging to the adsorbed chains. With this rinsed system, a new cycle of dipping-adsorption-rinsing steps can be started to simulate the adsorption of a new layer of oppositely charged oligoelectrolytes. In this case, this procedure was used to simulate the assembly of a film of four layers, $(PDADMAC/PSS)_2$. The measures of the structural parameters of the film were taken from independent simulation runs after the rinsing of every newly deposited layer: in this case, the rinsed system was dipped into a previously equilibrated solution containing only water and salt ions at a given concentration taken from three selected values, $C_{NaCl} = 0$ (pure water), $0.1M$ and $0.5M$. The dipped system was then relaxed by following the same simulation cycle described above, even though the final MD simulation step ran only for 100 ns. This assumes that the main changes during the relaxation of the swollen film correspond to the redistribution of water molecules and ions, which have a rather fast dynamics. Trajectory measures of the resulting swollen film were obtained from the last 50 ns of this final MD step, at a sampling rate of 0.1 ns. Finally, in order to obtain a minimal statistical sampling of the film structure, three independent runs of the whole simulation protocol were performed. Figure 1 shows a scheme of the different steps involved in this protocol, specifying the manual procedures, the MD runs and the simulated times corresponding to the latter.

In general, the computing cost associated to atomistic simulations is firstly determined by the amount of atoms in the simulation box. In our simulation protocol, this amount will grow with the number of already deposited layers. Here, we simulated systems within a range from approximately $3 \cdot 10^5$–$5 \cdot 10^5$ atoms. These quantities are relatively moderate and, therefore, do not require a big number of processors to reach a optimum parallel computing performance. In particular, we found that the use of 192 processors led to the best performance for the most demanding simulation runs, i.e., the ones corresponding to the long latest NVT step of the system relaxation cycle after each dipping preparation. During such runs, half of the used processors were reserved for the calculation of the electrostatic interactions with the PME method. Regarding processors load balancing, Gromacs usually manages it in a rather efficient way. In this case, the initial configurations of the dipped systems are quite far from a relaxed state, due to their manual preparation. Therefore, the Gromacs dynamic load balancing algorithm was challenged until the systems approached relaxation, reaching load imbalances of around 3 % at the beginning of these runs. After such initial issues, most of the simulation time was spent with imbalances below 1 %. Finally, the

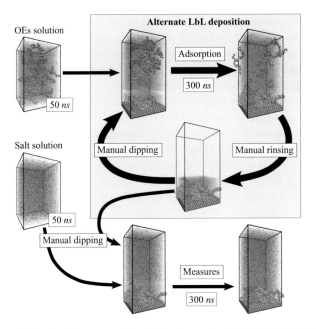

Fig. 1 Scheme of the simulation protocol used for the growth of the (PDADMAC/PSS)$_2$ OEMs. See the text for a detailed explanation

overall performance of the simulations was of approximately 12 ns per day. Taking into account all the simulation steps required by our protocol, the number of layers simulated and the independent runs carried out, the total computing cost of these simulations has been of the order of $1.6 \cdot 10^5$ cpu-h. Unfortunately, the sequential nature of the LbL growth process limited notably the parallelization of the simulations, since the relaxation runs of one given layer can not be carried out before all previous layers have been already obtained. This made the simulation of the assembly of the multilayer significantly time consuming, despite the relatively modest overall computing requirements.

4 Mesoscopic Properties of the Four Layers System

In this last Section we briefly present a selection of results on the structural properties of the $(PDADMAC/PSS)_2$ thin film. In particular, we show the measures of the main mesoscopic structural parameters obtained for the third and fourth layers (L_3 and L_4, respectively) and the three different added salt concentrations explored in our simulations, C_{NaCl}. The atomistic details of the film structure obtained from these simulations and their application to the development of an accurate coarse-grained model will be discussed elsewhere.

In Table 1, the measured values of the coverage fraction and the average film height, or average film thickness, are shown. The coverage fraction is simply the fraction of surface area of the substrate that is not exposed to vertical deposition trajectories. It is measured by simply discretizing the horizontal plane of the film surface into a square lattice of $n_x \times n_y$ regions, using the following criterium: a region is considered to be covered if there exists some atom belonging to the adsorbed oligoelectrolyte chains whose horizontal coordinates lie within such region. The coverage fraction is then the ratio between the total area of the covered regions over the total horizontal area of the substrate. Using the same horizontal discretization scheme for the horizontal positions of the oligoelectrolyte atoms, the free surface of the film is defined as the height profile, $h(x_i, y_i)$, obtained by assigning to each region the maximum height of the oligoelectrolyte atoms that it contains, if any. The average height of the film can then be calculated in two ways: the most straightforward estimation consists in simply taking the average value of the surface

Table 1 Coverage fraction and film thickness measured for the three and four layers system with the three selected added salt concentrations

	f_c		$\langle h \rangle$ (nm)		$\langle h^* \rangle$ (nm)	
C_{NaCl} (M)	L_3	L_4	L_3	L_4	L_3	L_4
0.0	0.71 ± 0.06	0.76 ± 0.07	2.6 ± 0.4	3.5 ± 0.3	3.7 ± 0.3	4.6 ± 0.2
0.1	0.72 ± 0.06	0.77 ± 0.08	2.6 ± 0.4	3.3 ± 0.2	3.6 ± 0.4	4.3 ± 0.2
0.5	0.72 ± 0.06	0.78 ± 0.07	2.6 ± 0.7	3.3 ± 0.3	3.6 ± 0.7	4.3 ± 0.3

height profile, $\langle h \rangle = (1/n_x n_y) \sum_{i,j} h(x_i, y_i)$. In the second method the average of the surface height profile is also calculated, but only the covered regions are taken into account. We denote this latter estimation as $\langle h^* \rangle = (1/n_x^* n_y^*) \sum_{i,j} h^*(x_i^*, y_i^*)$, where the star symbol indicates the exclusion of the uncovered regions. The results, obtained for a discretization lattice of $n_x \times n_y = 50 \times 50$ regions, show that, even the differences are relatively small compared to the error intervals, the coverage fraction and the film height grow with the number of layers, as is expected. In particular, the comparison of the estimated film thickness, $\langle h \rangle$, with experimental ellipsometry measurements has shown a rather good agreement [48, 49], confirming the validity of our simulation approach. The meaning of the second estimation, $\langle h^* \rangle$, will be explained below.

The second interesting indication provided by these results is related to the overall effect of the added salt concentration on the structure of the swollen film: we can observe that the coverage tends to increase and the thickness tends to decrease as the concentration grows. This indicates a flatter film structure for higher added salt concentrations, with a lower amount of water trapped within the oligoelectrolyte chains.

Finally, Table 2 shows the results for another important mesoscopic property of thin films: the surface roughness. Here, we also provide two different estimations of this parameter, based on the same approaches to the surface height profile used above, $h(x_i, y_i)$ and $h^*(x_i^*, y_i^*)$. The surface roughness is then taken as the average root mean square of such surface height profiles,

$$R_{\text{RMS}}(h) = \left(\frac{1}{n_x n_y} \sum_{i,j} [h(x_i, y_i) - \langle h \rangle]^2 \right)^{1/2}$$

and

$$R_{\text{RMS}}^*(h^*) = \left(\frac{1}{n_x^* n_y^*} \sum_{i,j} [h^*(x_i^*, y_i^*) - \langle h^* \rangle]^2 \right)^{1/2}.$$

In this case, the comparison between the latter estimation, R_{RMS}^*, and the corresponding experimental results, performed by means of AFM measurements, provides a better and also reasonably good agreement, supporting as well the validity of our simulation approach [48]. The reason to have a better agreement with AFM

Table 2 Root mean squared film surface roughness corresponding to the three and four layers system and the three selected added salt concentrations

C_{NaCl} (M)	$R_{\text{RMS}}(h)$ (nm)		$R_{\text{RMS}}^*(h^*)$ (nm)	
	L_3	L_4	L_3	L_4
0.0	2.0 ± 0.3	2.3 ± 0.3	1.4 ± 0.4	1.5 ± 0.2
0.1	2.0 ± 0.2	2.3 ± 0.2	1.3 ± 0.2	1.5 ± 0.2
0.5	1.9 ± 0.3	2.2 ± 0.4	1.3 ± 0.3	1.4 ± 0.2

measurements when the uncovered regions are disregarded can be attributed to the known effective averaging that such devices produce on the surface features that have a characteristic size smaller than their maximum resolution power. In Table 1, we have shown that the fraction of uncovered regions in the three and four layers films is lower than 0.3. Taking into account the dimensions of the substrate, this corresponds to uncovered surface areas with a characteristic length scale of the order of few nanometers. Since this scale is below the nominal resolution power of the AFM used in the reference experiments, it turned out that the contribution of the uncovered regions to the overall height profile was impossible to capture in such measurements. As a last remark, the evolution of the film roughness with the number of layers and the added salt concentration agrees with the conclusions pointed above: whereas the roughness grows with the amount of layers, the added salt tends to reduce it, as corresponds to flatter overall structures.

Acknowledgements We acknowledge the *Deutsche Forschungsgemeinschaft (DFG)* within the Priority Program SSP 1369 for its financial support to this research.

References

1. Ariga, K., Hill, J.P., Ji, Q.: Layer-by-layer assembly as a versatile bottom-up nanofabrication technique for exploratory research and realistic application. Phys. Chem. Chem. Phys. **9**(19), 2319–2340 (2007)
2. Arsenault, A.C., Halfyard, J., Wang, Z., Kitaev, V., Ozin, G.A., Manners, I., Mihi, A., Miguez, H.: Tailoring photonic crystals with nanometer-scale precision using polyelectrolyte multilayers. Langmuir **21**(2), 499–503 (2005)
3. Behrens, S.H., Grier, D.G.: The charge of glass and silica surfaces. J. Chem. Phys. **115**(14), 6716–6721 (2001)
4. Berendsen, H.J.C., Grigera, J.R., Straatsma, P.T.: The missing term in effective pair potentials. J. Phys. Chem. **91**, 6269–6271 (1987)
5. Block, S., Helm, C.A.: Single polyelectrolyte layers adsorbed at high salt conditions: Polyelectrolyte brush domains coexisting with flatly adsorbed chains. Macromolecules **42**(17), 6733–6740 (2009)
6. Bruening, M.L., Dotzauer, D.M., Jain, P., Ouyang, L., Baker, G.L.: Creation of functional membranes using polyelectrolyte multilayers and polymer brushes. Langmuir **24**(15), 7663–7673 (2008)
7. Cerdà, J.J., Qiao, B., Holm, C.: Modeling strategies for polyelectrolyte multilayers. Eur. Phys. J. Special Topics **177**, 129–148 (2009)
8. Cerdà, J.J., Qiao, B., Holm, C.: Understanding polyelectrolyte multilayers: an open challenge for simulations. Soft Matter **5**(22), 4412–4425 (2009)
9. Cerdà, J.J., Holm, C., Qiao, B.: Modeling the Structure and Dynamics of Polyelectrolyte Multilayers, chapter 5, 1st edn. pp. 121–166. Wiley, Hoboken, NJ (2012)
10. Chiarelli, P.A., Johal, M.S., Casson, J.L., Roberts, J.B., Robinson, J.M., Wang, H.-L.: Controlled fabrication of polyelectrolyte multilayer thin films using spin-assembly. Adv. Mater. **13**(15), 1167–1171 (2001). ISSN 1521–4095
11. Cho, J., Char, K., Hong, J.-D., Lee, K.-B.: Fabrication of highly ordered multilayer films using a spin self-assembly method. Adv. Mater. **13**(14), 1076–1078 (2001)
12. Darden, T., York, D., Pedersen, L.: Particle Mesh Ewald: An $N \log(N)$ method for Ewald sums in large systems. J. Chem. Phys. **98**, 10089–10092 (1993)

13. Datta, S., Cecil, C., Bhattacharyya, D.: Functionalized membranes by layer-by-layer assembly of polyelectrolytes and in situ polymerization of acrylic acid for applications in enzymatic catalysis. Ind. Eng. Chem. Res. **47**(14), 4586–4597 (2008)
14. Decher, G.: Fuzzy nanoassemblies: toward layered polymeric multicomposites. Science **277**(5330), 1232–1237 (1997)
15. Decher, G., Schlenoff, J.B. (eds.): Multilayer Thin Films: Sequential Assembly of Nanocomposite Materials, 2nd edn. Wiley, New York (2012)
16. Decher, G., Hong, J., Schmitt, J.: Buildup of ultrathin multilayer films by a self-assembly process: Iii. consecutively alternating adsorption of anionic and cationic polyelectrolytes on charged surfaces. Thin Solid Films **210–211**, 831–835 (1992)
17. Dubas, S.T., Schlenoff, J.B.: Factors controlling the growth of polyelectrolyte multilayers. Macromolecules **32**(24), 8153–8160 (1999)
18. Dubas, S.T., Schlenoff, J.B.: Swelling and smoothing of polyelectrolyte multilayers by salt. Langmuir **17**(25), 7725–7727 (2001)
19. Essmann, U., Perera, L., Berkowitz, M.L., Darden, T., Lee, H., Pedersen, L.: A smooth Particle Mesh Ewald method. J. Chem. Phys. **103**, 8577 (1995)
20. Garg, A., Heflin, J.R., Gibson, H.W., Davis, R.M.: Study of film structure and adsorption kinetics of polyelectrolyte multilayer films: effect of pH and polymer concentration. Langmuir **24**(19), 10887–10894 (2008)
21. Glinel, K., Moussa, A., Jonas, A.M., Laschewsky, A.: Influence of polyelectrolyte charge density on the formation of multilayers of strong polyelectrolytes at low ionic strength. Langmuir **18**(4), 1408–1412 (2002)
22. Gopinathan, A., Kim, Y.W., Gopinathan, A., Kim, Y.W.: Polymer translocation in crowded environments. Phys. Rev. Lett. **99**, 228106 (2007)
23. Gopinadhan, M., Ivanova, O., Ahrens, H., Günther, J.-U., Steitz, R., Helm, C.A.: The influence of secondary interactions during the formation of polyelectrolyte multilayers: layer thickness, bound water and layer interpenetration. J. Phys. Chem. B **111**(29), 8426–8434 (2007)
24. Gribova, V., Auzely-Velty, R., Picart, C.: Polyelectrolyte multilayer assemblies on materials surfaces: From cell adhesion to tissue engineering. Chem. Mater. **24**(5), 854–869 (2012)
25. Haitami, A.E.E., Martel, D., Ball, V., Nguyen, H.C., Gonthier, E., Labb e, P., Voegel, J., Schaaf, P., Senger, B., Boulmedais, F.: Effect of the supporting electrolyte anion on the thickness of PSS/PAH multilayer films and on their permeability to an electroactive probe. Langmuir **25**(4), 2282–2289 (2009)
26. Hess, B., Bekker, H., Berendsen, H.J.C., Fraaije, J.G.E.M.: LINCS: A linear constraint solver for molecular simulations. J. Comput. Chem. **18**, 1463–1472 (1997)
27. Hess, B., Holm, C., van der Vegt, N.: Modeling multi-body effects in ionic solutions with a concentration dependent dielectric permittivity. Phys. Rev. Lett. **96**, 147801 (2006)
28. Hess, B., Holm, C., van der Vegt, N.: Osmotic coefficients of atomistic NaCl (aq) force-fields. J. Chem. Phys. **124**, 164509 (2006)
29. Hess, B., Kutzner, C., van der Spoel, D., Lindahl, E.: GROMACS 4: algorithms for highly efficient, load-balanced, and scalable molecular simulation. J. Chem. Theory Comput. **4**(3), 435–447 (2008)
30. Ho, T.A., Argyris, D., Papavassiliou, D.V., Striolo, A.: Interfacial water on crystalline silica: a comparative molecular dynamics simulation study. Mol. Simul. **37**, 172–195 (2011)
31. Hoda, N., Larson, G.R.: Modeling the buildup of exponentially growing polyelectrolyte multilayer films. J. Phys. Chem. B **113**(13), 4232–4241 (2009)
32. Izvekov, S., Voth, G.A.: Multiscale coarse graining of liquid-state systems. J. Chem. Phys. **123**(13), 134105 (2005)
33. Jiang, L., Lu, F., Chang, Q., Liu, Y., Liu, H., Li, Y., Xu, W., Cui, G., Zhuang, J., Li, X.: Fabrication of ultrathin films with large third-order nonlinear optical properties. Chem. Phys. Chem. **6**(3), 481–486 (2005)
34. Jorgensen, W.L., Maxwell, D.S., Tirado-Rives, J.: Development and testing of the OPLS all-atom force field on conformational energetics and properties of organic liquids. J. Am. Chem. Soc., **118**(45), 11225–11236 (1996)

35. Khopade, A.J., Arulsudar, N., Khopade, S.A., Hartmann, J.:   Ultrathin antibiotic walled microcapsules. Biomacromolecules **6**(1), 229–234 (2005)
36. Kolasinska, M., Krastev, R., Gutberlet, T., Warszynski, P.:   Layer-by-layer deposition of polyelectrolytes. dipping versus spraying. Langmuir **25**(2), 1224–1232 (2009)
37. Kovacevic, D., van der Burgh, S., de Keizer, A., Stuart, M.A.C.: Kinetics of formation and dissolution of weak polyelectrolyte multilayers: role of salt and free polyions. Langmuir **18**(14), 5607–5612 (2002)
38. Lavalle, P., Gergely, C., Cuisinier, F.J.G., Decher, G., Schaaf, P., Voegel, J.C., Picart, C.: Comparison of the structure of polyelectrolyte multilayer films exhibiting a linear and an exponential growth regime: an in situ atomic force microscopy study. Macromolecules **35**(11), 4458–4465 (2002)
39. Lavalle, P., Picart, C., Mutterer, J., Gergely, C., Reiss, H., Voegel, J.-C., Senger, B., and Schaaf, P. Modeling the buildup of polyelectrolyte multilayer films having exponential growth. J. Phys. Chem. B **108**(2), 635–648 (2004)
40. Lyubartsev, A.P., Laaksonen, A.:   Calculation of effective interaction potentials from radial distribution functions: a reverse monte carlo approach. Phys. Rev. E **52**, 3730–3737 (1995)
41. Lyubartsev, A.P., Marčelja, S.:   Evaluation of effective ion-ion potentials in aqueous electrolytes. Phys. Rev. E **65**(4), 041202 (2002)
42. Malaisamy, R., Bruening, M.L.: High-flux nanofiltration membranes prepared by adsorption of multilayer polyelectrolyte membranes on polymeric supports. Langmuir **21**(23), 10587–10592 (2005)
43. Martínez, L., Andrade, R., Birgin, E.G., Martínez, J.M.: PACKMOL: a package for building initial configurations for molecular dynamics simulations. J. Comput. Chem. **30**, 2157–2164 (2009)
44. McAloney, R.A., Sinyor, M., Dudnik, V., Goh, M.C.: Atomic force microscopy studies of salt effects on polyelectrolyte multilayer film morphology. Langmuir **17**, 6655–6663 (2001)
45. McAloney, R.A., Dudnik, V., Goh, M.C.: Kinetics of salt-induced annealing of a polyelectrolyte multilayer film morphology. Langmuir **19**(9), 3947–3952 (2003)
46. Messina, R.: Polyelectrolyte multilayering on a charged planar surface. Macromolecules **37**(2), 621–629 (2004)
47. Messina, R., Holm, C., Kremer, K.: Polyelectrolyte adsorption and multilayering on charged colloidal particles. J. Polym. Sci. Part B: Polym. Phys. **42**, 3557 (2004)
48. Micciulla, S., Sánchez, P.A., Smiatek, J., Qiao, B., Sega, M., Laschewsky, A., Holm, C., von Klitzing, R.:   Layer-by-layer formation of oligoelectrolyte multilayers: a combined experimental and computational study. Soft Materials **12**, S14–S21 (2014)
49. Micciulla, S., Dodoo, S., Chevigny, C., Laschewsky, A., von Klitzing, R.:   Short versus long chain polyelectrolyte multilayers: a direct comparison of self-assembly and structural properties. Phys. Chem. Chem. Phys. **16**, 21988–21998 (2014)
50. Miyamoto, S., Kollman, A.P.:   Settle: an analytical version of the SHAKE and RATTLE algorithm for rigid water models. J. Comput. Chem. **13**, 952–962 (1992)
51. Müller, M., Meier-Haack, J., Schwarz, S., Buchhammer, H., Eichhorn, E., Janke, A., Kessler, B., Nagel, J., Oelmann, M., Reihs, T., Lunkwitz, K.: Polyelectrolyte multilayers and their interactions. J. Adhes. **80**(6), 521–547 (2004)
52. Nosé, S., Klein, M.L.: Constant pressure molecular dynamics for molecular systems. Mol. Phys. **50**(5), 1055–1076 (1983)
53. Parrinello, M., Rahman, A.:   Polymorphic transitions in single crystals: A new molecular dynamics method. J. Appl. Phys. **52**, 7182–7190 (1981)
54. Patel, P.A., Jeon, J., Mather, P.T., Dobrynin, A.V.:   Molecular dynamics simulations of layer-by-layer assembly of polyelectrolytes at charged surfaces: Effects of chain degree of polymerization and fraction of charged monomers. Langmuir **21**(13), 6113–6122 (2005)
55. Patel, P.A., Jeon, J., Mather, P.T., Dobrynin, A.V.:   Molecular dynamics simulations of multilayer polyelectrolyte films: effect of electrostatic and short-range interactions. Langmuir **22**(24), 9994–10002 (2006)

56. Picart, C., Mutterer, J., Richert, L., Luo, Y., Prestwich, G.D., Schaaf, P., Voegel, J.-C., Lavalle, P.: Molecular basis for the explanation of the exponential growth of polyelectrolyte multilayers. Proc. Natl. Acad. Sci. USA **99**(20), 12531–12535 (2002)
57. Qiao, B., Cerdà, J.J., Holm, C.: Poly(styrenesulfonate)-poly(diallyldimethylammonium) mixtures: toward the understanding of polyelectrolyte complexes and multilayers via atomistic simulations. Macromolecules **43**(18), 7828–7838 (2010)
58. Qiao, B., Cerdà, J.J., Holm, C.: Atomistic study of surface effects on polyelectrolyte adsorption: case study of a poly(styrenesulfonate) monolayer. Macromolecules **44**(6), 1707–1718 (2011)
59. Qiao, B., Sega, M., Holm, C.: An atomistic study of a poly(styrene sulfonate)/poly(diallyldimethylammonium) bilayer: the role of surface properties and charge reversal. Phys. Chem. Chem. Phys. **13**(36), 16336–16342 (2011)
60. Qiao, B.-F., Sega, M., Holm, C.: Properties of water in the interfacial region of a polyelectrolyte bilayer adsorbed onto a substrate studied by computer simulations. Phys. Chem. Chem. Phys. **14**, 11425–11432 (2012)
61. Reddy, G., Yethiraj, A.: Solvent effects in polyelectrolyte adsorption: computer simulations with explicit and implicit solvent. J. Chem. Phys. **132**(7), 074903 (2010)
62. Reith, D., Pütz, M., Müller-Plathe, F.: Deriving effective mesoscale potentials from atomistic simulations. J. Comput. Chem. **24**(13), 1624–1636 (2003)
63. Riegler, H., Essler, F.: Polyelectrolytes. 2. intrinsic or extrinsic charge compensation? quantitative charge analysis of PAH/PSS multilayers. Langmuir **18**(17), 6694–6698 (2002)
64. Roiter, Y., Trotsenko, O., Tokarev, V., Minko, S.: Single molecule experiments visualizing adsorbed polyelectrolyte molecules in the full range of mono- and divalent counterion concentrations. J. Am. Chem. Soc. **132**(39), 13660–13662 (2010)
65. Salomäki, M., Kankare, J.: Specific anion effect in swelling of polyelectrolyte multilayers. Macromolecules **41**(12), 4423–4428 (2008)
66. Salomaki, M., Laiho, T., Kankare, J.: Counteranion-controlled properties of polyelectrolyte multilayers. Macromolecules **37**(25), 9585–9590 (2004)
67. Schlenoff, J.B.: Retrospective on the future of polyelectrolyte multilayers. Langmuir **25**(24), 14007–14010 (2009)
68. Schlenoff, J.B., Dubas, S.T., Farhat, T.: Sprayed polyelectrolyte multilayers. Langmuir **16**(26), 9968–9969 (2000)
69. Schoeler, B., Poptoshev, E., Caruso, F.: Growth of multilayer films of fixed and variable charge density polyelectrolytes: effect of mutual charge and secondary interactions. Macromolecules **36**(14), 5258–5264 (2003)
70. Schönhoff, M.: Self-assembled polyelectrolyte multilayers. Curr. Opin. Colloid Interface Sci. **8**, 86–95 (2003)
71. Schönhoff, M., Ball, V., Bausch, A.R., Dejugnat, C., Delorme, N., Glinel, K., von Klitzing, R., Steitz, R.: Hydration and internal properties of polyelectrolyte multilayers. Colloids Surf. A **303**(1–2), 14–29 (2007)
72. Shin, Y., Roberts, J.E., Santore, M.M.: The relationship between polymer/substrate charge density and charge overcompensation by adsorbed polyelectrolyte layers. J. Colloid Interface Sci. **247**(1), 220–30 (2002)
73. Smiatek, J., Sega, M., Holm, C., Schiller, U.D., Schmid, F.: Mesoscopic simulations of the counterion-induced electro-osmotic flow: a comparative study. J. Chem. Phys. **130**(24), 244702 (2009)
74. Tanchak, O.M., Yager, K.G., Fritzsche, H., Harroun, T., Katsaras, J., Barrett, C.J.: Ion distribution in multilayers of weak polyelectrolytes: a neutron reflectometry study. J. Chem. Phys. **129**(8), 084901 (2008)
75. Thierry, B., Winnik, F.M., Merhi, Y., Tabrizian, M.: Nanocoatings onto arteries via layer-by-layer deposition: toward the in vivo repair of damaged blood vessels. J. Am. Chem. Soc. **125**(25), 7494–5 (2003)
76. Trau, D., Renneberg, R.: Encapsulation of glucose oxidase microparticles within a nanoscale layer-by-layer film: immobilization and biosensor applications. Biosens. Bioelectron. **18**(12), 1491–1499 (2003)

77. Tschöp, W., Kremer, K., Batoulis, J., Bürger, T., Hahn, O.: Simulation of polymer melts. ii. from coarse-grained models back to atomistic description. Acta Polymer. **49**, 75–79 (1998)
78. von Klitzing, R., Wong, J.E., Jäger, W., Steitz, R.: Short range interactions in polyelectrolyte multilayers. Curr. Opin. Colloid Interface Sci. **9**(1–2), 158–162 (2004)
79. Yeh, I.-C., Berkowitz, M.L.: Ewald summation for systems with slab geometry. J. Chem. Phys. **111**(7), 3155–3162 (1999)

Mechanochemistry of Cyclopropane Ring-Opening Reactions

Miriam Wollenhaupt, Martin Zoloff, and Dominik Marx

Abstract Since the end of the 1960th, the Woodward–Hoffmann rules have been a well-grounded and powerful tool to understand and predict pericyclic reactions. Recently, astonishing results on such reactions subject to mechanochemical activation by external forces have revealed reaction pathways at sufficiently large forces which are not expected from the Woodward–Hoffmann rules. These findings have started a controversy whether the "Woodward–Hoffmann rules are broken in mechanochemistry". Our study of ring opening of cyclopropane shows that the electronic structure underlying the dis- and conrotatory pathways, which are greatly distorted upon applying forces to an extent that eventually the "thermally forbidden" process becomes "mechanochemically allowed", does not change. It is rather the mechanical work that lowers the activation barrier and therefore promotes reaction pathways to products not expected from the Woodward–Hoffmann rules. A front cover and an additional cover profile article have been devoted to these findings (Wollenhaupt et al., Chem Phys Chem 16:1593–1597, 2015)

1 Scientific Background

The conventional way of starting chemical reactions—for example in an industrial process but also in a flask in the laboratory—is to use thermal, photochemical, or electrochemical activation [15]. These techniques have been applied for hundreds of years; taking for example the procedure of cooking, the effect of adding heat was efficiently used even though no one knew what was happening on a molecular level. Nowadays recently developed methods such as atomic force microscopy (AFM) [9] or sonochemical processes in an ultrasonic bath [10] provide the possibility of applying mechanical forces to single molecules while monitoring the stress and the resulting change in the molecules' structure. Therefore, it is now well appreciated that external mechanical forces can be applied to specific covalent bonds within molecular systems, defining the field of covalent mechanochemistry.

M. Wollenhaupt (✉) • M. Zoloff • D. Marx
Lehrstuhl für Theoretische Chemie, Ruhr-Universität Bochum, 44870 Bochum, Germany
e-mail: Miriam.Wollenhaupt@theochem.rub.de; Martin.Zoloff@theochem.rub.de;
Dominik.Marx@theochem.rub.de

© Springer International Publishing Switzerland 2016   229
W.E. Nagel et al. (eds.), *High Performance Computing in Science
and Engineering '15*, DOI 10.1007/978-3-319-24633-8_15

In the field of theoretical chemistry, codes for calculating geometries, energies, and even reaction pathways as a function of constant force rather than upon imposing structural constraints (i.e. isotensional versus isometric stretching, see Ref. [15] for a comprehensive discussion) have been implemented and successfully used. All this research is related to the field of (covalent) mechanochemistry, which deals with the influence of external mechanical forces on molecules and, in particular, on their reactions, see Refs. [2, 3, 15] for reviews.

As shown in Fig. 1, under mechanical forces, gem-dichlorocyclopropanes incorporated along the backbone of cis-polybutadiene exhibit an electrocyclic ring-opening accompanied with a chlorine migration [13]. Polymers with an equal distribution of cis- and trans-gDDC show an unexpected reaction probability.

Theoretical research on cis- and trans-1,1-dichloro-2,3-dimethylcyclopropane, which can be regarded as model systems to gDDC, has revealed some striking features, such as a lack of selectivity at high forces or unexpected branching ratios [4]. Figure 2 depicts the reaction sequences of this molecule together with the transitions among the various species. 1,1-dichloro-2,3-dimethylcyclopropane offers the opportunity to open its ring in a disrotatory and in a conrotatory way

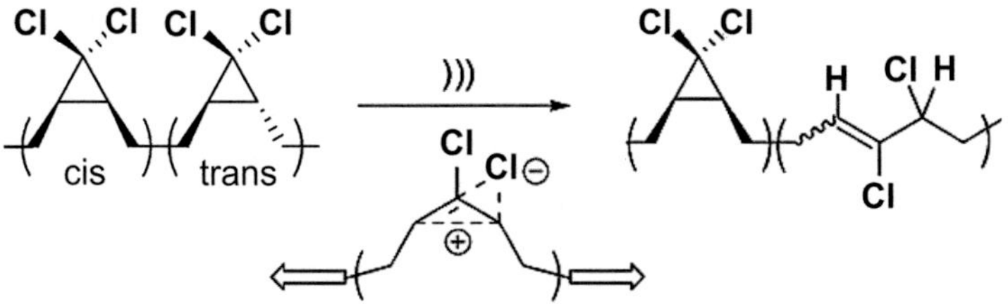

Fig. 1 Ring-opening reaction of cis- and trans-gem-dichlorocyclopropanes under sonication. The ultrasound produces collapsing bubbles that cause shear forces at the ends of the polymer chains. The chlorine migration is proposed to take place through a transition with a negatively charges chlorine ion and a positive charge remaining on the carbon ring. Parts of this picture are adapted from Ref. [13]

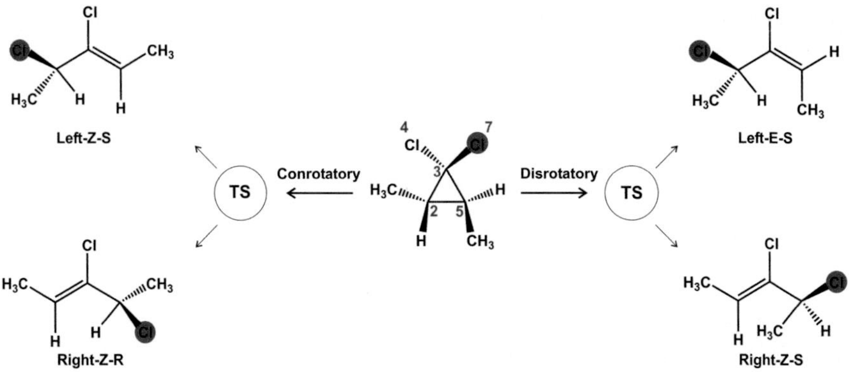

Fig. 2 Scheme of the reaction progress and the stereochemistry of the resulting products

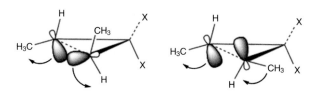

Fig. 3 Ring-opening mechanisms according to Woodward-Hoffmann rules. *Left*: Disrotatory. *Right*: Conrotatory

(see Fig. 3). The Woodward-Hoffmann rules [20, 21] predict the thermal opening of a 2π-electron system to proceed in a disrotatory fashion.

Starting in 1965, Woodward and Hoffmann established their unifying theory on the mechanisms of pericyclic reactions [20]. Building upon basic concepts of molecular orbital theory, they have derived powerful rules to classify them into symmetry-allowed and symmetry-forbidden reactions [11, 21]. Woodward and Hoffmann have been able to connect electronic structure with relative energetics for certain classes of chemical reactions in a qualitatively predictive manner. The famous Woodward-Hoffmann rules cover the field of thermal and photochemical reactions.

Recently, experimental [10] as well as theoretical [3, 5, 6, 12, 14, 16–18] studies have been performed on pericyclic reactions in combination with mechano-chemical activation. These investigations have featured results not expected from the Woodward-Hoffmann rules and have started a controversy about the "validity of the Woodward-Hoffmann rules". The key question is whether the application of an external force to a molecule qualitatively changes the electronic structure that underlies the novel reaction pathway, or if it simply changes relative energies due to the additional mechanical force term which generates mechanical work, and thus excess (potential) energy, as the reaction proceeds along the reaction coordinate.

The answer to this question might appear obvious at first sight: since the external force term does not enter the electronic Hamiltonian, there is no change of the electronic structure to be expected—thus rendering any further investigation obsolete. Yet, it is not as simple as that since external forces certainly do distort the energy landscape and thus reaction mechanisms to an extent that is comparable to what distinguishes Woodward-Hoffmann allowed from forbidden thermal reactions in the first place.

2   Results and Discussion

In this study, we have conducted geometry optimizations and calculations of intrinsic reaction coordinates (IRCs) within our isotensional formalism, [15, 16] i.e. we have explicitly applied constant collinear forces of opposite directions to the terminal carbon atoms of trans-1,1-dichloro-2,3-dimethylcyclopropane. Taking the transition state (TS) of the symmetry-allowed disrotatory ring opening at zero force, i.e. upon thermal activation, as a starting point, we have obtained a series of force-transformed TSs by successively increasing the force. At forces exceeding 1.6 nN, the mechanism is found to switch from dis- to conrotatory ring opening, which has been followed up to 3.0 nN and down to zero force [19].

The total energy along the IRCs of the dis- and conrotatory ring opening processes in the thermal limit case and at six representative forces is shown in Fig. 4. The IRCs are parameterized in terms of the Root Square Displacement (RSD) where, starting at the TS (where RSD $= 0$ Å), the RSD of a specific structure of the

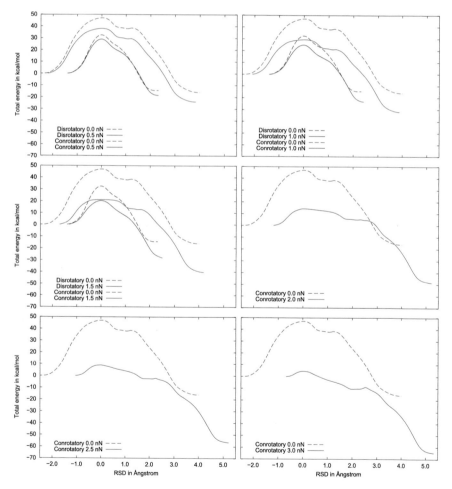

Fig. 4 Total isotensional energies including the work term along the disrotatory (*blue lines*) and conrotatory (*red lines*) ring opening pathways of trans-1,1-dichloro-2,3-dimethylcyclopropane relative to the respective reactant energies according to the IRCs optimized in the thermal limit (*dashed lines*) and at six representative forces (*solid lines*)

IRC is calculated as the accumulation of the structural displacements of the heavy atoms between adjacent structures along the IRC up to this point. Positive (negative) values indicate evolution toward the product (reactant).

The Woodward-Hoffmann allowed and forbidden thermal pathways corresponding to dis- and conrotatory ring openings, respectively, feature the following difference: The TS of the symmetry-forbidden conrotatory mechanism is higher in energy and the corresponding pathway is therefore disfavored. Comparing the thermal scenario with the IRCs at finite forces of each mechanism shows no fundamental changes in the shape of the underlying potential energy surface due to the application of an external force, except for a significant lowering of the activation energies (see Fig. 5) [19]. Activation energies have been calculated also at finite forces as usual, i.e. as the difference between energies of transition state and reactant.

These findings raise the question: Does the underlying electronic structure itself change significantly? To answer this question, we have analyzed the electronic structure along the IRCs as a function of constant external force by using the concepts of "atoms in molecules" (AIM) and "natural resonance theory" (NRT).

Within the AIM framework [1] of equilibrium structures, the electron density along the line (so called bond path) connecting two bonded atoms possesses a minimum at the bond critical point (BCP). The existence of a BCP correlates with the presence of a bond and the magnitude of the former is a measure of the strength of the latter. The direct comparison of the electron densities for the inspected bonds (see colored sketch in Fig. 6) in the thermal limit and at finite forces within one type of mechanism (either thermally Woodward-Hoffmann allowed or forbidden) shows

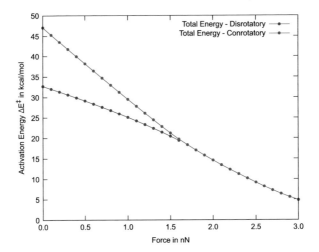

Fig. 5 Total isotensional activation energies including the work term of the disrotatory (*blue*) and conrotatory (*red*) ring opening reactions of trans-1,1-dichloro-2,3-dimethylcyclopropane as a function of the external mechanical force

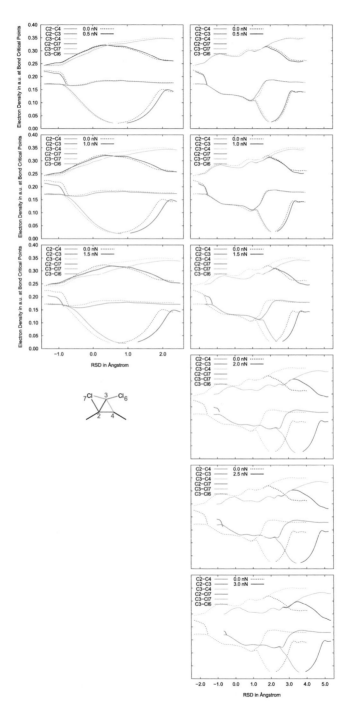

Fig. 6 Evolution of the electron density in atomic units at the line critical points of selected bonds as indicated along the disrotatory (*left*) and conrotatory (*right*) reaction pathways given by the respective IRCs (see Fig. 4) at zero (*dashed lines*) and finite forces (*solid lines*)

a very similar evolution along the IRCs all the way from reactants to products, which is evident from the close proximity of the dashed and corresponding solid lines. The topological AIM bonding analysis therefore shows no significant changes of the electronic structure of the disrotatory and the conrotatory pathways with increasing force [19].

As a complementary approach to the before mentioned topological analysis by AIM, we have employed the Lewis structure based NRT method. NRT analysis provides a localizing scheme to describe complicated wave functions in terms of a set of Lewis valence structures [7, 8]. We identified 39 important NRT resonance structures, which we have been able to condense into six main classes (see Fig. 7) [19].

The evolution of the contributions of the various NRT classes to the total wavefunction in terms of relative weights along the reaction pathway of disrotatory (left) and conrotatory (right) ring opening are depicted in Fig. 8 in the thermal limit (top) and at finite forces. Close to the reactant state, the wavefunction is mainly described by the reactant structures (orange), as expected. On the way to the TS, the contribution of the reactant structure decreases and the chlorine-detached structures (blue) start to play a bigger role. Around the disrotatory TSs, chlorine-migration (green) contributes significantly, whereas ring-opening structures (brown) come into play close to the conrotatory TSs instead. Upon moving toward the products, the relative weight of the product structures (red) increases until it eventually dominates. Very importantly, although the conrotatory process, which is the Woodward-Hoffmann forbidden thermal reaction, becomes the mechanochemically allowed process beyond a critical force of approximately 1.6 nN, it is characterized by the same distribution of NRT weights along the IRCs all the way from zero up to 3 nN, see Fig. 8 [19].

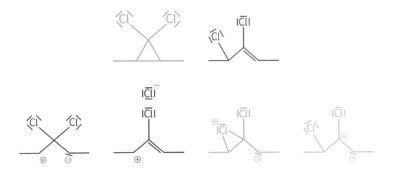

Fig. 7 The six classes of resonance structures used for NRT decomposition. They describe the course of the reaction: reactant structures (*orange*) and product structures (*red*), as well as ring-opening structures (*brown*), chlorine-detached structures (*blue*), chlorine-migration (*green*) and chlorine-reattached structures (*cyan*)

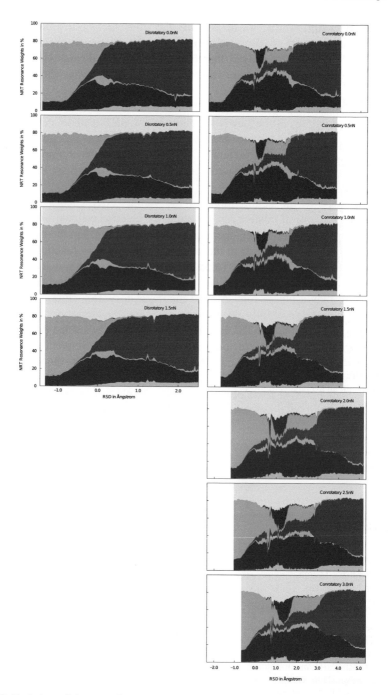

Fig. 8 Evolution of the natural resonance weights in percent along the disrotatory (*left*) and conrotatory (*right*) reaction pathways at zero and finite forces. The six classes of resonance structures used for NRT decomposition are sketched in Fig. 7 and define the color code of the histograms

3 Conclusions

In summary, we have analysed the Woodward-Hoffmann allowed and forbidden reaction pathways of trans-1,1-dichloro-2,3-dimethylcyclopropane as a function of an externally applied force. By using AIM and NRT methods, we are able to show that the electronic structure does not change significantly along both reaction pathways which are systematically distorted as a result of applying the mechanical force. It is rather the mechanical work term which distorts the potential energy surface and, therefore, changes the corresponding activation energies to eventually energetically favor the product that is not expected from the Woodward-Hoffmann rules [19]. The fundamental consequence is that the Woodward-Hoffmann rules, which are based on electronic structure considerations of certain classes of thermally and photochemically activated reactions, should not be considered to be transferable to the same reactions once mechanochemically activated. Instead, one needs to think of mechanochemical reactions to evolve on what we call "force-transformed potential energy surfaces", which is a concept that the Marx group introduced several years back [16] and put in a broad context [15] more recently.

Acknowledgements Partial financial support is provided by the DFG Koselleck Grant "Understanding Mechanochemistry" to Dominik Marx We wish to thank Przemyslaw Dopieralski and Martin Krupička for their contributions to this work. Computer resources have kindly been provided by HLRS Stuttgart (account ID 12982). Calculations were run using the distributed memory (MPI) parallelization scheme. In most runs 192 cores (6 nodes in the Hornet system) were used, with an average wall time of about 5.8 h (equivalent to ca. 1100 core-hours per run). A few runs were performed using 128 cores (4 nodes in Hornet) with an average wall time of about 9.1 h (ca. 1100 core-hours per run), whereas another group of runs were done using 256 cores (8 nodes in Hornet) with an average wall time of about 5.3 h (ca. 1300 core-hours). The requirement of RAM for each of these runs was on average 15 GB of total memory (2.5 GB per node for most runs). A typical run required approximately 1.5 GB of disk space for permanent data storage and an additional 4.5 GB for scratch data.

References

1. Bader, R.F.W.: Atoms in Molecules: A Quantum Theory. International Series of Monographs on Chemistry. Clarendon Press, Oxford (1990)
2. Beyer, M.K., Clausen-Schaumann, H.: Chem. Rev. **105**, 2921–2948 (2005)
3. Caruso, M.M., Davis, D.A., Shen, Q., Odom, S.A., Sottos, N.R., White, S.R., Moore, J.S.: Chem. Rev. **109**, 5755–5798 (2009)
4. Dopieralski, P., Ribas-Arino, J., Marx, D.: Angew. Chem. Int. Ed. **50**, 7105–7108 (2011)
5. Dopieralski, P., Anjukandi, P., Rückert, M., Shiga, M., Ribas-Arino, J., Marx, D.: J. Mater. Chem. **21**, 8309–8316 (2011)
6. Friedrichs, J., Lüßmann, M., Frank, I.: Chem. Phys. Chem. **11**, 3339–3342 (2010)
7. Glendening, E.D., Weinhold, F.: J. Comput. Chem. **19**, 593–609 (1998)
8. Glendening, E.D., Badenhoop, J.K., Weinhold, F.: J. Comput. Chem. **19**, 628–646 (1998)
9. Grandbois, M., Beyer, M., Rief, M., Clausen-Schaumann, H., Gaub, H.E.: Science **283**, 1727–1730 (1999)

10. Hickenboth, C.R., Moore, J.S., White, S.R., Sottos, N.R., Baudry, J., Wilson, S.R.: Nature **446**, 423–427 (2007)
11. Hoffmann, R., Woodward, R.B.: Acc. Chem. Res. **1**, 17–22 (1968)
12. Kochhar, G.S., Bailey, A., Mosey, N.J.: Angew. Chem. Int. Ed. **49**, 7452–7455 (2010)
13. Lenhardt, J.M., Black, A.L., Craig, S.L.: J. Am. Chem. Soc. **131**, 10818–10819 (2009)
14. Ong, M.T., Leiding, J., Tao, H., Virshup, A.M., Martínez, T.J.: J. Am. Chem. Soc. **131**, 6377–6379 (2009)
15. Ribas-Arino, J., Marx, D.: Chem. Rev. **112**, 5412–5487 (2012)
16. Ribas-Arino, J., Shiga, M., Marx, D.: Angew. Chem. **121**, 4254–4257 (2009)
17. Ribas-Arino, J., Shiga, M., Marx, D.: Chem. Eur. J. **15**, 13331–13335 (2009)
18. Ribas-Arino, J., Shiga, M., Marx, D.: J. Am. Chem. Soc. **132**, 10609–10614 (2010)
19. Wollenhaupt, M., Krupička, M., Marx, D.: Chem. Phys. Chem **16**, 1593–1597 (2015)
20. Woodward, R.B., Hoffmann, R.: J. Am. Chem. Soc. **87**, 395–397 (1965)
21. Woodward, R.B., Hoffmann, R.: Angew. Chem. Int. Ed. Engl. **8**, 781–853 (1969)

Part III
Bioinformatics

Willi Jäger

Modelling and simulations are important tools in linking processes on the level of atoms and molecules to the dynamics and structure of molecular complexes and building blocks for biological cells and organisms. Using the concepts and methods of molecular dynamics, computational biosciences as computational chemical and material sciences can determine how structures and functions are developing and acting in various environments. The approach used is only partially based on quantum-mechanical computations, but is mostly starting from a level of larger scale, since ab initio computations so far cannot handle the arising huge systems.

High-performance supercomputing is the key technology for molecular dynamics simulations to solve the challenging problems. In this volume the reports of the following two selected projects are presented:

1. *Computational modelling of a biocatalyst at a hydrophobic substrate interface*
   (Sven Benson, Jürgen Pleiss)
2. *Dynathor: Dynamics of the Complex of Cytochrome P450 and Cytochrome P450 Reductase in a Phospholipid Bilayer*
   (Xiaofeng Yu, Daria Kokh, Prajwal Nandekar, Ghulam Mustafa, Stefan Richter and Rebecca C. Wade)

Both contributions are dealing with molecular modelling and simulation of enzyme activities various environments, in particular at interfaces exercising a decisive influence on the processes. They are prototypical for a wider class of problems, which can only be tackled using advanced multiprocessor systems.

The first project is exploring the effects of solvents to protein structure and dynamics and their influence on enzyme stability, solubility, catalytic activity, substrate specificity and selectivity. A new approach to investigate the influx of

W. Jäger (✉)
Interdisziplinäres Zentrum für Wissenschaftliches Rechnen (IWR), Universität Heidelberg, Im Neuenheimer Feld 368, 69120 Heidelberg, Germany
e-mail: jaeger@iwr.uni-heidelberg.de

solvent molecules through the protein structure and into the active site cavity is proposed and implemented. In a case study the enzyme Candida Antarctica lipase B and the solvent water are investigated.

The second project developed a model and computational methods to tread the interaction of cytochrome P450 enzymes and their redox partner cytochrome P450 reductase in a phospholipid bilayer. Simulation studies on the HLRS CRAY systems contributed to a better quantitative understanding of the activity of these enzymes, which play important roles in drug metabolism, steroid biosynthesis and xenobiotic degradation. In humans, P450s are essential for the metabolism of 70–80 % of all drugs.

The presented evaluation of methods of random acceleration molecular dynamics is a methodical contribution of general interest.

The results obtained in these presented projects are major steps on the way to a deeper understanding of basic biochemical processes and to improvements in drug design and biotechnology.

Computational Modeling of a Biocatalyst at a Hydrophobic Substrate Interface

Sven Benson and Jürgen Pleiss

Abstract Solvent molecules play a crucial role in the function of proteins. The solvent flux method (SFM) was developed to comprehensively characterize the influx of solvent molecules from the solvent environment into the active site of a protein by molecular dynamics simulations. This was achieved by introducing a solvent concentration gradient and by reorienting and rescaling the velocity vector of all solvent molecules contained within a spherical volume enclosing the protein, thus inducing an accelerated solvent influx toward the active site. In addition to the detection of solvent access pathway within the protein structure, it is hereby possible to identify potential amino acid positions relevant to solvent-related enzyme engineering with high statistical significance. The method is particularly aimed at improving the reverse hydrolysis reaction rates in nonaqueous media. Candida antarctica lipase B (CALB) binds to a triglyceride-water interface with its substrate entrance channel oriented toward the hydrophobic substrate interface. The lipase-triglyceride-water system served as a model system for SFM to evaluate the influx of water molecules to the active site. As a proof of principle for SFM, a previously known water access pathway in CALB was identified as the primary water channel. In addition, a secondary water channel and two pathways for water access which contribute to water leakage between the protein and the triglyceride-water interface were identified.

1   Introduction

The first pioneering studies of an esterase in a non-aqueous solvent were performed 75 years ago, but it was not until the 1980s that the synthetic opportunities of enzymes in organic solvents were realized. Multiple studies confirmed that the catalytic activity sensitively depends on the water content. Water content and solvent polarity were identified as the major determinants of catalytic activity and stereoselectivity. It was suggested that the effect of organic solvents on an

S. Benson · J. Pleiss (✉)
Institute of Technical Biochemistry, University of Stuttgart, Allmandring 31, 70569 Stuttgart, Germany
e-mail: Juergen.Pleiss@itb.uni-stuttgart.de

Originally published in Journal of Chemical Theory and Computation
©American Chemical Society 2014. Reprint with permission by Springer
International Publishing Switzerland 2015, DOI 10.1007/978-3-319-24633-8_16

enzyme is primarily caused by the interactions between the solvent and the enzyme-bound water rather than with the enzyme itself, and that a low amount of water is required to provide sufficient conformational flexibility to the enzyme. To explore the molecular basis of enzyme stability, solubility, catalytic activity, substrate specificity, and selectivity, the effect of solvents to protein structure and dynamics was investigated by computational modelling. A deep understanding of protein packing, the structure of the active site, and the contribution of protein motions to function is prerequisite to a successful design of enzymes, the prediction of mutants, and the choice of appropriate solvents.

2   Enzymes at Substrate Interfaces

Water molecules are directly involved in many enzymatic reactions such as hydrolysis reactions catalyzed by esterase and lipases. While this natural degradative enzyme function is of substantial interest to a range of industrial applications, significant strides have been made during the last decades to shift the chemical equilibrium towards the reverse synthesis reaction by transferring hydrolases into non-natural organic solvent environments. The presence of residual hydration water is a prerequisite for enzyme activity in non-aqueous environments. Moreover, defining and maintaining dry conditions is difficult as well as costly to achieve on the industrial scale, which motivated Larsen [13] to propose an approach to suppress water as a nucleophile in CALB by rationally designing variants that diminish the influx of water to the active site cavity and thus dramatically increase transacylation rates for vinyl butyrate over hydrolysis rates in butanol-water mixtures. CALB catalyzes hydrolysis of triglycerides in its native physiological setting; the natural substrate triglyceride is water-insoluble and forms phase-separated interfaces in aqueous environments and thus CALB attaches to these interfaces in its active conformation, which is hereby oriented with its substrate channel towards the triglyceride phase [6]. The most obvious water access pathway is thus considered to be blocked by the triglyceride which forms a hydrophobic barrier. However, a constant influx of water to the active site cavity is a prerequisite for hydrolysis to occur. Therefore, Larsen [13] proposed the existence of a water channel, which was targeted for mutation and implicitly validated by experiment. Due to the difficulty of examining individual water molecules by experimental means, molecular dynamics (MD) studies have been conducted extensively to elucidate enzyme-solvent interactions under various conditions, which are comprehensively summarized in a recent review [14]. By analyzing hydration water and its influence on protein secondary structure flexibility, it may also be possible to indirectly identify water entrance channels [4]. In the present study, we propose an MD-based approach, the solvent flux method (SFM), which allows for a holistic characterization of the influx of solvent molecules through the protein structure and into the active site cavity. In this context, SFM was benchmarked for the influx of water molecules to the active site of CALB by comparing in silico predictions to the experimental results by Larsen [13].

3 Simulation Details

More than $1\mu s$ of MD simulations were conducted on a system consisting of 100,000 atoms using the Hermit cluster, a Cray XE6 system of the High Performance Computing Center (HLRS) in Stuttgart, Germany. Over 300 simulations, including benchmark and test runs, were performed on 256 cores in parallel. The GROMACS 4.5.3 [9, 22] software was used to model an NPT ensemble at 298.15 K and $1bar$, applying the leap-frog integrator [23] with a time step of $2fs$. The Berendsen thermo- and barostats were applied for their robustness and efficiency during equilibration. This was necessary since the SFM method introduces significant periodic rescaling and reorientation of water molecule velocity vectors close to the protein and thus drives the systems out of equilibrium. Water molecules with adjusted velocity vectors were loosely coupled with a relaxation time of 50 ps to prolong the velocity rescaling effect, while the remaining bulk water molecules, the triglyceride molecules, and the protein were coupled tightly in intervals of 0.1 ps. To ensure that the structural integrity of the system remained intact during simulation and to avoid protein denaturation or distortions in the protein-triglyceride interface, position restraints with force constants of $1000\,kJmol^{-1}nm^{-2}$ were applied to the protein backbone and the triglyceride molecules in all spatial dimensions. To avoid artifacts due to position restraints under NPT conditions, the reference coordinates of each molecule group's center of mass were rescaled periodically. Pressure coupling was applied separately to the x-y plane and in z-dimension for a more accurate pressure representation of the planar triglyceride layer [17] in the model system. Pressure coupling intervals were set to 100 ps for the adjusted water molecules and to 5 ps for the protein, triglyceride and bulk water molecules. The center of mass movements were removed every 100 simulation steps independently for all system components. Periodic boundary conditions were applied in all three dimensions. Hydrogen bonds were constrained with the LINCS algorithm [18]. Long-range electrostatics were calculated with the particle mesh Ewald method [5]. Lennard-Jones interactions were treated with a cutoff and capped at 1.2 nm. The OPLS-AA all-atom force field [11] was used to parameterize CALB. Structural information of CALB was obtained from the Protein Data Bank (PDB: 1TCA [21]). Triglyceride molecules were parameterized with the Berger lipid model [3] and the 1–4 interactions adapted with the half-ϵ double-pairlist method (Neale and Pomès, 2011, Combination rules for united-atom lipids and oplsaa proteins. http://www.pomeslab.com/files/lipidCombinationRules.pdf. Accessed 2 March 2011, Unpublished document) for consistency with the OPLS-AA force field. Water molecules were parameterized with the TIP3P [10] model.

4   Modeling of Candida antarctica Lipase B Attached to a Triglyceride Layer

A model system was created to mimic the conditions of the functionally active conformation of CALB at the triglyceride-water interface. As a prerequisite step, a random molecular arrangement of caproic (C6) triglyceride molecules and water molecules had to be equilibrated into a phase-separated triglyceride layer, in accordance to the approach proposed by Gruber [6]. The resulting triglyceride-water interface hereby represents a model for the surface of large-scale aggregates in triglyceride in water emulsions (Fig. 1). This model was deemed appropriate considering that the diameter of CALB ($\approx$5 nm) is three orders of magnitude smaller than experimentally determined triglyceride droplet diameters ($\approx$2.0 μm) [12]; hence the droplet surface curvature is expected to be negligible on the model scale. Furthermore, when considering that the Coulomb potential decays with $1/r$, a layer thickness of $\approx$7 nm should be sizeable enough to ensure that the most significant Coulomb contributions to the non-bonded potential between the protein and the triglyceride surface are accounted for. Moreover, the lamellar-like nanostructures observed for the triglyceride layer model are in agreement with previously published results on large-scale triglyceride aggregates [19] and our findings reported in a previous study [1]. The potential implication of these structural features on the mobility of water molecules are subject to a detailed follow-up study and will therefore not be elaborated on in this context. CALB was added to the triglyceride-water system and was attached to the layer with the active site entrance pointing towards the triglyceride phase, thus representing a model for the active binding

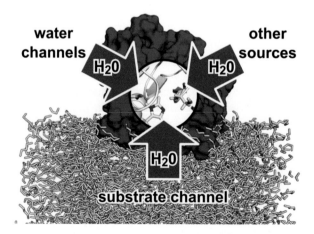

Fig. 1 The protein structure of Candida antarctica Lipase B (PDB: 1TCA) was adsorbed at a surface model of its natural substrate triglyceride (caproic (C6) fatty acid chain length). During equilibration, the protein immersed into the substrate layer, its main entrance being covered by the hydrophobic triglyceride molecule phase. This system served as the basis to model the water influx from potential sources. Reprinted with permission, copyright 2014 American Chemical Society[2]

configuration for triglyceride hydrolysis. The system was equilibrated for a further 500 ns, while monitoring protein adhesion and immersion depth (unpublished data).

5 Modeling the Water Influx into the Active Site of Candida antarctica Lipase B at a Triglyceride-Water Interface

The design and benchmarking of SFM was tightly coupled to the test case system of protein CALB attached to the triglyceride-water interface and using water as the influx solvent. Two major technical considerations were formative for SFM during the design phase. Firstly, it was considered necessary to differentiate solvent molecules that pass through a water access pathway into the active site cavity from other bulk solvent molecules. This was achieved in a recursive manner, by exploiting the fact that solvent molecules that are present within the active site cavity necessarily have to have passed through a water access pathway of the protein at a preceding point in time. It was thus possible to reconstruct solvent molecule pathways through the protein and therefore to detect general access pathways on a statistical basis. To identify such water molecules and to ensure that they had indeed passed into the active site cavity from the exterior, the water molecule closest to the reference atom at the intended influx site (here: oxygen atom of Ser105 of the catalytic triad of CALB) and within a spherical water removal cutoff of 0.35 nm was periodically removed from the simulation system after $\Delta t_{ITER} = 10$ ps (Fig. 2). Not only did this ensure that after a given number of iterations all solvent molecules previously present within the active site cavity were evacuated, it also introduced an increased influx of solvent molecules due to the solvent concentration gradient between the active site of the protein and the exterior solvent bulk. Moreover, the removal of water served as a mechanistic model for water consumption during the native CALB hydrolysis reaction. The induced solvent influx ties in with the second major technical consideration, which is the acceleration of conventional MD to rapidly overcome rate-limiting energy barriers specific to water influx and thus realize a significant data depth in a feasible duration of real-time. This was achieved by introducing a reorientation and rescaling of velocity vectors of water molecules that were located within a water velocity rescaling cutoff of 3 nm surrounding the active site reference, which was adjusted to incorporate the entirety of the protein and its first solvation shell. Within this cutoff, the velocity vectors of water molecules were periodically readjusted at constant time intervals Δt_{ITER}, simultaneously to the removal of water molecules. Velocity magnitudes of water molecules present within this cutoff were rescaled with a constant factor v_{MULT}, which effectively defines the strength of a periodic velocity pulse that is introduced to overcome the influx-rate-limiting energy barriers in a timely manner. The orientation of the new vectors was defined by assigning a fraction v_{OLD} of the respectively rescaled vector magnitudes to the unit vectors in the original direction, while a larger fraction $v_{CENT}=100\%-v_{OLD}$ was assigned to unit vectors pointing towards the active site reference (Fig. 2). Summation of these vector pairs yielded

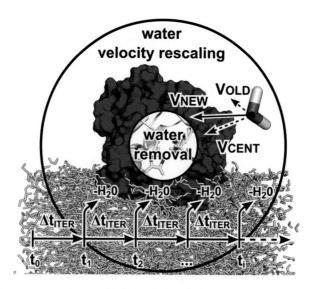

Fig. 2 Velocity vector magnitudes of water molecules found within the water velocity rescaling cutoff around an active site reference atom are rescaled by a factor of vMULT and then transformed to v_{NEW} after every iteration Δt_{ITER}. After every iteration Δt_{ITER}, a single water molecule is removed within the water removal cutoff radius. The direction of v_{NEW} is defined by X% of the rescaled magnitudes that are allocated to the original vector directions (v_{OLD}) and by Y% of the rescaled magnitudes that are allocated to the direction of the active site reference atom (v_{CENT}), whereby $X+Y = 100$ and $Y \gg X$. Reprinted with permission, copyright 2014 American Chemical Society[2]

the new velocity vectors v_{NEW}, which were used to update the initial conditions at the beginning of an iteration. Thereafter, temperatures were allowed to converge back towards equilibrium during Δt_{ITER}. It proved necessary to additionally fine-tune the duration of this effect by modulating the temperature coupling parameter τ_{ITER}, thus coupling the velocity-adjusted water molecules more loosely than other system components. The process of single water molecule removal, velocity readjustment and simulation continuation was repeated for a predefined number of iteration steps n_{ITER}. With ΔN_{WAT} denoting the sum of all water molecules removed during a full SFM run, it was thus possible to express the total water removal rate at the active site reference as $\Delta N_{WAT}/n_{ITER}$, $\Delta N_{WAT}/n_{ITER}=1$ signifying water removal at every iteration.

6 Applying the Solvent Flux Method to Candida antarctica Lipase B

For the case study of the enzyme Candida antarctica lipase B (CALB) with the solvent water the aforementioned SFM parameters were adjusted in a thorough benchmarking to both optimize the solvent flux $\Delta N_{WAT}/n_{ITER} = 1$ as well as maintain a stable simulation system. The feasibility of SFM analysis was then assessed by comparing simulation results with the successful experimental study of Larsen on the same system [13], wherein enzyme variants were engineered that interfere with water access and thus increase transacylation versus hydrolysis rates. Water influx towards the active site reference oxygen atom of Ser105 of CALB was induced by SFM, due to the water concentration gradient between the protein exterior and the active site, induced by the periodic removal of water molecules and the periodic introduction of a radial velocity pulse to the water molecules in close proximity to the protein. After an iteration $n_{EVAC} = 100$, it was ensured that all water molecules that were originally present in the active site cavity had been removed and thus any water molecules that were consecutively removed in $n_i > n_{EVAC}$ necessarily had to have passed through a water access pathway of the protein to reach the active site. The pathway of any water molecule that was thus removed during an iteration ni, with $n_{EVAC} < n_i < n_{ITER} = 500$, was reconstructed by superimposing the system coordinates in a least square fit of protein backbone atoms of all prior iterations $n_j < n_i$. Moreover, the contact frequency between the removed water molecules and all amino acids was evaluated on the basis of the reconstructed water pathways by counting the number of iterations during which a water molecule was found within the contact distance $r_{HSPOT} = 0.4\,nm$ to a specific amino acid. The SFM analysis was applied to three different initial conformations of the model system (Fig. 1) which were derived from a prior $100ns$ MD simulation under conventional equilibrium conditions without external forces or water removal, whereby the starting coordinates corresponded to the minimal, the maximal and an average RMSD value of the CALB protein backbone. For each initial conformation, SFM was performed 50 times for a total of 150 independent runs, each consisting of 500 iterations of $\Delta t_{ITER} = 10$ ps, amounting to a total of $750\,ns$ of accelerated SFM analysis. Contact frequencies were extracted from this data set, normalized and merged for a holistic and statistically significant representation of probable water access pathways during influx to the active site cavity of CALB.

7 Water Access Channels and Critical Amino Acid Positions Identified in Candida antarctica Lipase B by the Solvent Flux Method

Amino acid positions corresponding to increased contact frequencies were defined as hotspot areas H1-H9 (Fig. 3). Individual hotspot positions (Fig. 4) contained in hotspot areas were defined on the basis of their contact frequency, their relative position within the CALB structure and by visualizing reconstructed water influx pathways (Fig. 5). Hotspot positions were thereby predominantly found in close proximity to the triglyceride-water interface. Positions Thr42/Ser47 of H1 and Gln106 of H3 coincide with the mutational strategy on CALB of Larsen [13], who proposed and implicitly validated a water channel at these exact positions. Because

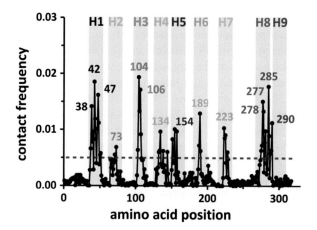

Fig. 3 Contact frequency between influx water and position-specific amino acid are presented as the number of observed water contacts relative to all evaluated system conformations of the SFM analysis (all potential contacts). Hotspot areas H1-H9 were defined as peaks greater than a contact frequency cutoff value (*red dotted line*) and were arbitrarily colored for reasons of easy distinction and reference in Figs. 4 and 5. Single positions that correspond to the highest values of the hotspot areas were labeled and defined as hotspot positions. Reprinted with permission, copyright 2014 American Chemical Society[2]

	pos.	AA	property	rel. freq.	d_act [Å]	positioning	assessment	chan
H1	38	PRO	hphob	0.71	5.7	periphery	adjacent to reported water channel	-
	42	THR	hphil	1.00	7.8	periphery	reported water channel	Larson
	47	SER	hphil	0.85	8.5	periphery	reported water channel	Larson
H2	73	LEU	hphob	0.27	10.8	surface	probable barrier, adjacent to Lys290	-
H3	104	TRP	hphob	0.96	4.8	active site	active site	-
	106	GLN	hphil	0.98	4.8	active site	reported water channel	Larson
H4	134	ASP	neg	0.54	11.5	periphery	potential secondary water entrance	S1
H5	154	VAL	hphob	0.56	12.0	periphery	entrance substrate channel	-
H6	189	ILE	hphob	0.96	9.4	surface	entrance substrate channel	-
H7	223	ASP	neg	0.65	12.7	surface	potential secondary water entrance	S2
H8	277	LEU	hphob	0.77	8.7	periphery	probable barrier, adj. to Asp223 and Ser47	
	278	LEU	hphob	0.88	6.9	periphery	entrance substrate channel	
	285	ILE	hphob	0.94	11.3	periphery	entrance substrate channel	
H9	290	LYS	pos	0.52	17.7	surface	potential secondary water entrance	S3

Fig. 4 Compiled hotspot areas H1-H9 obtained from 150 SFM runs analyzing water influx into CALB attached to a triglyceride-water interface depicted in Fig. 1. Data shown includes amino acid positions (pos.), amino acids (AA), amino acid properties in regards to charge and hydrophobicity and hydrophilicity (property), the relative contact frequency calculated by SFM (rel. freq.), the distance (d_{act}) of residue Cα carbon atoms relative to the oxygen reference atom of Ser105 at the active site, the spatial positioning of the amino acid in the protein structure (positioning), a general assessment of the individual hotspot positions (assessment), and a channel definition (channel), whereby the primary water channel (PWC) is the water channel reported by Larsen [13], the secondary water channel (SWC) a potential secondary water channel, and two interfacial water entrances (IWE1, IWE2) located at the protein-triglyceride interface that may contribute to the water influx via the substrate channel

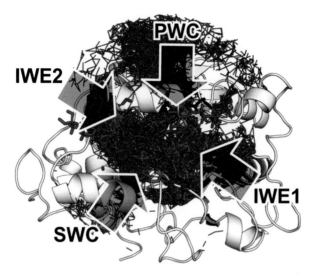

Fig. 5 Water influx pathways from five full SFM runs of 500 iterations length are superimposed onto the same depiction to illustrate potential water access pathways (*arrows*). Helices between the PWC and SWC channels and the triglyceride layer stop water "leakage" in between the protein and the interface, whereas IWE1 and IWE2 significantly contribute to the influx via the substrate channel, which is situated in the central part of the protein in both depictions. Reprinted with permission, copyright 2014 American Chemical Society[2]

it had the highest relative contact frequency (Fig. 4), we refer to this channel as the primary water channel (PWC). Three additional potential water access pathways with high relative contact frequencies were identified, the secondary water channel (SWC) at position Val134 of H4 and two interfacial water entrances IWE1 and IWE2 at position Asp223 of H7 and Lys290 of H9, respectively, which feature polar residues on the protein surface and are in immediate contact with the triglyceride interface. IWE1 and IWE2 are therefore considered to potentially contribute to the significant water influx through the substrate channel leading from the triglyceride interface to the active site (Fig. 5), which is lined by positions Val154 of H5, Ile189 of H6, and Leu278/Ile285 of H8. Other potential hotspot positions, such as Pro38 of H1, Leu73 of H2, and Leu277 of H8 were estimated to be potential barriers. Their increased contact frequency was attributed to their close proximity to a water access pathway or in the case of Trp104 of H3 to Ser105.

PWC was identified as the most frequented water access pathway (Fig. 4). Even residue Gln46, which was defined as the outer entrance in the study of Larsen [13], was resolved by SFM (Fig. 3). Water influx via PWC occurred in the loop region between $\alpha5$ and $\beta2$, according to the secondary structure numbering by Uppenberg [21]. Access to SWC occurred in the loop region between $\alpha5$ and $\beta5$, which is located in close proximity to the catalytically active Ser105 and is structurally slightly submerged relative to the protein surface, yet fully solvent accessible between $\alpha7$ and the loop region between $\beta6$ and $\beta7$. Access to entrance IWE1, which is located directly at the boundary between CALB and the triglyceride layer, occurred at the loop preceding $\alpha 9$, which was slightly denatured at its N-terminus due to the direct interaction with the triglyceride layer. IWE1 may thus contribute to the influx of water via the triglyceride layer and through the substrate channel, as water molecules were observed to "leak" in between the protein and the triglyceride layer (Fig. 5). Access to entrance IWE2 occurred in the loop region between $\beta 8$ and $\beta 9$ close to the C-terminus of CALB, which also directly borders the triglyceride-water interface and can thus potentially also contribute to the interfacial leakage.

8 The Solvent Flux Method in the Context of Hydrolase Engineering

Results from SFM suggest that the triglyceride layer, where the enzyme CALB is attached to in its catalytically active conformation, does not constitute a hydrophobic barrier that excludes water molecules from entering the enzyme via the buried substrate channel. On the contrary, significant interfacial leakage between the protein and the triglyceride interface was observed (Fig. 5) which is comparable in significance to the water influx via the known primary water channel (PWC) [13] and higher than the influx via the potential secondary water channel (SWC) identified by SFM (Fig. 4). The fact that during equilibration of the protein-triglyceride system the protein immersed into the triglyceride layer implies that the observed interfacial water leakage is not a consequence of incomplete protein adsorption in the model. The origin of the interfacial water leakage is instead accredited to the self-association structures of polar triglyceride moieties that are formed in aggregates at equilibrium [1]. Lamellar-like self-association structures might therefore serve as a polar microenvironment that can facilitate an influx of water molecules through the interface and into the active site via the substrate channel. In this regard, the polar amino acids of IWE1 and IWE2 are potential avenues for water access through the protein-triglyceride-water interface (Fig. 5). These considerations are subject of an ongoing study to clarify the involvement of triglyceride molecule association on interfacial water, where the quality of the interfacial model is placed under particular scrutiny. It was demonstrated that the solvent flux method presented in this study is able to successfully identify solvent access pathways in proteins, by revealing the primary water channel (PWC) and other key positions in CALB that may increase synthesis versus hydrolysis in non-aqueous media when appropriately mutated, as demonstrated by Larsen et al. [13]. Although SFM introduces artificial external forces to MD simulations by the periodic velocity adjustment of water molecules and may thus raise concerns about misleading biases, one must consider the limitations of conventional MD in sampling rare events such as the passage of a molecule through a channel, both in real and computational time. This is particularly difficult when seeking to simultaneously model multiple rare events. Due to the limitations of computational resources, exploring complex problems such as the solvent flux through an enzyme with significant statistics is currently not feasible by conventional MD approaches.

Therefore, increasing the sampling efficiency of MD by introducing biases to the potential function to overcome energy barriers has become a widely used strategy, such as accelerated MD (aMD) [8], replica-exchange MD (REMD) [24], steered MD (SMD) [15] or random acceleration MD (RAMD) [16]. The multitude of successful applications and insights attained by these methods is testament to their usefulness in studying biophysical effects that would otherwise be inaccessible. In this context, SFM offers the means to rapidly and comprehensively sample solvent flux behavior throughout a protein, which allows for an atomistic resolution of structural features and positions that are relevant to enzyme-solvent interactions. It is thereby unique in its approach of simultaneously modeling the accelerated influx of multiple solvent molecules that randomly come into contact with the protein from the solvent bulk, and to extract meaningful data in form of contact frequencies and compiled pathways of individual water molecules during influx. In this regard, other comparable methods are either less deliberate in overcoming specific energy barriers (aMD, REMD), are restricted to single molecules and predefined geometric reaction coordinates (SMD), or are more suited to resolving the efflux of single substrates (RAMD). SFM holds the potential of exploring solvent-enzyme interactions beyond the hydration-related context, which seems to be required, considering that identifying the role of residual water in facilitating enzyme activity in non-aqueous synthetic hydrolase applications has proven to be nontrivial [20]. Even studies on "dry" systems have shown that residual water is in fact not restricted to the hydration of proteins, but instead partitions into the solvent environment under equilibrium conditions [7]. Understanding the behavior of residual free water in reaction media may thus prove particularly useful for synthetic hydrolase applications in non-aqueous environments, specifically to eliminate water as a competing nucleophile. SFM in conjunction with the desired organic nucleophile may additionally benefit enzyme engineering in improving synthesis reactions by rational design. In principle, SFM can thereby aid in resolving the competition between synthesis and hydrolysis, which remains a fundamental problem in designing a particular hydrolase application and is difficult to overcome without a deeper mechanistic understanding of molecular solvent fluxes.

Acknowledgements We acknowledge the Cluster of Excellence in Simulation Technology (EXC 310/1) at the University of Stuttgart for financial support and the High Performance Computing Center Stuttgart (HLRS) for computational resources. This chapter is based on a previous publication by the authors [2]. Copyright 2014 American Chemical Society.

References

1. Benson, S.P., Pleiss, J.: Molecular dynamics simulations of self-emulsifying drug-delivery systems (SEDDS): influence of excipients on droplet nanostructure and drug localization. Langmuir **30**(28), 8471–8480 (2014a)
2. Benson, S.P., Pleiss, J.: Solvent flux method (SFM): a case study of water access to Candida antarctica lipase b. J. Chem. Theory Comput. **10**(11), 5206–5214 (2014b)

3. Berger, O., Edholm, O., Jähnig, F.: Molecular dynamics simulations of a fluid bilayer of dipalmitoylphosphatidylcholine at full hydration, constant pressure, and constant temperature. Biophys. J. **72**(5), 2002 (1997)
4. Bös, F., Pleiss, J.: Multiple molecular dynamics simulations of TEM β-lactamase: dynamics and water binding of the ω-loop. Biophys. J. **97**(9), 2550–2558 (2009)
5. Essmann, U., Perera, L., Berkowitz, M.L., Darden, T., Lee, H., Pedersen, L.G.: A smooth particle mesh Ewald method. J. Chem. Phys. **103**(19), 8577–8593 (1995)
6. Gruber, C.C., Pleiss, J.: Lipase b from Candida antarctica binds to hydrophobic substrate–water interfaces via hydrophobic anchors surrounding the active site entrance. J. Mol. Catal. B Enzym. **84**, 48–54 (2012)
7. Halling, P.J.: What can we learn by studying enzymes in non–aqueous media? Philos. Trans. R. Soc. B **359**(1448), 1287–1297 (2004)
8. Hamelberg, D., Mongan, J., McCammon, J.A.: Accelerated molecular dynamics: a promising and efficient simulation method for biomolecules. J. Chem. Phys. **120**(24), 11919–11929 (2004)
9. Hess, B., Kutzner, C., Van Der Spoel, D., Lindahl, E.: Gromacs 4: algorithms for highly efficient, load-balanced, and scalable molecular simulation. J. Chem. Theory Comput. **4**(3), 435–447 (2008)
10. Jorgensen, W.L., Chandrasekhar, J., Madura, J.D., Impey, R.W., Klein, M.L.: Comparison of simple potential functions for simulating liquid water. J. Chem. Phys. **79**(2), 926–935 (1983)
11. Jorgensen, W.L., Maxwell, D.S., Tirado-Rives, J.: Development and testing of the OPLS all-atom force field on conformational energetics and properties of organic liquids. J. Am. Chem. Soc. **118**(45), 11225–11236 (1996)
12. Jurado, E., Camacho, F., Luzón, G., Fernández-Serrano, M., García-Román, M.: Kinetic model for the enzymatic hydrolysis of tributyrin in o/w emulsions. Chem. Eng. Sci. **61**(15), 5010–5020 (2006)
13. Larsen, M.W., Zielinska, D.F., Martinelle, M., Hidalgo, A., Jensen, L.J., Bornscheuer, U.T., Hult, K.: Suppression of water as a nucleophile in Candida antarctica lipase b catalysis. Chem. Bio. Chem. **11**(6), 796–801 (2010). doi:10.1002/cbic.200900743. http://dx.doi.org/10.1002/cbic.200900743
14. Lousa, D., Baptista, A.M., Soares, C.M.: A molecular perspective on nonaqueous biocatalysis: contributions from simulation studies. Phys. Chem. Chem. Phys. **15**, 13723–13736 (2013). doi:10.1039/C3CP51761F. http://dx.doi.org/10.1039/C3CP51761F
15. Lu, H., Isralewitz B., Krammer, A., Vogel, V., Schulten, K.: Unfolding of Titin immunoglobulin domains by steered molecular dynamics simulation. Biophys. J. **75**(2), 662–671 (1998)
16. Lüdemann, S.K., Lounnas, V., Wade, R.C.: How do substrates enter and products exit the buried active site of cytochrome p450cam? 2. Steered molecular dynamics and adiabatic mapping of substrate pathways. J. Mol. Biol. **303**(5), 813–830 (2000)
17. Marrink, S., Mark, A.: Effect of undulations on surface tension in simulated bilayers. J. Phys. Chem. B **105**(26), 6122–6127 (2001)
18. Ryckaert, J.-P., Ciccotti, G., Berendsen, H.J.: Numerical integration of the cartesian equations of motion of a system with constraints: molecular dynamics of n-alkanes. J. Comput. Phys. **23**(3), 327–341 (1977)
19. Sum, A.K., Biddy, M.J., de Pablo, J.J., Tupy, M.J.: Predictive molecular model for the thermodynamic and transport properties of triacylglycerols. J. Phys. Chem. B **107**(51), 14443–14451 (2003)
20. Torres, S., Castro, G.R.: Non-aqueous biocatalysis in homogeneous solvent systems. Food Technol. Biotechnol. **42**(4), 271–277 (2004)
21. Uppenberg, J., Hansen, M.T., Patkar, S., Jones, T.A.: The sequence, crystal structure determination and refinement of two crystal forms of lipase b from Candida antarctica. Structure **2**(4), 293–308 (1994)
22. Van Der Spoel, D., Lindahl, E., Hess, B., Groenhof, G., Mark, A.E., Berendsen, H.J.: Gromacs: fast, flexible, and free. J. Comb. Chem. **26**(16), 1701–1718 (2005)

23. Wensink, E.J., Hoffmann, A.C., van Maaren, P.J., van der Spoel, D.: Dynamic properties of water/alcohol mixtures studied by computer simulation. J. Chem. Phys. **119**(14), 7308–7317 (2003)
24. Zhou, R.: Replica exchange molecular dynamics method for protein folding simulation. In: Protein Folding Protocols, pp. 205–223. Springer, New York (2006)

Dynathor: Dynamics of the Complex of Cytochrome P450 and Cytochrome P450 Reductase in a Phospholipid Bilayer

Xiaofeng Yu, Daria B. Kokh, Prajwal Nandekar, Ghulam Mustafa, Stefan Richter, and Rebecca C. Wade

Abstract The Dynathor project aims at understanding the interaction of cytochrome P450 (CYP, P450) enzymes and their redox partner, cytochrome P450 reductase (CPR), in a phospholipid bilayer. Through simulation studies on the HLRS CRAY systems (initially XE6, later on XC40), we investigated the interactions of models of membrane-bound P450s (CYP51 and CYP1A1) and CPR. A model of membrane-bound *T. brucei* CYP51 and the human CPR was successfully built and simulated for 217.5 ns. A model of human CYP1A1 and the human CPR in a phospholipid bilayer was also built and is being simulated. These models can be used as starting points to understand the selectivity of the interactions of CYPs with CPR in the native membrane-bound forms, and thus may aid drug discovery projects.

Furthermore, we evaluated the method of Random Acceleration Molecular Dynamics (RAMD) for use in calculating relative residence times of proteins with small molecules. This would allow the computation of relative off rates for drug molecules. As a small model system for this evaluation, we use the N-terminal domain of human heat shock protein 90 (HSP90) and a set of different ligands. The obtained RAMD simulated residence times show a clear correlation with experimental values.

X. Yu • D.B. Kokh • P. Nandekar • G. Mustafa • S. Richter (✉)
HITS gGmbH, Schloss-Wolfsbrunnenweg 35, 69118 Heidelberg, Germany
e-mail: mcmsoft@h-its.org; Stefan.Richter@h-its.org

R.C. Wade (✉)
HITS gGmbH, Schloss-Wolfsbrunnenweg 35, 69118 Heidelberg, Germany

Center for Molecular Biology (ZMBH), DKFZ-ZMBH Alliance, Heidelberg University, Im Neuenheimer Feld 282, 69120 Heidelberg, Germany

Interdisciplinary Center for Scientific Computing (IWR), Heidelberg University, Im Neuenheimer Feld 368, 69120 Heidelberg, Germany
e-mail: Rebecca.Wade@h-its.org

© Springer International Publishing Switzerland 2015
W.E. Nagel et al. (eds.), *High Performance Computing in Science and Engineering '15*, DOI 10.1007/978-3-319-24633-8_17

1 Technical Description

1.1 NAMD Simulations

NAMD is a software package for molecular dynamics simulations, provided by the University of Illinois [1–3]. Simulations where initially performed on the HLRS Hermit, a Cray XE6 computer [4]. Starting from November 2014, the new HLRS Hornet system, a Cray XC40 [5] computer was used. On Hermit, we used NAMD version 2.9, compiled using the Cray software stack. Simulations on Hornet used the newest NAMD version 2.10b2. This version switch was required since NAMD version 2.9 did not support the Cray Aries network environment implemented on the Hornet system. The Hornet system with NAMD 2.10b2 shows an approximate 2.4× increase in simulation performance at 1024 cores compared to the Hermit system using NAMD 2.9 (see Fig. 1). Calculations for the cytochrome P450 system were performed using 1024 cores.

1.2 RAMD Simulations

Random Acceleration Molecular Dynamics (RAMD) [6, 7] uses the NAMD software package to perform molecular dynamics simulations of two molecule systems bound in a complex. This method can be used to simulate the egress of a bound

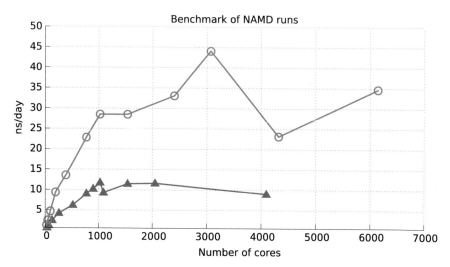

Fig. 1 Scaling of NAMD simulations. NAMD 2.9 simulations of CYP51-CPR complex in the membrane (390,422 atoms) on a Cray XE6 (*filled triangle*) compared to NAMD 2.10b2 simulations on a Cray XC40 (*open circle*). At 1024 cores, simulations on the Cray XC40 were 2.4 times faster using NAMD 2.10b2

ligand from a protein. The method relies on the application of random forces to guide a bound molecule out of the binding pocket. One problem of this random force application is that it disrupts simulation and requires all threads to synchronize. Therefore, so far, only a maximum of 48 cores was used for a single RAMD simulation. However, for calculations of off-rates for bound ligands, a large number of similar independent simulations needs to be carried out in the same system. These simulations can be run in parallel. Also new scripting techniques within NAMD [8] possibly will not interfere with the parallel performance. Since this would reduce the average runtime of a single NAMD run from currently several hours up to 20 h on hornet, the migration of the RAMD code would be beneficial (Fig. 2).

For the method evaluation we have used N-terminal domain (NTD) of human HSP90 (heat shock protein 90), for which experimental dissociation rates for a quite large set of compounds are available. We have carried out simulations for 11 ligands; four of them bound to the helical conformation and the rest bound to the loop conformation of the HSP90 binding site. For estimation of a relative residence time, 25 RAMD simulations for each ligand were performed. For evaluation of the sensitivity of ligand residence time to the magnitude of the force applied in RAMD simulations, we performed two set of simulations for each ligand using two different values of the acceleration force: 15 and 14 kcal mol^{-1} Å^{-1} (data not shown). The maximum length of each trajectory was 10 ns-5,000,000 steps, which takes about

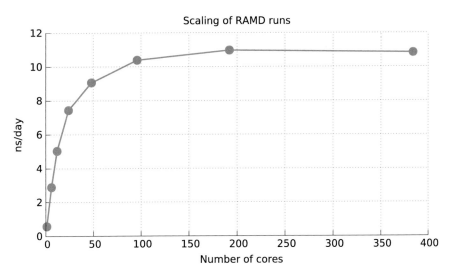

Fig. 2 Scaling of RAMD simulations of HSP90 with inhibitor (41,831 atoms). RAMD is an algorithm added to the NAMD software via TCL text commands. The nature of TCL text commands disrupts the parallel performance of NAMD of a single RAMD run. Although some of these commands can be used in a parallel environment, RAMD so far does not take advantage of this. New scripting techniques within NAMD [8] might overcome this limitation. A large number of RAMD simulations are required for calculating relative off-rates for ligands

20 h on 2 nodes. In practice, however, the most of ligands require much less time for dissociation, and thus most simulations are much shorter.

Effective RAMD residence times obtained for 11 ligands show a clear correlation with experimental values (data not shown) although the scaling factor is different for the two protein conformations simulated. The result was found to be quite robust with respect to variations of the magnitude of the force applied if the RAMD residence time was longer than hundreds of picoseconds.

2 Simulations of Membrane-Bound Complexes of P450 and CPR

2.1 Introduction

Cytochrome P450 (CYP, P450) enzymes play important roles in drug metabolism, steroid biosynthesis and xenobiotic degradation. In humans, P450s are essential for the metabolism of 70–80 % of all drugs. The catalytic function of P450s is monooxygenation, which is to add one oxygen atom to an organic bond (-RH). In the catalytic cycle of P450s, two electrons need to be transferred to the active site of P450s from their redox partners. For many P450s in eukaryotes, the redox partner is cytochrome P450 reductase (CPR). CPR consists of three domains: the FMN domain which contains the flavin mononucleotide (FMN), the FAD domain with the flavin adenine dinucleotide (FAD), and the NADP domain with the nicotinamide adenine dinucleotide phosphate (NADP). In eukaryotes, both proteins locate in the endoplasmic reticulum (ER) membrane with a trans-membrane helix anchor. The association of the globular domains of the two proteins is mainly driven by electrostatic interactions. The positively charged proximal side of P450 interacts with the negatively charged FMN domain of the CPR.

Many studies have been carried out for the soluble forms of P450s and their redox partners. The crystal structure of the complex of the P450BM3 CYP domain and its FMN domain (which, exceptionally for P450s, are in one polypeptide chain in P450BM3) gives direct evidence of how P450s may interact with CPR [9]. A model of the CYP2B4-cytochrome b5 complex was built using solid-state NMR data on CYP2B4 and the membrane-bound cytochrome b5 [10]. For the interactions of CYP2B4 with CPR, mutation studies have shown residues that affect binding. However, no crystal structure has been solved of complexes of P450s and CPR, and little is known about their interactions with the membrane or possible conformational changes of the proteins upon binding or during the catalytic cycle.

Understanding the P450-CPR interactions in the membrane is of great interest. First of all, it can help to investigate the catalytic mechanism of P450s in the native membrane-bound form. It can give insights into the association and dissociation of the two proteins in the membrane environment and how this varies with P450 sequence. Furthermore, the bound complex can be used to study ligand access and egress tunnels in P450s which can be influenced by the protein-protein interactions. Lastly, it can help to improve the prediction of drug metabolism by P450s and the design of drugs which bind to P450s.

2.2   Methods

Molecular dynamics (MD) simulations have been used to build models of different CYPs in the membrane. Our laboratory has established a systematic multiscale simulation protocol employing coarse-grain and atomic detail models, to model and simulate P450 in a 1-palmitoyl-2-oleoyl-sn-glycero-3-phosphocholine (POPC) bilayer and this has been used to model human CYP2C9 and *Trypanosoma brucei* CYP51 in a POPC bilayer [11, 12]. Other researchers have also recently built and simulated models of P450s in bilayers including for CYP2C9 [13], CYP3A4 [14, 15], CYP19 [16] and other human drug-metabolizing P450s [17]. Sündermann et al. [18] performed a simulation of a complex of CYP2D6 and the open form of CPR in a bilayer. This complex was however built manually and the time scale of the simulation (10 ns) was too short to investigate the dynamics of the equilibrated complex in the membrane.

Our in-house software, Simulation of Diffusional Association (SDA, http://mcm. h-its.org/sda7/) has provided a useful tool for predicting the structures of protein-protein encounter complexes by using Brownian dynamics simulations [19–21]. It is particularly useful for electrostatically driven complexes [22, 23]. Our predicted structure of the P450cam-putidaredoxin complex published in 2007 [24] is in good agreement with the later published crystal structure of the complex in 2013 [25, 26].

We have tested using SDA to predict the complex of soluble forms of CYP2B4 and CPR. The complex is in good agreement with experimental data and refinement of the docked encounter complex using molecular dynamics simulation improves agreement with experimental data [9, 27]. Thus, we used SDA to predict the complexes of CYP51 with CPR and of CYP1A1 with CPR. The membrane-bound models of CYP51 and CYP1A1 have been built and simulated by using the protocol described in [11]. The P450-CPR encounter complexes defined using SDA rigid-body docking were superimposed onto the membrane-bound models of CYP51 and CYP1A1. The missing trans-membrane helix of CPR was then be inserted into the membrane and the missing linker residues modeled to connect the trans-membrane helix and the FMN domain of the CPR in the P450-CPR complexes. These models of membrane-bound complexes of P450 and CPR provided starting structures for MD simulations at HLRS. The Amber force field ff99SB is used for the protein and the Amber gaff force field for the POPC lipids and the cofactors. The systems simulated contain roughly 1 million atoms each. The procedure for this study is shown in Fig. 3.

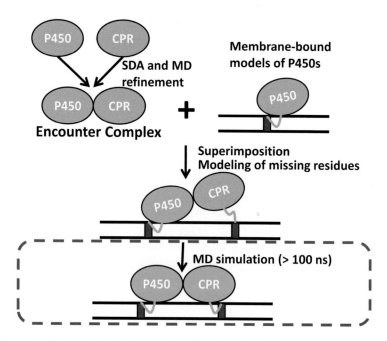

Fig. 3 Protocol to build and simulate membrane-bound models of CYP and CPR

2.3   *Results*

Using the protocol depicted in Fig. 3, we successfully built a membrane-bound model of the complex of *T. brucei* CYP51 and human CPR and simulated this model for 217.5 ns, which is 20 times longer than the recent simulation of a related system [18]. The starting structure and the equilibrated structure are shown in Fig. 4. In Fig. 4a, the membrane-bound *T. brucei* CYP51 was taken after MD simulation and the CPR was built onto it. In this model, the globular domain of CPR was relatively far away from the membrane, with the FMN domain interacting with the proximal side of CYP51 and the FAD domain extending away from the membrane. However, the FAD domain of CPR got much closer to the membrane in the simulation, although the distance between the center of mass of CPR and that of the membrane remained mostly similar during the simulation. The globular domain of CPR has a much larger distance to the membrane than that of CYP51 in the simulation. The reason can be that CPR is prohibited from getting much closer to the membrane because of its interaction with CYP51.

In the simulation of the membrane-bound complex of CYP51 and CPR, CYP51 maintained an orientation similar to the starting structure. The starting orientation of CYP51 in the membrane has been shown to be consistent with experimental data [12]. The 217.5 ns simulation of the model shows that the interaction of CYP51 and CPR is stable in the membrane-bound form.

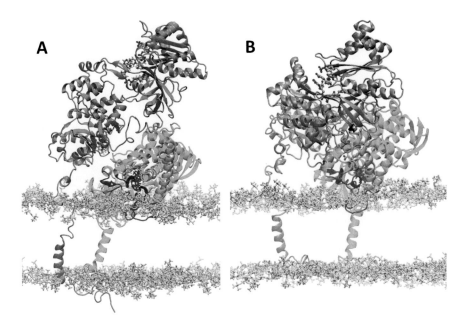

Fig. 4 (**a**) The starting structure and (**b**) the equilibrated structure of CYP51 (*green colored cartoons*) and CPR (*orange colored cartoons*) in membrane (sticks view)

The proposed protocol gives the possibility to build models of membrane-bound forms of different P450s and CPR. Having membrane-bound models of P450s and CPR can help to compare P450s, to understand selectivity of CPR for different P450s and thus to aid drug design. The model of the membrane-bound form of CYP51 and CPR provides a basis for understanding the behavior of both CYP51 and CPR in the complex on a lipid bilayer. This model may help to design drugs against parasitic CYP51s, taken their interaction with CPR into consideration.

A model of the membrane-bound complex of CYP1A1 and CPR was also built using the same protocol (Fig. 5a and b). The CYP1A1-CPR membrane bound complex was converted to coarse-grained model and 12 μs of MD simulation was run. During the simulation, the CYP1A1-CPR complex was stable in the membrane. The distance between CYP1A1 and FMN binding domain of CPR reduced from 32 to 29 Å and distance between FMN and HEME co-factors reduced from 24 to 19 Å (Fig. 5c). Thus, CYP1A1 and FMN domain of CPR came closer during the course of 12 μs of coarse-grain MD simulation.

The final structure obtained from coarse-grained MD simulation was converted to an all-atom model using the previously described protocol [11]. The all-atom structure of the CYP1A1-CPR complex in the membrane was then simulated for 15 ns using NAMD software. The RMSD of CYP1A1 and the FMN binding domain, showed a continuous increase during the simulation, indicating that it is not yet equilibrated. The length of the simulation will be increased to obtain a well converged structure of CYP1A1-CPR complex in membrane.

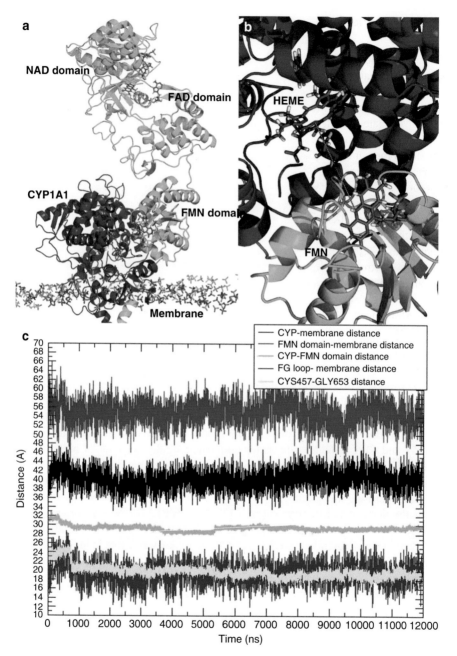

Fig. 5 (**a**) The orientation of the CYP1A1-CPR complex in the membrane, (**b**) the positions of the HEME and FMN co-factors in the CYP1A1-CPR complex. (**c**) the observed parameters characterizing the arrangement of the two proteins in the bilayer during the coarse-grained CYP1A1-CPR complex simulation

3   Future Plans

So far 900,000 CPU×h of the 4,000,000×h have been used. The remaining CPU time will primarily be used for NAMD production runs of CYP systems.

In future studies, we are interested in continuing to investigate the dynamics of the P450-CPR complex, especially in the membrane-bound form. Specifically, we will investigate how the two proteins interact with each other in the soluble form and, when the two proteins are in the membrane-bound form, how they orient in the membrane and how the membrane influences the association and dissociation of the two proteins and the structure of their complex. The studies will be carried out for complexes of three different P450s with CPR allowing for comparison of the effects of sequence differences on binding and identification of critical interactions and conformational changes. These studies will provide the basis for subsequent simulations to investigate ligand access and egress mechanism in the P450-CPR complexes. Furthermore, these studies will provide a step towards understanding how the electrons are transferred from the CPR to the P450 and how drugs are metabolized by this complex.

Acknowledgements We gratefully acknowledge the support of the Klaus Tschira Foundation, the German Academic Exchange Service (DAAD) for scholarships to Prajwal Nandekar and Ghulam Mustafa as well as the EU/EFPIA Innovative Medicines Initiative (IMI) Joint Undertaking, "Kinetics for Drug Discovery", K4DD (grant no. 115366). Finally, we would like to thank the HLRS for providing CPU time for this project.

References

1. Phillips, J.C., Braun, R., Wang, W., Gumbart, J., Tajkhorshid, E., Villa, E., Chipot, C., Skeel, R.D., Kalé, L., Schulten, K.: Scalable molecular dynamics with NAMD. J. Comput. Chem. **26**, 1781–1802 (2005)
2. Theoretical and U. o. I. a. U.-C. Computational Biophysics Group: NAMD 2.9 release notes. http://www.ks.uiuc.edu/Research/namd/2.9/notes.html. Online; Accessed 27 March 2015
3. Theoretical and U. o. I. a. U.-C. Computational Biophysics Group: NAMD 2.10b2 release notes. http://www.ks.uiuc.edu/Research/namd/2.10b2/notes.html. Online; Accessed 27 March 2015
4. HLRS: Cray XE6. https://wickie.hlrs.de/platforms/index.php/Cray_XE6. Online; Accessed 27 March 2015
5. HLRS: Cray XC40. https://wickie.hlrs.de/platforms/index.php/Cray_XC40. Online; Accessed 27 March 2015
6. Lüdemann, S.K., Lounnas, V., Wade, R.C.: How do substrates enter and products exit the buried active site of cytochrome P450cam? 1. Random expulsion molecular dynamics investigation of ligand access channels and mechanisms. J. Mol. Biol. **303**, 797–811 (2000)
7. Lüdemann, S.K., Lounnas, V., Wade, R.C.: How do substrates enter and products exit the buried active site of cytochrome P450cam? 2. Steered molecular dynamics and adiabatic mapping of substrate pathways. J. Mol. Biol. **303**, 813–830 (2000)
8. Phillips, J.C., Stone, J.E., Vandivort, K.L., Armstrong, T.G., Wozniak, J.M., Wilde, M., Schulten, K.: Petascale tcl with NAMD, VMD, and Swift/T, in Proceedings of the 1st First Workshop for High Performance Technical Computing in Dynamic Languages, HPTCDL '14, (Piscataway, NJ, USA), pp. 6–17. IEEE Press, New York (2014)

9. Sevrioukova, I.F., Li, H., Zhang, H., Peterson, J.A., Poulos, T.L.: Structure of a cytochrome P450-redox partner electron-transfer complex. Proc. Natl. Acad. Sci. U.S.A. **96**, 1863–1868 (1999)

10. Ahuja, S., Jahr, N., Im, S.-C., Vivekanandan, S., Popovych, N., Le Clair, S.V., Huang, R., Soong, R., Xu, J., Yamamoto, K., Nanga, R.P., Bridges, A., Waskell, L., Ramamoorthy, A.: A model of the membrane-bound cytochrome b5-cytochrome P450 complex from nmr and mutagenesis data. J. Biol. Chem. **288**, 22080–22095 (2013)

11. Cojocaru, V., Balali-Mood, K., Sansom, M.S.P., Wade, R.C.: Structure and dynamics of the membrane-bound cytochrome P450 2C9. PLoS. Comput. Biol. **7**, e1002152 (2011)

12. Yu, X., Cojocaru, V., Mustafa, G., Salo-Ahen, O.M.H., Lepesheva, G.I., Wade, R.C.: Dynamics of CYP51: implications for function and inhibitor design. J. Mol. Recognit. **28**, 59–73 (2015)

13. Berka, K., Hendrychová, T., Anzenbacher, P., Otyepka, M.: Membrane position of ibuprofen agrees with suggested access path entrance to cytochrome P450 2C9 active site. J. Phys. Chem. A **115**, 11248–11255 (2011)

14. Denisov, I.G., Shih, A.Y., Sligar, S.G.: Structural differences between soluble and membrane bound cytochrome P450s. J. Inorg. Biochem. **108**, 150–158 (2012)

15. Baylon, J.L., Lenov, I.L., Sligar, S.G., Tajkhorshid, E.: Characterizing the membrane-bound state of cytochrome P450 3A4: structure, depth of insertion, and orientation. J. Am. Chem. Soc. **135**, 8542–8551 (2013)

16. Sgrignani, J., Magistrato, A.: Influence of the membrane lipophilic environment on the structure and on the substrate access/egress routes of the human aromatase enzyme. A computational study. J. Chem Inf. Model. **52**, 1595–1606 (2012)

17. Berka, K., Paloncýová, M., Anzenbacher, P., Otyepka, M.: Behavior of human cytochromes P450 on lipid membranes. J. Phys. Chem. B **117**, 11556–11564 (2013)

18. Sündermann, A., Oostenbrink, C.: Molecular dynamics simulations give insight into the conformational change, complex formation, and electron transfer pathway for cytochrome P450 reductase. Protein Sci. **22**, 1183–1195 (2013)

19. Gabdoulline, R.R., Wade, R.C.: Brownian dynamics simulation of protein-protein diffusional encounter. Methods **14**, 329–341 (1998)

20. Gabdoulline, R.R., Wade, R.C.: On the contributions of diffusion and thermal activation to electron transfer between phormidium laminosum plastocyanin and cytochrome f: Brownian dynamics simulations with explicit modeling of nonpolar desolvation interactions and electron transfer events. J. Am. Chem. Soc. **131**, 9230–9238 (2009)

21. Motiejunas, D., Gabdoulline, R., Wang, T., Feldman-Salit, A., Johann, T., Winn, P.J., Wade, R.C.: Protein-protein docking by simulating the process of association subject to biochemical constraints. Proteins **71**, 1955–1969 (2008)

22. Sandikci, A., Gloge, F., Martinez, M., Mayer, M.P., Wade, R., Bukau, B., Kramer, G.: Dynamic enzyme docking to the ribosome coordinates N-terminal processing with polypeptide folding. Nat. Struct. Mol. Biol. **20**, 843–850 (2013)

23. Pachov, G.V., Gabdoulline, R.R., Wade, R.C.: On the structure and dynamics of the complex of the nucleosome and the linker histone. Nucleic Acids Res. **39**, 5255–5263 (2011)

24. Karyakin, A., Motiejunas, D., Wade, R.C., Jung, C.: FTIR studies of the redox partner interaction in cytochrome P450: the PDX-P450cam couple. Biochim. Biophys. Acta **1770**, 420–431 (2007).

25. Tripathi, S., Li, H., Poulos, T.L.: Structural basis for effector control and redox partner recognition in cytochrome P450. Science **340**, 1227–1230 (2013)

26. Hiruma, Y., Hass, M.A.S., Kikui, Y., Liu, W.-M., Ölmez, B., Skinner, S.P., Blok, A., Kloosterman, A., Koteishi, H., Löhr, F., Schwalbe, H., Nojiri, M., Ubbink, M.: The structure of the cytochrome P450cam-putidaredoxin complex determined by paramagnetic nmr spectroscopy and crystallography. J. Mol. Biol. **425**, 4353–4365 (2013)

27. Im, S.-C., Waskell, L.: The interaction of microsomal cytochrome P450 2B4 with its redox partners, cytochrome P450 reductase and cytochrome b(5). Arch Biochem. Biophys. **507**, 144–153 (2011)

Part IV
Reactive Flows

Dietmar Kröner

In this section about reactive flows we have two contributions, which are both supported by the Deutsche Forschungsgemeinschaft. The main focus of the first one concerns the improvement of the efficiency by using a multi-regional approach for a model problem and the second one the realistic simulation of rocket combustion for different flow regimes.

F. Zhang, T. Zirwes, P. Habisreuther, and H. Bockhorn study the "Numerical Simulation of Turbulent Combustion with a Multi-Regional Approach". In particular they use the Navier-Stokes equations (without transport for the species) in regions with no chemical reactions and the Navier-Stokes equations with a combustion model in regions where chemical reactions occur. The efficiency of this method is shown for a large eddy simulation of a model burner with a premixed flame. The obtained results are consistent with conventional single-regional computation results but the multi-regional approach turns out to be much faster and has a great potential to speed up further the simulation process. For the simulation the OpenFOAM code has been used.

In the second contribution "Numerical Simulations of Rocket Combustion Chambers on Massively Parallel Systems" by R. Keller, M. Seidl, M. Lempke, P. Gerlinger, and M. Aigner the flow in a rocket combustion chamber is considered. The big challenges are the different scales for the velocity, the pressure and the temperature. Due to pressure and temperature the fluid can be in different states (gas or liquid) and a phase transition can occur. The phases are separated by a sharp interface. The inhouse code TASCOM3D coupled with the particle tracking code SPRAYSIM is used for the numerical simulations. The liquid phase is assumed to be a dilute spray of discrete, non-interacting droplets. Multiple droplets are represented by computational parcels and for each parcel the equation of motion

D. Kröner (✉)
Abteilung für Angewandte Mathematik, Universität Freiburg, Hermann-Herder-Str. 10, 79104 Freiburg, Germany
e-mail: dietmar.kroener@mathematik.uni-freiburg.de

are solved. For increasing pressure and decreasing temperature the equations of state are adapted. For the initial data in the simulation experiments three different droplet size distributions are used. The values are taken from measurements. The computations have been performed on HERMIT and HORNET. The efficiency on HORNET is more than two times larger than on HERMIT.

Numerical Simulation of Turbulent Combustion with a Multi-Regional Approach

Feichi Zhang, Thorsten Zirwes, Peter Habisreuther, and Henning Bockhorn

Abstract The current work uses a multi-regional method for improving the computing performance of large-scale combustion simulations. In this manner, the solution of the isothermal flow within the burner is treated separately from the domain with combustion reaction. For the fresh gas flow within the nozzle only the Navier-Stokes equations for a non-reactive, fixed composition flow are solved, whereas the combustion model accounting for the chemical reactions is enabled in the ignition zone downstream of the burner. Because the chemistry solution takes a major part of the total computing time, the approach saves that part of execution time for the computing nodes located within the nozzle, where no chemical reaction occurs. In the present study, the potential of this methodology has been assessed by large eddy simulation (LES) of a model burner operated with a premixed methane/air flame. The multi-regional simulation showed consistent results with data obtained from the conventional single-regional computation. It however has been proven to be considerably faster than the comparable single-zonal LES, denoting an improved computing performance.

1 Introduction

For most industrial burners, the flow conditions generated within the nozzle are hardly assessable and cannot be forecasted from the general operation conditions such as mass flow rate or equivalence ratio. For example, there may be significant changes in cross-sections of the nozzle due to redirections, which result in complicated flow patterns like separations. Additionally, stabilization devices like swirlers are often installed within the nozzle, which lead to large, long-term vortical structures. These flow features are then continuously transported (through convection and dissipation) until they reach the reaction zone outside the burner and interact with the flame. Therefore, the resolution of the flow structures within the nozzle is of great importance for the chemistry modeling downstream. For this

F. Zhang (✉) • T. Zirwes • P. Habisreuther • H. Bockhorn
Division of Combustion Technology, Engler-Bunte-Institute, Karlsruhe Institute of Technology, Engler-Bunte-Ring 1, 76131 Karlsruhe, Germany
e-mail: feichi.zhang@kit.edu

© Springer International Publishing Switzerland 2016
W.E. Nagel et al. (eds.), *High Performance Computing in Science and Engineering '15*, DOI 10.1007/978-3-319-24633-8_18

reason, the computational grid has to cover a major part of the nozzle with a large number of computing nodes to ensure well resolved flow conditions that affect the reaction zones downstream.

For detailed combustion simulation, one transport equation has to be solved for each reacting species. Therewith, the reaction source term is highly non-linear in concentrations and temperature, and, hence, its evaluation is very CPU-intensive [1]. Moreover, thickness of the reaction zone and time scales of the chemical reactions have to be resolved. Consequently, the chemistry calculation for the simulation of a reactive flow requires a large amount of extra computing effort in comparison to non-reactive flows. For this reason, combustion models are generally used to simplify the reaction system and chemical source terms. Nonetheless, depending on complexity of the combustion model, the computational effort for the chemistry calculation could easily overwhelm the computing time required for solving the Navier-Stokes equations. For example, by using tabulated chemistry techniques with the flamelet (FLL) [2] or the flamelet generated manifold (FGM) [3] model, multi-dimensional interpolation has to be applied to obtain chemical scalars from a pre-calculated look-up table spanned by a number of control variables (e.g. progress variable or mixture fraction). An additional transport equation has to be solved for each of those scalars [1]. Therefore, in case that the chemistry modeling is computationally expensive and the proportion of computing cells located within the non-reactive zone is large, it is reasonable to omit the part of chemistry calculation for that region to avoid execution time for the unneeded chemistry calculations.

The objective of the current work is to apply a geometry-based multi-regional approach to reduce the overall computing time for combustion modeling. In this method, the wall-bounded flow region within the nozzle and the flame propagation zone downstream are geometrically decoupled at the exit plane of the nozzle. The segregated regions are then calculated by means of a non-reactive and a reactive flow simulation. Multi-regional modeling with the objective of reducing computational effort has already found a widespread use, for example, the coupling of LES (large eddy simulation) with RANS (Reynolds-averaged Navier-Stokes) models [4–6] or the hybrid CAA (computational aero-acoustics) method [7, 8] for far-field acoustic calculation. Compared to these available multi-zonal concepts, the novelty of the present approach is given by the multi-physics modeling of the flow and chemistry for different zones, or non-reactive and reactive handling, respectively.

2   Modeling Concepts

2.1   Combustion Modeling

A turbulent flame speed closure (TFC) combustion model is used in the present work to evaluate the turbulent mean rate of a global reaction progress variable [8, 9], which is combined with a tabulated chemistry technique to obtain chemical scalars.

The basic idea is to presume that the chemical reaction is intrinsically faster than the turbulent mixing (flamelet assumption: Damköhler number $Da > 1$), so that the entire flame can be considered as an ensemble of distinct, premixed reaction sheets with different stoichiometries. The mixing process is controlled by the mixture fraction $\tilde{\xi}$ and the subsequent chemical reaction by the progress variable $\tilde{\theta}$. The structures of the reaction-diffusion flame sheets, i.e., profiles of the chemical scalars along the flame axis are calculated with a solution of premixed flat flames (1D) in a pre-processing stage (e.g. with the CHEMKIN program package [10]). To take into account the effect of turbulence on the mixing, the pre-calculated flame profiles are averaged through a probability density function (PDF) of ξ, which has a presumed β-function shape determined by the mean value $\tilde{\xi}$ and variance $\widetilde{\xi''^2}$ of the mixture fraction [2]. Finally, all averaged (integrated over the PDF) quantities are gathered in a look-up table spanned by the three control parameters: $\tilde{\xi}$, $\widetilde{\xi''^2}$ and $\tilde{\theta}$.

Figure 1 shows a flow chart of the concept. In addition to the basic flow equations three additional transport equations are solved for the control parameters. The

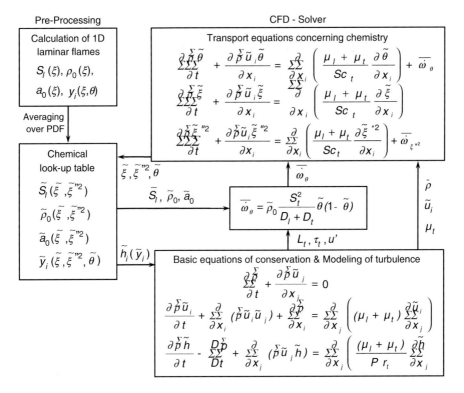

Fig. 1 Coupling of the CFD solution with tabulated chemistry

turbulent mean rate $\bar{\tilde{\omega}}_\theta$ of $\tilde{\theta}$ formation is calculated by means of the turbulent flame speed S_t [11]

$$\frac{S_t}{S_l} = 1 + \frac{u'}{S_l}\left(1 + Da^{-2}\right)^{-1/4} \tag{1}$$

which is a function of the turbulence parameters (turbulence intensity u' and turbulent length scale L_t) and the chemo-physical properties (laminar flame speed S_l, density ρ_0 and thermal diffusivity a_0 of the unburnt mixture). The turbulence parameters are derived from the turbulence modeling [12, 13]. The species mass fractions $\tilde{y}_i$ as well as S_l, ρ_0 and a_0 are searched by multi-dimensional interpolation through the look-up table via the control parameters, as shown in Fig. 1. The Dahmköhler number Da in Eq. (1) is evaluated by the ratio of the turbulent time scale $\tau_t = L_t/u'$ to the chemical time scale $\tau_c \propto a_0/S_l^2$. This concept has already been validated for different flame configurations [14, 15]. More details about this combustion model can be found in [8, 9].

2.2 Multi-Regional Solution Strategy

2.2.1 Basic Idea

Compared to a simulation without chemical reaction, the combustion modeling concept proposed in the last section will cause extra computing time due to the solution of additional transport equations for the control variables, including evaluation of the non-linear source terms therein. Moreover, a major part of computing time is attributed to the multi-dimensional interpolation used by the tabulated chemistry, which strongly depends on the size of the look-up table, i.e. the number of included species and the pre-defined values of the control variables. According to this, the basic idea of the multi-regional modeling is to separately apply flow (non-reactive) and combustion (reactive) simulation to different zones, which are connected to each other through an interface boundary. In case that the share of grid nodes located in the non-reactive zone is large, the approach can lead to a considerably reduced computational cost by saving time that would otherwise be needed by the chemistry model in non-reactive regions.

Ideally the reaction zone of combustion should be isolated entirely from the remaining non-reaction region, so that the maximal possible amount of computing time can be saved. However, due to the complex geometrical shape of the flame front caused by the turbulent flow, it is difficult to capture the joint face of reactive and non-reactive regions. On the other hand, the separation of these zones can easily be accomplished based on a fixed geometry, for example, burner interior or discharge of exhaust gas in connection with the combustion chamber. As illustrated in Fig. 2, the wall-bound fixed composition flow (e.g. fuel/air) in zone 1 is computed with LES, whereas the combustion model is enabled after the fresh gas has left the burner (zone

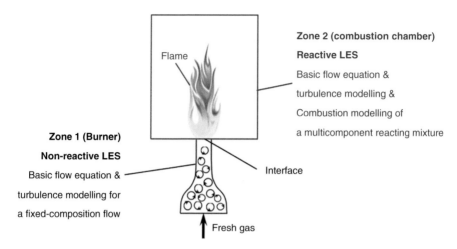

Fig. 2 Two-regional combustion simulation

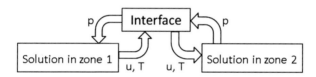

Fig. 3 Exchange of velocity, temperature and pressure through the interface boundary

2). Both regions are directly connected at the interface, which is colocated with the outlet plane of zone 1 (nozzle) and inlet plane of zone 2.

The solution process in this case consists of two separate runs in sequence, i.e., a non-reactive LES for the burner flow (zone 1) and a subsequent combustion LES downstream (zone 2). For the former computation, only the basic conservation equations for mass, momentum and energy are solved, whereas procedures used by combustion modeling are included for the latter run. At the interface, the calculated flow velocities and temperature at the outlet of zone 1 are directly mapped onto the inlet boundary of zone 2 and used there as inflow condition. The combustion simulation in zone 2 provides the unknown pressure field at the outlet of zone 1 for the non-reactive simulation as feedback. This two-way coupling is necessary due to compressibility effects of the fluid, so that pressure waves generated downstream in zone 2 can propagate against the bulk flow and be noticed at the exit plane of the burner. By doing so, the velocity field required for the inlet of zone 2 is interpolated from the previous non-reactive solution in zone 1 at the interface boundary, and vice versa for the pressure, as illustrated in Fig. 3. Note that due to the interpolation performed at the interface boundary, both the outlet of zone 1 and inlet of zone 2 may have different mesh topologies.

2.2.2 Performance Improvement

Assuming that the total number of grid nodes N_t is covered by N_1 cells for the first zone and N_2 for the second zone, i.e. $N_t = N_1 + N_2$; and the required computing time for a non-reactive LES and a combustion LES on the same grid is t_{flow} and t_{comb}, the theoretical speed-up factor S by using the two-regional approach compared to the conventional single-regional method can be derived:

$$S = \underbrace{\frac{N_1}{N_1 + N_2}}_{\alpha} \cdot \underbrace{\frac{t_{comb} - t_{flow}}{t_{comb}}}_{\eta} = \alpha \cdot \eta \qquad (2)$$

In Eq. (2), α indicates the portion of cells allocated to zone 1 and η the speed-up factor of a non-reactive LES compared to the corresponding combustion simulation. It is obvious that the hybrid method would be very beneficial if a large number of grid nodes are used for resolving the non-reactive zone 1 (in this case the interior burner flow region), leading to a large α value. In addition, highly sophisticated combustion models have found a widespread use. For example, in the conditional moment closure (CMC) [16, 17] and the detailed chemistry combustion model [18], transport equations have to be solved for all (conditioned) chemical species. Other modeling concepts use additional control parameters like the scalar dissipation rate or variance of the progress variable [2]. Moreover, a number of combustion related problems like pollutant emissions (e.g. NOx or soot) or thermal radiation may become an important issue [1] and need extra computing effort, leading to a large value of η. Consequently, the multi-regional method has a great potential to speed up the simulation process, particularly for engineering applications, where flows within complex burner systems have to be resolved appropriately. In addition, this approach reduces the memory overhead significantly, because less grid points for combustion modeling have to be taken into account.

One drawback of the multi-regional handling is attributed to the fact that the parallel performance is affected while using a large number of processors, because each individual simulation for zone 1 and zone 2 is less CPU-intensive compared to the single-zonal modeling, in particular for the non-reactive run in zone 1. More discussion about this issue is given in the following Sect. 3.4. Another drawback of the hybrid approach is that two individual simulations have to be prepared instead of one, which leads to additional manual operations, for example, the generation of two separate grids and two sets of settings for each simulation. On the other hand, the separate handling of different regions represents a beneficial feature too, because the grids for the non-reactive and reactive zone can be changed independently from each other, which allows flexible testing of grid and geometry effects for the different zones.

3 Application

3.1 Computational Method

The open source CFD code OpenFOAM [19] has been used in the current work to perform LES in a compressible formulation, which solves the filtered transport equations employing the finite volume method. The pressure-implicit split-operator (PISO) technique is applied for the pressure correction [20]. Discretization of the convective fluxes is based on a second order interpolation scheme. The diffusion terms are discretized with the central difference scheme of second order accuracy. The Smagorinsky model [12] has been used for modeling the subgrid scale Reynolds stresses. The turbulent Schmidt and Prandtl numbers have been modeled as constant to account for the subgrid scale scalar fluxes ($Sc_t = Pr_t = 0.7$). Moreover, a turbulence inflow generator [21] has been implemented into OpenFOAM, which provides transiently and spatially correlated velocity components at the inlet boundary. The non-reflecting boundary condition (NRBC) proposed by Poinsot and Lele [22], which is available in OpenFOAM, has been used for all opening boundaries like inlet or outlet, to avoid unphysical reflections of pressure waves at those boundaries.

The development of the two-regional solver is based on an extension of the standard application "chtMultiRegionFoam" in OpenFOAM for conjugate heat transfer between a solid region and a fluid region. Thereby, execution for the solid region has been replaced by LES modeling of chemically reacting flows. The combustion model proposed in Sect. 2 has been implemented into the new solver. The data transfer through the interface boundary is accomplished with a conservative interpolation algorithm between volume meshes by local Galerkin projection [23], which is available in the standard libraries of OpenFOAM.

3.2 Numerical Setups

In order to validate the multi-regional approach, it has been applied to simulate a model burner, which has been experimentally investigated at the Technische Universität Berlin [24]. The selected case is operated with a fuel-rich mixture of methane and air (equivalence ratio $\phi = 1.3$). The fresh gas is injected through a convergent nozzle into the atmosphere ($T_0 = 293$ K, $p_0 = 1$ atm) and then ignited outside. This results in a primary premixed flame at rich condition, which is followed by a diffusive post-combustion. The Reynolds number Re based on the diameter D of the nozzle at the exit plane and the bulk flow velocity is 7500.

The simulation domain consists of a part of the convergent nozzle (zone 1) and a large cylindrical domain downstream (zone 2: 23D in length and diameter), which are connected by the interface, as illustrated in Fig. 4. The computational grid consists of $N_t = N_1 + N_2 = 2.2 + 4.4 = 6.6$ million finite volumes with

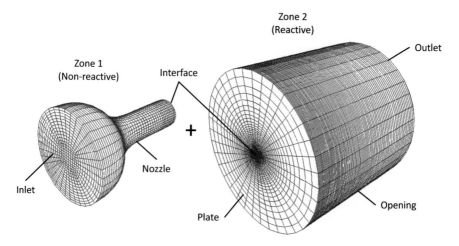

Fig. 4 Schematic drawing of grids and boundary conditions used for the 2-zonal LES

Table 1 List of boundary conditions

Zone 1			Zone 2			
Inlet	Nozzle	Interface		Plate	Opening	Outlet
u	Inflow generator	Non-slip	Zero-gradient	*Mapped from zone 1*	Non-slip	Zero-gradient	Zero-gradient
T	293 K	293 K	Zero-gradient	*Mapped from zone 1*	293 K	293 K	Zero-gradient
p	NRBC	Zero-gradient	*Mapped from zone 2*	NRBC	Zero-gradient	NRBC	NRBC

hexahedral shaped elements and is systematically refined along the burner wall and the shear layer up to the smallest length of $\Delta_{min} = 0.2$ mm. An overview of boundary conditions used in this case is given in Table 1. The flow velocities and turbulence parameters at the inlet of zone 1 are set according to LDA (laser doppler anemometry) measurement within the nozzle, which are used by the inflow-generator boundary condition. At the interface, velocity and temperature calculated in zone 1 are mapped to the inlet surface of zone 2, thereby providing necessary inlet conditions for the combustion simulation in zone 2.

For the non-reactive LES in zone 1, only a fixed-composition flow is considered, whereas a reacting mixture is used for zone 2, which also accounts for thermo-physical properties (e.g. heat capacities) of the chemical species. For the combustion modeling in zone 2 (see Sect. 2.1), the look-up table with pre-calculated chemical scalars has a size of $N_\xi \times N_{\xi''^2} \times N_\theta = 41 \times 11 \times 29$, and is generated by solving 1D premixed flames with the **CHEMKIN** code [10] and employing the GRI-30 reaction mechanism [25]. Although the GRI-30 mechanism contains more than 50 reacting species, only 14 species with relatively large mass fractions have been taken into account in order to save computing time.

The time step was set to $\Delta t = 25\,\mu s$, which allows the overall CFL (Courant-Friedrichs-Lewy) number to be below 0.5. After running the case for some flow-through times, statistical averaging and sampling of the flow variables has been conducted over a total simulation time of 3.6 s (144,000 time steps). The simulation has been carried out in parallel with 300 processors from the HPC platform ForHLR I [26] by consuming approx. 50,000 CPU hours. For comparison, a reference solution has been created with the conventional single-regional approach with the same boundary conditions, as well as the same mesh topology.

3.3 Simulation Results

Figure 5 shows instantaneous contours of the calculated velocity components and temperature in a plane passing through the centerline axis. The flame front is indicated by iso-contours of the progress variable $\tilde{\theta} = 0.1/0.3/0.5/0.7/0.9$ therein, which is corrugated and stretched by the turbulent flow. As Re is relatively low, large ring vortices or coherent flow structures are generated at the shear layer. Further downstream, these vortical structures become unstable and break down into pieces. The evolution of the flame is accompanied by these vortical motions. Their mutual interaction can be depicted by the external shape of the temperature contours in Fig. 5. The interface boundary located at the exit plane of the nozzle is indicated in

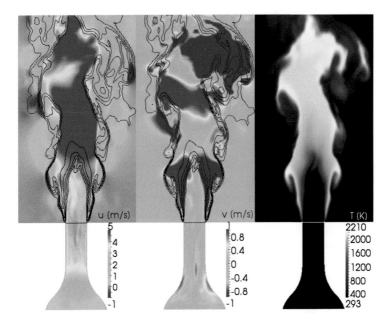

Fig. 5 Instantaneous contours of the streamiest and radial velocity as well as the temperature

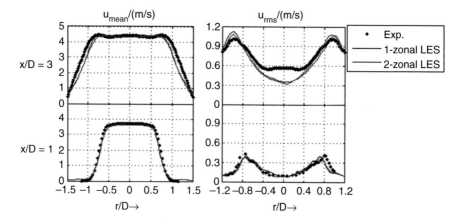

Fig. 6 Comparison of streamwise velocities calculated with the 1- and 2-zonal LES with the experiment: time mean values on the *left* and root mean square (rms) values on the *right*

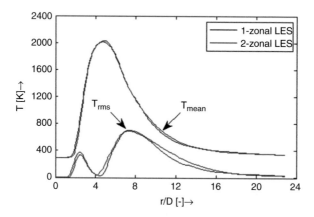

Fig. 7 Mean/rms temperature profiles along the centerline axis calculated with 1- and 2-zonal LES

each sub-plot of Fig. 5 by the horizontal line. A smooth transition of the velocity components through the interface can be identified, showing that the transmission of data from zone 1 to zone 2 through the interface works correctly.

Figure 6 compares the radial profiles of the time mean and root mean square (rms) streamwise velocity calculated by both 1- and 2-zonal LES together with the measured data. A quantitatively good agreement between the 1- and 2-zonal simulation has been achieved. Similar results have been found for the mean and rms temperature distributions along the centerline axis too, as shown in Fig. 7. Therefore, statistics of the flow field as well as flame properties (e.g. length and thickness) as obtained from the single-regional modeling can be reproduced well with the multi-regional simulation. This confirms the validity of the hybrid simulation for combustion modeling.

3.4   Performance

Figure 8 shows breakdowns of the execution time generated from different solution stages of a non-reactive and a reactive LES, which have been run with the same grid and boundary conditions as described in Sect. 3.2. The measurement was conducted in serial on the ForHLR I cluster (processor type: Intel Xeon E5-2670) [26] and for one single time loop. It is evident that the simulation without combustion reaction is faster than the reactive one by $\eta = 68\%$, which is mainly attributed to the solution of transport equations for the control variables as well as the multi-dimensional interpolation while searching species mass fractions from the look-up table (see Sect. 2.1). For the reacting case, the temperature is evaluated from specific enthalpies of the species by means of Newton's iterative method, which leads to further computational effort. The solutions of the Navier-Stokes equations and pressure correction equation in the combustion case are more time-consuming compared to the non-reactive case, because the chemical reaction causes large fluctuations of the fluid properties like density or viscosity, so that more iteration loops are required to obtain a converged solution. According to Eq. (2) and the measurement shown in Fig. 8, the theoretical speed-up factor by using the two-regional approach is calculated to

$$S = \alpha \cdot \eta = 0.33 \cdot 0.68 \approx 22.4\% \qquad (3)$$

where $\alpha = 0.33$ represents the proportion of grid cells assigned to zone 1 (non-reactive).

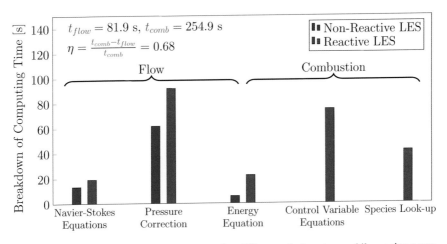

Fig. 8 Comparison of execution time consumed at different solution stages while running a non-reactive and a reactive LES

The method for parallel computing used by OpenFOAM is known as domain decomposition, where the computational grid associated with internal cell volumes and boundary patch elements are partitioned to a number of sub-domains and allocated to the same number of processors. To exploit the full capacity of the CPUs, each grid used for zone 1 and zone 2 has to be assigned with the same number of sub-domains or processors. Figure 9 on the left compares parallel performance of the single- and double-regional LES, where a better scale-up behavior" is observed for the former one. In this case, as the number of nodes per processor decreases with the number of CPU's used, the share of actual calculation time in comparison to that of multi-processor communications (e.g. data transfer, synchronization) decreases, which leads to a reduced efficiency of the parallel computing. This is especially the case for the non-reactive modeling in zone 1, as it is much less demanding in computing time than the combustion simulation in zone 2. The one-zonal LES is more CPU-intensive due to chemistry modeling for the whole domain, and therefore, a better parallel performance can be achieved.

This issue has a major impact on the speed-up behavior of the multi-regional simulation in comparison to the single-regional one. As indicated on the right hand side of Fig. 9, the speed-up factor S decreases with increased number of processors. The ideal value of $S > 20\%$ [see Eq. (3)] is only obtained by using less than 60 CPU cores. Nevertheless, an acceleration of the computing process by $S > 10\%$ has been achieved in parallel runs with up to 120 processors for the current case, which validates the performance improvement by using the proposed multi-regional method. Basically, better scalability can be accomplished with an increased node number for zone 1, because the non-reactive LES becomes more computationally intensive and can be run more efficiently in parallel mode. This also leads to an increased α and a better speed-up S of the multi-regional modeling.

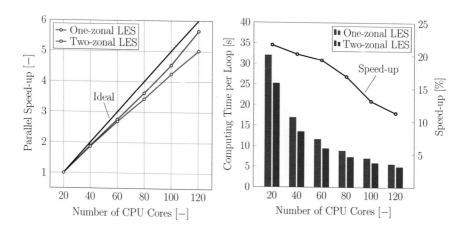

Fig. 9 Comparison of parallel performance of one- and two-zonal LES (*left*) and speed-up performance of the two-regional LES by using different numbers of CPU cores (*right*)

4   Summary and Outlook

A multi-regional concept in the framework of turbulent combustion modeling has been explored in the present work. It is based on the separate solution of the non-reactive region within the burner and the reaction zone downstream. One numerical example with LES of a model flame has been used to demonstrate the practicability of this approach. The multi-zonal simulation has been confirmed to be able to reproduce results of the same quality as obtained with the conventional single-regional simulation. However, the hybrid modeling is computationally more efficient compared to the 1-zonal run. The method may generate particular interest when applying the more fine-grained LES methodology instead of the commonly used RANS approach for industrial applications. Also, the multi-zonal approach for LES combustion modeling may be incorporated in general engineering CFD codes. In this case, millions of core hours may be saved for serial tests on prototypes, for example, if a series of operation conditions with different mass flow rates or equivalence ratios have to be tested. In addition, modifications of the grids for zone 1 (burner) and zone 2 (combustion chamber) can be conducted independently, which avoids the generation of computational grids for the whole system. The highly reliable results from LES can improve the predictability of instationary effects like combustion instabilities. Consequently, the time scale and cost required for the development of modern combustion devices may be reduced.

Acknowledgements The authors wish to acknowledge the financial support by the German Research Council (DFG) through the Research Unit DFG-BO693/27 "Combustion Noise" and the computing time from the ForHLR I cluster at SCC at KIT.

References

1. Poinsot, T., Veynante, D.: Theoretical and Numerical Combustion, 2nd edn. Edwards Inc., Philadelphia, PA (2005). ISBN 978-1-930217-10-2
2. Peters, N.: Turbulent Combustion. Cambridge University Press, Cambridge (2000)
3. van Oijen, J., Lammers, F., de Goey, L.: Modeling of complex premixed burner systems by using flamelet-generated manifolds. Combust. Flame **127**(3), 2124–2134 (2001)
4. Fröhlich, J., von Terzi, D.: Hybrid LES/RANS methods for the simulation of turbulent flows. Prog. Aerosp. Sci. **44**(5), 349–377 (2008)
5. Spalart, P.R.: Comments on the feasibility of LES for wing and on a hybrid RANS/LES approach. In: First ASOSR CONFERENCE on DNS/LES, Arlington, TX (1997)
6. Strelets, M.: Detached Eddy simulation of massively separated flows. In: AIAA 2001–0879 (2001)
7. Schwarz, A., Janicka, J. (eds.): Combustion Noise. Springer, Berlin/Heidelberg (2009). ISBN-10: 3642020372
8. Zhang, F., Habisreuther, P., Bockhorn, H., Nawroth, H., Paschereit, C.O.: On prediction of combustion generated noise with the turbulent heat release rate. Acta Acustica united Acustica **99**, 940–951 (2013). doi:10.3813/AAA.918673
9. Zhang, F.: Numerical modeling of noise generated by turbulent combustion. Dissertation, Karlsruhe Institute of Technology/Shaker Verlag, Germany (2014). ISBN: 978-3-8440-2508-8

10. Kee, J., Rupley, F., Miller, J.: Chemkin-II: a Fortran chemical kinetics package for the analysis of gas-phase chemical kinetics, Report No. SAND89-8009B, Sandia National Laboratories (1989)
11. Schmid, H., Habisreuther, P., Leuckel, W.: A model for calculating heat release in premixed turbulent flames. Combust. Flame **113**, 79–91 (1998)
12. Fröhlich, J.: Large Eddy Simulation Turbulenter Strömungen. Teubner Verlag, Stuttgart (2006)
13. Zhang, F., Habisreuther, P., Hettel, M., Bockhorn, H.: Modelling of a premixed swirl-stabilized flame using a turbulent flame speed closure model in LES. Flow Turbul. Combust. **82**, 537–551 (2009)
14. Zhang, F., Habisreuther, P., Bockhorn, H.: A unified TFC combustion model for numerical computation of turbulent gas flames. In: Nagel, W.E., Kröner, D.B., Resch, M.M. (eds.) High Performance Computing in Science and Engineering '12. Springer, Berlin/Heidelberg (2013). ISBN: 978-3-642-33374-3
15. Zhang, F., Habisreuther, P., Hettel, M., Bockhorn, H.: A newly developed unified turbulent flame speed closure (UTFC) combustion model for numerical simulation of turbulent gas flames. In: Proceedings of the 25, Deutscher Flammentag, pp. 177–182 (2011)
16. Bilger, R.: Conditional moment closure for turbulent reacting flow. Phys. Fluids **5**(2), 436–444 (1993)
17. Navarro-Martinez, S., Kronenburg, A.: LES-CMC simulations of a lifted methane flame. Proc. Combust. Inst. **32**, 1509–1516 (2009)
18. Domenico, M.D., Gerlinger, P., Noll, B.: Numerical simulations of confined, turbulent, lean, premixed flames using a detailed chemistry combustion model. In: Proceedings of ASME Turbo Expo 2011, GT2011-45520
19. OpenCFD Ltd.: OpenFOAM user guide, Version 2.3.0 (2014)
20. Ferziger, J., Perić, M.: Computational Methods for Fluid Dynamics. Springer, Berlin (2002)
21. Klein, M., Sadiki, A., Janicka, J.: A digital filter based generation of inflow data for spatially developing direct numerical or large eddy simulations. J. Comput. Phys. **286**, 652–665 (2003)
22. Poinsot, T., Lele, S.: Boundary conditions for direct simulation of compressible viscous flows. J. Comput. Phys. **101**, 104–129 (1992)
23. Farrell, P.E., Maddison, J.R.: Conservative interpolation between volume meshes by local Galerkin projection. Comput. Methods Appl. Mech. Eng. (2010). doi:10.1016/j.cma.2010.07.015
24. Nawroth, H., Paschereit, C.O., Zhang, F., Habisreuther, P., Bockhorn, H.: Flow investigation and acoustic measurements of an unconfined turbulent premixed jet flame. In: AIAA Paper 2013–2459 (2013)
25. Smith, G.P., Golden, D.M., Frenklach, M., Moriarty, N.W., Eiteneer, B., Goldenberg, M., Bowman, C.T., Hanson, R.K., Song, S., Gardiner, W.C., Jr., Lissianski, V.V., Qin, Z.: http://www.me.berkeley.edu/gri_mech/ (2000)
26. ForHLR user guide. Steinbuch Centre for Computing (SCC), Karlsruhe Institute of Technology (KIT). http://www.bwhpc-c5.de/wiki/index.php/ForHLR (2014)

Numerical Simulations of Rocket Combustion Chambers on Massively Parallel Systems

Roman Keller, Martin Seidl, Markus Lempke, Peter Gerlinger, and Manfred Aigner

Abstract The compressible, implicit combustion code TASCOM3D is used for the simulation of rocket combustion chambers. Coupled Euler-Lagrange simulations for a subcritical operated model rocket combustor at 5 bar pressure are performed. A quarter of the rectangular combustor is discretized for three-dimensional RANS simulations. Three spray simulations with different initial droplet size distributions are performed. Simulation results exhibit only limited dependency on the initial droplet size distribution. Finally, performance of TASCOM3D on HERMIT and HORNET is compared. A different scaling behavior on both machines with respect to local block size variations is observed.

1   Introduction

The in-house code TASCOM3D is used for numerical simulations of rocket combustion chambers. The main aspects to be considered for CFD simulations of rocket combustors are:

1. turbulence phenomena
2. combustion processes
3. thermodynamic and molecular transport properties
4. grid resolution and discretization

For an accurate simulation of rocket combustors it is essential to predict fluid properties and flow phenomena with sufficient accuracy. The fluid properties may differ significantly from an ideal gas behavior due to the extreme conditions in rocket engines. Pressures up to 100 bar and more and temperatures from below 100 K for the injected propellants up to about 4000 K within the reaction zone in the combustion chamber make these simulations very challenging. The fluids in the combustion chamber can be in different states of matter (gas-like or liquid-like) depending on the pressure and temperature. If a propellant is injected at cryogenic

R. Keller (✉) • M. Seidl • M. Lempke • P. Gerlinger • M. Aigner

Institut für Verbrennungstechnik der Luft- und Raumfahrt, Pfaffenwaldring 38-40, 70569 Stuttgart, Germany

e-mail: roman.keller@dlr.de

© Springer International Publishing Switzerland 2016

W.E. Nagel et al. (eds.), *High Performance Computing in Science and Engineering '15*, DOI 10.1007/978-3-319-24633-8_19

temperature and pressure below the thermodynamic critical pressure of the fluid, a discontinuous phase transition from liquid to gas will occur during heat up. The liquid and gaseous phase must be handled separately. In the present work, this is done with a Lagrangian approach for the liquid phase (tracking of the liquid droplets) and an Eulerian approach for the gaseous phase.

If injection temperature and/or pressure are above the critical temperature and/or pressure of the fluid, a continuous transition from liquid-like state to a gaseous state is observed. The liquid-like and gaseous phase can no longer be distinguished and a combined treatment is required. This is achieved within a single-fluid model based on real gas thermodynamics which is briefly introduced in Sect. 3.

This leads to three different types of simulations for rocket combustion chambers:

1. simulations with gaseous phase only
2. simulations with liquid and gaseous phase below the critical pressures
3. simulations with liquid-like and gaseous phase above the critical pressure

The first type requires a gas phase solver with ideal gas thermodynamics only. Although operation conditions can be above the critical pressure where the fluid is considered to be supercritical, the fluid behavior can be modeled with ideal gas thermodynamics. An example for such a simulation of a combustion chamber is the Pennstate Preburner Combustor [11]. These gas-only combustion chambers have practical significance for full-flow staged combustion (FFSC) cycles where both fluids enter the combustion chamber in a preburned gaseous state.

The second approach, simulation of chambers with liquid and gaseous phase at subcritical conditions (low to medium pressure), requires separate treatment for the liquid and gaseous phase. Simulations of a combustion chamber with coupled liquid and gaseous solvers are presented in [12] and in Sect. 4. These operation conditions occur in upper-stage combustion chambers or during the ignition phase while the chamber pressure is still low.

For simulations of cryogenic fluid injection at supercritical pressure, the fluid properties can deviate significantly from the ideal gas behavior. Thus real gas modeling is required. Fortunately, no phase transition occurs and the entire regime can be modeled in the same framework as a single phase. These operation conditions are present in most main stage rocket engines and therefore are very interesting for future research.

Due to the complexity of the employed models and the resulting high computing times, the utilization of high performance computing systems is inevitable. Consequently, it is crucial to examine the performance of the simulation codes on the used platforms. This was done for the Cray XE6 (HERMIT) in [7] and performance of the Cray XE6 and the Cray XC40 (HORNET) will be compared in this report.

A brief introduction of the codes is given in Sect. 2, followed by a brief introduction to real-gas thermodynamics in Sect. 3. Then, results for the M3 combustion chamber are shown in Sect. 4. Code performance on HORNET is compared with HERMIT in Sect. 5.

2 Numerical Method

The scientific in-house code TASCOM3D (Turbulent All Speed Combustion Multigrid Solver 3D) has been applied successfully during the last two decades to
simulate reacting and non-reacting super- and subsonic flows. Reacting flows are
described by solving the fully compressible Navier-Stokes, turbulence and species
transport equations. Additionally, an assumed PDF (probability density function) approach is available to take turbulence-chemistry-interaction into account.
The three-dimensional conservative form of the Reynolds-averaged Navier-Stokes
equations is given by

$$\frac{\partial \mathbf{Q}}{\partial t} + \frac{\partial (\mathbf{F} - \mathbf{F}_v)}{\partial x} + \frac{\partial (\mathbf{G} - \mathbf{G}_v)}{\partial y} + \frac{\partial (\mathbf{H} - \mathbf{H}_v)}{\partial z} = \mathbf{S}, \tag{1}$$

where

$$\mathbf{Q} = [\rho, \rho u, \rho v, \rho w, \rho E, \rho k, \rho \omega, \rho Y_i]^T, \ i = 1, 2, \ldots, N_k - 1. \tag{2}$$

The conservative variable vector $\mathbf{Q}$ consists of the density ρ, the velocity
components u, v and w, the total specific energy E, the turbulence variables k and ω
and the species mass fractions Y_i. N_k is the total number of species in the gaseous
phase. $\mathbf{F}$, $\mathbf{G}$, and $\mathbf{H}$ are the vectors specifying the inviscid fluxes in the x-, y- and z-
direction, respectively. $\mathbf{F}_v$, $\mathbf{G}_v$, and $\mathbf{H}_v$ are their viscous counterparts. The source
vector $\mathbf{S}$ includes terms from turbulence, chemistry and the liquid phase and is
given by

$$\mathbf{S} = [0, 0, 0, 0, S_E, S_k, S_\omega, S_{Y_i}]^T, \tag{3}$$

where S_E is the energy source term from the liquid phase, S_q and S_ω are the source
terms of the turbulence variables and S_{Y_i} the source terms of the species mass
fractions due to combustion. For turbulence closure the k-ω model of Wilcox [20] is
used. Additionally some other two-equation models are available like the q-ω model
of Coakley [1] and a Delayed Detached Eddy Simulation (DDES) approach [18].
The spatial discretization is performed on block structured grids based on a
finite volume scheme. For the reconstruction of the cell interface values, MLP^{ld}
(Multidimensional Limiting Process—low diffusion) [5] with up to sixth order
is used to prevent oscillations at discontinuities. MLP uses diagonal values to
improve the TVD (Total Variation Diminishing) limiting behavior [22]. Using these
interface values, the AUSM$^+$-up flux vector splitting [13] is employed to calculate
the inviscid fluxes. The unsteady set of Eq. (1) is solved with an implicit Lower-
Upper Symmetric Gauss-Seidel (LU-SGS) [6] algorithm. Furthermore, finite-rate
chemistry is treated in a fully coupled manner. The code is parallelized with
Message Passing Interface (MPI). More details concerning TASCOM3D may be
found in [4, 6, 19].

For multiphase simulations the flow solver TASCOM3D is coupled with the Lagrangian particle tracking code SPRAYSIM [8]. The liquid phase is assumed to be a dilute spray of discrete, non-interacting droplets. In order to limit the computational effort multiple droplets are represented by computational parcels. For each numerical parcel the equations of motion for mass, momentum, and energy

$$\frac{\partial \mathbf{Q}_d}{\partial t} = \mathbf{S}_d \tag{4}$$

are solved where the variable vector $\mathbf{Q}_d$ given by

$$\mathbf{Q}_d = [D_d, u_d, v_d, w_d, T_d]^T \tag{5}$$

includes the droplet diameter D_d, the velocities u_d, v_d, w_d, and the droplet temperature T_d. In addition to the provision of the required gas phase values to the Lagrange solver, the spray feedback is considered by regarding the accumulated spray source terms in the mass, momentum, turbulence, and species Eq. (1) of the gas phase solver (two-way coupling). For more details the interested reader is referred to Lempke et al. [10, 12].

3   Non-ideal Fluids in Rocket Combustors

3.1   Non-ideal Thermodynamics

For sufficiently low pressures and high temperatures, the thermodynamic relation between pressure, temperature, and density can accurately be described by the well-known ideal gas (IG) equation of state (EOS)

$$p = \rho \frac{R_u}{M_w} T \tag{6}$$

where p, T, M_w and R_u are the pressure, temperature, molecular weight of the mixture and the universal gas constant. However, with increasing pressure and decreasing temperature, deviations from this law are observed and cannot be neglected anymore for conditions like they occur in rocket combustors. This is due to the fact that the ideal gas law neglects the volume of the molecules and inter-molecular attractive forces. Both effects are important for high-density conditions. A large number of so-called 'real gas', better called 'real fluid', equations of state have been developed to account for these effects. One of the most famous and simplest is the Soave-Redlich-Kwong (SRK) EOS

$$p = \frac{\rho R_u T}{M_w - b\rho} - \frac{a}{M_w} \frac{\rho^2}{M_w + b\rho}. \tag{7}$$

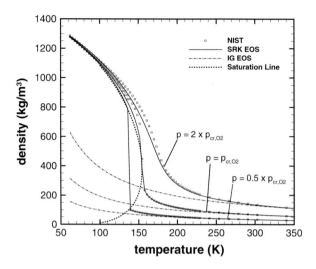

Fig. 1 Density of oxygen versus temperature for three different pressure levels

The parameters a (temperature dependent) and b for a mixture are obtained via mixing and combining rules from their pure species counterparts. The SRK is generally applicable to any pure fluid or mixture and continuously describes the p-ρ-T-relation for gases, liquids, and multi-phase regimes with remarkable accuracy over a wide range of thermodynamic states. For a more detailed description and a general introduction to real fluid properties, the interested reader is referred to textbooks on thermodynamics, e.g. [15]. Figure 1 shows the density-temperature relation for three different pressures for pure oxygen which is used as oxidizer in many rocket engines. Values from NIST database are plotted together with values predicted by the SRK EOS. Depending on the pressure level in the combustor and the injection temperatures, oxygen may be injected in a gas-like state (low density) or a liquid-like state (high density).

For chamber pressures below the thermodynamic critical pressure of oxygen at $p_{cr,O_2} = 50.43$ bar and cryogenic injection temperatures below the saturation temperature, the liquid oxygen (LOX) will undergo a discontinuous phase transition during heat up in the chamber. Surface tension between the liquid and gaseous phase leads to an abrupt and distinct separation of both phases. Associated flow phenomena are primary and secondary atomization of the LOX jet into small ligaments and droplet and their final evaporation into the gas phase.

In contrast, for pressures above the critical pressure or sufficiently high temperatures of the injected oxygen, only a single phase will occur and a continuous transition from the cool injection conditions to the hot reaction zone is observed.

For a consistent implementation of a real fluid EOS into a CFD code, it is important to use Eq. (7) in combination with fundamental thermodynamic relations for the calculation of other thermodynamic properties like enthalpy or speed of sound. For more details see for example [15, 21].

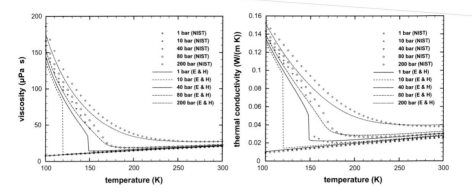

Fig. 2 Viscosity (*left*) and thermal conductivity (*right*) of oxygen versus temperature for various pressures

3.2 Non-ideal Molecular Transport Properties

In addition to deviations of thermodynamic properties from ideal gases, also deviations of molecular transport properties like viscosity, thermal conductivity, and diffusion coefficients from ideal gas values have to be considered. These are usually calculated from empirical models. For example, for non-polar fluids or fluid mixtures, the model of Ely and Hanley [2, 3] may be used for prediction of real fluid viscosities and thermal conductivities. Figure 2 shows a comparison of both properties for oxygen between values from NIST database and values obtained from the model of Ely and Hanley in combination with the use of the SRK EOS.

3.3 Non-ideal Flow Phenomena

Apart from thermodynamic and transport properties, certain flow phenomena may become non-negligible for conditions present in rocket engines. For example, the Soret effect (mass diffusion due to a temperature gradient) or the reciprocal Dufour effect (energy flux due to species concentration gradients) may become important in locally confined regions. Oefelein [14], however, observed that for injection of propellants with shear-coaxial injectors (typical for many rocket engines) their contribution may be neglected.

4 The M3 Combustion Chamber

The M3 Combustion Chamber is a single injector test chamber providing optical access for diagnostic techniques as well as investigations on spray disintegration and vaporization. It is operated at the DLR research center Lampoldshausen [17].

The operation conditions are comparable to upper-stage engine conditions like the HM7B or RL 10A. The chamber pressure for the simulated test case is 5 bar. The chamber has a length of 140 mm and a quadratic cross section of 60×60 mm. The subcritical liquid oxygen (LOX) is injected with 24 m/s. The mass flow rates of hydrogen and LOX are 5.5 and 39 g/s, respectively [17]. The test case has been investigated numerically by Schlotz [16] but only the rear part of the chamber was simulated. For the numerical setup of the present 3D-simulation, symmetry is exploited by discretizing only a quarter of the squared chamber to limit the number of grid cells to 1.5 million. The grid is refined close to the wall in order to achieve $y^+ < 1$ (Fig. 3). The length of the hydrogen injector is 20 mm and a fully developed turbulent flow is imposed at the inflow. The LOX spray is initialized using size and velocity distribution from the experimental investigation [17] assuming a Rosin-Rammler droplet size distribution. Adiabatic no-slip boundary conditions are applied at the chamber walls.

Figure 4 shows the temperature distribution on the symmetric planes as well as representative particle traces and corresponding droplet sizes. Three different initial droplet size distributions are simulated. All the distributions are extracted from the measured values of Sender et al. [17] and fitted to a Rosin-Rammler distribution. The droplet distributions were measured in the entire chamber, but not enough droplets were detected at the inflow to get sufficient statistical data for numerical boundary conditions. Thus the initial droplet size distribution must be extrapolated from the measured values within the chamber. For case one, the first quarter of the chamber is considered, case two considers all droplets up to half of the chamber length whereas case three takes all measured particles in the chamber into account.

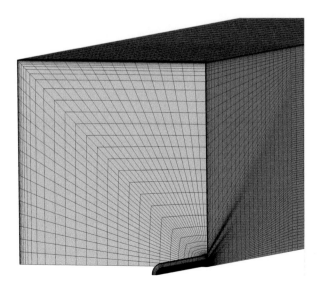

Fig. 3 Numerical grid for the M3 calculation (every second grid point is shown)

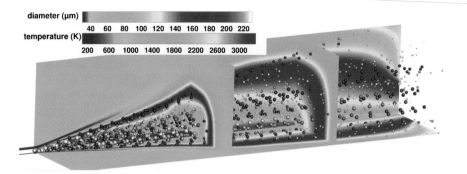

Fig. 4 Temperature distribution in the symmetry planes and two cross-sections along the chamber axis and particle positions with diameter

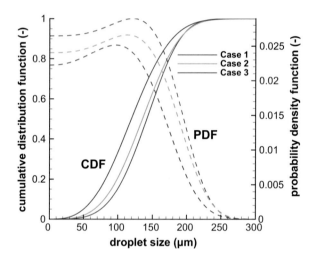

Fig. 5 CDF and PDF of the initial droplet distributions for the three simulation cases

It should be noted that the experimenter themselves stated that they are not very confident with their measured droplet sizes.

Figure 5 shows the probability density function (PDF) and the cumulative distribution function (CDF) of the assumed initial distributions. The different initial distributions are chosen to investigate the influence of the spray's initial conditions.

Figure 6 shows the temperature profiles for different positions along the x-axis for the investigated cases. A small influence of the different initial spray conditions on the temperature profiles is visible. The differences are in general more noticeable close to the symmetry plane. In contrast to most results from literature [9], the initial droplet size distribution has only a minor impact. While distributions extrapolated from only the measured values were tested, using idealized distributions from literature or other experiments could result in larger deviations. Due to the large uncertainties in the measured distributions this was not done.

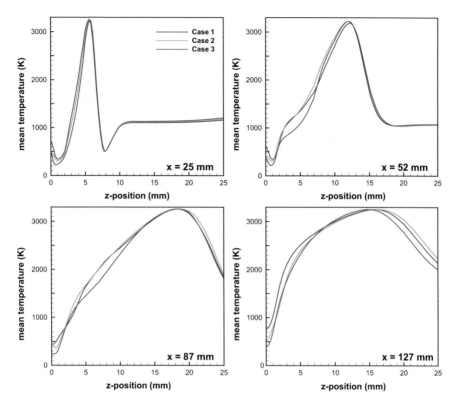

Fig. 6 Temperature profiles along the x-axis

In Fig. 7, the OH mole fraction is compared with a reverse Abel transform of the OH* emission. The different initial droplet distributions did not change the hydroxyl distribution significantly and therefore only case 3 is shown. The OH* emission shows strong asymmetric behavior which is not observed in the RANS simulation. The simulation predicts a thinner OH layer at the beginning of the chamber and the maximum values are visible at the end of the chamber. For the measured emission, however, the maximum value is reached shortly before the middle of the chamber. Both simulation and experiment show a broadening of the OH around one third of the chamber length. While the exact characteristics of the measured OH* emission do not match the simulation, the overall agreement is reasonable.

5   Performance Comparison of HERMIT and HORNET

In the last HLRS report, a comprehensive performance analysis for TASCOM3D and SPRAYSIM on the CRAY XE6 (HERMIT) was performed [7]. In this report, the speedup of TASCOM3D on the CRAY XC40 (HORNET) compared to HERMIT

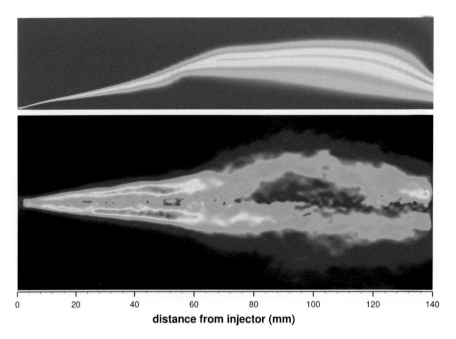

Fig. 7 Hydroxyl mole fraction in the symmetry plane (*top*) and the OH* emission image from the experiment (*bottom*)

Table 1 Performance comparison of TASCOM3D for HERMIT and HORNET

	2D (16 × 16)	2D (32 × 32)	Speedup 32/16	3D
HERMIT	210 s	856 s	4.08	587 s
HORNET	90 s	307 s	3.41	241 s
Speedup	2.3	2.79		2.44

is investigated. Table 1 shows calculation times for two-dimensional (blocksizes of 16 × 16 and 32 × 32 cells) and three-dimensional calculations on both systems. All calculations on HORNET show a speedup compared to the same calculations on HERMIT of at least a factor of two which is the factor for increased computing charge for HORNET compared to HERMIT due to faster CPUs. Thus calculations on HORNET are more cost efficient. The speedup on HORNET compared to HERMIT is more profound on larger blocksizes for the 2D calculations. A simple explanation for this may be larger cache sizes on HORNET's Haswell CPUs. Thus the decrease of the blocksize from 32 × 32 to 16 × 16, which quadruples the number of CPUs, is no longer efficient on HORNET like it is on HERMIT. These results suggest, that the investigation of the most efficient blocksizes for 2D and 3D calculations [7] should be repeated on HORNET.

6 Conclusion

A general overview of relevant simulation approaches for rocket combustion chamber was outlined. A new test case for liquid-gas coupling of SPRAYSIM and TASCOM3D was shown in this report. The real-gas thermodynamics required for combustion chambers operated at high-pressures was briefly introduced as a current and future topic in this area. The M3 combustion chamber shows only limited dependency in the temperature profiles on the initial droplet size distribution. The measured OH* distribution and the simulated hydroxyl mole fraction show a similar behavior. The transition from HERMIT to HORNET showed speedups larger than a factor of two for all cases. Since the cost for one CPU-hour on HORNET is twice as high as on HERMIT, this is desirable. On HERMIT, a reduction in blocksizes leads to increased efficiency for down to 16×16 blocksizes, this is no longer the case on HORNET. This implies that new optimal blocksizes for the 2D and 3D simulations should be investigated. Also, the limited dependency on the initial droplet distribution is in contrast with results from literature and should be investigated further using distributions with larger discrepancies. Furthermore, much better agreement with experimental results for the M3 combustion chamber is expected from unsteady (URANS or LES) simulations.

Acknowledgements The presented work was performed within the framework of the SFBTR 40 funded by the Deutsche Forschungsgemeinschaft (DFG). This support is greatly appreciated. All simulations were performed on the Cray XE6 (HERMIT) and XC40 (HORNET) cluster at the High Performance Computing Center Stuttgart (HLRS) under the grant number scrcomb. The authors wish to thank for the computing time and the technical support.

References

1. Coakley, T.J.: Turbulence modeling for high speed flows. In: AIAA Paper, No. 97–0436 (1992)
2. Ely, J.F., Hanley, J.M.: Prediction of transport properties. 1. Viscosity of fluids and mixtures. Ind. Eng. Chem. Fundam. **20**, 323–332 (1981)
3. Ely, J.F., Hanley, J.M.: Prediction of transport properties. 2. Thermal conductivity of pure fluids and mixtures. Ind. Eng. Chem. Fundam. **22**, 90–97 (1983)
4. Gerlinger, P.: Investigation of an assumed pdf approach for finite-rate chemistry. Combust. Sci. Technol. **175**(5), 841–872 (2003)
5. Gerlinger, P.: Multi-dimensional limiting for high-order schemes including turbulence and combustion. J. Comput. Phys. **231**, 2199–2228 (2012)
6. Gerlinger, P., Möbus, H., Brüggemann, D.: An implicit multigrid method for turbulent combustion. J. Comput. Phys. **167**(2), 247–276 (2001)
7. Keller, R., Lempke, M., Simsont, Y.H., Gerlinger, P., Aigner, M.: Parallelization and performance analysis of an implicit compressible combustion code for aerospace applications. In: High Performance Computing in Science and Engineering'14, pp. 251–266. Springer, Berlin (2015)
8. Le Clercq, P., Doué, N., Rachner, M., Aigner, M.: Validation of a multicomponent-fuel model for spray computations. In: AIAA Paper 2009-1188 (2009)
9. Lefebvre, A.H.: Atomization and Sprays. Taylor & Francis, New York (1989)

10. Lempke, M., Gerlinger, P., Rachner, M., Aigner, M.: Numerical simulation of subcritical model
    rocket combustors with single shear coaxial injectors. Sonderforschungsbereich/Transregio 40
    1, 201–212 (2010)
11. Lempke, M., Keller, R., Gerlinger, P.: Influence of spatial discretization and unsteadiness on
    the simulation of rocket combustors. Int. J. Numer. Methods Fluids **79**, 437–455 (2015)
12. Lempke, M., Gerlinger, P., Seidl, M.J., Aigner, M.: Unsteady High-Order Simulation of a
    Liquid Oxygen/Gaseous Hydrogen Rocket Combustor. J. Propuls. Power (2015). doi:10.2514/
    1.B35726
13. Liou, M.S.: A sequel to AUSM, part II : AUSM$^+$-up for all speeds. J. Comput. Phys. **214**(1),
    137–170 (2006)
14. Oefelein, J.C.: Large eddy simulation of a shear-coaxial LOX-H$_2$ jet at supercritical pressure.
    In: AIAA Paper 2002-4030 (2002)
15. Poling, B.E., Prausnitz, J.M., O'Connell, J.P.: The Properties of Gases and Liquids, 5th edn.
    McGraw-Hill, New York (2001)
16. Schlotz, D.: Modellierung laminarer und turbulenter Flüssig-Sauerstoff/Wasserstoff-
    Sprayflammen unter kryogenen Hochdruckbedingungen. Ph.D. thesis, Universität Stuttgart
    (2001)
17. Sender, J., Lecourt, R., Oschwald, M., Haidn, O.: Application of droplet-tracking-velocimetry
    to LOX/GH$_2$ coaxial-spray combustion with varying combustion chamber pressures. In:
    Annual Conference on Liquid Atomization and Spray Systems, 13th, Florence, Italy, 9–11
    July, 1997, ONERA, TP, 1997-139 (1997)
18. Spalart, P., Deck, S., Shur, M., Squires, K., Strelets, M., Travin, A.: A new version of detached-
    eddy simulation, resistant to ambiguous grid densities. Theor. Comput. Fluid Dyn. **20**(3),
    181–195 (2006)
19. Stoll, P., Gerlinger, P., Brüggemann, D.: Domain decomposition for an implicit LU-SGS
    scheme using overlapping grids. In: AIAA Paper, pp. 97–1869 (1997)
20. Wilcox, D.C.: Formulation of the k-ω turbulence model revisited. AIAA J. **46**(11), 2823–2838
    (2008)
21. Yang, V.: Liquid-propellant rocket engine injector dynamics and combustion processes at
    supercritical conditions. Technical Report, Department of Mechanical Engineering, The
    Pennsylvania State University (2004)
22. Yoon, S.H., Kim, C., Kim, K.H.: Multi-dimensional limiting process for three-dimensional
    flow physics analyses. J. Comput. Phys. **227**(12), 6001–6043 (2008)

Part V
Computational Fluid Dynamics

Ewald Krämer

As already in the previous years, a large number of high-class research in the field of Fluid Dynamics has been conducted on the supercomputers of the HLRS during the reporting period. This emphasises the indispensability of High Performance Computing (HPC) in general and of the computational resources provided by the HLRS in particular for leading edge research in this area. Thirty Four annual reports had been submitted and underwent a peer review process. Due to limited space, only 18 projects could be selected for publication in this book, which means that a number of high-qualified contributions could not be admitted. Even though the presented collection cannot entirely represent an area this vast, the selected papers demonstrate the state-of-the-art use of high-performance computing in Germany.

Again, the spectrum of projects ranges from Direct Numerical Simulations (DNS) over Large Eddy Simulations (LES) and hybrid methods (DES) to Reynolds Averaged Navier Stokes (RANS, URANS) simulations on the one hand, and from code development over fundamental research up to application oriented research on the other hand. In most of the cases, Finite-Volume codes have been applied, but also Lattice Boltzmann Methods (LBM) have been deployed for particle-containing flows. Significant progress has been made in the development of Discontinuous Galerkin (DG) methods, which have the potential to someday supersede the largely used Finite-Volume methods. In-house codes were used in the presented projects as well as commercial codes, the latter of which have notably improved with respect of scalability on massive parallel high-performance computers. All CFD-projects presented in this book were either run on the Cray XE6 Hermit, which offers a peak performance of more than 1 PFLOP/s, or on the Cray XC40 Hornet, which was installed in 2014 and offers almost the fourfold peak performance of its predecessor.

E. Krämer (✉)
Institut für Aerodynamik und Gasdynamik, Universität Stuttgart, Pfaffenwaldring 21, 70550 Stuttgart, Germany
e-mail: kraemer@iag.uni-stuttgart.de

The CFD section starts with five projects applying DNS for the investigation of fundamental flow physical problems. Sakai and Uhlmann from the Karlsruhe Institute of Technology, Institute for Hydromechanics, have studied the mechanism of turbulence induced secondary flow formation in a straight open duct with rectangular cross-section using an in-house pseudo-spectral DNS code. Due to the free surface, the flow in an open duct significantly distinguishes from a closed duct flow and the understanding of open duct flows is less established. The presented work provides an interesting new insight into the influence of a free surface on secondary flow near the upper corners of an open channel.

Chu and Laurien from the Institute of Nuclear Technology and Energy Systems, University of Stuttgart, have investigated the heat transfer problem arising in the cooling system of nuclear power plants or heavy-duty coolers. Direct numerical simulations of carbon dioxide flow under supercritical pressures in a heated vertical pipe including buoyancy effects have been performed using the open-source code OpenFOAM. The issue is that with a supercritical fluid as working fluid in a heated pipe, relaminarization may occur, which drastically reduces cooling efficiency. Various computational cases under different conditions considering forced convection and mixed convection and including upward and downward flow have been simulated and the changes in mean flow and turbulence statistics have been analysed in detail. The findings of this work can help develop new turbulence models for this kind of practical applications.

The mechanisms of primary break-up of a liquid swirling jet produced by a pressure swirl atomizer for spray generation have been studied by Galbiati, Ertl, Tonini, Cossali, and Weigand at the Institute of Aerospace Thermodynamics, University of Stuttgart. The simulations were performed on Hornet using the in-house code Free Surface 3D, which is based on the Volume-of-Fluid method. A detailed DNS study of the flow field and the droplet formation has been made, describing with high accuracy the interaction between the liquid jet and the environmental gas. In particular, the influences of different grid resolutions and inlet velocity boundary conditions have been evaluated using grids with up to 670 million cells. It could be found that for too a coarse resolution the break-up of the conical jet into droplets takes place too early and the performance is overestimated. Also the strong influence of the inlet velocity profile has been demonstrated.

The next three contributions describe the results of different projects running under the biannual "Call for Large-Scale Projects" of the Gauss Centre for Supercomputing (GCS). Projects considered in these calls require more than 35 million core hours per year. The first paper by Wilke and Sesterhenn, TU Berlin, Fachgebiet Numerische Fluiddynamik, summarizes the work performed within two separate projects, both dealing with a supersonic jet impinging on a flat plate. The one part is dedicated to the discrete tonal noise emitted by the impinging jet, whereas the other part refers to the heat transfer. In both cases, the used in-house code directly solves the governing Navier-Stokes equations in a characteristic pressure-velocity-entropy-formulation, and a dynamic mode decomposition (DMD) is applied to identify the dynamical behaviour which is the source both of the impinging jet noise and of the unsteadiness of the heat transfer at the plate. The simulations were performed on

Hermit and later in the year on Hornet, where an excellent scaling efficiency of the code could be demonstrated up to 16,384 cores.

Edelmann (now DLR Stuttgart) and Rist of the Institute of Aerodynamics and Gas Dynamics of the University of Stuttgart have investigated the influence of forward-facing steps on the disturbance amplification in compressible laminar boundary layer flows. They applied the in-house DNS-code NS3D for their simulations and used a so-called disturbance approach, i.e. a steady base flow was computed first and then an unsteady disturbance flow in a second step. Different base flow solutions were produced for a high subsonic and a low supersonic free stream Mach number (both of them being relevant for the flow along the wing of an aircraft in cruise flight) and various step heights, displaying fundamental differences in the flow, e.g. in regard to the streamwise pressure gradients and the existence or absence of laminar separation bubbles. The unsteady disturbance flow was initiated by a wave packet that contains all frequencies which become unstable in the considered area. It turned out that for step heights lower than a critical value, the additional amplification caused by the step can be predicted if it is possible to compute the steady base flow accurately enough for subsequent local linear stability analysis.

At the Institute of Aerodynamics of the RWTH Aachen, over the previous years a high-fidelity, massively parallelized flow solver using the MILES (monotone integrated LES) approach has been applied very successfully to various aerodynamic problems. The code has been run on locally refined Cartesian hierarchical meshes. In the present contribution, Cetin, Pogorelov, Lintermann, Cheng, Meinke, and Schröder report on their recent highly resolved simulations of a helicopter engine jet (with a simplified as well as with a complete engine nozzle with additional struts) and an axial fan at two different mass flow rates. The studies on the influence of the geometry variation and the different mass flow rates, respectively, were performed on grids with a cell number in the order of magnitude of 300 millions. However, for grid convergence tests, meshes with more than 1 billion cells were used. The code showed a very good scaling behaviour on Hornet. 40,032 cores for the jet case and even 91,872 cores for the fan case, i.e. nearly the complete system, were used for the speed-up tests. In the next step, the highly resolved LES results shall be used for determining the acoustic field using computational aero-acoustic methods.

LES has also been applied by Eberhardt und Hickel from the Institute of Aerodynamics and Fluid Mechanics, Technical University Munich, to study the mixing process downstream of a strut injector. This type of injector is intended to be used in the combustion chamber of a scramjet propulsion system, in which the combustion takes place at supersonic flow velocities and where fast mixing of fuel and air is requisite for efficient operation. In particular, the complex mixing process of a pilot injection for flame stabilization has been investigated and analyzed in detail. The results were achieved using the in-house implicit LES-code INCA with the Adaptive Local Deconvolution Method (ALDM) as subgrid-scale model. An almost linear speed-up has been achieved on Hermit with down to 30,000 grid cells per core.

The commercial code ANSYS CFX v16.0 has been used by Krappel, Ruprecht, and Riedelbauch from the Institute of Fluid Mechanics and Hydraulic Machinery,

University of Stuttgart, to perform challenging transient flow simulations in an operating Francis turbine in part load conditions. They chose two hybrid RANS-LES approaches, a scale resolving simulation using the SAS-SST model, and a Detached Eddy Simulation. The results reveal a rotating low pressure zone in the draft tube, the so-called vortex rope phenomenon, which leads to a very complex flow that to resolve numerically needs a large computational effort (up to 300 million grid points and up to 40,000 time steps for the considered 40 runner revolutions). A very comprehensive description of the scaling behaviour shows that the newer version of the code (CFX-v16.0) is superior to the former version v14.5, as the newer one can be run in the enhanced stability mode using CRAY-MPI instead of standard MPI. Also the multipass decomposition method developed for applications with rotating parts contributes to the improvement of the parallel performance. A comparison between Hermit and Hornet revealed that the simulations on Hornet run approximately three times faster than on Hermit for the same number of cores, but that the speed-up on Hermit is better than on Hornet for the present application.

For another type of turbine, a low head propeller turbine, numerical investigations of a full load operation point were performed at the same institute by Junginger and Riedelbauch. Again, ANSYS CFX v16.0 has been employed using the k-ω-SST turbulence model in RANS mode and the SAS-SST turbulence model for scale resolving simulations. Two major deviations in the results obtained with both turbulence models were found in the shape of the vortex rope that develops from a low pressure zone downstream the runner hub and in the time-averaged velocity profile in the draft tube. Although the SAS model is able to resolve smaller flow structures, it overpredicts the head losses in particular in the draft tube compared to the SST model and the experiments. Laser-Doppler-Velocimetry and pressure pulsation measurements are planned for the future for a better validation of the numerical results.

Biegger and Weigand from the Institute of Aerospace Thermodynamics in Stuttgart performed Detached Eddy Simulations for studying the flow and the heat transfer in a swirl tube. The open-source code OpenFOAM has been applied for this tasks and the results have been compared to own experimental data. The mean velocity profiles agree quite well, however, unfortunately comparisons for more challenging quantities like velocity fluctuations, wall stresses, wall heat fluxes or pressure losses are not shown. Results are discussed for Reynolds number 10,000 and three different swirl numbers. It was found that the averaged heat transfer is four times larger than for a smooth tube flow. The number of cells was 9 million for this case and 15 million for a higher Reynolds number. The parallel performance analyses on Hermit were performed for a 25 million cell grid, limiting the number of cores that provide a reasonable speed-up to 512 (i.e. 50,000 cells per core).

The high-fidelity simulation of the aerodynamics of wind turbines under real operating conditions and the determination of the unsteady loads resulting therefrom is a major research topic at the Institute of Aerodynamics and Gas Dynamics, University of Stuttgart. In the present article, Schulz, Fischer, Weihing, Lutz, and Krämer report on some of the latest results in this field. In the first part, the effect of yawed inflow on the loads and on the aerodynamics of a wind turbine is described, and the second part is dedicated to the control of wind turbine loads caused by the tower blockage by means of a trailing edge flap. For all simulations, the FLOWer code originally developed by the German Aerospace Center (DLR) was employed. For the first part, a DES approach on a 50 million cells grid has been used, whereas for the second task a URANS method has been applied. A major outcome of the first part of the investigations was the quantification of the reduction in blade loads and power output due to the yawed inflow, with the asymmetry of the load distribution over the rotor disk leading to high load fluctuations during one revolution. Also the wake deflection, which is highly important for wind park design and control, was determined. For the second part, it is shown that actively deflected flaps are able to decrease the tower induced loads fluctuations by 50 %. A comparison of URANS results to the results obtained with a newly implemented IDDES method shows a better representation of the turbine near wake and its vortex structures by the latter approach. For the FLOWer code, about 30,000 cells per core are a reasonable number on Hermit as well as on Hornet. Weak scaling tests on Hornet revealed a good scaling behaviour for this cell loading up to 4096 cores.

Latest results of the helicopter group at the same institute are presented by Kranzinger, Kowarsch, Schuff, Keßler, and Krämer in the next contribution. They report on a weak and strong scaling study which shows substantial improvements of the scalability of the FLOWer code by introducing a node-to-node MPI communication strategy. The study has been performed with 135 million grid cells and shows a nearly linear strong scaling on Hornet up to 8000 cores. Also the weak scaling was found to be very good for 270 million grid cells computed on 8016 cores. Results not shown in the paper revealed a further increase in parallel efficiency by using OpenMP in addition to the node-to-node based communication. Additionally, an overview of an extremely flexible CFD-CSD coupling interface based on radial basis functions is given which is able to handle structured, unstructured, and overset meshes without topology limitations and performance drawbacks. To demonstrate the capability of these new features, a full helicopter configuration using a mesh with 192 million grid cells has been investigated with one of the aims being a most accurate prediction of the blade vortex interaction, which is the main source of impulsive noise emission of a helicopter in descent flight.

Also the next contribution comes from the Institute of Aerodynamics and Gas Dynamics. It has been prepared by Mayer, Zimmermann, Wawrzinek, Lutz, and Krämer and deals with the effects of the flank shape of three-dimensional shock control bumps (SCBs) on the buffet behaviour of an aircraft wing during transonic flight. SCBs are known to be able to effectively reduce the wave drag of the wing at off-design conditions. However, very recently their potential to delay and alleviate transonic buffet was unveiled and is now subject to further research. The exact geometry of the bump significantly affects the local flow characteristics and, thus, the buffet behaviour. The present paper focuses specifically on the flank shape of a wedge-type SCB. The main outcome of the study is a bump geometry that simultaneously improves the aerodynamic performance as well as the buffet characteristics. Again, the FLOWer code was used in RANS mode. A parameter study was performed for different lengths and widths of the front and the rear flank and the effect of these geometry parameters on the bump performance as well as on the buffet behaviour has been analyzed. In preparation of future work concerning the optimization of the buffet alleviating bumps, where the use of the DLR TAU-code instead of the FLOWer-code is an option, performance tests for both codes have been performed on Hornet. In the present article, only the results of the scaling tests for the TAU-code are presented, as the FLOWer results have been described in detail in the last year's contribution. The tests have been very comprehensive and in particular address the performance of different compilers (CRAY, GNU, and INTEL) available on Hornet.

The flow solvers applied in the next three contributions are based on the Discontinuous Galerkin (DG) method, which may be considered as a combination of a finite-element scheme (with a continuous higher order polynomial in each grid cell) and a finite-volume scheme (allowing for discontinuities at the cell faces). DG methods are supposed to have the potential to some day replace the well-established finite-volume production codes, at least for certain applications. The working group of Munz at the Institute of Aerodynamics and Gas Dynamics in Stuttgart has built-up a DG-based high-order simulation framework, which has been tailored for high performance computing aiming at solving large scale problems on massively parallel architectures. In the first report, Atak, Beck, Bolemann, Flad, Frank, and Munz show results from this framework, which uses a very efficient variant of a DG formulation, the so-called spectral element method (DGSEM). Recent simulations for three test cases are presented. These are a DNS of an airfoil at Reynolds number 100,000, an LES based direct aero-acoustic simulation of a side-view mirror, and a DNS of a high-speed turbulent boundary layer flow. For the latter test case, close to 94,000 cores were used on Hornet. This simulation involved 1.5 billion degrees of freedom and represents the largest DG simulation published so far. The parallel performance of the used DGSEM solver FLEXI is excellent, as high-order DG methods in general offer superior weak and strong scaling capabilities if implemented appropriately.

Hoffmann, Sonntag, and Munz from the above mentioned working group together with Boblest and Ertl from the Visualization Center, University of Stuttgart, Hempert and Iben from Robert Bosch GmbH, Offenhäuser and Glass from the HLRS, and Sadlo from the Interdisciplinary Center for Scientific Computing, University of Heidelberg, report on a project that aims at performing highly resolved unsteady DG computations, in particular using the FLEXI code, for industry relevant problems. The focus of their contribution is twofold. First, the adaptation of the solver to industrial applications and the optimization of the parallel efficiency are described, and, second, a visualization tool specifically adapted to the properties of DG data is presented. This tool shall be combined with the solver to realize an in-situ visualization strategy in the near future. The industrial problem under consideration has been the injection of compressed natural gas under supersonic conditions into the intake manifold of an engine. The very different spatial and temporal scales occurring in this case make the numerical simulation exceptionally difficult and time consuming. Direct aero-acoustic simulations were performed to predict the noise emission directly from the simulation data. Validation of the solver was done in terms of the averaged velocity flow field and the sound pressure level showing a very good agreement to the available experimental data. A high scalability of the code was demonstrated for this application case up to 9600 cores. The visualization is based on the open source software ParaView. A reader plug-in fully parallelized using MPI has been developed that reads in the solution data together with the underlying mesh.

Also another working group at Stuttgart's Institute of Aerodynamics and Gas Dynamics is concerned with the discretization of the RANS equations by the Discontinuous Galerkin method. In their contribution, Wurst, Keßler, and Krämer describe the latest extensions of their in-house code SUNWinT. Their focus is on highly separated flows and on rotor flows. To this end, they have implemented a Detached Eddy Simulation method as well as a Chimera grid technique in order to be able to handle moving bodies. The report contains results for the turbulent flow around an airfoil at different angles of attack, for which they have used a near-body grid overlapping a background grid. The agreement between a single-grid solution on the one hand and experimental data on the other hand is excellent. For the instationary DES, a backward facing step was considered, using 141,000 hexahedral cells with 20 degrees of freedom per element. Also for this case, the agreement to experimental data and to numerical results obtained by Spalart et al. was very good.

The last two contributions in the CFD section of this book are based on the Lattice Boltzmann approach. The contribution of Xie, Günther, and Harting is about mesoscale simulations of anisotropic colloidal particles at fluid-fluid interfaces. The work has been conducted in cooperation between the Department of Applied Physics of the Eindhoven University of Technology and the Faculty of Science and Technology, Mesa+ Institute, University of Twente. The authors have used a Lattice Boltzmann method combined with molecular dynamics to simulate multi-component fluids and suspended particles. First, they have studied the ensemble of ellipsoidal particles at a spherical interface. Their findings provide relevant insight in the dynamics of emulsion formation, which is generally difficult to investigate experimentally. Second, they have investigated the behaviour of magnetic Janus particles adsorbed at fluid-fluid interfaces interacting with an external magnetic field. They can show that these particles are promising building blocks of reconfigurable and programmable self-assembled structures.

Masilamani, Klimach, and Roller from the Chair for Simulation Techniques and Scientific Computing at the University of Siegen present the latest results of their integrated simulations of electro-membrane processes for desalination of sea water. They have used a multi-species Lattice Boltzmann method to investigate the electrodialysis process. Here, sea water flows through a channel between ion exchange membranes which are kept at distance by spacer structures inside the channel. The ions are driven out of the channel and through the membranes by an external electrical field. The code had been developed with a strong focus on high computational performance. The report contains a very comprehensive study on the performance and scalability behaviour of the code on Hornet. For these analyses, 2048 compute nodes (i.e. 49,152 cores) were used, and problem sizes from 512 up to

68 billion elements were considered. As for the Hermit system, an excellent strong and weak scaling could be demonstrated on Hornet for single fluid simulations as well as for simulations with three species.

Not only the number, but also the large thematic variety of ambitious projects performed during the reporting period at the HLRS in the field of Computational Fluid Dynamics is again impressive. None of them could have been realized without access to leading edge HPC facilities. This demonstrates the high value as well as the indispensability of supercomputing in this area. The transition from the CRAY XE6 Hermit to the CRAY XC40 Hornet during the year 2014 at the HLRS could be managed by the individual projects without major difficulties. As could be expected, the run times for the simulations could mostly be reduced by a factor of two or even three, enabling the handling of even more complex (realistic) problems at even higher accuracy. Thanks have to be given to the staff of the HLRS and CRAY for their valuable support to the individual projects.

High-Resolution Numerical Analysis of Turbulent Flow in Straight Ducts with Rectangular Cross-Section

Yoshiyuki Sakai and Markus Uhlmann

Abstract Turbulent secondary motion of straight open duct flows with a rectangular cross-section was studied by means of direct numerical simulations, and the unique mean flow patterns were analysed with the aid of instantaneous coherent structure analysis for their Reynolds number dependence. Similar to the closed duct counterparts, it was found that the mean streamwise vorticity pattern is the statistical footprint of the most probable locations of the quasi-streamwise vortices. Furthermore, the existence of tightly-concentrated vortices with preferable rotational directions inside the mixed corners formed by no-/free-slip boundaries was observed. Such flow structures correspond to the side-wall high-speed streaks located directly on the free-slip plane, independently of Reynolds number, as well as the following low-speed streaks reside approximately 50 wall units away from the free-slip plane for friction Reynolds number larger than 200.

1 Introduction

Fluid flow in a straight duct with rectangular cross-section exhibits turbulence-induced secondary motion of small amplitude (few percent of the bulk velocity), but with large consequences for momentum, heat and mass transport; hence the difficulties in experimental measurements and engineering significance co-exist.

Much of the previous attention was paid upon the closed duct configuration with the square cross-section at marginal to moderate Reynolds numbers (e.g. [7, 9], where high-resolution direct numerical simulations were performed up to bulk Reynolds number of 3500). As a consequence, understanding in Reynolds number dependence up to a point where the flow exhibits a clear scale separation between near-wall structures and outer-scale structures still needs to be established. Furthermore, thorough numerical investigations in aspect ratio dependence covering a range over where the flow structures start to be detached from the side-walls,

Y. Sakai (✉) • M. Uhlmann
Karlsruhe Institute of Technology, Karlsruhe, Germany
e-mail: yoshiyuki.sakai@kit.edu; markus.uhlmann@kit.edu

© Springer International Publishing Switzerland 2016
W.E. Nagel et al. (eds.), *High Performance Computing in Science and Engineering '15*, DOI 10.1007/978-3-319-24633-8_20

to the point where the side-wall influence vanishes at the duct centre need to be achieved. Rigorous understanding in such phenomenon will, for instance, serve as a theoretical backbone of wind/water-tunnel design in fluid labs, where the side-wall effects need to be negligible at the measurement windows.

The corresponding open duct flow—featuring a free surface—is characterized by a distinct secondary flow pattern, leading to such practically important effects as the so-called "dip phenomenon": the maximum of average streamwise velocity is not found at the surface of a river, but somewhat below. Despite such practical importance, understanding of open duct flows is less established than the closed-duct counterpart.

In the present work, we investigate the mechanism of secondary flow formation in open duct flows. Here, free-surface deformation is neglected by enforcing a free-slip boundary condition on the top boundary. In the case of negligible surface tension, this hypothesis amounts to requiring the Froude number to be sufficiently small such that gravity effectively suppresses any significant deformation of the surface. Particular emphasis in our analysis is placed upon the dynamics of coherent structures and the consequences for Reynolds number and aspect-ratio scaling.

2 Numerical Methods

2.1 Algorithm and Implementation

For the purpose of current studies, the pseudo-spectral DNS code previously used in the studies in closed square duct flows by Uhlmann et al. [9], Pinelli et al. [7] and Sekimoto et al. [8] has been extended to incorporate the free-slip boundary condition. The code integrates the Navier-Stokes equations by expanding flow variables in terms of truncated Fourier series in the streamwise direction on equidistant grid points, while Chebyshev polynomials are used in the two cross-stream directions on collocated Chebyshev–Gauss–Lobatto points. A fractional step method is employed in order to decouple the momentum equations from the continuity constraint. The temporal integration is based on the Crank–Nicolson scheme for the viscous terms and a three-step low-storage Runge–Kutta method for the nonlinear terms. The 2D Helmholtz and Poisson problems for each Fourier mode are solved by a fast diagonalisation technique. The code is MPI parallel and the parallelisation of the pseudo-spectral algorithm is achieved by a global data transpose strategy. A schematic of the data decomposition by slices is shown in Fig. 1 and a cyclic communication pattern, which avoids communication cascade and used for global data transpose, is illustrated in Fig. 2. The cost of such communication is $2 \times$ (number of parallel process $- 1$) MPI send/receive calls per data transpose per MPI process. Asynchronous communications are used to

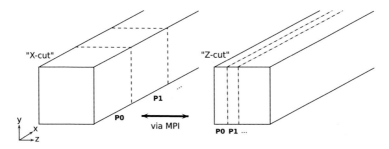

Fig. 1 "Slice" data decomposition strategy used for parallelisation of our DNS code

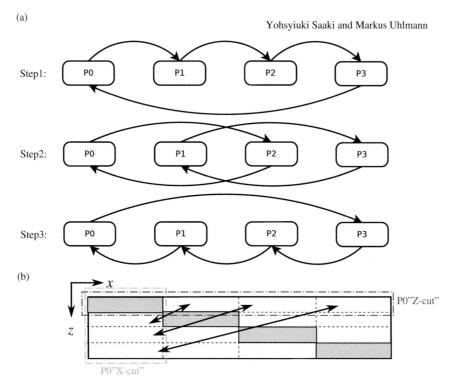

Fig. 2 (**a**) A cyclic all-to-all communication pattern used for data transpose between "X-cut" and "Z-cut" data decompositions (cf. Fig. 1), illustrated by an example of parallel execution model with 4 MPI processes. Each "P*" element represents individual MPI rank, while arrows show data flow by MPI communications. (**b**) Top view of the above example's domain. Sub-domains surrounded by solid lines (coloured in *light gray*) are the overlapped regions between "X-cut" and "Z-cut" of each MPI process, hence no communication is required. Any other areas are required to be communicated to other MPI processes. An example of such communications for the MPI process "P0" is illustrated

minimise the communication overhead. It should be noted that this parallelisation strategy is standard in spectral methods applied to plane channel flows.

2.2 Code Performance

In order to demonstrate the efficiency of our DNS code, we have carried out strong scaling tests on CRAY XC40 HORNET at HLRS (cf. Table 1, Fig. 3). The results show that for the chosen problem size (approx. 10^8 modes), the code maintains a good parallel efficiency up to several hundred cores. With roughly 10^6 modes per processor core we obtain the desired efficiency of $\geq 65\%$ over a range of system sizes, at a cost of approx. 3 s per full Runge-Kutta timestep.

Table 1 Strong scaling tests with the modal resolution of $M_x = 3075$, $M_y = 33$, $M_z = 1025$

nproc	Walltime/ timestep (sec)	Speed-up	Para. efficiency (%)
1	427.76	1.00	100
5	99.19	4.31	86
25	22.73	18.82	75.28
41	15.54	27.53	67.15
205	3.12	137.10	66.89
1025	2.21	193.56	19

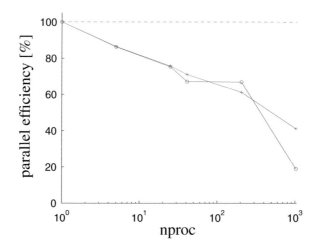

Fig. 3 Parallel efficiency for two numbers of Chebyshev modes in y-direction: $M_y = 33$, *open circle* in *blue* ; $M_y = 67$, *plus symbol* in *red*. The number of modes in the x-/z-direction are fixed at $M_x = 3075$ and $M_z = 1025$ respectively. The runs were performed on Cray XC40 HORNET at HLRS

2.3   Simulations

Wall-bounded turbulent flows—such as channel flow and boundary layers—suffer from severe near-wall grid resolution requirements as the Reynolds number increases. In the duct geometry, such requirement is even more demanding since the side-wall boundary layers also need to be resolved. Two kinds of Reynolds numbers need to be introduced for convenience hereafter, namely: bulk Reynolds number ($Re_b = U_b h/\nu$, formed with the bulk velocity U_b, the duct semi-(full-) height $h(H)$ and the kinematic viscosity ν); and friction Reynolds number ($Re_\tau = u_\tau h/\nu$ or $Re_\tau = u_\tau H/\nu$, where u_τ is the friction velocity defined as $u_\tau = \sqrt{\tau_w/\rho}$ with the wall shear stress τ_w averaged in time and space on no-slip walls, and the constant density ρ). Please note that we always use h to normalise length scales in the bulk unit in the closed duct configuration, while H is used in the open duct configuration, unless stated otherwise. In the current studies, we chose the grid resolutions to satisfy $\Delta x^+ \leq 15.0$; $\Delta y_{max}^+, \Delta z_{max}^+ \leq 4.0$, where the superscript '+' stands for wall units: $l^+ = l/\delta_\nu$ with $\delta_\nu = \nu/u_\tau$ the viscous length scale. The corresponding $\Delta y(z)_{min}^+$ values remained to be smaller than 0.06. CFL number is maintained to be always under 0.3. Numerical box size of our simulations has a fixed streamwise length $L_x/h = 4\pi$ for closed ducts, and $L_x/H = 8\pi$ for open ducts respectively. The box size in y-direction is also fixed at $L_y/h = 2$ and $L_y/H = 1$, while the size in z-direction is varied according to the aspect ratio ($A = W/h$ for closed ducts, $A = W/H$ for open ducts) of our interests. Please refer to Fig. 4 for a schematic of the domain dimensions. It is also essential to note that the statistical data are averaged over a sufficiently long time interval which can be estimated as approximately 10,000 bulk time units (a bulk time unit is defined as h/U_b, or H/U_b).

More than 30 separate simulations have been performed so far to cover a wide range of parameter space necessary for the current studies (cf. Fig. 5). The total amount of the generated data is approximately 15 TB, including several

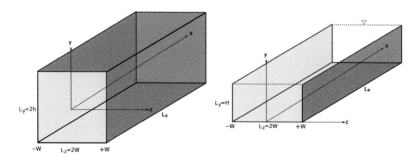

Fig. 4 Coordinates system and geometry of: (**a**) closed; and (**b**) open duct

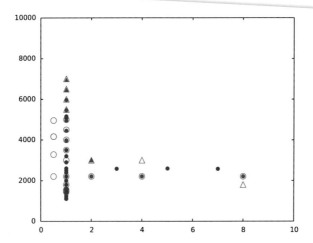

Fig. 5 Parameter map of duct DNS simulations. The *red symbols* represent the simulations for the current study, while the *blue symbols* represent the previous studies by: [1, 3, 7, 9, 10] for closed duct; [4] for open duct. *Open* and *filled symbols* represent open and closed duct configurations respectively, while the shape of the symbols shows whether those runs' statistics are fully converged (circle), or not (triangle)

Table 2 Representative examples of simulations. In the "Type" column, "o" indicates open duct, while "c" indicates closed duct

Type	A	Re_b	$[M_x,M_y,M_z]$	Machine	nproc	Walltime
o	1.0	2205	[256,97,193]	HERMIT	32	1 month
o	2.0	2205	[256,97,385]	HERMIT	64	1.5 months
o	4.0	2205	[256,97,769]	HERMIT	64	3.5 months
o	8.0	2205	[256,97,1153]	HORNET	48	3 months
o	1.0	3000	[384,97,193]	HORNET	48	1 month
o	1.0	5000	[512,129,257]	HORNET	96	1.5 month
o	1.0	6500	[768,193,385]	HORNET	96	6 months
c	2.0	2205	[193,129,257]	HERMIT	32	1.5 month
c	4.0	2205	[193,129,513]	HERMIT	64	2 months
c	8.0	2205	[193,129,1025]	HERMIT	64	4 months
c	1.0	4000	[512,257,257]	HORNET	96	5 months
c	1.0	6000	[768,385,385]	HORNET	120	12 months

thousands instantaneous flow fields necessary for scientific visualisations as well as coherent structure analysis and its statistics discussed in the following section. Some representative simulation configurations are summarised in Table 2 including spatial resolution, number of processors and typical walltime to complete such simulations.

As it is mentioned in Sect. 2.2, our DNS code maintains a good parallel efficiency up to several hundred cores. However, since queueing time for reserving such number of cores was inconveniently long, we have utilised only up to 120 CPUs in order to maximise the throughput of our simulations. In total 3.45851 million core-hours of computational time was granted for Cray XE6 HERMIT (used: 100 %); while computational time of 6.359745 million core-hours was granted for Cray XC40 HORNET (used: 53 %). The estimated percentage of usage for HORNET at the end of the current project period (31st May 2015) at the current rate of throughput is around 72 %.

3 Results and Discussions

3.1 Coherent Structure Analysis on Open Duct Flows

We have identified the centres of vortical structures by the technique proposed by Kida and Miura [5] for the open duct case at bulk Reynolds number $Re_b = 2205$ with the aspect ratio set at unity. It was found that the mean streamwise vorticity pattern in the turbulent open duct flows is the statistical footprint of the most probable locations of the quasi-streamwise vortices, similarly to the corresponding closed duct cases (cf. Fig. 6, also [7, 9]). There is, however, a significant difference between the open and the closed duct statistics, which is the tightly-concentrated vortices with preferable rotational directions that exist in the mixed-boundary corners. Our results show that those vortices persist in the mixed-boundary corners much more likely than anywhere else in the duct domain. By considering streaks as byproducts of those near-wall vortices, our finding is consistent with the existence of statistically highly concentrated near-free-surface low-speed streaks found experimentally by Grega et al. [2] (cf. their Fig. 12).

3.2 Reynolds Number Dependence in Open Duct Flows

Gavrilakis [1] demonstrated with their direct numerical simulations of square closed duct flow ($Re_b = 2205$) that there is an ambiguity on the selection of the normalisation velocity scale for the near-wall dynamics. The ambiguity is caused by the fact that, unlike canonical plane channel flows, the average wall friction has a variation along the duct perimeter due to the mean secondary flow. Gavrilakis [1] highlighted this aspect of duct flows by studying the mean streamwise velocity profile along the wall bisector in logarithmic scale, and argued that the local friction

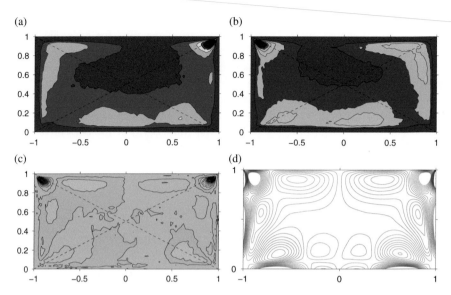

Fig. 6 Probability of occurrence of vortex centres for the open duct case with $Re_b = 2205$, detected by the technique proposed by Kida and Miura [5]. (**a**) vortices with *positive* streamwise vorticity; (**b**) *negative* streamwise vorticity; and (**c**) the difference between (**a**) and (**b**). The iso-contours indicate 0.1(0.1)0.9 times the maximum values (except (**c**) where $-0.9(0.1)0.9$ times the maximum absolute value is used instead. Negative values correspond to vortices with negative vorticity). The statistics in (**a**)–(**c**) were accumulated from 1000 instantaneous snapshots over a time interval of $721.5H/U_b$. (**d**) Mean streamwise vorticity contours indicate $-0.9(0.1)0.9$ times the maximum absolute value where *red* and *blue lines* correspond to positive and negative values, respectively. The vorticity field was averaged over a time interval of $8000H/U_b$

Table 3 Numerical set-up of the open duct simulations discussed in Sect. 3.2

A	Re_b	Re_τ	$(Re_\tau)_{bisector}$	$[M_x, M_y, M_z]$
1.0	1500	104	113	[256,97,193]
1.0	2205	150	170	[256,97,193]
1.0	3000	197	218	[384,97,193]
1.0	5000	309	330	[512,129,257]

velocity should be used for the quantities within the viscous sublayer, while the global friction velocity is more appropriate for the quantity near the duct centre. However, as it was pointed out by Huser and Biringen [3], Gavrilakis' data also show a closer fit to the law of the wall ($u^+ = \frac{1}{0.41}\log y^+ + 5.5$) in the logarithmic region if the local friction velocity is used throughout for normalisation. Please also note that [6] experimentally determined a Reynolds number-independent logarithmic relation ($u^+ = \frac{1}{0.412}\log y^+ + 5.29$) from their open duct measurements.

We analysed four of our open duct simulations with Reynolds numbers $Re_b = 1500, 2205, 3000, 5000$ and aspect ratio $A = 1$ (cf. Table 3 for the configurations) and the results agree well with such observation except the lowest Reynolds number case, whose logarithmic region does not exist (cf. Fig. 7). Furthermore, it was

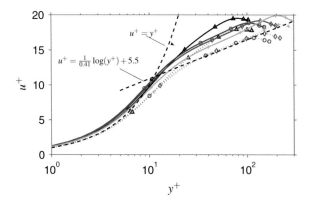

Fig. 7 Mean streamwise velocity in logarithmic scale. *Symbols* indicate the bulk Reynolds numbers: *triangle*, $Re_b = 1500$; *open circle in blue*, $Re_b = 2205$; *diamond symbol in red*, $Re_b = 3000$; *plus symbol in green*, $Re_b = 5000$. *Solid lines* indicate the values normalised by the global friction velocity u_τ, while *dotted lines* indicate the values normalised by the local friction velocity $(u_\tau)_{bisector}$

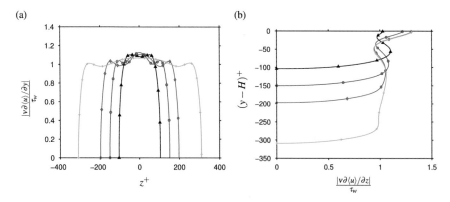

Fig. 8 Mean local wall shear stress normalised by the average over the whole no-slip walls, along (**a**) bottom wall and (**b**) side walls (the origin has been translated to the corner). *Symbols* indicate the bulk Reynolds numbers: *triangle*, $Re_b = 1500$; *open circle in blue*, $Re_b = 2205$; *diamond symbol in red*, $Re_b = 3000$; *plus symbol in green*, $Re_b = 5000$

observed that the discrepancy between the two scalings reduce as Reynolds number increases. This phenomenon can be explained by studying the distribution of the mean wall shear stress along the duct bottom wall shown in Fig. 8a. It can be seen that by increasing Reynolds number, the local variation of the bottom wall shear stress becomes smaller and develops a channel-like plateau around the bottom wall bisector. Furthermore, examining the number of extrema of the bottom wall shear stress profile shown in Fig. 8a shows that the $Re_b = 1500$ case exhibits a three velocity streaks state (high-, low-, and high-speed streaks arrangement, as discussed in [7]), while the $Re_b = 2205$ case's profile indicates a five streaks state. The highest Reynolds case, whose bottom-wall shear profile has a plateau around the wall bisector, can theoretically host up to thirteen streaks. However, at this high Reynolds number, the middle-wall streaks travel more freely and randomly in contrast to the quasi-permanent high-speed streaks residing next to the bottom wall corners (mean distance from the closest side wall $\approx$ 50 wall units [7]), thus less significant signatures result in the mean wall-shear profile.

In contrast to the bottom wall statistics behaving similarly to the closed duct counterparts, the open duct side walls host distinctive features especially near the corners formed by no-/free-slip boundaries. Such differences are apparent, for example, by examining the wall shear stress along the side walls shown in Fig. 8b. Here, we observe the signatures of existence of the nearest high-speed streak from the free-slip plane to appear directly on the plane. The positions of the following low-speed streaks appear to move away from the free-slip plane as Reynolds number increases, then settle down at approximately 50 wall units (i.e. $d^+ \approx 50$) for $Re_b \geq 3000$ with $L_y^+ = L_y/\delta_v \geq 200$.

The time evolutions of the instantaneous low-speed streak locations, determined as the location of the minimum wall shear stress, are plotted for the three Reynolds numbers in Fig. 9. Please note that data from $Re_b = 1450$ had to be used here instead of $Re_b = 1500$, due to the availability of such large number of instantaneous data. It can be observed that the amplitude of the streaks' lateral movement are much more restricted and therefore their paths are significantly more distinguishable than the ones from the bottom wall, which appear similar to the ones previously studied for closed ducts (cf. [7] their Fig. 6). Regarding the side-walls, the corresponding probability density function of the distance from the free-slip plane is shown in Fig. 10, illustrating that the locations of their peaks are actually at similar distances at all Reynolds numbers ($d^+ \approx 25$), but the tails of the distribution develop away from the free-slip plane as Reynolds number is increased, which brings the averaged location of the low-speed streaks away from the plane up to ≈ 50 wall units for high enough Reynolds numbers.

4 Conclusions and Outlook

We have performed a series of direct numerical simulations both in the open and the closed duct configurations with variable Reynolds number and aspect ratio.

Our coherent structure analysis revealed that, likewise in turbulent closed duct flows, the mean streamwise vorticity pattern in turbulent open duct flows is the statistical footprint of the most probable locations of the quasi-streamwise vortices. The same analysis also revealed the existence of tightly-concentrated vortices with preferable rotational directions in the corners formed by no-/free-slip boundaries.

Those in-corner vortices and their corresponding low-speed streaks are later demonstrated to be the main players of distinctive features on the duct side walls, such as the unique wall shear stress profile whose peak laying directly on the free-slip boundary.

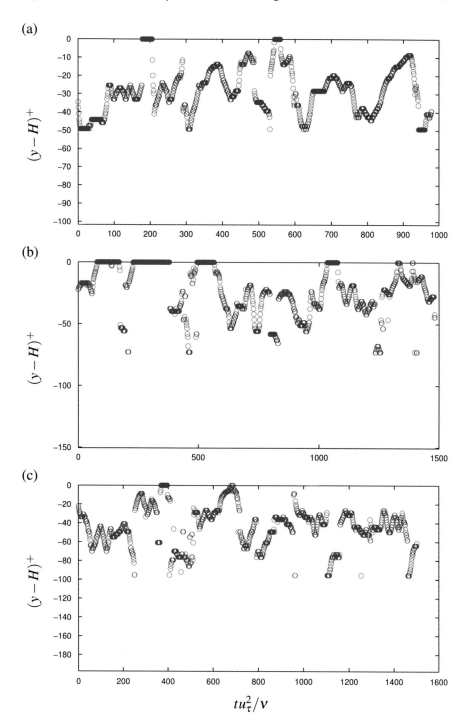

Fig. 9 Time evolution of position of minimum of the wall shear stress at $z/H = 1$, $x/H = 0$. (**a**) $Re_b = 1450$; (**b**) $Re_b = 2205$; (**c**) $Re_b = 3000$

Fig. 10 Probability density function of low speed streak distance from the free-slip plane (d^+), computed considering the instantaneous location of the minimum value of wall skin friction. *Symbols* indicate the bulk Reynolds numbers: *triangle, $Re_b = 1450$; open circle in blue, $Re_b = 2205$; diamond symbol in red, $Re_b = 3000$*

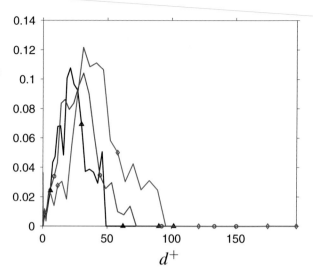

Results regarding other aspects of turbulent duct flows that are not reported here, including variable aspect ratio effects on open/closed duct flows, as well as closed duct flows with Reynolds numbers significantly higher than previous studies, will be discussed in details in other forms of publication in near-future elsewhere.

Overall, the outcomes of our current project at HLRS has been fruitful. However, computing high Reynolds number duct flows with high aspect ratio is still a challenge even with the aid of sophisticated HPC facilities such as Cray XC40 HORNET. Alongside the severe spatio-temporal resolution requirements mentioned earlier, the main difficulties arise from the limited flexibility of Chebyshev-Gauss-Labatto point distribution used in spatial discretisation in the cross-stream directions, which let us tweak only one parameter, namely number of grid points for each direction. As a consequence, near-wall regions are in general over-resolved in order to maintain adequate resolution around the duct core region, where Chevyshev-Gauss-Lobatto points have the coarsest distribution, which carries severe timestep size restriction thus slower computations. To cope with such challenges by obtaining better grid-design freedom while maintaining spectral accuracy, our DNS code is currently under a major upgrade with a use of spectral element spatial discretisation method in the cross-stream directions. The upgraded code would not only enable us to compute higher Reynolds number/aspect ratio duct flows efficiently, but also to incorporate wall-roughness elements as well as angled duct side walls within our simulations, which are of great interest to civil/environmental engineering applications.

References

1. Gavrilakis, S.: Numerical simulation of low-Reynolds-number turbulent flow through a straight square duct. J. Fluid Mech. **244**(1), 101 (1992)
2. Grega, L.M., Wei, T., Leighton, R.I., Neves, J.C.: Turbulent mixed-boundary flow in a corner formed by a solid wall and a free surface. J. Fluid Mech. **294**(1), 17–46 (1995)
3. Huser, A., Biringen, S.: Direct numerical simulation of turbulent flow in a square duct. J. Fluid Mech. **257**, 65–95 (1992)
4. Joung, Y., Choi, S.-U.: Direct numerical simulation of low Reynolds number flows in an open-channel with sidewalls. Int. J. Numer. Methods Fluids (2) **62**, 854–874 (2010)
5. Kida, S., Miura, H.: Swirl condition in low-pressure vortices. J. Phys. Soc. Jpn. **67**(7), 2166–2169 (1998)
6. Nezu, I., Rodi, W.: Open-channel flow measurements with a laser Doppler anemometer. J. Hydraul. Eng. **112**(5), 335–355 (1986)
7. Pinelli, A., Uhlmann, M., Sekimoto, A., Kawahara, G.: Reynolds number dependence of mean flow structure in square duct turbulence. J. Fluid Mech. **644**, 107 (2010)
8. Sekimoto, A., Kawahara, G., Sekiyama, K., Uhlmann, M., Pinelli, A.: Turbulence- and buoyancy-driven secondary flow in a horizontal square duct heated from below. Phys. Fluids **23**(7), 075103 (2011)
9. Uhlmann, M., Pinelli, A., Kawahara, G., Sekimoto, A.: Marginally turbulent flow in a square duct. J. Fluid Mech. **588**(2000), 153–162 (2007)
10. Vinuesa, R., Noorani, A., Lozano-Durán, A., Khoury, G.E., Schlatter, P., Fischer, P.F., Nagib, H.M.: Aspect ratio effects in turbulent duct flows studied through direct numerical simulation. J. Turbul. **00**(00), 1–28 (2014)

Investigation of Convective Heat Transfer to Supercritical Carbon Dioxide with Direct Numerical Simulation

Xu Chu and Eckart Laurien

Abstract For fluids at supercritical pressure, the phase change from liquid to gas does not exist. In the meanwhile, the fluid properties change drastically in a narrow temperature range. With supercritical fluid as working fluid in a heated pipe, heat transfer deterioration and recovery have been observed, which correspond to the turbulent flow relaminarization and recovery. Direct Numerical Simulation (DNS) of supercritical carbon dioxide flow in a heated vertical circular pipe at a pressure of 8 MPa is developed with the open-source code OpenFOAM in the present study. Forced convection cases and mixed convection including upward and downward flow have been considered in the simulation. The change of the mean flow and turbulence statistics has been analysed in detail. In the forced convection, flow turbulence is attenuated due to acceleration from thermal expansion, which leads to a peak of the wall temperature. Buoyancy has a stronger impact to the flow. In the upward flow, the average streamwise velocity distribution turns into an M-shape profile because of the "external" effect of buoyancy. Besides that, negative buoyancy production caused by the density variation reduces the Reynolds shear stress to almost zero, which means that the flow is relaminarized. Further downstream turbulence is recovered. This behaviour of flow turbulence is confirmed by visualization of turbulent streaks and vortex structures. The observation of the flow turbulence of this can help to develop advanced turbulence models for applications in nuclear or conventional energy generation technologies.

1 Introduction

For fluids at supercritical pressure, a phase change from liquid to gas does not exist as in sub-critical flows. When in an isobaric process the fluid temperature increases from below to above the pseudo-critical temperature (T_{pc}), the density (ρ), the thermal conductivity (κ) and the dynamic viscosity (μ) decrease drastically, and

X. Chu (✉) • E. Laurien
Institute of Nuclear Technology and Energy Systems, University of Stuttgart, Pfaffenwaldring 31, 70569 Stuttgart, Germany
e-mail: xu.chu@ike.uni-stuttgart.de; eckart.laurien@ike.uni-stuttgart.de

© Springer International Publishing Switzerland 2016
W.E. Nagel et al. (eds.), *High Performance Computing in Science and Engineering '15*, DOI 10.1007/978-3-319-24633-8_21

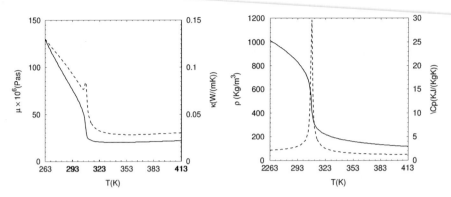

Fig. 1 Variation of thermo-physical properties viscosity μ, heat capacity C_p, heat conductivity κ and density ρ of CO_2 as a function of the temperature at a supercritical pressure of $P_0 = 80$ bar

the specific heat capacity (C_p) shows a high peak (Fig. 1). These strong variations in their properties make supercritical fluid useful in many industrial applications, such as extraction, cleaning and as a refrigerant in air-conditioning and refrigerating systems [1]. The Supercritical Water-Cooled Reactor is considered as a Generation IV nuclear reactor concept, which directly uses supercritical water as the working fluid, has attracted much attention since the 1990s because of its high efficiency, compact size, and reduced complexity [2].

However, deteriorated heat transfer has been found in various experimental studies [3–6] for heated flows at supercritical pressure. In these works, the heat transfer deterioration has been investigated under different flow conditions: mass flow rate, heat flux, pipe geometry, flow direction, etc. Comprehensive reviews on the topic of heat transfer to fluids at supercritical pressure have been given by Duffey and Pioro [7] and Yoo [8]. Besides experiments, numerical approaches using CFD codes have also been carried out to study heat transfer of fluid at supercritical pressure. Turbulence modelling using different two-equation models is the common method for a first approach in most of the early studies. He et al. [9] did a comprehensive investigation of Low-Reynolds turbulence models using an in-house code. He claims that Low-Reynolds number turbulence models, whose damping functions are based on variables readily responding to buoyancy and flow acceleration significantly over-predict flow laminarization and therefore heat transfer deterioration. Other RANS simulations have been carried out by Cheng et al. [10], Palko and Anglart [11], and Yang et al. [12]. Results from these works suggest that, even in some cases, simulations have good agreement with experiments, but in general the current turbulent models are not reliable for solving the heat transfer problem of supercritical fluids.

Obviously, a fundamental understanding of heat transfer mechanisms through DNS/LES is necessary for further development of turbulence modeling. A well-know DNS work is accomplished by Bae et al. [13, 14]. He investigated the turbulent heat transfer of supercritical fluids with direct numerical simulation at inlet Reynolds number $Re_0 = 5400$ for a circular pipe and $Re_0 = 9000$ for an annular pipe. For circular pipe flow, several cases with different diameters, different wall heat fluxes and flow directions have been performed and compared. In a recent work, Nemati et al. [15] rerun some cases of Bae and focused on the averaging artifact called the Jensen inequality. Through the present study, it is aimed to use OpenFOAM to develop a DNS of turbulent heat transfer of supercritical fluids for $Re_0 = 5400$, verify the method with other DNS studies, and analyse the turbulence data. In the future, it is planned to make a direct comparison between DNS and experiment results at higher Reynolds numbers.

2 Computational Details

2.1 Governing Equations and Numerical Method

In the present work, Direct Numerical Simulation of heated pipe flow with CO_2 at supercritical pressure is performed on the basis of the Navier-Stokes Equations in a Cartesian coordinate system $x_{j,j=1,2,3}$ using the low-Mach-number assumption, where the acoustic interactions and compressibility effects are eliminated ((1)–(3)). The state variables are the velocity components $U_{j,j=1,2,3}$ and the enthalpy h. We can determine all thermodynamic properties density ρ, dynamic viscosity μ, specific heat capacity at constant pressure C_p and thermal conductivity κ, as functions of the enthalpy at a reference pressure P_0, see (4). The property data are obtained from the NIST web chemistry book [16].

$$\frac{\partial \rho}{\partial t} + \frac{\partial(\rho U_j)}{\partial x_j} = 0 \tag{1}$$

$$\frac{\partial \rho U_i}{\partial t} + \frac{\partial(\rho U_i U_j)}{\partial x_j} = -\frac{\partial P}{\partial x_i} + \frac{\partial}{\partial x_j}(\mu(\frac{\partial U_i}{\partial x_j} + \frac{\partial U_j}{\partial x_i})) \pm \rho g \delta_{i1} \tag{2}$$

$$\frac{\partial \rho h}{\partial t} + \frac{\partial(\rho U_j h)}{\partial x_j} = \frac{\partial}{\partial x_j}(\kappa \frac{\partial T}{\partial x_j}) \tag{3}$$

$$h = h(P_0, T), T = T(P_0, h), \rho = \rho(P_0, h), \mu = \mu(P_0, h),$$
$$C_p = C_p(P_0, h), \kappa = \kappa(P_0, h) \tag{4}$$

The model equations (1–4) are integrated using the open source program Open-FOAM 2.2.2 [17], which is a finite volume CFD code written in C++. OpenFOAM offers standard solvers for various physical problems and the flexibility to edit and create new solvers and functions at the same time. The velocity-pressure coupling is handled with the PISO (Pressure implicit with splitting of operator) algorithm [17]. For the temporal discretization the second-order implicit differencing scheme is applied. The spatial discretization method of the convection term in the momentum equation is central differencing. The upwind three-point interpolation QUICK-scheme [18] is used for the convection term in the energy equation to reduce oscillations due to sharp gradients. Such methods are also used in other DNS/LES turbulent heat transfer simulations of supercritical fluids [13, 15, 19].

The geometry of the pipe is presented in Fig. 2. The mesh consists of a structured O-grid with five blocks. The resolution is listed in Table 1, converted from Cartesian to Cylindrical Coordinates under the condition that the total number of cells is approximately the same. An inflow generator with a mesh resolution of $115 \times 120 \times 240$ along the radial (r), circumferential (θ) and axial direction (z) is used to generate fully developed turbulent flow at the inlet. For the heated main pipe, the mesh resolution is equivalent to $115 \times 120 \times 1440$. A uniform grid spacing is used in the axial direction, while the radial grid is refined close to the wall, which corresponds to a dimensionless grid resolution of $0.16(wall) < \Delta r^+ < 1.84(center)$, $(R\Delta\Theta)^+ \approx 9.6$, $\Delta z^+ = 7.6$ in wall units, based on constant-temperature properties at the inlet.

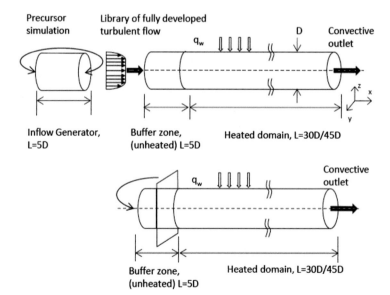

Fig. 2 Velocity inlet boundary conditions, *upper*: library method, *down*: mapping method

Table 1 Mesh resolution of current DNS and other authors

Number of cells	Radial	Circumferential	Axial	Δr^+	$(R\Delta\theta)^+$	Δz^+
Bae et al.	69	129	769	0.18	9.6	14.6
Nemati et al.	120	128	1536	0.17	8	8.8
Present	115	120	1440	0.16	7.6	9.6

2.2 Inflow Turbulence

Figure 2 shows the integration domain of heated pipe flow in the simulations. The pipe consists of two sections. The first section (buffer zone) has a length of 5 diameters and no wall heat flux. Turbulent flow should develop here before entering the heated domain. This eliminates any entry effects. In the second section, the wall is heated with constant heat flux. The pipe length is 30 or 45 diameters to ensure the observation of flow relaminarization and recovery. The outlet boundary condition for the velocity is the convective boundary condition from Orlanski [20], as seen in (5), where could be any scalar variable or the velocity vector and U_c is the convective velocity of the outflow.

$$\frac{\partial \phi}{\partial t} + U_c \frac{\partial(\phi)}{\partial x} = 0 \qquad (5)$$

The velocity inlet boundary condition of LES/DNS should be treated carefully. In many cases, the fluid behaviour within the domain is determined largely by the inlet behaviour. The inflow generator should generate the turbulence structure within a short pipe length. Besides, simplicity to implement and modify should also be considered. Tabor et al. [21] wrote a review of inlet boundary condition for LES/DNS based on OpenFOAM, including precursor simulation methods, internal mapping methods and different synthesis turbulent inflow generator. He concluded that although the synthesis inflow generator is easy to specify parameters of the turbulence, such as length scales or turbulent energy levels, it is inherently inaccurate, and requires long enough inlet development section. Precursor simulation methods produces true turbulence data and thus are inherently more accurate, however, can be cumbersome to modify to generate the required state of turbulence. The internal mapping method, also referred as recycling-and-rescaling method, is able to generate turbulent flow within relatively short distance. Besides, it is simple to adapt it to wide range of simulation conditions. The library method is often used in similar works by other authors [13, 15] and also in early works of current simulations. In order to run cases with different conditions (pipe diameter, inlet velocity) efficiently, the mapping inlet boundary condition is used instead. Details of two inlet boundary conditions will be given in the following.

In the current study, two different approaches for inlet boundary condition have been implemented and validated: the library method from precursor simulation (denoted as library method) and the inlet mapping method (denoted as mapping method), as shown in Fig. 2. In the library method a precursor simulation is performed with the same dimensionless resolution as the main pipe but cyclic boundary conditions. The data are stored as a library. The inlet velocity of the main pipe is read from that library for every time step. For the mapping method, the velocity inlet boundary condition is mapped from a cutting plane downstream. In each time step the total mass flux at the inlet is scaled to match a required value. To ensure the quality of the inflow turbulence before entering the heated domain, the

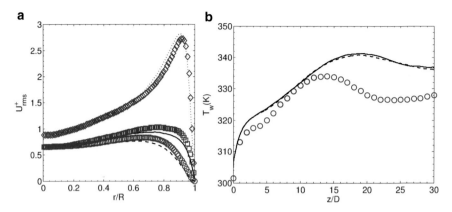

Fig. 3 Validation of the mapping inlet boundary condition, *left*: dimensionless velocity fluctuation, *lines*: current DNS, *symbols*: DNS results from Eggels et al. [22], *from top to bottom* $U_{z,rms}$, $U_{\theta,rms}$, $U_{r,rms}$; *right*: averaged wall temperature in one of simulation case 22U with two different boundary conditions

dimensionless velocity fluctuations in radial, circumferential and axial directions are validated with the DNS study of Eggels et al. [22] in Fig. 3 left. The dimensionless velocity fluctuation matches the reference data satisfying well. Besides that, we also simulate one of our cases with these two inflow boundary conditions and demonstrate the averaged wall temperature along the pipe in the right picture of Fig. 3. It is clear that substituting the library inflow generator with the mapping inflow generator would barely impact the heat transfer character of the flow.

2.3 Code Validation

OpenFOAM is designed as a general-purpose CFD code rather than a specially designed DNS code. It has been used for Direct Numerical Simulation before [23]. We still find it necessary to validate the code with experiments or other reliable DNS studies. Due to the character of supercritical fluids, experiments with detailed flow information, e.g. velocity and temperature in the pipe, have not been performed at low Reynolds numbers. Therefore, we choose the experimental work by Shehata and McEligot [24] for validation, which is turbulent pipe flow with heated air. In this flow the thermo-physical properties vary with temperature as the result of heating, which could be compared with fluids at supercritical pressure. The inlet Reynolds number Re_0 of the chosen case Run635 is 6000, which means that we can use the same mesh resolution as introduced above. The pipe diameter is $D = 27.4$ mm, the inlet pressure and flow temperature are $P_0 = 0.1$ MPa, $T_0 = 25\,^{\circ}$C, the wall heat flux is fixed to $q_w = 4.11$ kW/m^2. The geometry and boundary conditions are the same as introduced in this section. This experimental case is a forced convection dominant case, which means the buoyancy effect is relatively small. But because of the flow acceleration due to thermal expansion, the turbulent flow is relaminarized in the heated pipe. The averaged wall temperature along the pipe and the flow profile in the pipe are shown in Figs. 4 and 5.

In Fig. 4, the average wall temperature result from DNS rises under the effect of wall heat flux, which matches excellently with the experiment data. In Fig. 5, the averaged axial velocity profile and averaged flow temperature profile on different cutting planes of the pipe are also very close to the measured profiles.

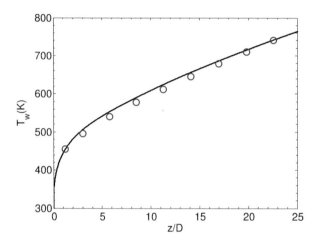

Fig. 4 Average wall temperature T_w of case Run635 from Shehata and McEligot [24] (*symbol*) and the DNS result (*line*)

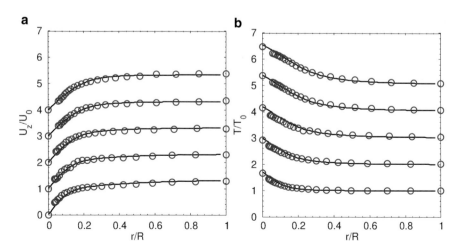

Fig. 5 Average velocity profile U_z/U_0 (*left*) and average flow temperature profile T/T_0 (*right*), DNS results in *line*, experiments in *symbol*, *from bottom to top* are $z = 3.2D$, $z = 8.7D$, $z = 14.2D$, $z = 19.9D$, $z = 24.5D$

Table 2 Simulation conditions, $Re_0 = 5400$, $T_0 = 301.15$ K, $P_0 = 8$ MPa

Case	Reference	Type	Direction	D (mm)	q_w (kW/m²)	Max. $Gr/Re^{2.7} \times 10^4$	Max. $Kv \times 10^7$
1F	Bae, Nemati	Forced	–	1	61.6	0	7.2
1U	Bae, Nemati	Mixed	Up	1	61.6	0.78	7.2
2F	–	Forced	–	2	15.4	0	3.6
2U	–	Mixed	Up	2	15.4	5.5	3.6
2D	–	Mixed	Down	2	15.4	5.3	3.6
22F	–	Forced	–	2	15.4	0	7.2
22U	Bae, Nemati	Mixed	Up	2	30.8	6.2	7.2
22D	Bae	Mixed	Down	2	30.8	5.6	7.2

2.4 Simulation Conditions

In this chapter, simulation conditions of heated turbulent vertical pipe flow at supercritical pressure will be introduced as listed in Table 2. To ensure the mesh resolution of DNS, inlet Reynolds number of all cases is set to $Re_0 = 5400$, which based on the properties at the constant flow temperature 301.15 K and the pressure 8 MPa. Some of the cases correspond to the studies of Bae et al. [13] and Nemati et al. [15]. Both forced convection cases (case 1F, 2F and 22F) and mixed convection cases (all other cases) are simulated through the control of the buoyancy term in the momentum equation. For mixed convection, the influence of the flow direction due to buoyancy could be examined through the comparison of upward and downward flow. Setting up the pipe diameter to 1 mm (case 1F and 1U) and 2 mm (all other cases) could help to investigate the influence of buoyancy, because the Grashof number ($Gr = g\beta_0 q_w R^4/(v_0^2 k_0)$) is proportional to the third power of pipe diameter. Therefore, $Gr/Re^{2.7}$, which presents the influence of buoyancy, is one order higher in 2 mm pipe cases than 1 mm pipe of the mixed convection cases. The dimensionless number $Kv = (v_b/U_b^2)/(dU_b/dx)$ introduced by Shehata and McEligot [24] represents the effect of flow acceleration (if $Kv > 10^{-5}$).

3 Results and Discussion

In the flow statistics below, we define the mean quantities with Reynolds- and Favre averaging, where $\bar{\phi}$ is the Reynolds average of any quantity and $\tilde{\phi} = \frac{\overline{\rho\phi}}{\bar{\rho}}$ is the mass-weighted (Favre) average. The corresponding fluctuations are denoted with $\phi' = \phi - \bar{\phi}$ and $\phi'' = \phi - \tilde{\phi}$.

The velocity components and the enthalpy are Favre averaged, whereas pressure, density and transport properties are Reynolds averaged.

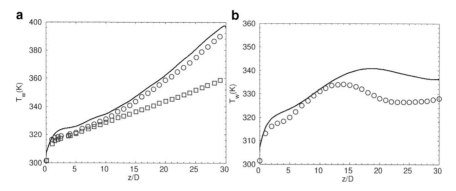

Fig. 6 $T_w(K)$ of the present DNS (*line*) with results by Bae et al. (*open circle*), Nemati et al. (*open square*), *left*: Case 1U, *right*: Case 22U

3.1 Heat Transfer Character

As mentioned in Sect. 2.4, some cases of the present DNS have the same conditions as DNS studies of other authors, in order to validate the code. This could also be meaningful for setting a data base for turbulence model development. In Fig. 6, the averaged wall temperature of two cases are compared with Bae et al. [13] and Nemati et al. [15]. As for the case 1U, current DNS is closer to Bae, but apparently higher than the results from Nemati. In the case 22U, both the heat transfer deterioration and recovery have been captured in current DNS and in Bae DNS. But the wall temperature is generally higher than Bae after about 15D downstream direction of the pipe. This discrepancy has also been mentioned by Nemati et al. [15]. All three codes and numerical methods have been validated with the same experiment from Shehata and McEligot [24], and the agreement is coincidently excellent. Obviously, the special thermo-physical properties of supercritical fluid bring extra difficulty to solve this problem numerically. Low Mach N-S equations are implemented in all three cases. An obvious difference is the property data base, NIST and PROPATH. But still we cannot confirm that this is the reason of discrepancy. More cooperation is needed for setting up a trustable DNS data base.

3.2 Flow Statistics

Mixed convection cases 2U (upward flow) and 2D (downward flow) will be detailed analysed in this chapter. In these cases, the simulation conditions (mass flux, heat flux, pipe diameter) are the same. Figure 7 shows the averaged wall temperature along the pipe in the three cases. In the downward case, the wall temperature rises only slightly downstream of a peak value at about $z = 5D$. A steep rise of the wall temperature and a slow recovery are captured in the upward case 2U. These

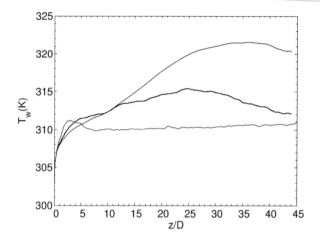

Fig. 7 Averaged wall temperature $T_w(K)$ of Case 2F (*black*) 2U (*blue*) and 2D (*red*)

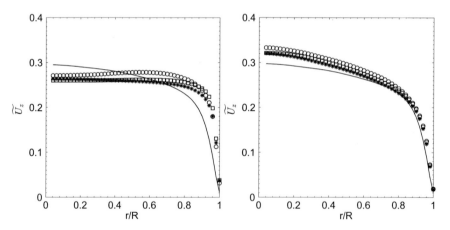

Fig. 8 Mean velocity profiles $\widetilde{U}_z$ (m/s) of 2U (*left*) and 2D (*right*)

mechanisms are generally denoted as heat transfer deterioration and recovery. The range of the wall temperature in the forced convection flow (2F) is between that of the upward and downward flow. Interestingly, a wall temperature peak is identified at about $z = 25D$ even in the forced convection case. In the forced convection cases with smaller pipe diameter and shorter pipe [13, 15], this peak is not observed, because it is expected to be located further downstream.

Figure 8 shows the Favre averaged velocity profile at different cutting planes along the downstream direction. In the upward flow (2U), the enormous density difference of the supercritical fluid and the gravity lead to a buoyancy effect. The buoyancy accelerates the lower-density fluid near the wall and decelerates high-density fluid near the pipe centreline. This distorts the mean velocity profile to form an M-shape. This effect of turbulence to the mean velocity profile caused by

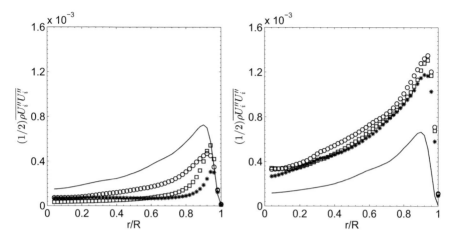

Fig. 9 Turbulent kinetic energy (TKE) $(1/2)\overline{\rho U_i'' U_i''}$ (kg/(ms^2)) of Case 2U (*left*) and 2D (*right*)

local flow acceleration or deceleration is called "external" or "indirect" effect. In downward flow (2D), buoyancy shows a opposite effect, which is an acceleration in the bulk and deceleration near the wall.

The turbulent kinetic energy (TKE) per unit volume $(1/2)\overline{\rho U_i'' U_i''}$ is the sum of the three Reynolds-stress components and indicates the intensity of turbulence Fig. 9). The downward flow, where $g(\partial\rho/\partial z) < 0$, is referred to as "unstable" in terms of the density stratification: buoyancy delivers a positive contribution to the turbulence. Therefore, the TKE of downward flow (2D) increases monotonically along the flow direction. For upward flow (2U), a stable density case $g(\partial\rho/\partial z) > 0$, buoyancy has a damping effect on turbulence initially and then turns to amplification. This influence is demonstrated in the TKE of case 2U. In the forced-convection flow case 2F, where the mean wall temperature increases, the TKE goes down, which means the turbulence intensity is depressed even without the influence of buoyancy. But as for the turbulent kinetic energy, it is clear that the turbulence intensity is more influenced by buoyancy (cases 2U and 2F) than the variable properties and thermal expansion (case 2F).

In Fig. 10 profiles of the Reynolds shear stress $\overline{\rho U_z'' U_r''}$ are presented. The forced convective heat transfer case 2F shows a similar peak value in all locations, but small depression could still be found at 10D and 20D, where the wall temperature rises. After that, at location 30D and 40D, it rises over the value at the beginning 0D. In the upward flow case 2U $\overline{\rho U_z'' U_r''}$ is strongly influenced by buoyancy, it drops to almost zero at 30D, where strong turbulent heat transfer deterioration is detected, and turns to negative at 40D. On the other side, the downward flow case 2D has a monotonically increasing Reynolds shear stress due to the positive effect of buoyancy to turbulence. Because of the limitation of pages, we will not go further analyze of more turbulence statistics like turbulence heat flux $\overline{\rho U_z'' h''}$,

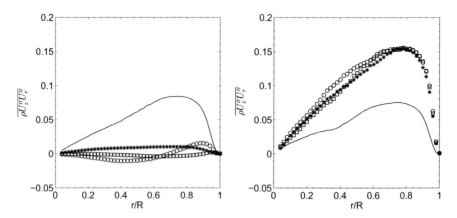

Fig. 10 Reynolds shear stress $\overline{\rho U_z'' U_r''}$ (kg/(ms^2)) of Case 2U (*left*) and 2D (*right*)

$\overline{\rho U_r'' h''}$, buoyancy production term $(gR/U_0^2)\overline{\rho' U_z'}$, which will be demonstrated in our further publications.

3.3 Visualization

Fluid motion on different cutting planes of the upward flow case 2U is visualized in Fig. 11, from the laminar sub-layer ($y_{ref}^+ = 5$), to the pipe centre line ($y_{ref}^+ = 150$). In the figure, the instantaneous contour is plotted, in which "warm" colours represent high speed streaks and "cold" colours represent low speed streaks. In a typical wall-bounded turbulent flow, the majority of the turbulence production occurs in the buffer region [25]. The coherent structures are believed to carry a large part of the turbulent kinetic energy and maintain the flow turbulence. At the inlet of a pipe, flow motion corresponds to a typical turbulent pipe flow. In the top figure of this column, $y_{ref}^+ = 5$ indicates that the cutting plane is located in the viscous sub-layer of the pipe according to law of the wall, where viscous shear dominates over turbulent shear. Therefore, most area of the plane is occupied by a uniform velocity. Most high- and low speed streaks can be found in the following three figures ($y_{ref}^+ = 15, y_{ref}^+ = 30, y_{ref}^+ = 50$), which belong to the buffer layer of the flow. Near the pipe centre ($y_{ref}^+ = 100, y_{ref}^+ = 150$), the streaks appear less frequently. In the second column, the flow is relaminarized because of the buoyancy effect. We observe that the turbulent-energy carrying streaks in the buffer layer disappear, especially at location $y_{ref}^+ = 30$ and $y_{ref}^+ = 50$. This explains the lower turbulent kinetic energy level in Fig. 9. In the third column, streaks reappear in the buffer layer. This visualization shows the fluid pattern at different locations of the pipe and could explain us the reason of flow relaminarization and recovery.

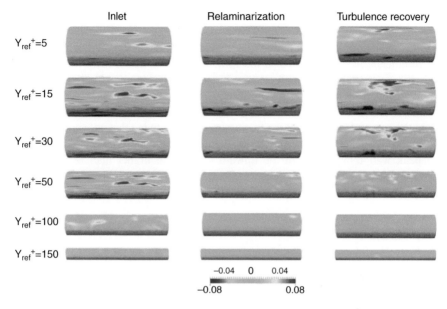

Fig. 11 Turbulent streaks of upward flow case 2U, at different wall distance y^+, at the position of inlet, flow relaminarized and flow recovered

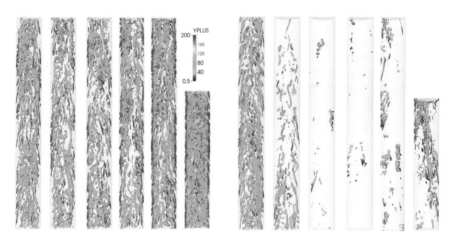

Fig. 12 Vortex structure according to lambda-2 criterium of 2F (*left*) and 2U (*right*)

In Fig. 12, the vortex structure of the flow is captured with the lambda-2 criterium and the threshold is set to $\lambda = -2 \times 10^4$ to obtain an optimal isosurface for visualization. The vortex structure is coloured by the dimensionless wall distance y_{ref}^+. Both, the forced-convection case (left) and the upward-flow mixed-convection case (right) are presented here. For convenience, the long pipe is decomposed to six parts as marked under the pictures. In the forced convection case, the decreased flow turbulence due to acceleration could be identified in the vicinity of $z = 20D$, which corresponds to the low level of turbulent kinetic energy in Fig. 10. After that, the density of the vortex rises to a higher level than at the inlet, which means the turbulence is reconstructed in the flow. And in the upward flow on the right side, the vortex is strongly affected by buoyancy. The vortex structure could be barely observed in the third and fourth part of the pipe, which means the flow is relaminarized. It is followed by a reproduction of buoyancy in the later section. From the comparison between the two cases, it is concluded that buoyancy impacts the convective heat transfer of supercritical fluid stronger than acceleration. It causes the peak wall temperature in the upward case to be higher than in the forced convection case. This has also been mentioned in Sect. 2.4 by analysing the indexes for buoyancy $Gr/Re^{2.7}$ and acceleration Kv.

4   Computational Performance

Parallel computational performance will be discussed in this chapter. The hardware utilized for computation is Hornet located at the High-Performance Computing Center Stuttgart (HLRS, Stuttgart). Hornet is a Cray XC40 system that consists of 3944 compute nodes. Each node has two Intel Haswell processors (E5-2680 v3, 12 cores) and 128 GB memory, and the nodes are interconnected by a Cray Aries network with a Dragonfly topology. This amounts to a total of 94,656 cores and a theoretical peak performance of 3.8 PFlops.

Data points in Fig. 13 show the speedup, defined as the wall-clock time on p cores divided by the wall-clock time on 96 cores, and parallel efficiency of OpenFOAM in the current study. In this case, the geometry consists of a total cell number of

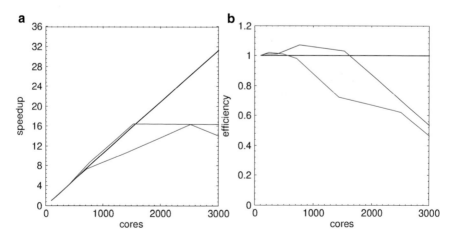

Fig. 13 Speedup and efficiency of 22 Mio. cells DNS case on Hornet, ideal (*black line*), PCG solver (*blue line*), GAMG solver (*red line*)

22×10^6. The mesh is decomposed with the Scotch algorithm [17]. The parallel computation is performed with up to about 3000 cores. The largest fraction of the total run-time to solve Eqs. (1)–(4) is the solution of the Poisson equation (pressure equation) within the PISO algorithm. Therefore, PCG (preconditioned conjugate gradient solver) and GAMG (generalised geometric-algebraic multi-grid solver) are compared as iterative linear solvers for the Poisson equation. In the test case, the speedup is calculated from a measurement of the wall clock time of 100 time steps. The first 100 time steps after the simulation-begin are ignored in order to avoid possible start-up fluctuations. In the current real-field test case, OpenFOAM shows excellent speedup up to 1536 cores with the PCG algorithm, although our the total mesh size is not that huge. In this case, each CPU core has only a decomposed-domain size of about 14,000 cells. Interestingly, there is also a slightly super-linear speed-up of 107 % at 768 cores and 103 % at 1536 cores. After that, the parallel efficiency drops down immediately to about 50 % at 3072 cores. With the GAMG linear solver, scalability could not be exceeded with PCG. The efficiency starts to decrease after 768 cores. Also it should be noticed that parallel performance of GAMG solver strongly depends on the detailed setup of the solver, like the number of pre- and postsweeps.

5   Conclusions

Direct Numerical Simulation of supercritical carbon dioxide flow in heated circular pipe is developed with OpenFOAM in the present study. The inlet Reynolds number of turbulent pipe flow is $Re_0 = 5400$ to fulfil the dimensionless resolution requirement. First, the numerical solver and method are validated with an experimental case of heated-air pipe flow. In this forced convection-dominant case, the flow is relaminarized due to acceleration from thermal expansion. The results of our DNS calculation agree very well with the experimental data.

Second, various computational cases with different pipe diameter, mass flux, and heat flux, leading to forced convection and mixed convection cases have been simulated. Both upward and downward flow directions are considered. In the forced convection case, a decreasing turbulence kinetic energy has been observed, which is due to flow acceleration from thermal expansion. In upward flow of mixed convection, relaminarization occurred followed by turbulence recovery in downstream direction because of the influence of buoyancy. In this process, buoyancy reduces the Reynolds shear stress to almost zero and then changes its sign to negative. The relaminarization of turbulent flow leads to heat transfer deterioration, as it is observed from the current DNS and also experiments. For the downward-flow case, buoyancy has an opposite effect, which means increasing the flow turbulence and as a result, the convective heat transfer is promoted. The turbulence statistic of our DNS may aid in the development of new turbulence models for practical applications in the future.

Acknowledgements The research presented in this paper is supported by Forschungsinstitut für Kerntechnik und Energiewandlung e.V. The authors would like to thank the kind support from HLRS and Cray team.

References

1. Brunner, G.: Applications of supercritical fluids. Annu. Rev. Chem. Biomol. Eng. **1**(1), 321–342 (2010)
2. Dostal, V., Driscoll, M.J., Hejzlar, P.: A Supercritical Carbon Dioxide Cycle for Next Generation Nuclear Reactors. Massachusetts Institute of Technology, Cambridge (2004). http://hdl.handle.net/1721.1/17746
3. Shiralkar, B.S., Griffith, P.: The Deterioration in Heat Transfer to Fluids at Supercritical Pressure and High Heat Fluxes. MIT, Cambridge (1968)
4. Bae, Y.-Y., Kim, H.-Y.: Convective heat transfer to CO_2 at a supercritical pressure flowing vertically upward in tubes and an annular channel. Exp. Thermal Fluid Sci. **33**(2), 329–339 (2009)
5. Licht, J., Anderson, M., Corradini, M.: Heat transfer to water at supercritical pressures in a circular and square annular flow geometry. Int. J. Heat Fluid Flow **29**(1), 156–166 (2008)
6. Li, Z.-H., Jiang, P.-X., Zhao, C.-R., Zhang, Y.: Experimental investigation of convection heat transfer of CO_2 at supercritical pressures in a vertical circular tube. Exp. Thermal Fluid Sci. **34**(8), 1162–1171 (2010)
7. Duffey, R.B., Pioro, I.L.: Experimental heat transfer of supercritical carbon dioxide flowing inside channels (survey). Nucl. Eng. Des. **235**(8), 913–924 (2005)
8. Yoo, J.Y.: The turbulent flows of supercritical fluids with heat transfer. Annu. Rev. Fluid Mech. **45**(1), 495–525 (2013)
9. He, S., Kim, W.S., Bae, J.H.: Assessment of performance of turbulence models in predicting supercritical pressure heat transfer in a vertical tube. Int. J. Heat Mass Transf. **51**(19–20), 4659–4675 (2008)
10. Cheng, X., Kuang, B., Yang, Y.H.: Numerical analysis of heat transfer in supercritical water cooled flow channels. Nucl. Eng. Des. **237**(3), 240–252 (2007)
11. Palko, D., Anglart, H.: Theoretical and numerical study of heat transfer deterioration in high performance light water reactor. Sci. Technol. Nucl. Ins. **2008**(2), 1–5 (2008)
12. Yang, J., Oka, Y., Ishiwatari, Y., Liu, J., Yoo, J.: Numerical investigation of heat transfer in upward flows of supercritical water in circular tubes and tight fuel rod bundles. Nucl. Eng. Des. **237**(4), 420–430 (2007)
13. Bae, J.H., Yoo, J.Y., Choi, H.: Direct numerical simulation of turbulent supercritical flows with heat transfer. Phys. Fluids **17**(10), 105104 (2005)
14. Bae, J.H., Yoo, J.Y., McEligot, D.M.: Direct numerical simulation of heated CO_2 flows at supercritical pressure in a vertical annulus at Re = 8900. Phys. Fluids **20**(5), 055108 (2008)
15. Nemati, H., Patel, A., Boersma, B.J., Pecnik, R.: Mean statistics of a heated turbulent pipe flow at supercritical pressure. Int. J. Heat Mass Transf. **83**, 741–752 (2015)
16. NIST (The National Institute of Standards and Technology): NIST Chemistry WebBook (2015). http://webbook.nist.gov/chemistry/
17. OpenFOAM Foundation: OpenFOAM User Guide (2013). http://foam.sourceforge.net/docs/Guides-a4/UserGuide.pdf
18. Leonard, B.P.: A stable and accurate convective modelling procedure based on quadratic upstream interpolation. Comput. Methods Appl. Mech. Eng. **19**(1), 59–98 (1979)
19. Ničeno, B., Sharabi, M.: Large eddy simulation of turbulent heat transfer at supercritical pressures. Nucl. Eng. Des. **261**, 44–55 (2013)
20. Orlanski, I.: A simple boundary condition for unbounded hyperbolic flows. J. Comput. Phys. **21**(3), 251–269 (1976)

21. Tabor, G.R., Baba-Ahmadi, M.H.: Inlet conditions for large eddy simulation: a review. Comput. Fluids **39**(4), 553–567 (2010)
22. Eggels, J.G.M., Unger, F., Weiss, M.H., Westerweel, J., Adrian, R.J., Friedrich, R., Nieuwstadt, F.T.M.: Fully developed turbulent pipe flow: a comparison between direct numerical simulation and experiment. J. Fluid Mech. **268**, 175–210 (1994). doi:10.1017/S002211209400131X. http://journals.cambridge.org/article_S002211209400131X
23. Komen, E., Shams, A., Camilo, L., Koren, B.: Quasi-DNS capabilities of openfoam for different mesh types. Comput. Fluids **96**, 87–104 (2014)
24. Shehata, A.M., McEligot, D.M.: Turbulence Structure in the Viscous Layer of Strongly Heated Gas Flows. Lockheed Idaho Technologies Co., Idaho Falls (1995). http://www.osti.gov/scitech/servlets/purl/578698-bqeLBG/webviewable/
25. Pope, S.B.: Turbulent Flows. Cambridge University Press, Cambridge (2000)

DNS Investigation of the Primary Breakup in a Conical Swirled Jet

Claudio Galbiati, Moritz Ertl, Simona Tonini, G. Elvio Cossali, and Bernhard Weigand

Abstract The mechanisms of primary break-up of a jet produced by a Pressure Swirl Atomizer for aircraft use is investigated. A grid study has been done together with the inlet boundary adopted to setup the simulation. The simulations are undertaken using the in-house multiphase code FS3D based on the volume of Fluid (VOF) method. All simulations were carried out on the Cray XC40 at the High Performance Computing Center Stuttgart (HLRS).

1   Introduction

In many industrial fields it is necessary to break-up a liquid jet to create a spray. This is connected to the definition of the spray, where the liquid surface is increased to improve heat and mass exchange. The pressure swirl atomizer (PSA) is one of the most commonly used injector thanks to its simple construction and high reliability in operation. It is widely used, from spray-drying, cooling, and painting, sprinklers and fire suppression systems. Various tangential ports feed liquid to the internal chamber creating a vortex inside the atomizer: the liquid rotates and, due to the centrifugal force, it is displaced near the wall [8]. Therefore, the swirl motion creates a depression within the internal region causing entrainment at the surface in contact with the gas. The jet, which is obtained at the exit section of the atomizer, is conical, and is swirled: that two conditions lead to the phenomenon that the rupture of the liquid jet is not clearly understood. Its interaction with the external gas induces instabilities on the jet surface which grows until the liquid breaks-up into ligaments and, then, into droplets. This process is called primary break-up, and it is followed by a secondary one, that is caused by different phenomena (e.g. aerodynamic forces, evaporation, coalescence, impact between droplets). Despite the description of the main flow field is quite well documented by experiments, the atomization process

M. Ertl (✉) • B. Weigand
Institut für Thermodynamik der Luft- und Raumfahrt, Universität Stuttgart, Pfaffenwaldring 31, 70569 Stuttgart, Germany
e-mail: moritz.ertl@itlr.uni-stuttgart.de

C. Galbiati • S. Tonini • G.E. Cossali
University of Bergamo, Bergamo, Italy

© Springer International Publishing Switzerland 2016
W.E. Nagel et al. (eds.), *High Performance Computing in Science and Engineering '15*, DOI 10.1007/978-3-319-24633-8_22

continues to be a difficult point to define, even more for the primary break-up that could be described only by theoretical models. One possible way to describe this phenomenon is to perform numerical simulation which are able to define with high accuracy the interaction between the liquid jet and the environmental gas, and moreover the pinching of the liquid from the main jet. For that reasons, the present work is aimed at describing the atomization process by Direct Numerical Simulation (DNS) using the in-house code Free Surface 3D, adopting different grid sizes and inlet boundaries, for the jet injected from a pressure swirl atomizer for aircraft engines. It is important, to comprehend the behaviour of the jet by means of the following questions: What droplet sizes are formed by the break-up? What is the influence of the droplets on the generated spray? Can we use the knowledge of these parameter, to better control the production of pollutants like NOx after combustion?

2 Numerical Method

Free Surface 3D (FS3D) is a direct numerical simulation code for incompressible multiphase flows based on the volume of fluid (VOF) method. The code has been developed at the Institute of Aerospace Thermodynamics (ITLR), University of Stuttgart for more than 20 years. It is used for investigations in several scientific areas concerning multiphase flows including droplet deformation [12], droplet wall interaction [13], droplet film interaction [4], droplet collision [14] and bubbles [16]. The code has also been expanded to solve for heat and mass transfer in order to simulate evaporating droplets [6, 15]. FS3D is also used by the Mathematical Modeling and Analysis group at the Center for Smart Interface, Technical University Darmstadt, which has added for example the capability to simulate chemical reactions [1].

FS3D solves the incompressible Navier-Stokes equations for the mass and the momentum conservation:

$$\frac{\partial \rho}{\partial t} + \nabla \cdot (\rho \mathbf{u}) = 0, \tag{1}$$

$$\frac{\partial (\rho \mathbf{u})}{\partial t} + \nabla \cdot [(\rho \mathbf{u}) \otimes \mathbf{u}] = -\nabla p + \nabla \cdot \mathbf{S} + \rho \mathbf{g} + \mathbf{f}_\gamma. \tag{2}$$

Here $\mathbf{u}$ denotes the velocity vector, t the time, ρ the density and p the pressure. Furthermore $\mathbf{S}$ is the viscosity stress tensor, which has the form

$$\mathbf{S} = \mu \left(\nabla \mathbf{u} + (\nabla \mathbf{u})^T \right). \tag{3}$$

The surface tension is calculated as a volume force and is included as the term $\mathbf{f}_\gamma$. The energy equation and a diffusion calculation are also implemented and can be utilized for temperature dependent problems. The treatment of a purely fluid mechanics problems such as our jet break-up case doesn't require them, as the diffusion is negligible compared to the convection and the conservation equations are uncoupled from the energy equation due to the incompressible formulation.

2.1 Simulation of Multiphase Flows with VOF and PLIC

The VOF method for multiphase flows proposed by Hirt and Nichols [7] introduces a new field variable f which represents the volume fraction of a fluid in each computational cell in order to capture the interface. It is used to identify the location of the interface in the computational domain and to determine the curvature. The variable f can have the following values:

$$f(\mathbf{x}, t) = \begin{cases} 0 & \text{outside the liquid phase,} \\ (0, 1) & \text{at the interface,} \\ 1 & \text{inside the liquid phase.} \end{cases}$$

The density is calculated by the following equation

$$\rho(\mathbf{x}, t) = \rho_g + (\rho_l - \rho_g)f(\mathbf{x}, t), \tag{4}$$

where the subscript l denotes the liquid phase and g the gaseous phase. An additional equation for the transport of the variable f is introduced

$$\frac{\partial f}{\partial t} + \nabla \cdot (\mathbf{u}f) = 0. \tag{5}$$

The flux for the above Eq. (5) is calculated using the method of piecewise linear interface reconstruction (PLIC). This idea of combining the VOF method with the PLIC method was first introduced by Rider and Kothe [11]. The use of the PLIC reconstruction becomes necessary due to the f variable only representing the amount of fluid inside a cell as can be seen in Fig. 1a. Without an accurate information about the spatial distribution it is impossible to precisely determine how much liquid and how much gas is transported across a cell boundary. In order to capture the interface the normal vector $\mathbf{n}_\gamma$ of the interface is calculated from the gradient of the f-field using information from the 26 surrounding cells. A plane with the normal vector $\mathbf{n}_\gamma$ is placed inside the cell and its position is iteratively calculated to encase the volume fraction provided by the VOF variable f inside the cell. A 2D example of a PLIC reconstructed interface is shown in Fig. 1b. The f flux across a cell boundary is then calculated with the reconstructed interface in the finite volume formulation.

FS3D is vectorized for usage on the Nec SX9 vector computer and is parallelized with MPI and OpenMPI. The code is validated and has good performance on the Cray XC40 Hornet supercomputer of the HLRS Stuttgart [10].

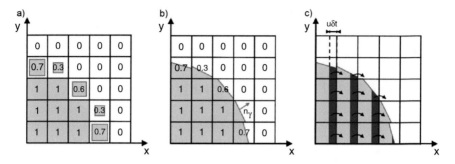

Fig. 1 (**a**) f-Field without interface information; (**b**) interface reconstruction with the PLIC-method; (**c**) calculation of the f-flux from a PLIC reconstructed interface

3   Numerical Setup

This work is aimed at simulating the near nozzle break-up behaviour of a jet emerging from a pressure swirl atomizer. The jet has an annular cross-sectional shape, where the internal and the external regions are occupied by the environmental gas. Moreover, in the internal part the pressure attains lower values than in the external region due to a need to balance the centrifugal force acting on the liquid, which leads to the ambient gas moving into the injector. For this reason, the inlet boundary is composed by an annulus shaped nozzle, where liquid is injected into the domain, and a circle inside, where the environmental gas is flowing out (see picture (a) in Fig. 3).

3.1   Inlet Velocity Profile

To setup correctly the simulation it is necessary to impose a swirl motion at the liquid that enters into the domain: with this aim, two new inlet boundaries have been implemented into FS3D. The first is able to impose a constant velocity profile for all velocity components in cylindrical coordinates: axial u_{axial}, swirl u_{swirl} and radial u_{radial}. The second inlet boundary that has been implemented is able to set into the inlet area a realistic velocity profile of the flow coming out from the nozzle. To understand the influence of the inlet velocity profile, both the boundaries have been used to simulate the behaviour of the jet for one real operating condition. Figure 2 shows the velocity profile for each velocity components for the test under investigation: the continuous line represents the velocity profile extracted from a previous Large-Eddies Simulation (LES) of the internal flow field. The circles over these lines represent the discretization adopted to set the input boundary. The dashed red line show the division of the inlet profile adopted for the liquid (for large value of the radius r), and for the gas (for small value of r).

Fig. 2 Radial distribution of the velocity components at the nozzle exit over the nozzle radius for the case under investigation

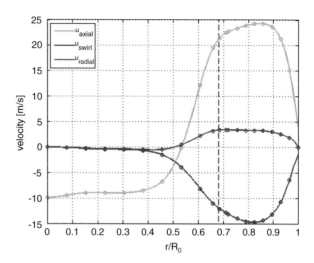

Table 1 Averaged inlet velocity components for each phase

	u_{axial} [m/s]	u_{swirl} [m/s]	u_{radial} [m/s]
Liquid	19.05	14.27	2.34
Gas	-0.18	7.60	1.38

Starting from the realistic velocity profile for each components estimated from LES, the average value has been then calculated and reported here in Table 1: these values are used to setup the constant inlet boundary.

For the liquid flow, the turbulent intensity $Tu = \frac{\sqrt{\overline{u'^2}}}{U_0}$ is defined as the ratio of the averaged magnitude of velocity fluctuations and the mean velocity magnitude $\overline{U_0} = 23.85$ m/s, and for the case under investigation Tu is equal to 13 %. The Reynolds number is equal to $Re \approx 1700$, as defined in Eq. (6), where $2h = 103\,\mu$m is the liquid lamella thickness while ρ_l and μ_l are the liquid density and dynamic viscosity,

$$Re = \frac{\rho_l \overline{U_0} 2h}{\mu_l}. \tag{6}$$

The ambient gas considered is air at a density $\rho_g = 13.27$ kg/m^3, while the fuel adopted is JET-A1, which has the density, kinematic viscosity and surface tension: $\rho_l = 801$ kg/m^3, $\mu_l = 1.29 \cdot 10^{-2}$ Pa s, and $\sigma = 2.23 \cdot 10^{-2}$ N/m. During the simulation all of these properties remain constant.

3.2   Description of the Domain

A rectangular domain has been adopted: the inlet boundaries are located at the center of the left side domain, while the remaining parts of the left side are set as no-slip wall. A continuous (Neumann) boundary condition is chosen for the other sides of the domain as can be seen in Fig.3. The dimension are as follows: 10.5 times the nozzle diameter $D = 625\,\mu m$ along both the directions of the nozzle plane (y and z) and 7.5D in the axial direction of the jet (x). The annular inlet boundary has a thickness equal to the initial liquid film thickness of the jet $2h$.

In the present work, a grid study has been carried out to see the influence of the grid on the prediction of the jet break-up. For that reason four grids have been generated with different Kolmogorov factors KF as reported in Table 2 by imposing a constant profile. This parameter represents the ratio between the size of the cells L_c and the smallest dissipative length scale λ_k according to the Kolmogorov's theory of the universal equilibrium,

$$KF = \frac{L_c}{\lambda_k} \in [1, 10]. \tag{7}$$

To perform a DNS calculation, the value of this parameter must be lower than 10 [5]. The value of λ_k is defined in Eq. (8), where the turbulent length scale L_t has

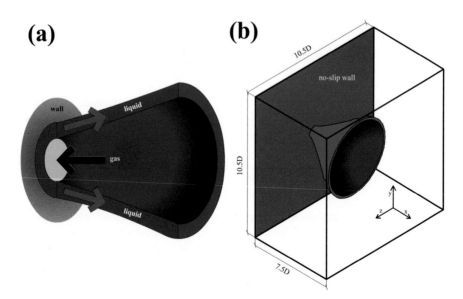

Fig. 3 (a) Topology of the inlet boundaries showing the annular nozzle injecting the liquid and the backflow region in the center surrounded by a no-slip wall. (b) A view of the domain dimension and boundaries—the transparent sides are continuous (Neumann) boundaries

Table 2 Kolmogorov factor and corresponding dimension of the grid. For each grid the type of inlet boundary profile is given

KF	Total amount of cells	Constant profile	Realistic profile	Number of nodes
6.56	6.71e+08	X		86
7.65	5.37e+08	X		86
8.19	3.03e+08	X	X	86
10.93	1.00e+08	X		22

been set equal to $\frac{2h}{10}$,

$$\lambda_k = \frac{L_t}{Re_t^{3/4}}, \text{ with } \quad Re_t = \frac{\rho_l \sqrt{u'^2} L_t}{\mu_l}. \tag{8}$$

4   Results

In this section the three-dimensional (3D) jet structure of ligaments and droplets will be discussed to provide a first impression of the jet behaviour by changing the grid size.

4.1   Grid Study

This kind of analysis has been made to see the influence of the grid size on the final prediction of the jet structure and the subsequent rupture into droplets. In Fig. 4 it is possible to observe the 3D jet surface for the four simulations under investigation at 0.1 ms after the start of the simulation, together with the internal flow field. It can been seen, if the grid becomes coarser, the annular ligaments that after the first break-up have different sizes and numbers: for $KF = 6.56$, it is possible to observe that there are three ligaments, with one closer to tip of the jet that starts to break-up in a smaller structure, while if KF increases the number of ligaments decrease to one (for the case at $KF = 10.93$ with a larger thickness).

Also in Fig. 4 it is possible to observe the velocity distribution of the air, which shows at the tip of the jet the presence of two counter rotating vortices that induce an additional aerodynamic disturbance on the jet surface and especially at the tip of the jet.

Figure 5 shows the results obtained from the simulation at 0.3 ms after the start of the simulation: at this point the jet is already ruptured and a primary spray is generated.

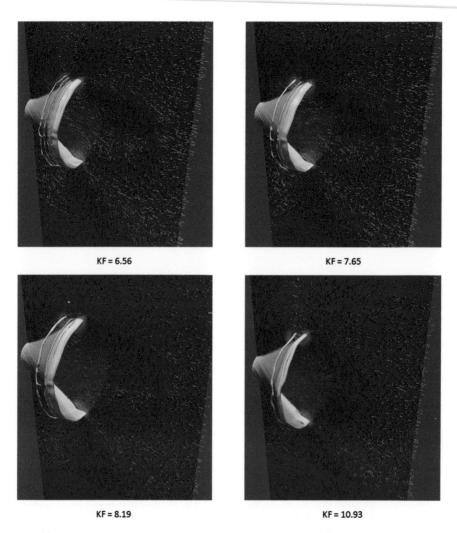

Fig. 4 Comparison of the jet at $t = 0.1$ ms for the four different grids investigated sorted by Kolmogorov factor KF. The liquid surface is displayed in *white*. The *central slice* shows the velocity magnitude and the direction of the velocity vectors is display as *orange arrow glyphs*

For the finer grid it is possible to observe the presence of the liquid film more forward after the injection point than for the other cases, where there are more ligaments, and droplets, for the coarser cases.

To quantify the effect of the grid on the break-up process, the break-up length of the jet l_b, following the methodology described by A. Dumachel et al. [2] has been estimated. Table 3 shows the results of this analysis. It is possible to observe that increasing the Kolmogorov factor leads to a faster break-up of the jet, due

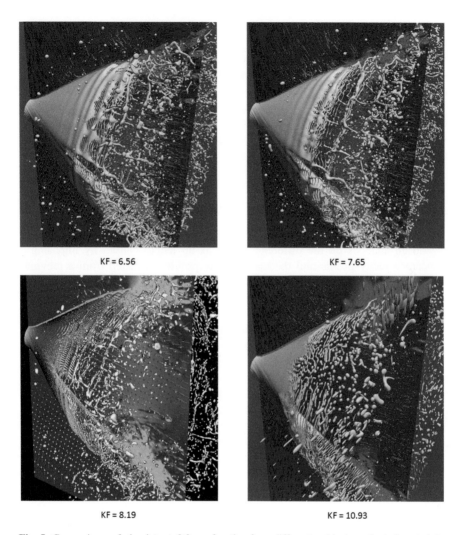

KF = 6.56

KF = 7.65

KF = 8.19

KF = 10.93

Fig. 5 Comparison of the jet at 0.3 ms for the four different grids investigated sorted by Kolmogorov factor KF. The liquid surface is displayed in *white*. The *central slice* shows the velocity magnitude and the direction of the velocity vectors is display as *orange arrow glyphs*

Table 3 Normalized break-up length of the swirled jet at different grid size

KF	$\frac{\bar{l}_b}{D}$
6.56	3.63
7.65	3.47
8.19	3.25
10.93	2.06

to the faster thinning of the liquid film connected to a worse description of the flow. However, if there is an increment of the break-up length by increasing the Kolmogorov factor, the results obtained show that they are not so different for $KF < 10$, while for $KF > 10$ the simulation is not able to reproduce correctly the behaviour of the jet.

From this analysis it has been established that there isn't an appreciable effect on the estimation of the break-up length for the cases where $KF < 10$. Therefore the successive analysis has been done using the grid with $KF = 8.19$ which has a lower number of elements and thus the calculations can be performed faster.

4.2 Comparison with Different Inlet Profiles

We do an analysis on the difference in break-up length caused by imposing a constant velocity profile and the real velocity profile, extracted from a previous internal nozzle flow simulation. Figure 6 shows the results from both the simulations at 0.1, 0.2 and 0.3 ms after the start. As it can be observed, for the two cases the development of the jet seems to be similar but with a different shape at the tip: the case with a realistic profile imposed tends to thinning faster than the other cases, where the film liquid moves deeper into the domain before it starts to break-up. That is connected to the higher value of shear stress generated from the realistic profile inside the liquid film. Indeed the normalized value of the break-up length over the nozzle diameter is 2.59 for the realistic inlet velocity profile and 3.25 by imposing a constant inlet velocity. Moreover, it is possible to observe that for the realistic profile, the spray angle is larger than the other one, which instead penetrates more. This suggest that imposing an inlet velocity profile with the same averaged properties doesn't lead to the same behaviour of the jet. In order to correctly simulate the spray it is necessary to set realistic inlet profiles as boundary conditions.

5 Performance

The general performance of FS3D on a computer cluster has been discussed in previous HLRS reports [9, 17]. However these analysis have been done for the previous systems Nec SX9 and Cray XE6 "Hermit". We will therefore now provide a performance analysis of the Code FS3D on the Cray XC40 "Hornet" system. As stated in Sect. 2, FS3D is parallelized with MPI. In order to provide meaningful results, we use different realistic simulation cases for our investigation. For the investigation of smaller grids, which can still be calculated on one single node, we use the case of a single oscillating droplet. This case is used for example in the investigation of Non-Newtonian fluids [3]. The case has a resolution of $256^3 \approx 16$ million cells. All small calculations have been done for 1 h walltime. The cycles per minute were then calculated from the total amount of cycles, the amount of node

Constant Profile   Real Profile

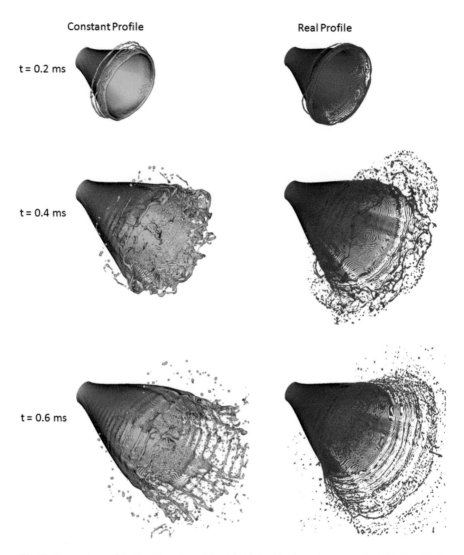

t = 0.2 ms

t = 0.4 ms

t = 0.6 ms

Fig. 6 Comparison of the liquid surface of the jet obtained by imposing a constant inlet profile (in *green*) and the realistic inlet profile (in *blue*) at different times after injection

and the utime value stated in the log file,

$$Cycles/\min = \frac{\#Cycles}{\frac{utime}{\#Nodes}}. \tag{9}$$

Calculations have been done on one, two and four nodes with one, two, four, eight and sixteen processors per node. The results have been visualized in Fig. 7. It can be observed, that the performance of FS3D for a small amount of processors is very

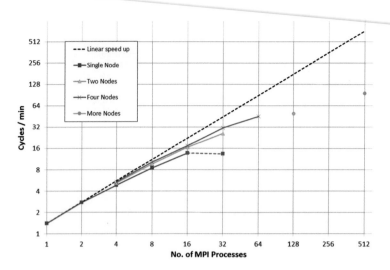

Fig. 7 Performance for a droplet oscillation with 16 millions cells: cycles per minutes over different amounts of processors and nodes

good. The increase in calculation speed is very close to the linear speed up to 32 processors. Afterwards the performance is still increasing, but at a visibly lower rate compared to the linear speed up. The simulation with 128 and 512 processors have been calculated with 16 processors per node on 8 and 32 nodes respectively. It can furthermore be observed that a distribution of the work onto more nodes performs slightly better, then the same amount of processors on less nodes.

For the investigation of large simulations we use the swirl jet described in Sect. 3.2 with a resolution of $512 \cdot 768 \cdot 768 \approx 303$ million cells. Since the utime values for these cases could not be obtained from the log file we use the somewhat less accurate walltime to calculate the cycles per minute. The results of the performance analysis are shown in Fig. 8. For this analysis we did each calculation twice—once with hyperthreading and once without. The reference case with 64 processors was calculated on 4 nodes with 16 processors each. This was the smallest possible amount of nodes to provide the necessary memory. The other cases were calculated with 24 processor per node and with 48 processors per node for hyperthreading. The performance shows a good increase for configurations with up to 512 processors. For 1024 processors only a small increase can be obtained. For 4096 processors the communication demands outweigh the gains in computational speed. A solution might be to use the MPI parallelization between the nodes in combination with OpenMP on the nodes. A first try with such a hybrid parallelization with three MPI process per node and eight OMP processors per MPI process however provided no improvement. A project to improve FS3D in that respect is under way with the help from the Cray team.

The graph also shows a better performance for every simulation without hyperthreading. This is not unexpected since hyperthreading puts two MPI processes on

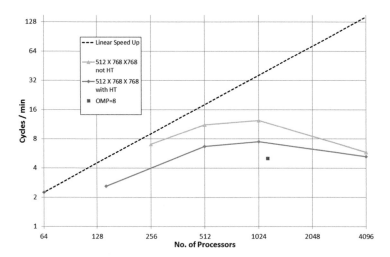

Fig. 8 Performance of a jet break-up case uses 303 million cells: cycles per minutes over different amounts of processors. *Blue diamonds* show the cases with hyperthreading

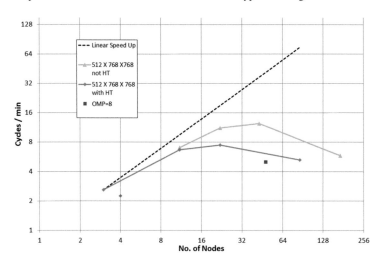

Fig. 9 Performance of a jet break-up case uses 303 million cells: cycles per minutes over different amounts of nodes. *Blue diamonds* show the cases with hyperthreading

each core and therefore provides less bandwidth. Furthermore, simulations without hyperthreading use double the amount of nodes and the performance analysis of the small cases has already shown, that the use of additional nodes provides performance improvements. In order to obtain a better understanding of the cost of hyperthreading we plot for this case the performance per node. The results are shown in Fig. 9. The linear speedup is calculated from the case with the fewest nodes.

The diagram shows, that the cases without hyperthreading perform worse than the cases without even on a per node comparison. FS3D therefore should not be used with the hyperthreading option.

6 Conclusion

A grid study for the simulation of a swirling jet has been carried out together with an analysis of the influence of the inflow velocity boundary condition. The simulation have been done using the ITLR inhouse multiphase CFD solver FS3D. For the grid study computational domains with Kolmogorov factors of 10.93, 8.19, 7.65, 6.56 have been calculated and the dimensionless jet break-up length has been calculated. The break-up length is dependent on the grid for all above mentioned simulations, but the deviations become unacceptable for Kolmogorov factors above 10. Therefore the case with $KF = 8.19$ was selected as the optimum trade off between numerical accuracy and computational cost. A strong influence of the velocity profiles has been demonstrated and therefore the realistic profile has been selected for subsequent simulations. A performance analysis has shown good results for computations with up to 512 processors. Simulations with 1024 processors still provide an acceptable increase in performance. For a larger number of processors, the performance actually decreases. To solve this problem and allow for efficient use of even larger amounts of processors, a move to a hybrid MPI/OpenMP parallelization is under way. An analysis of hyperthreading has shown that no efficient reduction in node usage can be achieved with FS3D.

Acknowledgements The authors greatly appreciate the *High Performance Computing Center Stuttgart* (HLRS) for its support and supply of computational time on the Cray XE6 and Cray XC40 platforms under the Grant No. FS3D/11142. The authors also greatly acknowledge financial support of this project from DFG for the priority program SPP 1423 "Prozess Spray".

References

1. Bothe, D., Kröger, M., Warnecke, H.J.: A VOF-based conservative method for the simulation of reactive mass transfer from rising bubbles. Fluid Dyn. Mater. Process. **7**(3), 303–316 (2011)
2. Dumouchel, C., Cousin, J., Triballier, K.: Experimental analysis of liquid-gas interface at low Weber number: interface length and fractal dimension. Exp. Fluids **39**(8), 651–666 (2005)
3. Ertl, M., Roth, N., Brenn, G., Gomaa, H., Weigand, B.: Simulations and experiments on shape oscillations of newtonian and non-newtonian liquid droplets. In: ILASS 2013, p. 7 (2013)
4. Gomaa, H., Stotz, I., Sievers, M., Lamanna, G., Weigand, B.: Preliminary investigation on diesel droplet impact on oil wallfilms in diesel engines. In: ILASS – Europe 2011, 24th European Conference on Liquid Atomization and Spray Systems, Estoril, September 2011 (2011)
5. Gorokhovski, M., Herrmann, M.: Modeling primary atomization. Annu. Rev. Fluid Mech. **40**, 343–366 (2008)

6. Hase, M., Weigand, B.: A numerical model for 3D transient evaporation processes based on the volume-of-fluid method. In: ICHMT International Symposium on Advances in Computational, pp. 1–23 (2004)
7. Hirt, C.W., Nichols, B.D.: Volume of fluid (VOF) method for the dynamics of free boundaries. J. Comput. Phys. **39**(1), 201–225 (1981). doi:10.1016/0021-9991(81)90145-5
8. Lefebvre, A.H.: Atomization and Sprays. Hemisphere, New York (1989)
9. Rauschenberger, P., Birkefeld, A., Reitzle, M., Meister, C., Weigand, B.: A parallelized method for direct numerical simulations of rigid particles in multiphase flow. In: High Performance Computing in Science and Engineering '14, pp. 321–334. Springer, Berlin (2014)
10. Rauschenberger, P., Schlottke, J., Weigand, B.: A computation technique for rigid particle flows in an Eulerian framework using the multiphase DNS code FS3D. In: High Performance Computing in Science and Engineering '11 Transactions of the High Performance Computing Center, Stuttgart (HLRS) (2011)
11. Rider, W.J., Kothe, D.B.: Reconstructing volume tracking. J. Comput. Phys. **141**(2), 112–152 (1998). doi:10.1006/jcph.1998.5906
12. Rieber, M., Graf, F., Hase, M., Roth, N., Weigand, B.: Numerical simulation of moving spherical and strongly deformed droplets. In: Proceedings ILASS-Europe, pp. 1–6 (2000)
13. Roth, N., Schlottke, J., Urban, J., Weigand, B.: Simulations of droplet impact on cold wall without wetting. In: ILASS, pp. 1–7 (2008)
14. Roth, N., Gomaa, H., Weigand, B.: Droplet collisions at high weber numbers: experiments and numerical simulations. In: Proceedings of the DIPSI Workshop 2010 on Droplet Impact Phenomena & Spray Investigation, Bergamo (2010)
15. Schlottke, J., Rauschenberger, P., Weigand, B., Ma, C., Bothe, D.: Volume of fluid direct numerical simulation of heat and mass transfer using sharp temperature and concentration fields. In: ILASS - Europe 2011, 24th European Conference on Liquid Atomization and Spray Systems, Estoril. http://www.ilass.uci.edu/ (2011)
16. Weking, H., Schlottke, J., Boger, M., Munz, C.D., Weigand, B.: DNS of rising bubbles using VOF and balanced force surface tension. In: High Performance Computing on Vector Systems. Springer, Berlin, Heidelberg, New York (2010)
17. Zhu, C., Ertl, M., Meister, C., Rauschenberger, P ., Birkefeld, A., Weigand, B.: Direct numerical simulation of inelastic Non-Newtonian jet breakup. In: High Performance Computing in Science and Engineering '13, pp. 321–335. Springer, Berlin (2013)

Numerical Simulation of Subsonic and Supersonic Impinging Jets

Robert Wilke and Jörn Sesterhenn

Abstract This report concentrates on fully turbulent confined round impinging jets with focus on heat transfer and the source mechanism of the impinging tones. Direct numerical simulations were performed with Reynolds numbers of $Re = 3300$ (subsonic and supersonic) and $Re = 8000$ (subsonic) using grid sizes of $512 \times 512 \times 512$ respectively $1024 \times 1024 \times 1024$ points. The transient flow field is analysed using a dynamic mode decomposition (DMD). It is shown that there is a dominant frequency with which the heat transfer at the impinging plate fluctuates. The corresponding structures are the vortex rings developing in the shear layer of the free jet region of the impinging jet. The same structures are together with the standoff shock responsible for the discrete tones referred to as impinging tones emitted by supersonic impinging jets.

1 Introduction

Subsonic impinging jets provide an effective cooling method for various applications such as the cooling of turbine blades of aircraft. No less important is the usage as rocket engine or vertical and/or short take-off and landing (V/STOL) aircraft aero engine. Those jets are typically operated under pressure ratios high enough to allow a supersonic flow. The operation is characterised by the emission of destructive loud noise that can cause material fatigue and deafness.

Both, heat transfer and jet noise are investigated using direct numerical simulations (DNS). Each project is described in an own section (Sects. 3 and 4) after the common numerical framework is explained (Sect. 2).

R. Wilke (✉) • J. Sesterhenn
Technische Universität Berlin, Fachgebiet Numerische Fluiddynamik, Müller-Breslau-Str. 11, 10623 Berlin, Germany
e-mail: robert.wilke@tnt.tu-berlin.de; joern.sesterhenn@tu-berlin.de; http://www.cfd.tu-berlin.de/

© Springer International Publishing Switzerland 2016
W.E. Nagel et al. (eds.), *High Performance Computing in Science and Engineering '15*, DOI 10.1007/978-3-319-24633-8_23

2 Numerical Method

2.1 Discretisation

The governing Navier-Stokes equations are formulated in a characteristic pressure-velocity-entropy-formulation, as described by Sesterhenn [25] and solved directly numerically. This formulation has advantages in the fields of boundary conditions, parallelization and space discretisation. No turbulence modelling is required since the smallest scales of turbulent motion are resolved. The spatial discretisation uses 6th order compact central schemes of Lele [16] for the diffusive terms and compact 5th order upwind finite differences of Adams et al. [1] for the convective terms. To advance in time a 4th order Runge-Kutta scheme is applied. In order to avoid Gibbs oscillations in the vicinity of the standoff shock (present at supersonic impinging jets) an adaptive shock-capturing filter developed by Bogey et al. [2] that automatically detects shocks is used.

2.2 Computation of Statistical Variables

The impinging jet at $Re = 8000$ has to be resolved with more than one billion grid points in order to achieve an adequate spacial resolution of the Kolmogorov scales. Storing one time step with the necessary five variables (pressure, velocity (x,y,z) and entropy) requires 41 GB of storage. It is easily noticeable that it is not sensible today to store thousands of time steps so as to do statistical analysis as post-processing. Therefore the code was restructured in 2014 and allows now the computation of statistical variables e.g. mean values, variances or complicated budget terms on-the-fly. This is done in four steps. In the first step the simulation runs until the influence of the start e.g. the starting of the impinging jet is faded away and the flow reaches a settled or periodic state. Afterwards, mean values of the required variables and terms are computed. In the third step, fluctuations of expressions are computed. In the final step only operation on mean values obtained in the second and third step are performed. Applying this strategy, the required storage is reduced to a fraction, but one has to be very carefully since additional quantities may not be computable after the simulation was conducted.

2.3 High Performance Computing

Investigating physics by means of direct numerical simulation requires huge computing capacity, which can only be provided by the most powerful high performance computers that are available nowadays. The Kolomgorov scales that need to be resolved lead at high Reynolds numbers of $Re = 8000$ to capacities in the order of much more than ten million core hours per computation. The load is partitioned between a huge number of processes, e.g. 8192 or 16,384. Each process solves the Navier-Stokes equations for a fractional part of the computational domain (block). This approach is referred to as domain decomposition, see [7]. In order to calculate derivatives, information from the adjacent blocks are needed. Therefore the decomposed domain is rearranged so that each process receives grid lines that span the entire domain in the particular direction. The total number of grid points per process remains constant and is typically either 32^3 or 64^3. Figure 1 exemplary shows the transformation from the original decomposition to the decomposition used for the calculation of derivatives in x-direction. The required inter-process communication is managed via MPI libraries.

The code has been successfully ported from CRAY XE6 (Hermit) to CRAY XC40 (Hornet). Figure 2 shows nearly perfect linear scaling up to 16,384 cores

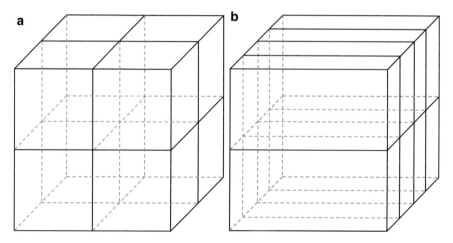

Fig. 1 Domain decomposition of a three-dimensional domain. (**a**) Original decomposition; (**b**) Transformated decomposition for the calculation of derivatives in x-direction

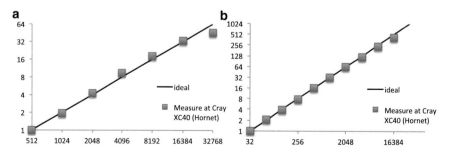

Fig. 2 Strong and weak scaling of the code on CRAY XC40 (Hornet) (**a**) Strong scaling; simulations run with 1024^3 grid points; (**b**) Weak scaling; simulations run with 64^3 grid points per core

for the given problem. Using auto-vectorisation, the efficiency with 16,384 cores is 102 % (strong) respectively 83 % (weak) enables to efficiently perform computations on Hornet with 16,384 cores. Grids with 512^3 (1024^3) points are typically parallelized on $16^3 = 4096$ ($32 \times 16 \times 16 = 8192$ or $32 \times 32 \times 16 = 16,384$) cores. The preferred wall time interval is 24 h.

2.4 Computational Domain

The present simulation is conducted on a numerical grid of size $512 \times 512 \times 512$ ($Re = 3300$) respectively $1024 \times 1024 \times 1024$ ($Re = 8000$) points for the computational domain sized $12D \times 5D \times 12D$, where D is the inlet diameter, see Fig. 3. A confined impinging jet is characterised by the presence of two walls, the impinging plate and the orifice plate. The grid is refined in those wall-adjacent regions in order to ascertain a maximum value in time and space of the dimensionless wall distance y^+ of the closest grid point to the wall smaller than one for both plates. For the x- and z-direction a slight symmetric grid stretching is applied which refines the shear layer of the jet. The refinements use hyperbolic tangent respectively hyperbolic sin functions resulting in a change of the mesh spacing lower than 1 % for all directions and cases. Table 1 shows the physical parameters of the simulation.

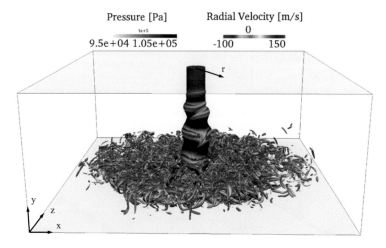

Fig. 3 Computational domain with iso-surfaces at $Ma = 0.2$ coloured with pressure and at $Q = 10^5$ m^2 s^{-4} coloured with radial velocity. $Re = 8000$; $p_o/p_\infty = 1.5$

Table 1 Physical parameters of the simulation

p_o/p_∞	p_∞ (Pa)	Ma	T_o (K)	T_W (K)	Re	Pr	κ	R [J/(kg K)]
1.5	10^5	0.784	293.15	373.15	3300	0.71	1.4	287
1.5	10^5	0.784	293.15	373.15	8000	0.71	1.4	287
2.15	10^5	1.106	293.15	373.15	3300	0.71	1.4	287

$p_o, p_\infty, T_o, T_W, Re, Pr, \kappa, R$ denote the total pressure, ambient pressure, total temperature, wall temperature, Reynolds number, Prandtl number, ratio of specific heats and the specific gas constant

2.5 Boundary Conditions

The computational domain is delimited by four non-reflecting boundary conditions, one isothermal wall which is the impinging plate and one boundary consisting of an isothermal wall and the inlet. The walls are fully acoustically reflective. The location of the nozzle is defined using a hyperbolic tangent profile with a disturbed thin laminar annular shear layer as described in [31].

A sponge region is applied for the outlet area $r/D > 5$, that smoothly forces the values of pressure, velocity and entropy to reference values. This destroys vortices before leaving the computational domain. The reference values at the outlet were obtained by a preliminary large eddy simulation of a greater domain.

2.6 Kolmogorov Scales

The scales of turbulent motion span a huge range from the size of the domain to the smallest energy dissipating ones. Since the turbulent kinetic energy is transferred downwards to smaller and smaller scales, the smallest ones have to be resolved by the numerical grid in order to obtain a reliable solution of the turbulent flow. They are given with the kinematic viscosity ν and dissipation rate ϵ by $l_\eta \approx \left(\frac{\nu^3}{\epsilon}\right)^{\frac{1}{4}}$ and are valid for isotropic turbulence, that occurs at sufficiently high Reynolds numbers. The Reynolds numbers of 3300 and 8000 can be considered as low in that context or *not sufficiently high* to obtain isotopic turbulence. Therefore the Kolmogorov microscales provide a conservative clue.

The ratio of the mesh width to the Kolmogorov length scale $r_K = \left(h_x h_y h_z\right)^{1/3}/l_\eta$ reaches a maximum value of 1.6 ($Re = 3300$, subsonic) respectively 2.1 ($Re = 3300$, supersonic) at the lower wall. For supersonic turbulent boundary layers, Pirozzoli et al. [18] showed that the typical size of small-scale eddies is about $5..6l_\eta$. A strongly different behaviour for a boundary layer of high subsonic Mach number is not expected. The maximum ratio in the area of the free jet is $r_K = 1.3$ ($Re = 3300$, subsonic) respectively $r_K = 1.5$ ($Re = 3300$, supersonic) and of the

wall jet is $r_K = 1.5$ ($Re = 3300$, subsonic)/$r_K = 1.5$ ($Re = 3300$, supersonic). At
the standoff shock a value of $r_K = 2.7$ occurs and marks the global maximum.

2.7 Boundary Layer

In addition to the criterion due to turbulent motion of the jet, the boundary layer
including the viscous sub-layer also has to be resolved appropriately in order to
achieve reliable results of the heat transfer at the impinging plate. The maximum
dimensionless wall distance y^+ of the present simulations $Re = 3300$ and ($Re =
8000$) occurs at $r/D = 0.46$ and reaches a value of $y^+ = 0.64$ ($y^+ = 0.58$). The
minimum number of points in the viscous sub-layer $y^+ \leq 5$ is seven (eight) for the
entire domain. Figure 4 shows the velocity- and temperature boundary layer profile
for different distances from the stagnation point. u^+ and T^+ are the dimensionless
radial velocity and the dimensionless temperature.

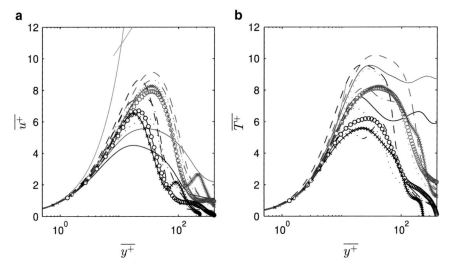

Fig. 4 Mean boundary layer profiles for different radial positions. *Black: $Re = 3300$; red: $Re =
8000$. Thick line: $r/D = 0.5$, Dashed line: $r/D = 1.0$; Dotted line: $r/D = 1.5$; Dashdotted line:
$r/D = 2.0$; line with circles: $r/D = 3.0$; line with asterisks: $r/D = 4.5$; Thin line: $u^+ = y^+$,
$u^+ = \ln(y^+)/0.41 + 5.1$ respectively $T^+ = Pr y^+$.* (**a**) Dimensionless Velocity; (**b**) Dimensionless
Temperature

2.8 Dynamic Mode Decomposition

A dynamic mode decomposition (DMD) is used to relate coherent structures to the heat transfer characteristics of subsonic impinging jets (Sect. 3) and to tonal noise of the supersonic impinging jets (Sect. 4). The Dynamic Mode Decomposition (DMD) as described by Schmid and Sesterhenn [21–23] extracts dynamic information out of a sequence of snapshots for a specific time interval that are either generated experimentally or numerically. The temporal dynamics of the flow is approximated by a linear snapshot to snapshot operator. The dominant eigenfunctions of this evolution operator (companion matrix) form a set of dynamically relevant modal structures. The so called dynamic modes accurately describe the motion of the fluid and can be interpreted as a generalisation of global stability modes. The mathematical background as well as the algorithm are given in [21, 22].

3 Subsonic Jet Impingement Heat Transfer

3.1 Introduction

Impinging jets and have been studied for decades. General information including schematic illustrations of the flow fields as well as distributions of local Nusselt numbers for plenty of different geometrical configurations and Reynolds numbers Re can be found in several reviews, such as [12, 13, 29, 30] based on experimental and numerical results. Since experiments cannot provide all quantities of the entire flow domain spacially and temporally well resolved, the understanding of the turbulent flow field requires simulations. Most existing publications of numerical nature use either turbulence modelling for the closure of the Reynolds-averaged Navier-Stokes (RANS) equations, e.g. [33], or large eddy simulation (LES), e.g. [5]. Almost all available direct numerical simulations (DNS) are either two-dimensional, e.g. [4], or do not exhibit an appropriate spatial resolution in the three-dimensional case, e.g. [8]. Recent investigations come from Dairay et al. [6]. He conducted a DNS of a round impinging jet with a nozzle to plate distance of $h/D = 2$ and focused on the secondary maximum of the heat transfer distribution and the connection to elongated structures.

This study deals with direct numerical simulations of a turbulent round impinging jets with $h/D = 5$ and Reynolds numbers of $Re = 3300$ and $Re = 8000$. Both impinging jets are fully turbulent since the critical Reynolds number of the case is 3000 [12].

3.2 Results

3.2.1 Influence of The Reynolds Number

The results out of the strategy described in Sect. 2 are already available for the simulation at $Re = 3300$. The third step of the greater simulation is not completely converged at this time. In this section, we present some results of the first and second phase. Figure 5 shows mean distributions of the radial velocity and the temperature for different radial positions. As expected, both simulations have a similar mean flow field with the distinction, that the boundary layer thickness decreases with an increasing Reynolds number.

The mean Nusselt number

$$Nu(r) = -\frac{D}{\Delta T}\frac{\partial T}{\partial y}(r)\Big|_W \tag{1}$$

is shown in Fig. 6a and scaled with $Re^{0.555}$ Fig. 6b. The scaling law was linear interpolated from the laws for $h/D = 4$ and $h/D = 6$ found by Lee [15]. There are plenty of those laws determined from experiments. All of them are similar but the coefficients differ based on the set-up (nozzle shape, nozzle-to-plate distance,..). Regarding the scaled distributions, the Nusselt number profile of the simulations at $Re = 3300$ and $(Re = 8000)$ feature a local maximum which is at once global, close to the axis of the jet at $r/D = 0.18$ $(r/D = 0.23)$. The secondary maximum of the $Re = 3300$ case is located at $r/D = 1..1.4$. The $Re = 8000$ case does not have this secondary maximum, but it is visible that the slope in this area is decreased.

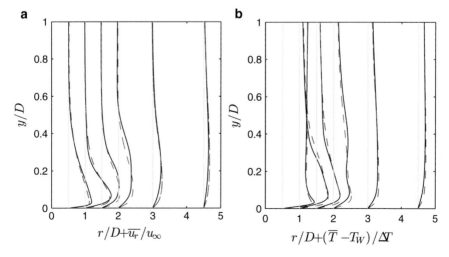

Fig. 5 Mean distributions for different radial positions r/D *solid line*: $Re = 3300$, *red dashed line*: $Re = 8000$. (**a**) Radial velocity; (**b**) Temperature difference $T - T_o$ relative to $\Delta T = T_W - T_o$

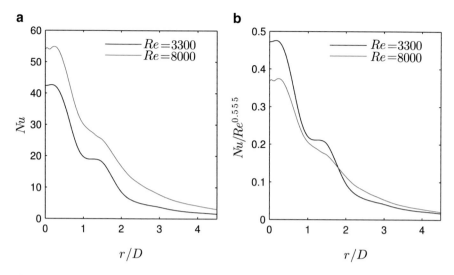

Fig. 6 Mean Nusselt number. (**a**) Mean Nusselt number; (**b**) Mean Nusselt number scaled with $Re^{0.555}$

A detailed analysis of the origins of this differences can and will be performed when all statistical results are available. However, the two maxima are connected to the primary vortex rings impinging on the plate close to the jet axis (primary maximum) respectively to a vortex pair (secondary maximum) consisting of the primary and a counter-rotating secondary vortex that moves from the stagnation point in radial direction until it breaks down into smaller structures [32]. Buchlin[3] reported for experimental data at much higher Reynolds numbers that $Re = 60000$ that the strength of both maxima decrease and the inner one moves to the stagnation point with increased nozzle to plate distances h/D and decreased Reynolds number. The simulation results show a reversed trend for the considered Reynolds numbers (Fig. 6).

3.2.2 The Dynamic Modes of Jet Impingement Heat Transfer

The aim of the performed DMD in this project is to find the modes corresponding to heat transfer characteristics of the impinging jet. Therefore a time series of the Nusselt number ($Re = 3300$) at the impinging plate is analysed beforehand of the DMD. The region $r/D = 1.2$ was chosen since it is located in the secondary maximum of the Nusselt number as illustrated in Fig. 6. The time series of an arbitrary point on the circle is shown in Fig. 7. The periodical behaviour of the Nusselt number is obvious. In order to determine the appropriate frequency range to be resolved with the DMD, a Fourier transformation of the time series around the circle $r/D = 1.2$ was conducted and than averaged. Figure 8 shows the spectrum of the signal whose amplitude A is measured in decibel with a reference value of

Fig. 7 Time series of the
Nusselt number on an
arbitrary point at $r/D = 1.2$,
$Re = 3300$

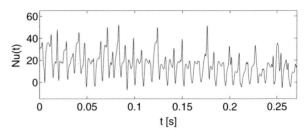

Fig. 8 Circumferential
averaged FFT of the time
series of the Nusselt number
at $r/D = 1.2$, $Re = 3300$

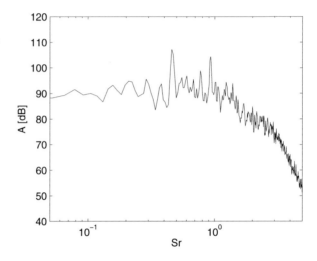

2×10^{-5}. The spectrum exhibits two peaks with a Strouhal number of $Sr = 0.46$
and $Sr = 0.92$ corresponding to the first harmonic. Considering the dominant
frequencies, a time span of 0.271 s represented by 120 snapshots and one out of
four grid points ($128 \times 128 \times 128$) was used for the DMD.

Figure 9 shows the eigenvalue spectrum of the DMD. The imaginary part
of the eigenvalues λ_i is equal to the dimensionless frequency of the dynamic
structure whereas the real part quantifies the damping ($Re(\lambda_i) < 0$) or growing
($Re(\lambda_i) > 0$) of the mode. The mode at $(0,0)$ represents the mean flow field since
it has no frequency and is not damped. Pairs of complex conjugate eigenvectors
and eigenvalues belong together to a single dynamic mode. The identified relevant
frequencies of the time series of the Nusselt number $Sr = 0.46$ and $Sr = 0.92$ are
marked additionally to the $(0,0)$ eigenvalue in the figure.

The corresponding structures are illustrated in Fig. 10. The first column shows
the Nusselt number at the impinging plate and Q (second invariant of the velocity
gradient tensor) representing vortical structures on a cut through the jet axis. In
the second column iso-surfaces of Q are snapped. The impinging plate represents
the Nusselt Number again. The first row corresponds to the frequency $Sr = 0.46$
whereas the second one corresponds to $Sr = 0.92$. The superposition of both modes
is shown in the third row. The DMD results show that the vortex rings developing in

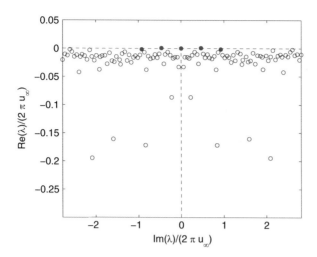

Fig. 9 Eigenvalue spectrum of the DMD, $Re = 3300$, marked frequencies: 0, ± 0.46 and ± 0.92

the shear layer are a superposition of a larger and a smaller structure. The larger structure corresponds to the frequency $Sr = 0.46$ and is the dominant one in the shear layer of the free jet region. The smaller one oscillates with the double frequency $Sr = 0.92$ and becomes important in the wall jet region. Both modes are axis-symmetric.

The DMD results are in agreement with the previously described relation between the movement of the vortical structures of the impinging jet and the dynamics of the Nusselt number on the impinging plate. The DMD delivers the dominant frequencies that are connected to the structures. This is an important step for heat transfer enhancement. Janetzke [14] showed that a pulsating inlet condition leads to an increase in heat transfer efficiency of more than 30 % for Strouhal numbers in the range around $Sr = 0.9$. The reason behind the existence of this efficient frequency range remained unclear. Following the DMD results, it can be supposed that a pulsating inlet with the eigenfrequency of the impinging jet leads to an increase of the heat transfer efficiency. This theory will be verified in future work.

3.3   Conclusion

This report presents a dynamic mode decomposition (DMD) of data obtained from a direct numerical simulation of a subsonic confined impinging jet at $Re = 3300$. The spectral analysis of the time series of Nusselt number at the impinging plate features two dominant frequencies at $Sr = 0.46$ and $Sr = 0.92$. The same frequencies were found in the eigenvalue spectrum of the DMD. The corresponding structures match the primary and secondary vortex rings of the transient flow field of the impinging

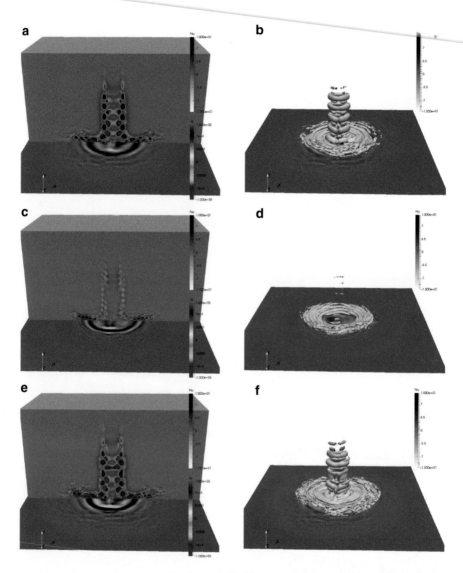

Fig. 10 Dynamic modes representing the periodical heat transfer characteristics of the impinging jet. The impinging plate is coloured with the Nusselt number. *First column*: Q (second invariant of the velocity gradient tensor) on a cut through the jet axis. *Second column*: iso-surfaces of Q. (**a**) $Sr = 0.46$; (**b**) $Sr = 0.46$; (**c**) $Sr = 0.92$; (**d**) $Sr = 0.92$; (**e**) $Sr = 0.46$ and $Sr = 0.92$; (**f**) $Sr = 0.46$ and $Sr = 0.92$

jet. Given the connection between the dominant frequencies and the unsteady heat transfer at the impinging plate, it can be supposed that a pulsating inlet with these dominant frequencies leads to an increase of the heat transfer efficiency of jet impingement.

4 Supersonic Impinging Jet Noise

4.1 Introduction

Compressible impinging jets are characterised by the appearance of immensely loud tonal noise. This noise can cause deafness and material fatigue. Until now the source mechanism is not completely understood. The discrete tones are an additional noise source to the ones occurring in the free jet.

In order to distinguish the impinging tones, a brief summary of the free jet noise sources following Tam [27] and Schulze [24] is given. For subsonic jets the only source is the turbulent mixing noise that is caused by large and small turbulent structures in the mixing layer of the jet. As soon as the nozzle pressure ratio (NPR) reaches a value to allow a supersonic flow, broadband shock-associated noise and screech noise appear as well in the case of under-expanded jets. Broadband shock-associated noise is caused by the interaction between downstream propagating large scale structures and the quasi periodic shock cell structure of the under-expanded jet.

Screech noise is a mark for discrete tones that are generated by a feedback mechanism. This source has to be distinguished carefully from the impinging tones, since both emit very strong tonal noise at similar frequencies. The screech noise feedback mechanism is shown in Fig. 11a. Vortical structures develop in the shear layer and grow while they are convected downstream. When the structures reach the fourth or fifth shock cell, they interact with those and emit strong acoustic waves that propagate upstream. These reach the nozzle lip or upper plate and excite the shear layer of the jet which leads to new instability waves and the close of the feedback mechanism. Based on the work of Powell [20] and Panda [17] five acoustic modes, labelled A-E, exist. They occur associated with the screech noise and therefore are

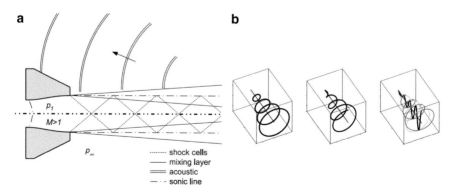

Fig. 11 The screech feedback loop and its associated acoustic modes of the free jet. (**a**) The screech feedback loop. Taken from [24]; (**b**) Instability modes of the free jet. *Left*: toroidal (axis symmetric), *middle*: helical, *right*: flapping (superposition of two counter-rotating helical modes). Adapted from [24]

referred to as screech modes. Depending on the nozzle pressure ratio, the free jet can feature varicose (toroidal) modes: A (A1, A2), helical modes: C or flapping modes that are a superposition of two counter-rotating helical modes: B, D. The shape of mode E is unknown. The mode shapes are sketched in Fig. 11b.

Supersonic impinging jet noise has been experimentally investigated for decades. However the source mechanism of the impinging tones is still not completely understood. First of all one has to distinguish between screech and impinging tones. Since the screech noise source is located between the rear edge of the third and the fifth shock cell, the distance between the nozzle/upper plate and the impinging plate has to be at least five diameters ($h/D \geq 5$) in order to allow the generation of screech tones [26]. Therefore a large proportion of the research was concentrated on very close plates and high pressure ratios ($NPR > 3$) [26]. It is generally agreed that a feedback mechanisms similar to the screech feedback loop is responsible for the appearance of impinging tones [9–11, 19]. However competing theories exist with respect to the close of the loop. Ho and Nosseir [11] named the primary vortices impinging on the wall as a possible link in the feedback chain. On the contrary Powell [19] and Henderson [10] suggested that standoff shock oscillations can have an important role in the feedback mechanism. A recent numerical investigation was performed by Uzun et al. [28]. He conducted a large eddy simulation with a plate distance of five diameters and a Mach number of 1.5 and found using a DMD coherent axis symmetric structures corresponding to the dominant tone at $Sr \approx 0.33$. A delimitation to screech and the missing mechanism in the feedback loop is not given.

In the present investigation a direct numerical simulation at $Re = 3300, NPR = 2.15, h/D = 5$ is carried out in order to identify the missing chain link in the impinging tone feedback mechanism by means of a dynamic mode decomposition (DMD). A delimitation to screech is given.

4.2 Results

4.2.1 Characterisation of the Discrete Tones of the Supersonic Impinging Jet: Distinction to Screech

The aim of this section is to show that the discrete tones occurring in the noise spectrum of the supersonic impinging jet are not related to screech. As explained, having a plate distance large enough, so that the relevant shock cells (number three to five) fit into the domain, it is possible that screech noise is radiated by the impinging jet. Figure 12 shows the pressure and axial velocity profile along the jet axis of the impinging jet. The profile is compared to one obtained from a free jet simulation. The for screech required five shock cells of the impinging jet are nearly undisturbed from the impinging plate. The shock cell spacing does not differ between the two cases. Only the amplitude of the displacement of the impinging jet profile is slightly decreased. Approaching the impinging plate closer than about

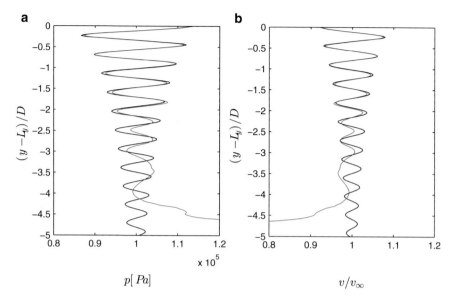

Fig. 12 (**a**) Pressure and (**b**) axial velocity profiles along the jet axis of the free *dashed line* and the impinging jet *red solid line* with equal physical parameters

two diameters the impinging jet shock cells are stronger and stronger influenced and disappear. The profiles lead to the conclusion that if screech occurs in the impinging jet noise spectrum it would have to have the exact same frequency as in the free jet case, since the relevant shock cells are even spaced.

Figure 13 shows the noise spectrum of the impinging and the free jet. The impinging jet features a discrete tone at a Strouhal number of $Sr = 0.353$ accompanied by the first harmonic whereas the free jet tone is at $Sr = 0.375$ and is missing its harmonics. In order to prove and strengthen the difference of the tones of the two cases, a dynamic mode decomposition using $128 \times 256 \times 128$ points for either simulations was carried out and compared. Figure 14 shows the eigenvalue spectra of the impinging and the free jet. They detect that the screech frequency of the free jet at $Sr = 0.375$ is a superposition of two frequencies at $Sr = 0.353$ and $Sr = 0.399$. On the contrary, the tonal noise of the impinging jet is associated to one mode only. Figure 15 shows the corresponding structures by isosurfaces of the pressure. The impinging jet mode is axis-symmetric however both free jet modes as well as the superposition are helical and even orientated.

The comparison suggests that the tonal noise of the impinging jet is an impinging tone and not associated with screech, since the dominant frequency differs from the one of the free jet despite equal spacings of the relevant shock cells. In addition the corresponding modes have different shapes: single axis-symmetric respectively helical being a superposition of two helical ones (free jet).

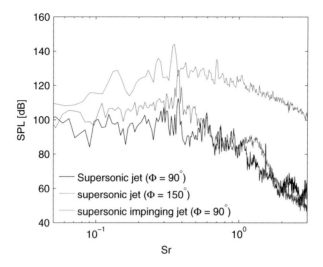

Fig. 13 Noise spectra of the impinging and the free jet. Reference pressure: 2×10^{-5} Pa

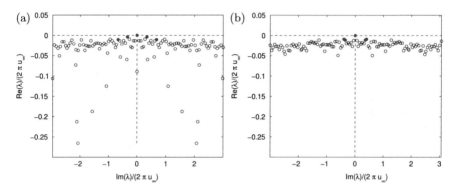

Fig. 14 Eigenvalue spectra of the impinging and free jet, obtained from a DMD. (**a**) Eigenvalue spectrum of the impinging jet, mean value and corresponding frequencies to tonal noise at $Sr = \pm 0.353$ and the first harmonic are marked; (**b**) Eigenvalue spectrum of the free jet, mean value and corresponding frequencies to tonal noise at $Sr = \pm 0.375$ are marked

4.2.2 The Origin of the Impinging Tones

As described before, possible origins of the impinging tones that trigger the feedback mechanism are (a) the primary vortices developing in the shear layer that impinge on the plate and (b) a standoff shock oscillation. Figure 16 shows a sequence representing one period of three dimensional structures of the impinging tone mode ($Sr = 0.353$, without the first harmonic). Vortical structures are detected by Q isosurfaces $Q = 1 \times 10^5 \, \text{m}^2 \, \text{s}^{-2}$ and coloured green. The standoff shock is detected by isosurfaces of the pressure gradient in y (wall normal) direction $\frac{\partial p}{\partial y} = -2 \times 10^{-5}$ (blue) and $\frac{\partial p}{\partial y} = -2 \times 10^5$ (red). The impinging plate is coloured

(a) (b)

(c) (d)

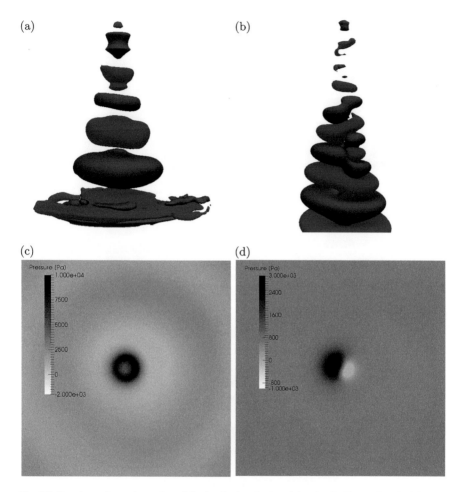

Fig. 15 Dominant dynamic modes of the impinging (**a,c**) and free jet (**b,d**) associated with tonal noise. The three dimensional structures are isosurfaces of the pressure (*first row*). The *second row* shows the pressure on a cut normal to the jet axis 2 respectively six diameters from the upper plate

with the pressure. It is visible that the impinging tone mode is related to both the primary vortices of the shear layer and the standoff shock oscillation.

The relation between the structures and the acoustic emission can be seen in Fig. 17 that shows snapshots of the flow field at three consecutive time steps (each in one row). The development of the ring vortices in the shear layer as well as the standoff shock can be seen in the first column where Q is depicted. In the final picture (g) the ring vortex reaches the standoff shock. The corresponding velocity divergence is imaged in (i) and shows strong acoustic waves that are radiated and have their origin in the crossing point of the vortex and the shock. This gives evidence that neither the vortices impinging on the plate nor the oscillation itself of the standoff shock is the origin of the impinging tone. The violent interaction

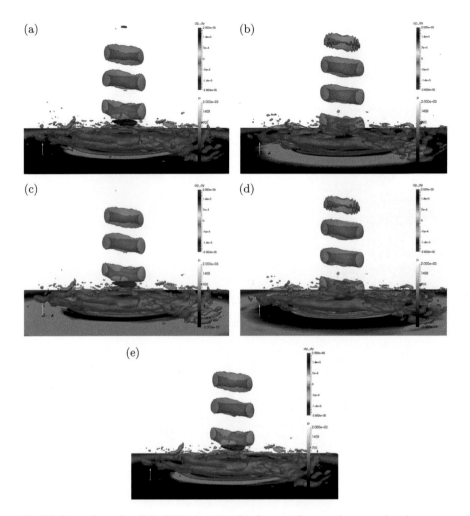

Fig. 16 Dynamic mode of the impinging tone for five equally spaced consecutive time steps (a)–(e) representing one impinging tone cycle. Vortical structures visualised at $Q = 1 \times 10^5 \, \mathrm{m^2 \, s^{-2}}$, shock structure visualised at $\frac{\partial p}{\partial y} = \pm - 2 \times 10^5 \, \mathrm{Pa \, m^{-1}}$, impinging plate coloured with the pressure

between both players leads to those strong acoustic emissions referred to as impinging tones.

4.3 Conclusion

A DNS of an under-expanded impinging jet at $Re = 3300, NPR = 2.15, h/D = 5$ was performed using a grid of 512^3 points. The noise spectrum features a discrete tone at $Sr = 0.353$ and its first harmonic. A comparison to a free jet with equal

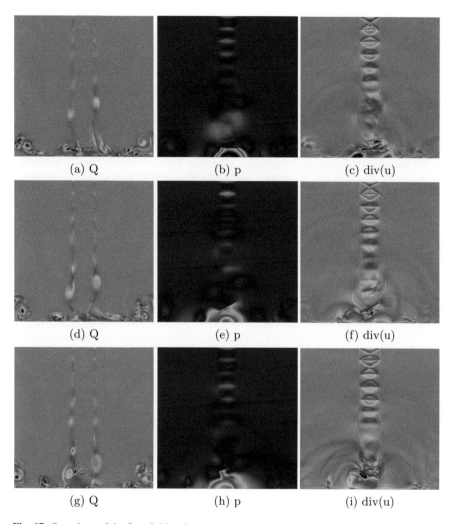

Fig. 17 Snapshots of the flow field at three consecutive time steps (each in one row) showing the interaction between ring vortices and the standoff shock. *First column*: *Q*, *second column*: pressure and *third column*: divergence of the velocity

physical parameters distinguished the tonal noise from screech. The missing chain link in the feedback mechanism of the impinging tones was identified as being the primary vortex-standoff shock-interaction.

Acknowledgements The simulations were performed on the national supercomputer Cray XE6 (Hermit) and Cray XC40 (Hornet) at the High Performance Computing Center Stuttgart (HLRS) under the grant numbers GCS-NOIJ/12993 and GCS-ARSI/44027.

The authors gratefully acknowledge support by the Deutsche Forschungsgemeinschaft (DFG) as part of collaborative research center SFB 1029 "Substantial efficiency increase in gas turbines through direct use of coupled unsteady combustion and flow dynamics".

References

1. Adams, N.A., Shariff, K.: A high-resolution hybrid compact-ENO scheme for shock-turbulence interaction problems. J. Comput. Phys. **127**, S.27–S.51 (1996). http://dx.doi.org/10.1006/jcph.1996.0156. doi:10.1006/jcph.1996.0156
2. Bogey, C., de Cacqueray, N., Bailly, C.: A shock-capturing methodology based on adaptive spatial filtering for high-order non-linear computations. J. Comput. Phys. **228**(5), 1447–1465 (2009). http://dx.doi.org/http://dx.doi.org/10.1016/j.jcp.2008.10.042. doi:http://dx.doi.org/10.1016/j.jcp.2008.10.042. ISSN 0021–9991
3. Buchlin, J.: Convective heat transfer in impinging-gas-jet arrangements. J. Appl. Fluid Mech. **4**(3), 137–149 (2011)
4. Chung, Y.M., Luo, K.H.: Unsteady heat transfer analysis of an impinging jet. J. Heat Transf. **124**(6), 1039–1048 (2002). http://dx.doi.org/10.1115/1.1469522. ISBN 0022–1481
5. Cziesla, T., Biswas, G., Chattopadhyay, H., Mitra, N.: Large-eddy simulation of flow and heat transfer in an impinging slot jet. Int. J. Heat Fluid Flow **22**(5), 500–508 (2001). http://dx.doi.org/http://dx.doi.org/10.1016/S0142-727X(01)00105-9. doi:http://dx.doi.org/10.1016/S0142--727X(01)00105--9. ISSN 0142–727X
6. Dairay, T., Fortuné, V., Lamballais, E., Brizzi, L.-E.: Direct numerical simulation of a turbulent jet impinging on a heated wall. J. Fluid Mech. **764**, 362–394 (2015). http://dx.doi.org/10.1017/jfm.2014.715. doi:10.1017/jfm.2014.715. ISSN 1469–7645
7. Eidson, T.M., Erlebacher, G.: Implementation of a fully balanced periodic tridiagonal solver on a parallel distributed memory architecture. Concurrency Pract. Experience **7**(4), S.273–S.302 (1995)
8. Hattori, H., Nagano, Y.: Direct numerical simulation of turbulent heat transfer in plane impinging jet. Int. J. Heat Fluid Flow **25**(5), 749–758 (2004) http://dx.doi.org/http://dx.doi.org/10.1016/j.ijheatfluidflow.2004.05.004. doi:http://dx.doi.org/10.1016/j.ijheatfluidflow.2004.05.004. ISSN 0142–727X. Selected papers from the 4th International Symposium on Turbulence Heat and Mass Transfer
9. Henderson, B.: The connection between sound production and jet structure of the supersonic impinging jet. J. Acoust. Soc. Am. **111**(2), S.735–S.747 (2002). http://dx.doi.org/http://dx.doi.org/10.1121/1.1436069. doi:http://dx.doi.org/10.1121/1.1436069
10. Henderson, B., Powell, A.: Experiments concerning tones produced by an axisymmetric choked jet impinging on flat plates. J. Sound Vib. **168**(2), S.307–S.326 (1993). http://dx.doi.org/http://dx.doi.org/10.1006/jsvi.1993.1375. doi:http://dx.doi.org/10.1006/jsvi.1993.1375. ISSN 0022–460X
11. Ho, C.-M., Nosseir, N.S.: Dynamics of an impinging jet. Part 1. The feedback phenomenon. J. Fluid Mech. **105**(4), S.119–S.142 (1981). http://dx.doi.org/10.1017/S0022112081003133. doi:10.1017/S0022112081003133. ISSN 1469–7645
12. Hrycak, P.: Heat Transfer from Impinging Jets. A Literature Review. New Jersey Institute of Technology, Newark, NJ (1981). Forschungsbericht
13. Jambunathan, K., Lai, E., Moss, M., Button, B.: A review of heat transfer data for single circular jet impingement. Int. J. Heat Fluid Flow **13**(2), S.106–S.115 (1992). http://dx.doi.org/http://dx.doi.org/10.1016/0142-727X(92)90017-4. doi:http://dx.doi.org/10.1016/0142--727X(92)90017--4
14. Janetzke, T.: Experimentelle Untersuchungen zur Effizienzsteigerung von Prallkühlkonfigurationen durch dynamische Ringwirbel hoher Amplitude, TU Berlin, Dissertation (2010)

15. Jungho Lee, S.-J.L.: Stagnation region heat transfer of a turbulent axisymmetric jet impingement. Exp. Heat Transfer **12**(2), 137–156 (1999). http://dx.doi.org/10.1080/089161599269753. doi:10.1080/089161599269753
16. Lele, S.K.: Compact finite difference schemes with spectral-like resolution. J. Comput. Phys. **103**(1), 16–42 (1992). http://dx.doi.org/10.1016/0021-9991(92)90324-R. doi:10.1016/0021–9991(92)90324-R
17. Panda, J.: Shock oscillation in underexpanded screeching jets. J. Fluid Mech. **363**, S.173–S.198 (1998). http://dx.doi.org/10.1017/S0022112098008842. doi:10.1017/S0022112098008842. ISSN 1469–7645
18. Pirozzoli, S., Bernardini, M., Grasso, F.: Characterization of coherent vortical structures in a supersonic turbulent boundary layer. J. Fluid Mech. **613**, 205–231 (2008). http://dx.doi.org/10.1017/S0022112008003005. doi:10.1017/S0022112008003005. ISSN 1469–7645
19. Powell, A.: The sound-producing oscillations of round underexpanded jets impinging on normal plates. J. Acoust. Soc. Am. **83**, S.515–S.533 (1988)
20. Powell, A., Umeda, Y., Ishii, R.: Observations of the oscillation modes of choked circular jets. J. Acoust. Soc. Am. **92**(5), S.2823–S.2836 (1992). http://dx.doi.org/http://dx.doi.org/10.1121/1.404398. doi:http://dx.doi.org/10.1121/1.404398
21. Schmid, P.J.: Dynamic mode decomposition of numerical and experimental data. J. Fluid Mech. **656**, 5–28 (2010). http://dx.doi.org/10.1017/S0022112010001217. doi:10.1017/S0022112010001217. ISSN 1469–7645
22. Schmid, P.: Application of the dynamic mode decomposition to experimental data. **50**(4), 1123–1130 (2011). http://dx.doi.org/10.1007/s00348-010-0911-3. doi:10.1007/s00348–010–0911–3. ISBN 0723–4864
23. Schmid, P.J., Sesterhenn, J.L.: Dynamic mode decomposition of numerical and experimental data. In: 61st APS meeting of American Physical Society, San Antonio, p. S.208 (2008)
24. Schulze, J.: Adjoint based jet-noise minimization, TU Berlin, Dissertation (2013)
25. Sesterhenn, J.L.: A characteristic–type formulation of the Navier–Stokes equations for high order upwind schemes. Comput. Fluids **30**(1), S.37–S.67 (2001)
26. Sinibaldi, G., Lacagnina, G., Marino, L., Romano, G.P.: Aeroacoustics and aerodynamics of impinging supersonic jets: analysis of the screech tones. Phys. Fluids (1994-present) **25**(8) (2013). http://dx.doi.org/http://dx.doi.org/10.1063/1.4819333. doi:http://dx.doi.org/10.1063/1.4819333
27. Tam, C.K.W.: Supersonic jet noise. Annu. Rev. Fluid Mech. **27**(1), 17–43 (1995). http://dx.doi.org/10.1146/annurev.fl.27.010195.000313. doi:10.1146/annurev.fl.27.010195.000313
28. Uzun, A., Kumar, R., Hussaini, M.Y., Alvi, F.S.: Simulation of tonal noise generation by supersonic impinging jets. AIAA J. **51**(7), S.1593–S.1611 (2013). http://dx.doi.org/10.2514/1.J051839. doi:10.2514/1.J051839
29. Viskanta, R.: Heat transfer to impinging isothermal gas and flame jets. Exp. Thermal Fluid Sci. **6**(2), S.111–S.134 (1993). http://dx.doi.org/http://dx.doi.org/10.1016/0894-1777(93)90022-B. doi:http://dx.doi.org/10.1016/0894--1777(93)90022--B
30. Weigand, B., Spring, S.: Multiple jet impingement - a review. Heat Transf. Res. **42**(2), S.101–S.142 (2011). ISSN 1064–2285
31. Wilke, R., Sesterhenn, J.L.: Direct numerical simulation of heat transfer of a round subsonic impinging jet. In: Notes on Numerical Fluid Mechanics and Multidisciplinary Design Bd. 127, pp. S.147–S.159. Springer, Berlin (2014)
32. Wilke, R., Sesterhenn, J.L.: Numerical simulation of impinging jets. In: High Performance Computing in Science and Engineering '14, pp. S.275–S.287. Springer, Berlin (2015)
33. Zuckerman, N., Lior, N.: Impingement heat transfer: correlations and numerical modeling. J. Heat Transf. **127**(5), 544–552 (2005). http://dx.doi.org/10.1115/1.1861921. ISBN 0022–1481

On the Impact of Forward-Facing Steps on Disturbance Amplification in Boundary-Layer Flows

Christopher Edelmann and Ulrich Rist

Abstract Using our DNS code *NS3D* boundary-layer flows over smooth walls and walls containing a forward-facing step have been simulated using a so-called disturbance formulation, i.e., by first computing a steady two-dimensional base flow and then an unsteady disturbance flow. The influence of step height and free-stream Mach number on streamwise pressure gradient and the occurrence or absence of laminar separation bubbles are shown for some base flows. A fundamental difference between subsonic and a supersonic flow over the same step height is that the flow directly after the step exhibits an adverse pressure gradient with laminar separation in the subsonic case but due to an expansion of the supersonic flow around the corner, a favourable pressure gradient without separation in the supersonic case. The unsteady disturbance flow is initialised by a wave packet in the upstream part of the integration domain that contains all frequencies which become unstable in the considered area. Amplification factors, so called N-factors, are then extracted from the frequency spectra of the disturbance flow for quantification of the impact of step parameters on disturbance amplification. Our results show that the observed differences between different cases are fully due to the local flow properties of the base flow, i.e. streamwise pressure gradient and Mach number. Three-dimensional simulation results for a supersonic case with a large step height where the N-factor approach is no longer applicable are presented towards the end. Self-excited three-dimensional disturbances appear in this case, probably due to a three-dimensional global instability.

C. Edelmann
DLR, Pfaffenwaldring 38, 70569 Stuttgart, Germany
e-mail: Christopher.Edelmann@dlr.de

U. Rist (✉)
IAG, University of Stuttgart, Pfaffenwaldring 21, 70550 Stuttgart, Germany
e-mail: rist@iag.uni-stuttgart.de

© Springer International Publishing Switzerland 2016
W.E. Nagel et al. (eds.), *High Performance Computing in Science and Engineering '15*, DOI 10.1007/978-3-319-24633-8_24

1 Introduction

Laminar flow is important for drag and hence fuel reduction, but laminar flow is only possible if the surface obeys certain smoothness parameters. Important generic surface roughness shapes are steps and gaps which may occur at the joint of two adjacent surface panels. Previous investigations, like [3, 12, 13, 19], tried to parameterise the detrimental influence of such surface protuberances on the growth factor of small-amplitude disturbances, so-called Tollmien-Schlichting waves. These waves are the precursors of laminar-turbulent transition in a flow with low turbulence levels in the free stream. Previous investigations measured the impact of surface imperfections by a so-called ΔN factor that quantifies the additional amplification in a flow with surface imperfection relative to the flow over the corresponding smooth surface. The present project uses Direct Numerical Simulation (DNS) on a High-Performance Computer (HPC) to investigate the effects of Mach number, pressure gradient, step position and step height on ΔN for forward-facing steps.

The paper is organized as follows. The numerical method is presented in Sect. 2 which includes discussion of implementation aspects on the HPC. This is followed by exemplary results in Sect. 3, a discussion of the limits of the present approach in Sect. 3.3 and conclusions in Sect. 4.

2 Numerical Method

A sketch of the integration domain for a flat plate with a forward-facing (i.e. against the flow direction) step is shown in Fig. 1. The x-axis is normal to the leading edge, parallel to the wall and pointing in streamwise direction. The y-axis points in the wall-normal direction, whereas the z-axis is aligned with the spanwise direction. The computational domain extends in the x-direction from x_0 to x_E, the location of the step is given by x_S, where all x positions are measured as distances from the leading edge of the plate. The height of the step is given by H. The coordinate $x_{D,M}$ denotes the position of a disturbance strip where the stability of the steady laminar base flow will be probed by small-amplitude disturbances (see further down). Most cases studied are two-dimensional and thus only an x-y plane needs to be considered. In the three-dimensional case, the flow field is assumed to be periodic in the spanwise direction with the spanwise wave length λ_z. In the following, all spatial dimensions will be given in non-dimensional form as Reynolds numbers.

In Fig. 2, a sketch of the domain decomposition with forward-facing step is shown. A minimum of three domains (red domains labelled 1–3) is necessary to discretise the step geometry. To increase the performance by further parallelising the simulation (see Sect. 2.2), the basic domains can be split into smaller sub-domains. An example is given by the blue sub-domains in Fig. 2.

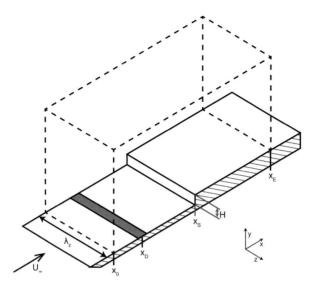

Fig. 1 Sketch of computational domain with forward-facing step with step height H at x_S on a flat plate

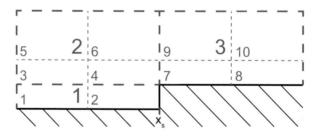

Fig. 2 Sketch of domain decomposition with step geometry

The present DNS code *NS3D* was originally written by Babucke [2] and is continuously enhanced as part of the ongoing work on transition research at the IAG. In the following, the most important aspects for the present study will be explained. A more detailed insight can be found in the thesis by Babucke [2].

NS3D offers a multitude of different finite-difference schemes for spatial discretisation in x- and y- direction. These schemes can be classified in explicit, compact and sub-domain compact schemes, cf. [8, 9]. Hereby, sub-domain compact means that in every sub-domain a compact scheme is used, whereas the coupling between sub-domains is accomplished using an explicit scheme. For the present study, explicit as well as compact and sub-domain compact differences are used because of parallelisation issues (see Sect. 2.2). The numerical properties of the different schemes are chosen to be as similar as possible.

As non-linear terms generate higher harmonics, damping of under-resolved wavelengths is necessary. Thus, alternating up- and downwind-biased finite-differences are applied for convective terms. The compact finite differences used for convective terms are of 6th-order accuracy.

Time integration is performed using the classical fourth-order accurate Runge-Kutta scheme. It consists of four steps at two equidistant time levels. After each sub-step, the direction of the alternating up- and downwind-biased finite-differences changes [10]. A preferred direction is avoided by changing the alternation of the differences after each timestep, using all possible combinations in x-, y- and z-[1] direction.

2.1 Selective Frequency Damping

For stability analyses of flows, a steady baseflow solution is needed. As the code *NS3D* is a time-accurate high-order code with low numerical dissipation, it is impossible to reach this solution if the flow under consideration is globally unstable. Thus, selective frequency damping (SFD) was implemented in the code. With SFD, a temporal filter is used to calculate a solution which successively suppresses temporal fluctuations. The method is based on the work by Åkervik et al. [1] which again elaborates an idea by Pruett [15]. Parameters which need to be defined by the user are the filter width Δ and the control coefficient χ. Details of the method can be find in Åkervik et al. [1].

2.2 Parallelisation

Three different high performance computing systems were used for the present numerical simulations. At the beginning, two vector machines, the NEC SX-8 and SX-9 were available at the HLRS. Later on, these machines were replaced by a massive-parallel system, the Cray XE6 "Hermit". Both vector machines have very similar requirements for parallelisation, but for the massive-parallel system some new problems arose.

For both vector and massive-parallel systems, the domain in the x-y plane is divided into sub-domains with data exchange using the *Message Passing Interface* (*MPI*) [11]. This domain decomposition in the x-y-directions not only serves for parallelisation purposes but also allows more complex geometrical configurations

---

[1] When finite differences are used in the z-direction.

like the step geometry for the present study (see Fig. 2). In the z-direction, loops are parallelised utilizing a shared memory parallelisation.

A typical configuration on vector machines would be for example for a two-dimensional configuration a decomposition in eight domains using eight CPUs running on one SX-8 node or for a three-dimensional computation a decomposition in eight domains each running on one SX-8 node, using 64 CPUs in total. On the vector machines, compact finite differences are used. The resulting tridiagonal system is solved using a pipelined Thomas algorithm (see Babucke [2], pp. 48–51 and Povitsky [14]), leading to an efficient parallelisation for the typical amount of sub-domains (and thus MPI-processes) used on the vector machines.

As the new massive-parallel "Hermit" system has a much higher peak performance than the vector machines, another strategy is used to achieve highest performance. Instead of relatively few, very powerful CPUs, a great number of "standard" CPUs, similar to those used in personal computers, are built together in one big system. The large-grain sub-domain configuration used for the vector machines is no longer sufficient as the computational time for a given problem would increase severely. Therefore, for the present forward-facing step problem, the x-y-plane is typically split into up to more than 1000 sub-domains. In this scenario, the pipelined Thomas algorithm of the compact scheme performs poorly as huge dead-lock times arise when processes are waiting for other processes to calculate data they need. An illustration of this behaviour is given in Table 1 for a typical two-dimensional run which was run over 20,000 time steps using a domain that contains 448 subdomains in streamwise direction x plus 320 subdomains in wall-normal-direction y, i.e., 811,008 grid points. The performance advantage of the explicit finite differences compared to the other two alternatives is quite obvious from the data given in the table: 103 s instead of 352, i.e., a speed-up factor of 3.4. The reason for this speed-up is that data exchange is necessary only once per Runge-Kutta timestep such that each MPI-process performs the calculations for its sub-domain independent from the other processes.

Typical resource requirements and performance data for three-dimensional simulations are as follows: For a case with 786,432 grid points per spanwise Fourier mode, 42 Fourier modes and 129 collocation points in spanwise direction 6144 cores have been used using simultaneous OpenMP and MPI parallelisation. The execution time in this case is 0.3 s/time step.

Table 1 Comparison of execution times using different spatial discretisation schemes

	Compact	Sub-domain compact	Explicit
CPU time (s)	352	153	103

3   Results

Many cases with different step heights, step positions and free-stream conditions (identified by the Mach number Ma_∞) have been investigated. Each case starts with the computation of a two-dimensional steady base flow. If this fails (e.g., if SFD is not applied) the according case cannot be investigated further with the present approach (see Sect. 3.3).

Baseflow results for subsonic and supersonic cases are presented in Sect. 3.1, followed by a description of the N-factor method to quantify the results in Sect. 3.2 and a discussion of a case with a large step height H in Sect. 3.3.

3.1   Base Flows

Laminar separation bubbles are the most influential flow phenomena in all base flows. Depending on flow parameters, either only one separation bubble in front of the step or two zones of separated flow, one in front and one on top of the step are found for subsonic cases.

Figure 3 shows streamlines, pressure contours, the region of reversed flow and representative velocity profiles in the area around the step for high subsonic Mach number and a small step height ($Ma_\infty = 0.8$, $Re_S = 2{,}450{,}000$, $Re_H = 1320$), where Re_S and Re_H are the Reynolds numbers based on step position (x_S) and step height (H), respectively. Here, only one very small separation bubble is found in front of the step. Streamlines near the step are tightened but follow the contour smoothly. The pressure rises in front of the step and then decreases as the flow is accelerated around it.

The profiles of the streamwise velocity component u at four streamwise positions are shown for two grid resolutions. Refined grid means a grid refinement in both the x- and y-directions by a factor of two. Results for both resolutions are found to be in very good accordance.

The boundary layer depicted in these profiles starts with a typical laminar boundary-layer profile far upstream of the step and develops an inflection point in front of the step. The third position at $Re_x = 2{,}450{,}320$ is immediately past the step. Here, a small sub-boundary-layer starting at the step corner can be seen, whereas the outer profile remains nearly unchanged. At $Re_x = 2{,}800{,}000$ the sub-boundary-layer has vanished and the flat-plate boundary layer is recovered.

Figure 4 presents a case with higher step but otherwise identical parameters ($Ma_\infty = 0.8$, $Re_S = 2{,}450{,}000$, $Re_H = 2640$). Now, two zones of separated flow are observed. The first one is much bigger than the second, which is in particular much flatter. Compared to the first case, the first separation bubble is much longer, with the region of separated flow being roughly ten times longer, whereas the step is only twice the height in absolute values and approximately four times higher when both cases are compared using local flow properties at the step. Pressure contours

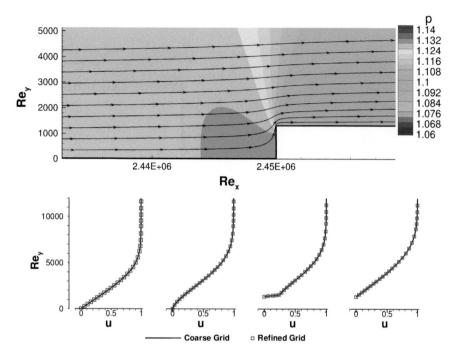

Fig. 3 Baseflow near the step for $Ma_\infty = 0.8$, $Re_S = 2,450,000$, $Re_H = 1320$. (**a**) Streamlines and pressure contours. Region of reversed flow visualized by *dashed line*. y-direction stretched by factor 2. (**b**) Profiles of streamwise velocity u at four streamwise positions $Re_x = 2,000,000$, $Re_x = 2,445,000$, $Re_x = 2,450,320$, and $Re_x = 2,800,000$, respectively

show that the pressure in front of the step reaches higher values compared to the case with only one separation bubble. Additionally, a distinct low-pressure zone is observed on top of the step, after which the pressure rises again.

The profiles of streamwise velocity u are shown for two grid resolutions and at the same four streamwise positions as before. Whereas profiles far upstream and downstream of the step are nearly identical to those from the case with only one separation bubble, differences around the step are visible. In front of the step, the flow separation can be seen. Only marginally negative velocities are observed. Downstream of the step, the sub-boundary-layer is now bigger than in the previous case. A very small region of separated flow is existent but barely visible.

Figure 5 compares the wall pressure along the streamwise coordinate for both cases. Upstream of the step, the pressure equals the level of the flat plate. It remains nearly unchanged until approximately $Re_x = 2,000,000$, where it rises until the step location. Here, an immediate pressure drop occurs. After the step, the wall pressure increases until the level of the flat plate solution is again asymptotically reached.

When the flow at the step reaches supersonic conditions, which would be a typical situation for an airliner wing, the base-flow topology changes markedly. Figure 6 presents results for $Ma_\infty = 1.06$, $Re_S = 2,300,000$, $Re_H = 2640$

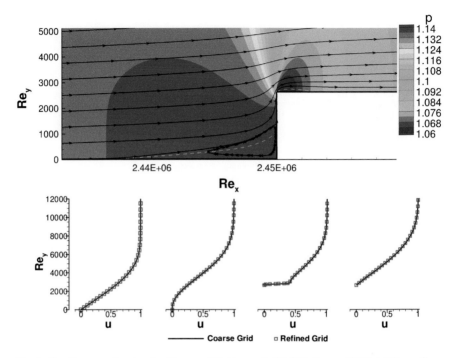

Fig. 4 Baseflow near the step for $Ma_\infty = 0.8$, $Re_S = 2,450,000$, $Re_H = 2640$. (**a**) Streamlines and pressure contours. Region of reversed flow visualized by *dashed line*. y-direction stretched by factor 2. (**b**) Profiles of streamwise velocity u at four streamwise positions $Re_x = 2,000,000$, $Re_x = 2,445,000$, $Re_x = 2,450,320$, and $Re_x = 2,800,000$, respectively

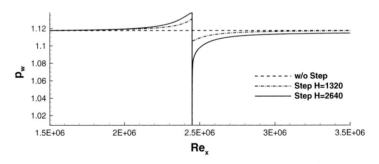

Fig. 5 Comparison of wall pressure p_w along x for $Ma_\infty = 0.8$, $Re_S = 2,450,000$, $Re_H = 2640$ and various step heights H

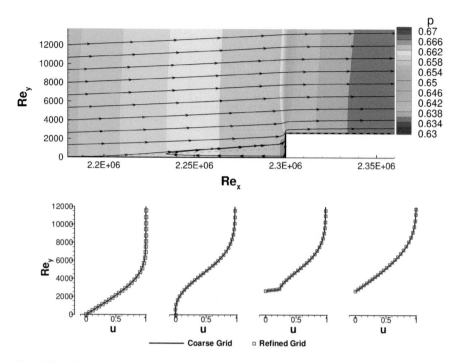

Fig. 6 Baseflow near the step for $Ma_\infty = 1.06$, $Re_S = 2{,}300{,}000$, $Re_H = 2640$. (**a**) Streamlines and pressure contours. Region of reversed flow visualized by *dashed line*. *y*-direction stretched by factor 5. (**b**) Profiles of streamwise velocity *u* at four streamwise positions $Re_x = 1{,}800{,}000$, $Re_x = 2{,}250{,}000$, $Re_x = 2{,}300{,}320$, and $Re_x = 2{,}800{,}000$

in the same way as for the subsonic cases before. The relative step height in terms of boundary-layer momentum thickness is the same as for the case shown in Fig. 4 which exhibits two separation bubbles in subsonic flow. Now, only one separation zone is observed. Compared to the subsonic case, this separation bubble is approximately eight times longer. The rise of the separation streamline from the separation point at the wall towards the step is very slow and constant, such that the separation zone acts like a smooth ramp in front of the step. Thus, flow deflection around the step is much smoother than in the subsonic case, allowing the flow to follow the wall contour after the step without further separation. Due to the upward flow deflection in front of the step (because of the separation bubble) the pressure increases. Afterwards the flow expands around the step, leading to a pressure drop in the right half of the domain.

The velocity profiles at four different streamwise locations show again a very good grid convergence. Far upstream of the step the nearly unaltered boundary-layer profile of the similarity solution can be seen. Inside the separation zone only marginal negative velocities are observed. The third profile is directly on top of the step. Here, like in the subsonic case, the oncoming profile is cut-off and a new sub-boundary-layer is observed. However, this sub-boundary-layer is smaller

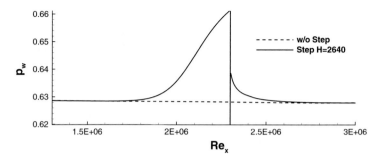

Fig. 7 Wall pressure p_w along x for $Ma_\infty = 1.06$, $Re_S = 2{,}300{,}000$, $Re_H = 2640$

in the supersonic case because of the smoother flow deflection around the step. Far downstream, the boundary-layer profile of the flat plate is recovered as in all cases shown here.

The wall pressure distribution in streamwise direction x is illustrated in Fig. 7. Starting from the nearly constant pressure of the flat plate, the pressure rises (with inverse curvature compared to the subsonic case) until the step location. Here, a sudden pressure drop occurs, followed by a steep rise and a slow decay until a constant value is reached again. This decay stands in contrast to the progression of the wall pressure in the subsonic cases, where the wall pressure increases after the step.

For supersonic cases, three different steady flow regimes are observed when conducting two-dimensional simulations. In the first regime, one very long zone of separated flow is found in front of the step. For increasing step height, while all other parameters stay constant, secondary separation zones within the separation bubble are visible. When the step height is further increased, a second region of separated flow on top of the step is formed. However, it is found that cases within both the latter aforementioned regimes are naturally three-dimensionally unsteady, thus a purely two-dimensional analysis seems unphysical. The flow characteristics of these cases are therefore described in Sect. 3.3, where three-dimensional base flows are discussed.

3.2 Disturbance Excitation and N-factor Computations

A wave packet is introduced into the steady baseflow by blowing and suction over a finite span of time at the wall within the disturbance strip introduced in Fig. 1.

A multitude of different disturbance excitations were tested. Usually the best results are obtained when not a single Dirac-pulse is used but the amplitudes of the blowing/suction are first smoothly increased and subsequently decreased in time. By turning the disturbance strip on and off, sound waves are emitted which travel in up- and downstream direction. These sound waves interact with the step

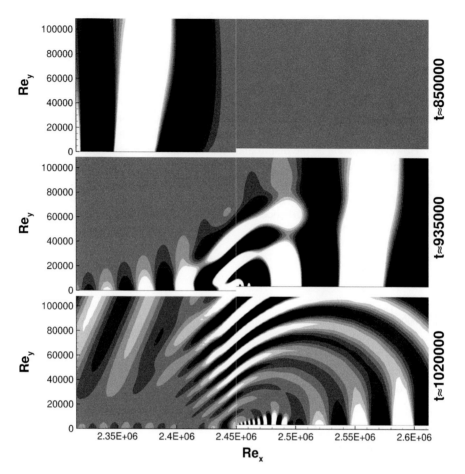

Fig. 8 Sound waves at step for $Ma_\infty = 0.8$, $Re_S = 2,450,000$, $Re_H = 3520$. Six contour levels of p' between $-5 \cdot 10^{-11}$ and $5 \cdot 10^{-11}$ shown

as well. A certain optimization with respect to using multiple frequencies for the blowing/suction was necessary to reduce the amplitudes of these unavoidable sound waves [4]. Nevertheless, the unsteady, instantaneous flow field is rather complex as can be seen using three snapshots of the pressure disturbances p' in Fig. 8.

Two main phenomena are observed. Firstly, a part of the sound wave is reflected with the main direction pointing upstream at an angle of 45°. Secondly, a new wave packet emerges at the step inside the boundary layer. In the region directly after the step, a very small dominant wavelength of approximately $Re_x = 6000$ width and a small propagation speed of $c \approx 0.2$ is found, which leads to a frequency in the range of the dominant frequency of the sound waves. Wavelengths as well as amplitudes of the wave packet in that region very near the step increase rapidly. However, an accurate Fourier analysis is not possible as the wave packet is superposed by the sound waves. Slightly further downstream, the analysis becomes possible as the sound waves have passed.

The generation of acoustics at the disturbance strip and the thereby induced excitation of a small wave packet at the step due to receptivity is only a side effect of the chosen disturbance method. The main purpose remains the generation of a wave packet in the boundary layer where all frequencies of interest are existent with sufficient amplitude such that their amplitude development can be studied. At the beginning, the wave packet develops in the same manner as on a flat plate without step. However, as soon as the disturbance reaches the region where, caused by the step, the pressure deviates from the pressure without step, the development of the wave packet differs from the development without step.

The spatial growth of the wave packet is illustrated in Fig. 9 using contours of the streamwise velocity disturbance (u') for three different time steps in the region

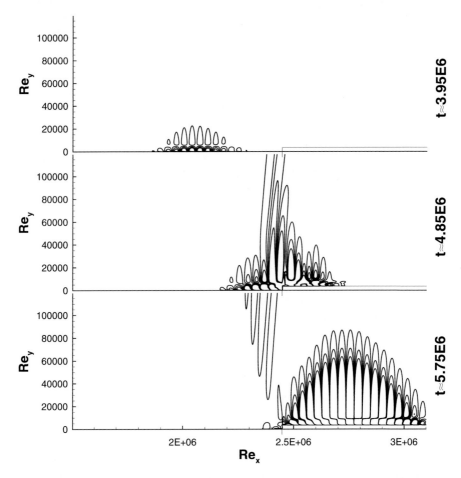

Fig. 9 Contours of u' for three different timesteps. Main wave packet $Ma_\infty = 0.8$, $Re_S = 2,450,000$, $Re_H = 3520$. y-direction stretched by factor 5

around the step. At the step, the appearance of the wave packet changes as some irregular structures in the near-wall region are observed. Behind the step, the wave packet is again similar to the wave packet in front of the step but with much higher amplitude. Additionally, it can be seen that the wave packet induces sound waves at the step which are mainly directed in upstream direction.

For quantification of the influence of steps so-called N-factors are used. This term comes from the e^N-method proposed by van Ingen [18] and Smith and Gamberoni [17] for prediction of laminar-turbulent transition in boundary-layer flows. The N-factor is just a measure of the maximal possible streamwise disturbance growth. Originally, it would be calculated from linear stability theory. However, here we extract it from the results of disturbance-flow calculations with our code *NS3D* as follows.

First, a Fourier analysis of the time signals at each grid point is performed to obtain the disturbance amplitude field $\Upsilon(x, y, f)$, where x and y are the spatial coordinates, and f is the frequency. Additional details like the elimination of the sound waves from the Fourier analysis are not presented here, they are discussed in [4]. A maximum search in wall-normal direction y is then performed at every streamwise station x and for all frequencies f to yield $A(x, f) = \max_y(\Upsilon)$. Next, we define $A_0(f)$ as the amplitude where the growth curve $A(x, f)$ for each frequency f enters the unstable region, i.e., starts to grow. From this we obtain normalised amplitude growth-curves $n = \ln(A(x, f)/A_0(f))$ like those which are plotted in Fig. 10. The envelope curve $N(x)$ which is indicated by the red line in the figure is the so-called N-factor curve then.

Quantification of the influence of surface roughness is thereafter performed by comparing the N-factor curves for the flow over the smooth wall with the results for the flow over the roughness, see Fig. 11. It turns out that the difference ΔN between the two is not a constant but rather a function of x.

Figure 12 presents a comparison of N-factor curves for two Mach numbers, $Ma = 0.6$ and $Ma = 1.06$ but for the same relative step height $Re_H = 2640$. The first difference between the two is the stronger disturbance growth in the flat-plate boundary layer in the subsonic case that leads to generally higher N-factors

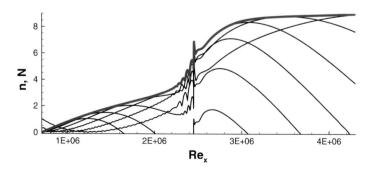

Fig. 10 Evaluation of N-factor curve from amplitudes of simulation with optimized disturbance excitation

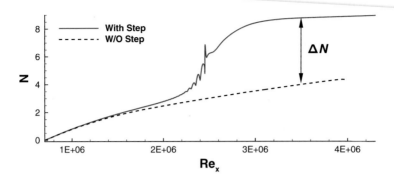

Fig. 11 Definition of additional amplification ΔN

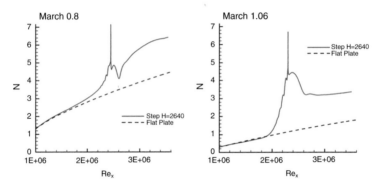

Fig. 12 Comparison of N-factor curves at different Mach number but same relative step height

there. The next difference is the different qualitative behaviour in the presence of the step. Two regions of amplification occur in the subsonic case and only one in the supersonic. Both cases show a ΔN after the step that can be quantified from the figure: $\Delta N \approx 2$ in the subsonic and an approximately constant $\Delta N \approx 1.6$ downstream of the step in the supersonic case.

Another striking qualitative difference between bot cases is that the subsonic wave packet continues to grow *after* the step whereas its amplitude is *reduced* in the supersonic case. However, in light of the above discussion of the streamwise pressure gradient, together with the occurrence (or absence) of a second separation bubble, this is not surprising but fully consistent with the observed qualitative differences in the two base flows. Adverse pressure gradients (i.e., streamwise pressure increase) and boundary-layer separation leads to increased disturbance amplification, whereas flow expansion and streamwise pressure decrease, reduce the disturbance amplitudes. Comparisons with linear stability theory have also confirmed these results in a quantitative manner. Furthermore, all N-factor results of the present research are published in [4–6]. They shall not be repeated here. Instead, we present some more DNS results that go beyond the applicability of linear stability theory and the N-factor method in the next section.

3.3   Limits of the Present Approach: High Steps in the Supersonic Regime

The influence of steps on laminar-turbulent transition has been represented by a local additional ΔN-factor in the present study. Basis for this approach is the simulation of a steady, two-dimensional baseflow (see Sect. 3.1). The time-accurate DNS-code $NS3D$ allows simulation of steady base flows for subsequent N-factor calculations as shown above. However, when the height of the step exceeds a certain value, new phenomena arise. These steps are in the following called "high steps". Differences are again found between sub- and supersonic cases, but we shall limit our presentation here to the supersonic case at $Ma_\infty = 1.06$.

The required three-dimensional simulations are computationally extremely expensive. Thus, only relatively few cases with high steps were simulated and only a quite short description of the physical phenomena follows.

For flows at $Ma_\infty = 1.06$ and the step located at $Re_S = 2,450,000$, a steady baseflow is reached for step heights up to $Re_H = 7000$ if a two-dimensional simulation is conducted. At $Re_H = 9000$, the simulation crashed after a while. A grid refinement study would be necessary to clarify the situation in this case.

The baseflow for a case with a step height of $Re_H = 6000$, located at $Re_S = 2,450,000$ and $Ma_\infty = 1.06$ is visualized in Fig. 13. Compared to Fig. 6, used for the description of supersonic base flows in Sect. 3.1, some differences are observable. First, the separation bubble in front of the step is even bigger and has a relative length of $L_1/H \approx 51$. Second, a very small separation bubble on top of the step emerges. However, this separation bubble is very small. The relative length of the second bubble is only $L_2/H \approx 1/6$ and the height of the region with negative u-velocity is approximately $1/6H$. Furthermore, a secondary separation zone is observed within the separated zone in front of the step.

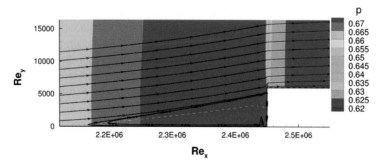

Fig. 13 Baseflow around step geometry for $Ma_\infty = 1.06$ with $Re_S = 2,450,000$ and $Re_H = 6000$. Region of reversed flow visualized by *dashed line*. y-direction stretched by factor 10

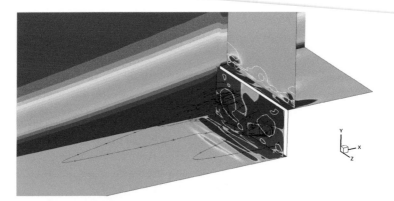

Fig. 14 Snapshot of three-dimensional flow around step geometry for $Ma_\infty = 1.06$ with $Re_S = 2,450,000$ and $Re_H = 6000$. y- and z-directions stretched by factor 10. x-plane: w-contours with 21 contour levels between $-5 \cdot 10^{-4}$ and $5 \cdot 10^{-4}$. λ_2 (vortex identification criterion [7]) isolines in *white*. y-plane: Contours of ω_x at the walls. Twenty contour levels between $-1 \cdot 10^{-6}$ and $1 \cdot 10^{-6}$. z-plane: u-contours with 21 contour levels between 0 and 1

This flow is now used as a baseflow for a three-dimensional simulation without introduced disturbances. It is found that a spanwise (w) velocity component emerges in front of the step and grows in time. As soon as the w-component exceeds $w \approx 0.001$, the unsteady motion becomes visible in the other variables as well. Soon thereafter, the secondary separation bubble disappears. Saturation of the w-component seems to be reached at $w \approx 0.05$ which is found directly in front of the step. However, in the time simulated, neither a steady nor a truly periodic state was achieved.

A snapshot of the flow field at the last time step simulated is shown in Fig. 14. Directly in front of the step, the separation bubble is highly three-dimensional with all three velocity components being in the same order of magnitude. In the shown plane just in front of the step, two main vortices can be seen. These are of extremely short spatial extent as they are not visible after the step and the separation bubble in front of the step gets increasingly two-dimensional in the upstream direction. This fact is further illustrated by the streamline remaining two-dimensional in the upstream region and becoming chaotic, three-dimensional directly in front of the step. On top of the step, some weak streaky structures are visible. However, the maximum w-velocity in the shown plane after the step is only $w \approx 0.001$ compared to $w \approx 0.02$ in the displayed plane in front of the step. Together with a strong u-component the flow is again nearly two-dimensional after the step.

The described state is only a snapshot and it is not clear whether a periodic or a steady state will be reached if the simulation is continued. A comparable behaviour is observed for a step height of $Re_H = 5000$. Here, the secondary separation bubble within the separated zone in front of the step is found as well, whereas the separation on top of the step is missing. However, in this case, the w-velocity grows much slower and even though the simulation is run for twice as long, the maximum w-component has only reached $w \approx 0.015$.

To check that the three-dimensional behaviour is not encountered in the earlier supersonic cases used for the N-factor calculation, an additional full three-dimensional simulation has been conducted for one of those. In that case, the w-velocity stayed within the computational error with a maximum of $w \approx 5 \cdot 10^{-15}$.

4 Conclusions

The existing HPC-DNS code *NS3D* has been used to compute amplification factors (N-factors) for boundary-layer flows over smooth walls and walls with steps. Compressible cases with high subsonic and low supersonic Mach numbers have been investigated. Many interesting and unexpected differences between the supersonic and subsonic flow conditions have been detected. The most prominent is the occurrence of different streamwise pressure gradients and flow fields with different laminar separation bubbles. As these have a predictable influence on the growth of disturbances which are precursors of laminar-turbulent transition, it turns out that the additional amplification ΔN caused by a step relative to the smooth case can be predicted if it is possible to compute the steady base accurate enough for subsequent local linear stability analysis.

However, if the step height exceeds some critical value, new phenomena occur which need to be studied further. Here, the observed unsteady flow occurring for high steps at $Ma_\infty = 1.06$ was briefly described but the flow in this case is far from being understood. The simulation was neither run long enough nor was the influence of spanwise wave length studied. Furthermore, a two-dimensional linear stability study as conducted by Robinet [16] for separation bubbles induced by shock-wave/boundary-layer interaction is recommended.

Acknowledgements The simulations have been performed on the national supercomputers of the High Performance Computing Centre Stuttgart (HLRS) under grant number GCS-Lamt/44026.

References

1. Åkervik, E., Brandt, L., Henningson, D.S., Hoepffner, J., Marxen, O., Schlatter, P.: Steady solutions of the Navier-Stokes equations by selective frequency damping. Phys. Fluids **18**(6), 68–102 (2006)
2. Babucke, A.: Direct numerical simulation of noise-generation mechanisms in the mixing layer of a jet. Ph.D. thesis, Universität Stuttgart (2009)
3. Crouch, J.D., Kosorygin, V.S., Ng, L.L.: Modeling the effects of steps on boundary-layer transition. In: Govindarajan, R. (ed.) IUTAM Symposium on Laminar-Turbulent Transition. Fluid Mechanics and its Applications, vol. 78, pp. 37–44. Springer, Netherlands (2006)

4. Edelmann, C.: Influence of forward-facing steps on laminar-turbulent transition. Ph.D. thesis, University of Stuttgart (2015)
5. Edelmann, C., Rist, U.: Impact of forward-facing steps on laminar-turbulent transition in subsonic flows. In: Dillmann, A., Heller, G., Krämer, E., Kreplin, H.-P., Nitsche, W., Rist, U. (eds.) New Results in Numerical and Experimental Fluid Mechanics IX. Notes on Numerical Fluid Mechanics and Multidisciplinary Design, vol. 124, pp. 155–162. Springer, Berlin (2014)
6. Edelmann, C.A., Rist, U.: Impact of forward-facing steps on laminar-turbulent transition in transonic flows. AIAA J. **53**(9), 2504–2511 (2015). doi:10.2514/1.J053529
7. Jeong, J., Hussain, F.: On the identification of a vortex. J. Fluid Mech. **285**, 69–94 (1995)
8. Keller, M., Kloker, M.J.: Direct numerical simulations of film cooling in a supersonic boundary-layer flow on massively-parallel supercomputers. In: Resch, M.M., Bez, W., Focht, E., Kobaysahi, H., Kovalenko, Y. (eds.) Sustained Simulation Performance 2013: Proceedings of the Joint Workshop on Sustained Simulation Performance, University of Stuttgart (HLRS) and Tohoku University, pp. 107–128. Springer, Berlin (2013)
9. Keller, M., Kloker, M.J.: DNS of effusion cooling in a supersonic boundary-layer flow: influence of turbulence. In: Proceedings of the 44th AIAA Thermophysics Conference, AIAA Paper 2013–2897 (2013)
10. Kloker, M.: A robust high-resolution split-type compact fd-scheme for spatial direct numerical simulation of boundary-layer transition. Appl. Sci. Res. **59**(4), 353–377 (1998)
11. Mpi documents (2013). http://www.mpi-forum.org/docs/docs.html
12. Nenni, J.P., Gluyas, G.L.: Aerodynamic design and analysis of an LFC surface. Astronaut. Aeronaut. **4**, 52–57 (1966)
13. Perraud, J., Arnal, D., Séraudie, A., Tran, D.: Laminar-turbulent transition on aerodynamic surfaces with imperfections. In: RTO AVT, Prague (2004)
14. Povitsky, A.: Parallelization of the pipelined Thomas algorithm. Technical Report (1998)
15. Pruett, C.D., Gatski, T.B., Grosch, C.E., Thacker, W.D.: The temporally filtered Navier–Stokes equations: properties of the residual stress. Phys. Fluids **15**(8), 2127–2140 (2003)
16. Robinet, J.-Ch.: Bifurcations in shock-wave/laminar-boundary-layer interaction: global instability approach. J. Fluid Mech. **579**(1), 85–112 (2007)
17. Smith, A.M.O., Gamberoni, N.: Transition, pressure gradient and stability theory. Technical Report ES-26388, El Segundo, CA. Douglas Aircraft Company, El Segundo Division (1956)
18. van Ingen, J.L.: A suggested semi-empirical method for the calculation of the boundary layer transition region. Technical Report, Delft University of Technology (1956)
19. Wang, Y., Gaster, M.: Effect of surface steps on boundary layer transition. Exp. Fluids **39**(4), 679–686 (2005)

Large-Scale Simulations of a Non-generic Helicopter Engine Nozzle and a Ducted Axial Fan

Mehmet Onur Cetin, Alexej Pogorelov, Andreas Lintermann, Hsun-Jen Cheng, Matthias Meinke, and Wolfgang Schröder

Abstract Large-eddy simulations (LESs) of a helicopter engine jet and an axial fan are performed by using locally refined Cartesian hierarchical meshes. For the computations a high-fidelity, massively parallelized solver for compressible flow is used. To verify the numerical method, a coaxial hot round jet is computed and the results are compared to reference data. The analysis is complemented by a grid convergence study for both applications, i.e., for the helicopter engine jet and the axial fan. For the helicopter engine jet, additional computations have been performed for two different nozzle geometries, i.e., a simplified nozzle geometry that is consisting of a center body and divergent outer annular channel, and a complete engine nozzle geometry with four additional struts were used. The presence of the struts results in a different potential core break-down and turbulence intensity. Furthermore, for the axial fan configuration, computations have been performed at two different volume flow rates. The reduction of the volume flow rate results in an interaction of the tip-gap vortex with the neighboring blade which leads to a higher turbulent kinetic energy near and inside the tip-gap region.

1 Introduction

The prediction and reduction of noise generated by turbulent flows has become one of the major tasks of todays aircraft development and is also one of the key goals in European aircraft policy. Compared to the year 2000 the perceived noise level of flying aircraft should to be reduced by 65 % until the year 2050. To comply with

M.O. Cetin (✉) • A. Pogorelov • A. Lintermann • H.J. Cheng
Institute of Aerodynamics, RWTH Aachen University, Wüllnerstr. 5a, 52062 Aachen, Germany
e-mail: office@aia.rwth-aachen.de

M. Meinke • W. Schröder
Institute of Aerodynamics, RWTH Aachen University, Wüllnerstr. 5a, 52062 Aachen, Germany

Forschungszentrum Jülich, 52425 Jülich, Germany

JARA – High-Performance Computing, 52062 Aachen, Germany
e-mail: office@aia.rwth-aachen.de

© Springer International Publishing Switzerland 2016
W.E. Nagel et al. (eds.), *High Performance Computing in Science and Engineering '15*, DOI 10.1007/978-3-319-24633-8_25

new noise level regulations, reliable, efficient and accurate aeroacoustic predictions are required, i.e., for low noise design of technical devices such as helicopter engine nozzles or axial fans.

In jet and fan flows as investigated in this paper, the acoustic field is dependent on the unsteady turbulent flow field. That is, the reliability of the acoustic field prediction is strictly related to the accuracy of the flow field solution. However, the accurate prediction of the turbulent flow field requires extensive computing resources, i.e., the solution of the turbulent flow field necessitates a scale resolving LES that requires a high mesh resolution to cover the major part of the turbulence energy spectrum.

Codes used in industry mainly rely on computationally efficient solutions of the Reynolds averaged Navier-Stokes equations (RANS), they, however, only provide solutions which are time averaged over all turbulent scales. In contrast, details of the turbulent flow field obtained by an LES allow a thorough analysis of the physical mechanisms responsible for the noise generation and therefore provide valuable information for design modifications resulting in reduced noise emission.

With the recent substantial growth of the computing technology, numerical simulation methods such as LES is possible for the flow field prediction of many simplified practical applications. Vast number of studies have been done in the past to numerically simulate for instance, jets with simplified nozzle geometries under laboratory conditions [1–3]. Moreover, few authors applied LES to tip leakage flows. Investigations have been performed, e.g., by You et al. [4–7] for a linear cascade with a moving end wall and Boudet et al. [8] for a single airfoil tip-clearance flow focusing on the noise generation. Real applications however, often have more complicated geometries and higher Reynolds numbers. Application-relevant Reynolds numbers, i.e., $Re > 10^5$ require large computational meshes. Additionally, the inclusion of geometries to the computational domain, e.g., a jet nozzle or an axial fan, requires additional local refinement of the mesh to avoid any unphysical flow behavior like boundary layer separations at the wall due to an underresolved turbulence spectrum. To overcome these difficulties and to analyze the grid dependence on the flow field, highly resolved LESs are performed in this study.

This paper is structured as follows. First, the numerical methods are presented in Sect. 2 and validated in Sect. 3. Subsequently, the results of nozzle-jet and axial fan simulations are discussed in Sects. 4 and 5. Computational features and scalability analysis are given in Sect. 6. Finally, some conclusion is outlined in Sect. 7.

2 Numerical Method

An LES model based on a finite volume method is used to simulate the compressible unsteady turbulent flow by solving the Navier-Stokes equations. For the LES an implicit grid filter is assumed and the monotone integrated LES (MILES) approach [9] is adopted, i.e., the dissipative part of the truncation error of the numerical method is assumed to mimic the dissipation of the non-resolved subgrid scale stresses. This solution method has been validated and successfully used, e.g., in [10, 11]. The governing equations are spatially discretized by using the modified advection upstream splitting method (AUSM) [12]. The cell center gradients are computed using a second-order accurate least-squares reconstruction scheme [13], i.e., the overall spatial approximation is second-order accurate. For stability reasons, small cut-cells are treated using an interpolation and flux-redistribution method developed by Schneiders et al. [14]. A second order 5-stage Runge-Kutta method is used for the temporal integration. A massively parallel grid generator is used to create a computational hierarchical Cartesian mesh featuring local refinement [15]. The interested reader is referred to [12] for the details of the numerical methods, i.e., the discretization and computation of the viscid and inviscid fluxes.

3 Validation Test

To verify the solution procedure, a simulation of a coaxial hot round jet at a Mach number of $M = \frac{u_s}{a_s} = 0.9$ and a Reynolds number of $Re_D = \frac{\rho_s u_s D}{\mu} = 4 \times 10^5$ are performed, where u_s is the secondary jet inflow velocity, a_s is the sound speed, D is the jet diameter, ρ is the density and μ is the dynamic viscosity. The results are compared to the findings in [1]. A mesh with 24 million cells is used for the simulation. The computation and sampling time are chosen long enough to fulfill the statistical convergence, i.e., 5000 LES snapshots are used to average the flow field.

To introduce turbulence, the jet is forced with artificial instability modes, that are introduced in the shear layers of the jet [16], where a hyperbolic tangential velocity profile is prescribed. The velocity distribution at the inlet section reads [1]

$$u_0(r) = \frac{u_s}{2}\left(1 + \tanh\frac{r_s - r}{2\delta_\theta}\right) + \frac{u_p - u_s}{2}\left(1 + \tanh\frac{r_p - r}{2\delta_\theta}\right). \tag{1}$$

where u_s is the secondary jet velocity, u_p is the primary jet velocity, r_s is the secondary jet radius, r_p is the primary jet radius and δ_θ is the momentum thickness which is chosen to be $0.05r_s$ [1]. The Inflow density distribution is defined by the Crocco-Buseman relation and the ideal gas law is used [1]. The flow parameters of the corresponding jet are shown in Table 1. Figure 1 illustrates the mean density contours and instantaneous vorticity field. To discuss the mean flow quantitatively, the streamwise profiles of the axial velocity and the jet half-width velocity are compared in Fig. 2. In agreement with the reference result, the velocity decay begins

Table 1 Simulation parameters (subscript 'p' is primary and 's' is secondary)

$Re = \rho_s u_s D/\mu_s$	$M = u_s/a_s$	u_s/u_p	T_s/T_∞	T_s/T_p	r_s/r_p
400,000	0.9	0.9	1.0	0.37	2

(a) 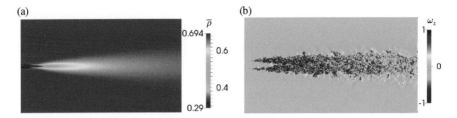 (b)

Fig. 1 (**a**) Mean density contours, and (**b**) instantaneous vorticity contours

(a) 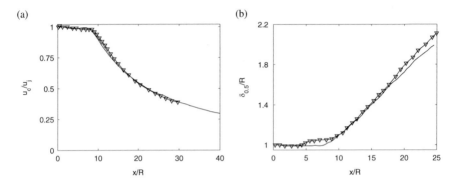 (b)

Fig. 2 Streamwise profiles of the, (**a**) axial velocity decay, and (**b**) jet half-width velocity for: (—) current simulation, (∇) Koh et al. [1]

(a) 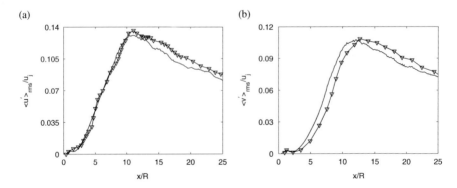 (b)

Fig. 3 Streamwise profiles of the, (**a**) rms fluctuating axial velocity, and (**b**) rms fluctuated radial velocity for: (—) current simulation, (∇) Koh et al. [1]

at the end of the potential core at about $x \approx 9R$, where $R = D/2$ is the jet inlet radius in Fig. 2a. Moreover, as expected, the evolution of the jet half-width in Fig. 2b evidences a linear growth begins almost at the end of the potential core. The comparison shows a good agreement in the interval $0 < x/R < 18$. However, a slightly different jet spreading rate is observed for $x/R > 18$. Streamwise profiles of rms-axial and rms-radial velocity fluctuations are displayed in Fig. 3. Both profiles show a good agreement with the reference data. A minor shift is detected on both profiles, where a slightly different potential core break down influences the Reynolds stress distribution of the jet. Overall, the comparisons of the first and second moment streamwise profiles show convincing agreement with the reference study.

4 Helicopter Engine Jet

In this section, simulation results of round jets emanating from a non-generic nozzle are presented. To investigate the grid dependence and influence of the nozzle geometry to the jet development two variants of the nozzle geometries are used. The Reynolds number is $Re_D = 7.5 \times 10^5$ based on the jet inlet diameter D, and the Mach number is $M = 0.341$.

4.1 Grid Convergence Test

The grid convergence study is done for two different meshes using the same geometry, i.e., case a and b. These meshes possess 0.329×10^9 and 1.097×10^9 cells, respectively. To ensure the statistical convergence 2250 LES snapshots for the case a and 3000 LES snapshots for the case b are used to average the flow field. A zoomed view on the mesh of case a is shown in Fig. 4. The computational specifications are given in Table 2. The nozzle geometry is included to the computational domain explicitly, where the inlet of the nozzle is located downstream of the last turbine stage.

Operating conditions are set at the inlet boundary that were taken from the measurements of a full-scale turbo-shaft engine [17]. Isotropic synthetic turbulence is injected at the inlet plane with approx. 10 % turbulence intensity [18]. For the outflow and lateral boundaries of the jet domain, static pressure is kept constant and other variables are extrapolated from the internal domain. To damp the numerical

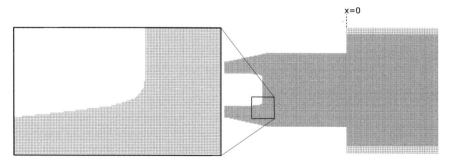

Fig. 4 Cartesian mesh view of the nozzle, *right*: resolution of a coarse mesh with 41 million cells, *left*: enlargement of a certain area for the case *a*

Table 2 Nozzle jet simulation parameters

	Re	M	Mesh cells	Simulation time
Jet case a	750, 000	0.341	329×10^6	480 D/u
Jet case b	750, 000	0.341	1097×10^6	440 D/u
Jet case c	750, 000	0.341	328×10^6	480 D/u

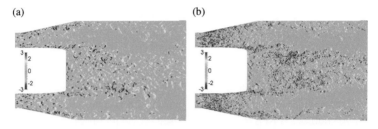

(a)   (b)

Fig. 5 Instantaneous axial velocity gradients in the rear part of the nozzle, (**a**) case a, (**b**) case b

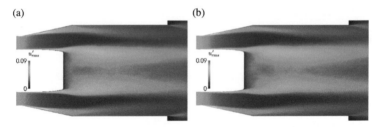

(a)   (b)

Fig. 6 Mean rms fluctuating axial velocity contours in the rear part of the nozzle, (**a**) case a, (**b**) case b

reflections at the boundaries, sponge layers are prescribed [19]. At the nozzle-wall a no-slip condition with a zero pressure and density gradient is applied.

The contours of the axial velocity gradients in streamwise direction are shown in Fig. 5. From the gradients distribution, it is obvious that the mesh resolution plays an important role in the formation of the turbulent structures in the rear part of the nozzle. To discuss the mean flow, a comparison based on the rms fluctuating axial velocity contours is displayed in Fig. 6. A juxtaposition of the centerline velocity decay is shown in Fig. 7a, where R is the nozzle-exit radius and u_j inflow velocity. An earlier potential core break-down is detected for the case b. However, downstream of the core, a similar velocity decay is observed.

Figure 7b illustrates the radial profile of the axial velocity at the nozzle-exit. The nozzle-exit profile is formed by the wake flow shedding from the center body, i.e., the velocity increases for $0 < r/R < 0.6$ and decreases for $r/R \geq 0.6$. Velocity profiles for case a and b are however, alike. At the nozzle wall however, a slight difference is observed due to the different boundary layer thicknesses. Streamwise

(a)   (b)

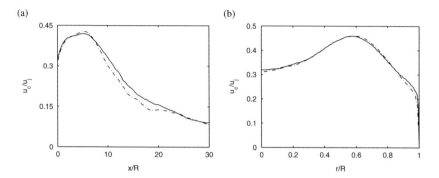

Fig. 7 (**a**) Streamwise profile of the mean centerline velocity, and (**b**) radial profile of the axial velocity at the nozzle-exit $x/R = 0$ for: (—) mesh case a, (-·-) mesh case b

(a)   (b)

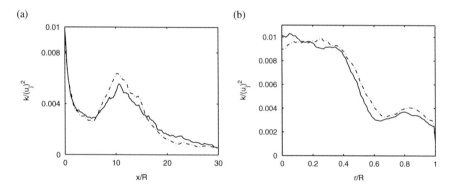

Fig. 8 (**a**) Streamwise profile of the turbulent kinetic energy, and (**b**) radial profile of the turbulent kinetic energy at the nozzle-exit $x/R = 0$ for: (—) case a, (-·-) case b

and radial profiles of the turbulent kinetic energy defined as

$$k = \frac{1}{2}\left(\overline{u'^2} + \overline{v'^2} + \overline{w'^2}\right). \tag{2}$$

are shown in Fig. 8. Both comparisons show that the behavior and the magnitude of the turbulent kinetic energy distribution almost coincides for both cases a and b. To summarize, from the grid dependence study it is obvious that a mesh resolution of $\mathcal{O}(300 \times 10^6)$ cells (case a), is fine enough to resolve important features of turbulence. This mesh resolution can therefore be regarded as sufficient.

4.2 Analysis of Two Variants of the Nozzle Geometry

In the following, two variants of the helicopter engine nozzle geometry are analyzed by simulation, i.e., first a simplified nozzle geometry (case a) which consist of a divergent annular nozzle and a center body, and a complete nozzle geometry (case c) with four additional struts which are supporting the center body inside the nozzle are used. A perspective view of the nozzle geometries is illustrated in Fig. 9. To visualize the turbulent jet flow field, the instantaneous 3D Q-criterion [20] of the jet of case a is illustrated in Fig. 10. The mean flow field is discussed in Fig. 11. The centerline velocity decay comparison in Fig. 11a evidences that case c has an appreciably earlier potential core break down than case a which is due to the enhanced turbulence generation originating from the struts. Furthermore, the comparison of the streamwise profile of the rms fluctuating axial velocity comparison in Fig. 11b shows that the presence of the struts are diminishing the turbulent intensity at the potential core, indicating potentially different sound generation mechanisms for the different jets. Radial profiles of the axial velocity at the nozzle-exit are illustrated in Fig. 12a. An almost constant velocity profile

(a)   (b)

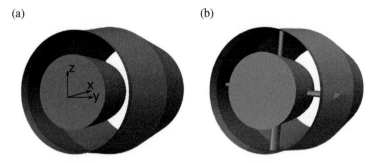

Fig. 9 Nozzle geometries of the (**a**) case a, and (**b**) case c

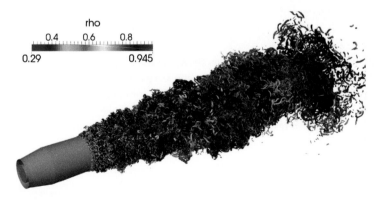

Fig. 10 Perspective view of contours of the Q-criterion color coded with density

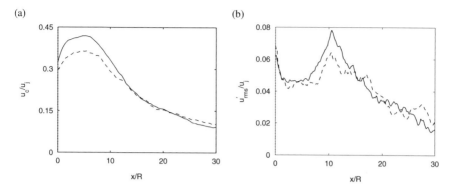

Fig. 11 Streamwise profile of the (**a**) axial velocity, and (**b**) rms fluctuated axial velocity for: (—) case a, (−−) case c

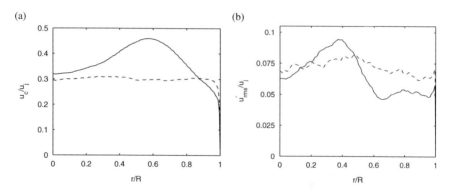

Fig. 12 Nozzle-exit $x/R = 0$ radial profile of the (**a**) axial velocity, and (**b**) rms fluctuated axial velocity for: (—) case a, (−−) case c

for case c is observed. Note that since the nozzle-exit velocity distribution is not axisymmetric, the profile behavior is highly sensitive to the angle of the radius at the nozzle-exit. However, the profile of case a increases for $0 < r/R < 0.6$ and dramatically decreases for $r/R \geq 0.6$. Similarly, case c possesses an almost constant rms fluctuating axial velocity distribution, while case a possesses a velocity deficit for $0.4 < r/R < 0.6$ (see Fig. 12b). A detailed analysis of the geometry effect and non-axisymmetric flow on the jet development is carried out in [21, 22].

5 Axial Fan

In the following section, a rotating axial fan is investigated by LES. The fan configuration has five twisted blades. To reduce the computational costs only one out of five blades is simulated by solving the Navier-Stokes equations in the rotating frame of reference with periodic boundary conditions in the azimuthal direction as discussed by Pogorelov et. al [23]. First, the sensitivity of the computational results on the grid resolution is analyzed using two different meshes for a fixed operating point defined by the flow coefficient $\phi = \frac{4\dot{V}}{\pi^2 D_o^3 n} = 0.165$. Afterwards, the impact of the flow rate on the flow field with special focus on the tip-gap region is discussed. Therefore, computations at two volume flow rates, i.e., $\phi = 0.165$ and $\phi = 0.195$ are conducted. The diameter of the outer casing wall is $D_o = 300$ mm and the inner diameter of the hub is $D_i = 135$ mm. The rotational speed is $n = 3000$ rpm and the gap between the blade tip and the outer casing wall is $s/D_o = 0.01$ for all computations. The Reynolds number based on the relative velocity of the outer casing wall is $Re = \pi D_o^2 n / \nu = 9.36 \times 10^5$, where ν is the kinematic viscosity.

5.1 Grid Convergence Test

Two different meshes are used for the computations, i.e, case d with 250 million cells and case e with 1 billion cells. Computational resources for both cases are summarized in Table 3. The simulations were conducted on 7992 computing cores for mesh case d and 31,992 computing cores for mesh case e. The time step for both cases is $\Delta t = 1.936 \times 10^{-5} \frac{1}{\pi n}$ and the number of time steps corresponding to four full rotations of the rotor is 0.64×10^6 resulting in a wall clock time of approx. 250 h for each computation. After two full rotations 2000 samples of instantaneous data were collected for statistical analysis.

Figure 13 illustrates a coarse example of the computational grid for the 72° fan section, where Fig. 13a shows the overall computational grid and Fig. 13b highlights the resolution and refinement of the grid in wall regions.

Figure 14 depicts the experimental operating line of the fan showing the pressure coefficient $\Psi = \Delta p / (\frac{\pi^2}{2} \rho D_o^2 n^2)$ as a function of the flow coefficient Φ, where $\Delta p = p_{stat,out} - p_{0,in}$, i.e., the difference of the static pressure at the outlet $p_{stat,out}$ and the stagnation pressure at the inlet $p_{0,in}$ of the fan. As shown, both numerical results agree well with the experimental data.

To give an impression of the overall flow field, Fig. 15a depicts the instantaneous contours of the Q-criterion [20] visualizing the vortical structures. Several physical phenomena are evident, e.g., a development of a passage vortex on the blade root initiating a transition of the incoming boundary layer, a turbulent wake behind the trailing edge of the blade, a transition region on the blades suction side and a jet-like tip-gap vortex due to the leakage flow through the tip-gap generating a turbulent wake region. The flow field inside the tip-gap is caused by the pressure difference

Table 3 Axial fan simulation parameters		# cores	Mesh cells	Disc space
	Case d	7992	250×10^6	20 TB
	Case e	31,992	1000×10^6	80 TB

(a)   (b)

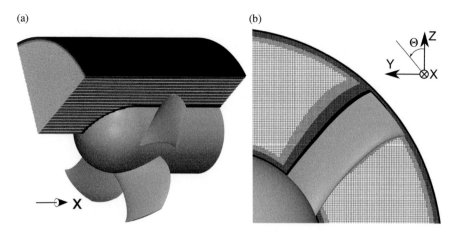

Fig. 13 Coarse example of the Cartesian mesh resolving one out of five blades of the axial fan;
(**a**) axial fan geometry and the overall mesh; (**b**) detailed view of the mesh resolution around the
blade

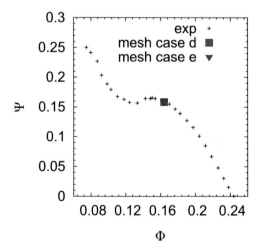

Fig. 14 Operating map for a constant tip-gap size of $s/D_0 = 0.01$ showing the pressure coefficient
$\Psi = \frac{\Delta p}{(\frac{\pi^2}{2}\rho D_o^2 n^2)}$, where $\Delta p = p_{stat,out} - p_{0,in}$, i.e., the difference of the static pressure at the outlet
$p_{stat,out}$ and the stagnation pressure at the inlet $p_{0,in}$ of the fan, versus the flow coefficient $\Phi = \frac{4\dot{V}}{\pi^2 D_o^3 n}$; experimental data from [24]

between the pressure and the suction side as illustrated by Fig. 15b which shows the
instantaneous contours of the pressure coefficient $C_p = \frac{2(p-p_{in})}{\rho(\pi D_o n)^2}$ in the gap region,
where p_{in} is the mean inlet pressure. Low pressure regions are observed inside the
tip-gap due to separations on the surface of the blade tip and upstream of the blade
due to the tip-gap vortex.

(a)   (b)

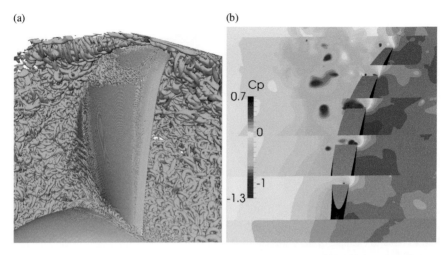

Fig. 15 Instantaneous contours of (**a**) the Q-criterion showing the vortical structures around the blade and (**b**) the pressure coefficient showing low pressure regions caused by the tip-gap vortex and the separation vortices inside the tip-gap for mesh case e

(a)   (b)

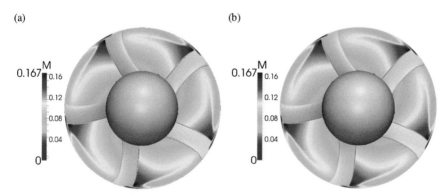

Fig. 16 Men relative Mach number contours at a constant axial location $x/D_o = 0.617$; (**a**) mesh case d and (**b**) mesh case e

To analyze the impact of the grid resolution Fig. 16 shows the time-averaged relative Mach number contours at a constant axial location $x/D_o = 0.617$ for mesh case d and e. A marginal impact of the grid resolutions on the Mach number contours in Fig. 16 is observed. For both mesh cases, the Mach number increases in the radial direction and drops in the wake generated by the tip-gap vortex. A maximum Mach number is observed near the tip-gap region.

To further quantify the impact of the mesh, Figs. 17 and 18 show axial distributions of the pressure coefficient and the turbulent kinetic energy at 80 % span and two circumferential locations, i.e., $\Theta = -45°$ and $\Theta = -35°$. Upstream of the blade, the pressure coefficients in Fig. 17 shows a smooth drop due to the suction

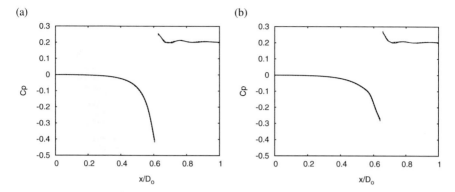

Fig. 17 (**a**) Axial distribution of the pressure coefficient at $\Theta = -45°$ and 80% span; (**b**) Axial distribution of the pressure coefficient at $\Theta = -35°$ and 80% span; mesh case d (—), mesh case e (-·-)

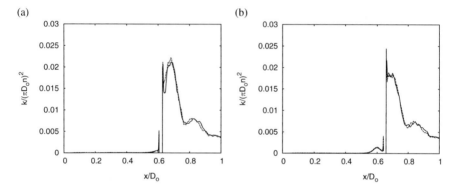

Fig. 18 (**a**) Axial distribution of the turbulent kinetic energy at $\Theta = -45°$ and 80% span; (**b**) Axial distribution of the turbulent kinetic energy at $\Theta = -35°$ and 80% span; mesh case e (—), mesh case e (-·-)

region. Larger values are observed on the pressure side of the blade, as already observed in Fig. 15b. The impact of the mesh on the pressure coefficient is negligibly small and the curves almost perfectly match for both locations. The turbulent kinetic energy distribution in Fig. 18 shows high values on the pressure side of the blade due to the turbulent wake generated by tip-gap vortex of the neighboring blade. Only a small impact of the grid resolution at both location its observed.

5.2 Volume Flow Rate Variation

In this section the impact of the flow coefficient on the flow field near the tip-gap region is demonstrated for mesh case d. Figures 19 and 20 show the time-averaged

(a)   (b)

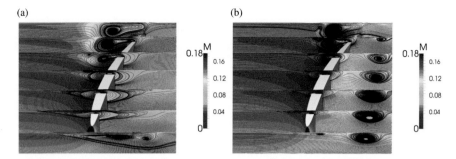

Fig. 19 Time-averaged contours of Mach number contours and streamlines of projected velocity vector in different radial planes from $\theta = -30°$ to $\theta = -60°$, for $\phi = 0.165$ (*left*) and $\phi = 0.195$ (*right*). (**a**) $\phi = 0.165$. (**b**) $\phi = 0.195$

(a)   (b)

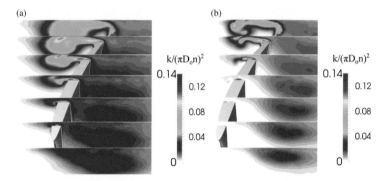

Fig. 20 Turbulent kinetic energy contours in several radial planes from $\theta = -30°$ to $\theta = -60°$, for $\phi = 0.165$ (*left*) and $\phi = 0.195$ (*right*). (**a**) $\phi = 0.165$. (**b**) $\phi = 0.195$

Mach number contours including streamlines of the projected velocity vector and the turbulent kinetic energy contours in several radial planes from $\theta = -30°$ to $\theta = -60°$, for $\phi = 0.165$ and $\phi = 0.195$. For $\phi = 0.195$, the tip-gap vortex passes by the leading edge of the neighboring blade without any interaction. A small counter-rotating separation vortex appears near the leading edge, which, however, has a very low turbulent kinetic energy such that a marginal effect compared to the wake, which impinges upon the blade for the lower volume flow rate, is observed. For $\phi = 0.165$, the tip-gap vortex, which has a larger diameter and is more turbulent, spreads in the axial direction and breaks up in two vortices, where the left vortex is fed by the right vortex and grows. Subsequently, the turbulent left vortex interacts with the leading edge of the blade, which causes strong fluctuations near the gap region extending further upstream compared to $\phi = 0.195$. This results in a larger tip-gap vortex with a higher turbulent kinetic energy, which is created due to the back flow caused by the adverse pressure gradient, supporting the turbulent transition on the suction side. For both volume flow rates, a counter rotating separation vortex is observed next to the tip-gap vortex, which disappears earlier for $\phi = 0.165$.

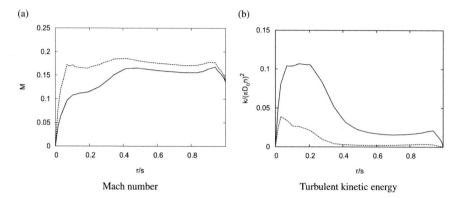

(a)

(b)

Mach number

Turbulent kinetic energy

Fig. 21 Radial distribution of Mach number (**a**) and turbulent kinetic energy (**b**) inside the tip-gap at $x/D_o = 0.617$ and $\theta = -45°$, for $\phi = 0.165$ (—) and $\phi = 0.195$ (— —)

In addition, Fig. 21 depicts the radial distribution of the Mach number and the turbulent kinetic energy inside the tip-gap at $x/D_o = 0.617$ and $\theta = -45°$. The results for $\phi = 0.165$ show a lower Mach number and a higher turbulent kinetic energy inside the tip-gap.

6   Computational Specifications and Scalability Analysis

The simulations were carried out on the CRAY XC40 at HLRS Stuttgart, containing two socket nodes with 12 cores at 2.5 GHz. Each node is equipped with 128 GB of RAM, i.e., each core has 5.33 GB of memory available for the computation. For the scaling experiments a maximum number of grid points for both cases on the order of 1.0×10^9 cells has been used. The scaling test for the fan case has been performed on 228–3828 nodes (i.e., a total number of 5472–91,872 CPUs has been used). For the helicopter engine jet case 418–1668 nodes (10,008–40,032 CPUs) have been used. Details of the scaling experiments for both cases are given in Tables 4 and 5. Up to 183×10^3 cells per domain can be used for the computations owing to the high memory capacity of each nodes. The scaling results are shown in Fig. 22 which indicate a good speed-up for both configurations. Almost the full machine of the

Table 4 Strong scaling speed-up measurements in the fan case

Cores	Cells/domain	Speedup
5472	183×10^3	1.0
10,968	91×10^3	1.87
21,936	45×10^3	3.23
32,880	30×10^3	4.53
91,872	10×10^3	14.49

Table 5 Strong scaling
speed-up measurements in the
jet case

Cores	Cells/domain	Speedup
10,008	100×10^3	1.0
20,016	50×10^3	1.68
40,032	25×10^3	2.94

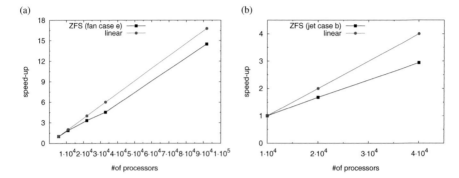

Fig. 22 Scaling test based on simulations of (**a**) ducted axial fan, and (**b**) helicopter engine jet

Hornet system is used for the fan case, and almost half of the machine is used for the jet case.

7 Conclusion

In this study, LESs simulations have been performed for a helicopter engine jet and a ducted axial fan. To verify the numerical solution method, a coaxial circular hot jet has been computed and the results were compared to reference data. The comparison showed a convincing agreement and hence the applied numerical method can be regarded as reliable. Large-scale computations for a turbulent jet configuration were conducted for a grid convergence test. Hierarchical Cartesian meshes featuring local refinement with a number of cells of 0.329×10^9 and 1.097×10^9 were used. Analogously, a grid dependence study was performed for an axial fan configuration using meshes with a number of cells of 0.25×10^9 and 1×10^9. The analysis showed that $\mathcal{O}(0.3 \times 10^9)$ cells are sufficient to resolve all relevant turbulent structures. Furthermore, computations have been performed for different nozzle geometries, i.e., a simplified nozzle geometry (case a) that consists of a center body and a divergent outer annular channel, and a complete engine nozzle geometry (case c) with 4 additional struts. The presence of the struts results in a different potential core break-down and turbulence intensity at the jet field.

Finally, computations have been performed at two different volume flow rates $\phi = 0.165$ and $\phi = 0.195$ for the axial fan configuration. A periodic boundary condition was used in the azimuthal direction to reduce the computational costs. The smaller volume flow rate results in an interaction of the tip-gap vortex with the neighboring blade. The interaction leads to a higher turbulent kinetic energy near and inside the tip-gap region.

A code speed-up analysis for the two cases showed a good scalability. For the axial fan configuration almost the complete machine could be used, while for the jet case almost half of the machine was used. For a better understanding of the unsteady flow phenomena, further analysis of the flow field need to be performed. Results of the LES will be used to determine the acoustic field using computational aeroacoustics (CAA) methods in future studies.

Acknowledgements The research has received funding from the European Community's Seventh Framework Programme (FP7, 2007–2013), PEOPLE programme under the grant agreement No. FP7-290042 (COPAGT project) as well as by the German Federal Ministry of Economics and Technology via the "Arbeitsgemeinschaft industrieller Forschungsvereinigungen Otto von Guericke e.V." (AiF) and the "Forschungsvereinigung Luft- und Trocknungstechnik e.V." (FLT) under the grant no. 177747N (L238). The authors gratefully acknowledge for provision of supercomputing time and technical support granted the High Performance Computing Center Stuttgart (HLRS).

References

1. Koh, S.R., Schröder, W., Meinke, M.: Comput. Fluids **78**, 24 (2013)
2. Bogey, C., Bailly, C.: AIAA J. **43**(5), 1000 (2005)
3. Freund, J.B.: J. Fluid Mech. **438**, 277 (2001)
4. You, D., Wang, M., Moin, P.: AIAA Paper 2002-0981 (2002)
5. You, D., Wang, M., Moin, P.: AIAA J. **42**(2), 271–279, (2004)
6. You, D., Wang, M., Moin, P., Mittal, R.: Phys. Fluids **18**(10), 105102 (2006)
7. You, D., Wang, M., Mittal, R., Moin, P.: AIAA Paper 2003-0838 (2003)
8. Boudet, J., Caro, J., Jacob, M.: AIAA Paper 2010-3978 (2010)
9. Boris, J.P., Grinstein, F.F., Oran, E.S., Kolbe, R.L.: Fluid Dyn. Res. **10**(4–6), 199 (1992)
10. Konopka, M., Meinke, M., Schröder, W.: AIAA J. **50**(10), 2102–2114 (2012)
11. Alkishriwi, N., Meinke, M., Schröder, W.: Comput. Fluids **37**(7), 786–792 (2008)
12. Meinke, M., Schröder, W., Krause, E., Rister, T.: Comput. Fluids **31**(4), 695 (2002)
13. Hartmann, D., Meinke, M., Schröder, W.: Comput. Fluids **37**(9), 1103 (2008)
14. Schneiders, L., Hartmann, D., Meinke, M., Schröder, W.: J. Comput. Phys. **235**, 786 (2013)
15. Lintermann, A., Schlimpert, S., Grimmen, J.H., Günther, C., Meinke, M., Schröder, W.: Comput. Methods Appl. Mech. Eng. **277**, 131 (2014)
16. Bogey, C., Bailly, C., Juvé, D.: Theor. Comput. Fluid Dyn. **16**(4), 273 (2003)
17. Pardowitz, B., Tapken, U., Knobloch, K., Bake, F., Bouty, E., Davis, I., Bennett, G.: 20th AIAA/CEAS Aeroacoustics Conference (2014)
18. Kunnen, R.P., Siewert, C., Meinke, M., Schröder, W., Beheng, K.D.: Atmos. Res. **127**, 8–21 (2013)
19. Freund, J.B.: AIAA J. **35**(4), 740–742 (1997)
20. Jeong, J., Hussain, F.: On the identification of a vortex. J. Fluid Mech. **285**, 69–94 (1995)
21. Cetin, M.O., Schröder, W., Meinke, M.: 19th DGLR/STAB Conference (2014)
22. Cetin, M.O., Meinke, M., Schröder, W.: ICWSF Conference (2015)
23. Pogorelov, A., Meinke, M., Schröder, W., Kessler, R.: AIAA Paper 2015-1979 (2015)
24. Zhu, T., Carolus, T.H.: ASME Paper GT2013-94100 (2013)

Large-Eddy Simulation of a Scramjet Strut Injector with Pilot Injection

Sebastian Eberhardt and Stefan Hickel

Abstract A scramjet is an air breathing jet engine for hypersonic flight, in which combustion takes place at supersonic flow velocity. Fast mixing of fuel and the compressed air flow is thus essential for the efficient operation. Because the penetration depth of fuel injection perpendicular to the main flow direction is very small, strut injectors that are positioned directly within the supersonic core flow are usually used in large chambers. We performed large-eddy simulations for a generic strut-injector geometry. The main objective of this paper is the analysis of the pilot injection for flame stabilisation.

1 Introduction

In a scramjet combustor, efficient and fast mixing of injected fuel with the surrounding airflow is essential to enable combustion during the very short residence time of the reactants in the combustion chamber. The penetration depth of fuel injection perpendicular to the main flow direction is very small. Therefore strut injectors are used in large chambers, which are positioned within the supersonic core flow. Depending on the flight Mach number, the conditions in a supersonic combustion chamber allow for autoignition of the fuel. The governing quantity for this process is the static temperature of the gas in the mixing region. If the flight Mach number is small, the temperature in the combustion chamber is also low, and not sufficient to ignite the fuel. Another scenario where autoignition may not be possible, depending on the setup, are ground experiments. To ignite the fuel, a pilot

S. Eberhardt
Institute of Aerodynamics and Fluid Mechanics, Technische Universität München,
Boltzmannstr. 15, 85748 Garching, Germany
e-mail: sebastian.eberhardt@tum.de

S. Hickel (✉)
Institute of Aerodynamics and Fluid Mechanics, Technische Universität München,
Boltzmannstr. 15, 85748 Garching, Germany

Faculty of Aerospace Engineering, Aerodynamics Group, Technische Universiteit Delft,
P.O. Box 5058, 2600 GB Delft, The Netherlands
e-mail: S.Hickel@tudelft.nl; sh@tum.de

© Springer International Publishing Switzerland 2016
W.E. Nagel et al. (eds.), *High Performance Computing in Science
and Engineering '15*, DOI 10.1007/978-3-319-24633-8_26

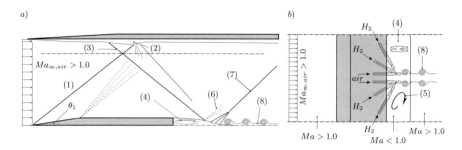

Fig. 1 Sketch of the strut injector pilot injection flow from the *side* (**a**) and *top* (**b**)

flame is then used. We present large-eddy simulation results for such a pilot flame. The strut-injector geometry is based on an experimental facilities at the Institute for Flight Propulsion of Technische Universität München. Fuel and oxidizer for the pilot flame are provided by several injections from the injector base. The main injection is switched off here to isolate effects belonging to the pilot flame. Results for the main injection of the same injector can be found in Eberhardt and Hickel [1].

Figure 1 provides an overview of the flow field: a primary shock (1) is generated by the sharp leading edge of the injector. This shock causes separation of the boundary-layer (2) and a separation shock (3) at the top wall. This reflected shock enters the recirculation region (4) following the strut's base where the shock is reflected as an expansion (6). Hydrogen and air is injected through six injectors at the base of the strut. A second, horizontal, recirculation (5) is created by the injection jets, which are oriented in such a way that the three jets of one side cross in a single point. Following the expansion, the recirculation collapses, creating a re-compression shock (7). In the wake of the injector a turbulent shear- and mixing-zone (8) is formed.

2   Experimental Setup

The combustion chamber used in the experiment was described by Fuhrmann [2] and is sketched in Fig. 2. It consists of a short section with parallel walls and a strut injector located in the middle of the channel. The strut injector has identical dimensions as the one investigated in [1]. Following the section with constant cross section is a diverging part with an opening angle of 5°. Fuel is injected through perpendicular nozzles on the strut injector's top and bottom. Optionally, hydrogen wall injectors in the divergent part of the chamber can be activated during operation. For the pilot injection, a total of six injections at the base of the injector are used. The two middle ones are parallel to the main flow direction and inject air, the other four are sloped in such a way that the hydrogen jets impinge on the air jets to enhance mixing. All pilot injectors have a diameter of $D = 0.5$ mm. On the top wall of the chamber are bores for pressure taps to measure the wall pressure. The walls are made of copper with active water cooling to enable continuous operation without

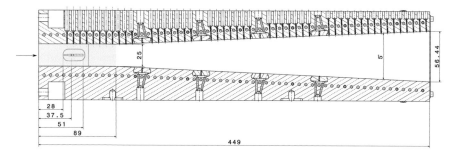

Fig. 2 Technical sketch as provided by Fuhrmann [2] of the experimental combustion chamber downstream from the laval nozzle. Highlighted in *blue* is the simulated section

overheating of the walls. The section with parallel walls measures $L_x = 89$ mm in x-direction, $L_y = 25$ mm in y-direction and $L_z = 27$ mm z-direction.

The air flow through the combustion chamber has a Mach number of $Ma_\infty = 2.15$ with a static temperature of $T_\infty = 509.2$ K. The injection fluids both have a total temperature of $T_0 = 300$ K. The total pressure can be calculated from the mass flow rate measured in the experiment. These mass flows are 0.25 g/s at both air injections and 22.5 mg/s hydrogen through each of the four hydrogen injectors. The resulting pressure values are $p_{0,inj,air} = 5.456 \cdot 10^5$ Pa for the air injectors and $p_{0,inj,H_2} = 1.857 \cdot 10^5$ Pa for the hydrogen injection.

3 Numerical Method and Computational Details

We solve the compressible multi-species Navier-Stokes equations (see [3], e.g.) with a finite-volume method on an adaptive Cartesian grid. The subgrid-scale (SGS) turbulence model is provided by the Adaptive Local Deconvolution Method (ALDM) of Hickel et al. [4, 5], which follows an implicit LES approach. ALDM is implemented for Cartesian collocated grids and used to discretize the convective terms of the Navier-Stokes equations. The diffusive terms are discretized by second order centered differences. The fully conservative cut-element immersed boundary technique of Örley et al. [6] is employed to represent the strut injector. For time-integration the explicit third order accurate Runge-Kutta scheme of Gottlieb and Shu [7] is used.

The LES in this study were performed using the flow solver INCA,[1] which is written in FORTRAN 2003. It uses a classical block-structured grid topology. For blocks which are not part of the same process, non-blocking communication according to the MPI standard[2] is employed, otherwise values are copied directly.

---

[1] http://www.inca-cfd.org.

[2] http://www.mpi-forum.org.

Three ghost-cell layers need to be exchanged as INCA uses discretization schemes that operate on six-cell stencils. Additional OpenMP directives allow a hybrid OpenMP/MPI parallelization, where several OpenMP threads share the workload assigned to one MPI process. The grid blocks can be distributed on an arbitrary number of MPI tasks, where the local blocks are stored in a linked list. Several load balancing strategies can be used to distribute the grid blocks: either INCA's internal partitioner that aims at balancing the computational load (total no. of cells per task) while minimizing the communication load (total no. of communicated data between tasks), or the more sophisticated METIS[3] library are possible choices.

To evaluate the scaling capabilities of INCA we performed a strong-scaling study on Hermit. The test problem is chosen to be a compressible Taylor-Green vortex and only one OpenMP thread per MPI process is used. We consider this the worst-case scenario. The computational domain is a cube with periodic boundary conditions in all directions. The domain is discretized by $N = 256^3$ cells with homogeneous grid spacing and the cube has an edge length of 2π on all sides. The domain is divided into subdomains of identical size according to the number of cores used, starting with 32 cores up to 2048. The number of cells per core, ghost cells per core, and ratio of ghost cells to cells is given in Table 1. Figure 3 shows the strong scaling

Table 1 Grid statistics for strong scaling of INCA

No. of cores	Cells/core	Ghost cells/core	Ghost cells/cells (%)
32	524,288	132,312	25.2
64	262,144	80,856	30.8
128	131,072	55,128	42.1
256	65,536	35,544	54.2
512	32,768	22,104	67.5
1024	16,384	15,384	93.9
2048	8192	10,200	124.5

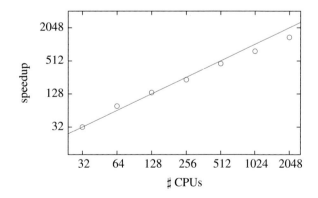

Fig. 3 Strong scaling of INCA on Hermit: speed-up on up to 64 nodes

[3] http://glaros.dtc.umn.edu.

on Hermit. For two and three nodes (64 and 128 cores), we observe a super-linear scaling that we attribute to optimized cache usage for these processor configurations. Up to the maximum investigated number of nodes, we observe good strong scaling properties of INCA. We conclude that an almost linear speed-up is possible by decomposing a computational grid into grid blocks with approximately 30,000 cells or more.

4   Numerical Setup

Figure 2 shows the computational domain and its position within the combustion chamber, which is the section with parallel walls and the strut injector. The inlet and outlet are supersonic boundary conditions. The four side walls are modeled as symmetries which corresponds to frictionless walls. These outer walls are not resolved to reduce computational costs and because the main focus of the study is the mixing zone in the wake of the strut. The injector surface and injection tubes are modeled as adiabatic walls by the immersed boundary technique [6]. Figure 4 displays a technical drawing of the experimental strut interior, (a) as given by Fuhrmann [2], and the wall created with the immersed boundary, (b), used in the simulations. The positions where injection boundary conditions are applied are pointed out as well. The boundary conditions applied for the injections is a total temperature and total pressure boundary condition.

For the calculation of the total pressure, the assumption (based on the Fanno flow) is made that the flow at the injection tube exit is critical. With this assumption

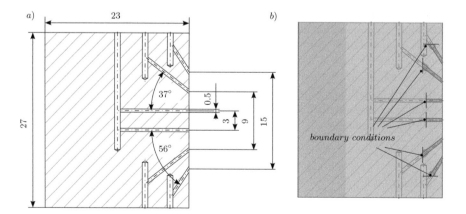

Fig. 4 (**a**) Technical drawing [2] of the experiments injector. (**b**) Representation of the injector geometry for the simulation, displayed by the immersed boundary wall as an overlay over the technical drawing

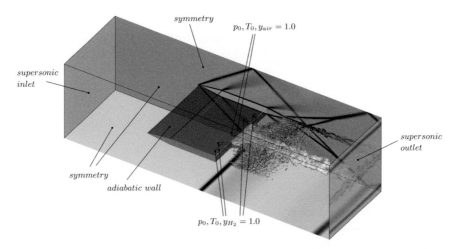

Fig. 5 Overview of the computational domain (including visualization of the flow) with applied boundary conditions. The origin of the computational frame of reference is the center of the injector's sharp leading edge

and the fact that the mass flow rate has a maximum at the point of critical flow, the total pressure for the boundary condition can be calculated:

$$p_0 = \frac{\dot{m}}{A} \frac{1}{\psi} \sqrt{\frac{RT_0}{2}}, \ \psi = \left(\frac{2}{\gamma + 1}\right)^{\frac{1}{\gamma - 1}} \cdot \sqrt{\frac{\gamma}{\gamma - 1}}. \tag{1}$$

We like to add that this functional only depends on the gas, the total temperature of the gas, the mass flow rate and the geometry of the tube. A summary of all boundary conditions used in this simulation is provided by Fig. 5.

The computational grid for this simulation consists of a total of 42 million cells with refinements around the injector wall on the outside and on the inside of the injection tubes. To capture the mixing zone, the wake of the strut is refined as well over a distance of 19 mm, which is almost one injector length. The grid has a homogeneous point distribution in z-direction. Figure 6 shows the grid topology and every fifth grid line.

5   Results

First we want to discuss the overall flow field around the strut injector. Figure 7 displays the instantaneous density gradient magnitude on a slice normal to the z-axis at a location $z = -0.0015$ m, which is the center of one of the air injections. The primary shocks from the injector leading edge are reflected at the top and bottom symmetries. They enter the recirculation region and increase the size of the subsonic

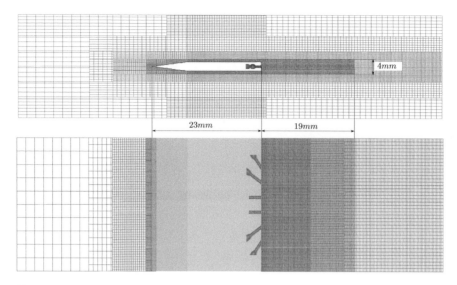

Fig. 6 Computational grid used for the simulation, displaying every fifth cell on a $z = 0.0$ (*top*) and a $y = 0.0$ (*bottom*) plane and an overlay of the immersed boundary

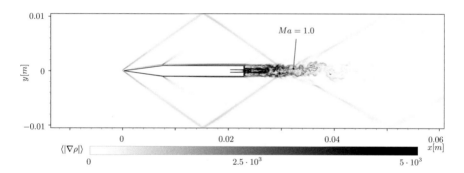

Fig. 7 Instantaneous density gradient magnitude on a $z = -0.0015\,\text{m}$ slice. This position cuts through one air injection at its center

region behind the injector base. From the base the injected supersonic air jet is visible which enters the recirculation. Following the recirculation re-compression socks are visible, as well as an unsteady, periodic shedding along the direction of the y-axis. The sonic line is added in Figs. 7, 8 and 9 to visualize the supersonic character of the injection jets and the region where the main flow re-accelerates to supersonic speed. Figure 8 displays a slice through the center of the strut at $y = 0$ slice. The Figure contains a graphical representation of the strut injector walls. Positions (*a*) through (*f*) mark the slice positions plotted in Fig. 9. The six injected jets and their super sonic region are clearly visible, with the two outer ones on each side impinging on the central jets. This creates a mixing region with strong

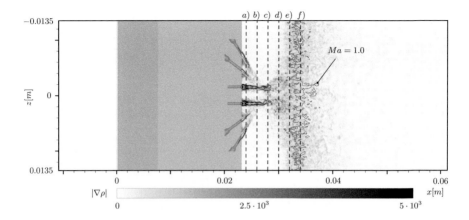

Fig. 8 Instantaneous density gradient magnitude on a $y = 0$ m slice including a $Ma = 1.0$ line and a geometric representation of the strut. Positions (**a**) through (**f**) mark the positions of slices in Figs. 9 and 15

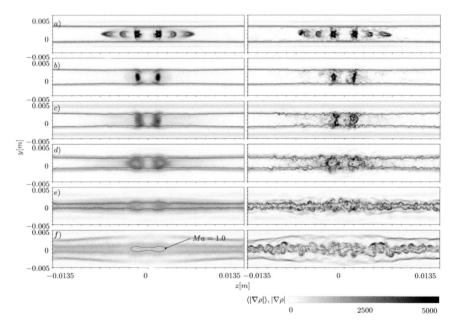

Fig. 9 Time averaged (*left*) and instantaneous (*right*) plots of the density gradient magnitude on slices with constant x-coordinate including a $Ma = 1.0$ line. The locations of the individual slices are marked in Fig. 8

turbulence within the subsonic recirculation behind the strut, which ensures a good mixture. The region $x \approx 0.034$ m indicates the re-acceleration to supersonic speed with presence of strong turbulence. The turbulence originates from the borders of the recirculation region, which collapses in this region. The same flow features are visible in the instantaneous density gradient magnitude on the right hand side of Fig. 9. Close to the strut the degree of turbulence is low and increases with larger distance. The inner part around the jets is also turbulent near the injector, as they enter the recirculation with high speeds. Turbulence is further enhanced by the hydrogen jets impinging on the air jets although they are subsonic at the impingement point. In plot (c) the incoming shock from the top channel wall is visible and in plot (f) the re-compression shock, both as a horizontal dark bar. The air jets have the longest supersonic core, as they become subsonic only shortly before the main flow re-accelerates to supersonic speed. As the injected mass has to be accelerated in addition to the main flow, the subsonic region behind the air jets has the largest extend in x-direction as can be seen in the bottom left plot of Fig. 9. This figure also shows that, due to the added mass, the recirculation region in the wake becomes thicker in y-direction in the center of the flow.

The highest Mach number in the entire flow field is reached in the core of the air jets. Even though the Mach number there is higher than in the hydrogen jets, the absolute velocity in the hydrogen is higher due to the high speed of sound of the gas. The maximum Mach number in the air jets is around 2.65 and 1.4 in the hydrogen jets, as can be seen in Fig. 10. The sonic line in the same figure also shows a clear retardation in the x-direction following the air jets confirming the previous observation.

To analyze the mixing behavior of the injector, first the turbulence quantities variance of the velocity components, see Fig. 11, and turbulent kinetic energy, see Fig. 12, are analyzed. They are displayed in the figures on a slice at $y = 0$ m. These quantities show the edges of the injected jets to be the most turbulent part in the close wake of the injector. The air jets have the strongest variance in u direction due to the x-axis parallel injection, see Fig. 11a. Accordingly, the outer hydrogen jets create the largest fluctuations in w direction (z-axis) due to their large angle, see Fig. 11a. The largest turbulence intensity on the slice, however, is visible a short distance

Fig. 10 Mach number on a $y = 0$ m slice in the wake of the strut including a $Ma = 1.0$ line

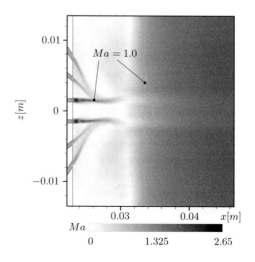

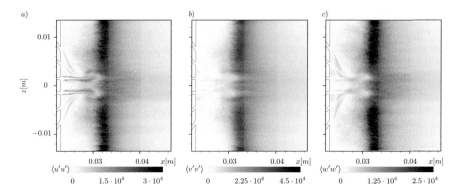

Fig. 11 Variance of the velocity fluctuations (**a**) $\langle u'u' \rangle$, (**b**) $\langle v'v' \rangle$ and (**c**) $\langle w'w' \rangle$ in all three directions on a $y = 0$ m slice

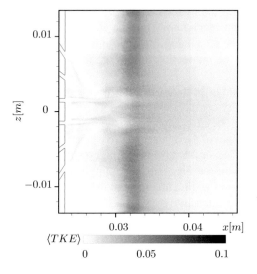

Fig. 12 Turbulent kinetic energy on a $y = 0$ m slice

downstream, where the flow re-accelerates to supersonic speed. This is again the region where the turbulence from the upper and lower border of the recirculation merge in the center of the flow. The shockwaves reflected from the upper and lower combustor walls hit the recirculation upstream of the maximum of turbulence and its trace can not be seen on the center slice.

The last of the turbulence statistic quantities considered here is the turbulent transport in the three directions, Fig. 13. It visualizes the mass transported by turbulence independent of the mass fractions of the mixing species. Again, the largest values of turbulent transport are visible in the x-direction (a) and z-direction (c). As expected, the vertical transport is the smallest on the symmetry plane. In

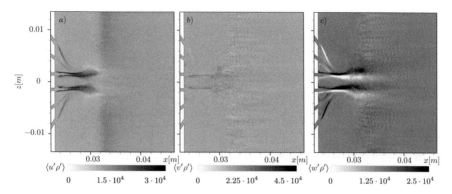

Fig. 13 Turbulent transport in all three directions on a $y = 0$ m slice

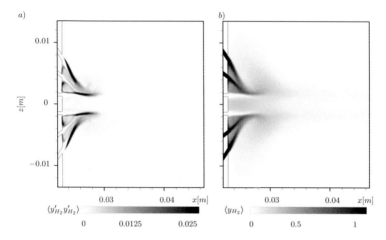

Fig. 14 Variance of the hydrogen mass fraction and hydrogen mean mass fraction on a $y = 0$ m slice

contrast to the variances discussed earlier, the region of highest turbulence is less pronounced in the turbulent transport. The region with the strongest transport again are the borders of the air jets, and the outer borders of the hydrogen jets.

The variance of the hydrogen mass fraction, Fig. 14a, supports the result of the velocity fluctuations as it shows maxima at the air jet boundaries where the hydrogen impinges, as well as on the outer boundary of the hydrogen jets. Furthermore, the figure shows that the six injection jets create two separated mixing zones, one on each side, with the most mixing on the outer boundary of the respective air jet. The mass fraction vanishes quickly because of the hydrogen's low density, and so there is no visible rise in mass-fraction variance in the region of high turbulence following the mixing. The mass fraction is plotted in Fig. 14b where the four jets are clearly visible. The region where the mass fraction spreads in z-direction at $x \approx 0.03$ m is caused by the horizontal recirculation. Following the region at $x \approx 0.032$ m the mass fraction is barely visible as it mixes with the air.

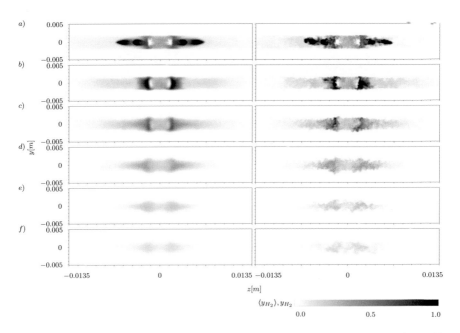

Fig. 15 Time averaged and instantaneous plots of the hydrogen mass fraction on slices with constant x-coordinate. The locations of the individual slices are marked in Fig. 8

A similar observation is made in Fig. 15. The plot shows the mean (left) and instantaneous (right) hydrogen mass fraction on $x = const.$ slices at the locations (a) through (f) as indicated in Fig. 8. The main regions of high mass fraction are located left and right of the air jets, which are clearly visible as white regions in the slices close to the injector. Further downstream the distribution becomes more homogeneous as the gases mix and the value of the H_2 mass fraction rapidly drops.

For hydrogen mixing the volume fraction gives additional information on the distribution and does not decay as quickly as the mass fraction. This can be seen in Figs. 16 and 17, where the mean volume fraction is shown. The first figure shows the density gradient magnitude in white to black contours and the volume fraction in white to blue at a position $z = -0.0015\,\text{m}$. The volume fraction is visible over the entire length of the domain following the injection and is spreading in spanwise direction further downstream. The region where the shocks from the channel walls hit the recirculation area shows a thickening due to the imposed upstream pressure gradient on the subsonic recirculation. Figure 17 displays the volume fraction distribution ranging from pure hydrogen on the one side of the color scale to pure air on the other side of the scale. Streamlines show the horizontal recirculation which enhances the mixing and is created by the hydrogen jets.

One of the goals of this study is to evaluate the pilot injection with respect to combustion and flame stabilization. The hydrogen volume fraction allows for a straight forward detection of the stoichiometric surface, which is a first

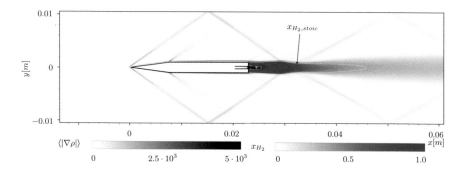

Fig. 16 Time averaged density gradient magnitude (*grey*) and hydrogen volume fraction (*blue*) on a $z = -0.0015$ m slice. *Yellow line* displays the stoichiometric line for hydrogen–air combustion

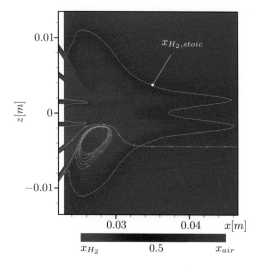

Fig. 17 Pure hydrogen volume fraction (*red*) to pure air volume fraction (*blue*) on a $y = 0$ m slice. *Yellow line* displays the stoichiometric line for hydrogen–air combustion

coarse approximation of the flame front location. The stoichiometric iso-line $x_{H_2,stoich} = 0.296$ for the combustion of hydrogen fuel with air has been added to Figs. 16 and 17 to give a rough estimate of where the main reaction zone will be. The large area of the stoichiometric surface indicates that the six hole pilot injector is a rather good solution for enabling ignition and for flame stabilization. Additionally, the subsonic recirculation region will have a higher temperature than the surrounding supersonic flow and the impinging oblique shock waves improve this even further.

Acknowledgements The support of this research within the Research Training Group "Acro
Thermodynamic Design of a Scramjet Propulsion System for Future Space Transportation
Systems" 1095/2 by the Deutsche Forschungsgemeinschaft (DFG) and the High Performance
Computing Centre Stuttgart (HLRS) is greatly acknowledged.

References

1. Eberhardt, S., Hickel, S.: Large-eddy simulation of a central strut injector with perpendicular
   injection for Scramjet applications. In: proceedings of 5th European Conference for Aerospace
   Sciences EUCASS (2013)
2. Fuhrmann, S., Hupfer, A., Kau, H.-P.: Investigations on multi-stage supersonic combustion in a
   model combustor. AIAA 2011–2332 (2011)
3. Tritschler, V.K., Zubel, M., Hickel, S., Adams, N.A.: Evolution of length scales and statistics
   of Richtmyer-Meshkov instability from direct numerical simulations. Phys. Rev. E 90, 063001
   (2014). doi:10.1103/PhysRevE.90.063001
4. Hickel, S., Adams, N.A., Domaradzki, J.A.: An adaptive local deconvolution method for implicit
   LES. J. Comput. Phys. (2006). doi:10.1016/j.jcp.2005.08.017
5. Hickel, S., Egerer, C.P., Larsson, J.: Subgrid-scale modeling for implicit Large Eddy Simulation
   of compressible flows and shock turbulence interaction. Phys. Fluids 26, 106101 (2014).
   doi:10.1063/1.4898641
6. Örley, F., Pasquariello, V., Hickel, S., Adams, N.A.: Cut-element based immersed boundary
   method for moving geometries in compressible liquid flows with cavitation. J. Comput. Phys.
   283, 1–22 (2015). doi:10.1016/j.jcp.2014.11.028
7. Gottlieb, S., Shu, C.-W.: Total variation diminishing Runge-Kutta schemes. Math. Comput. **67**,
   73–85 (1998)

Turbulence Resolving Flow Simulations of a Francis Turbine with a Commercial CFD Code

Timo Krappel, Albert Ruprecht, and Stefan Riedelbauch

Abstract Transient flow simulations of a Francis turbine in part load conditions using hybrid RANS-LES turbulence models are presented. In the draft tube a rotating low pressure zone—the vortex rope phenomenon—arises, which leads to very complex flow phenomena. A detailed resolution of the flow in space and time is leading to large computational effort.

For the flow simulations the commercial CFD code Ansys CFX version 16.0 is used, as the parallel performance especially for this application has improved. Two hybrid RANS-LES turbulence models, the Scale Adaptive Simulation (SAS) and Detached Eddy Simulation (DES) are applied. To have a very good resolution of turbulence, meshes in the range of 50 to 300 million mesh nodes are investigated running on up to more than 2000 cores in parallel.

1 Introduction

The operation of Francis turbines in part load conditions leads to complex flow phenomena especially in the draft tube. The flow field at the runner outlet has a high swirling component leading to a stagnant region in the core of the flow and around this the vortex rope phenomenon evolves. The rotating vortex rope can lead to severe structural damages in hydraulic machines. Therefore, a better understanding by detailed numerical analysis can help in the design process.

Preliminary studies reveal [7, 9] that it is necessary to simulate the turbulence, in contrast to RANS turbulence models, where the turbulence is modelled. As pure LES simulations have computational costs being too high for such a high Reynolds number application, two hybrid RANS-LES turbulence are investigated in this paper. These models are the Scale Adaptive Simulation (SAS) and Detached Eddy Simulation (DES) approaches.

T. Krappel (✉) • A. Ruprecht • S. Riedelbauch
Institute of Fluid Mechanics and Hydraulic Machinery, Pfaffenwaldring 10, 70550 Stuttgart, Germany
e-mail: timo.krappel@ihs.uni-stuttgart.de

© Springer International Publishing Switzerland 2016
W.E. Nagel et al. (eds.), *High Performance Computing in Science and Engineering '15*, DOI 10.1007/978-3-319-24633-8_27

421

2   Numerical Methods

The flow simulations of the Francis turbine presented in this paper are performed with the commercial CFD code Ansys CFX version 16.0. This CFD codes is able to handle the rotation of the turbine runner and to couple different meshes by an general-grid-interface. The finite-volume method is used for discretisation, while the volumes of discretisation are built around the cell nodes.

2.1   Domain Decomposition

For domain decomposition an optimised method for rotor-stator-interface is used, called multipass partitioning. This method uses a special multipass algorithm to optimise the partition boundaries. Circumferentially-banded partitions adjacent to each transient rotor stator interface are generated ensuring that interface nodes remain in the same partition as the two domains slide relative to each other [1].

2.2   Turbulence Modeling

2.2.1   Scale Adaptive Simulation

The SAS approach enables the unsteady SST RANS turbulence model [10] to operate in SRS (Scale Resolving Simulation) mode [14] meaning that it can work like a LES turbulence model. This is achieved by introducing a new quantity, namely Q_{SAS}, into the ω-equation of the SST model [3, 4, 12]. Q_{SAS} is defined as:

$$Q_{SAS} = \max\left[\rho \zeta_2 \kappa S^2 \left(\frac{L}{L_{vK}}\right)^2 - C\frac{2\rho k}{\sigma_\Phi} \max\left(\frac{|\nabla\omega|^2}{\omega^2}, \frac{|\nabla k|^2}{k^2}\right), 0\right], \qquad (1)$$

containing the turbulent length scale L and the von Karman length scale L_{vK}. Based on the theory of Rotta [17], L_{vK} describes the second derivative of the velocity field. The manipulation of the ω-equation leads to a reduction of the turbulent eddy viscosity. The test case of decay isotropic turbulence [2] shows not enough damping at the high wave-number limit to dissipate the energy at the smallest scales [12]. One way to limit the eddy viscosity is the following formulation:

$$\mu_t = \max(\mu_t^{SAS}, \mu_t^{LES}) \qquad (2)$$

For the calculation of the LES eddy viscosity the WALE model is used [15]. This model does not need wall damping functions.

2.2.2 Delayed Detached Eddy Simulation

The detached eddy simulation based on the SST turbulence model [13] replaces the dissipation term in the k-equation as follows [20]:

$$D_k = \beta^* k\omega = \frac{k^{3/2}}{l_t} . \tag{3}$$

The length scales are according to e.g. [18]:

$$l_{RANS} = \frac{\sqrt{k}}{\beta^* \omega} \qquad l_{LES} = C_{DES}\Delta \tag{4}$$

$$C_{DES} = (1 - F_1)C_{DES,k-\epsilon} + F_1 C_{DES,k-\omega} \tag{5}$$

whereas $C_{DES,k-\epsilon} = 0.61$ and $C_{DES,k-\omega} = 0.80$ are used as DES-coefficients [13], depending on the RANS-SST constant F_1. For the DES formulation the length scale is:

$$l_t = l_{DES} = \min(l_{RANS}, l_{LES}) \tag{6}$$

This formulation has the deficiency of the "modelled stress depletion" (MSD) that occurs in the boundary layer if the mesh is not fine enough for LES-resolution and the modelled Reynolds stresses are reduced. In order to activate the LES behaviour later for such grids, a "delay" function has been introduced [19]:

$$f_d \equiv 1 - tanh((8r_d)^3) \tag{7}$$

which is developed to be 1 in LES- and 0 in RANS-regions, with:

$$r_d \equiv \frac{\nu_t + \nu}{\sqrt{U_{i,j}U_{i,j}} \, \kappa^2 d^2} \tag{8}$$

The length scale is therefore:

$$l_t = l_{DDES} = (1 - f_d)l_{RANS} + f_d \, l_{LES} \tag{9}$$

2.3 Temporal and Spatial Discretisation

For temporal discretisation a second order backward Euler scheme and for spatial discretisation a bounded second order central differencing scheme [5] is used. For the turbulence quantities a bounded second order backward Euler scheme is applied for the temporal discretisation and a first order scheme for the spatial discretisation [11].

3 Francis Turbine Case

3.1 Computational Setup

The discretised geometry displaying the mesh with 150 million nodes can be seen
in Fig. 1. The mesh consists of four different domains coupled with GGI (general
grid interface). These domains are for the components of the spiral casing, stay and
guide vanes, runner and draft tube with expansion tank (in streamwise direction).
The number of stay and guide vanes is 23 and 24 each and the runner consists of 13
blades.

   At the inlet of the spiral casing steady boundary condition profiles are applied
with a mass flow corresponding to the part load operating point. The profiles are
resulting from a separate pipe flow simulation.

   Due to the awareness of preliminary studies [7, 9], where meshes with 16M
(million) and 40M nodes were used (draft tube y^+: 16 and 9), larger meshes are
used in this study. For the 50M mesh the boundary layer is resolved to the laminar
region ($y^+ = 1$) and the draft tube is refined at the end of the vortex rope, where the
large structures decay to small turbulent structures. Furthermore, two larger meshes
with 150M and 300M nodes are investigated (see Table 1), also having a resolved
boundary layer to the laminar region ($y^+ = 1$). All components of the machine have
a finer resolution with a higher concentration of nodes in the runner and especially
in the draft tube.

Fig. 1 Visualisation of the computational domain and mesh for the 150M

Table 1 Description of different grid sizes for different domains in million nodes and the
corresponding time steps

Name	Spiral casing	Stay and guide vanes	Runner	Draft tube	Total	#Time steps/rev	Δt in °/time step
50M	4.54	10.18	13.47	22.14	50.33	720	0.5
150M	7.29	17.95	29.90	98.81	153.95	840	0.43
300M	11.84	27.92	54.98	211.62	306.36	1000	0.36

Table 2 Description of wall adjacent elements normalised in dimensionless units for different domains; x^+ in spanwise and z^+ in streamwise direction

Name	Runner		Draft tube cone		Draft tube diffuser	
	x^+	z^+	x^+	z^+	x^+	z^+
50M	247	141	355	196	443	277
150M	178	101	174	109	252	171
300M	140	80	130	87	196	131

According to the definition of the wall adjacent node in dimensionless units:

$$y^+ = y\frac{u_\tau}{\nu} \tag{10}$$

with y the wall distance of the first node, u_τ the friction velocity and ν the kinematic viscosity, the corresponding parameters for the other directions of the wall adjacent element in streamwise x^+ and spanwise z^+ direction are defined as follows:

$$x^+ = x\frac{u_\tau}{\nu} \qquad z^+ = z\frac{u_\tau}{\nu} \tag{11}$$

For a wall-resolved LES a resolution of $x^+ \cong 50 - 150$ and $z^+ \cong 15 - 40$ would be required [16]. The 50M mesh is well resolved in terms of RANS (see Table 2). The two steps of mesh refinement, the 150M and the 300M mesh, have an even better resolution of the boundary layer. Especially the 300M mesh is quite close to an adequate, but still away from a really wall-resolved LES mesh.

The chosen time steps for the different meshes are listed in Table 1. It is essential to have a Courant number smaller or equal one to resolve turbulent structures. Therefore, the number of time steps per runner revolution is between 720 and 1000 leading to time steps between 0.5° and 0.36° runner rotation. In the area between guide vanes and runner, where the highest flow velocities appear, the Courant number rises—locally restricted—up to values slightly above one. The Courant number in the rest of the computational domain, and especially in the draft tube, is strictly below one.

3.2 Global Machine Data

First of all, the predicted torque for different mesh types is compared. The 150M simulation—featuring a boundary layer resolution of $y^+ = 1$—shows a torque, which is 1.3 % lower than the 16M simulation and 1.1 % lower than the 40M simulation. The predicted torque of the 50M simulation, with the same boundary layer resolution as the 150M simulation, is only marginally higher compared to the largest mesh. This indicates the necessity of a resolved boundary layer to predict the torque correctly.

The hydraulic losses mainly differ for the runner and the draft tube component. The simulation with the 150M mesh differs by 0.2 % of total hydraulic head from the coarse mesh simulations. The results of the coarse mesh simulations are very similar.

3.3 Flow Analysis

In the draft tube cone a flow analysis is done for time-averaged velocity components (see Fig. 2). The simulation time for all configurations equals 40 runner revolutions. Therefore, only the 50M and 150M simulations can be post-processed, as for the 300M simulations no statistical meaningful data is available at the moment of writing this paper.

In the draft tube cone, there is no backflow region in the core of the flow with the finer meshes in contrast to the coarse mesh simulation (16M) (see Fig. 2 top, left). At higher radius the axial velocity is somewhat lower with the finer meshes. This might be due to the better resolution of the boundary layer. The 50M and the 150M are quite similar for the axial velocity component. The swirling component is also higher with finer meshes in contrast to the coarse mesh (see Fig. 2 top, right). The

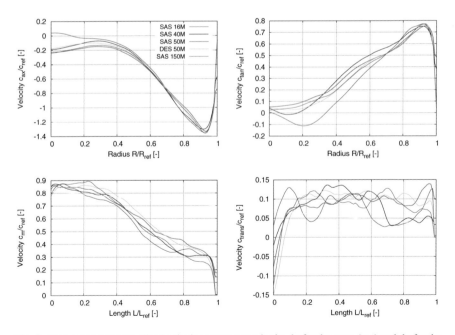

Fig. 2 Time-averaged normalised velocity components in the draft tube cone (*top*) and draft tube diffuser (*bottom*) on *horizontal lines* each; legend is the same for all figures

result of the finest mesh (150M) shows an offset to higher values for the tangential velocity component compared to the 50M simulation.

In the draft tube diffuser, the meridional velocity component is the lowest for the coarse mesh simulation (see Fig. 2 bottom, left). The results for the finer meshes show higher values, instead. A similar trend for the different configurations is also visible for the vertical, transversal velocity component (see Fig. 2 bottom, right).

3.4 Vortex Rope

For the evaluation of the pressure fluctuations in the draft tube cone, which are mainly caused by the rotating vortex rope, the wall pressure signal is used. The reason for the pressure fluctuation at the wall is mainly caused by the vortex rope, but there is also an influence resulting from the runner blade wakes. The first mode of the vortex rope has a frequency of around 1/3 of the runner frequency and the runner blade wakes first mode is—according to the number of blades—13 times the runner frequency. These frequencies can be seen according to the time signal (see Fig. 3) and the FFT signal (see Fig. 4) for different simulation configurations.

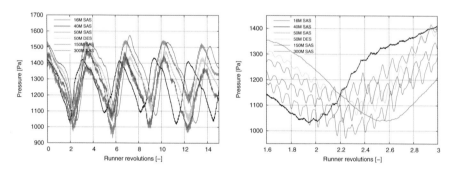

Fig. 3 Time signal of the wall pressure in the draft tube cone for different time ranges

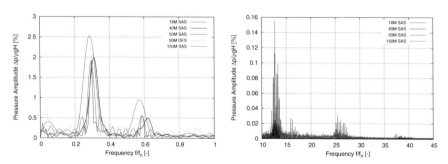

Fig. 4 FFT signal of the wall pressure in the draft tube cone for different frequency ranges

The coarse mesh simulation predicts a noticeable higher pressure fluctuation amplitude with a slightly smaller frequency for the vortex rope mode than the fine mesh simulations. The amplitudes of the other simulations are quite similar, whereas the frequencies are slightly different. The influence of the runner blade wakes is not visible for the 16M and very small for 40M simulation as the boundary layer is not well resolved in both cases. The pressure signal of the other simulations with finer meshes clearly show the runner blade wakes and in particular the 150M and 300M simulations. The 300M simulation even predicts smallest fluctuations, as it can be seen in the pressure signal. As the data base for the 300M simulation is too small at the moment of writing this paper, no FFT analysis is done.

3.5 Turbulence Evaluation

One goal of this study is to simulate as much turbulence as possible with a hybrid RANS-LES turbulence model in LES manner, instead to model the turbulence in RANS manner. Firstly, this is evaluated by the amount of eddy viscosity that is introduced by the turbulence model into the Navier-Stokes equations. The eddy viscosity ratio (turbulent eddy viscosity/dynamic viscosity) is in the draft tube in the order of 1000 for RANS simulations [8]. An indicator for activated LES-behaviour is an eddy viscosity ratio of around one order below. The eddy viscosity in the draft tube can be seen in Fig. 5 (left). The coarse mesh simulation (16M) has a distinct region of higher values of viscosity ratio between the end of the cone, mainly in the elbow and the beginning of the diffuser. The mesh size in this region is less extended—leading to smaller eddy viscosity values—for the 40M and 50M simulation, whereas the 50M mesh is optimised in this region, where the vortex rope decays and more turbulence appears. The difference between the SAS and DES turbulence model for the 50M simulation is that the DES seems to slightly further reduce the eddy viscosity in the elbow region. Close to the floor of the diffuser, the DES model shows some higher values of eddy viscosity. A further grid refinement (150M) leads to values of around 50 in the elbow, with some small areas with values around 100. The finest grid (300M) with almost 212 million grid nodes in the draft tube shows a further reduction of the eddy viscosity in the elbow, so that only small areas with values of around 50 appear.

The second way to determine the resolution of turbulent structures is a visualisation using the Q-criterion being the second invariant of the velocity gradient tensor:

$$Q = 0.5 \left(\Omega^2 - S^2 \right) \tag{12}$$

where S and Ω are the symmetric and asymmetric components of the velocity gradient tensor [6].

As shown in Fig. 5 (right), the coarse mesh simulation (16M) mainly predicts large structures, like the vortex rope in the cone and its decay in the elbow. Due to the general mesh refinement, the 40M simulation shows some more resolved structures.

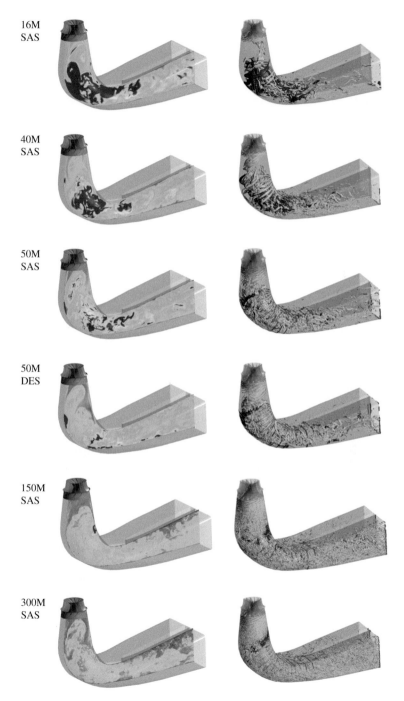

Fig. 5 Cutting plane in the draft tube coloured with the eddy viscosity ratio between 0 and 100 (*left*) and isosurface of the velocity invariant Q of 250 coloured with the same eddy viscosity ratio (*right*) (except for the 150M and 300M simulation with a ratio between 0 and 50)

The result of the 50M mesh with a improved mesh refinement of the boundary layer,
draft cone and elbow results in more resolved turbulent structures downstream of
the runner, namely the runner blade wakes and a more detailed prediction of the
turbulent structures in the elbow. A difference between the turbulence models SAS
and DES is not visible for this mesh. The further refinement of the draft tube mesh
generally leads to a resolution of more turbulent structures throughout the whole
draft tube. Close to the vortex rope and in the diffuser typical LES-structures are
visible.

4   Computational Resources

All flow simulations were performed on the Cray XE6 Hermit and Cray XC40
Hornet installed at the HLRS Stuttgart (and some tests on Cray XE6 Hera). The Cray
XE6 Hermit architecture contains 3552 compute nodes with AMD®Interlagos®
processors with 16 cores, 32 GB memory and Gemini interconnect. The Cray
XC40 Hornet has 3944 compute nodes with Intel®Xeon® processors with 12 cores,
128 GB memory and Aries interconnect. For the pre-processing steps interpolation
and partitioning, special nodes with large memory are necessary, as the required
memory for the 150M and 300M mesh is roughly in the range between 128 GB and
512 GB.

Between CFX-v14.5 and CFX-v16.0 quite a lot improvements have been made
regarding parallel performance and especially for the applications of Francis turbine
flow simulations. One difference between these versions is that the newer one can
be used in ESM-mode (enhanced scalability mode) using CRAY-MPI, instead of
standard MPI. As a result, the parallel performance is quite well for the 50M mesh
up to around 1000 cores on Cray XE6 in contrast to the performance of the older
version (Fig. 6 top, left). Another improvement can be seen through the multipass
decomposition method (Sect. 2.1) developed for applications with rotating parts and
GGI (general grid interface) (Fig. 6 top, right).

A comparison of the performance on different architectures, Cray XE6 and Cray
XC40, can be seen in Fig. 6 middle row. The speed up on Cray XE6 is better than
on Cray XC40. A reason for this could be the much smaller simulation time needed
by the solver itself due to the faster architecture. Assuming that if the amount of
communication is roughly the same, this part gets a higher weight per time step
respectively in the speed up curve. On the other hand, the simulation time for five
time steps is much lower on Cray XC40 than on Cray XE6 (see Fig. 6 middle, right).
For smaller core numbers of around 256 the simulation is more than a factor of three
faster and around 1000 cores it is more than a factor of 2.5 faster.

For the larger mesh with 150 million nodes (150M) the code shows good
performance up to 1728 cores and adequate performance up to 2304 cores (Fig. 6
bottom, left) on Cray XC40 Hornet. The largest mesh with 300 million nodes
(300M) performs quite well up to 2048 cores (Fig. 6 bottom, right) on Cray XE6
Hera (courtesy of Friedemann Unger, Ansys Germany).

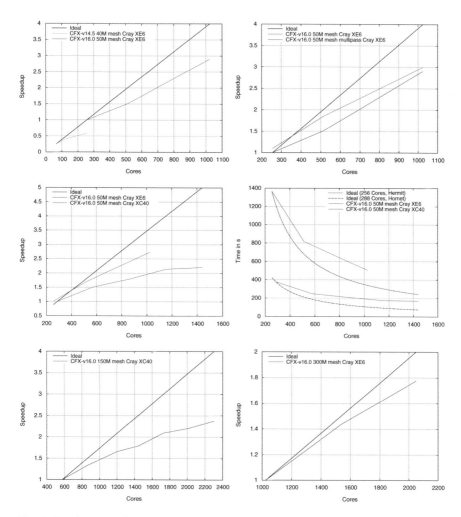

Fig. 6 Speed up tests for the transient flow simulations of the Francis turbine using the SAS turbulence model

About 80 runner revolutions are necessary as simulation time. The first 20 revolutions are needed to get rid of the starting oscillations and the second 60 revolutions are used for flow field evaluation like time-averaged velocity and Reynolds stress components. That number translates into a simulation time of 32 days for the 50M mesh on 864 cores, of 73 days for the 150M mesh on 1440 cores and of 85 days for the 300M mesh on 2016 cores. These pure computational times increase by a factor of around two to five due to queuing time caused by the loading of the cluster.

5   Conclusion

Simulations of the turbulent flow of a Francis turbine at part load operating
conditions were performed using the commercial CFD code Ansys CFX version
16.0. Two commonly used hybrid RANS-LES turbulence models were used for
different mesh sizes of in total 50, 150 and 300 million mesh nodes. It could be
shown that an increased mesh resolution predicts more turbulent structures in the
draft tube. The finer meshes show a LES-like behaviour, despite the high Reynolds
number. Due to the better boundary layer resolution, the runner blade wakes can be
determined in the draft tube cone.

An intensive performance study for different mesh sizes and platforms showed
the clearly improved parallel performance of the current version of Ansys CFX. The
code scales up to 2000 cores and more for the larger mesh.

Those types of transient flow simulations using hybrid RANS-LES turbulence models require large computational resources. Long computational times
are necessary to get time-averaged results for the mean velocity and especially
averaged turbulent quantities like the Reynolds stresses. In the current status of the
simulations, only the mean velocity for the 50M and 150M simulations could be
analysed. Further investigation will follow as well as for the 300M simulation.

Acknowledgements The research leading to the results presented in this paper is part of a
common research project of the Institute of Fluid Mechanics and Hydraulic Machinery, Voith
Hydro Holding GmbH & Co. KG and Ansys Germany GmbH.

References

1. Ansys CFX-Solver, Release 16.0: CFX-Solver Manager User's Guide (2015)
2. Comte-Bellot, G., Corrsin, S.: Simple Eulerian time correlation of full- and narrow-band
   velocity signals in grid-generated, isotropic turbulence. J. Fluid Mech. **48**(Part 2), 273–337
   (1971)
3. Egorov, Y., Menter, F.R.: Development and application of SST-SAS turbulence model in the
   DESIDER project. In: Advances in Hybrid RANS-LES Modelling. Notes on Numerical Fluid
   Mechanics and Multidisciplinary Design, vol. 97, pp. 261–270. Springer, Heidelberg (2008)
4. Egorov, Y., Menter, F.R., Cokljat, D.: The scale-adaptive simulation method for unsteady
   turbulent flow predictions. Part 2: application to aerodynamic flows. J. Flow Turbul. Combust.
   85(1), 139–165 (2010)
5. Jasak, H., Weller, H.G., Gosman, A.D.: High resolution NVD differencing scheme for
   arbitrarily unstructured meshes. Int. J. Numer. Methods Fluids **31**, 431–449 (1999)
6. Jeong, J., Hussain, F.: On the identification of a vortex. J. Fluid Mech. **285**, 69–94 (1995)
7. Krappel, T., Ruprecht, A., Riedelbauch, S.: Flow simulation of a Francis turbine using the SAS
   turbulence model. In: High Performance Computing in Science and Engineering '13. Springer,
   Berlin (2013)
8. Krappel, T., Ruprecht, A., Riedelbauch, S., Jester-Zuerker, R., Jung, A.: Investigation of
   Francis turbine part load instabilities using flow simulations with a Hybrid RANS-LES
   turbulence model, In: 27th IAHR Symposium on Hydraulic Machinery and Systems, Montreal
   (2014)

9. Krappel, T., Ruprecht, A., Riedelbauch, S.: Flow simulations of a Francis turbines using hybrid RANS-LES turbulence models. In: High Performance Computing in Science and Engineering '14. Springer, Berlin (2015)
10. Menter, F.R.: Two-equation eddy-viscosity turbulence models for engineering applications. AIAA J. **32**(8), 269–289 (1994)
11. Menter, F.R.: Best Practice: Scale-Resolving Simulations in ANSYS CFD Version 1.0 ANSYS Germany GmbH (April 2012)
12. Menter, F.R., Egorov, Y.: The scale-adaptive simulation method for unsteady turbulent flow predictions. Part 1: theory and model description. J. Flow Turbul. Combust. **85**(1), 113–138 (2010)
13. Menter, F.R., Kuntz, M., Langtry, R.: Ten years of industrial experience with the SST turbulence model. In: Turbulence, Heat and Mass Transfer, vol. 4, pp. 625–632. Begell House, New York (2003)
14. Menter, F.R., Schütze, J., Gritskevich, M.: Global vs. zonal approaches in hybrid RANS-LES turbulence modelling. In: Progress in Hybrid RANS-LES Modelling. Notes on Numerical Fluid Mechanics and Multidisciplinary Design, vol. 117, pp. 15–28. Springer, Heidelberg (2012)
15. Nicoud, F., Ducros, F.: Subgrid-scale stress modelling based on the square of the velocity gradient tensor. Flow Turbul. Combust. **62**, 183–200 (1999)
16. Piomelli, U., Chasnov, J.: Large-eddy simulations: theory and applications. In: Turbulence and Transition Modelling. Kluwer Academic, Dordrecht (1996)
17. Rotta, J.C.: Turbulente Strömungen. BG Teubner, Stuttgart (1972)
18. Shur, M.L., Spalart, P.R., Strelets, M.K., Travin, A.K.: A hybrid RANS-LES approach with delayed-DES and wall-modelled LES capabilities. Int. J. Heat Fluid Flow **29**, 1638–1649 (2008)
19. Spalart, P.R., Deck, S., Shur, M.L., Squires, K.D., Strelets, M.Kh., Travin, A.: A new version of detached-eddy simulation, resistant to ambiguous grid densities. Theor. Comput. Fluid Dyn. **20**, 181–195 (2006)
20. Travin, A., Shur, M., Strelets, M., Spalart, P.R.: Physical and numerical upgrades in the detached-eddy simulation of complex turbulent flows. In: Advances in LES of Complex Flows. Fluid Mechanics and Its Applications, vol. 65, pp 239–254. Kluwer Academic, Dordrecht (2002)

Numerical Investigation of a Full Load Operation Point for a Low Head Propeller Turbine

Bernd Junginger and Stefan Riedelbauch

Abstract The performance of low head turbines heavily depends on the draft tube. In the design process of hydraulic machines the geometry is often simplified, meaning gaps between runner and shroud and, if occurring, gaps at the trailing edge of the guide vanes are often neglected. The gap flow can however lead to a stabilization of the draft tube flow. In order to investigate this, a numerical analysis of a 4-blade runner with tip clearance is carried out. For the investigated operating point a full load vortex in the draft tube develops. The numerical results are evaluated against experimental measurements of integral quantities head, torque and discharge according to IEC 60 193 standard. Additionally to the investigations of the integral quantities, an evaluation of the vortex rope shape, turbulence quantities and velocity profiles in the draft tube are compared for the different numerical approaches.

1 Introduction

The European Water Framework Directive advises all member states of the European Union to achieve an ecologically good status of all flowing water [2]. Hence a revitalization of unused dams and weirs all over Europe are moving back into the focus of the energy providers. The investigated 4-blade propeller turbine could be adapted to fulfill the standards of the European Water Framework Directive.

The energy mix in Germany in the year 2014 consists of about 26.2 % of renewable energy. The quota of energy gained from hydro power is around 4 % of the total energy mix. This means that about 15 % of the renewable energy is gained from hydro power [1]. In contrary to photovoltaic and wind energy, hydro power has the advantage of providing a predictable energy output. Water power plants are able to compensate fluctuations of the power supply system generated from photovoltaic

B. Junginger (✉) • S. Riedelbauch
Institute of Fluid Mechanics and Hydraulic Machinery, Pfaffenwaldring 10, 70550 Stuttgart,
Germany
e-mail: junginger@ihs.uni-stuttgart.de

© Springer International Publishing Switzerland 2016   435
W.E. Nagel et al. (eds.), *High Performance Computing in Science
and Engineering '15*, DOI 10.1007/978-3-319-24633-8_28

and wind energy. As a result water turbines are operated more and more in off-design conditions to balance fluctuations of the electric grid.

The performance of low head turbines like Kaplan and propeller turbines, which are generally used for these types of applications, heavily depend on the draft tube flow. Simplifications of the geometry model, which are often made for the design process, can lead to a falsification of the flow field and, thus, to an unreliable performance prediction of the machine. Typical simplifications are the willful neglect of gaps and the circumferential averaging at mixing plane interfaces between stationary and rotating components. The gap flow between runner and shroud in particular can influence the flow field in the draft tube. For off design operation points like full load and part load vortex ropes typically develop in the draft tube.

In this paper an operation point with the characteristic values: $n'_1 = n'_{1opt}$ and $Q'_1 = 1.13\,Q'_{1opt}$ is analyzed. At the investigated operation point a full load vortex is generated in the draft tube. These vortices originate downstream the runner hub due to the existing velocity distributions. The full load vortex can arise in a symmetrical or an asymmetrical shape and in some cases the vortex is pulsating [9].

For the investigated operating point there is no gap at the trailing edge of the guide vanes. The normalized runner gap size analyzed in this paper $\tau = 2$ which is defined as:

$$\tau = \frac{s\,c_{cl}}{D_{ru}} \tag{1}$$

Eq. (1) contains the runner diameter of the turbine D_{ru}, the runner gap size s and a constant for the axial test rig at the Institute of Fluid Mechanics and Hydraulic Machinery c_{cl}. The measured runner gap size in the experiment is $\tau = 1$. Due to the cylindrical contour of the shroud at the runner, the gap size will increase to about the double size while operating the machine.

2   Numerical Setup

The flow simulations of the turbine are performed with the commercial CFD code Ansys CFX version 16.0. Due to the possible asymmetrical shape of the full load vortex the focus of this paper lies on the transient analysis of the draft tube flow field. The entire geometry without simplifications is modeled for all presented simulations [8]. The hydraulic contour of the investigated propeller turbine is illustrated in Fig. 1.

A k-ω-SST as well as a SAS-SST turbulence model are compared in this paper. The numerical results are compared with experimental measurements performed in the laboratory of the Institute of Fluid Mechanic and Hydraulic Machinery in Stuttgart. The model turbine is installed in the closed loop test rig. The measurements of head, torque and discharge in general fulfill the standard of the

Fig. 1 Hydraulic contour of the propeller turbine guide: vanes = *green*, runner = *red*, and draft tube = *blue*

Table 1 Number of elements and averaged y^+ for the turbine parts for the investigated grid

Turbine part	Elements	y^+
Guide vanes	5.2M	1.4
Runner	11.9M	4.8
Draft tube with expansion tank	14.3M	1.5
Total	31.3M	

IEC 60 193, which defines the standard of model acceptance tests for hydraulic machines [5].

A second order backward Euler scheme is used for the temporal discretisation. For spatial discretisation a bounded second order central differencing scheme is applied while using the SAS turbulence model [6]. A second order backward Euler scheme is applied for temporal discretisation of the turbulence quantities and a first order scheme is used for the spatial discretisation [11].

The mesh evaluated in this paper has medium grid density. During the project three different grid densities are analyzed. Strong scaling tests of a coarse grid, which consists of approximately 10 million elements (10M), is presented in Sect. 5. The focus of the analysis of this paper lies, however on a medium sized grid of about 31 million (30M) elements. The number of elements used for generating the 30M mesh are listed in Table 1. The runner mesh with the tip clearance and the draft tube mesh are finer than the guide vane grid. About 45 % of the elements are placed in the draft tube. The high grid density in the draft tube is used to get a better resolution of e.g. the full load vortex and if possible vortices induced by the tip clearance of the runner. For an upcoming more refined grid an additional refinement of the draft tube mesh is planned. The number of elements planed for the fine grid is approximately 45–50 million. A Courant number smaller than one is aspired to resolve the turbulent structures.

438

B. Junginger and S. Riedelbauch

3 Turbulence Modeling

For the numerical investigations two turbulence models are used. One of them is the k-ω-SST turbulence model, which represents the standard in turbo machinery. The other one is the SAS-SST turbulence model. The k-ω-SST model is a classical RANS model whereas the SAS-SST model is a hybrid RANS-LES turbulence model.

3.1 k-ω-SST

The k-ω-SST model represents a RANS model and is using the Boussinesq hypothesis to solve the turbulence quantities of the flow. The set of equations needs closure, meaning a sufficient number of equations for all unknowns is necessary. In the unknowns also include the Reynolds-Stress tensor resulting from the averaging procedure. The k-ω-SST model by Menter is a two-equation eddy viscosity turbulence model [10]. The Boussinesq eddy viscosity assumption is the basis for all two-equation turbulence models. It implies that the Reynolds stress tensor τ_{ij} is proportional to the mean strain rate tensor S_{ij}, which can be written:

$$- \rho \overline{u_i' u_j'} = \mu_t \left(\frac{\partial U_i}{\partial x_j} + \frac{\partial U_j}{\partial x_i} - \frac{2}{3} \frac{\partial U_k}{\partial x_k} \delta_{ij} \right) - \frac{2}{3} \rho k \delta_{ij} \tag{2}$$

The k-ω-SST model combines the advantages of the k-ω and the k-ϵ turbulence models. As a result, no additional damping function is necessary for the viscous sublayer due to the usage of the k-ω formulation. The SST-model is able to switch to a k-ϵ formulation in the inner flow region. With the use of a blending function the common k-ω problem of being too sensitive to the inlet free-stream turbulence properties can be avoided.

3.2 SAS-SST

Smaller scales of turbulent eddies in the flow field can be resolved by applying the SAS-SST turbulence model. The SAS-SST approach enables unsteady RANS to operate in Scale Resolving Simulation (SRS) mode [13]. An additional source term Q_{SAS} is introduced in the transport equation for the turbulence eddy frequency ω of the RANS SST model described in Eq. (3) [3, 4, 12].

$$\frac{\partial \rho \omega}{\partial t} + \frac{\partial}{\partial x_j} \left(\rho U_j \omega \right) = \alpha \frac{\omega}{k} P_k - \rho \beta \omega^2 + Q_{SAS} +$$

$$+ \frac{\partial}{\partial x_j} \left[\left(\mu + \frac{\mu_t}{\sigma_\omega} \right) \frac{\omega}{x_j} \right] + (1 - F_1) \frac{2\rho}{\sigma_{\omega 2}} \frac{1}{\omega} \frac{\partial k}{\partial x_j} \frac{\partial \omega}{\partial x_j} \tag{3}$$

The additional source term Q_{SAS} is defined as

$$Q_{SAS} = \max \left[\rho \zeta_2 \kappa S^2 \left(\frac{L}{L_{vK}} \right)^2 - C \frac{2\rho\kappa}{\sigma_\Phi} \max \left(\frac{1}{\omega^2} \frac{\partial\omega}{\partial x_j} \frac{\partial\omega}{\partial x_j}, \frac{1}{k^2} \frac{\partial k}{\partial x_j} \frac{\partial k}{\partial x_j} \right), 0 \right] \quad (4)$$

containing the Karman length scale L_{vK} and the turbulent length scale L. The Karman length scale is defined as

$$L_{vK} = \kappa \left| \frac{\overline{U}'}{\overline{U}''} \right|$$

$$\overline{U}'' = \sqrt{\frac{\partial^2 \overline{U}_i}{\partial x_k^2} \frac{\partial^2 \overline{U}_i}{\partial x_j^2}}, \quad \overline{U}' = S = \sqrt{2S_{ij}S_{ij}}, \quad S_{ij} = \frac{1}{2} \left(\frac{\partial \overline{U}_i}{\partial x_j} + \frac{\partial \overline{U}_j}{\partial x_i} \right) \quad (5)$$

and is the essential quantity for the SAS-SST model to switch in SRS mode [14]. Based on the information provided by the Karman length scale a dynamically adjustment of the resolved structures for the SAS model is possible, resulting in a LES like behavior. The reduced eddy viscosity enables the generation of smaller structures, leading in a turbulence cascade down, to grid limit. A limiter similar to the Smagorinsky LES model is introduced to the turbulence eddy viscosity due to the insufficient damping of the SAS-SST model for small turbulent scales at grid limit [3].

The ability of the SAS-SST model to operate in RANS mode if the grid density and the time step are not fine enough is one advantage in comparison to DES-type models [15]. For these models the mesh resolution and the time step are a critical criteria.

4 Results

Results of different numerical approaches are compared to integral quantities measured in the laboratory of the Institute of Fluid Mechanics and Hydraulic Machinery. The shape of the full load vortex and the velocity profiles in the cone and diffuser of the draft tube are analyzed for the different numerical approaches. Additional turbulence quantities in the draft tube are evaluated.

Upon achieving a periodic flow behavior the time averaging of selected variables is started. In this paper a time-averaging over 50 runner revolutions is performed for all carried out simulations.

	Head $\frac{H_{cfd}-H_{exp}}{H_{exp}}$ [%]	Torque $\frac{T_{cfd}-T_{exp}}{T_{exp}}$ [%]
SST	2.6	2.6
SAS	4.2	2.6

Table 2 Normalized integral quantities, head and torque

4.1 Integral Quantities

The integral quantities head and torque are compared with experimental results, while the discharge is a steady state inlet boundary condition. All calculated quantities are normalized with experimental results. The deviations of the computational to the experimental results are shown in Table 2. Differences in the results can be seen in head as well as in torque. All applied turbulence models overestimate the results of the measurement. The head losses in the guide vane domain are about 0.9 % of the total head of the turbine for both investigated turbulence models. In the runner small deviations between the k-ω-SST and the SAS-SST model can be observed. The losses calculated by the k-ω-SST model are about 0.6 % higher than those of the SAS-SST model. A significant deviation in the losses occurs in the draft tube domain. The calculated head losses of the SAS-SST model are about 2.7 % higher than by simulating with a k-ω-SST model. The torque generated at the runner blades is at the same level for both investigated turbulence models.

Owing to the specifications of the IEC 60 193 the total measurement error occurring in the experiment for the measured quantities head, torque and discharge is on a very small level. As a result of the overprediction of head and torque the estimated gap width for the computation is either to small or the losses produced by both investigated turbulence models, in particular the SAS-SST model, are too high. Further investigations are necessary to verify the numerical approaches. Additional experimental measurements are desirable to get more validation parameters.

4.2 Flow Analysis

The velocity profiles of time-averaged meridional, circumferential and radial velocity components are analyzed at two positions of the draft tube. The first velocity profile is evaluated in the draft tube cone close to the start of the draft tube, and the second one close to the draft tube exit (see Fig. 3). The velocity profiles at the evaluation lines in the draft tube cone and diffuser are illustrated in Fig. 2. In particular in the cone (see Fig. 3a), where the full load vortex develops, differences in the velocity profiles between the k-ω-SST and the SAS-SST model can be observed. A significantly larger stagnation region develops in the middle of the flow field, behind the runner hub, for the SAS-SST model compared to the k-ω-SST model. The maximum meridional velocity is similar for both turbulence models. Differences in the circumferential component of the velocity can be seen in the middle of the flow field. The circumferential component is decreasing to a minimum

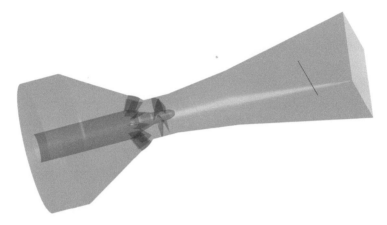

Fig. 2 Evaluation of velocity profiles in the cone (*yellow line*) and diffuser of the draft tube (*red line*)

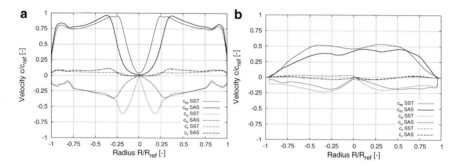

Fig. 3 Time-averaged velocity components in the draft tube. (**a**) Cone (**b**) Diffuser

level at the same radial position as seen in the meridional velocity component. The maximum circumferential velocity is higher for the k-ω-SST model at the edge of the stagnation region. Expect for the stagnation region only minor deviations of the circumferential velocity component can be observed for the two turbulence models. The same characteristic can be noticed for the radial component. The stagnation region indicate the size of the full load vortex, which arises from the hub of the runner.

Flow separations develop in the middle of the draft tube resulting from the rectangular shape at the end of the diffuser and the transition between cone and diffuser. In the time-averaged velocity profiles at the evaluation spot in the diffuser differences between the turbulence models can be observed. However, a clear characteristic can not be detected.

It can be confirmed that averaging over 50 runner revolutions in the draft tube cone is sufficient. In the draft tube diffuser a longer averaging procedure seems to be demandable. Laser-Doppler-Velocimetry measurements are planed in the future to validate the computational results.

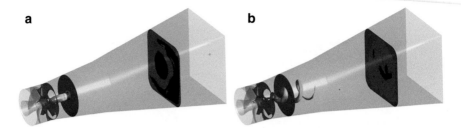

Fig. 4 Isosurface of pressure (5000 Pa), colored by the viscosity ratio 0–2000, planes in the evaluation spots for the velocity profile. (**a**) SST and (**b**) SAS

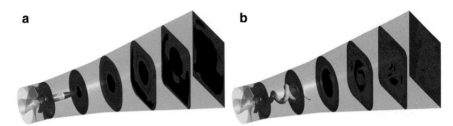

Fig. 5 Isosurface of pressure (5000 Pa), colored by the viscosity ratio 0–2000 at various planes in the draft tube. (**a**) SST and (**b**) SAS

4.3 Vortex Rope

The full vortex rope develops from a low pressure zone downstream the runner hub and is visualized with a pressure isosurface in Fig. 4. The shape of the vortex rope differs strongly between the two numerical approaches. While the vortex rope has a straight shape using a k-ω-SST turbulence model, a quite strong deviation of the shape can be noticed applying a SAS-SST model for turbulence modeling. The vortex rope is developing a shape resembling a corkscrew, when using the SAS-SST model. The effects of the different shape of the vortex rope are illustrated at several evaluation planes in the draft tube (see Fig. 5). The viscosity ratio is plotted on selected planes in the draft tube. The region of the low pressure zone in the center of the draft tube is clearly smaller for straight vortex rope compared to the corkscrew shaped vortex rope. As a result of the asymmetric shape of the vortex rope, when applying the SAS-SST model, more flow separations in the draft tube diffuser are developing than by applying a k-ω-SST turbulence model. This transient phenomena can also be observed in the velocity profiles in the diffuser, plotted in Fig. 3b, even though the velocity components are time-averaged over 50 runner revolutions. Optical evaluations of the vortex rope at the model turbine indicate that the shape of the vortex is similar to the shape computed with the SAS-SST turbulence model. Further experimental measurements using a high speed camera are planed in order to verify the numerical results.

a   b

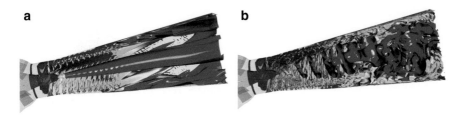

Fig. 6 Isosurface of velocity invariant Q = 1 colored by the viscosity ratio 0–500. (**a**) SST and (**b**) SAS

4.4 Turbulence Evaluation

The flow structures resolved by the k-ω-SST simulation are larger than those resolved by the SAS-SST model. The SAS-SST model is able to resolve smaller flow structures, as can be seen in Fig. 6. There are several options to resolve the flow structures. In this paper two options are illustrated. One is the velocity invariant Q (Q = $0.5(\Omega^2$-$S^2)$) and the other one is the viscosity ratio which can be used as an indicator for the LES-content of the turbulence model (see also Fig. 5) [7]. Higher viscosity ratio numbers produce a higher damping of the flow structures. The largest flow structures for both turbulence models occur downstream the hub in the middle of the draft tube, where the full load vortex is positioned. Both turbulence models are capable of resolving small turbulent structures in the draft tube cone with the exception of the inner flow region downstream the runner hub. The small turbulent structures induced by the vortex rope can not be resolved by the k-ω-SST model and only with some limitations by the SAS-SST model. In the draft tube diffuser the flow separations and the variety of small vortices, especially in the corners of the draft tube, lead to a decay of larger turbulent structures. The decay is indicated by higher values of the turbulence eddy viscosity. A mesh refinement in the draft tube, which is planed with the 45M grid, is necessary to achieve a better resolution of the turbulent structures.

5 Speed-Up

For the flow simulations the high performance cluster Cray XE6 (Hermit) and Cray XC40 (Hornet) are used. Both clusters are installed at the HLRS Stuttgart. The transfer from Hermit to Hornet was done in the end of the year 2014. Hence, two speed up tests are performed, one on the Hermit system and another on the Hornet system. On Hermit the analysis is carried out for a mesh with approximately 10 million elements using the version 14.5 of the flow solver Ansys CFX. In the 10M mesh the tip clearance at the runner is neglected. Further results of the coarse 10M mesh are not discussed in this paper. A speed-up of the 10M mesh is performed

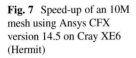
Fig. 7 Speed-up of an 10M
mesh using Ansys CFX
version 14.5 on Cray XE6
(Hermit)

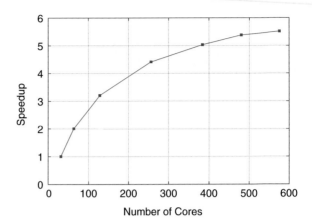

on Hornet. Therefore, Ansys CFX version 15.0 and version 16.0 are used. Finally, a speed-up of the discussed computational grid with approximately 30 million elements is carried out. A minimum of 40 runner revolutions are necessary to get fully time-averaged velocity profiles in the draft tube cone. For the draft tube diffuser the required minimum number of runner revolutions for fully time-averaged velocity profiles increases up to at least 50.

5.1 Hermit

The results of the strong scaling analysis performed on the Cray Hermit are illustrated in Fig. 7. The parallel performance of the code, while using more than 256 cores, is decreasing away from the ideal parallel performance. Therefore the computations on Hermit were performed on 256 cores.

5.2 Hornet

After the installation of the Cray XC40 at the HLRS a strong scaling test with the same grid density was performed on the Hornet system. A comparison between the installed CFX versions 15.0 and 16.0 was carried out, illustrated in Fig. 8a. Moreover, a speed up test was performed with the medium grid density of 30M using CFX version 16.0 (see Fig. 8b). The parallel performance with version 16.0 is clearly improved compared to version 15.0. The number of cores that can be used before a significant decrease of the parallel performance occurs has more than doubled from 192 cores to 576 cores. A rather good parallel performance is displayed by performing a strong scaling analysis for the medium grid size. The decline of the speedup curves appears by around 768 cores.

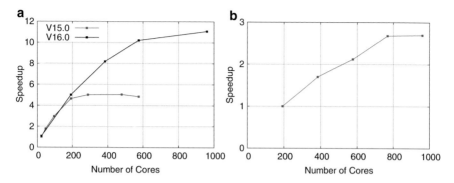

Fig. 8 Speed-up on Cray XC40 (Hornet). (**a**) Speed-up of an 10M mesh using Ansys CFX version 15.0 and V16.0 on Cray XC40. (**b**) Speed-up of an 30M mesh using Ansys CFX version 16.0 on Cray XC40

The computation time is reduced about the factor of 2 comparing the performance between the Cray XC40 and the Cray XE6. With the new version of the flow solver Ansys CFX a further increase in the performance can be noted. The increase in performance when using 576 cores is about factor 2 between version 15.0 and version 16.0. The parallel performance of version 16.0 is in particular improved when more higher core numbers are used. The plateau of version 15.0 in the speed up is reached by 384 cores for the investigated 10M grid. The parallel performance of version 16.0 is still excellent at this point. A decrease in speed up can be observed if the simulation is running on 960 cores.

For the computation of the propeller turbine 576 cores are used. This enables the computation of approximately 2 runner revolutions in 24 h. The minimum simulation time needed to achieve acceptable time-averaged results for the entire machine is at least 30 computation days.

6   Conclusion and Outlook

Simulations of a full load operation point are performed with a medium grid size of approximately 30 million elements applying two different turbulence models. The effect of the different numerical approaches on the flow field, especially in the draft tube, can be observed in the shape of the vortex rope which is differing from straight vortex for the k-ω-SST model and a vortex in the shape of a corkscrew for the SAS-SST model. The shapes of the vortex lead to different sizes of the stagnation region, which can be seen in the time-averaged velocity profiles evaluated in the draft tube cone. Influences of the vortex rope can be observed on the meridional, the circumferential and the radial velocity component. At the second evaluation spot in the draft tube diffuser the effects of the vortex rope is smoothed out. Due to flow separations at transition from cone to diffuser a time averaging of more than 50 runner revolutions seems necessary. The numerical results overpredict the experimental results for the evaluated quantities head and torque independent of the applied turbulence model. Differences between the numerical models occur only in the head but not in the torque. The SAS-SST model generates significantly larger head losses in the draft tube than the k-ω-SST model. The losses of the other turbine components are compared to the differences of the head losses in the draft tube on a similar level. Thy hybrid SAS-SST turbulence model is capable of resolving smaller turbulent structures than the k-ω-SST model. Both turbulence models are limited in the resolution of the turbulent flow structures downstream the runner hub in the middle of the draft tube cone and the draft tube diffuser. An increase of the mesh density is an option to enable smaller turbulent structures to be resolved.

A strong scaling analysis is performed on the Cray XE6 and the Cray XC40. With the installation of the Cray XC40 (Hornet) the performance for the carried out simulation has increased. The simulation time needed for the computation of a 10M grid on the Cray XC40 (Hornet) has been significantly reduced with factor 2.5 compared to the Cray XE6 (Hermit). An additional increase is provided by version 16.0 of Ansys CFX which improves the performance especially for higher core numbers used for the simulation.

Due to the lack of experimental results, additional investigations on the model turbine are projected to validate the numerical data. Measurements using a Laser-Doppler-Velocimetry are planed in the draft tube to verify the velocity profiles. Moreover, videos recorded with a high speed camera are planed to visualize the shape of the full load vortex. A simulation of the full load operation point with a refined grid is planed to see the effects of the refinement on the evaluated quantities. Additional simulations with varying gap sizes are planed to investigate the effects of the runner gap on the flow field in the draft tube.

References

1. Bundesverband der Energie- und Wasserwirtschaft e.V.: BDEW aktualisiert Angaben zum Erzeugungsmix 2014: Erneuerbare Energien erzeugen immer mehr Strom. https://www.bdew.de/internet.nsf/id/20150306-pi-erneuerbare-energien-erzeugen-mehr-strom-de 21.April 2015 (2014)
2. Commission of the European Communities: Towards sustainable water management in the European Union—first stage in the implementation of the Water Framework Directive. In: 2000/60/ECCOM 2007, 128, pp. 1–13 (2000)
3. Egorov, Y., Menter, F.R.: Development and application of SST-SAS turbulence model the DESIDER project. In: Advances in Hybrid RANS-LES Modelling. Notes on Numerical Fluid Mechanics and Multidisciplinary Design, vol. 97, pp. 261–270 (2008)
4. Egorov, Y., Menter, F.R., Cokljat, D.: The scale-adaptive simulation method for unsteady turbulent flow predictions. Part 2: application to aerodynamic flows. J. Flow Turbul. Combust. 85(1), 139–165 (2010)
5. International Electrical Commission. (ed.): International standard IEC 60193 Second Edition 1999–11. Hydraulic Turbines, Storage Pumps and Pump-Turbines—Model Acceptance Tests, Geneva (1999)

6. Jasak, H., Weller, H.G., Gosman, A.D.: High resolution NVD differencing scheme for arbitrarily unstructured meshes. Int. J. Numer. Methods Fluids **31**, 431–449 (1999)
7. Jeong, J., Hussain, F.: On the identification of a vortex. J. Fluid Mech. **285**, 69–94 (1995)
8. Jošt, D., Škerlavaj, A., Lipej, A.: Numerical flow simulation and efficiency prediction for axial turbines by advanced turbulence models. In: 26th IAHR Symposium on Hydraulic Machinery and Systems, Beijing (2012)
9. Kirschner. O.: Experimentelle Untersuchung des Wirbelzopfes im geraden Saugrohr einer Modellpumpturbine. Dissertation, IHS-Mitteilung 32, University of Stuttgart (2011)
10. Menter, F.R.: Two-equation eddy-viscosity turbulence models for engineering applications. AIAA-J. **32**(8), 269–289 (1994)
11. Menter, F.R.: Best Practise: Scale-Resolving Simulations in ANSYS CFD Version 1.0. ANSYS Germany, Darmstadt (2012)
12. Menter, F.R., Egorov, Y.: The scale-adaptive simulation method for unsteady turbulent flow predictions. Part 1: theory and model description. J. Flow Turbul. Combust. **85**(1), 113–138 (2010)
13. Menter, F.R., Schütze, J., Gritskevich, M.: Global vs. zonal approaches in hybrid RANS-LES turbulence modelling. In: Progress in Hybrid RANS-LES Modelling. Notes on Numerical Fluid Mechanics and Multidisciplinary Design, vol. 117, pp. 15–28, http://link.springer.com/chapter/10.1007%2F978-3-642-31818-4_2 (2012)
14. Rotta, J.C.: Turbulente Strömungen. BG Teubner, Stuttgart (1972)
15. Shur, M.L., Spalart, P.R., Strelets, M.K., Travin, A.K. A hybrid RANS-LES approach with delayed-DES and wall-modelled LES capabilities. Int. J. Heat Fluid Flow **29**, 1638–1649 (2008)

Detached Eddy Simulation of Flow and Heat Transfer in Swirl Tubes

Christoph Biegger and Bernhard Weigand

Abstract Detached Eddy Simulations (DES) are carried out to predict the flow structure and the heat transfer of swirling flows in tubes. Such swirl tubes are a promising method for heat transfer enhancement compared to the flow in a smooth tube. However, the physics in such a swirl tube is far from being fully understood. The numerical results in terms of velocity and Nusselt number are compared to own experimental data. DES and experiments agree well for the mean velocity profile and the numerics give insight in the turbulent vortex structure in a swirl tube. The axial velocity is characterized by a backflow in the tube center due to the induced strong swirl. The occurrence of this backflow region depends on the swirl strength and its influence on the heat transfer is still under research. Therefore, we simulate swirl tubes with different swirl numbers for the same mass flow and consequently same Reynolds number. Furthermore, we show a speed-up comparison for the parallelization with OpenFOAM and provide details about the computational performance of a DES on the new Cray XC40.

1 Introduction

Over the last decades the turbine entry temperature in gas turbines increased well above the melting temperature of the blade material. Due to these high thermal loads, more and more complex cooling techniques are necessary for the internal turbine blade cooling for a reliable long term operation. A promising and efficient method for leading edge turbine blade cooling is a swirl tube displayed in Fig. 1. A swirl tube consists of one or more tangential inlets, which induce a highly 3D swirling flow. The increased turbulence and high velocities near the wall enhance the heat transfer. However, the complex flow and the heat transfer in such a system are far from being fully understood. In literature a number of experimental investigations [1–4] can be found dealing with swirling flows in tubes, but numerical investigations are very rare.

C. Biegger (✉) • B. Weigand
Institute of Aerospace Thermodynamics (ITLR), University of Stuttgart, Pfaffenwaldring 31, 50569 Stuttgart, Germany
e-mail: christoph.biegger@itlr.uni-stuttgart.de

© Springer International Publishing Switzerland 2016
W.E. Nagel et al. (eds.), *High Performance Computing in Science and Engineering '15*, DOI 10.1007/978-3-319-24633-8_29

Fig. 1 Turbine blade sketch
with leading edge swirl tube
[5]

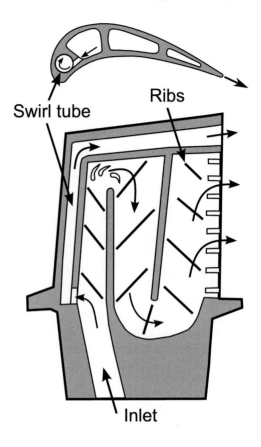

Ribs

Swirl tube

Inlet

In this paper, we investigated numerically the flow field and the heat transfer in swirl tubes via Detached Eddy Simulation (DES) using the open source code OpenFOAM [6]. The DES described by Spalart et al. [7] solves the near wall region by using the Reynolds-Averaged Navier-Stokes (RANS) equations, whereas the free stream region is simulated via a Large Eddy Simulation (LES) [8]. The main advantages of coupling LES and RANS is the drastic reduction of computational costs compared to full LES and that a DES provides more reasonable results than RANS especially in complex 3D problems. Here we used the Spalart-Allmaras turbulence model [9] for closing the Navier-Stokes equations. We solve the energy equation by using the turbulent Prandtl number [10] concept. In addition, we validated the numerical method simulating a turbulent channel flow with constant but different wall temperatures and compared it with DNS data [11].

The numerical swirl tube results are compared with own experimental data obtained at the Institute of Aerospace Thermodynamics at the University of Stuttgart (ITLR). The velocity field was studied via Particle Image Velocimetry (PIV) [12] and the cooling capability was investigated by measuring the surface temperature and the heat transfer coefficient applying the established transient technique using Thermochromic Liquid Crystals (TLC) [13].

2   Numerical Setup

The open source code OpenFOAM is used to simulate swirl tubes via DES, which solves the near wall region with RANS equations and the free stream region via LES [8]. The Spalart-Allmaras turbulence model [9] is used for closing the Navier-Stokes equation. The numerical setup is validated simulating a turbulent channel flow with constant but different wall temperatures and compared to DNS data from Iida and Kasagi [11]. The velocities and temperature profiles as well as the fluctuations show a good agreement [5].

The time averaged (RANS) and filtered (LES) equations show a structural similarity and read for the time averaged velocity $\langle u_i \rangle$ or the filtered velocity $\bar{u}$:

$$\frac{\partial \langle u_i \rangle}{\partial t} + \nabla(\langle u_i \rangle \langle u_j \rangle) = -\nabla \langle p \rangle + \nabla(v \nabla \langle u_i \rangle) - \nabla \tau_{ij}^{RANS}. \tag{1}$$

$$\frac{\partial \bar{u}_i}{\partial t} + \nabla(\bar{u}_i \bar{u}_j) = -\nabla \bar{p} + \nabla(v \nabla \bar{u}_i) - \nabla \tau_{ij}^{LES}. \tag{2}$$

Here v is the kinematic viscosity and p the density-divided pressure. The turbulent stresses on the right-hand side are modeled using the eddy viscosity concept with a turbulent viscosity v_t:

$$\tau_{ij} = v_t \nabla S_{ij}. \tag{3}$$

Here S_{ij} is the strain rate tensor. The turbulent viscosity $v_t = \tilde{v} f_{v1}$ is then simulated by a single transport equation proposed by Spalart and Allmaras [9]:

$$\frac{D\tilde{v}}{Dt} = c_{b1}\tilde{S}\tilde{v} + \frac{1}{\sigma}\left[\nabla((v + \tilde{v})\nabla\tilde{v}) + c_{b2}(\nabla\tilde{v})^2\right] - c_{w1}f_w\left(\frac{\tilde{v}}{\tilde{d}}\right)^2. \tag{4}$$

Here the variables σ, c_{b1}, c_{b2} and c_{w1} are model constants. The last term represents a destruction term for the modified viscosity $\tilde{v}$ depending on the DES limiter $\tilde{d} = \min(d, C_{DES}\Delta)$. Here d is the distance to the nearest wall, C_{DES} a constant and Δ the grid spacing. The limiter switches between RANS near the wall and LES in the free stream region, so the Spalart-Allmaras model act as a turbulence-viscosity model where $d \ll \Delta$ and as a subgrid-scale model where

$d \gg \Delta$. The used grid spacing is set to the maximum cell size of all directions $\Delta = \max(\Delta x, \Delta y, \Delta z)$.

The energy equation for the time averaged temperature $\langle T \rangle$ or the filtered temperature $\overline{T}$ reads

$$\frac{\partial \langle T \rangle}{\partial t} + \langle u_j \rangle \nabla \langle T \rangle = \nabla(\Gamma_{\mathit{eff}} \, \nabla \langle T \rangle). \tag{5}$$

$$\frac{\partial \overline{T}}{\partial t} + \overline{u}_j \nabla \overline{T} = \nabla(\Gamma_{\mathit{eff}} \, \nabla \overline{T}) \, . \tag{6}$$

Here Γ_{eff} is the effective thermal diffusivity and implies the molecular diffusivity $\Gamma = \nu/Pr$ and the turbulent diffusivity $\Gamma_t = \nu_t/Pr_t$ using the turbulent Prandtl number concept with $Pr_t = 0.85$ [10].

The computational domain shown in Fig. 2 refers to the experimental setup at the ITLR and represents a generic model of a turbine blade leading edge swirl tube. The swirl tube (2) has an inner diameter of $D = 50\,\mathrm{mm}$, an axial length of $L = 20D$ and consists of an inlet section (1) with two $180°$ displaced tangential inlet ducts. According to the experiment an additional outlet tube (3) and plenum (4) are simulated to minimize the influence of the outlet boundary condition and to assure a stable simulation.

The investigated Reynolds numbers based on the mean axial velocity and the tube diameter are $Re = \overline{V}_z D/\nu = 10{,}000$ and $20{,}000$ with the kinematic viscosity ν. The representative Reynolds number for such flows in real gas turbine cooling channels is around $20{,}000$. So the here investigated Reynolds numbers are of physical and engineering interest. The geometrical swirl number used to characterize the swirling flow is based on the inlet duct area and the tube cross-section: $S_{geo} = (D/2 - h/2)A_{tube}/(D/2\,A_{inlet}) = 5.3$.

We performed DES calculations with the finite-volume code OpenFOAM [6]. The Pressure-Implicit Split-Operator (PISO) algorithm is used as pressure corrector for the momentum equations. For the time discretization a second-order backward differencing scheme is applied. The viscous and convective fluxes are approximated with a second-order accurate central differences scheme. The computational meshes

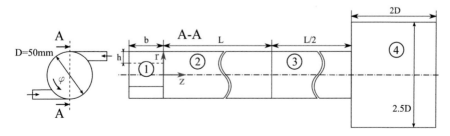

Fig. 2 Computational domain with two inlets (1), swirl tube (2), outlet tube (3) and outlet plenum (4)

for the two Reynolds numbers are hexahedral O-grids with a total mesh size of 9 and 15 million cells, respectively. A cross-section of the swirl tube mesh for $Re = 10,000$ is shown in Fig. 3. The wall is resolved providing a dimensionless wall distance of $y_1^+ < 1.5$ for the first node near the wall. As a reference scale we used the Kolmogorov length scale, which can be approximated by $\eta = D\,Re_D^{-3/4}$ with the diameter D as the characteristic length [14]. The time step is adjusted with a maximum Courant number of 0.9 mostly in the inlet part. The simulation is run for $3\,L/\overline{V}_z$, before starting averaging for $15\,L/\overline{V}_z$. A summary of the two meshes regarding the number of cells, the Kolmogorov length scale and the first wall and center cell size is listed in Table 1.

No-slip boundary conditions were applied for all walls. A parabolic velocity and uniform temperature distribution ($T_{in} = 333\,K$) was prescribed at the inflow boundary. A turbulent flow is created by the interaction of the flow with the curved tube walls. The wall temperature is kept constant at $T_{wall} = 293\,K$. At the outlet, a fixed pressure value is set and zero gradient boundary conditions are applied for all other variables.

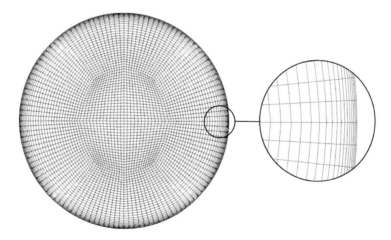

Fig. 3 Cross-section of the hexahedral tube mesh and a detailed view of the wall resolution [5]

Table 1 Summary of number of cells, Kolmogorov length scale η and used wall and center cell sizes for the computational meshes

Mesh	Cells	Length scale η [m]	Δy_{wall} [m]	$(\Delta x, \Delta y, \Delta z)_{center}$ [m]
$Re = 10,000$	$9 \cdot 10^6$	$5 \cdot 10^{-5}$	$3 \cdot 10^{-5}$	$9.4 \cdot 10^{-4}, 11.8 \cdot 10^{-4}, 11.0 \cdot 10^{-4}$
$Re = 20,000$	$15 \cdot 10^6$	$3 \cdot 10^{-5}$	$1.5 \cdot 10^{-5}$	$8.3 \cdot 10^{-4}, 10.6 \cdot 10^{-4}, \ 9.6 \cdot 10^{-4}$

3   Results

In this section we discuss the flow field, the velocity fluctuations, the helical vortex structure and the wall shear stress in the swirl tube for the swirl number $S = 5.3$ and the Reynolds number $Re = 10,000$. Furthermore, we present the heat transfer and the pressure loss and analyze the thermal performance to a smooth tube flow.

3.1   Flow Field

The non-dimensional circumferential and axial velocities in the swirl tube are shown in Fig. 4 compared to experimental data at seven axial positions z/D. Additionally, contour plots of the DES visualize the flow field. Since a large tangential velocity is imposed into the tube a large circumferential velocity V_ϕ can be observed, which is

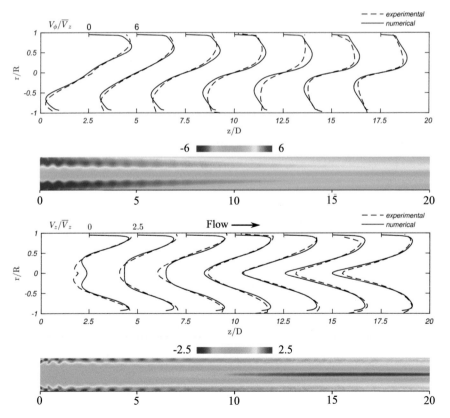

Fig. 4 Mean circumferential (*top*) and axial velocity (*bottom*) in the swirl tube for the swirl number $S = 5.3$ and $Re = 10,000$

more than two times larger than the axial velocity V_z. Close to the inlet at $z/D = 2.5$ the profile is characterized by a solid body vortex ($V_\phi = \Omega r$) with the maximum velocity in the wall region. Further downstream the maximum shifts to the tube center and the profile can be described by a potential vortex ($V_\phi = \Gamma/r$). Here r is the radius, Ω the angular velocity and Γ the circulation. The axial velocity shows its maximum velocity in the near wall region and a backflow in the tube center. The swirling flow is strong enough that this backflow occurs over the whole tube length. Both velocities show a good agreement to the experimental data.

The axial, radial and tangential velocity fluctuations in the tube midplane are shown in Fig. 5 at the same positions as the velocities. All three fluctuations are in the same order of magnitude. In the tube center they are increasing towards the end of the tube especially after $z/D = 10$. This correlates with the strong backflow on the tube axis. In the inlet region at $z/D = 2.5$ two symmetric maxima appear on both sides in the border region especially for the radial component. This elucidates a strong radial mixing and fluid exchange, which might be responsible for the high heat transfer close to the inlet.

The dominant vortex structure in the swirl tube is evaluated using the Q-criterion ($Q = 1/2\,(D_{ij}^2 - S_{ij}^2)$) and shown in Fig. 6. S_{ij} and D_{ij} are the symmetric and antisymmetric component of the velocity gradient tensor. Occurring vortices are

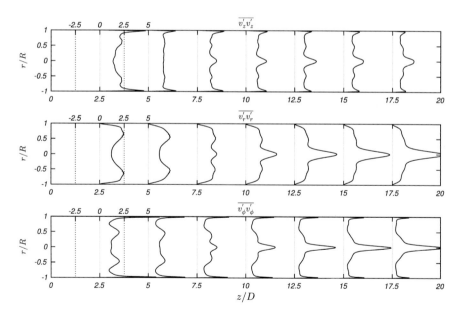

Fig. 5 Velocity fluctuations in the swirl tube for the swirl number $S = 5.3$ and $Re = 10,000$

Fig. 6 Helical vortex in the swirl tube for the swirl number $S = 5.3$ and $Re = 10,000$, iso-surfaces of $Q = 1$, *color* = axial velocity [5]

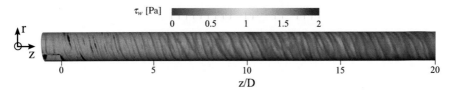

Fig. 7 Instantaneous wall shear stress for the swirl number $S = 5.3$ and $Re = 10,000$

shown with iso-surfaces of $Q > 0$, where the vorticity dominates the shear rate. The flow structure is characterized by a double helical vortex which enhances the turbulent momentum transfer in the wall region and causes a high heat transfer in the swirl tube compared to an axial tube flow.

3.2 Wall Shear Stress

The wall shear stress in fluids along a boundary is defined as $\tau_w = \mu \, (\partial u / \partial n)|_w$ and plotted in Fig. 7 for the here presented simulation. Here μ is the dynamic viscosity, u the velocity parallel to the wall and n the normal distance to the wall. The instantaneous snapshot clearly shows the swirling stripes of higher and lower velocities near the wall. Obviously the highest shear stresses occur close to the inlet and decrease slowly towards the tube end.

3.3 Heat Transfer

The circumferential averaged local Nusselt number

$$Nu = \frac{-\frac{\partial T}{\partial n}|_{wall} \, D}{T_{wall} - T_{fluid}},\tag{7}$$

is illustrated in Fig. 8 normalized with Nu_0 for a smooth tube flow. Here the Dittus-Boelter correlation is used given by $Nu_0 = 0.023 \, Re^{0.8} \, Pr^n$ with $n = 0.3$ if $T_{wall} < T_{fluid}$ [15]. Due to the large velocity near the wall high heat transfer coefficients are obtained up to eight times larger than the one for a smooth tube.

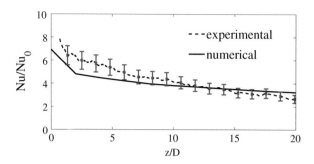

Fig. 8 Normalized local Nusselt number for the swirl number $S = 5.3$ and $Re = 10,000$ [5]

Table 2 Summary of pressure loss Δp, friction factor f, normalized friction factor f/f_0 and averaged heat transfer results in terms of Nusselt number $\overline{Nu}$, normalized Nusselt number $\overline{Nu}/Nu_0$ and thermal performance parameter $(\overline{Nu}/Nu_0)/(f/f_0)^{(1/3)}$

Simulation	Δp [Pa]	f	f/f_0	$\overline{Nu}$	$\overline{Nu}/Nu_0$	$(\overline{Nu}/Nu_0)/(f/f_0)^{(1/3)}$
$S = 5.3$, $Re = 10,000$	118.2	0.989	31.26	124.7	3.79	1.20

With decreasing swirl towards the end of the tube the Nusselt numbers decrease as well but are still larger than for an axial tube flow. The globally averaged Nusselt number is around four times larger than the one in a smooth tube. The circumferential velocity in Fig. 4 with its large value in the wall region and the enhanced turbulence are the major mechanisms for the high heat transfer in the swirl tube. Comparing experiment and DES near the inlet, the simulation slightly underestimates the heat transfer coefficients, but further downstream both show similar values in the uncertainty range of the experiment ($\pm 13\%$).

3.4 Pressure Loss

For turbine blade cooling the pressure loss is important to compare the performance of different cooling devices. In Table 2 the pressure loss along the tube is listed together with the friction factor $f = \Delta p/(1/2\ \rho \overline{V_z^2})\ D/L$. For comparison the friction factor is normalized with the correlation by Blasius for a smooth tube flow given by $f_0 = 0.3164\ Re_D^{-0.25}$. Additionally, the circumferential average Nusselt number is given and also normalized with the Dittus-Boelter correlation for a smooth tube flow. The high heat transfer enhancement of around 3.8 necessary for the cooling of high thermal load components is paid by a friction factor of more than 30. On the other hand the thermal performance parameter $(\overline{Nu}/Nu_0)/(f/f_0)^{(1/3)}$ is 1.2, which shows an appropriate cooling performance compared to a smooth tube flow.

4 Cray Performance Analysis

We investigated the parallel computing performance of OpenFOAM (Version 2.2.1) on the Cray XE6 (hermit) simulating a DES at a Reynolds number of $Re = 10,000$. The used mesh was a structured O-grid with 25.6 million cells analog to the shown mesh in Fig. 3. For the speed-up test we used 64 up to 4096 parallel cores which represents $400,000$ and 6250 cells per core (cpc), respectively. The performed DES was a transient, incompressible simulation with an additional scalar transport equation for the temperature as already explained in Sect. 2.

The obtained speed-up is shown against an ideal speed-up in Fig. 9. All simulations with up to 4096 cores are compared to the simulation with 64 cores. Up to a usage of 512 cores (50,000 cpc) we recognize a nearly linear speed-up which shows the known good parallelization of OpenFOAM. For 1024 (25,000 cpc) and 2048 cores (12,500 cpc) the performance gain is smaller but still with an acceptable speed-up. Only with 3072 cores (8333 cpc) or less than around 10,000 cells per core the speed-up is decreasing, so the communication and the data transfer between the cores take longer than the solution of the equation system. The comparison showed that OpenFOAM is suitable for using a high performance cluster such as hermit or hornet available at the High Performance Computing Center Stuttgart (HLRS).

Furthermore, we are currently running a DES at a higher Reynolds number of $Re = 20,000$ on the new Cray XC40 (hornet) and two simulations with lower swirl numbers $S = 1.0$ and 0.75. The simulation differences regarding grid size, time step and CPUh between $Re = 10,000$ and $20,000$ and for the different swirl numbers $S = 5.3, 1.0$ and 0.75 are listed in Table 3. If we compare the two Reynolds number cases for $S = 5.3$, the simulation for $Re = 20,000$ has a factor of 1.67 larger mesh and a factor of 7.67 lower time step. This results at least in a 12.8 times higher

Fig. 9 OpenFOAM speed-up on the Cray XE6 for a DES test simulation with 25.6 million cells

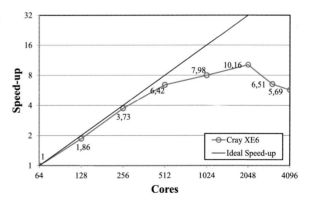

Table 3 Simulation parameters of the latest DES (cpc = cells per core)

	Cells	Time step	Cores	CPUh
DES $Re = 10,000\ S = 5.3$	$9 \cdot 10^6$	$2.3 \cdot 10^{-6}$	512 on XE6: 17,500 cpc	$300,000$
DES $Re = 10,000\ S = 1.0$	$9 \cdot 10^6$	$4.8 \cdot 10^{-6}$	600 on XC40: 15,000 cpc	Running
DES $Re = 10,000\ S = 0.75$	$9 \cdot 10^6$	$6.0 \cdot 10^{-6}$	600 on XC40: 15,000 cpc	Running
DES $Re = 20,000\ S = 5.3$	$15 \cdot 10^6$	$3.0 \cdot 10^{-7}$	1200 on XC40: 12,500 cpc	$1,500,000$

computational time, which would be around 3,800,000 CPUh on the Cray XE6. So we clearly see that the new Cray XC40 is at least 2.5 times faster than the previous Cray XE6 using OpenFOAM for an incompressible DES case. The two simulations with lower swirl numbers are still running, so just some first results will be discussed in the next section.

5 Current and Future Work

In a swirl tube the axial velocity is characterized by a backflow in the tube center due to the induced strong swirl as already discussed in the flow field section. The occurrence of this backflow region depends on the swirl strength and its influence on the heat transfer is still under research. Kobiela [16] analytically estimated a limitation swirl number $S_{limit} = 0.8$ based on an angular momentum balance and a linear axial velocity profile $V_z = V_{z,wall}\ r/R$. Therefore, we simulate two swirl tubes with lower swirl numbers $S = 1.0$ and 0.75 but with the same mass flow and consequently same Reynolds number than the high swirl number case ($S = 5.3$). The structured O-grid is the same as for $S = 5.3$ and the respective simulation time steps are listed in Table 3.

Figure 10 shows the instantaneous axial velocity for all three investigated swirl numbers $S = 5.3$, 1.0 and 0.75 for $Re = 10,000$. A zero axial velocity is indicated by the white line and the black color represents negative velocity and therefore backflow regions. For the highest swirl number a large backflow region over the whole tube occurs which we already discussed. For a swirl number of $S = 1.0$, which means the inlet duct area is equal to the tube cross-section, just a few backflow spots occur. Simulating a swirl number of $S = 0.75$ and with it further reducing the angular momentum more or less no backflow can be observed. This means the analytical limitation swirl number of $S_{limit} = 0.8$ could be confirmed with the here performed simulations. More detailed investigations on the flow structure and the heat transfer are planned in the future.

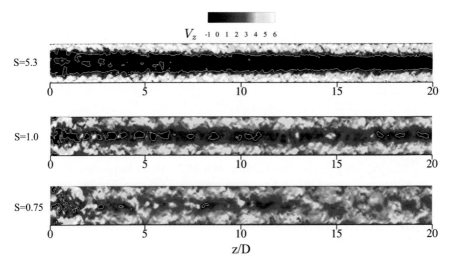

Fig. 10 Instantaneous axial velocity in the swirl tube for different swirl numbers $S = 5.3, 1.0$ and 0.75 at $Re = 10,000$; the *white line* represents zero axial velocity

6 Conclusion

We numerically investigated the flow and the heat transfer in a swirl tube via DES on the Cray XE6 and XC40 at the HLRS and compared it to own experimental results. We presented the axial and circumferential mean velocity field, the velocity fluctuations, the helical vortex structure and the wall shear stress in the swirl tube. Additionally, we discussed the heat transfer, the pressure loss and the thermal performance parameter compared to a smooth tube flow. Numerical and experimental results agree well for the mean velocity profile. The heat transfer shows deviations in the inlet region, but agrees well close to the outlet. The averaged heat transfer is around four times larger than the one in a smooth tube flow, which is paid by a friction factor of 31 again compared to a smooth tube flow. On the other hand the thermal performance parameter is 1.2, which shows that a swirl tube is well applicable for cooling of high thermal load components such as turbine blades.

Furthermore, the speed-up comparison simulating a DES using OpenFOAM indicates that a significant speed-up can be achieved up to around 12,500 cells per core. The next step will be to finish the current simulations with the lower swirl numbers $S = 1.0$ and 0.75 to investigate the flow structure and the heat transfer in more detail. This will help to understand the formation of the backflow region in a swirl tube.

Acknowledgements The authors greatly thanks the High Performance Computing Center Stuttgart (HLRS) for support and supply of computational time on the Cray XE6 and Cray XC40 platforms. The authors also greatly appreciate the funding of this project by the Deutsche Forschungsgemeinschaft (DFG) under the Grant No. WE 2549/25-2.

References

1. Chang, F., Dhir, V.K.: Mechanisms of heat transfer enhancement and slow decay of swirl in tubes using tangential injection. Int. J. Heat Fluid Flow **16**(2), 78–87 (1995)
2. Glezer, B., Moon, H.K., Kerrebrock, J., Bons, J., Guenette, G.: Heat transfer in a rotating radial channel with swirling internal flow. ASME Paper No. 98-GT-214 (1998)
3. Ligrani, P.M., Hedlund, C.R., Babinchak, B.T., Thambu, R., Moon, H.-K., Glezer, B.: Flow phenomena in swirl chambers. Exp. Fluids **24**, 254–264 (1998)
4. Hedlund, C.R., Ligrani, P.M.: Local swirl chamber heat transfer and flow structure at different Reynolds numbers. J. Turbomach. **122**, 375–385 (2000)
5. Biegger, C., Sotgiu, C., Weigand, B.: Numerical investigation of flow and heat transfer in a swirl tube. Int. J. Thermal Sci. **96**, 319–330 (2015). doi:10.1016/j.ijthermalsci.2014.12.001
6. Jasak, H., Jemcov, A., Tukovic, Z.: OpenFOAM: A C++ library for complex physics simulations. In: International Workshop on Coupled Methods in Numerical Dynamics, pp. 1–20 (2007)
7. Spalart, P.R., Deck, S., Shur, M.L., Squires, K.D., Strelets, M.Kh., Travin, A.: A new version of Detached Eddy Simulation, resistant to ambiguous grid densities. Theor. Comput. Fluid Dyn. **20**, 181–195 (2006)
8. Fröhlich, J., von Terzi, D.: Hybrid LES/RANS methods for the simulation of turbulent flows. Prog. Aerosp. Sci. **44**, 349–377 (2008)
9. Spalart, P.R., Allmaras, S.R.: A one-equation turbulence model for aerodynamic flows. La recherche Aérospatiale **1**, 5–21 (1994)
10. Kays, W.M.: Turbulent Prandtl number—where are we? J. Heat Transf. **116**, 284–295 (1992)
11. Iida, O., Kasagi, N.: Heat Transfer of Fully Developed Turbulent Channel Flow with Iso-Thermal Walls. Turbulence and Heat Transfer Laboratory, University of Tokyo. http://thtlab.jp/index-orig.html (2001)
12. Biegger, C., Cabitza, A., Weigand, B.: Three Components- and Tomographic-PIV Measurements of a Cyclone Cooling Flow in a Swirl Tube. ASME Turbo Expo GT2013-94424 (2013)
13. Biegger, C., Weigand, B.: Heat Transfer Measurements in a Swirl Chamber Using the Transient Liquid Crystal Technique. 15th International Heat Transfer Conference, IHTC15-9231 (2014)
14. Pope, S.B.: Turbulent Flows. Cambridge University Press, Cambridge (2000)
15. Dittus, F.W., Boelter, L.M.K.: University of California Publications of Engineering, vol. 2, p. 443. University of California, Berkeley (1930)
16. Kobiela, B.: Wärmeübertragung in einer Zyklonkühlkammer einer Gasturbinenschaufel. Doctoral thesis, University of Stuttgart (2014)

Evaluation and Control of Loads on Wind Turbines under Different Operating Conditions by Means of CFD

Christoph Schulz, Annette Fischer, Pascal Weihing, Thorsten Lutz, and Ewald Krämer

Abstract This article presents results from Computational Fluid Dynamics (CFD) simulations of wind turbines performed within the project *WEALoads*. The project is devoted to the unsteady load response of wind turbines under realistic environmental conditions and to mechanisms to control these loads. An extract of the latest research done on these topics is presented in this article. Based on previous studies of wind turbines in complex terrain done by the authors the effects of the terrain on the load response is tried to break down to several single events. Some of these are the shear of the wind profile, turbulence intensity of the inflow, inclined inflow or yawed inflow. In this article the effect of yawed inflow on the loads and aerodynamics of a wind turbine is described in further detail. The second part of the studies is dedicated to the control of wind turbine loads which is one of the key factors in the current turbine design processes. An example to reduce load fluctuations caused by tower blockage is shown. This can be seen as a prove of concept for the method applied. All of the simulations were performed using the flow solver *FLOWer* from DLR (German Aerospace Center). For the studies of the first part a *Detached Eddy Simulations (DES)* approach was used whereas for the second part *URANS* methods were applied. Afterwards, a newly implemented DES method giving a more realistic prediction of the flow field around the turbine and consequently the loads is described in more detail. The general outcome of the article is that a load reduction under yawed inflow can be observed as well as a wake deflection which is of high importance in case of wind park development and control. Moreover, a bridge between the yaw results, the load reduction method and the new high fidelity *DES* methods is build to give an outlook to future works.

C. Schulz (✉) • A. Fischer • P. Weihing • T. Lutz • E. Krämer
Institute of Aerodynamics and Gas Dynamics, Pfaffenwaldring 21, 70569 Stuttgart, Germany
e-mail: schulz@iag.uni-stuttgart.de; weihing@iag.uni-stuttgart.de; fischer@iag.uni-stuttgart.de; lutz@iag.uni-stuttgart.de; kraemer@iag.uni-stuttgart.de

© Springer International Publishing Switzerland 2016
W.E. Nagel et al. (eds.), *High Performance Computing in Science and Engineering '15*, DOI 10.1007/978-3-319-24633-8_30

463

1 Introduction

There is no doubt the wind energy sector has become one of the most important in the renewable energy market in recent years. Several studies have been performed to optimise single turbines or wind farms regarding the reduction of maintenance costs, loads and power output increase among others [2, 8]. For the development of reliable turbines requiring only little maintenance, it is essential to accurately predict the aerodynamic loads on the blades. Especially, changes in the local angle of attack over one rotation of the blade cause fatigue loading even in an undisturbed inflow. It is quite obvious that for the purpose of more effective and reliable turbines it is needed to get a deeper understanding of the aerodynamics of the turbine and its response to realistic inflow conditions. Furthermore, a need of methods to control these loads is getting into focus [11]. The installation of these load control concepts is often expensive and time consuming. As a consequence, a lot of manufactures do not take the effort of tests on real wind turbines. In all described cases numerical schemes give more freedom of development and make it possible to perform studies impossible in field tests.

These are just some points why within the framework of wind turbine development CFD is becoming increasingly important. The rise of high–performance computing capacities is a another driving factor for these developments. In contrast to the widely used engineering tools, often based on blade element momentum theory, solving the Navier–Stokes equations enables the most accurate approach. Therefore, it is needed to directly represent all relevant wind turbine components [16, 20, 28].

Weihing et al. [29] showed the response of wind turbines in complex terrain and of a turbine sited in a half wake situation. The latter is a typical case of a wind park. The first is a growing interest in Germany as more and more onshore sites get into the focus of the wind energy industry. A problem of these investigations was the overlap of different single effects. As a consequence, the impact of one single factor could not be determined in detail. Hence, further studies on single effects regarding the inflow have been performed. Some of these are the shear of the wind profile, turbulence intensity of the inflow, inclined inflow or yawed inflow. In this article results of yawed inflow on a generic 2.4 MW turbine are presented which are of relevance for a single turbine because of the reduction of load and power output as the projected rotor area decreases under yawed conditions [18, 22]. Additionally, this study is also of major importance for wind park planning, as the wake of a turbine under yawed inflow is deflected in relation to the inflow direction because of the varying induction over one rotation. This effect can be used in wind park control to optimise the overall power output by reducing the power output of a single turbine. Studies of these effects have been performed by [22] on a model wind turbine and by [18] on field data. A second emphasis of the article is the control of loads of a turbine. For this study an academic test turbine from the KIC–OFFWINDTECH project was picked as many reference data are available for the calculation case chosen [3, 4]. To reduce the loads for the blade passing the tower the concept of active trailing edge flaps is applied. Active trailing edge flaps change the flap angle in relation to the azimuthal position and as a consequence camber and the aerodynamic properties of the airfoil [1, 13]. By means of this the load fluctuations during the tower passage are reduced.

For all presented simulations the CFD solver *FLOWer* [12] by *DLR (German Aerospace Center)* is used. First of all, a general description of the code and the numerical setups are given before some improvements of the numerical methods are shown. Afterwards, the results of the yawed inflow cases and the load reduction approach by means of active trailing edge flaps are presented.

2 Numerical Setup and Computational Details

2.1 Simulation of Yawed Inflow (CaseBase & CaseYaw)

The computational domain of this test case consists of different meshes which are overlapped using the overset grid technique as also described by [11, 19–22, 28, 29]. The background domain spans 500 m × 400 m × 475 m. At the inlet as well as on the sides and the top of the background domain farfield boundary conditions are chosen. The ground is represented by a no–slip boundary condition. Having a hub height of 95 m and a rotor diameter of 109 m the turbine is tilted by 5°, preconed by 2° and pitched by 4.41°. At the chosen operational point the turbine has a rotor frequency of 12.81 RPM. The inflow velocity is 11.01 m/s and the turbine is rotated 30° around its z–axis for *CaseYaw* to create yawed flow conditions (Fig. 1). In the reference test case *CaseBase* the yaw angle γ is equal to zero and the rotor plane orthogonal to the inflow. All relevant turbine components like the blades, tower, hub and the nacelle are represented as full model. They have been meshed with fully resolved boundary layers ensuring $y^+ = 1$ of the first cell. All components are meshed separately and overlapped within *FLOWer* using the overset grid technique according to [5]. In total, the complete set-up is discretized by

Fig. 1 Illustration of the yawed inflow with a yaw angle γ=30°

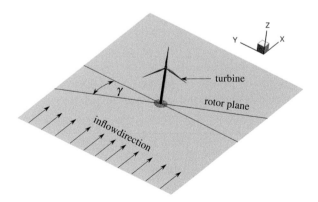

approx. 50 mio. cells. As spatial discretization a second order central differencing scheme JST [10], whereas for temporal discretization an implicit dual time stepping scheme [9] is used with a time-step equivalent to 2.5° azimuthal blade movement. Moreover, the simulation uses an Detached-Eddy Simulation approach as developed by Spalart [25, 26] in combination with the *Spalart–Allmaras turbulence model with Edwards modification* [7]. In order to obtain converged results for the evaluation 26 revolutions have been performed before data were extracted for two revolutions. For reasons of comparison the same simulation has been performed without yaw using all other parameters similar.

2.2 Simulations of the Load Control Case (CaseControl)

The modified NREL 5 MW wind turbine, as used in the KIC–OFFWINDTECH project [3, 4], was chosen as reference turbine for the present investigations. The turbine has a rotor diameter of 126 m and a hub height of 90 m. The precone angle is -2.5°, the rotor is tilted by 5° and the rotor frequency is 11.7 RPM for the chosen operation point. More information about the turbine can be found in [15].

In the test case a uniform inflow with an inflow velocity of 11.3 m/s is used and the rotor plane is perpendicular to the inflow. In the KIC–OFFWINDTECH project this case is called B1.2 [3]. The computational domain spans $520 \text{ m} \times 400 \text{ m} \times 400 \text{ m}$ and the background mesh has approximately 10 mio. cells. The ground is represented by a slip boundary condition, whereas all other boundary conditions are realized as farfield conditions. The blade is discretized with 2.25 mio. cells and a fully resolved boundary layer, ensuring $y^+ = 1$ of the first cell. Again, all parts of the turbine are meshed separately and overlapped in *FLOWer*. The whole setup includes 27.26 mio. cells. In the present case, the turbulence was modelled with the *Wilcox k- ω* turbulence model. The spatial and temporal discretization is the same as for *CaseBase* and *CaseYaw*, the time-step corresponds to 1.5° azimuthal blade movement, with a maximum of 40 inner iterations. 20 revolutions of the turbine were performed, extracting the data for the 20th revolution as reference for the later *CaseControl* evaluations. During the 21st revolution, trailing edge flaps were deflected in order to minimize the fatigue loads caused by the tower blockage. For present investigations, one independent flap per blade, covering 10 % of the chord length and located from 70 % up to 90 % radius was installed at the modified NREL 5 MW wind turbine. In several iteration steps, the movement of the flaps was improved to reduce the effect of the tower blockage as much as possible (Fig. 2).

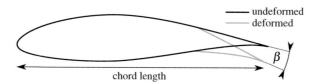

Fig. 2 Nomenclature of an airfoil equipped with trailing edge flap. The flap covers 30 % of the chord length and the flap angle β is equal to 12° for reasons of clarity

2.3 Numerical Methods

High fidelity simulation of wind turbines, aiming to improve the understanding of the interaction of atmospheric turbulence and load response, require methods that allow the resolution of the dominant load affecting eddies. Since wall resolved *LES* is still unfeasible for Reynolds numbers of full scale wind turbines, a widely used overcome is the so called *Detached Eddy Simulation (DES)* approach which is a family of hybrid RANS/LES methods that resolves the energy carrying eddies far from walls but treats the boundary layer using a classical RANS model. In *WEALoads* most computations [20, 28, 29] have been performed using *DES97* [26] or *DelayedDES (DDES)* [25]. In order to increase resolved turbulence activity near the wall, e.g. in presence of ambient turbulence, Shur et al. [23] proposed the *Improved Delayed DES (IDDES)* method which incorporates the functionality of *wall modelled LES (WMLES)* in the strategy of *DDES*. In principal, the method requires a near-wall modification of the *LES* filter width Δ and a more rapid transition between *RANS* and *LES* length scales compared to *DES97* or *DDES*. The filter width which enters L_{LES} is modified to

$$\Delta = \min\left[\max\left(C_w d_w, C_w h_{max}, h_{wn}\right), h_{max}\right], \tag{1}$$

where d_w is the wall distance, h_{max} is the maximum extent of the cell, h_{hwn} is the extent in wall normal direction and C_w (= 0.15) is an empirical constant. The modified *DES* length scale reads

$$L_{IDDES} = \tilde{f}_d(1 + f_e)L_{RANS} + (1 - \tilde{f}_d)L_{LES}. \tag{2}$$

The function $\tilde{f}_{dt}$ blends the *DDES* and *WMLES* branches of the model

$$\tilde{f}_d = \max\left[(1 - f_{dt}), f_b\right]. \tag{3}$$

The basic weighting function for the *WMLES* branch is f_b, which depends on the ratio d_w/h_{max} and indicates (for $f_b = 0$) whether the grid is fine enough to resolve boundary-layer dominant eddies.

$$f_b = \min\left[2\exp(-9\alpha^2), 1.0\right], \quad \alpha = 0.25 - d_w/h_{max} \tag{4}$$

The term $(1 - f_{dt})$ is a sensor for turbulent content in the flow. It is calculated from

$$f_{dt} = 1 - \tanh\left[(8r_{dt})^3\right] \tag{5}$$

$$r_{dt} = \frac{\nu_t}{\kappa^2 d_w^2 \max\left(\sqrt{\frac{\partial U_i}{\partial x_j}\frac{\partial U_i}{\partial x_j}}, 10^{-10}\right)} \tag{6}$$

In the presence of resolved turbulence, r_{dt} goes to zero, resulting f_{dt} getting close to 1 and therefore $(1 - f_{dt})$ close to zero. As a result, the model uses the *WMLES* branch. The elevating function f_e in Eq. (2) is active at the *RANS-LES* interface and is instrumental in combating the log-layer mismatch (details see [23]). *IDDES* has been implemented in *FLOWer* for *Spalart–Allmaras* [24] and *Menter SST* [17] turbulence models.

2.4 Simulation Clusters

Most of the simulations in the year 2014 have been performed at the *High Performance Computing Center Stuttgart (HLRS)* with *AMD Interlagos* processors on *Cray XE6*. With the begin of the test phase of *Cray XC40* in 2014 this new machine with its *Intel(R) Xeon(R)* processors has been used as well. *CaseBase* and *CaseYaw* used 1600 processors, with the option to use only every second processor to increase memory and speed, whereas *CaseControl* was computed on 288 processors using every processor. In total, *CaseBase* and *CaseYaw* simulations required approx. 220,000 CPUh. For *CaseControl* approx. 1440 CPUh per revolution were needed. It is important to mention that for the surrounding study of different inflow conditions various of theses simulations had to be performed, but are not shown here.

In order to check the latest developments in the *FLOWer*-code and to get reliable information about its suitability regarding massively parallel computations a weak scaling test has been performed on *Cray XC40* similar to the one done by [29] on *Cray XE6*. Other scaling tests using the *FLOWer*-code can be found in [6].

For the weak scaling test *FLOWer* was compiled with *ifort* on *Cray XC40*. The problem size was increased from 128 blocks up to 4096, keeping the cell loading on each block constant to 32^3. The time for 1000 iterations was measured for each run and normalized with the value obtained for 128 cores. In Fig. 3 the efficiency vs. the number of used cores is plotted. With an efficiency of 0.77 on 4096 cores, *FLOWer* shows acceptable performance to apply highly resolved hybrid RANS/LES simulations on *Cray XC40* as described in Sects. 2.3 and later in 3.3.

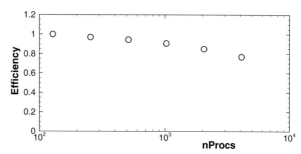

Fig. 3 Results from weak scaling tests: Efficiency of *FLOWer* on *Cray XC40* using *ifort* fortran compiler and a constant cell loading of 32^3 for each MPI process (normalized with the time per iteration on 128 cores)

3   Results

3.1   Yawed Inflow

In this section the results of *CaseYaw* will be described and compared to *CaseBase* to highlight the impact of yawed inflow on wind turbine aerodynamics. At first, a closer look at the flow field shall give a basic impression of the different situations. Figure 4 shows a top view of a plane extracted at hub height of the turbine. Clearly visible is the near wake of the turbine by the reduced downstream velocity. Looking more into detail various effects of the yawed inflow can be detected. First of all, the deflection of the wake shall be investigated by looking at the wake center line in Fig. 4b. This line has been determined as the connection of the mid–points between two vortices of the same age. It becomes obvious that the center line is rotated compared to the inflow direction towards the side of the downwind blade about $\chi_\tau=2.2°$ caused by the changing angle of attack over one revolution. This leads to a variation of the induced velocities in radial and azimuthal direction. Under consideration of the Biot–Savart law the induction at the downwind blade is larger than for the upwind blade. This effect is strengthened by the fact that the induced velocities are pointing in the axial direction of the turbine and not towards the inflow. The simulated wake deflection can be compared to analytical models given by [27] which determine the skew angle $\chi = \chi_\tau + \gamma$ using the following relation:

$$tan(\chi) = \frac{sin\,(\gamma) - a \cdot tan\,(0.5\chi)}{cos\,(\gamma) - a} \tag{7}$$

The value of the axial induction factor can be calculated from the free stream velocity u_{ref} and the velocity in the rotor plane u_2 as follows:

$$a = \frac{u_{ref} - u_2}{u_{ref}} \tag{8}$$

Applying Eq. (8) to the simulation results leads to an axial induction factor of 0.14 giving according to Eq. (7) $\chi = 32.33°$ which is close to the value of $32.2°$ seen in the flow field of the simulation in Fig. 4. In fact, this means that the wake of turbine

(a)   (b)

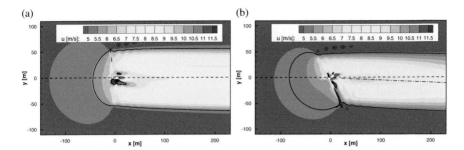

Fig. 4 Flow field for the baseline and yaw case. The solid black line indicates a velocity equal to 95 % of the freestream velocity and shows the edge of the wake downstream the turbine. The dash dotted line symbolizes the wake center line. (**a**) CaseBase (**b**) CaseYaw

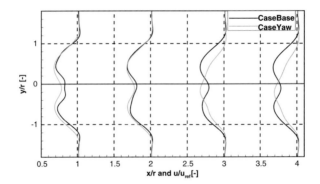

Fig. 5 Wake deficit of *CaseBase* and *CaseYaw* at different downstream positions. The distance in x is normalized by the rotor radius and the wind speed by the free stream velocity u_{ref}. The different velocity profiles are shifted by the corresponding downstream distance

under yawed conditions is stronger deflected than the one of a non–yawed turbine. A phenomenon which for example can be used for wind park control. Another interesting aspect is the widening of the wake. This can be seen in the flow field of Fig. 4 as well as in the wake deficit at different positions downstream depicted in Fig. 5. The wake widens symmetrically at $x/r = 2$ to about $y/r = \pm 1.20$ and further downstream at $x/r = 4$ to $y/r = \pm 1.30$ leading to a wake widening over distance, also called wake expansion rate, of 0.1. In comparison to this the wake of *CaseYaw* develops differently, not only because of the deflection but also regarding wake widening in total. At $x/r = 2$ it has expanded to $y/r = {}^{+1.20}_{-1.25}$ and narrows afterwards at $x/r = 4$ to $y/r = {}^{+1.00}_{-1.10}$. A reason for this can also be seen in the basic structure of the wake deficit. In *CaseYaw* the wake recovers more quickly and the mean velocity inside the wake at $x/r \geq 3$ is higher than in *CaseBase* leading to the described narrowing. Using a coordinate system related to the turbine the wake development can be visualized like in Fig. 6. In *CaseBase* the wake 150 m downstream the rotor plane has developed almost in line with the turbine x–axis, meaning no movement in y–direction is visible whereas in the yaw case a strong movement of the wake

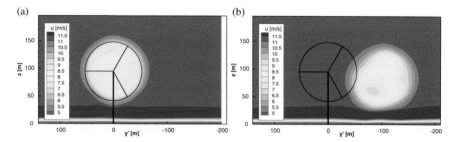

Fig. 6 Wake three rotor radii downstream the rotor plane. *Solid black line* indicating the turbine. The coordinate systems have been rotated about the yaw angle to better illustrate the skew angle impact. Data extracted 150 m downstream the rotor plane. (**a**) CaseBase (**b**) CaseYaw

towards negative y–values is prompting. The wake for *CaseYaw* has moved about 100 m in the mentioned direction and seems to be more deformed from a circular shape than in *CaseBase*. Taking some integral data into account the evaluation of the normalized axial blade force Fx_{norm} and normalized power P_{norm} shown in Figs. 7 and 8 is of relevance. The normalization of each quantity was done by the corresponding mean value of *CaseBase* determined over one rotation:

$$Fx_{norm}(Azimuth) = \frac{Fx(Azimuth)}{Fx_{Base}} \; ; \; P_{norm}(Azimuth) = \frac{P(Azimuth)}{P_{Base}} \tag{9}$$

For *CaseBase* the tower impact can be seen in an area between 150° and 215° in a load reduction of about 11 percentage points. Moreover, the loads at the side for the upward moving blade (270° side) are larger than for the opposite side. This means that the loads are shifted to the left side in Fig. 7. An effect caused by the tilt angle of the turbine and the resulting change of the angle of attack over one rotation. Similar observations have been described by [16, 20, 21]. In the power output the latter effect can not be seen, whereas the tower impact is clearly visible in the areas around 60°, 180° and 300° by a dip in the graph of about four percentage points. Comparing these results to the grey graph in the two plots showing the corresponding data for *CaseYaw* major differences can be detected. Firstly, a closer look at the normalized axial blade force shows a massive decrease of the load level. The mean load over one rotation is reduced about 20 percentage points compared to *CaseBase*. Moreover, the loads are shifted in general towards the right and in special to the top right corner of the plot, meaning the sector between 0° and 90° azimuthal angle. This change of the situation is caused by different overlaying effects. First of all, the general load reduction is a consequence of the reduced projected area of the turbine regarding the inflow. This aspect shall be discussed later on when it comes to power output. The movement of the loads to the top results from the same fact as the shift of the loads to the left in *CaseBase*. A yawed inflow produces comparable angle of attack and relative wind speed differences between 180° and 360° azimuth as a tilt of the rotor between 90° and 270° azimuth. The local angle of attack at the blade

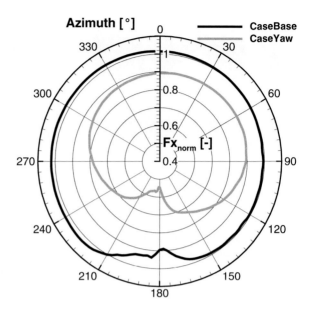

Fig. 7 Comparison of the normalized axial blade loads Fx_{norm} for yawed and non-yawed inflow over the azimuth

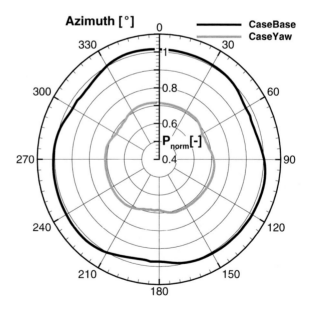

Fig. 8 Comparison of the normalized power output P_{norm} for yawed and non–yawed inflow over the azimuth

at 180° is smaller than under axial inflow conditions and the relative velocity at the blade consequently higher. Depending on which effect is dominant for the loads of the turbine these are increased or reduced, for this turbine the latter is true. Taking global inflow angles into consideration explains the mentioned shift of the loads towards the right side of Fig. 7 [20]. For the right side of the turbine, the effective area of the blade that faces the inflow is bigger than on the left side. The blade is more orthogonal to the flow as a result of the turbine cone angle in combination with the yawed inflow. As a consequence, the loads are higher on this side.

Keeping the view on the global flow field also explains the power reduction visible in Fig. 8 which is caused by the decreasing projected area of the rotor plane in relation to the flow as well as by the already described changes in aerodynamics e.g. the angle of attack variation. Overall, the power output under yawed conditions is reduced by 29 % compared to the mean power output of *CaseBase* while the impact of the tower blockage remains clearly visible. Schepers [18] mentions a cosine relation between power output and yaw angle, as the projected area of the rotor plane depends on the cosine of the yaw angle:

$$P = P_{CaseBase} \cdot cos^x(\gamma) \tag{10}$$

Power again is related to the third power of the wind speed, which would make $x = 3$ a first hand assumption. Nonetheless, as described by [18] the value of x varies depending on the turbine. For the analysed case, a value of 2.38 can be determined which can be validated by the results of the not shown simulations performed .

In total, it can be seen that yawed inflow impacts the loads and power output of a wind turbine as well as the near wake because of the changed aerodynamics compared to a axial inflow. Yaw can be used to maximize wind park power output by reducing a single turbines power output, but getting the downstream turbines out of the wake of the upstream ones. How these positions inside the wake can influence turbine behaviour has been shown in previous studies [29]. To improve the estimations made for downstream turbines a more realistic wake and wake propagation is needed. Therefore, different reconstruction themes as well as new developments as in Sect. 2.3 described are tested. Moreover, it becomes obvious that yawed inflow leads to high load fluctuations over one rotation of the turbine. To reduce these impact and the fatigue of the turbine advanced load control concepts are needed. One approach to handle this is explained in Sect. 3.2.

3.2 Load Control

Yawed inflow leads to unsymmetrical load fluctuations at turbines. But not only turbines with yawed inflow or in the wake of another rotor experience load variations, as the tower of each rotor affects the turbine itself. These load fluctuations lead to high fatigue loads, which shall be limited in order to reduce the costs of

Fig. 9 Comparison of the
thrust for the reference case
without flap deflection and
for two different deflections.
Only the area of interest
around the tower passage is
displayed

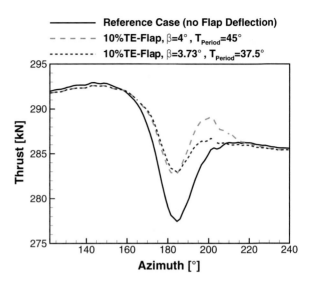

energy and increase the competitiveness of wind energy towards other sources of
energy.

Figure 9 shows the potential of the load alleviation system to reduce the effects
of the tower blockage. Two different flap deflections along with the reference case
are shown for the area of interest around the tower passage at 180° azimuth. The
azimuth angle represents the position of the blade, whereby an azimuth angle of 0°
indicates the upright position of the blade and 180° means, that the blade is in front
of the tower. As the flap is only deflected close to the tower the thrust over the rest of
the rotation is not influenced. In the area of interest, reaching from approx. 160° to
210°, the blade aerodynamics and consequently its thrust is influenced by the tower
blockage, leading to a drop of the thrust of about 5 %. The reference case shows the
20th revolution of the original KIC–OFFWINDTECH setup, whereas the two other
graphs show the 21st revolution, where flaps were integrated in the setup. A positive
flap angle β leads to a higher cambering of the airfoil (Fig. 2). In the present case,
the flap deflection is a function of time and described by

$$\beta(t) = \beta_m + \Delta\beta \cdot cos\left(2\pi \cdot \frac{1}{T_{Period}} \cdot t\right). \tag{11}$$

The parameter β_m is the mean deflection angle over the deflection period and $\Delta\beta$ the
maximum amplitude of β. In Fig. 9 the value of β specifies the maximum deflection
angle and T_{Period} the duration of one deflection period. In both cases, the deflection
starts at azimuth=166.5°. In the first case, the deflection ends after T_{Period}=45°. It
can be seen, that the flap movement reduces the load fluctuations caused by the
tower blockage by 50 %, but only up to azimuth=195°. Afterwards, the deflection
causes an overshoot. This is a result of the time shift between flap deflection and the
reaction of the blade to the change of the cambering. In order to prevent this, the

deflection angle and T_{Period} were reduced, which leads to the same reduction of the thrust drop without the overshoot.

In a next step, other types of flaps, like e.g. coupled leading and trailing edge flaps and their potential for load reduction can be investigated [14]. Moreover, the effect of flaps on load variations, caused by yawed inflow or upstream turbines, can be investigated, leading to an extension of the scope of possible applications for load alleviation systems.

3.3 Improvement of Numerical Methods

The newly implemented *IDDES* method has been applied to a simulation of an isolated rotor with focus on flow separation close to the hub. Each blade was discretized by $400 \times 224 \times 140$ cells in spanwise, circumferential and wall normal direction, respectively. The overall domain consisted of 120 Mio. cells. The simulations were run on *Cray XC40* using 4096 cores. Preliminary results are presented in Fig. 10, where vortex structures are compared for *URANS* and *IDDES* simulations by means of the $\Lambda_2 (= -5e^{-7})$ criterion. It is evident, that different turbulence modelling predicts qualitatively different wakes in the inner portion of the rotor, where the flow is separated. For *URANS*, the wake is characterized by two distinct hub vortices. The outer hub vortex is generated at the outermost reattachment line, where the profile has still airfoil shape and the flow direction is led by the trailing edge. The helix of the inner vortex develops in the transition region to a circular profile. Since the local ratio of circumferential velocity and the wind velocity is very small, close to the hub, the pitch of the inner helix is higher compared to the outer one. When applying the *IDDES* method and entering the *LES* mode, the length scale in the destruction term of the background *RANS* model

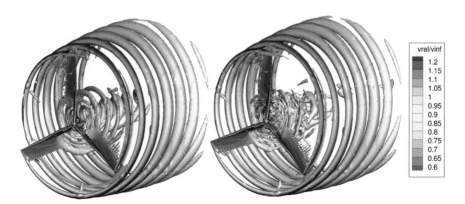

Fig. 10 Vortex vizualisation for an isolated rotor by the $\Lambda_2 = -5e^{-7}$ isosurface. Contour levels indicate relative velocity magnitudes. *Left*: *URANS*, *right*: *IDDES*

is reduced. As a consequence, a *Smagorinsky*–model like behavior is achieved in regions of separated flow. Hence, in the right plot of Fig. 10, the coherent structures break down to different scales until they are dissipated at the filter width of the grid. The qualitative difference in resolving the flow field observed in Fig. 10 is expected to be transferable to the prediction of the blade loads. Analyses, not shown here indicate significantly higher load fluctuations predicted by *IDDES* compared to quasi converged loads obtained by *URANS*.

4 Conclusions

The presented article showed different aspects of wind turbine simulations under realistic conditions. In order to gain a deeper understanding for complex flow and for the control of aerodynamic response to those the events of a realistic atmospheric inflow were separated and investigated in detail on their own. An extract of the results of these analyses were shown. Taking the example of yawed inflow conditions a overall power reduction of the turbine could be investigated as well as a change of the blade load behaviour leading to high load fluctuations per rotation of the turbine and to a change of fatigue. Closely related to these effects occurring from e.g. the change of the angle of attack a distortion of the wake of the turbine was presented. Overall, the flow situation of yaw is quite likely in reality. Sometimes unwanted as inflow directions change and sometimes wanted to optimize wind farm power generation. The results of the yaw impact on the wake are e.g. needed for these. Regarding the impact on the loads a bridge between these simulations and a case study to control the loads of a turbine by active trailing edge flaps was build. In the approach of load control in detail different flap deflections were investigated in order to reduce the impact of the tower blockage on the fatigue load of a wind turbine. In the final part of the article newly implemented *IDDES* resulting in better representation of the turbine near wake and its vortex structures. At the same time this new approach highlights the demand for high–performance clusters as an enormous increase of computational effort for future calculations can be expected. Overall regarding complexity a step back compared to [29] was done to get better understanding of the occurring aerodynamic effects at wind turbines and to develop new methods inside the field of wind turbine simulations in atmospheric environments and the control of their loads. The future step will be the combination of all parts shown in this article into a single simulation trying to cover as much as possible of reality.

Acknowledgements The authors gratefully acknowledge the *High Performance Computing Center Stuttgart* for providing computational resources.

References

1. Andersen, P., Barlas, T., Buhl, T.: 2d numerical comparison of trailing edge flaps –upwind wp1b3. Technical report, Technical University of Delft (2001)
2. Barthelmie, R.J., Jensen, L.E.: Evaluation of wind farm efficiency and wind turbine wakes at the nysted offshore wind farm. Wind Energy **13**(6), 573–586 (2010)
3. Bekiropoulos, D., Lutz, T., Baltazar, J., Lehmkuhl, O., Glodic, N.: D2013-3.1: Comparison of benchmark results from cfd-simulation. Deliverable report, KIC-OFFWINDTECH (2013)
4. Bekiropoulos, D., Rieß, D., Lutz, T., Krämer, E., Matha, D., Werner, M., Cheng, P.W.: Simulation of unsteady aerodynamic effects on floating offshore wind turbines. DEWEK (2012)
5. Benek, J.A., Steger, J.L., Dougherty, F.C., Buning, P.G.: In: Chimera A Grid-Embedding Technique (1986)
6. Busch, E., Wurst, M., Keßler, M., Krämer, E.: Computational aeroacoustics with higher order methods. In: Nagel, W.E., Kröner, D.H., Resch, M.M. (eds.) High Performance Computing in Science and Engineering 12, pp. 239–253. Springer, Berlin/Heidelberg (2013)
7. Edwards, J.R., Chandra, S.: Comparison of eddy viscosity-transport turbulence models for three-dimensional, shock-separated flowfields. AIAA J. **34**(4), 756–763 (1996)
8. Goit, J.P., Meyers, J.: Optimal control of energy extraction in wind-farm boundary layers. J. Fluid Mech. **768**(4), 5–50 (2015)
9. Jameson, A.: Time dependent calculations using multigrid, with applications to unsteady flows past airfoils and wings. AIAA Pap. **1596**, 1991 (1991)
10. Jameson, A., Schmidt, W., Turkel, E. et al.: Numerical solutions of the euler equations by finite volume methods using runge-kutta time-stepping schemes. AIAA Pap. **1259**, 1981 (1981)
11. Jost, E., Fischer, A., Lutz, T., Krämer, E.: Cfd studies of a 10 mw wind turbine equipped with active trailing edge flaps. In: 10th EAWE Ph.D. Seminar on Wind Energy in Europe, Orlans, pp. 51–54, 28–31 October 2014
12. Kroll, N., Rossow, C.-C., Becker, K., Thiele, F.: The megaflow project. Aerosp. Sci. Technol. **4**(4), 223–237 (2000)
13. Lackner, M.A., van Kuik, G.: A comparison of smart rotor control approaches using trailing edge flaps and individual pitch control. Wind Energy **13**(2-3), 117–134 (2010)
14. Lambie, B.: Aeroelastic investigation of a wind turbine airfoil with self-adaptive camber. PhD thesis, Technical University of Darmstadt (2011)
15. Matha, D., Schuon, F., Lutz, T.: Baseline fowt definition v4. Deliverable report d3.1, KIC-OFFWINDTECH
16. Meister, K.: Numerische Untersuchung zum aerodynamischen und aeroelastischen Verhalten einer Windenergieanlage bei turbulenter atmosphärischer Zuströmung. PhD thesis, p. 246. Shaker Verlag (2015). ISBN: 978-3-8440-3962-7
17. Menter, F.R.: Two-equation eddy-viscosity turbulence models for engineering applications. AIAA J. **32**(8), 1598–1605 (1994)
18. Schepers, J.: Engineering models in wind energy aerodynamics. Ph.D. thesis, TU Delft (2012)
19. Schulz, C.: Cfd of wind turbines in complex terrain. In: IEA R&D Wind Task 11 - Topical Expert Meeting Challenges on Wind Energy Deployment in Complex Terrain (2013)
20. Schulz, C., Klein, L., Weihing, P., Lutz, T. et al.: Cfd studies on wind turbines in complex terrain under atmospheric inflow conditions. J. Phys.: Conf. Ser. **524**, 012134 (2014)
21. Schulz, C., Lutz, T., Krämer, E.: Simulation of wind turbines in complex terrain by means of direct cfd. In: 10th EAWE PhD Seminar on Wind Energy in Europe, Orlans, pp. 51–54, 28–31 October 2014
22. Schulz, C., Meister, K., Lutz, T., Krämer, E.: Investigations on the wake development of the mexico rotor considering different inflow conditions. In: Contributions to the 19th STAB/DGLR Symposium Munich, Germany 2014, Notes on Numerical Fluid Mechanics and Multidisciplinary Design. STAB, Springer, New York (2014) (under review)

23. Shur, M.L., Spalart, P.R., Strelets, M.K., Travin, A.K.: A hybrid rans-les approach with delayed-des and wall-modelled les capabilities. Int. J. Heat Fluid Flow **29**(6), 1638–1649 (2008)
24. Spalart, P.R., Allmaras, S.R.: A one-equation turbulence model for aerodynamic flows. 30th Aerospace Sciences Meeting and Exhibit. http://arc.aiaa.org/doi/abs/10.2514/6.1992-439 (1992)
25. Spalart, P.R., Deck, S., Shur, M., Squires, K., Strelets, M.K., Travin, A.: A new version of detached-eddy simulation, resistant to ambiguous grid densities. Theor. Comput. Fluid Dyn. **20**(3), 181–195 (2006)
26. Spalart, P.R., Jou, W., Strelets, M., Allmaras, S.: Comments of feasibility of les for wings, and on a hybrid rans/les approach (1997)
27. Tsalicoglou, C., Jafari, S., Chokani, N., Abhari, R.S.: Rans computations of mexico rotor in uniform and yawed inflow. J. Eng. Gas Turbines Power **136**(1), 011202 (2014)
28. Weihing, P., Meister, K., Schulz, C., Lutz, T. et al.: Cfd simulations on interference effects between offshore wind turbines. J. Phys.: Conf. Ser. **524**, 012143 (2014)
29. Weihing, P., Schulz, C., Lutz, T., Krämer, E.: Cfd performance analyses of wind turbines operating in complex environments. In: Nagel, W.E., Kröner, D.H., Resch, M.M. (eds.) High Performance Computing in Science and Engineering 14, pp. 403–415. Springer International Publishing, New York (2015)

Advances in Parallelization and High-Fidelity Simulation of Helicopter Phenomena

Patrick P. Kranzinger, Ulrich Kowarsch, Matthias Schuff, Manuel Keßler, and Ewald Krämer

Abstract A weak and strong scaling study is presented which shows substantial improvements to the scalability of the CFD solver FLOWer by introducing a node-to-node MPI communication strategy. Furthermore, an overview of an extremely flexible and reusable CFD-CSD coupling interface is giving. It is able to handle unstructured, structured, and overset meshes without topology limitations and performance drawbacks. Finally, using these new capabilities a full helicopter configuration is investigated with regard to its aeroacoustic noise emission.

1 Introduction

The recent increase of computational power available at HLRS allows increasing the case size of aerodynamic helicopter investigations accordingly. In combination with newly implemented higher order methods, the numerical exploration of helicopter specific phenomena like noise emissions, tail shake or dynamic stall, as well as in-ground effect flights becomes possible [1, 2].

However, the performance increase of computing clusters mainly resulted from increasing the number of available computing cores, instead of significantly increasing the per-core performance itself. This burdens great demands on scalability on the used CFD codes. Substantial improvements of the scalability of the CFD code FLOWer by using intra-node shared memory for communication purposes will be shown.

In addition, a new CFD-CSD coupling library implementation will be presented, which enables the usage of the increased computational power for a large variety of future studies, requiring mesh topology independent loads and moments evaluation, e.g. for applying automated optimizers, mesh-to-mesh interpolation and parameter and time dependent mesh deformation. It is completely reusable within any structured and unstructured CFD code, and offers a new level of grid and application flexibility.

P.P. Kranzinger (✉) • U. Kowarsch • M. Schuff • M. Keßler • E. Krämer
Institut für Aerodynamik und Gasdynamik, Universität Stuttgart, Pfaffenwaldring 21, 70569 Stuttgart, Germany
e-mail: kranzinger@iag.uni-stuttgart.de

© Springer International Publishing Switzerland 2016
W.E. Nagel et al. (eds.), *High Performance Computing in Science and Engineering '15*, DOI 10.1007/978-3-319-24633-8_31

With these new options available, an earlier H145 isolated rotor investigation was extended to a full helicopter configuration. The influence of the fuselage to the aeroacoustic noise emission, in terms of e. g. fuselage shading, reflection and diffraction effects, will be investigated. Furthermore, for acoustic investigation the anti-torque Fenestron system is taken into account as well.

2   Numerical Methods

2.1   Computational Fluid Dynamics (CFD)

For high-fidelity aerodynamic investigations, the block structured finite volume Reynolds-averaged Navier-Stokes (RANS) CFD code FLOWer, initially developed by the German Aerospace Center (DLR) [3], is used. For the closure of the RANS equations various types of turbulence models are available. For helicopter and contra rotating open rotor investigations, the Wilcox k-ω turbulence model [4] with a fully turbulent far field flow state has shown best robustness and accuracy. The time discretization is achieved by integrating the governing differential equations with the implicit dual time-stepping approach according to Jameson [5]. The consideration of grid motions using an Arbitrary Lagrangian Eulerian approach enables the code for helicopter flow simulation. In addition, the Chimera technique for over-set grids simplifies the meshing of complex helicopter geometries like rotor-fuselage configurations including relative grid movements. To consider the effects of fluid-structure interaction on the rotor blade, the mesh is deformed to a given structural deformation of the blade in each time step. The efficiency of the computation is achieved by a multi-block structure of the grid to enable parallel computing.

Within the last years, the CFD solver was extended by IAG [6] with different methods of fifth order spatial WENO schemes to guarantee a detailed conservation of the flow field and especially vortices. Besides the basic WENO scheme according to Jiang [7], an improved order preserving WENO scheme near discontinuities denoted as WENO-Z according to Borges [8] is available. The compact reconstruction WENO (CRWENO) scheme [9] also available in FLOWer improves the spectral resolution with a higher efficiency than the basic WENO scheme. Irrespectively of the WENO scheme in use, the base reconstruction at the cell boundaries results in a Riemann problem, which is solved using the upwind HLLC scheme of Toro [10]. The viscous fluxes are solved with conventional central differences of second order accuracy.

2.2 Flight Mechanics (FM) and Structural Dynamics (CSD)

For helicopter applications, a proper reproduction of the flight state including the aero-elasticity of the rotor blades is mandatory. Especially in forward flight, blade elasticity influences the aerodynamic behavior and force generation substantially.

At IAG, the structural deformation of the rotor blades is modeled using the Comprehensive Analytical Model of Rotorcraft Aerodynamics and Dynamics (CAMRAD II) code [11] as part of a weak coupling scheme: The CAMRAD II code provides solutions for the blade deformation and flight kinematics modeling the rotor blades as Euler-Bernoulli beams with isotropic material and elastic axis. For aerodynamics load estimations a low-fidelity aerodynamics model based on lifting line theory and two dimensional steady airfoil data tables is used. The initial deformation and trim-angle values provided are used for performing a CFD based aerodynamics simulation, providing load results of high-fidelity. By correcting the internal low-fidelity loads evaluation of CAMRAD II with the CFD results, the CSD internal aerodynamics are successively replaced by the CFD solution, leading to blade dynamics based on CFD loads with deformation and deflection calculated with CSD.

In order to fit the specified global forces and moments, in case of an isolated rotor trim the collective and two cyclic control angles are determined with a fixed rotor orientation, known as the wind-tunnel trim. For a complete helicopter, three additional degrees of freedom are taken into account for the spatial fuselage orientation and the tail rotor thrust. This approach is designated as a free-flight trim.

3 Tool Chain Improvements

3.1 Advanced CFD-CSD Coupling Implementation

For more than a decade now, the performance of a single computing core has been nearly stagnating. Present day's high performance computing clusters offer 100,000 cores and more instead. Thus, efficient parallelization is a key skill for being able to use the power of current and future computation clusters.

Due to restrictions on the mesh topology when using the former CFD-CSD coupling implementation, the preparation of rotor blade meshes in consideration of efficient parallelization was strictly limited [12]. After completing the block splitting process, some blocks remained, whose geometric dimensions could not be further shrunk. Increasing the spatial discretization would lead to a higher number of grid nodes per block. Hence, the power of next generations of supercomputers could not be used adequately.

In addition, up to now resulting line loads and integral sectional loads were also evaluated based on the same mesh topology depending scheme. Geometric changes e.g. of the twist angle distribution or due to parameter variations describing the base shapes of rotor blades, cause a variation of the load evaluation points, which entails uncertainties when comparing results of parameter studies. Common loads for overset grids could not be evaluated considering the overlapping area appropriately, which severely constrained, the usage of Chimera setups for modeling rotor blades, a promising method for increasing the mesh density at the blade tip locally.

To overcome these limitations and because of the ongoing development of the unstructured Discontinuous Galerkin code SUNWinT [13] and the free wake method FIRST, a mesh deformation algorithm and a load integration algorithm were developed. Both algorithms are implemented as separate object oriented C++ libraries. The key design goals were reusability on current and future codes, a software design which allows easy extensions for future applications, no limitations concerning mesh topologies, and attaining the performance of the former FLOWer internal implementation.

For achieving these aims, radial basis functions (RBF) have been selected as backbone for the deformation algorithm. The load integration algorithm uses a three-dimensional spatial search tree (octree) for evaluating overlapping regions and computing sectional and line loads.

Both algorithms are able to handle overlapping Chimera structures and structured as well as unstructured meshes, respecting relative motions (e.g. if slotted leading and trailing edge flaps are modeled) [14].

3.1.1 Deformation

Radial Basis Function Interpolation

RBF can be used for interpolation of discrete data in an n-dimensional space. The deformation of a three-dimensional CFD mesh can thus be interpreted as an interpolation of the known discrete deformation of the surface in the surrounding area.

RBFs are real-valued functions whose value only depend on a certain distance [15], so that

$$\phi(\mathbf{x}, \mathbf{d}) = \phi(\|\mathbf{x} - \mathbf{d}\|)$$

where $\mathbf{x}$ is the evaluation point and $\mathbf{d}$ the point of reference. These functions can be used for interpolating data by defining a function for each data point and summing them up. When deforming volume meshes, the data point is a point on a surface and the data to be interpolated is its displacement in the deformed state. By adding a constant and linear polynomial to this sum, offsets and linear parts of the data are represented directly analytically. When handling CFD meshes, translation and rotation in space can be interpreted as x, y, z-offset and a linear function of x, y, z.

The displacement of an arbitrary point $\mathbf{p}$ may be described as

$$s(\mathbf{p})_{x,y,z} = \sum_{i=1}^{n} \left(\alpha_{i\ x,y,z} \phi \left(\|\mathbf{p} - \mathbf{p}_{surf_i}\| \right) \right) + q(\mathbf{p})_{x,y,z}$$

where α_i is a weighting factor for each RBF and $q(\mathbf{p})$ is an arbitrary linear polynomial [16]:

$$q(\mathbf{p})_{x,y,z} = [\beta_0\ \beta_1\ \beta_2\ \beta_3]_{x,y,z} \cdot \begin{bmatrix} 1 \\ \mathbf{p}_x \\ \mathbf{p}_y \\ \mathbf{p}_z \end{bmatrix}$$

The requirement is that there is a number of n reference points—or *surface points*—$\mathbf{p}_{surf}$, that have a known deformation.

Thus, the displacement of these surface points must exactly map with the $s(\mathbf{p}_{surf})$ such that

$$s(\mathbf{p}_{surf_i})_{x,y,z} = \mathbf{d}_{b_i\ x,y,z}$$

with $\mathbf{d}_{b_i}$ being the known displacement of the deformed surface points.

As averaged rigid body motions should be solely covered by the polynomial part $q(\mathbf{p})$, the requirement

$$\sum_{i=1}^{n} \alpha_i \mathbf{P}_{b_i} = 0$$

with

$$\mathbf{P}_{b_i} = \begin{bmatrix} 1 & \mathbf{p}_{surf_i\ x} & \mathbf{p}_{surf_i\ y} & \mathbf{p}_{surf_i\ z} \end{bmatrix} \tag{1}$$

has to be fulfilled [17].

Eventually, an equation system can be built and the coefficients for the basis functions (α) and the rigid body motions (β) can be computed by inverting the resulting dense matrix $\mathbf{M}_{complete}$:

$$\begin{bmatrix} \mathbf{d}_{b\ x,y,z} \\ 0 \end{bmatrix} = \underbrace{\begin{bmatrix} \mathbf{M}_{bb} & \mathbf{P}_b \\ \mathbf{P}_b^T & 0 \end{bmatrix}}_{M_{complete}} \begin{bmatrix} \alpha \\ \beta \end{bmatrix}$$

where $\mathbf{M}_{bb}$ with row i and column j is the evaluation of the basis functions for the reference points between each other $\phi(\mathbf{p}_{surf_i}, \mathbf{p}_{surf_j})$ and $\mathbf{P}_b$ is an $n \times 4$ matrix with the row i given as in (1).

As $\mathbf{M}_{complete}$ contains no information of the actual displacement, but only information of the distances of each node to every other in non-deformed state, its inversion needs only to be executed once as long as the non-deformed surfaces will not change.

Detailed description of RBFs can be found in [15]. A detailed study of different RBFs for deforming volume meshes has been done by [17].

Performance

The performance of a RBF based deformation algorithm depends on two things. Firstly, the inversion of a densely populated $(n_{surf} + 4) \times (n_{surf} + 4)$ matrix. As a rule, this is a $\mathcal{O}(n_{surf})^3$ problem and can be solved by LU decomposition or Gaussian elimination. Secondly, applying each RBF coefficient to each volume grid node, where the number of radial basis function coefficients is equal to the number of selected surface mesh points. Hence, the complexity for each time step is $(n_{surf} \cdot n_{volumemesh})$. As n_{surf} can be kept mostly constant with increasing case size (as shown in [14]) the resulting deformation algorithm complexity is $\mathcal{O}(n_{volumemesh})$.

For computational setups with about 35,000 mesh points per MPI process, representing an average per process workload for the midterm future, the wall time required for applying the deformation is < 1 s, which is comparable to the performance of the original FLOWer internal algorithm. Increasing the case size will not affect the wall time required for deforming the meshes, as the number of mesh points per MPI process is kept constant.

Validation

A trim validation with an isolated five bladed rotor at a flight speed of 125 kts has been conducted. As reference, the former Hermite interpolation deformation was used [12].

In Fig. 1 the pitch angles are plotted against the iteration steps for both trim runs. The convergence of these angles is almost the same, with some neglectable discrepancy which is due to a slightly different twist of the elastic rotor blade

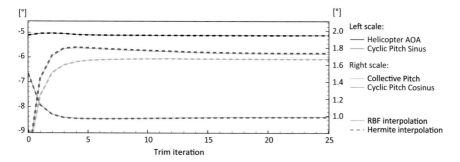

Fig. 1 Angles of pitch and helicopter's angle of attack (AOA) during trim iterations

Fig. 2 Deformed near surface meshes of strongly coupled tail boom in conjunction with a weakly coupled main rotor

because of the misinterpretation of the beam angles in the former algorithm (as mentioned in [14]).

Further Applications

The implementation allows the usage of different information sources to deform various components of the complete simulated CFD environment separately. Exemplarily, the deformation of a ground boundary layer to represent ground roughness for in-ground-effect forward flight simulations, or strongly coupled tailboom deformation in combination with weakly coupled structural rotor dynamics (cf. Fig. 2), have become possible.

3.1.2 Load Integration

The load integration algorithm collects the complete surfaces of all simulation components as specified from the CFD solver, disassembles all cell faces to triangles, and sorts the resulting triangles by their center of area into an octree. For a helicopter rotor blade, the resulting octree is exemplarily shown in Fig. 3.

For respecting overset areas of surfaces, an algorithm was implemented based on the octree data structure, which allows finding all overset objects with a complexity of $n \log(n)$. The actual overlap is computed by projecting two possible overlap candidates to a common plane. The overlapping area is identified by a geometric comparison and summed up for each triangle separately. The actual overlap is later on respected by defining a specific overlap factor f_o for each element, which is

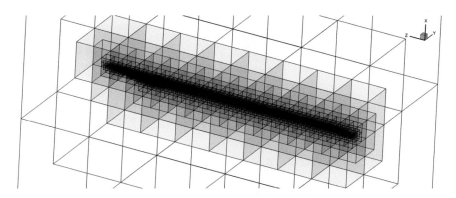

Fig. 3 Visualization of resulting octree of a helicopter rotor blade

always taken into account when computing loads and moments:

$$f_o = 1 - 0.5 \cdot \frac{A_{overlapping}}{A_{complete}}$$

Hence, for a complete overlap the factor results to 0.5; for no overlap it is 1.0. Table 1 shows validation results for overlapping rotor blade surfaces.

Line loads are now mesh topology independently evaluated at defining planes, cutting all components selected for load integration. Using the octree, all triangular faces cut by the plane are identified. All segments resulting from cutting the faces by the plane build up together a two-dimensional outline of the cut structure on the cutting plane. Line loads are now calculated by summing up the length of the resulting segments l_i multiplied with the specific overlap factor $f_{o, i}$, and the three-

Table 1 Global loads and moments for partly, and completely overlapping rotor blade surfaces in relation to stand-alone blade surface

	No overlap	Partly overlapping (‰)	Completely overlapping (‰)
ΔF_x	Reference	1.19	0.68
ΔF_y	Reference	1.03	0.73
ΔF_z	Reference	1.46	0.73
ΔM_x	Reference	0.90	0.63
ΔM_y	Reference	1.02	0.63
ΔM_z	Reference	1.10	0.55

dimensional force vector of all cut triangles normalized by their surface areas $\mathbf{f}_i/A_i$:

$$\mathbf{f} = \sum_{i=1}^{n_{cut}} l_i \cdot f_{o,i} \cdot \frac{\mathbf{f}_i}{A_i}$$

Resulting moments are evaluated the same way relating to a global or sectional reference point. Therefore, the distance vector from the reference point to each segment center $\mathbf{d}_i$ is calculated and taken into account

$$\mathbf{m} = \sum_{i=1}^{n_{cut}} l_i \cdot f_{o,i} \cdot \frac{\mathbf{f}_i}{A_i} \times \mathbf{d}_i.$$

Integral sectional loads, demanded e.g. by CAMRAD II, are also evaluated by defining cutting planes. However, all triangles between two cutting planes are taken into account with their complete three-dimensional surfaces. The cut triangles are split into a part, which lies in the specific section, and a part outside of the specific section.

In all cases, the resulting algorithm has a complexity of $\mathcal{O}\left(n_{surfpts} \cdot log(n_{surfpts})\right)$. For current computation setups with up to 400 million volume mesh cells and an appropriate surface discretization, the wall time required for computing all loads required for post-processing and CFD-CSD coupling is below 100 ms (Fig. 4).

3.2   Improvements in Code Scaling

The suitability of the FLOWer code has been demonstrated in the past in several successfully performed highly parallel simulations, which have been presented at the HLRS annual user workshop [6]. The CFD code is parallelized by the distribution of the overall grid cells into work packages for each MPI rank defined by mesh blocks. Each MPI rank runs on a physical CPU core, whereby all available cores per node are used. At each block boundary, data exchange in terms of an MPI message has to be performed to proceed the global solution. This leads to a large number of MPI messages. When increasing the number of MPI processes, the wall time required for exchanging the blocks' boundary data became dominant. It was found, that the performance limitations were not caused by the total size of data exchanged during the rank-to-rank communication, but the number of MPI messages itself was limiting the performance.

This issue was overcome by introducing intra-node shared memory with one MPI process managing all MPI-ranks' boundary exchanges on the node. Thereby, a transition from core-to-core to node-to-node communication was achieved. Intra-node boundary communication is directly performed over the MPI-decoupled shared memory. In addition, OpenMP directives were implemented for a hybrid

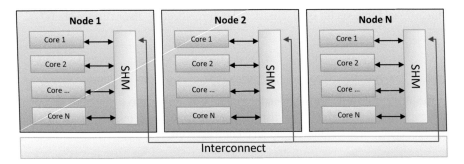

Fig. 4 Methodology of advanced MPI communication on a SMP system

parallelization approach, where memory demands prohibit the utilization of all cores in a node for MPI ranks.

A pure MPI scaling study was performed on the Hermit, and on the new Hornet system to evaluate the current efficiency of the CFD code FLOWer. A reference simulation with 135 million grid cells served as basis, which represents an average simulation size for the forthcoming research topics. The FLOWer code was compiled using the ifort respectively the crayftn compiler available on the Hornet, respectively on the Hermit user environment.

As basis served the numerical setup used for complete helicopter simula-tions featuring i.e. full viscous flux computation incl. turbulence modeling, and convergence acceleration based on the multigrid method. Running the reference simulation on the Hermit using 4.096 cores, computation of one subiteration takes 53.58 μs/grid point. Using the updated MPI inter-process communication this value could be reduced to 44.72 μs/grid point. In typical helicopter simulations 30–60 subiterations lead to numerical convergence within a time step.

3.2.1 Strong Scaling

The usual core count for the strong scaling simulation size is in the range between 2,000 and 6,000 cores depending on the available computational resources and time constraints (cf. Fig. 5a). The strong scaling shows highly satisfying results with only minor deviations from ideal scaling.

3.2.2 Weak Scaling

In case of the weak scaling the test run using 4,008 cores equals the strong scaling simulation case with 135 million grid cells. Slight unexpected variations are observed with a speed-up over 1.0 with increasing core amount (cf. Fig. 5b). However, this variation is within the usual scattering of node performance and intra-node loading, which is observed to vary in the range of ±10 %. Even the 270 million

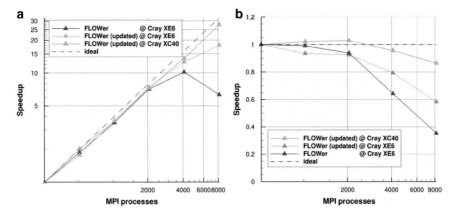

Fig. 5 Pure MPI FLOWer scaling study on HLRS systems. (**a**) Strong scaling (**b**)Weak scaling

cells test case using 8016 cores shows reasonable performance with ∼85 % of the reference computational speed with a total memory consumption of 4.06 TB.

Studies not listed here have shown efficient scaling using OpenMP in addition to the node-to-node based communication up to 32,000 cores.

4   Complete Helicopter Simulation

With the higher order methods and the improved parallelization, a further step in the investigation of fluid-structure interaction phenomena can be launched. Especially simulations of aeroacoustic noise benefit from a detailed preservation of pressure disturbances to map interaction effects.

For an engineering-oriented investigation an H145 helicopter configuration of Airbus Helicopters is investigated. To evaluate possible further improvements, the demand for a highly detailed investigation of the aerodynamics is certainly challenging.

4.1   Setup and Simulation

Figure 6 shows the helicopter surface and the mesh components highlighted with different colors. Overall 59 separate meshes, partly generated with an automated mesh generator [18], are required to reproduce the helicopter surface including their different relative movements, e.g. the rotor-blades, rotor-head and Fenestron tail-rotor system. Edges mark the block splitting within the individual meshes, to enable an efficient parallelization of the simulation on high performance systems. A maximum of 38,000 cells in each grid unit is not exceeded, which defines

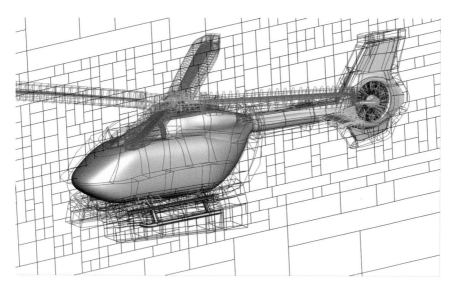

Fig. 6 Mesh blocking of the body grids

Table 2 Grid components of complete helicopter CFD setup

Component	No. of blocks	No. of cells (mio.)
Background	6563	106.3
Main rotor blade	4× 230	4× 6.9
Blade root	4× 56	4×1.0
Rotor hub system	723	9.2
Fuselage	1365	26.7
Skid system	375	7.1
Fenestron® stator + rotor	670	11.9
Total	10876	191.7

the parallelization limits of the setup. The overall mesh size is 192 million grid cells, which represents a high-fidelity simulation in the field of helicopter CFD investigations. The different grid components are listed in Table 2. Focus is set on vortex convection and mapping of vortex-structure interactions using the higher order methods. Since the method works best for Cartesian meshes, the body meshes are embedded in a Cartesian background mesh with refinements in the vicinity of the helicopter's surface. The usage of hanging grid nodes enables a coarsening towards the far field to ensure a reasonable grid size. With this approach more than 50 % of the overall cells are located in the Cartesian background mesh to guarantee best higher order results.

Different numerical methods to improve the physical behavior of the helicopter are taken into account. Besides aeroelasticity of the rotor blades, the mass flow through the engines is considered by a prescribed pressure at the inlet and exhaust.

The exhaust flux is furthermore prescribed with the average exhaust temperature of this helicopter type. Especially the mass flux influences the fuselage wake significantly in terms of occurrence and extent of separation areas.

As presented in previous HLRS annual reports of the HELISIM project, a free-flight trim technique has been established, which has been applied for this investigation [19].

The convergence is mainly dependent on physical processes in the flow field. As numerical convergence is already achieved after a few time steps, physical convergence is the limiting factor. To achieve physical convergence of the flow field starting vortices must have convected out of vicinity of the helicopter. For this case, three complete main rotor revolutions were required until all aerodynamic structures caused by startup effects have left the area of concern. Physical convergence of the complete CSD-CFD-coupled model has been achieved after eight trim iterations. For each trim iteration one complete rotor revolution has been computed with a time step of 0.5°. For better acoustic analysis, three further revolutions with a shorter time step of 0.25° has been computed after achieving convergence of the complete model.

4.2 Aerodynamic Results

Using the higher order methods ensures a highly preserved vortex structure around the helicopter (cf. Fig. 7). Besides the preservation of the blade vortex interaction (BVI) relevant main rotor blade tip vortices, the turbulent wakes of the fuselage and

Fig. 7 λ_2-visualization of the flow field around the H145 in a BVI relevant descent flight with pressure contours on the surface. Hot areas caused by the engine exhaust are marked in blue

Fig. 8 Azimuthal (time) derivative of the sectional lift coefficient $(c_l/d\Psi)$

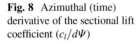
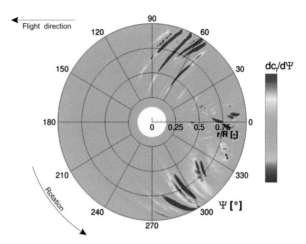

especially of the skids are visible. The area of high temperatures are marked in blue, which result from the engine exhaust. The detailed preservation of the wake ensures a realistic behavior of the Fenestron tail rotor operation in terms of provided thrust and therefore required control angles.

One of the primary drivers causing aeroacoustic impulsive noise emission by the rotor blade is the time rate of load change. Therefore, the azimuth angle was taken as the independent variable and the azimuthal derivative was formed. Figure 8 shows this azimuthal derivative of the sectional lift coefficient c_l in polar coordinates for one rotor revolution. The BVI events are clearly visible on the advancing blade side as well as on the retreating blade side in terms of high lift fluctuations. In case of the advancing side, four strong interaction events are seen in the area between $\Psi=40°$ and $90°$. All interactions show a strong gradient with short interactions periods. Taking the timing of the events into account, the first and third event show a nearly coherent interaction over the blade span at the same time. These are first indications of events with high noise emission due to a coherent interference of the caused pressure disturbances.

Focusing on the retreating side, high load fluctuations are found in the azimuth range between $\Psi=270°$ and $330°$, mostly symmetric to the advancing side. The interaction events show a stretched temporal influence compared to the advancing side, where the events are limited to short periods. A higher variation of the level of influence between BVI events is present compared to the advancing side, where all events have comparable magnitudes.

A more detailed investigation of a previous isolated rotor simulation is found in [2]. The analysis of the aeroacoustic noise emission of the full helicopter configuration is published in [20].

5 Conclusion

Generalized, high performance mesh deformation and load evaluation libraries have been developed and validated, that are suitable for structured, unstructured and overlapping meshes. A major achievement is the flexibility, re-usability and performance in highly distributed environments. Validation results show excellent agreement with former implementations. Further applications, e.g. deforming ground boundary layers for simulating in-ground-effect forward flights above rough terrain, or for analyzing ground roughness influence to wind turbines, and deformation of slotted and elastic flaps and slats, are in development.

The scaling investigation shows the high efficiency of the current FLOWer version operating on the HLRS Hornet system. These recent upgrades of the code's architecture in terms of MPI process data exchange allow the fulfillment of all demands of todays' high performance computing clusters in terms of parallelization. From our point of view, currently no additional updates are required to increase the code performance.

Furthermore, the present paper shows significant advantages using the higher order WENO reconstruction method newly implemented by IAG into the CFD code FLOWer. Especially in combination with the HLLC Riemann solver a significantly higher vortex conservation is achieved. Even if a higher computational effort is required, a computation with as much as 70 % less grid cells compared to a *2nd* order scheme still shows a more precise solution. The presented application to a helicopter forward flight situation shows obvious benefits of the higher accuracy of the fifth order scheme. Remarkable results are achieved in the computation of the acoustic noise emission being the outcome of a more detailed vortex topology and a good preservation of vortex strength.

Acknowledgements We greatly acknowledge the provision of supercomputing time and technical support by the High Performance Computing Center Stuttgart (HLRS) for our project HELISIM.

References

1. Kutz, B.M., Günther, T., Rumpf, A., Kuhn, A.: Numerical examination of a model rotor in brownout conditions. In: Proceedings of the American Helicopter Society, 70th Annual Forum, Montreal, May 2014
2. Kowarsch, U., Lippert, D., Schneider, S., Keßler, M., Krämer, E.: Aeroacoustic Simulation of an EC145T2 Rotor in descent flight. In: 71th American Helicopter Society Annual Forum, Virginia, May 2015
3. Kroll, N., Eisfeld, B., and Bleeke, H.: The Navier-Stokes code FLOWer. Notes Numer. Fluid Mech. **71**, 58–71 (1999)
4. Wilcox, D.: Multiscale model for turbulent flows. AIAA J. **26**(11), 1311–1320 (1988)
5. Jameson, A.: Time dependent calculations using multigrid, with applications to unsteady flows past airfoils and wings. In: AIAA 10th Computational Fluid Dynamics Conference, Honolulu (1991)

6. Kowarsch, U., Oehrle, C., Hollands, M., Keßler, M., Krämer, E.: Computation of helicopter phenomena using a higher order method. In: High Performance Computing in Science and Engineering '13. pp. 423–438. Springer, Berlin (2013)
7. Jiang, G.-S., Shu, C.-W.: Efficient Implementation of Weighted ENO Schemes. J. Comput. Phys. **126**, 202–228 (1996)
8. Borges, R., Carmona, M., Costa, B., Don, W.S.: An improved weighted essentially non-oscillatory scheme for hyperbolic conservation laws. J. Comput. Phys. **227**(6), 3191–3211 (2008)
9. Ghosh, D., Baeder, J.: Weighted non-linear compact schemes for the direct numerical simulation of compressible, turbulent flows. J. Sci. Comput. **61**(1), 61–89 (2014)
10. Toro, E.F.: Riemann Solvers and Numerical Methods for Fluid Dynamics. Springer, Berlin (1997)
11. Johnson, W.: CAMRAD II Comprehensive analytical model of rotorcraft aerodynamics and dynamics, 4th edn. Johnson Aeronautics, Palo Alto (2009)
12. Dietz, M.: Simulation der Umströmung von Hubschrauberkonfigurationen unter Berücksichtigung von Strömungs-Struktur-Kopplung und Trimmung, Ph.D. thesis, Institut für Aerodynamik und Gasdynamik, Universität Stuttgart (2009)
13. Wurst, M., Keßler, M., Krämer, E.: Detached eddy simulation using the discontinuous galerkin method. In: Dillmann, A., Heller, G., Krämer, E., Kreplin, H.-P., Nitsche, W., Rist, U. (eds.) New Results in Numerical and Experimental Fluid Mechanics IX. Notes on Numerical Fluid Mechanics and Multidisciplinary Design, vol. 124, pp. 435–442. Springer International Publishing, New York (2014)
14. Schuff, M., Kranzinger, P.P., Keßler, M., Krämer, E.: Advanced CFD-CSD Coupling: Generalized, High Performant, Radial Basis Function Based Volume Mesh Deformation Algorithm for Structured, Unstructured and Overlapping Meshes, 40th European Rotorcraft Forum, Southampton (2014)
15. Buhmann, M.D.: Radial Basis Functions: Theory and Implementations, 8th edn. Cambridge University Press, Cambridge (2008)
16. Beckert, A., Wendland, H.: Multivariate interpolation for fluid-structure-interaction problems using radial basis functions. Aerosp. Sci. Technol. **5**(2), 125–134 (2001)
17. De Boer, A., Van der Schoot, M., and Bijl, H.: Mesh deformation based on radial basis function interpolation. Comput. Struct. **85**(11), 784–795 (2007)
18. Kranzinger, P.P., Hollands, M., Keßler, M., Wagner, S., Krämer, E.: Generation and Verification of Meshes Used in Automated Process Chains to Optimize Rotor Blades, Paper AIAA-2012-1260, 50th AIAA Aerospace Sciences Conference, Nashville, TN (2012)
19. Embacher, M., Keßler, M., Dietz, M., and Krämer, E.: Coupled CFD-simulation of a helicopter in free flight trim. In: Proceedings of the American Helicopter Society 66th Annual Forum, Phoenix, 11–13 May 2010
20. Kowarsch, C., U. Öhrle, Schneider, S., Keßler, M., Krämer, E.: Aeroacoustic Simulation of a Complete EC145T2 Helicopter in Descent Flight, 41th European Rotorcraft Forum, Munich (2015)

Numerical Study of Three-Dimensional Shock Control Bump Flank Effects on Buffet Behavior

R. Mayer, D. Zimmermann, K. Wawrzinek, T. Lutz, and E. Krämer

Abstract Originally developed as a flow control device Shock Control Bumps (SCB) reduce wave drag of an aircraft wing at off-design in transonic speed effectively. Recently, another field of application for such bumps has been studied, namely the delay and alleviation of buffet, an unsteady shock motion due to continuous flow separation and re-attachment at the rear part of the airfoil. In principle the idea of buffet alleviation is the use of SCB as a sort of 'smart' vortex generator. Considerable effort has been undertaken to link geometrical bump features to buffet affecting flow characteristics. In this paper a parametric study on the influence of flank shape of a three-dimensional wedge-shaped SCB on its performance and buffet behavior is presented. It has been found that performance as well as buffet behavior can be improved by optimization of the bump flanks. The study shows that length of front and rear flank should be increased up to given constraints (e.g. flaps on a wing or inserts for a wind tunnel model) and a narrow front and wide rear flank increase $c_{L,max}$ and damp lift oscillations at buffet onset.

1 Introduction

First introduced by Ashill, Fulker and Shires [1] in 1992, Shock Control Bumps (SCB) are a passive method of flow control in transonic flow regime with the objective of reducing wave drag. Accurately designed, wave drag of an airfoil resp. a wing is negligible at design conditions. However, increasing angle of attack or Mach number leads to a strong increase of wave drag, referred to as transonic drag rise. This drag rise normally reduces flight performance significantly. Therefore, SCBs can help to improve performance at airfoil/wing off-design conditions (see e.g. [13]).

Recent studies focus on another field of application for SCBs, namely the alleviation of the buffet phenomenon. Buffet limits flight Mach number and maximum lift of an airliner. It is characterized by a continuously separation and

R. Mayer (✉) • D. Zimmermann • K. Wawrzinek • T. Lutz • E. Krämer
Institute of Aerodynamics and Gas Dynamics, University of Stuttgart, Stuttgart, Germany
e-mail: r.mayer@iag.uni-stuttgart.de; zimmermann@iag.uni-stuttgart.de;
wawrzinek@iag.uni-stuttgart.de; lutz@iag.uni-stuttgart.de; kraemer@iag.uni-stuttgart.de

© Springer International Publishing Switzerland 2016   495
W.E. Nagel et al. (eds.), *High Performance Computing in Science
and Engineering '15*, DOI 10.1007/978-3-319-24633-8_32

re-attachment of the boundary layer, resulting in shock and lift oscillations. It has been shown [2, 3] that SCBs can have a double benefit by reducing wave drag at off-design conditions and increasing lift coefficient at which buffet indicating lift oscillations first occur. In general, the intention is to use SCBs as a novel class of 'smart' vortex generators and thereby introducing stream-wise vortices in the flow. These vortices act as a sort of boundary layer control and prevent the flow downstream of the bump from separating. The precise mechanism by which the vorticity is generated by the bump is still unsolved and subject of current research (e.g. [3, 6, 10]). Recent publication [7] indicates that the span-wise pressure gradient present on the front flank of a SCB is strongly related to the vorticity generated by the bump. In addition, the rear flank design is correlated to flow separation present on the SCB. Bruce and Colliss [5] state that a wider tail can be beneficial at off-design conditions by reducing re-acceleration over the crest to obtain a reduction in the extent of local separation around the SCB crest region.

This paper presents a parametric study analyzing the effect of front and rear flank shapes of a generic wedge-shaped three-dimensional SCB on performance and buffet behavior of the airfoil. First, the numerical methods used as well as the baseline airfoil and bump design are presented. Then, the effect of flank shape on different parameters like aerodynamic efficiency, maximum lift coefficient and lift oscillations at buffet onset are discussed in detail. Finally, trends for the flank shape of buffet alleviating bumps are derived.

2 Numerical Methods

2.1 Flow Solver

FLOWer [15] is a block-structured RANS solver developed by German Aerospace Center (DLR) to meet the demands of aircraft aerodynamics. In order to determine a suitable setting for buffet simulations, this flow solver has been validated using the well documented OAT15A airfoil. The buffet characteristics of this airfoil have been analyzed excessively in wind tunnel experiments [12], providing a valuable database for buffet calibration of flow solver settings. Experimental buffet onset has been found at an angle of attack (AoA) of $\alpha = 3.25°$ with a buffet frequency of $f_{buffet} \approx 70\,Hz$. Three-dimensional (U)RANS simulations have been carried out with the settings discussed in Sect. 2.2 to analyze the buffet behavior of the OAT15A airfoil numerically. Out of a range of different turbulence models tested (Spalart-Allmaras (SA), SA-Edwards, SA-salsa, SST-k-ω and LEA-k-ω) the *Strain Adaptive Formulation of Spalart-Allmaras One-Equation Model* (SA-salsa) performed best, finding buffet onset (here defined as the lowest angle of attack at which URANS simulations capture lift oscillations) at $\alpha = 3.2°$ with an accuracy of $\Delta\alpha = 0.1°$. The buffet frequency was computed to $f_{buffet} \approx 73\,Hz$ which is in excellent agreement with the experimental results.

2.2 Computational Set-up

For all flow simulations, the cell-centered approach has been applied. The computations have been carried out with a central scheme (according to Jameson) for spatial discretization. Temporal discretization is achieved with a central Runge-Kutta scheme. For convergence acceleration, a 3-stage sawtooth-V-cycle multi-grid has been applied and the residual has been smoothed with variable coefficients according to Swanson. For CFL number varying between 1.0 and 7.5, computations have been carried out up to a residual of less than $1.0e^{-6}$. Based on the results from Sect. 2.1, 1-equation turbulence model SA-salsa has been used.

In order to prevent laminar shock-boundary-layer interactions in the numerical simulations and thereby reduce numerical robustness considerably, transition is tripped at 35 % of chord, just upstream of the most upstream shock location in case of buffet.

For URANS simulations, a dual time stepping scheme with a time step size of $5.7e^{-5}$ s has been applied, resolving an estimated buffet frequency of 70 Hz with approximately 250 time steps as recommended by several authors [3, 17]. The mean values as well as buffet characteristics have been determined by analyzing eight buffet cycles in fully established flow.

2.3 Grid

All simulations presented in this paper base on structured H-C type, multi-block meshes with boundary layer refinement which have been generated script-based, ensuring consistently high quality meshes for each SCB configuration and thereby reducing the grid influence on the solution to a minimum. A grid convergence study for those meshes can be found in [14]. Nevertheless, a short overview of the mesh topology shall be given here.

For the three-dimensional simulations the grid around the airfoil (see Sect. 3.1) has been extruded 30 % of chord length in span-wise direction. The far-field extends to 50 chord lengths from the airfoil. A boundary layer refinement of the grid leads to a first cell height of $y_1^+ \approx 0.8$. For better physical agreement with experimental results the scripts for automated mesh generation from [14] have been adapted to blunt trailing edge airfoils. In total, the meshes consist of approximately 6.0 million cells with 352 cells around the airfoil, 32 cells on the trailing edge and 64 cells in the wake in stream-wise direction. In accordance to [14] the number of span-wise cells has been set to 120.

3   Baseline Geometry

All calculations presented in this paper base on the OALT25 airfoil, further discussed in Sect. 3.1. Corresponding to the design condition of this airfoil, all simulations have been carried out for $Re = 13e^6$ and $Ma = 0.73$. Stagnation temperature has been set to $T_{st} = 300$ K.

3.1   Airfoil

As baseline airfoil OALT25 developed by ONERA has been used. Reference condition of this airfoil are $Ma = 0.73$, $Re = 13e^6$ and $AoA = 1.0°$. With a relative thickness of 12.18 % and a trailing edge thickness of 0.5 % of chord, this transonic airfoil has been designed for delayed transition. In the S3Ch wind tunnel at ONERA, the boundary layer remains laminar up to the shock at around 60 % of chord.

3.2   Shock Control Bump

In accordance to previous studies by Bogdanski [3], who analyzed the buffet behavior of SCBs designed for drag reduction at high lift coefficients, a Wedge-SCB has been optimized in height and position for minimum drag coefficient as a first step towards a buffet alleviating bump. Figure 1 shows the optimized Wedge-SCB on the suction side of the airfoil placed near the shock foot. The bump parameterization allows for detailed adaption to the flow condition but has also a rather simple shape which is essential for parametric studies on geometrical effects. The upstream flank, the crest plateau as well as the rear flank are plain. The side flanks are modeled by a 4th order polynomial using the base point (intersection between bump and airfoil), the top point (intersection between side flank and mid-section), the side flank angle at the base point as well as the curvature (4th derivative of polynomial) at the top point at each stream-wise position. The height of the bump refers to the crest plateau of the bump.

Using a SIMPLEX algorithm, the bump parameters have been optimized for minimum drag coefficient. The bump design point (BDP) has been chosen carefully based on an elaborate analysis of the baseline airfoil. At $\alpha_{BDP} = 1.7°$ respectively $c_{L,BDP} \approx 0.8$, MSES simulations [9] indicate that wave drag contributes with around 17 % to total drag, offering sufficient potential for drag reduction by SCB. In addition, an analysis of shock location for an AoA range between airfoil design point and buffet onset revealed a downstream movement of the shock with increasing AoA up to $\alpha \approx 2.5°$. For higher AoA the shock moves upstream again. Since bump position in relation to shock location affects the performance of the bump

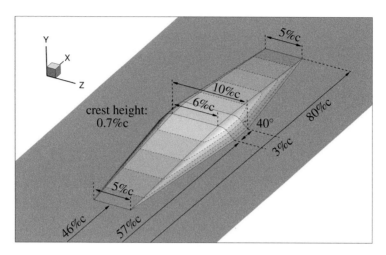

Fig. 1 Baseline Wedge-SCB with parametrization

significantly, the design AoA of the bump has been chosen to have the same shock location as at buffet onset.

At bump design condition, this bump reduces drag by 8.3 % leading to an improved aerodynamic efficiency of 9.2 %. A well known disadvantage of SCBs is their performance deterioration at bump off-design conditions. At design condition of the airfoil drag is increased by 20.7 %, reducing aerodynamic efficiency by 19.1 %.

In contrast to studies by Bogdanski [3], this baseline SCB does not improve the buffet behavior of the OALT25. The reason is a significant difference between highest AoA ($\alpha = 2.4°$) at which RANS simulations converged and buffet onset AoA ($\alpha = 3.4°$), not allowing for bump optimization close to buffet onset. Obviously, URANS simulations would be more appropriate for buffet alleviating bumps but these simulations are to costly for three-dimensional bump optimization.

4 Influence of Flank Shape

As found by other researchers [5, 6] the flanks of the bump affect the buffet behavior. Colliss [7] found that the vorticity generated by the bump strongly depends on the maximum span-wise pressure gradient on the front flank of the bump. Thinking of buffet alleviating bumps as a kind of 'smart' vortex generators it seems to be beneficial to maximize generated vorticity in order to strengthen the boundary layer downstream of the bump and thereby delaying separation. In addition, the rear flank design has found to be correlated to flow separation on the bump [5]. Since any kind of flow separation weakens the boundary layer downstream it can be assumed that a minimization of flow separation on the bump will help to delay buffet onset.

Table 1 Parameter space for flank variation

Setup	FFL (%c)	FFW (%c)	RFL (%c)	RFW (%c)	FFLf	FFWf	RFLf	RFWf
Baseline	11	5	20	5	1.0	1.0	1.0	1.0
FF1	5.5	2.5	20	5	0.5	0.5	1.0	1.0
FF2	5.5	10	20	5	0.5	2.0	1.0	1.0
FF3	5.5	15	20	5	0.5	3.0	1.0	1.0
FF4	11	2.5	20	5	1.0	0.5	1.0	1.0
FF5	11	10	20	5	1.0	2.0	1.0	1.0
FF6	11	15	20	5	1.0	3.0	1.0	1.0
RF1	11	5	10	2.5	1.0	1.0	0.5	0.5
RF2	11	5	10	10	1.0	1.0	0.5	2.0
RF3	11	5	10	15	1.0	1.0	0.5	3.0
RF4	11	5	20	2.5	1.0	1.0	1.0	0.5
RF5	11	5	20	10	1.0	1.0	1.0	2.0
RF6	11	5	20	15	1.0	1.0	1.0	3.0

FFL(f) front flank length (factor), *FFW(f)* front flank width (factor), *RFL(f)* rear flank length (factor), *RFW(f)* rear flank width (factor)

The precise mechanism by which vorticity is generated by the bump is still unsolved and subject of current research (e.g. [3, 6, 10]). The objective of the present paper is to link geometry parameters (such as flank length and width) to aerodynamic performance and buffet behavior of the bump.

In total, six different front flank (FF) shapes and six different rear flank (RF) shapes have been analyzed. Table 1 gives an overview of the parameter space. It has to be noted that only one flank (front or rear) was modified at a time.

4.1 Flank Parameter

As mentioned in Sect. 4 the maximum span-wise pressure gradient on the front flank of the bump as well as flow separation on the bump seem to be related to the buffet behavior. Therefore, the effect of flank shape on those two parameters shall be analyzed first.

Figure 2 shows the maximum span-wise pressure gradient extracted from a slice at stream-wise mid of front flank. The contour gives a good tendency for the effect of front and rear flank shape on this parameter. In general, a short and very wide front flank produces large pressure gradients. Compared to the baseline bump (blue square), the pressure gradient can be nearly sextupled. The shape of the rear flank has obviously no effect on the pressure gradient on the front flank.

Flow separation is determined by analyzing the skin friction coefficient c_f. For two-dimensional simulations $c_{fx} < 0$ is a sufficient criterion. In case of three-dimensional simulations the situation is more complex since cross flow has to be

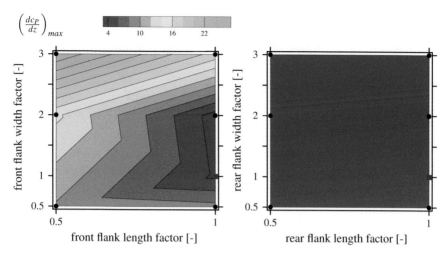

Fig. 2 Maximum span-wise pressure gradient on front flank at $Ma = 0.73$ and $AoA = 1.7°$ (bump design condition)

considered. For this reason Illi [11] used the 'combined' skin friction coefficient $c_{f\Phi}$.

$$c_{f\Phi} = \frac{2}{\pi} \tan^{-1} \left(sgn(c_{fx}) \cdot \frac{\sqrt{c_{fx}^2 + c_{fy}^2}}{|c_{fz}|} \right) \quad (1)$$

As a measure of the direction of surface skin friction, $c_{f\Phi}$ equals 1 for attached flows and -1 for separated flows. The range between -1 and 1 indicates dominating cross flows. In order to evaluate the percentage of separated flow on the bump, $c_{f\Phi}$ has been averaged in an area between 35 % and 80 % of chord in stream-wise direction over the entire span of 30 % of chord of the wing section.

As found by other authors [5] the rear flank has a stronger effect on flow separation compared to the front flank, see Fig. 3. Generally, a longer flank reduces stream-wise pressure gradients and thereby prevents the boundary layer from early separation. Considering the flank widths, wider bumps are beneficial with respect to flow separation.

4.2 Bump Performance

Performance of the bump is assessed by evaluating aerodynamic efficiency. Figure 4 shows the effect of flank shape on aerodynamic efficiency at bump design condition and Fig. 5 at airfoil design condition, respectively. First of all, it is interesting to

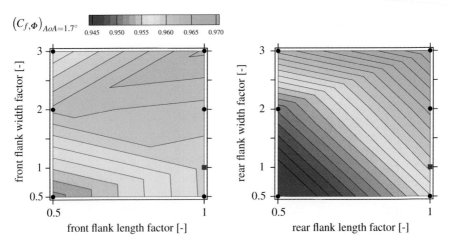

Fig. 3 Skin friction coefficient $C_{f,\Phi}$ at $Ma = 0.73$ and $AoA = 1.7°$ (bump design condition)

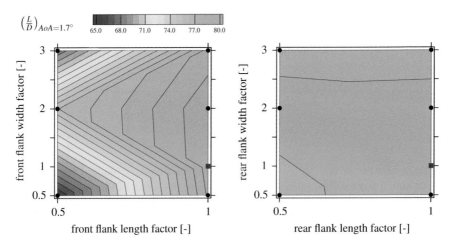

Fig. 4 Lift-to-drag ratio at $Ma = 0.73$ and $AoA = 1.7°$ (bump design condition)

see that the rear flank width does not affect aerodynamic efficiency at bump design condition whereas it affects performance at bump off-design conditions. The main reason for that is the positioning of the bump in relation to the shock position. At bump design condition the shock foot is on the front flank close to the crest. At bump off-design condition a double-shock systems establishes. The downstream/second shock occurs at the rear flank. Here, a wider flank further re-accelerates the flow and therefore increases shock strength, leading to a performance deterioration.

As written in Sect. 3.2 the bump has only been optimized in height and position. For all other parameters (like flank width) default parameters have been used. For

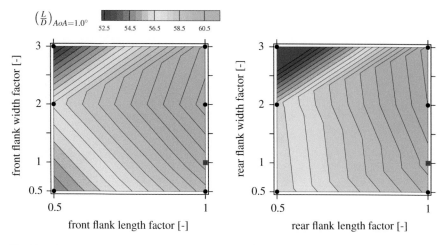

Fig. 5 Lift-to-drag ratio at $Ma = 0.73$ and $AoA = 1.0°$ (airfoil design condition)

this reason there is a slight increase of aerodynamic efficiency for wider front flanks compared to the baseline bump.

As a general trend, long front and rear flanks are beneficial with respect to aerodynamic efficiency. Concerning the width there seems to be an optimum at widths close to or equal to the width of the mid-section of the bump (here 10 %c).

4.3 Buffet Behaviour

In order to assess the buffet behavior, two flow characteristics have been analyzed. From an industrial point of view, maximum lift coefficient (Fig. 6) and strength of lift oscillations at buffet onset (Fig. 7) are most important. For analysis of lift oscillations, eight buffet cycles in a fully established flow field have been analyzed.

In general, long front flanks provide a higher maximum lift coefficient. This corresponds to the findings from Sect. 4.2 that an increased flank length is beneficial for the boundary layer state. Considering flank width there is a clear trend towards narrow front and wide rear flanks. One explanation might be an induced 'extra' lift, generated by the vortices emerging from the side flanks of the bump.

Comparable to maximum lift coefficient, longer flanks seem to reduce the root mean square of lift oscillations at buffet onset, see Fig. 7. However, the main parameter affecting lift oscillations seems to be the flank width. Here, rather extreme front flank widths perform better. This correlates to the results of the span-wise pressure gradient, verifying the findings of [7]. Considering rear flank width, wider bump tails reduce lift oscillations at buffet onset.

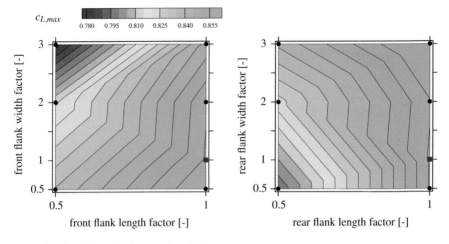

Fig. 6 Maximal lift coefficient at $Ma = 0.73$

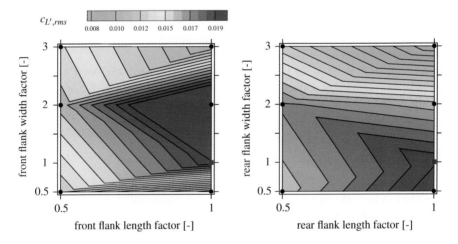

Fig. 7 RMS of lift oscillations at buffet onset ($Ma = 0.73$)

5   Trends for Optimal Flank Shape

Considering the results presented in Sect. 4 trends for optimal flank shape of buffet
alleviating bumps can be derived. Overall, reducing the length of the flanks worsens
both, performance and buffet behavior. Purely optimizing the bump for good buffet
behavior ($c_{L,max}$, $c_{L',rms}$) leads to a narrow front and a wide rear flank shape. In
contrast to that most desirable would be a 'double benefit bump', alleviating buffet
and improving performance at airfoil/wing off-design conditions. Including the
effect of flank shape on aerodynamic efficiency (see Sect. 4.2), the 'optimum'

flank shape slightly changes. Especially a 'too narrow' front flank deteriorates the performance.

Concluding, performance and especially buffet behavior of the baseline bump can be improved by modified flank shapes. Both, front and rear flank have only to be adapted in width, not in length. In fact, a further extension in stream-wise direction ('longer flanks') seems to be beneficial. But normally the bumps' length is a highly restricted parameter (e.g. by flaps, spoiler, bump inserts for wind tunnel measurements) not allowing for considerably increase of flank length. The front flank width should by slightly reduced by a factor of $0.5 < FFWf < 1$. Width of the rear flank should be increased up to the boundary of the analyzed parameter space ($RFWf \simeq 3$).

In addition, a comparison with flank parameters in Sect. 4.1 shows that both, maximum span-wise pressure gradient on the front flank as well as low separation on the bump shape are suitable design parameter for buffet bump optimization.

6 Computational Resources

In the near future several buffet alleviating bumps will be analyzed with higher order methods (e.g. DES) resulting in considerably larger meshes. It has not been decided yet about the flow solver to be used. Besides the structured flow solver FLOWer which has been used in the present study, the unstructured flow solver TAU (also developed by DLR) is up for discussion. The advantages of this solver are its excessive validation for buffet simulations (e.g. [11]) and significantly more implemented methods.

For numerical costly simulations, performance of the flow solver on the high performance cluster is very important for minimization of computational effort. Therefore, a scaling test on the CRAY XC40 was performed for both flow solvers to ensure an optimum utilization of resources. In this paper, the test results of the scaling tests applying the flow solver TAU release version 2014.2.0 are presented. A detailed description of the respective scaling test applying the structured flow solver FLOWer can be found in [16]. The scaling tests includes both, a weak scaling test and a strong scaling test, and are based on a generic test case. These test cases consist of a cubic mesh containing 33.5 million cells (33.9 million points) for the strong scaling test and 32,768 cells per core for the weak scaling test, respectively. The boundaries of the cube are set to far field conditions. For preprocessing of the TAU test cases, the native partitioner, included in the TAU software package, was applied. The usage of the new parallel initial partitioner was switched off and the old one was applied, which computes the partitions in a parallel matter but stores initially the entire grid on process 0 [8]. For all simulations, a central differences scheme was used, including a second order scheme for spatial discretization and a LUSGS scheme for backward Euler-time integration. The two-equation model k-ω was applied for turbulence modeling.

Three different compiler settings were tested to compile the TAU Code on the CRAY XC40: CRAY 8.3.8, GNU 4.9.2 and INTEL 15.0.2.164. Based on results of prior scaling tests on the CRAY XE6, the compiler PGI was excluded from the present scaling test due to its poor performance results [4]. For each compiler setting, the respective NetCDF version 4.3.2 and MPICH2 version 7.1.2 was included.

During all scaling tests, the so-called 'Real-time' of the solver TAU was logged to achieve scalable results. 'Real-time' is defined as the time from the first to the last iteration of the process and includes time the process spends blocked in case if it is waiting for other processes to complete. The initialization process and the write out time are not taken into account. For all tests, a number of 1000 Iterations have been calculated.

Starting with the weak scaling test, this test consists of several mesh sizes and process numbers in such a way that every process treats 32,768 cells, independent of the process count. A range from 128 to 3072 domains was tested. First of all, wall-clock time of the preprocessing tools 'subgrids' and 'preprocessing' of the TAU code were gathered and are presented in Fig. 8a and b, respectively. This data was collected for the CRAY-compiled version of the TAU code and partly complemented by the other two compiler versions. Both figures indicate a nearly linear increase of the logarithmic time plotted against the logarithmic number of cores. There are no major differences observable between the three tested compiler versions.

In the second place, the 'real-time' of the solver TAU was measured for all compiler version installed. To minimize any dependency on the allocated cores for the respective job, all tests are repeated three times. The results are given in Fig. 9a–c for each compiler version, respectively. Here, the 'real-time' for 1000 iterations is plotted against the number of cores.

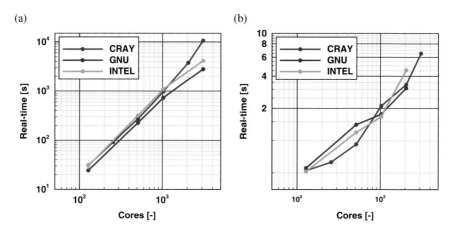

Fig. 8 'Real'-time of the preprocessing tools; CRAY-compiled version of TAU version 2014.2.0. (**a**) 'Subgrids' tool. (**b**) 'Preprocessing' tool

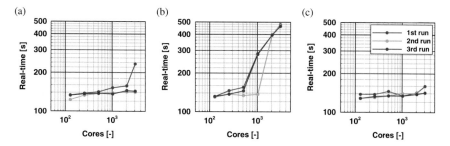

Fig. 9 'Real'-time results for different compilers; TAU version 2014.2.0. (**a**) CRAY-compiled version. (**b**) GNU-compiled version. (**c**) INTEL-compiled version

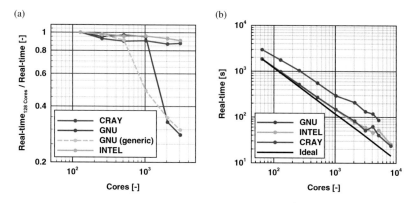

Fig. 10 CRAY-/GNU-/INTEL-compiled version of TAU code version 2014.2.0. (**a**) Weak scaling test; Smallest 'Real'-time results out of three test runs. (**b**) Strong scaling test

All versions of the code indicate differences in some points dependent on the allocated compute nodes. An extreme value of approx. 65 % deviation was found for the CRAY-compiled version and 3072 nodes. A second extreme value of approx. 105 % deviation is observable for the GNU-compiled version and 1024 nodes.

To compare the scaling behavior of all three compilers, the smallest values out of the three mentioned test runs are plotted against the number of nodes in Fig. 10a. All measured data points are normalized with the test result of 128 nodes, respectively. Analyzing the chart, the scaling behavior of the GNU-compiled version of TAU indicates major differences towards the CRAY-/INTEL-compiled code versions for high numbers of compute nodes (2048/3072). Those results are not an isolated case. As shown in Fig. 9b, all runs indicate a major increase of 'real-time' for 2048 and 3072 nodes. Even a second measurement series applying a GNU-compiled code version with slightly different compiler flags validates the presented results of the GNU-compiled code. Besides the mentioned observations, the INTEL compiler shows the best scaling efficiency followed by the CRAY compiler. The GNU compiler indicates values almost as good as the INTEL compiler, but only in the range of 128 till 1024 nodes. Apart from the extreme values given by the

GNU compiler at high node numbers, the worst scaling factor in this particular test amounts to 87.6 %.

For the strong scaling test, on the other hand, the mesh of 33.5 mio cells was divided into a range of different domain numbers. The first test results are obtained using 64 cores, which equals 2,097,152 cells per domain. Under certain circumstances, a core number higher than 5120 caused problems using the parallel in-code partitioner presumably because of the cell number per domain is less than 7000. This issue might be exacerbated when using high-order multi-grid schemes, which further reduce the cell number per domain. Furthermore, queuing duration is many times higher than simulation time, making this approach inefficient. Figure 10b depicts the 'Real-time' plotted against the number of cores for the three compilers. The TAU Code shows good linear behavior within a wide range of cores for all versions, followed by a deviation in the results of the INTEL and CRAY compiler. In the linear region the results for INTEL and CRAY overlap. For the CRAY compiler the deviation starts at 3072 cores while the INTEL compiler remains linear up to 4096 cores. Those results were reproducible in three separate runs. Interestingly, for a core number of 5120 domains, the results seem to meet the linear behavior again. The GNU results show a similar gradient but experience a time offset compared to the other compilers, meaning it takes longer to complete for any number of cores. The ideal results for the CRAY compiler depicted in Fig. 10b need parallel translation for the GNU results. The gradients of the results slightly differ from the ideal results, presumably due to processing overhead required for each additional core. This means, increasing the number of cores reduces the simulation time but increases the overall numerical costs. In order to get the most economic results a compromise between overall numerical cost and simulation time has to be found.

7 Conclusions

Originally designed for (wave) drag reduction in transonic flight, it has been shown that shock control bumps (SCB) can also alleviate buffet, an unsteady shock motion due to continuous flow separation and re-attachment at the rear part of the airfoil. Analyzing the effect of flank shape of a three-dimensional wedge-like SCB on performance and buffet behavior, this paper presents a parametric study on different flank configurations. Reference configuration is the OALT25 airfoil equipped with a performance optimized SCB (using RANS simulations), reducing (wave) drag at its design point considerably. By modifying front and rear flank shape of the bump, URANS simulations predict an improved buffet behavior. It was found that longer flanks are beneficial for both, performance at airfoil off-design condition and buffet behavior. Concerning width of the flanks, a narrow front and wide rear flank improve maximum lift coefficient and alleviate lift oscillations at buffet onset. Important for up-coming bump optimization, it was found that both, maximum span-wise pressure gradient as well as flow separation on the bump are suitable design objectives.

Acknowledgements This study is undertaken within the European-Russian project BUTERFLI, supported by the European FP7 program (FP7-AAT-2013.8-1-RTD-RUSSIA) under Grant Agreement no. 605605.

References

1. Ashill, P., Fulker, J., Shires, A.: A novel technique for controlling shock strength of laminar-flow aerofoil sections. In: Proceedings: 1st European Forum on Laminar Flow Technology (1992)
2. Bogdanski, S.: Numerische Untersuchungen zum stoßinduzierten Buffet. Verlag Dr. Hut, München (2014)
3. Bogdanski, S., Nübler, K., Lutz, T., Krämer, E.: Numerical investigation of the influence of shock control bumps on the buffet characteristics of transonic airfoils. In: New Results in Numerical and Experimental Fluid Mechanics IX, 124 edn., Springer International Publishing (2014)
4. Bogdanski, S., Gansel, P., Lutz, L., Krämer, E.: Impact of 3d shock control bumps on transonic buffet. In: High Performance Computing in Science and Engineering '14, pp. 447–461. Springer International Publishing (2015)
5. Bruce, P.J.K., Colliss, S.P.: Review of research into shock control bumps. Shock Waves **25**(5), 451–471 (2015)
6. Bruce, P.J.K., Colliss, S.P., Babinsky, H.: Three-dimensional shock control bumps: effects of geometry. In: 52nd AIAA Aerospace Sciences Meeting, pp. 1–17 (2014)
7. Colliss, S.P., Babinsky, H., Nübler, K., Lutz, T.: Vortical structures on three-dimensional shock control bumps.In: 51st AIAA Aerospace Sciences Meeting including the New Horizons Forum and Aerospace Exposition, January (2013)
8. DLR. Tau-code userguide 2014.2.0. 16.09.2014
9. Drela, M., Giles, M.B.: Viscous-inviscid analysis of transonic and low Reynolds number airfoils. AIAA J. **25**(10), 1347–1355 (1987)
10. Eastwood, J.P., Jarrett, J.P.: Toward designing with three-dimensional bumps for lift/drag improvement and buffet alleviation. AIAA J. **50**, 2882–2898 (2012)
11. Illi, S.A., Lutz, T.: Transonic tail buffet simulations on the ATRA research aircraft. Comput. Flight Testing **123**, 273–287 (2013)
12. Jacquin, L., Molton, L., Deck, S., Maury, B., Soulevant, D.: Experimental study of shock oscillation over a transonic supercritical profile. AIAA J. **47**, 1985–1994 (2009)
13. König, B., Pätzold, M., Lutz, T., Krämer, E., Rosemann, H., Richter, K., Uhlemann, H.: Numerical and experimental validation of three-dimensional shock control bumps. J. Aircr. **46**(2), 675–682 (2009)
14. Nuebler, K., Colliss, S.P., Lutz, T., Babinsky, H.: Numerical and experimental examination of shock control bump flow physics. In: High Performance Computing in Science and Engineering 2012, pp. 333–349. Springer, Berlin (2013)
15. Rossow, C., Kroll, N., Schwamborn, D.: The MEGAFLOW project - numerical flow simulation for aircraft. In: Progress in Industrial Mathematics at ECMI 2004, pp. 3–33, 8th edn. Springer, Berlin/Heidelberg (2006)
16. Schulz, C., Fischer, A., Weihing, P., Lutz, T., Krämer, E.: Evaluation and control of loads on wind turbines under different operating conditions by means of cfd. High Performance Computing in Science and Engineering '15 (2015, submitted)
17. Soda, A.: Numerical investigation of unsteady transonic shock/boundary-layer interaction for aeronautical applications. Ph.D. thesis, Technical University of Aachen (2006)

High Fidelity Scale-Resolving Computational Fluid Dynamics Using the High Order Discontinuous Galerkin Spectral Element Method

Muhammed Atak, Andrea Beck, Thomas Bolemann, David Flad, Hannes Frank, and Claus-Dieter Munz

Abstract In this report we give an overview of our high-order simulations of turbulent flows carried out on the HLRS systems. The simulation framework is built around a highly scalable solver based on the discontinuous Galerkin spectral element method (DGSEM). It has been designed to support large scale simulations on massively parallel architectures and at the same time enabling the use of complex geometries with unstructured, nonconforming meshes. We are thus capable of fully exploiting the performance of HLRS Cray XE6 (Hermit) and XC40 (Hornet) systems not just for academic benchmark problems but also industrial applications. We exemplify the capabilities of our framework at three recent simulations, where we have performed direct numerical and large eddy simulations of turbulent compressible flows. The test cases include a high-speed turbulent boundary layer flow utilizing close to 94,000 physical cores, a DNS of a NACA 0012 airfoil at $Re = 100,000$ and direct aeroacoustic simulations of a close-to-production car mirror at $Re_c = 100,000$.

1 Introduction

The focus of our activities within this project is the development of a numerical framework, which combines the high accuracy of academical codes with the flexibility necessary to tackle complex industrial applications. Our target applications are large eddy simulation (LES) and direct numerical simulation (DNS) and require large amounts of computational resources, which can only be provided by optimally utilizing massively parallel systems. Therefore, each component of the framework has been designed to contribute to a high parallel efficiency. To achieve these

M. Atak (✉) • A. Beck • T. Bolemann • D. Flad • H. Frank • C.-D. Munz
Institute of Aerodynamics and Gas Dynamics, Universität Stuttgart, Pfaffenwaldring 21, 70569 Stuttgart, Germany
e-mail: atak@iag.uni-stuttgart.de; beck@iag.uni-stuttgart.de; bolemann@iag.uni-stuttgart.de; flad@iag.uni-stuttgart.de; frank@iag.uni-stuttgart.de; munz@iag.uni-stuttgart.de

© Springer International Publishing Switzerland 2016
W.E. Nagel et al. (eds.), *High Performance Computing in Science and Engineering '15*, DOI 10.1007/978-3-319-24633-8_33

goals we have chosen a variant of the discontinuous Galerkin (DG) methods called the discontinuous Galerkin spectral element method (DGSEM) as the frameworks foundation. As in Finite Elements, DG schemes use a polynomial approximation of the solution within a grid cell, but do not require continuity at the cell boundaries. Instead they rely on Finite Volume like flux functions for the coupling between grid cells, making them more robust in the presence of strong solution gradients than classical Finite Element schemes. High-order DG methods in general are known for their superior scale resolving capabilities, making them well suited for LES and DNS [6] as these require the resolution of a broad range of scales. Furthermore, previous investigations have shown that DG methods can be implemented efficiently on parallel architectures and scale up to more then 100,000 physical cores. This can be attributed to the local nature of DG schemes, as only surface data needs to be communicated.

The DGSEM-based CFD solver FLEXI, which is used for the simulations, is the most recent development of IAG's DG-based codes. Stemming from the Spectral Element Method, DGSEM is considered as one of the most efficient DG formulations. It uses a tensor-product nodal basis to represent the solution and collocates integration and interpolation points. The operator can thus be implemented using a favorable dimension-by-dimension structure and supports very high polynomial degrees. The methods' restriction to hexahedral (but unstructured) meshes may introduce difficulties for complex geometries, but are easily circumvented by element splitting strategies and the use of non-conforming meshes.

We demonstrate the accuracy, versatility and performance of our framework at the example of three test cases, an airfoil, a car mirror and a flat plate boundary layer. The first case, addresses a highly accurate DNS of a NACA 0012 airfoil and serves as preparation step for the car mirror simulations. Here, a very high polynomial degree of $N = 11$ has been employed. In the second test case, a LES for direct aeroacoustics of a close-to-production car side-view mirror has been conducted, showing that our framework is well applicable in an industrial setting with complex geometries. The final case, a DNS of a supersonic turbulent boundary layer, proves that FLEXI is able to scale seamlessly from a few thousand cores up to the whole HLRS Cray XC40 Hornet system and that our postprocessing toolchain can efficiently handle this amount of data. This simulation involved 1.5 billion degrees of freedom (DOF) and represents the biggest DG simulation described in literature. We note that all simulations have been performed in an off-design setting concerning the optimal number of degrees of freedom per core.

2 An Overview of the FLEXI Framework

The CFD framework of our research group has been developed for the simulation of unsteady 3D turbulent flows, characterized by a broad range of turbulent scales. Earlier simulations of turbulent flows at low Reynolds numbers using the FLEXI framework showed very promising results [4]. Due to the framework's maturity after

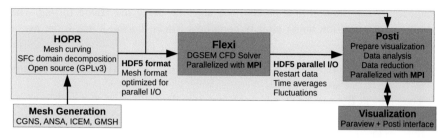

Fig. 1 Framework and workflow for a DGSEM simulation. *Blue*: serial, *red*: parallel

ongoing development and the encouraging results, we raised the bar and applied it to more complex flow phenomena in terms of higher Reynolds numbers and geometric complexity as well as even larger scale simulations.

The single components of the framework and the workflow are depicted in Fig. 1. The frameworks' three main components are the mesh preprocessor HOPR, generating high-order meshes, which are required for DG methods, the massively parallel DGSEM-based CFD solver FLEXI and the parallel postprocessing toolkit POSTI.

2.1 Pre- and Postprocessing

By design, high-order DG methods have a very high resolution within each cell, especially when operating at high polynomial degrees. It is therefore possible to use much coarser meshes compared to lower-order schemes, which is however in conflict with a sufficient geometry representation. Bassi and Rebay [3] have shown that in the case of curved boundaries a high-order DG discretization yields low-order accurate results when using straight-sided meshes. DG methods thus require a high-order boundary approximation to retain their accuracy in the presence of curved boundaries.

We have developed the standalone high-order preprocessor HOPR[1] to generate these high-order meshes by using input data from standard, linear mesh generators. HOPR has been recently made open source under the GPLv3 license. Several strategies for the element curving are implemented, e.g. using additional data obtained by supersampling the curved boundaries or a wall-normal approach, see [7] for a complete overview. Though the current DG solver FLEXI uses only hexahedra, our mesh preprocessor is also capable of generating hybrid curved meshes. In addition, HOPR also provides a domain decomposition prior to the simulation by sorting all elements along a unique space-filling curve. The elements are then in a non-overlapping fashion stored in a contiguous array, together with

---

[1] Available as open source under GPLv3 license at http://www.hopr-project.org.

connectivity information in a parallel readable file using the HDF5 format. The domain decomposition being already contained inside the mesh has some major advantages:

- The domain decomposition is directly available to all tools without requiring any additional libraries.
- The element list can be simply split into equally sized parts and each process reads its own set of contiguous non-overlapping data from the file. This massively reduces the file system load, which is especially beneficial for a large number of processes performing simultaneous I/O.
- As the solution is also stored according to the sorting in a single file, the simulations can be simply run, restarted and postprocessed on any number of processes without any need to convert or gather files.

We employ a Hilbert curve for the space-filling curve element sorting due to its superior clustering properties compared to other curves [9]. The main advantage of this approach embraces that proximate three-dimensional points are also proximate on the one-dimensional curve, which is an important property for domain decomposition. While specialized tools like METIS may provide a slightly better domain decomposition, we benefit from the fact, that for high-order DG methods a process has typically far less then 100, sometimes even as few as 1–2 single elements.

The domain decomposition obtained from the space filling curve approach of HOPR is depicted in Fig. 2 at the example of an unstructured hexahedral mesh. It shows the distribution of the grid cells on 128 and 4096 processes, each subdomain is shrinked towards its barycenter for visual purposes. The communication graph depicts for each domain its communication partners.

Postprocessing the simulation data for large datasets has become a major task in the last years. Like in mesh generation, DG, as a relatively young numerical technique, lacks the support of specialized visualization tools. DG solution data cannot be processed directly using standard tools like Tecplot or Paraview as those require linear solutions inside a cell, the DG solution, however, is a piecewise polynomial. We thus use a supersampling strategy and decompose a DG cell into multiple linear cells, which can be viewed using standard tools. However, the visualization is only a small part of the actual postprocessing procedure. Spatial averaging and computing correlations from time-averaged flow data as well as evaluating time accurate data logged each time step are only some examples for the very distinct tasks. Therefore our postprocessing toolkit POSTI has been designed to be easily adaptable to changing requirements. We employ a stripped-down modularized version of our solver as a basis, providing building blocks for I/O support, readers for geometry or solution data and interpolation and integration features. On top of this, we build separate tools, each specialized on single tasks. The resource consuming parts of our postprocessing toolkit POSTI feature an efficient parallelization. A recent and significant improvement of the framework is the development of a POSTI interface for Paraview. This enables the user to directly visualize FLEXI state files in Paraview without intermediate steps and provides Paraview access to POSTI features.

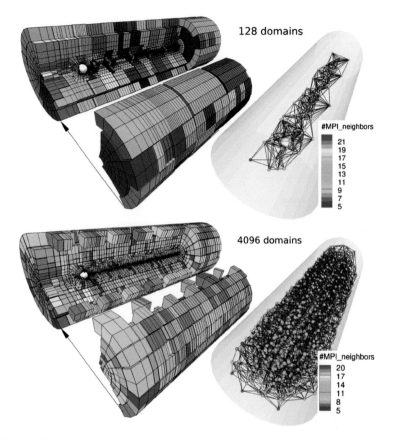

Fig. 2 Domain decomposition and communication graph for the unstructured mesh with 128 (*top*) and 4096 (*bottom*) domains

2.2 Parallel Performance

In this section, we present a parallel performance analysis of the DGSEM solver FLEXI. The strong scaling tests have been conducted on the new HLRS Cray XC40 Hornet cluster for a range of 48–12,288 physical cores, each test running for approximately 10 min. Figure 3 depicts two setups we have investigated: in the first setup, we employ a constant polynomial degree of $N = 5$ (216 DOFs/element) and use three different meshes from 768 to 12,288 elements. In the second setup we use a fixed mesh with 12,228 elements and benchmark 6 polynomial degrees from $N = 3$–9 (64–1000 DOFs/element). In both cases we do not apply multithreading and start at less then 100 processes. We constantly double the number of processes until we reach the limit one single element per process. For all cases we can observe superlinear scaling over a wide range of processes and scaling only degrades at one

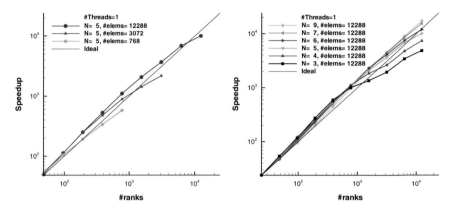

Fig. 3 Comparison of the strong scaling of the solver FLEXI without multithreading for varying number of elements and constant polynomial degree (*left*) and constant number of elements and varying polynomial degrees (*right*)

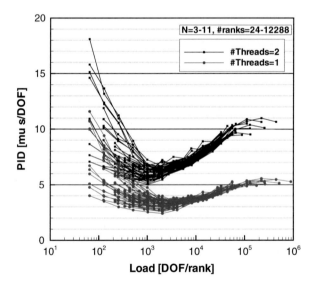

Fig. 4 Performance index (PID) with/without multithreading for polynomial degrees $N = 3–11$ and 24–12,228 processes

element per core. Note, that for the higher polynomial degrees $N = 7$ and $N = 9$, the scaling does not degrade even in the single element case.

In a third setup depicted in Fig. 4 we investigate solvers' performance index (PID), which is a convenient measure to compare different simulation setups: it represents the CPU-time required to update one degree of freedom within one time step, and is computed from the overall CPU time, the number of timesteps and degrees of freedom of the simulation,

$$\text{PID} = \frac{\text{Wall-clock-time \#cores}}{\text{\#DOF \#Timesteps}}.$$

The PID boils down the contradicting effects of scaling, compiler optimization and CPU cache on the performance to a single efficiency measure, showing three regions. In the leftmost part in Fig. 4 is the latency-dominated region, where we observe a very high PID. Here, communication latency hiding by doing inner work has no effect due to the low load per core ratio. The rightmost region, however, which is characterized by a high PID and high loads, is dominated by the memory-bandwidth of the nodes. In the central "sweet-spot" region the load is small enough to fit into the CPU-cache while still ensuring sufficient latency hiding. Furthermore, we note that in our case multithreading had no beneficial effect on the solver performance.

3   DNS of a NACA 0012 Airfoil at Re=100,000

In this project, we have computed a direct numerical simulation of a NACA 0012 airfoil at $Re = 100,000$ and an angle of attack of $2°$. This challenging flow case serves two purposes:

- The acoustic feedback mechanism described in Sect. 4 needs to be investigated further. To better understand the physical mechanisms, in particular in combination with turbulence, we have chosen a problem setup that generates a laminar vortex shedding and thus tonal noise. A comparison with literature [1, 8, 12] strongly suggest the existence of the feedback mechanism described for the car mirror case. Data analysis is currently underway to determine whether the feedback mechanism can also be identified for this case.
- The development of numerical schemes and modeling approaches for turbulent flows requires fully resolved DNS simulations as reference data. In particular for higher Reynolds numbers, complete spatial and temporal data sets are rarely available for comparison. This DNS computation provides a challenging test case for the LES models, as it incorporates fully laminar as well as turbulent flow regions and their interaction in the wake.

The NACA 0012 airfoil is located at the center of a C-shaped domain with radius $R = 100c$, where c denotes the airfoil chord. The chord is at an angle of attack of $2°$ to the free-stream flow. The spanwise extension is $0.1c$. A rectangular trip region of height $0.003c$ at position $x = 0.05c$ is located near the leading edge to force transition on the upper half. Periodic boundary conditions are applied on the spanwise faces, isothermal wall conditions on the geometry and free-stream condition on the outer boundaries. The Reynolds number was set to $Re = 100,000$, the Mach number to $M = 0.3$. The domain is discretized with 71,500 hexahedral elements with a local tensor-product polynomial ansatz of degree 11, resulting in 123.5 million DOFs. The non-linear fluxes are represented by a tensor-product ansatz of degree 13, leading to 196.2 million flux integration points. Figure 5 shows a spanwise slice of the computational grid.

The simulation was started from a free-stream initial state, and required about 25 convection times $T^* = c/U_\infty$ to achieve stable vortex shedding. Figure 6 depicts the temporal development of the lift and drag coefficients after the initial transient. All computations were run on the HLRS Cray XC40 Hornet, using 14,300 processes.

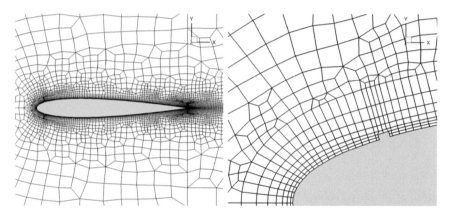

Fig. 5 *Left:* Computational grid for DNS of NACA 0012. *Right:* Zoom on upper trip region

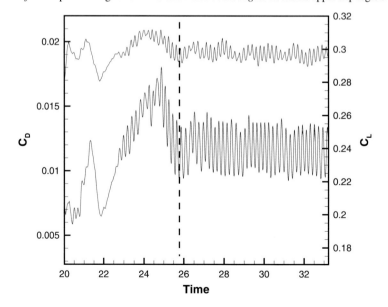

Fig. 6 Evolution of lift and drag coefficients over convective time units T^*. *Dashed vertical line* denotes onset of stable shedding

Table 1 summarizes the computational costs of this simulation. We have chosen the number of processes to achieve a balance between an optimal PID and queue waiting time, therefore, the core load is not optimal as for the case presented in Table 2.

Figures 7 and 8 show some preliminary results from the DNS computation. As expected, the trip generates rapid transition and results in a fully turbulent boundary layer on the upper airfoil surface. Due to the negative pressure gradient, the lower boundary layer remains laminar, even across the trip, and shows characteristic two-dimensional rollers near the trailing edge, which suggests the occurrence of tonal noise.

Table 1 Computational details for the DNS of NACA 0012 airfoil case

N	n_{DOF}	Procs	$PID[\mu s]$	Runtime$/T^*$
11	$1.235 \cdot 10^8$	14,300	8.2	23 min

Table 2 Computational setup for the NACA0012 airfoil case

N	n_{DOF}	Procs	$PID[\mu s]$	Runtime$/T^*$
7	$43.6 \cdot 10^6$	3,288	6.0	266.7 min

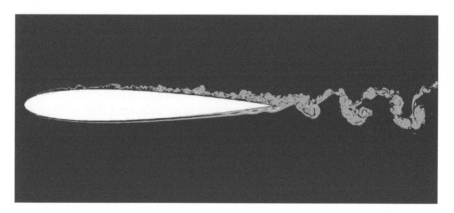

Fig. 7 Instantaneous vorticity magnitude at $t = 29$ and $z = 0.05c$

Fig. 8 Visualization of the vortical structures by the Q-criterion (Q=100). Tripped boundary layer on *top* surface becomes fully turbulent, while two-dimensional structures on lower side indicate laminar flow

As a next step, the DNS computation will be continued to gather sufficient statistical data for the analysis of the turbulent and acoustic quantities. It will then serve as a benchmark for the validation and assessment of LES modeling approaches.

4 Direct Aeroacoustic Simulation for the Analysis of Acoustic Feedback Mechanisms at a Side-View Mirror

For the acoustic optimization in automotive engineering the reduction of acoustic sources leading to narrow band noise is essential. As broad band noise sources such as engine noise or noise on road were significantly reduced, aeroacoustic noise sources leading to tonal noise are perceived as especially annoying by occupants. Particularly tonal noise at the side-view mirror is often found during development processes and needs to be removed in costly wind tunnel testing. To this end, the underlying mechanisms are not well understood. This work contributes to the understanding of the mechanism by comparing experimental and numerical data, both obtained at IAG. For the numerical analysis the fully compressible Navier-Stokes equations are solved time accurately using our in-house discontinuous Galerkin spectral element solver FLEXI. Using high order schemes allows for the direct simulation of acoustic within a hydrodynamic simulation, due to their excellent dissipation and dispersion properties. The simulation is scale-resolving in the sense of a large eddy simulation (LES). In comparison to state-of-the-art hybrid acoustic methods, this allows for the simulation of acoustic feedback mechanisms.

The idea of acoustic feedback leading to tonal noise is well established for airfoils at medium Reynolds number flows ($\mathcal{O}(10^5) - \mathcal{O}(10^6)$) and was investigated experimentally [1, 12] and numerically [8]. When there is a laminar flow almost up to the trailing edge, laminar separation is likely to occur. Within the separated shear layer instabilities are excited, rolling up as vortices downstream. The interaction of vortices with the trailing edge leads to sound emission and the upstream traveling acoustic wave may excite boundary layer instabilities of the same frequency. This mechanism is only self-maintaining for such frequencies with negligible overall phase difference. This criterion leads to the typical ladder-like structure of the acoustic spectrum when plotted over inflow velocity. The acoustic spectra of the side view mirror under consideration show a similar behavior. In addition, the absence of tonal peaks in the spectrum, when turbulators are applied to the surface further hint at the presence of acoustic feedback.

Prior to the study of the side view mirror, the simulation method was validated by an airfoil simulation described in [8]. The flow around a NACA 0012 airfoil is simulated fully scale-resolving (DNS) in two dimensions. The Reynolds number with respect to the chord length is $Re_c = 100,000$, the Mach number $M = 0.4$ and the angle of attack is $0°$. Vortex shedding downstream of the separation lead to several discrete spectral components. The perturbation ansatz also used in [8], examining only a single isolated wave package of vanishing amplitude, definitely demonstrates the existence of a feedback mechanism in the simulation. Figure 9 shows the non-dimensional power spectral density of DNS and perturbation simulation evaluated two chords above the trailing edge. In both cases several narrow band peaks are observed for which the frequencies collapse.

Analogous to the experiments, described in detail in [13], we consider an isolated side view mirror mounted on the floor. The setup and the computational mesh on the surface can be seen in Fig. 10. The inflow velocity is $U_\infty = 100\,\text{km/h}$ corresponding to a Reynolds number of $Re_S \approx 184{,}000$, with respect to the mirror inner surface length $S = 0.1\,\text{m}$, and Mach number $M = 0.082$. The computational domain has 17 million degrees of freedom, using a polynomial degree $N = 7$.

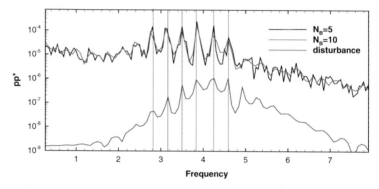

Fig. 9 Power spectral density of emitted acoustic field for DNS (*upper curves*) and perturbation calculation (*lower curve*)

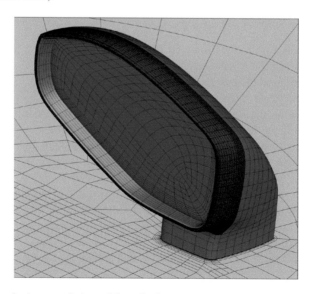

Fig. 10 Isolated mirror on windtunnel floor. Surface mesh with refinement near the trailing edge

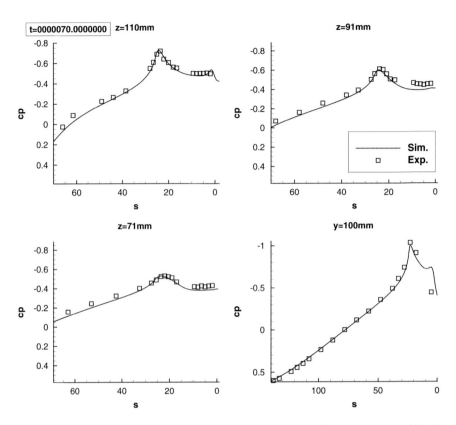

Fig. 11 Pressure distribution on the inner ($z = 110$, $z = 91$, $z = 71$) and *top* ($y = 100$ mm) mirror surface

The computation was carried out on HLRS Cray XC40 using 3288 processes. The simulation was run for 70 convective time units ($T^* = S/U_\infty$), where the last 40 time units where used to analyze the results.

Figure 11 shows the surface pressure distribution for three positions on the inner surface and one on the top surface of the side view mirror, where s is the wall-tangential coordinate up to the trailing edge. Excellent agreement of simulation and experimental data is observed at all positions. The pressure distribution shows that the flow is strongly accelerated up to an edge upstream of the trailing edge. At the edge the pressure peaks and the flow downstream is unable to follow the strong pressure rise. Consequently the boundary layer separates, characterized by the flat pressure distribution observed in Fig. 11 after the peak. The boundary layer separation is visualized in Fig. 12 by means of the time averaged velocity field obtained from Particle Image Velocimetry (PIV) and LES results. Again, there is an excellent agreement between experiments and simulation for the hydrodynamic quantities.

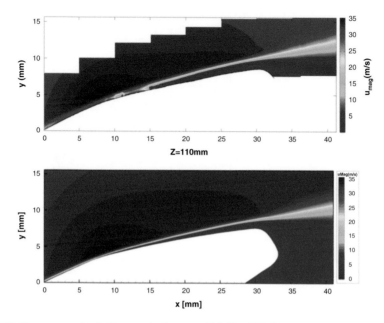

Fig. 12 Time averaged velocity magnitude at $z = 110$. *Top*: PIV, *bottom*: simulation

Fig. 13 Visualization of vortex structures by means of isosurfaces of instantaneous y-velocity fluctuations with $v' = \pm 0.01 U_\infty$. In proximity to the side surface, the y-velocity is aligned with the wall-normal direction

From the velocity field, turbulent re-attachment of the boundary layer can be ruled out, therefore all prerequisites for the feedback loop are given. The evaluation of the corresponding RMS-fluctuations (not plotted) shows the growth of perturbations along the separated shear layer. A qualitative impression of the acoustic source region at the mirror inner side is given in Fig. 13. Coherent vortices are identified by

Fig. 14 Akustisches
Spektrum bei Position 1,
$z = 300$ mm

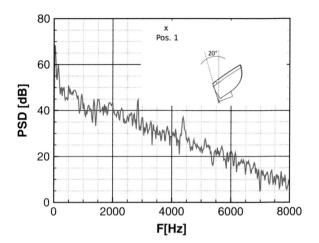

velocity fluctuation isosurfaces. The interaction of those vortices (or rollers) with the
trailing edge at an almost constant frequency results in distinct tonal noise. Similar
structures are also found at the mirror top surface. The power spectral density (PSD)
of the acoustic pressure in Fig. 14 shows two dominant tonal peaks. The frequencies
are at 2860 and 4370 Hz. A comparison to the hydrodynamic pressure spectra
reveals that the source of the low frequency is located at the mirror inner surface, the
higher frequency source at the mirror top surface. This is in good agreement with the
experiments. Despite the excellent agreement of hydrodynamic quantities and the
location of the sources, the measured frequencies deviate significantly from those
predicted by the simulation. The measured frequency at the top surface is 5100 Hz
and at the inner surface 3600 Hz. Up to now, there is no conclusive identification for
the deviation.

   In summary, we showed that the tonal noise generation mechanism observed at
an automotive side view mirror is closely related to acoustic feedback loops known
from airfoils or wing sections. The noise generation was successfully reproduced
within a direct acoustic simulation, resolving all relevant scales in time and space.
Such simulations are only possible and meaningful when using high order, highly
efficient methods, such as our DGSEM solver FLEXI. The calculation also shows
the advantage compared to state-of-the-art hybrid acoustic-hydrodynamic methods,
which cannot account for acoustic feedback in general. Table 2 summarizes the
computational details of the side view mirror simulation, run on HLRS Cray XC40
Hornet.

5 The Multicore Challenge: DNS of a Spatially-Developing Supersonic Turbulent Boundary Layer up to High Reynolds Numbers

The payload ratio and the economic effectiveness of space transportation systems are important characteristic factors for today's aerospace projects. These factors can even decide whether a project will be realized or discarded. In this context, air-breathing supersonic and hypersonic aircrafts may become a key technology for the future of space flight as they could significantly reduce the payload costs. Particularly scramjets—a ramjet engine where the combustion occurs at supersonic velocities—are considered as a favorable alternative for tomorrow's space missions. In contrast to conventional transportation technologies, which need to carry tons of oxygen supplies, the scramjet inhales the atmospheric air to gain the oxygen required for combustion. Thus, the aircraft becomes lighter, faster and eventually more economic since the payload can be drastically increased. The incoming air will not be compressed by moving parts like compressors, but through a series of shock waves generated by the specific shape of the intake and the high flight velocity, hence, the intake plays a decisive role for air-breathing vehicles. Furthermore, as the supply of compressed air is of fundamental importance for the subsequent efficient combustion of the fuel-air mixture to produce thrust, the intake also determines the operability limits of the whole system. The intake flow itself is characterized by laminar and turbulent boundary layers and their interaction with shock waves, yielding a three-dimensional unsteady complex flow pattern. Beyond that, due to interactions with shocks the boundary layer may experience intense heat loads leading to serious aircraft damages [2]. Hence, the investigation of the turbulent boundary layers are crucial not only for new cooling concepts, but also to ensure the structural integrity of such supersonic/hypersonic air-breathing propulsion systems.

As mentioned before, the flow within the intake (particularly at the upper part) is dominated by laminar and turbulent boundary layers emerging along the walls. Thereby, the initially laminar boundary layer becomes turbulent further downstream of the intake. The phenomenon describing the changeover from laminar state to turbulent is referred to as transition and depicts one of the main flow features encountered in the intake. During the operation state of the scramjet, however, the occurring shock waves traverse through the spatially-developing boundary layer until a quasi-equilibrium state between the shock strength and the boundary layer will be reached. Thus, in order to generate a reliable database for further complex studies, e.g. shock wave/boundary layer interactions (SWBLI) at different impingement locations, and also to demonstrate the high potential of DG schemes for compressible turbulent wall-bounded flows, we will show results of a DNS of a compressible spatially-developing flat-plate turbulent boundary layer up to a momentum thickness based Reynolds number of $Re_\theta = 3878$. We note that, the simulation of a turbulent boundary layer along the flat-plate represents the base study for canonical SWBLIs as it can be transferred into a SWBLI study by imposing shock waves via appropriate boundary conditions.

The free-stream Mach number, temperature and pressure of the supersonic boundary layer are given by $M_\infty = 2.67$, $T_\infty = 568\,K$ and $p_\infty = 14{,}890\,Pa$, respectively. Air was treated as a non-reacting, calorically perfect gas with constant Prandtl number $Pr = 0.71$ and with a constant specific heat ratio $\kappa = c_p/c_v = 1.4$. Sutherland's law was used to take the temperature dependency of the dynamic viscosity μ into account. The Reynolds number at the leading edge of the plate was set to $Re = 1.156 \cdot 10^5$ and isothermal no-slip conditions were applied at the wall. The wall temperature was equal to the adiabatic wall temperature $T_w = T_{ad} = 1242\,K$. Furthermore, sponge zones were used at the outflow regions to suppress any reflections from the outflow boundaries. To trigger the laminar-turbulent transition, periodic disturbances—given by amplitude and phase distributions along the inflow boundary—were added to the initial Blasius solution. In this work, we superimposed 5 discrete disturbances (determined by the eigenfunctions from linear stability theory) with a disturbance amplitude of $A = 0.02$ which yield a maximum RMS velocity of 0.43 % of the free-stream velocity. As for compressible flows the most unstable modes are obliquely traveling disturbance waves, we introduced two single oblique waves with a fundamental spanwise wave number of $\gamma_0 = 21$ to rapidly reach a turbulent state. The DNS was performed with the DGSEM code FLEXI using a polynomial degree of $N = 5$. The mesh consisted of $1500 \times 100 \times 45$ elements in x-, y- and z-direction resulting in 1.458 billion DOFs. To the best of the authors' knowledge, this simulation is the biggest computation within the DG community.

Due to the excellent scaling attributes of the solver, we were able to cope with the demanding costs by exploiting the whole computational resources on the HLRS Cray XC40 Hornet in an efficient way and run the DNS with up to 93,840 physical cores. An overview of the performance index (PID) for different numbers of processors used for this study is given in Table 3. The overview particularly reveals that the PID stays constant up to the limit of available physical cores, which proves that the solver achieves an ideal speedup without any performance losses even at running simulations on the whole machine.

The non-dimensional simulation time was $T = 4.25T^*$ whereas $T^* = 2L/U_\infty$ labels the characteristic flowthrough time with plate length L and free-stream velocity U_∞. Figure 15 shows the instantaneous distribution of the vortices along the plate at the final simulation time visualized by the λ_2-criterion and colored by the streamwise velocity component. Here, we can clearly see how the oblique disturbance waves introduced in the inflow generate hairpin vortices which in turn breakdown into turbulence further downstream.

The streamwise development of the time- and spanwise-averaged skin friction coefficient C_f is shown in Fig. 16. Whilst the leading edge of the plate always maintains its laminar state, the flow experiences transition where a sudden increase of C_f is observable around $x_{tr} \approx 3.5$ ($Re_{x_{tr}} = 3.5 \cdot 10^5$). After the overshoot in skin-friction, which is typical for transition, the C_f-profile in the turbulent regime, however, agrees very well with the skin-friction correlation according to the reference temperature approach [14].

Table 3 Performance index for varying numbers of processes on the HLRS Cray XC40 Hornet

Procs	12,000	24,000	93,840
Grid	$1500 \times 100 \times 45$	$1500 \times 100 \times 45$	$1500 \times 100 \times 45$
n_{DOF}	$1.498 \cdot 10^9$	$1.498 \cdot 10^9$	$1.498 \cdot 10^9$
n_{DOF}/Procs	121,500	60,750	15,537
PID [μs]	6	6	6

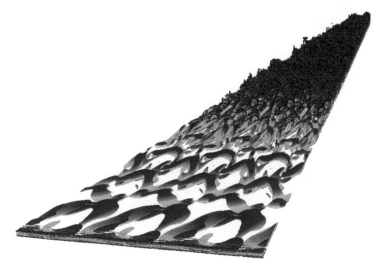

Fig. 15 λ_2-visualization of the turbulent structures along the flat-plate colored by the streamwise velocity component u

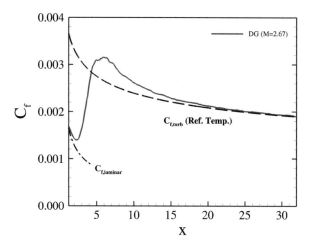

Fig. 16 Development of the skin friction coefficient C_f along the plate length

The Van Driest transformation of the streamwise velocity u accounts for compressibility effects and enables the comparison with incompressible scaling laws [14]. Figure 17 compiles the transformed velocity profiles u_{vD}^+ at different downstream positions along the plate. Here, we can clearly observe that the results perfectly match both the theoretical incompressible law of the wall ($u^+ = y^+$ and $u^+ = 2.44 \cdot ln(y^+) + 5.2$) and well-established DNS studies of spatially-developing turbulent boundary layers by Bernardini and Pirozzoli [5] and Schlatter and Örlü [11]. In Fig. 17 we also notice that the higher the momentum thickness based Reynolds number, the longer the velocity profile follows the incompressible log-law.

Furthermore, in Fig. 18 we compared the Reynolds-stresses, again, at different streamwise locations with previous DNS data and also with experimental PIV measurements of Piponniau et al. [10]. The Reynolds-stresses were rescaled according to Markovin's density scaling [14] to account for the varying density. Figure 18 exhibits that the results are in an excellent agreement with other DNS studies as

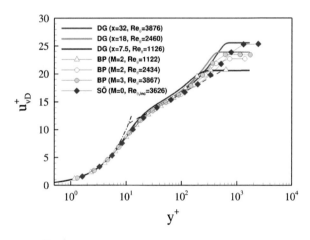

Fig. 17 Van Driest transformed velocity profiles u_{vD}^+ at different streamwise locations compared with DNS data by Bernardini and Pirozzoli (BP) [5] and Schlatter and Örlü (SÖ) [11]

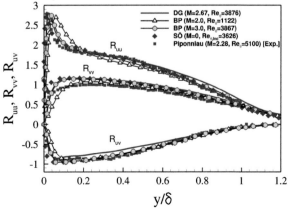

Fig. 18 Density weighted Reynolds-stresses at different streamwise locations compared with DNS data by Bernardini and Pirozzoli (BP) [5], Schlatter and Örlü (SÖ) [11] and experimental data by Piponniau et al. [10]

Table 4 Computational setup for the compressible supersonic turbulent boundary layer case

N	n_{DOF}	Processors	$PID[\mu s]$	Runtime/T^*
5	$1.458 \cdot 10^9$	93,840	6	23.41 h

well as with experimental data. A summary of the computational details is given in Table 4.

With this study, we laid the foundation of subsequent SWBLI studies, in which we are interested in assessing the impact of different shock impingement locations on the flow field and thus to gain a deeper insight into the complex physics involved. In this context, we will apply the FV subcell shock capturing which is already incorporated in our DG framework and enables the robust approximation of shock waves.

6 Summary and Outlook

In this report, we summarize the ongoing development in numerical schemes, parallelization efforts and toolchain improvements for our HPC simulation framework and provide details for large scale application cases.

We demonstrate that the excellent scaling results transfer from the HLRS Cray XE6 Hermit to the new XC40 Hornet, and our postprocessing toolchain is able to handle massive amounts of data. The efficiency of our simulation framework allows us to conduct large scale simulations with 10^7 to 10^9 spatial degrees of freedom, using up to 93,840 processes on the XC40 and 14,000 on the XE6. The presented test cases show that our framework is versatile and mature enough to conduct industrially relevant simulations of complex engineering flows, as well es very large scale computations of canonical test cases for basic research.

In the future, the capabilities of the framework will be extended to retain efficiency beyond 10^5 processes. We will apply our framework to direct numerical simulations of previously unfeasible and expensive computations of multiscale problems, both in basic research as well as in applied engineering.

Acknowledgements The research presented in this paper was supported in parts by the Deutsche Forschungsgemeinschaft (DFG), the Boysen Stiftung and the Audi AG. We truly appreciate the ongoing kind support by HLRS and Cray in Stuttgart.

References

1. Arbey, H., Bataille, J.: Noise generated by airfoil profiles placed in a uniform laminar flow. J. Fluid Mech. **134**, 33–47 (1983)

2. Babinsky, H., Harvey, J.K.: Shock Wave-Boundary-Layer Interactions. Cambridge University Press, Cambridge (2011)
3. Bassi, F., Rebay, S.: High-order accurate discontinuous finite element solution of the 2D Euler equations. J. Comput. Phys. **138**(2), 251–285 (1997)
4. Beck, A.D., Bolemann, T., Flad, D., Frank, H., Gassner, G.J., Hindenlang, F., Munz, C.-D.: High order discontinuous Galerkin spectral element methods for transitional and turbulent flow simulations. Int. J. Numer. Methods Fluids **76**(8), 522–548 (2014)
5. Bernardini, M., Pirozzoli, S.: Wall pressure fluctuations beneath supersonic turbulent boundary layers. Phys. Fluids **23**(8), 085102 (2011)
6. Gassner, G.J., Beck, A.D.: On the accuracy of high-order discretizations for underresolved turbulence simulations. Theor. Comput. Fluid Dyn. **27**(3-4), 221–237 (2013)
7. Hindenlang, F., Bolemann, T., Munz, C.-D.: Mesh curving techniques for high order discontinuous galerkin simulations. In: Kroll, N., Hirsch, C., Bassi, F., Johnston, C., Hillewaert, K. (eds.) IDIHOM: Industrialization of High-Order Methods-A Top-Down Approach, vol. 128, pp. 133–152. Springer International Publishing AG, Cham (2015)
8. Jones, L.E., Sandberg, R.D.: Numerical analysis of tonal airfoil self-noise and acoustic feedback-loops. J. Sound Vib. **330**(25), 6137–6152 (2011)
9. Moon, B., Jagadish, H., Faloutsos, C., Saltz, J.: Analysis of the clustering properties of the Hilbert space-filling curve. IEEE Trans. Knowl. Data Eng. **13**(1), 124–141 (2001)
10. Piponniau, S., Dussauge, J.P., Debieve, F.F., Dupont, P.: A simple model for low-frequency unsteadiness in shock-induced separation. J. Fluid Mech. **629**, 87–108 (2009)
11. Schlatter, P., "Orl"u, R.: Assessment of direct numerical simulation data of turbulent boundary layers. J. Fluid Mech. **659**, 116–126 (2010)
12. Tam, C.: Discrete tones of isolated airfoils. J. Acoust. Soc. Am. **55**, 1173–1177 (1974)
13. Werner, M., Würz, W., Krämer, E.: Experimental investigations of tonal noise on a vehicle side mirror. In: Notes on Numerical Fluid Mechanics and Multidisciplinary Design, Munich, Germany. Springer 2014, 2014, submitted
14. White, F.M.: Viscous Fluid Flow. McGraw-Hill, New York (1991)

Toward a Discontinuous Galerkin Fluid Dynamics Framework for Industrial Applications

Sebastian Boblest, Fabian Hempert, Malte Hoffmann, Philipp Offenhäuser, Matthias Sonntag, Filip Sadlo, Colin W. Glass, Claus-Dieter Munz, Thomas Ertl, and Uwe Iben

Abstract For many years, discontinuous Galerkin (DG) methods have been proving their value as highly efficient, very well scalable high-order methods for computational fluid dynamics (CFD) calculations. However, they have so far mainly been applied in the academic environment and the step toward an application in industry is still waited for. In this article, we report on our project that aims at creating a comprehensive CFD software that makes highly resolved unsteady industrial DG calculations an option. First, our focus lies on the adaptation of the solver itself to industrial problems and the optimization of the parallelization efficiency. Second, we present a visualization tool specifically tailored to the properties of DG data that will be combined with the solver to obtain an in-situ visualization strategy within our project in the near future.

M. Hoffmann (✉) • M. Sonntag • C.-D. Munz
Institute for Aerodynamics and Gas dynamics, University of Stuttgart, Pfaffenwaldring 21, 70569 Stuttgart, Germany
e-mail: hoffmann@iag.uni-stuttgart.de; sonntag@iag.uni-stuttgart.de; munz@iag.uni-stuttgart.de

S. Boblest • T. Ertl
Visualization Research Center, University of Stuttgart, Allmandring 19, 70569 Stuttgart, Germany
e-mail: sebastian.boblest@visus.uni-stuttgart.de; thomas.ertl@vis.uni-stuttgart.de

F. Hempert • U. Iben
Robert Bosch GmbH, Robert-Bosch-Campus 1, 71272 Renningen, Germany
e-mail: Fabian.Hempert@de.bosch.com; Uwe.Iben@de.bosch.com

P. Offenhäuser • C.W. Glass
High Performance Computing Center, University of Stuttgart, Nobelstrasse 19, 70569 Stuttgart, Germany
e-mail: offenhaeuser@hlrs.de; glass@hlrs.de

F. Sadlo
Interdisciplinary Center for Scientific Computing, Heidelberg University, Im Neuenheimer Feld 368, 69120 Heidelberg, Germany
e-mail: filip.sadlo@iwr.uni-heidelberg.de

© Springer International Publishing Switzerland 2016   531
W.E. Nagel et al. (eds.), *High Performance Computing in Science and Engineering '15*, DOI 10.1007/978-3-319-24633-8_34

1 Introduction

Numerical simulations of fluid flow play a very important role in the development and optimization of industrial products, for example engine components. It is highly desirable to have efficient and accurate high-order methods at hand for such computations. A method that shall be applicable in an industrial environment has to meet several requirements. It must be accurate and reliable at the same time, such that numerical simulations of a wide range of different physical situations can be handled. Further, it has to scale well with the number of processors used, because with modern super computers like the Cray XC40 the computational resources grow rapidly. Finally, techniques for the postprocessing and visualization of the huge amounts of resulting data must be available.

Discontinuous Galerkin (DG) methods are a very promising class of accurate and very well parallelizable high-order numerical schemes for computational fluid dynamics (CFD) simulations. For decades, they have been proving their value in such computations, however, so far mainly in the academic environment. The goal of our project is to develop a code package that encloses both a DG solver code and a postprocessing strategy for the resulting data that is applicable in industrial applications.

Over the last years, a big topic in the automotive industry was the reduction of carbon dioxide emissions. One strategy to reduce emissions, is the use of bi-fuel engines. In our project, we exemplarily study the injection of compressed natural gas into the intake manifold. The gas exits the injector at supersonic conditions and produces jets which are under-expanded. To investigate the noise emissions over the entire speed range in the intake manifold, understanding the details of the fluid flow behind the injector is necessary. The very different spatial and temporal scales that play a role under these conditions, make the numerical simulation exceptionally difficult and computationally expensive.

We discuss how our CFD code *Flexi* performs in such situations and show how our post-processing tool handles the resulting data.

2 Discontinuous Galerkin Spectral Element Method

Our code is based on the discontinuous Galerkin spectral element method (DG SEM). Details of this method are well described in the literature, see for example [4], so we restrict ourselves to a short overview.

2.1   Fundamentals of DG SEM

Starting point of DG SEM are the compressible Navier-Stokes equations expressed in conservative form

$$u_t(x) + \nabla_x \cdot F(u(x), \nabla_x u(x)) = 0 \quad \forall x \in \Omega. \tag{1}$$

Here, u is a vector containing the conservative variables, F contains the physical fluxes, and Ω is the computational domain. Within our code, the domain is split into hexahedral grid cells, where we explicitly allow curved faces to be able to efficiently treat complex geometries.

Each of these grid cells is mapped to the reference element $E = [-1, 1]^3$, where the actual computation is performed. This transformation yields

$$J(\xi)u_t(t, \xi) + \nabla_\xi \cdot \mathcal{F}(u(t, \xi), \nabla_x u(t, \xi)) = 0, \tag{2}$$

where $J(\xi)$ is the Jacobian of the mapping, $\mathcal{F}$ are the transformed fluxes, and the coordinates of the reference volume are $\xi = (\xi^1, \xi^2, \xi^3)^\top$. In the DG context, the conservation law (2) is multiplied with a test function Φ. To obtain the so-called weak formulation, the resulting equation is integrated over the reference element E. Here, integration by parts is applied to the second integral:

$$\int_E Ju_t \Phi \, d\xi + \int_{\partial E} \left(\mathcal{F}^* \cdot n\right) \Phi \, dS_\xi - \int_E \mathcal{F} \cdot \nabla_\xi \Phi \, d\xi = 0, \tag{3}$$

where n is the normal vector of the reference element E.

A polynomial tensor product basis of degree N is used to approximate the solution in the reference element in each spatial direction as

$$u(\xi) = \sum_{i,j,k=1}^{N+1} \hat{u}_{ijk} \psi_{ijk}(\xi) \quad \text{with } \psi_{ijk} = l_i(\xi^1)l_j(\xi^2)l_k(\xi^3), \tag{4}$$

where $\hat{u}_{ijk}$ are the nodal degrees of freedom (DOFs), and $l_i(\xi)$ are the one-dimensional Lagrange interpolation polynomials defined by the set of Gauss nodes $\{\xi_i\}_{i=1}^{N+1}$. For a computation with an order of accuracy A, it is advisable to use a mesh with at least equally many nodes per spatial direction per element. Within such a mesh, we have polynomials of degree $N = A - 1$ in each spatial direction. Each element then has $(N + 1)^3$ nodes, i.e., the simple case of eight Gauss nodes per cell corresponds to $N = 1$, of course. DG SEM uses a collocation technique, where the integration in the DG discretization is solved by Gauss quadrature using these Gauss nodes. Furthermore, DG SEM uses the same ansatz and test functions $\Phi = \psi$.

In the DG method, adjacent elements are coupled by fluxes through their surfaces. Adjacent elements have different states at the interface because the solution is discontinuous over element boundaries. A Riemann solver is used to approximate the boundary flux $\mathcal{F}^*$ with the state u of an element and the state u^+ of the neighbor element. This step is the only part of the algorithm, where data from two elements is required at the same time. Because of this low interdependence of elements, only the data of the faces at MPI borders have to be communicated between processors. This is the main ingredient for the very high scalability of this method. The parallel efficiency is further increased by computing the volume integral [Term 3 in Eq. (3)] during the communication process required for the surface integral [Term 2 in Eq. (3)], which reduces latency to a minimum. The parallel efficiency of the DG SEM implementation for strong scaling was further demonstrated by Hindenlang et al. [4].

The evaluation of the volume integral and surface integral determines the time derivative $\partial \hat{u}_{ijk}/\partial t$ for each DOF. Afterward, the time integration is done with an explicit third-order accurate Runge-Kutta scheme.

2.2 Shock Detection and Shock Capturing

The injection of natural gases into the intake manifold takes place at high pressures and therefore shocks occur during this process. DG SEM cannot resolve these shocks without generating unphysical oscillations. A local reduction of the polynomial degree in the vicinity of shocks or strong gradients has been used in several approaches to avoid this problem. However, this leads to a local decrease of the accuracy and resolution. To keep the resolution, some local mesh refinement around the shock locations has to be applied. Our code uses a shock capturing method that is described in detail by Sonntag and Munz [10]. In this approach, the number of degrees of freedom is kept fixed. Within the DG grid cell, a number of sub-cells are introduced, for which a second-order finite volume scheme is applied. To have the same number of degrees of freedom, the sub-cells are simply constructed around the Gauss points. Hence, the whole procedure does not change the structure of the solution: The values at the Gauss points in the grid cell may be interpreted either as values of the DG polynomial or as finite volume sub-cell values. In principle, it is a replacement of the evaluation of the volume integral and combines the decrease of the order of the approximation with a local grid refinement. No new degrees of freedom are generated with this approach, which keeps the memory layout left unchanged and allows for a straightforward implementation.

The detection of shocks and therewith the correct choice for the discretization of the time derivative is achieved with indicators. In the context of DG, a sensor proposed by Persson and Peraire [6] is often used, which, however, is only capable of detecting oscillating polynomials due to under-resolved scales and cannot separately identify physical shocks. Ducros et al. [1], on the other hand, proposed a general shock indicator. We use a combination of both sensors. This allows to separate the instabilities introduced by the physical shocks from the oscillations in the polynomial due to under-resolved scales. Consequently, we can detect all DG elements where one of these problems occurs and keep the calculation stable.

3 Simulation Case

We apply our code to investigate the flow behavior of a natural gas injector (NGI) from Bosch as shown in Fig. 1a. This injector is installed in several in-production cars, e.g., Volkswagen Caddy, Touran, and Passat, among others [8]. It is used for gas injection into an intake manifold. We focus on the external flow behind the injector to gain a better understanding of the flow behavior and its noise emissions. Thereby, we apply a direct aeroacoustic simulation, which allows us to calculate the sound pressure level (SPL) directly from the simulation data. The compressed natural gas exits the four kidney shaped orifices at supersonic conditions and under-expanded supersonic jets form. The distance of the centers of two opposite orifices is $D = 3.42 \times 10^{-3}$ m. The non-homogeneous flow at the orifices has, on average, a normal velocity $w = 505\,\mathrm{m\,s^{-1}}$, a pressure $p = 244$ kPa, a density $\rho = 1.97\,\mathrm{kg\,m^{-3}}$ and a Mach number $Ma = 1.21$. The simulations are performed with accuracy order $A = 5$, i.e., $N = 4$, on 164,000 elements, which leads to $164{,}000 \cdot 5^3 \approx 2 \times 10^7$ degrees of freedom. The surface mesh of the external injector is shown in Fig. 1b. A more detailed description of the geometry and the boundaries is given by Kraus et al. [5].

Since the main focus of this test case is to predict noise emission, we need to avoid non-physical reflection of outwards traveling waves. Therefore, we apply damping zones at the outlets [2] with a moving average flow field [7].

To give an impression of the flow field, Fig. 2 illustrates the flow contours for the axial velocity. From left to right a snapshot of the axial velocity, the averaged flow field and the fluctuations of the axial velocity are depicted. Clearly, the flow is highly turbulent. At about $z/D > 5$, the averaged axial velocity is similar to what could be expected for a subsonic circular jet. The fluctuations of the axial velocity show a very dynamic region close to the third and fourth shocks, which is in agreement with the presumed origin of a vast amount of the noise emissions.

(a)   (b)

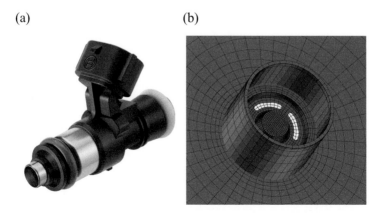

Fig. 1 NGI with the injector outlet in the *bottom left* (**a**) and the surface mesh of the injector outlet (**b**)

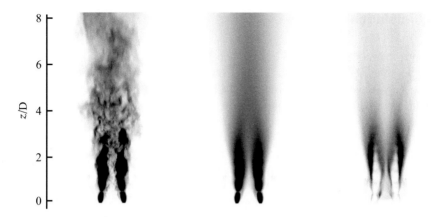

Fig. 2 Snapshot (*left*), temporal averaged (*middle*), and fluctuations (*right*) contours for axial velocity (levels from 0 to $700\,\mathrm{m\,s^{-1}}$ for the velocity and $0\,\mathrm{m^2\,s^{-2}}$ to $3 \times 10^5\,\mathrm{m^2\,s^{-2}}$ for the fluctuations; color scales from *white to black*)

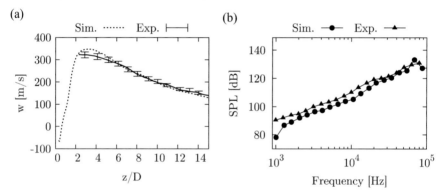

Fig. 3 Comparison of simulation results with experimental data for the axial velocity along the *center line* (**a**) and the SPL at 80° of the *center line* (**b**). Data reproduced from [3]

To validate the proposed simulation framework, we use experimental data for the averaged velocity flow field [9] and the SPL measurements [5]. Figure 3a shows the simulation results for the average axial velocity along the center line compared to the measurements. The simulation agrees well with the experimental data, only the velocity predicted in the range 5 to 10 z/D is marginally too high. The SPL was calculated at different observation points. The results for an observation point in radial direction from the injector are shown in Fig. 3b. The simulation is capable of predicting the noise emission with only little deviations from the experimental data, with the exception of low frequencies.

The high resolution of the simulation allows for further insights, which are inaccessible with less sophisticated methods. A focus of the framework is the use of massively parallel systems, with an emphasis on high scalability. This allows for an efficient integration into the engineering work-flow. Figure 4 presents the speedup

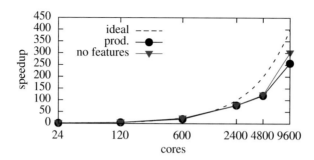

Fig. 4 Speedup comparison with respect to the run-time on 24 cores

of the full production code and a comparison with only the core functionality and the boundary conditions activated. For the latter, no gas injection occurred so no shock-capturing is needed. Only for computations on 9600 cores the disabling of features causes a relevant increase of the speedup, with a 65 % speedup for the production code, and a 75 % speedup with features disabled. This difference is mainly caused by the shock-capturing. An aeroacoustic simulation of 1 ms in physical time can be performed in under 1 h on 9600 cores and at a cost of less than 8000 core hours.

4 Visualization

An accurate and efficient CFD solver is only really applicable if it is supplemented with an equally accurate and efficient visualization tool. As we focus on industrial applications in our project, a further criterion is to choose a visualization approach that is widely used and has already proven its value both within the scientific and the industrial communities. Hence, the visualization within our project is based on the open source software ParaView [12].

We developed a reader plugin for this purpose, which reads in the solution data together with the underlying mesh. The visualization itself then uses the techniques implemented in ParaView. Our reader is fully parallelized using MPI. We use the server-client visualization concept of ParaView, where the parallelized server runs on the Cray XC40. Here, all calculations are performed. The user works on a desktop computer (client) and connects to the server.

4.1 Smoothing of Solution Data

Discontinuous Galerkin data have properties that make their visualization nontrivial. They are represented as piecewise polynomials within each computational cell, and hence their evaluation is numerically costly. While approaches exist that evaluate these polynomials directly without resampling [11], we chose a resampling strategy with variable resolution N_S due to performance reasons. Here, we subdivide each DG element into N_S^3 sub-cells. N_S can be chosen independently of the polynomial degree N, but of course all details contained in the data can only be revealed when $N_S \geq N$. We typically choose $N_S \approx (3/2)N$ to produce high quality output and lower N_S for quick overviews.

In a simple implementation emphasizing performance, we load the individual DG elements in ParaView and ignore the topology of the mesh, i.e., we do not provide the information how these cells are connected. Note that for the DG algorithm itself, neighbor relations are only relevant with respect to shared faces because fluxes through these faces must be computed. Other possible neighbor relations however, i.e., only one common corner or two common corners, that is a shared edge, are irrelevant for the CFD solver.

For the visualization however, this causes problems because of the discontinuities of the DG data. Some visualization techniques, such as volume rendering, also work well in this case. However, isosurfaces in particular suffer heavily from the discontinuities. Hence, to make such methods applicable for our data, we implemented an efficient method to smooth the DG data. To do that, we establish mesh connectivity for all resampling resolutions N_S and let ParaView only use one data value on each node that is shared by several elements. Figure 5 gives an overview of our method.

In a first step, we determine the element-element relations, i.e., the type of neighborhood relation for all elements. This information is implicitly part of the mesh files, where the unique nodes for each element are stored. Note that we can gain information about this relationship from the corner nodes alone, irrespective of the mesh order, see Fig. 5a. With our mesh structure, we have three kinds of neighborhood types, i.e., a common corner, a common edge (two common corners), or a common face (four common corners). This information is independent of the resampling resolution N_S, so we have to compute it only once a mesh is loaded.

The second step is to identify common nodes of neighboring elements. Of course, if two elements only share a corner, they have a single node in common independent of N_S, but for a common edge or face, we get $N_S + 1$ or $(N_S + 1)^2$ common points, respectively (Fig. 5b). This step has to be repeated each time N_S is changed.

Note that we do not establish connectivity over different MPI domains but only on a per-processor level. If we wanted to create a mesh that is connected everywhere to smooth the solution globally, we would have to introduce massive communication in MPI with negative impact on performance. As a consequence, our isosurfaces can still exhibit some artifacts even after connectivity has been established.

Despite this minor limitation however, establishing connectivity and smoothing the DG data leads to a significant improvement of the visual perception for all resampling resolutions N_S. To exemplify this, we use data from the simulation of the natural gas injector described in Sect. 3 computed on a mesh with 134,783 DG

(a) The three kinds of neighbor relations: common corner, common edge
and common face (left to right).

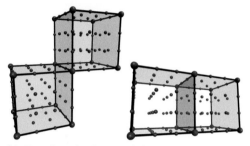

(b) For shared edges and faces, the number of
common points depends on the resampling resolution N_S.

Fig. 5 Computation of mesh connectivity. Our mesh structure allows for three types of neighbor-hood relations independent of the mesh order (**a**). We compute the relations between all elements in a first step. With a common edge, two elements share $N_S + 1$ points, with a common face they share $(N_S + 1)^2$ points (**b**) (In this Figure $N_S = 3$). In a second step we identify these identical points and hence establish connectivity within the mesh

elements and $N = 4$, and use different typical techniques to visualize them (Figs. 6, 7, 8). First, we depict the pressure on a slice through the center of the natural gas injector, where brighter colors correspond to higher pressure. Further, we show velocity streamlines, which are colored according to the velocity magnitude (high values are blue, low values are red), and finally we display a density isosurface colored according to the temperature (dark and bright colors correspond to low and high values, respectively).

In Fig. 6, we compare the visual impression for this test case with $N_S = 1$. While the pressure slice and the velocity streamlines are relatively unaffected by the discontinuities of the DG data, the difference in the quality of isosurfaces is obvious. Of course, with this low resolution, we can only get a rough impression of the data that was computed with a $N = 4$ simulation. But still, with connectivity,

Fig. 6 Influence of data smoothing on the visualization for low-resolution data sampling. The figures show examples of a density isosurface generated for $N_S = 1$ without (**a**) and with (**b**) smoothed DG data

(a)

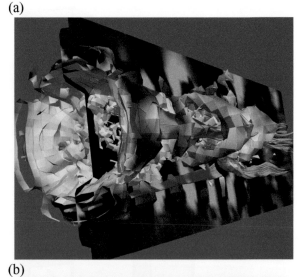

(b)

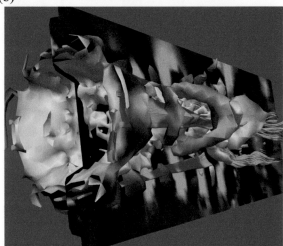

we get smooth, connected isosurfaces.

Figure 7 shows the same visualization with $N_S = 6$. Much finer details are now visible and the poor quality of the isosurfaces is less obvious but the zoomed in view in Fig. 8 still shows that the perception of the isosurfaces is highly improved with smoothed data.

Fig. 7 Influence of data
smoothing on the
visualization for
high-resolution data
sampling. The figures show
examples of a density
isosurface generated for
$N_S = 6$ without (**a**) and with
(**b**) smoothed data. See also
Fig. 8 for a zoomed view

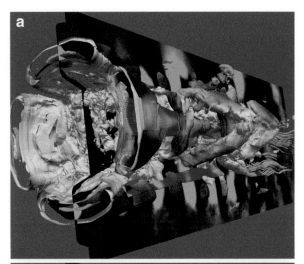

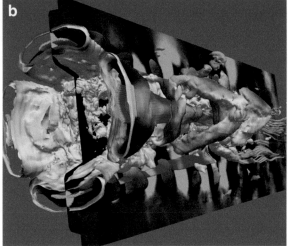

4.2   Scaling Efficiency of the Reader Plugin

To test the performance of our reader plugin, we again use a mesh of the natural gas
injector, but with a finer resolution of 1,059,328 DG elements and $N = 4$. We test
how fast our plugin can read in the mesh and five solution variables for each DOF,
and how fast it establishes mesh connectivity. We test this read-in process for the
resampling resolutions $N_S = 1$–7 and on 24, 48, 72, and 96 cores. This selection of
different values for N_S and for the number of cores corresponds to a huge range of
computational complexities for the different cases. Explicitly, visualizing the case
$N_S = 1$ on 96 cores means that each one gets approximately 1.1×10^4 cells of the

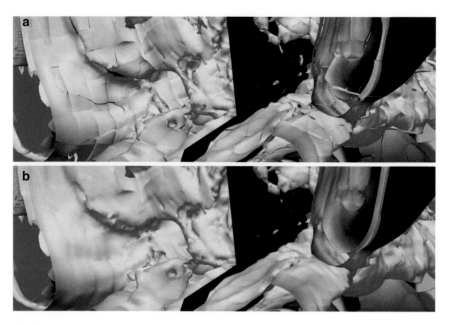

Fig. 8 Zoomed in view of Fig. 7 without (**a**) and with (**b**) smoothed DG data

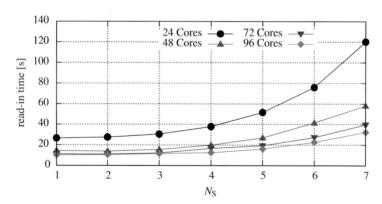

Fig. 9 Read-in time for a mesh with $N = 4$ and 1,059,328 DG elements together with a solution containing seven variables as a function of the resampling resolution N_S for different numbers of processors

refined mesh, but with $N_S = 7$ on 24 cores we have approximately 1.5×10^7 cells on each core.

Figure 9 gives the total read-in times as a function of N_S for the four different sets of cores. With 72 and 96 cores, the read-in process completes in less than 40 s even for the very high resolution $N_S = 7$.

Table 1 Ratio of the read-in time used to establish mesh connectivity (in percent)

# Cores	$N_S = 1$	$N_S = 2$	$N_S = 3$	$N_S = 4$	$N_S = 5$	$N_S = 6$	$N_S = 7$
	Computation of neighborhood relations						
24	79	77	69	56	41	28	17
48	38	39	35	28	20	13	9
72	23	23	21	15	13	9	6
96	14	14	13	12	9	6	5
	Identification of identical points						
24	1	4	11	24	41	56	64
48	1	3	10	23	38	50	66
72	1	3	9	18	35	50	63
96	1	2	7	18	32	45	58

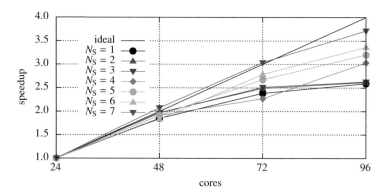

Fig. 10 Speedup efficiency of the reader plugin for the same situation as in Fig. 9

Table 1 shows the ratios of the read-in process that is required for this step for different resampling resolutions and numbers of cores. Of course, in absolute numbers, the computation of neighborhood relations is independent of N_S (approximately 21 s, 5.5 s, 2.5 s, and 1.5 s on 24, 48, 72, and 96 cores, respectively).

Further, establishing mesh connectivity scales with more than 100 % efficiency when increasing the number of cores, because not only the number of elements per core, but also the number of neighbor relations that have to be checked, decreases. This fact is also reflected in the speedup data that are shown in Fig. 10. Especially for high N_S, we get an excellent speedup, because establishing mesh connectivity then takes a relevant part of the complete read-in time.

5 Conclusion and Outlook

We presented an overview of our DG SEM fluid dynamics framework that aims at industrial applications. We showed that our CFD code *Flexi* is well adapted to modern multi-core supercomputers and presented an equally efficient visualization tool based on the open source visualization software ParaView.

One challenge of modern CFD simulations is the huge amount of data that is produced. In the future, we will deal with this challenge by further interlocking the CFD solver and the visualization tool. We will combine the computational step with the postprocessing in an in-situ visualization approach, where indicators similar to those described in Sect. 2.2 detect situations in the simulation that need closer inspection and trigger the visualization. This approach will then be applied to cavitation phenomena that can occur in a number of hydraulic systems. This way we will be able to obtain a better understanding of this phenomena, especially concerning the damage they can cause on, e.g., engine components, which is vital for the future development of such devices.

Acknowledgements This work is supported by the Federal Ministry of Education and Research (BMBF) within the HPC III project HONK "Industrialization of high-resolution numerical analysis of complex flow phenomena in hydraulic systems".

References

1. Ducros, F., Ferrand, V., Nicoud, F., Weber, C., Darracq, D., Gacherieu, C., Poinsot, T.: Large-eddy simulation of the shock/turbulence interaction. J. Comput. Phys. **152**(2), 517–549 (1999)
2. Flad, D.G., Frank, H.M., Beck, A.D., Munz, C.-D.: A discontinuous Galerkin spectral element method for the direct numerical simulation of aeroacoustics. In: Proceedings of the American Institute of Aeronautics and Astronautics (2014)
3. Hempert, F., Hoffmann, M., Iben, U., Munz, C.-D.: On the simulation of industrial gas dynamic applications with the discontinuous Galerkin spectral element method. In: Proceedings of the 12th International Symposium on Experimental and Computational Aerothermodynamics of Internal Flows (2015)
4. Hindenlang, F., Gassner, G., Altmann, C., Beck, A., Staudenmaier, M., Munz, C.-D.: Explicit discontinuous Galerkin methods for unsteady problems. Comput. Fluids **61**, 86–93 (2012)
5. Kraus, T., Hindenlang, F., Harlacher, D.F., Munz, C.-D., Roller, S.: Direct aeroacoustic simulation of near field noise during a gas injection process with a discontinuous Galerkin approach. In: Proceedings of the 33rd AIAA Aeroacoustics Conference (2012)
6. Persson, P.-O., Peraire, J.: Sub-cell shock capturing for discontinuous Galerkin methods. In: Proceedings of the American Institute of Aeronautics and Astronautics, vol. 112 (2006)
7. Pruett, C., Thomas, B., Grosch, C., Gatski, T.: A temporal approximate deconvolution model for large-eddy simulation. Phys. Fluids **18**(2), 8104 (2006)
8. Robert, B.G.: One million natural gas injection valves produced, Press Release (2010)
9. Schmidt, A.: Experimentelle Untersuchung einer Gasströmung durch ein CNG-Injektorventil mittels Particle-Image-Velocimetry (PIV). Master's thesis, Institute of Mechanics of the University of Kassel (2012)

10. Sonntag, M., Munz, C.-D.: Shock capturing for discontinuous Galerkin methods using finite volume subcells. In: Finite Volumes for Complex Applications VII-Elliptic, Parabolic and Hyperbolic Problems. Springer Proceedings in Mathematics & Statistics, vol. 78, pp. 945–953. Springer, Berlin (2014)
11. Üffinger, M., Frey, S., Ertl, T.: Interactive high-quality visualization of higher-order finite elements. Comput. Graphics Forum **29**(2), 337–346 (2010)
12. Utkarsh A.: The ParaView Guide. Kitware Inc. www.kitware.com (2015)

A High-Order Discontinuous Galerkin CFD Solver for Turbulent Flows

Michael Wurst, Manuel Keßler, and Ewald Krämer

Abstract The Discontinuous Galerkin method is used for the discretisation of the Reynolds-Averaged-Navier-Stokes equations. It is a high-order method in space reducing the amount of cells for calculations compared to standard CFD solvers. We are planning to use the method for two different kind of flow types: highly separated flows and rotor flows. For the first, we implemented a Detached Eddy Simulation method. For rotor flows we started to implement a Chimera grid technique so that we are able to handle moving bodies.

1 Introduction

The Discontinuous Galerkin (DG) method was first used by Reed and Hill [1] for neutron transport and it was first applied for conservation laws by Cockburn and Shu [2]. In recent years the method was extended by Bassi et al. for the Navier-Stokes equations [3] and Reynolds-Averaged-Navier-Stokes (RANS) equations [4]. Lübon [5] first applied the method for Detached Eddy Simulations (DES). In fluid mechanics the DG method combines ideas from finite volume (FV) discretisation techniques as well as from finite element (FE) discretisation techniques. A typical FE feature of the DG method is that the representation of the solution in a cell is given by a polynomial approximation, whose accuracy can easily improved by increasing its polynomial order. However, the solution between cells is discontinuous, thus the solution of this Riemann problem requires approximate Riemann solvers known from FV methods.

The Chimera method was first introduced for the Euler equations by Benek et al. [6]. Since then the method has been applied to many problems in fluid mechanics. It has basically two advantages compared to a single grid approach. The separate creation of a body-fitted grid and the background grid allows a meshing process, which would be more complicated with a single mesh or even impossible for some geometries, e.g. when using a structured solver. A first implementation of

M. Wurst (✉) • M. Keßler • E. Krämer
Institut für Aerodynamik und Gasdynamik, Universität Stuttgart, Pfaffenwaldring 21, 70569 Stuttgart, Germany
e-mail: wurst@iag.uni-stuttgart.de

© Springer International Publishing Switzerland 2016
W.E. Nagel et al. (eds.), *High Performance Computing in Science and Engineering '15*, DOI 10.1007/978-3-319-24633-8_35

the Chimera method for the DG method was done by Galbraith et al. for inviscid [7] and viscous flow [8].

2　Numerical Formulation

The basis for the numerical description of industrial technical flows are the RANS equations employing the Spalart-Allmaras (SA) turbulence model [9] as turbulence model. They are given by:

$$\frac{\partial \mathbf{u}}{\partial t} + \nabla \cdot (\mathscr{F}_{inv}(\mathbf{u}) - \mathscr{F}_{vis}(\mathbf{u}, \nabla \mathbf{u})) = \mathscr{S}(\mathbf{u}, \nabla \mathbf{u}). \tag{1}$$

The vector $\mathbf{u}$ is the vector of the conservative variables, $\mathscr{F}_{inv} = (F^x_{inv}, F^y_{inv}, F^z_{inv})$ and $\mathscr{F}_{vis} = (F^x_{vis}, F^y_{vis}, F^z_{vis})$ are the vectors of the inviscid and viscous fluxes:

$$U = \begin{pmatrix} \rho \\ \rho u_i \\ \rho E \\ \rho \tilde{v} \end{pmatrix}, F^j_{inv} = \begin{pmatrix} \rho u_j \\ \rho u_i u_j + p\delta_{ij} \\ u_j(\rho E + p) \\ \rho \tilde{v} u_j \end{pmatrix}, F^j_{vis} = \begin{pmatrix} 0 \\ \tau_{ij} \\ \tau_{ji} u_i + q_j \\ \frac{1}{\sigma}(\mu + \rho \tilde{v})\frac{\partial \tilde{v}}{\partial x_j} \end{pmatrix}, \tag{2}$$

where ρ is the fluid density, (u_1, u_2, u_3) the velocity vector, $\tilde{v}$ a viscosity like variable, p the pressure and E the specific total energy. The pressure p and the specific energy E are connected by the equation of state for an ideal gas

$$p = (\gamma - 1)(\rho E - \frac{1}{2}\rho(u_1^2 + u_2^2 + u_3^2)), \tag{3}$$

where $\gamma = \frac{c_p}{c_v} = 1.4$ is the ratio of specific heats. The viscous stress tensor τ_{ij} and the heat flux q_i include both contributions from the Reynolds averaging via a turbulent eddy viscosity μ_t and a turbulent thermal conductivity λ_t:

$$\tau_{ij} = (\mu + \mu_t)\left(\frac{\partial u_i}{\partial x_j} + \frac{\partial u_j}{\partial x_i} - \frac{2}{3}\frac{\partial u_k}{\partial x_k}\delta_{ij}\right) - \frac{2}{3}\delta_{ij}\rho k, \tag{4}$$

$$q_j = -(\lambda + \lambda_t)\frac{\partial T}{\partial x_j}. \tag{5}$$

Both the laminar and turbulent thermal conductivity can be connected to the laminar viscosity and the eddy viscosity, respectively, by

$$\lambda = \frac{\mu c_p}{Pr}, \qquad \lambda_t = \frac{\mu_t c_p}{Pr_t} \tag{6}$$

with a laminar Prandtl number $Pr = 0.72$ and turbulent Prandtl number $Pr = 0.9$. The turbulent eddy viscosity is then determined from the SA model as

$$\mu_t = \rho \tilde{v} f_{v1} \tag{7}$$

with a calibration function f_{v1}. The source term $\mathscr{S}$ in Eq. (1) is only present in the SA equation and serves as a term to model turbulent production and destruction in the RANS equations:

$$S = \frac{1}{\sigma} \rho c_{b2} \frac{\partial \tilde{v}}{\partial x_i} \frac{\partial \tilde{v}}{\partial x_i} + c_{b1} \rho \tilde{S} \tilde{v}$$
$$- c_{w1} f_w \frac{1}{\rho} \left(\frac{\rho \tilde{v}}{d} \right)^2 \tag{8}$$

In the source term the distance to the wall d is a decisive part to model the production and destruction. In the classical DES model from Spalart [10] d is replaced by

$$l_{DES} = \min(d, C_{DES}\Delta) . \tag{9}$$

Here Δ is a measure of the grid spacing (longest cell edge) and $C_{DES} = 0.65$ a constant to calibrate the model in order to operate the model in RANS mode close to the wall and in LES mode away from the wall. A disadvantage of the method is clearly that the model depends on the applied mesh, e.g. it can happen that the change from the RANS to the LES model is too early in the boundary layer. A possible improvement is the use of the DDES model [11], which also includes information about the flow field in the calculation of the turbulent length scale:

$$l_{DDES} = d - f_d \max(0, d - C_{DES}\Delta) \tag{10}$$

with

$$f_d = 1 - \tanh([8r_d]^3) \tag{11}$$

and

$$r_d = \frac{v_t + v}{\sqrt{u_{i,j} u_{i,j} \kappa^2 d^2}} . \tag{12}$$

Both DES models were developed for the simulation with second-order FV methods. As the DG method is normally used as a high-order method using polynomial basis functions of the order $p > 1$, this higher resolution inside a cell should be represented in a SA based DES model. Thus, the adaptation of the cell width Δ with the polynomial degree p is a possible solution

$$\Delta := \frac{\text{cell width } \Delta}{\text{polynomial degree } p} . \tag{13}$$

3 Discontinuous Galerkin Discretisation

3.1 Spatial Discretisation

The Navier-Stokes equations (1) are discretised with a standard Discontinuous Galerkin method. Multiplying the equation with an arbitrary test function $\mathbf{v}$, integrating over the domain Ω and integrating by parts of the divergence leads to the weak form of the Navier-Stokes equation:

$$
\int_\Omega \mathbf{v}_h \cdot \frac{\partial \mathbf{u}_h}{\partial t} \, d\Omega + \oint_{\partial\Omega} \mathbf{v}_h \left(\mathscr{F}_i(\mathbf{u}_h) - \mathscr{F}_v(\mathbf{u}_h, \nabla \mathbf{u}_h) \right) \cdot \mathbf{n} \, d\sigma
$$
$$
- \int_\Omega \nabla \mathbf{v}_h \cdot \left(\mathscr{F}_i(\mathbf{u}_h) - \mathscr{F}_v(\mathbf{u}_h, \nabla \mathbf{u}_h) \right) \, d\Omega = \int_\Omega \mathbf{v}_h \mathscr{S}. \tag{14}
$$

The Riemann problem in the surface integral is tackled with a HLL Riemann solver for the inviscid fluxes. For the gradients in the viscous flux terms either the BR1 scheme [3] or the BR2 scheme can be applied [12].

3.2 Temporal Discretisation

The temporal discretisation is handled either with an explicit or an implicit discretisation. For unsteady calculations we use standard explicit Runge-Kutta methods up to fourth order and for steady calculations an implicit Backward Euler method. The DG discretisation in Eq. (14) can be written in a compact form

$$
\mathbf{M} \frac{d\mathbf{U}}{dt} = \mathbf{R}(\mathbf{U}), \tag{15}
$$

with $\mathbf{U}$ as the global solution vector and $\mathbf{M}$ as the global mass matrix. Discretising this with an Euler backward scheme leads to:

$$
\mathbf{M} \frac{\mathbf{U}^{n+1} - \mathbf{U}^{n+1}}{\Delta t} = \mathbf{R}(\mathbf{U}^{n+1}). \tag{16}
$$

For the solution of the nonlinear system we linearise the system with a Newton method with only one Newton iteration per time-step:

$$
\left(\frac{\mathbf{M}}{\Delta t} - \frac{\partial \mathbf{R}(\mathbf{U}^n)}{\partial \mathbf{U}} \right) \Delta \mathbf{U} = \mathbf{A} \Delta \mathbf{U} = \mathbf{R}(\mathbf{U}^n). \tag{17}
$$

The linearisation of the residual $\mathbf{R}$ is done analytically and the resulting linear system is solved with a GMRES solver preconditioned with an ILU(0) method. For steady computations we apply a local-time stepping scheme with a constant CFL number for all cells in the domain. The CFL number varies according to the change of the L_2-norm of the residual vector:

$$\text{CFL}^{n+1} = \text{CFL}^0 \frac{\|\mathbf{R}(\mathbf{U}^0)\|_{L_2}}{\|\mathbf{R}(\mathbf{U}^n)\|_{L_2}}. \tag{18}$$

4 Chimera Method for the DG Discretisation

4.1 Basic Scheme

In the following the basic Chimera scheme is explained on the example of a two-dimensional cylinder in Fig. 1. In general, there are three different kinds of cells in the background grid and the near-body grid. They are distinguished with an additional vector I_{Blank}, which has the size of the number of cells. First of all, there are cells which do not lie in the overlapping region of the two grids. They are called active cells and the vector I_{Blank} will have a value of one for these cells (green cells in Fig. 1).

The second kind of cells are in the background mesh in the area where the body lies (blue cells). These cells are identified within the process of hole cutting. The process can be very critical for a successful Chimera method. For simulations with a movement of the body the hole cutting has to be done in every time-step and should be obviously implemented efficiently. For the stationary, static case which is treated here, we choose explicitly specified geometric forms as circles and rectangles for

Fig. 1 Typical Chimera setup

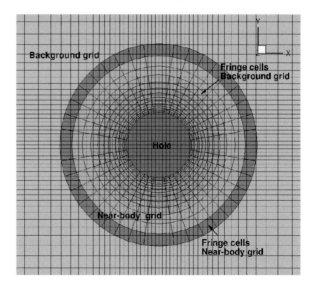

the form of the hole. The cells inside the hole are inactive cells and the vector I_{Blank} equals zero.

The third kind of cells are the fringe cells (red cells). They are important for the coupling of both grids and can be found in the background as well as the near-body grid. In the background grid they are the cells which are positioned next to the hole cells, and in the near-body grid they are the cells at the outer boundary of the grid. It is important that both lines of fringe cells form a closed surface without any gaps, which e.g. for the background grid would mean that it is possible that fluid flows into the hole, or vice versa. As the BR2 scheme needs one neighbor cell for the calculation of gradients, one row of fringe cells is sufficient, leading also to a compact Chimera stencil compared to high-order finite difference or finite volume schemes. Fringe cells have an I_{Blank} value of two. The cells between the fringe cells in the background and the near-body grid, respectively, are active cells. In the small overlap region of active cells the flow solution will be double valued. However, this area is commonly well enough resolved in simulations, so the differences should be small.

4.2 Definition of Interpolation Operators

For the coupling of the two grids we use a discrete L_2 projection to determine the modal coefficients of the basis functions. This has the advantage that by definition the error of the approximation becomes minimal in the L_2 sense. This projection satisfies the following property for the solution $\mathbf{u}_{FC}$ in the fringe cells

$$\int_\Omega \phi_i \mathbf{u}_{FC}^h \, d\Omega = \int_\Omega \phi_i \mathbf{u}_{IG}^h \, d\Omega, \tag{19}$$

where ϕ_i denote the basis functions in the fringe cells, and $\mathbf{u}_{FC}$ is defined as:

$$\mathbf{u}_{FC}^h = \sum_{n=1}^{n_{DOF}} \mathbf{u}_{FC,i} \phi_i. \tag{20}$$

The vector $\mathbf{u}_{IG}^h$ is the solution in the other, for the interpolation necessary grid. We approximate the integral containing $\mathbf{u}_{IG}^h$ by Gauss integration

$$\int_\Omega \phi_i \mathbf{u}_{IG}^h \, d\Omega = \sum_{k=1}^{n_{GP}} \phi_i(x_k) \mathbf{u}_{IG}^h(x_k) \omega_k, \tag{21}$$

leading to the following compact form for the determination of the unknown modal coefficients $\mathbf{u}_{FC,i}$

$$\mathbf{M} \mathbf{u}_{FC} = \mathbf{b}. \tag{22}$$

Fig. 2 Interpolation process

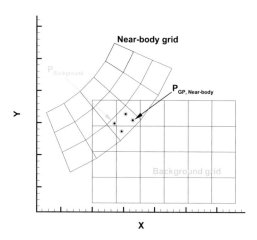

Here **M** is as the mass matrix of the fringe cell and **b** the sum in Eq. (21). In summary, that means, that the solution $\mathbf{u}_{IG}^h$ has to be evaluated at all Gauss points x_k. In the present Chimera method those Gauss points are defined in the fringe cell, however, the solution is obtained from those points evaluated in the corresponding grid. This can be seen in Fig. 2 for a generic case: For the Gauss points (black points) in the near-body grid exist points in the background grid (gray points). In general, these points do not lie in the same cell in the background grid, but are distributed over several cells. Additionally, it is not sufficient to find the correct partner cell, we also have to find the local coordinates (ξ, η) to evaluate the solution in the background. The transformation is usually given as $\mathbf{x} = \mathbf{x}(\xi, \eta)$, however, we need the inverse transformation. As this is sometimes difficult to get analytically (e.g. for curved elements), we use a Newton iteration to find the unit coordinates after we have identified the correct cell for the interpolation. As long as the Jacobian of the transformation stays positive, which is necessary anyway, the Newton iteration will converge, usually in very few steps.

4.3 Modification to the Time Integration

The implicit time integration can be adapted with two strategies for the integration of Chimera boundaries: explicit Chimera boundaries and implicit Chimera boundaries. In case of explicit Chimera boundaries the following scheme is applied:

1. Project the solution to the Chimera cells ($I_{Blank} = 2$) in both the background grid and the Chimera grid according to Eq. (22).
2. Adapt the linear system for these cells, so that a "no solution update" in the part of the linear system can be determined:

$$\mathbf{I}\Delta\mathbf{U} = 0, \tag{23}$$

which means that the RHS side is zeroed and the LHS is changed to the unit matrix $\mathbf{I}$ for interpolated and cut-out cells.

The projection is done for both the Chimera and the background grid with the solution vector $\mathbf{U}^n$. However, also for those cells a solution at t_{n+1} has to be determined, which means that additional iterations over the Newton algorithm have to be performed until the solution at the fringe cells does not change anymore. We usually use a fixed number of iterations per pseudo time-step, in the following results section we employ one, and alternatively three iterations.

In case of implicit Chimera boundaries the discrete projection is linearized as well, which means that the dependencies of the fringe cells on their donor cells are taken into account in the linear system. Linearising equation (22) leads to:

$$\left(\mathbf{I} - \mathbf{M}^{-1} \frac{\partial \mathbf{b}(\mathbf{U}^n)}{\partial U}\right) \Delta \mathbf{U} = \mathbf{M}^{-1} \mathbf{b} - \mathbf{U}^n. \tag{24}$$

5 Numerical Results

5.1 Chimera Method

Here we present results from [13], where we assessed the correct implementation and robustness of the Chimera method. We summarize two test cases, a simulation with the method of manufactured solutions and a turbulent simulation of a NACA0012 airfoil with different angles of attack. Both flows have a steady solution and are calculated with different polynomial orders. For the NACA0012 airfoil we performed the simulations with order sequencing: to achieve a fast convergence at high-order discretizations we start with a P1 solution and increase the order, so that the steady solution of order n is used as a restart solution for order $n + 1$. For the local time-stepping we start with $CFL^0 = 1$ at P1 and increase it up to $CFL^0 = 100$ for higher orders. The GMRES solver for the solution of the linear system uses typically 120–240 Krylov space vectors and 240–480 maximum iterations to solve the system with a residual tolerance of 10^{-10}.

5.2 Manufactured Solutions

The method of manufactured solutions is applied to investigate the discretization error of the implemented Chimera method. It needs a given analytical solution as a forcing term for the Navier-Stokes equations so that this solution satisfies the Navier Stokes equations. We choose the following analytical form for density, velocity and

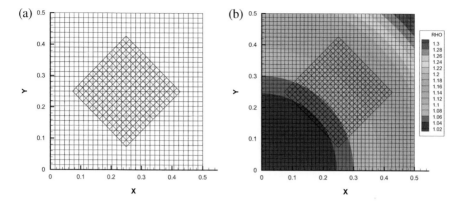

Fig. 3 Chimera grid (**a**) and density distribution (**b**) for the manufactured solutions

total energy (compare [14]):

$$\rho = 0.5(\sin(x^2 + y^2) + 1.5),$$
$$u = \sin(x^2 + y^2) + 0.5,$$
$$v = 0.1(\cos(x^2 + y^2) + 0.5),$$
$$E_t = 0.5(\cos(x^2 + y^2) + 1.5). \tag{25}$$

The problem is solved on a quadratic computational domain with length 0.5. The single grid uses an equidistant grid with N elements in each direction. For the Chimera solution we add another smaller grid with length 0.25 inside the domain, which has the same element size for its mesh but was rotated (compare Fig. 3a). The simulation is performed for different gridsizes and polynomial orders and for two different Reynolds numbers $Re = 100$ and $Re = 1000$. The resulting solution is plotted exemplarily for the density in Fig. 3b. The discretization error can then be seen in Fig. 4. The order of accuracy reaches for both grids and both Reynolds numbers the design order of the DG scheme. For the Chimera version the discretization error is somewhat higher but it achieves the same convergence order as the single grid version.

5.3 Turbulent Flow Around a NACA0012 Airfoil

A NACA0012 airfoil is used for the validation of the Chimera method for turbulent flows. The airfoil is examined at $Re = 2.88 \cdot 10^6$ with $Ma_\infty = 0.15$ with different angles of attack $\alpha = 0°, 10°, 15°$. The flow is assumed as fully turbulent with $\tilde{v}_\infty / v_\infty = 3$ on the farfield boundary for the SA turbulence model. The farfield itself is located 100 chords away from the airfoil. The grid has 2688 cells in the

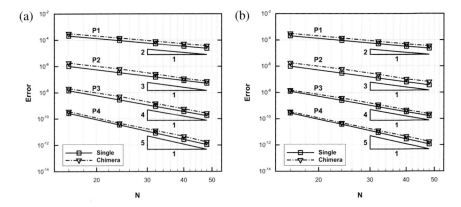

Fig. 4 Spatial order of accuracy for the method of manufactured solutions at different Reynolds numbers. (**a**) $Re = 100$. (**b**) $Re = 1000$

background grid and 2024 in the near-body grid. It is usually good for the method and has been paid attention to that the background grid has nearly the same grid features as the near-body grid, which means that the fringe cells and the cells from which they are interpolated should have approximately the same size. The y^+-value in the first cell row close to the wall was around 10 for both versions, a value which is typical for the DG method, as the high-order allows a proper resolution of the entire laminar sublayer within the first cell. The single grid version is a typical C-grid with 2862 cells. All cases were run up to sixth order. Experimental reference data for this case is available with pressure distributions from wind tunnel measurements from Gregory and O'Reilly [15].

The P1 solution overestimates the suction peak, leading to a too high lift coefficient. The agreement between the single grid version and the Chimera grid version is excellent. The distribution of the pressure coefficient for all angles of attack is the same (compare Fig. 5a–c). As a consequence, the differences in the forces are also negligible (compare Tables 1, 2, 3, 4, 5). Small differences in both simulations can be seen in the contour of $\tilde{\nu}$ in Fig. 7c and d. Here the contour lines of the Chimera grid simulation (red lines) and the single grid simulation (black lines) are shown in one plot. Before and at the trailing edge of the airfoil the lines are in good overlap. However, in the wake the lines differ slightly, but the influence on global forces and pressure distribution is very small.

Concerning the use of explicit or implicit Chimera boundaries differences exists. This can be seen in Fig. 6a–d exemplarily for the $\alpha = 10°$ case. The variant with three Chimera boundary iterations per Newton iteration needs as many time-steps as the implicit Chimera boundary version or the single grid version. However, it still needs three times more linear solves as the implicit Chimera boundary version or the single grid version. The explicit version with one Chimera boundary iteration performs quite well. The convergence rate at the beginning of each order

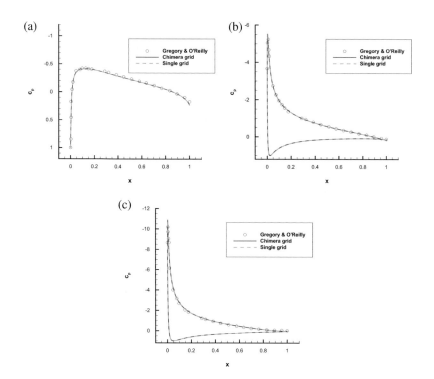

Fig. 5 Pressure coefficient for different angles of attack comparing single grid solution and Chimera grid solution for P5. (**a**) $\alpha = 0°$. (**b**) $\alpha = 10°$. (**c**) $\alpha = 15°$

Table 1 Drag coefficient for the NACA0012 airfoil ($\alpha = 0°$)

	P1	P2	P3	P4	P5
Single grid	0.010095	0.0091221	0.0091408	0.0091305	0.0091391
Imp. Chimera grid	0.010380	0.0091097	0.0091637	0.0091339	0.0091390

Table 2 Drag coefficient for the NACA0012 airfoil ($\alpha = 10°$)

	P1	P2	P3	P4	P5
Single grid	0.010366	0.013669	0.014048	0.013995	0.014014
Imp. Chimera BC	0.011063	0.013780	0.014071	0.013999	0.014014

Table 3 Lift coefficient for the NACA0012 airfoil ($\alpha = 10°$)

	P1	P2	P3	P4	P5
Single grid	1.17654	1.07999	1.07530	1.07515	1.07487
Imp. Chimera BC	1.18161	1.07928	1.07561	1.07552	1.07513

Table 4 Drag coefficient for the NACA0012 airfoil ($\alpha = 15°$)

	P1	P2	P3	P4	P5
Single grid	0.014395	0.022070	0.024435	0.024445	0.024424
Imp. Chimera grid	0.015645	0.022253	0.024478	0.024479	0.024440

Table 5 Lift coefficient for the NACA0012 airfoil ($\alpha = 15°$)

	P1	P2	P3	P4	P5
Single grid	1.77141	1.54365	1.51856	1.51716	1.51674
Imp. Chimera grid	1.76420	1.54230	1.51921	1.51789	1.51731

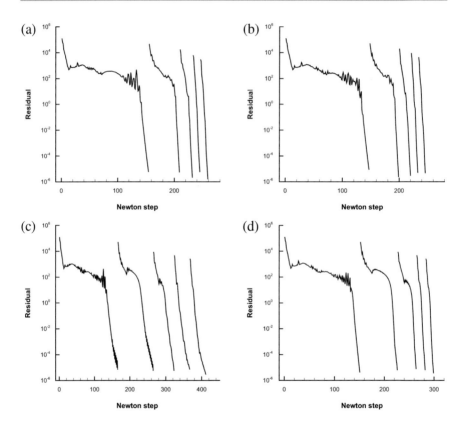

Fig. 6 Convergence history for the NACA0012 airfoil at $\alpha = 10°$ for P1,..., P5. (**a**) Single grid. (**b**) Implicit Chimera BC. (**c**) Explicit Chimera BC one Chimera BC iteration per Newton iteration. (**d**) Explicit Chimera BC three Chimera BC iteration per Newton iteration

is competitive, but in the end the rate degenerates. Altogether less linear solves as in the case with three iterations are performed. However, the implicit Chimera boundary is to be preferred (Fig. 7).

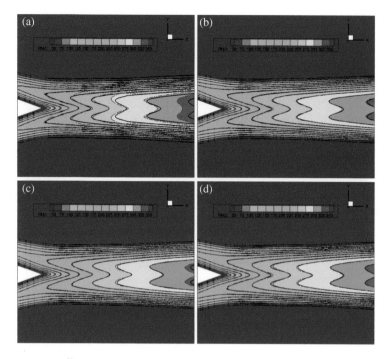

Fig. 7 Contour of $\tilde{\nu}$ for the NACA0012 airfoil (*Black lines*: single grid, *red lines*: Chimera grid, y-Axis with larger scale). (**a**) P2. (**b**) P3. (**c**) P4. (**d**) P5

5.4 Instationary DES Flows

5.4.1 DES Results

The simulation of a backward-facing step at $Re = 28{,}000$ is used for the comparison of the DES and DDES model [16]. The simulation is based on the experiment by Eaton and Vogel [17], numerical reference data is available in Spalart et al. [11]. A key property for the correct simulation of this kind of simulation is the definition of an inlet profile to achieve a correct boundary layer thickness close to the step. For that reason an inlet profile based on a power-law is imposed at the domain inlet. The spanwise width of the domain is two step heights (including 20 cells) employing periodic boundary conditions. The mesh for the simulation consist of only 141,000 hexahedral cells, probably too small for a simulation with a FV method. The simulation is carried out with fourth-order accurate elements consisting of 20 degree of freedoms (DOFs) per element. In Fig. 8a and b the mean friction coefficient at the lower wall behind the step is plotted in comparison to the experimental and numerical reference data. Temporal averaging was done for a period of seven flow through periods of the domain behind the step and after that the flow was also averaged in spanwise direction. It can be clearly seen that the

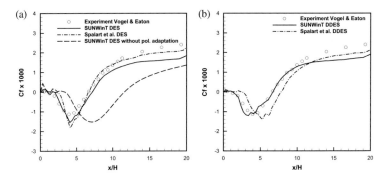

Fig. 8 Mean Friction coefficient at the lower wall behind the step employing the (**a**) DES (**b**) DDES model

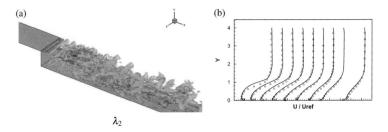

Fig. 9 (**a**) $\lambda_2 = -0.01$ colored with vorticity magnitude and (**b**) velocity profiles behind the step for the DES calculation

adaptation of the cell width is successful in this case. There is some difference to the reference data, which may be due to different inlet conditions. Spalart et al. give no details about their inlet velocity profile, but results have been found to be sensitive to the given profile.

In Fig. 9a the λ_2 vortex criterion is used for the identification of the vortices behind the step. Directly behind the step, the free shear layer begins to roll up to form essentially two dimensional vortices. After about 1 step height these vortices burst and the flow becomes three-dimensional. In the experiment the flow reattaches at 6.7 step heights behind the step. Here, the flow reattaches at $7.3\,h$ in the DES case and at $6.8\,h$ in the DDES case. Velocity profiles presented in Fig. 9b also show a good agreement with the experiments with only small deviations in the upper part of the profiles.

6 Computational Aspects

The good parallel performance of our solver up to 16,384 cores was shown before in [18] for explicit calculations. For implicit calculations we can expect the same behaviour for the assembly of the matrix as there is no additional communication

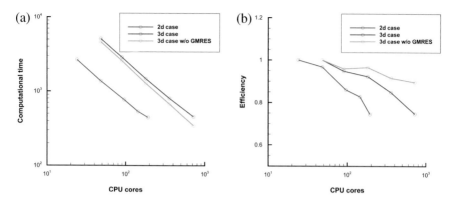

Fig. 10 Strong Scaling (**a**) and parallel efficiency (**b**)

compared to a residual evaluation. However, the parallel performance should a bit decrease for the GMRES linear system solver. This can be seen in Fig. 10: the parallel efficiency increases if the time for the linear solve is excluded. The reason for that might be that the performance of the GMRES method depends on a good preconditioning which on one core is done with an ILU method. In parallel this is not possible and a block ILU method must be chosen which is an ILU method on a every core. This leads to a higher amount of GMRES iterations decreasing the parallel performance. We rely here on the GMRES implementation of the trilinos package [19] which is available preinstalled at the Cray XC40 system at HLRS.

7   Conclusion

The Discontinuous Galerkin method was used for the discretisation of the RANS equations with DES capabilities. The DES capabilities were validated with a simulation of a backward facing step. Additionally, a Chimera method was implemented which will be used for rotor flows in the future. Until now, the method is able to achieve the design order of the DG scheme as shown in a simulation with the method of manufactured solutions. A first step towards turbulent calculations was done with the simulation of a NACA0012 airfoil at different angles of attack. Here, the Chimera grid simulations agreed well with single grid simulations without decreasing the robustness.

Acknowledgements   We greatly acknowledge the provision of supercomputing time and technical support by the High Performance Computing Center Stuttgart (HLRS) for our project DGDES.

References

1. Reed, W., Hill, T.: Triangular mesh methods for the neutron transport equation. Technical Report, Los Alamos Scientific Laboratory (1973)
2. Cockburn, B., Shu, C.: TVB Runge-Kutta local projection discontinuous Galerkin finite element method for conservation law II: general framework. Math. Comput. **52**(186), 411–435 (1989)
3. Bassi, F., Rebay, S.: A high-order accurate discontinuous finite element method for the numerical solution of the compressible Navier-Stokes equations. J. Comput. Phys. **131**, 267–279 (1997)
4. Bassi, F., Crivellini, A., Rebay, S., Savini, M.: Discontinuous Galerkin solution of the Reynolds-averaged Navier-Stokes and $k - \omega$ turbulence model equations. Comput. Fluids **34**, 507–540 (2005)
5. Lübon, C.: Turbulenzmodellierung und Detached Eddy Simulation mit einem Discontinuous Galerkin Verfahren von hoher Ordnung. PhD thesis, Universität Stuttgart (2009)
6. Benek, J., Steger, J., Dougherty, F.: A flexible grid embedding technique with application to the Euler equations. In: Fluid Dynamics and Co-located Conferences. American Institute of Aeronautics and Astronautics, July 1983
7. Galbraith, M., Orkwis, P., Benek, J.: Extending the Discontinuous Galerkin scheme to the Chimera overset method. In: Fluid Dynamics and Co-located Conferences. American Institute of Aeronautics and Astronautics, June 2011
8. Galbraith, M., Orkwis, P., Benek, J.: Discontinuous Galerkin scheme applied to Chimera overset viscous meshes on curved geometries. In: Fluid Dynamics and Co-located Conferences. American Institute of Aeronautics and Astronautics, June 2012
9. Spalart, P., Allmaras, S.: A one-equation turbulence model for aerodynamic flows. La Recherche Aerospatiale **1**, 5–21 (1994)
10. Spalart, P., Jou, W.H., Strelets, M., Allmaras, S.: Comments on the feasibility of LES for wings, and on a hybrid RANS/LES approach. In: Advances in DNS/LES (1997)
11. Spalart, P., Deck, S., Shur, M., Squires, K., Strelets, M., Travin, A.: A new version of detached-eddy simulation, resistant to ambiguous grid densities. Theor. Comput. Fluid Dyn. **20**(3), 181–195 (2006)
12. Bassi, F., Rebay, S., Mariotti, G., Pedinotti, S., Savini, M.: A high-order accurate discontinuous finite element method for inviscid and viscous turbomachinery flows. In: Decuypere, R.G.D. (ed.) 2nd European Conference on Turbomachinery Fluid Dynamics and Thermodynamics, Antwerpen, Belgium: Technologisch Instituut, pp. 99–108 (1997)
13. Wurst, M., Keßler, M., Krämer, E.: A high-order discontinuous Galerkin chimera method for laminar and turbulent flows. In: AIAA SciTech. American Institute of Aeronautics and Astronautics, Jan 2015
14. Glasby, R., Burgess, N., Anderson, W., Wang, L., Mavriplis, D., Allmaras, S.: Comparison of SU/PG and DG Finite-Element techniques for the compressible Navier-Stokes equations on anisotropic unstructured meshes. In: 51st AIAA Aerospace Sciences Meeting, AIAA 2013–0691 (2013)
15. Gregory, N., O'Reilly, C.: Low speed aerodynamic characteristics of NACA0012 aerofoil section, including the effects of upper surface roughness simulating hoar frost. Technical Report, Aeronautical Research Council (1970)
16. Wurst, M., Keßler, M., Krämer, E.: Aerodynamic and acoustic analysis of an extruded airfoil with a trailing edge device using detached eddy simulation with a discontinuous Galerkin method. In: Fluid Dynamics and Co-located Conferences. American Institute of Aeronautics and Astronautics, June 2013
17. Vogel, J., Eaton, J.: Combined heat transfer and fluid dynamic measurements downstreams of a backward facing step. J. Heat Transf. **107**, 922–929 (1985)

18. Busch, E., Wurst, M., Keßler, M., Kräämer, E.: Computational aeroacoustics with higher order methods. In: Nagel, W.E., Kröner, D.H., Resch, M.M. (eds.) High Performance Computing in Science and Engineering '12, pp. 239–253. Springer, Berlin (2013)
19. Heroux, M.A., Bartlett, R.A., Howle, V.E., Hoekstra, R.J., Hu, J.J., Kolda, T.G., Lehoucq, R.B., Long, K.R., Pawlowski, R.P., Phipps, E.T., Salinger, A.G., Thornquist, H.K., Tuminaro, R.S., Willenbring, J.M., Williams, A., Stanley, K.S.: An overview of the trilinos project. ACM Trans. Math. Softw. **31**(3), 397–423 (2005)

Mesoscale Simulations of Anisotropic Particles at Fluid-Fluid Interfaces

Qingguang Xie, Florian Günther, and Jens Harting

Abstract Anisotropic colloidal particles have attracted great interest in industrial applications, such as particle-stabilized emulsions and fabrication of self-assembled complex materials. However, our understanding of the fundamental properties of such systems is still limited. We combine the lattice Boltzmann method and molecular dynamics to simulate multicomponent fluids and suspended particles. We review two examples of our recent research on anisotropic particles. First, we study the ensemble of ellipsoidal particles at a spherical interface. The capillary interactions between the particles cause a local reordering on very long timescales leading to a continuous change in the interface configuration and to an increase of interfacial area. This effect can be utilized to counteract the thermodynamic instability of particle stabilized emulsions and thus offers the possibility to produce emulsions with exceptional stability. Second, we investigate the behaviour of magnetic Janus particles adsorbed at fluid-fluid interfaces interacting with an external magnetic field. We demonstrate that the strength of resulting capillary interactions can be tuned by altering the external field strength, opening up the possibility to create novel, reconfigurable materials.

1 Introduction

Anisotropic colloidal particles have drawn special attention for their potential in material science, such as particle-stabilized emulsions and the fabrication of self-assembled complex materials. The particles can strongly adsorb at the interfaces

Q. Xie • F. Günther
Department of Applied Physics, Eindhoven University of Technology, P.O. Box 513, NL-5600MB Eindhoven, The Netherlands

J. Harting (✉)
Department of Applied Physics, Eindhoven University of Technology, P.O. Box 513, NL-5600MB Eindhoven, The Netherlands

Faculty of Science and Technology, Mesa+ Institute, University of Twente, NL-7500AE Enschede, The Netherlands
e-mail: j.harting@tue.nl

© Springer International Publishing Switzerland 2016

W.E. Nagel et al. (eds.), *High Performance Computing in Science and Engineering '15*, DOI 10.1007/978-3-319-24633-8_36

between two immiscible fluids and as such stabilize an emulsion or allow the formation of 2D structures.

The stability of emulsions depends on several parameters like particle coverage at the interfaces and the wettability of the particles. It was found that the particle coverage at the interfaces is the most important parameter [1]. The colloidal particles act in a similar way as surfactants. In both cases the free energy of the interface is reduced. However, the fluid-fluid interfacial tension is not being modified by particles [2]. There are many kinds of particles/colloid types which can stabilize an emulsion. I.e., next to spheres [3, 4], the colloidal particles can also be of more complex nature and include anisotropic shapes [5], magnetic interactions [6, 7], or anisotropic Janus style properties [8].

The influence of the particle shape on the stabilization of Pickering emulsions was studied experimentally with prolate and oblate ellipsoids, e.g. in [9]. As the degree of the particle anisotropy increases, the effective coverage area increases. In this way they are more efficient stabilizers for emulsions than spherical particles. Furthermore, the rheological properties of the emulsion vary with changing aspect ratio because the coverage of the fluid interfaces and the capillary interactions differ.

In [10–14] the adsorption of a single particle at a flat interface is studied in absence of external fields. The stable configuration for elongated ellipsoids is the orientation parallel to the interface [10]. This state minimizes the free energy of the particle at the interface by reducing the interfacial area [11, 13, 14]. A more complex particle shape, e.g. like the super-ellipsoidal hematite particle [15], might allow several equilibrium orientations.

Furthermore, if particles are adsorbed at an interface they generally deform it. This deformation can be caused for example by particle anisotropy [16], external forces such as gravity or electromagnetic forces acting on the particles [17, 18], or non-constant interface curvature [19]. This deformation leads to capillary interactions between the particles. In case of ellipsoids at a flat interface it is a quadrupolar potential [20], which leads to spatial ordering [21].

Due to the short timescales and limited optical accessibility, the dynamics of the formation of emulsions has only found limited attendance so far [22]. In our first study case, we investigate the influence of the geometrical anisotropy and rotational degrees of freedom of ellipsoidal particles on the time development of fluid domain sizes in particle-stabilized emulsions as published in [23, 24].

Secondly, we show that Janus particles are promising building blocks of reconfigurable and programmable self-assembled structures. Janus particles are characterized by anisotropic surface chemical (e.g. wetting or catalytic) or physical (e.g. optical, electric, or magnetic) properties at well-defined areas on the particle. For symmetric Janus particles composed of hydrophobic and hydrophilic hemispheres, the equilibrium contact angle is 90° since each hemisphere immerses in its favourable fluid, and the interface remains flat [25]. However, due to surface roughness [26], anisotropic shape [27, 28], or the influence of external forces, Janus particles can tilt with respect to the interface. In a tilted orientation, the fluid-fluid interface around the Janus particle deforms in a dipolar fashion in order to fulfil boundary conditions stipulated by Young's equation [29]. Assuming small interface deformations, the particle-induced deformations obey $\nabla^2 h = 0$, where h is the interface height. This equation can be solved using a multipolar analysis, analagous to 2D electrostatics [30]. These particle induced deformations can cause particles to attract and repel in specific orientations, making them a useful tool for controlling the behaviour of particles at interfaces in a straightforward manner.

Previous investigations into capillary interactions between particles at fluid-fluid interfaces have focussed mainly around two themes: particle weight-induced deformations, which lead to monopolar interactions between particles and are responsible for e.g. the Cheerios effect [31]; and surface roughness or shape anisotropy induced deformations, which lead to quadrupolar interactions between particles [32] and are responsible for e.g. the suppression of the coffee ring effect [33]. With respect to Janus particles, Brugarolas et al. [34] showed that quadrupolar capillary interactions induced by surface roughness can be used to form fractal-like structures of Janus nanoparticle-shelled bubbles. However, a significant limitation of the above mentioned capillary interactions is that they are not dynamically tunable because they depend on the particle properties alone.

Davies et al. [35] recently found a way of creating dynamically tunable dipolar capillary interactions between magnetic ellipsoidal particles adsorbed at an interface under the influence of an external magnetic field. The structures that form depend on the dipole-field coupling, which can be controlled dynamically [36]. However, it is also desirable to create tunable capillary interactions between spherical particles without relying on particle shape anisotropy. Therefore, we show how to create tunable dipolar capillary interactions using spherical Janus particles adsorbed at fluid-fluid interfaces [37].

This current report is organised as follows. We present our hybrid molecular dynamics-lattice Boltzmann simulation method in Sect. 2. Section 2.3 contains our simulation results on anisotropic particles at fluid interfaces. Finally, Sect. 4 concludes the report.

2 Simulation Method

2.1 The Multicomponent Lattice Boltzmann Method

We use the lattice Boltzmann method (LBM) to simulate the motion of the two fluid components required for our applications. The LBM is a local mesoscopic algorithm, allowing for efficient parallel implementations, and has demonstrated itself as a powerful tool for numerical simulations of fluid flows [38, 39]. It has been extended to allow the simulation of, for example, multiphase/multicomponent fluids [40, 41] and suspensions of particles of arbitrary shape and wettability [42–45].

We implement the pseudopotential multicomponent LBM method of Shan and Chen [40] with a D3Q19 lattice [46] and review some relevant details in the following [23, 43, 44, 47, 48]. Two fluid components are modelled by following the evolution of each distribution function discretized in space and time according to the lattice Boltzmann equation:

$$f_i^c(\mathbf{x} + \mathbf{c}_i \Delta t, t + \Delta t) = f_i^c(\mathbf{x}, t) - \frac{\Delta t}{\tau^c}[f_i^c(\mathbf{x}, t) - f_i^{eq}(\rho^c(\mathbf{x}, t), \mathbf{u}^c(\mathbf{x}, t))], \qquad (1)$$

where $i = 1, \ldots, 19$, $f_i^c(\mathbf{x}, t)$ are the single-particle distribution functions for fluid component $c = 1$ or 2, $\mathbf{c}_i$ is the discrete velocity in ith direction, and τ^c is the relaxation time for component c. The macroscopic densities and velocities are defined as $\rho^c(\mathbf{x}, t) = \rho_0 \sum_i f_i^c(\mathbf{x}, t)$, where ρ_0 is a reference density, and $\mathbf{u}^c(\mathbf{x}, t) = \sum_i f_i^c(\mathbf{x}, t)\mathbf{c}_i / \rho^c(\mathbf{x}, t)$, respectively. Here, $f_i^{eq}(\rho^c(\mathbf{x}, t), \mathbf{u}^c(\mathbf{x}, t))$ is a third-order equilibrium distribution function. When sufficient lattice symmetry is guaranteed, the Navier-Stokes equations can be recovered from Eq. (1) on appropriate length and time scales [38]. For convenience we choose the lattice constant Δx, the timestep Δt, the unit mass ρ_0 and the relaxation time τ^c to be unity. The latter leads to a kinematic viscosity $\nu^c = \frac{1}{6}$ in lattice units.

In the Shan-Chen multicomponent model a mean-field interaction force

$$\mathbf{F}_C^c(\mathbf{x}, t) = -\Psi^c(\mathbf{x}, t) \sum_{c'} g_{cc'} \sum_{\mathbf{x}'} \Psi^{c'}(\mathbf{x}', t)(\mathbf{x}' - \mathbf{x}) \qquad (2)$$

is introduced between fluid components c and c' [40], in which $\mathbf{x}'$ denote the nearest neighbours of lattice site $\mathbf{x}$ and $g_{cc'}$ is a coupling constant determining the surface tension. $\Psi^c(\mathbf{x}, t)$ is an "effective mass", which is chosen as

$$\Psi^c(\mathbf{x}, t) \equiv \Psi(\rho^c(\mathbf{x}, t)) = 1 - e^{-\rho^c(\mathbf{x}, t)}. \qquad (3)$$

This force is then applied to the component c by adding a shift $\Delta\mathbf{u}^c(\mathbf{x}, t) = \frac{\tau^c \mathbf{F}_C^c(\mathbf{x}, t)}{\rho^c(\mathbf{x}, t)}$ to the velocity $\mathbf{u}^c(\mathbf{x}, t)$ in the equilibrium distribution. The Shan-Chen LB method is a diffuse interface method, resulting in an interface width of $\approx 5\Delta x$ [47].

2.2 Colloidal Particles

The trajectories of the colloidal particles are updated using a leap-frog integrator. The particle is discretized on the fluid lattice and coupled to the fluid species by means of a modified bounce-back boundary condition as pioneered by Ladd and Aidun [42, 49] and extended to multicomponent flows by Jansen and Harting [43].

Hydrodynamics leads to a lubrication force between the particles. This force is reproduced automatically by the simulation for sufficiently large particle separations. If the distance between the particles is so small that no free lattice point exists between them this reproduction fails. We apply a lubrication correction when the distance of the two particle is less than one lattice constant, where the force resulted from lubrication interaction is not correctly produced in simulation. The lubrication force (including the correction) already reduces the probability that the particles come closely together and overlap. For the few cases where the particles still would overlap we introduce the direct potential between the particles which is assumed to be a Hertz potential [50].

The outer shell of the particle is filled with a "virtual" fluid with the density

$$\rho^1_{\text{virt}}(\mathbf{x}, t) = \overline{\rho}^1(\mathbf{x}, t) + |\Delta\rho|, \tag{4}$$

$$\rho^2_{\text{virt}}(\mathbf{x}, t) = \overline{\rho}^2(\mathbf{x}, t) - |\Delta\rho|, \tag{5}$$

where $\overline{\rho}^1(\mathbf{x}, t)$ and $\overline{\rho}^2(\mathbf{x}, t)$ are the average of the density of neighbouring fluid nodes for component 1 and 2, respectively. The parameter $\Delta\rho$ is called the "particle colour" and dictates the contact angle of the particle. A particle colour $\Delta\rho = 0$ corresponds to a contact angle of $\theta = 90°$, i.e. a neutrally wetting particle. In order to simulate a Janus particle, we set different particle colours in well defined surface areas corresponding to the different hemispheres of the particle.

2.3 Implementation

We use our simulation code *LB3D* for the simulations presented in this report. *LB3D* combines a multicomponent lattice Boltzmann solver with a molecular dynamics module for suspended solid particles. The code has shown exceptional scaling behaviour and excellent performance in the past. It was the work horse for several previous projects at HLRS [23, 24, 39, 44, 47, 51, 52] and its performance has been documented in several previous reports.

3 Results

3.1 Ellipsoidal Particles at Spherical Fluid Interfaces

In [44] the time development of emulsions stabilized by ellipsoidal particles is studied. During the emulsion formation an additional time scale in the domain formation can be observed which is not present in systems stabilized by spherical particles. In order to understand this additional time development we investigate the following model systems: a single particle at a flat interface as well as a particle ensemble at flat and spherical interfaces.

Here, we focus on the particle ensemble at the interface of a spherical droplet. Pickering emulsions usually consist of (approximately) spherical droplets and a bijel has an even more complicated structure of curved interface. The simplest realization of a curved interface is a single droplet and as such this is studied in this section.

The simulated system is periodic and each period has a size of $L_S = 256$ lattice units. The droplet radius and the number of adsorbed particles are chosen to be $R_D = 0.3 L_S \approx 76.8$ and 600, respectively. At the beginning of the simulation the particles are placed orthogonal to the local interface tangential plane. Then, the particles flip to an orientation parallel to this tangential plane. This state is shown in Fig. 1a after $2 \cdot 10^5$ timesteps.

A preliminary comparison between flat and spherical interfaces has already been given in our previous contribution [24]. The time development of S is shown in Fig. 11(a) in [24]. It has been found that the influence of the interface curvature on the flipping process is larger than the influence of the particle coverage. The time required for the particles to flip is about a factor two smaller in the case of the curved interface.

Here, we investigate the pair correlation function $g(r)$ for the particle ensemble. Figure 1b shows g for $\chi_I \approx 0.27$ at three different times. After 10^4 timesteps it is still close to the correlation function of the initial condition. After 10^5 timesteps some changes can be seen. The first peak is reduced but the second peak is more pronounced. There is no substantial change between $1 \cdot 10^5$ and $2 \cdot 10^5$ timesteps. Compared to the state at 10^5 timesteps the correlation function shows pronounced peaks at longer distances from the particle (about $6 R_p$). The particles mostly reorder during the first 10^5 timesteps since at later times only minor changes in the particle order can be observed. Similar to the case of flat interfaces that was discussed in the previous section, the particle ensemble forms domains where the particles are ordered in a nematic fashion. The peaks in the correlation function are more pronounced in the case of droplets than in the case of a flat interface. The reason is given by the capillary interactions between the particles which are much stronger in the case of curved interfaces. In particular, non-zero capillary interactions persist between spheroids even in the case of neutrally wetting particles.

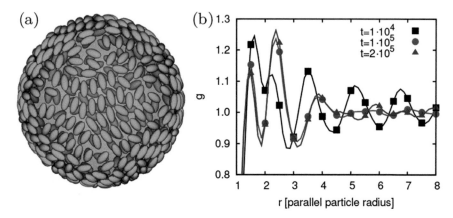

Fig. 1 (**a**) Snapshot of a particle ensemble at a spherical interface after $2 \cdot 10^5$ timesteps. (**b**) Time development of the correlation function $g(r)$ for particles adsorbed at a spherical interface

The time development of g at the droplet as discussed in this section differs from the behavior in the case of a flat interface. For the droplet, g arrives at its final structure after about 10^5 timesteps whereas at the flat interface about four times as many steps are required. In addition, for flat interfaces, g only shows one or two peaks (depending on χ_l), while for the particle covered droplet five peaks are found due to a larger range of ordering of the particles. This is a result of the stronger capillary interactions between the particles due to the interface curvature.

We can understand one of the additional timescales with the behavior of the ellipsoidal particles at a single droplet. The particles reorder and it can be shown that this leads to a small change of the shape of the droplet which is (almost) exactly spherical in the beginning [53]. A change of the interface shape caused by reordering of anisotropic particles leads to a change of the average domain size. The reordering of particle ensembles at flat as well as spherical interfaces takes of the order of 10^5 timesteps. This reordering takes place in idealized systems with constant interfaces which do not change their shape considerably. In real emulsions, however, the interface geometry changes substantially during their formation. For example, two droplets of a Pickering emulsion can coalesce. After this unification the particle ordering starts anew. This explains the fact that the additional timescale we find for emulsions is of the order of several 10^6 timesteps.

3.2 Janus Particles at Fluid Interfaces

In this section we follow Ref. [37] and present theoretical models describing the behaviour of magnetic Janus particles adsorbed at fluid-fluid interfaces interacting with an external magnetic field. Using numerical simulations, we test the model predictions and demonstrate that the magnetic Janus particles deform the interface which leads to tunable capillary interactions.

3.2.1 Theoretical Results

We consider a spherical Janus particle composed of apolar and polar hemispheres adsorbed at a fluid-fluid interface, as illustrated in Fig. 2a. The two hemispheres have opposite wettability, represented by the three-phase contact angles $\theta_A = 90° + \beta$ and $\theta_P = 90° - \beta$, respectively, where β represents the amphiphilicity of the particle. A larger β value corresponds to a greater degree of particle amphiphilicity. In its equilibrium state, the Janus particle takes an upright orientation with respect to the interface with its two hemispheres totally immersed in their favourable phases, as shown in Fig. 2a.

The free energy of the particle in its equilibrium configuration is

$$E_{\text{int}} = \gamma_{12}A_{12}^{\text{int}} + \gamma_{a1}A_{a1}^{\text{int}} + \gamma_{p2}A_{p2}^{\text{int}}, \tag{6}$$

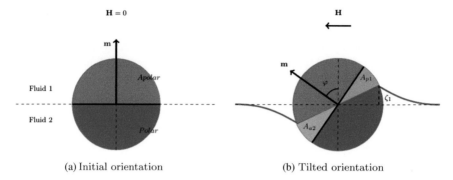

(a) Initial orientation   (b) Tilted orientation

Fig. 2 A single Janus particle adsorbed at a fluid-fluid interface in its equilibrium orientation (**a**) and in a tilted orientation (**b**). The Janus particle consists of apolar and polar hemispheres. It's magnetic dipole moment **m** is orthogonal to the Janus boundary, and the external magnetic field, **H**, is aligned parallel with the interface. The tilt angle φ is given by the angle between the particle's dipole moment and the undeformed interface normal. A_{a2} and A_{p1} are the surface areas of the apolar hemisphere immersed in fluid 2 and the polar hemisphere immersed in fluid 1, respectively. The *bold red line* represents the deformed interface and ζ_1 is the maximal interface height at the contact line (from [37])

where γ_{ij} are the interface tensions between phases i and j and A_{ij} are the contact surface areas between phases i and j, where $i,j = \{1:$ fluid, $2:$ fluid, $a:$ apolar, $p:$ polar$\}$. For a symmetric amphiphilic spherical particle, the apolar and polar surface areas are equal $A_{a1} = A_{p2} = 2\pi R^2$.

Due to the horizontal magnetic field, **H**, the particle experiences a torque $\tau = \mathbf{m} \times \mathbf{H}$ that causes the particle dipole axis to align with the field. However, interfacial tension resists the rotation causing the particle to take a tilted orientation with respect to the interface for a given dipole-field strength $B = |\mathbf{m}||\mathbf{H}|$ (see Fig. 2b). The tilt angle φ is defined as the angle between particle dipole-moment and the undeformed interface normal (i.e. the y-axis). The interface deforms around the particle so that each fluid is in contact with a larger area of its favourable particle surface (Fig. 2b). This interface deformation increases the fluid-fluid interface area and decreases the surface of each hemisphere in contact with its unfavourable fluid. The free energy of the system is reduced in total due to the dominant contribution of particle-fluid interface energies [29]. The free energy of a Janus particle in a tilted orientation can be written as

$$E_{\text{tilt}} = \gamma_{12}A_{12}^{\text{tilt}} + \gamma_{a1}A_{a1}^{\text{tilt}} + \gamma_{p2}A_{p2}^{\text{tilt}} + \gamma_{a2}A_{a2} + \gamma_{p1}A_{p1} + B\sin\varphi. \tag{7}$$

The free energy difference between the tilted orientation state and the initial state is

$$\begin{aligned}
\Delta E &= E_{\text{tilt}} - E_{\text{int}} \\
&= \gamma_{12}\left(A_{12}^{\text{tilt}} - A_{12}^{\text{int}}\right) + \gamma_{a1}\left(A_{a1}^{\text{tilt}} - A_{a1}^{\text{int}}\right) + \gamma_{p2}\left(A_{p2}^{\text{tilt}} - A_{p2}^{\text{int}}\right) \\
&\quad + \gamma_{a2}A_{a2} + \gamma_{p1}A_{p1} + B\sin\varphi.
\end{aligned} \tag{8}$$

Under the assumption of a flat interface [27], Eq. (8) finally reduces to

$$\Delta E = 4\varphi R^2 \gamma_{12} \sin \beta + B \sin \varphi. \tag{9}$$

There is no exact analytical expression for the free energy of a tilted Janus particle at an interface that includes interface deformations, due to the difficulty in modelling the shape of the interface and position of the contact line. However, in the limit of small interface deformations [30], we derive such an analytical expression for the free energy.

$$\Delta E = \frac{\pi}{2}\gamma_{12}\zeta_1^2 + 4(R^2\varphi - R\zeta_1)\gamma_{12} \sin \beta + B \sin \varphi, \tag{10}$$

in which ζ_1 is the maximal interface height at the contact line. We discuss the predictions of our models and compare them with our simulation results in the following section.

3.2.2 Simulation Results and Comparison to Theory

We numerically calculate the free energy of a single Janus particle adsorbed at a fluid-fluid interface using lattice Boltzmann simulations. We choose a system size $L = 128 \times 64 \times 128$ to eliminate finite size effects. The particle with radius $R = 10$ is placed at the interface. Our simulations are capable of capturing interface deformations fully without making any assumptions about the magnitude of the deformations or stipulating any particle-fluid boundary conditions.

To obtain the free energy of the Janus particle as a function of tilt angle, the total surface area of the deformed fluid-fluid interface and the corresponding particle surfaces have to be determined [29, 35, 54]. We utilise the ability to easily measure the force applied to the particle by the fluids using the lattice Boltzmann method. We first determine the contribution to the free energy neglecting the dipole-field contributions by integrating the torque on the particle as the particle rotates quasi-statically,

$$\Delta E = \int_0^{\varphi_{\text{tilt}}} \tau_\varphi d\varphi, \tag{11}$$

where φ_{tilt} is the tilt angle. We rotate the particle on the interface until it reaches the desired angle, then fix the position of the particle and allow the system to equilibrate. The remaining torque on the particle has its origin from the resistance applied to the particle from the fluid-fluid interface. By fitting the torque τ_φ with a hyperbolic tangent function and integrating the fitted function numerically, we obtain the free energy. In order to calculate the free energy of our small interface deformation based model [Eq. (10)], we measure the corresponding maximal height of the contact line

ζ_1 as a function of tilt angle. The height of the contact line increases linearly for small tilt angles, and reaches a plateau for large angles.

Figure 3a depicts a comparison of the free energy models that we developed in Sect. 3.2.1 with lattice Boltzmann simulations. The undeformed interface model [Eq. (9)] predicts that the energy varies linearly with the tilt angle for all amphiphilicities, $\beta = 21°$, $30°$, and $39°$. The model further predicts that the free energy increases as the amphiphilicity increases from $\beta = 21°$ to $\beta = 39°$ for any given φ. The analytical model which includes small interface deformations [Eq. (10)] shows interesting qualitative behaviour: For small tilt angles, $\varphi < 30°$, the energy varies quadratically with the angle for all amphiphilicities. Furthermore, for tilt angles $\varphi > 45°$, the energy varies linearly with the angle, due to the fact that for large angles, ζ_1 is constant. When comparing the two models with our simulation data (symbols), we find that the small deformation model captures the qualitative features of the data extremely well. In addition, the model quantitatively agrees with the numerical results for tilt angles $\varphi < 30°$ and for all amphiphilicities. As the tilt angle of the particles increases the quantitative deviation between model and data becomes more significant. However, for small amphiphilicities $\beta = 21°$ the a good agreement is found.

We now measure the tilt angle as a function of the dipole-field strength by minimizing the free energies in Eqs. (9) and (10) with respect to the tilt angle, as shown in Fig. 3b. For large dipole-field strengths $B > 1.5 A_p \gamma_{12}$, both analytical models perform well and predict the numerically measured tilt angles for all particle amphiphilicities β.

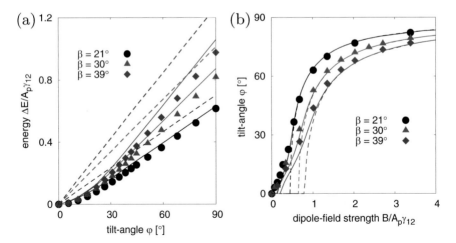

Fig. 3 (a) Free energy as a function of tilt angle φ and (b) φ as a function of dipole-field strength for different amphiphilicities $\beta = 21°$ (*circles*), $\beta = 30°$ (*triangles*) and $\beta = 39°$ (*diamonds*), which are calculated using the analytical model that excludes interface deformations (Eq. (9), *dashed lines*), our analytical model which assumes small interface deformations (Eq. (10), *solid lines*) and simulation results (symbols) (from [37])

(a)   (b)

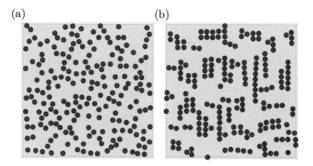

Fig. 4 (**a**) Janus particles are initially randomly distributed at the interface, (**b**) the particles reorder to form two-dimensional network-like structures driven by capillary forces. The system size is $L = 256 \times 96 \times 256$

For small dipole-field strengths, the tilt angle predicted by the undeformed interface model Eq. (9) shows large deviations from the simulations. These deviations increase with increasing amphiphilicity. In addition, the undeformed interface model predicts a large critical dipole-field strength at which the particle begins to rotate, which we do not observe in the simulations. This critical dipole-field strength increases with increasing particle amphiphilicity.

The small deformation model [Eq. (10)] shows significant improvement at predicting the tilt angle as compared to the undeformed interface model. The predicted tilt angle shows good agreement with simulation data for weak dipole-field strengths. In particular the model predicts a much smaller, though still finite, critical dipole-field strength, agreeing well with the numerical simulations. We conclude that interface deformations strongly affect the orientation of tilted Janus particles adsorbed at fluid-fluid interfaces.

The Janus particle deforms the interface in a dipolar fashion: a symmetric depression and rise on opposite sides of the particle. Capillary interactions arise if multiple particles are adsorbed at the interface. Figure 4 shows that the particles reorder to form two-dimensional network-like structures, driven by capillary forces [55]. We can tune the strength of capillary interactions by varying the dipole-field strength. These unique dipolar capillary interactions may be used to assemble particles into novel materials at a fluid-fluid interface.

4   Conclusions

We reviewed two recent contributions [23, 37] where we have applied a combined lattice Boltzmann and molecular dynamics method to simulate the dynamics of anisotropic particles at fluid interfaces. The first simulation aims to investigate the ensembles of ellipsoidal particles at a spherical interface. We quantified the reordering of the particles. Our findings provide relevant insight in the dynamics of emulsion formation which is generally difficult to investigate experimentally due to the required high temporal resolution of the measurement method and limited optical transparency of the experimental system. Anisotropic particles provide properties which might allow the generation of emulsions that are stable on substantially longer timescales. Secondly, we studied the behaviour of a magnetic spherical amphiphilic Janus particle adsorbed at a fluid-fluid interface in an external magnetic field directed parallel to the interface. The particles tilt under the influence of the magnetic field and thus deform the interface. We demonstrated that the magnitude of these deformations can be dynamically varied, and therefore the capillary interactions between monolayers of such particles are tunable, and we suggest that this may enable the production of novel, reconfigurable materials.

The simulations presented in this report can be seen as building blocks for large scale simulations of particle stabilized emulsions as presented in [23, 48]. They have been performed to obtain a fundamental understanding of the underlying physical principles. While the system sizes of these simulations were comparably small and allowed to run on as few as $\approx$ 100 cores, the simulations had to run for several million timesteps in order to fully equilibrate and a large number of simulations was required to cover the full parameter space.

Acknowledgements Financial support is greatly acknowledged from NWO/STW (Vidi grant 10787 of J. Harting and STW project 13291). We thank the High Performance Computing Center Stuttgart for the allocation of computing time on Hermit and Hornet.

References

1. Fan, H., Striolo, A.: Soft Matter **8**, 9533 (2012)
2. Frijters, S., Günther, F., Harting, J.: Soft Matter **8**(24), 6542 (2012)
3. He, Y., Yu, X.: Mater. Lett. **61**(10), 2071 (2007)
4. Aveyard, R., Binks, B.P., Clint, J.H.: Adv. Colloid Interf. Sci. **100–102**(0), 503 (2003)
5. Kalashnikova, I., Bizot, H., Bertoncini, P., Cathala, B., Capron, I.: Soft Matter **9**, 952 (2013)
6. Kim, E., Stratford, K., Cates, M.E.: Langmuir **26**(11), 7928 (2010)
7. Melle, S., Lask, M., Fuller, G.G.: Langmuir **21**(6), 2158 (2005)
8. Binks, B.P., Fletcher, P.D.I.: Langmuir **17**, 4708 (2001)
9. Madivala, B., Vandebril, S., Fransaer, J., Vermant, J.: Soft Matter **5**, 1717 (2009)
10. Günther, F., Janoschek, F., Frijters, S., Harting, J.: Comput. Fluids **80**, 184 (2013)
11. de Graaf, J., Dijkstra, M., van Roij, R.: J. Chem. Phys. **132**(16), 164902 (2010)
12. Dong, L., Johnson, D.T.: Langmuir **21**, 3838 (2005)
13. Bresme, F., Oettel, M.: J. Phys. Condens. Matter **19**(41), 413101 (2007)
14. Faraudo, J., Bresme, F.: J. Chem. Phys. **18**, 6518 (2003)
15. Morgan, A.R., Ballard, N., Rochford, L.A., Nurumbetov, G., Skelhon, T.S., Bon, S.A.F.: Soft Matter **9**, 487 (2013)
16. Lehle, H., Noruzifar, E., Oettel, M.: Eur. Phys. J. E: Soft Matter Biol. Phys. **26**, 151 (2008)
17. Bleibel, J., Dietrich, S., Domínguez, A., Oettel, M.: Phys. Rev. Lett. **107**, 128302 (2011)
18. Bleibel, J., Domínguez, A., Günther, F., Harting, J., Oettel, M.: Soft Matter **10**, 2945 (2014)
19. Zeng, C., Brau, F., Davidovitch, B., Dinsmore, A.D.: Soft Matter **8**, 8582 (2012)
20. Botto, L., Lewandowski, E.P., Cavallaro, M., Stebe, K.J.: Soft Matter **8**, 9957 (2012)
21. Madivala, B., Fransaer, J., Vermant, J.: Langmuir **25**, 2718 (2009)

22. Dai, L.L., Tarimala, S., Wu, C.Y., Guttula, S., Wu, J.: Scanning **30**, 87 (2008)
23. Günther, F., Frijters, S., Harting, J.: Soft Matter **10**, 4977 (2014)
24. Krüger, T., Frijters, S., Günther, F., Kaoui, B., Harting, J.: Eur. Phys. J. Spec. Top. **222**, 177 (2013)
25. Ondarçuhu, T., Fabre, P., Raphaël, E., Veyssié, M.: J. Phys. **51**, 1527 (1990)
26. Adams, D.J., Adams, S., Melrose, J., Weaver, A.C.: Colloid Surf. A **317**, 360 (2008)
27. Park, B.J., Lee, D.: ACS Nano **6**, 782 (2012)
28. Park, B.J., Lee, D.: Soft Matter **8**, 7690 (2012)
29. Rezvantalab, H., Shojaei-Zadeh, S.: Soft Matter **9**, 3640 (2013)
30. Stamou, D., Duschl, C., Johannsmann, D.: Phys. Rev. E **62**, 5263 (2000)
31. Dominic, V., Mahadevan, L.: Am. J. Phys. **73**, 817 (2005)
32. Loudet, J.C., Alsayed, A.M., Zhang, J., Yodh, A.G.: Phys. Rev. Lett. **94**, 018301 (2005)
33. Yunker, P.J., Still, T., Lohr, M.A., Yodh, A.G.: Nature **476**, 308 (2011)
34. Brugarolas, T., Park, B.J., Lee, M.H., Lee, D.: Adv. Funct. Mater. **21**, 3924 (2011)
35. Davies, G.B., Krüger, T., Coveney, P.V., Harting, J., Bresme, F.: Soft Matter **10**, 6742 (2014)
36. Davies, G.B., Krüger, T., Coveney, P.V., Harting, J., Bresme, F.: Adv. Mater. **26**, 6715 (2014)
37. Xie, Q., Davies, G., Günther, F., Harting, J.: Soft Matter **11**, 3581 (2015)
38. Succi, S.: The Lattice Boltzmann Equation for Fluid Dynamics and Beyond. Oxford University Press, Oxford (2001)
39. Harting, J., Harvey, M., Chin, J., Venturoli, M., Coveney, P.V.: Philos. Trans. R. Soc. Lond. A **363**, 1895 (2005)
40. Shan, X., Chen, H.: Phys. Rev. E **47**, 1815 (1993)
41. Cappelli, S., Xie, Q., Harting, J., Jong, A.M., Prins, M.W.J.: Dynamic wetting: status and prospective of single particle based experiments and simulations. N. Biotechnol. **32**, 420–432 (2015)
42. Ladd, A.J.C., Verberg, R.: J. Stat. Phys. **104**, 1191 (2001)
43. Jansen, F., Harting, J.: Phys. Rev. E **83**, 046707 (2011)
44. Günther, F., Janoschek, F., Frijters, S., Harting, J.: Comput. Fluids **80**, 184 (2013)
45. Janoschek, F., Toschi, F., Harting, J.: Phys. Rev. E **82**, 056710 (2010)
46. Qian, Y.H., D'Humières, D., Lallemand, P.: Europhys. Lett. **17**, 479 (1992)
47. Frijters, S., Günther, F., Harting, J.: Soft Matter **8**, 6542 (2012)
48. Frijters, S., Günther, F., Harting, J.: Phys. Rev. E **90**, 042307 (2014)
49. Aidun, C.K., Lu, Y., Ding, E.-J.: J. Fluid Mech. **373**, 287 (1998)
50. Hertz, H.: J. für reine und angewandte Math. **92**, 156 (1881)
51. Groen, D., Henrich, O., Janoschek, F., Coveney, P.V., Harting, J.: In: Wolfgang, F.B.M. (ed.) Jülich Blue Gene/P Extreme Scaling Workshop 2011. Jülich Supercomputing Centre, 52425 Jülich, Germany, Apr 2011. FZJ-JSC-IB-2011-02. http://juser.fz-juelich.de/record/15866/files/ib-2011-02.pdf
52. Schmieschek, S., Narváez Salazar, A., Harting, J.: In: Resch, M., Nagel, W., Kröner, D. (eds.) High Performance Computing in Science and Engineering '12, p. 39. Springer, Berlin (2013)
53. Kim, E., Stratford, K., Adhikari, R., Cates, M.E.: Langmuir **24**, 6549 (2008)
54. Dasgupta, S., Katava, M., Faraj, M., Auth, T., Gompper, G.: Langmuir **30**, 11873 (2014)
55. Xie, Q., Davies, G., Harting, J.: (2015, in preparation)

Highly Efficient Integrated Simulation of Electro-Membrane Processes for Desalination of Sea Water

Kannan Masilamani, Harald Klimach, and Sabine Roller

Abstract Sea water desalination is an important technology to provide fresh drinking water. However, traditional processes require large amounts of energy, and there is ongoing research for new strategies with less energy consumption. In the *HISEEM* project, we investigated the electrodialysis process in this context. For this process, sea water flows through a channel between ion exchange membranes, which are kept apart by structure, filling the complete channel. Ions are driven out of the channel and through the membranes by an external electric field. The large *Hornet* system in Stuttgart enabled us to perform detailed simulations of the overall setup, including the flow through a complex geometry, interactions with the membranes, driving forces by the electric field and interactions of the species in the solvent. To achieve this, we make use of the (multi-species) lattice Boltzmann method (LBM). Besides results from those simulations, we present performance and scalability of our implementation on the *Cray XC40* system *Hornet* at the *HLRS* in Stuttgart.

1 Introduction

Supply with clean drinking water is considered one of the essential issues for mankind today. Desalination of sea water offers the option to make use of the vast water reservoirs provided by the oceans. However, the desalination requires expensive energy and reducing the energy consumption is one of the main research topics in this domain. Electrodialysis is a promising candidate to master this process with low energy consumption. The process makes use of selective ion exchange membranes and an electric field to drive ions out of the sea water. Due to the multitude of physical effects interacting in the electrodialysis, it is not possible to understand the system analytically in every detail. Also measurements are difficult to conduct in the complete setting without disturbing the process too much.

K. Masilamani (✉) • H. Klimach • S. Roller
Simulationstechnik und wissenschaftliches Rechnen, University of Siegen, Hölderlinstr.3, 57076 Siegen, Germany
e-mail: kannan.masilamani@uni-siegen.de; harald.klimach@uni-siegen.de; sabine.roller@uni-siegen.de

© Springer International Publishing Switzerland 2016
W.E. Nagel et al. (eds.), *High Performance Computing in Science and Engineering '15*, DOI 10.1007/978-3-319-24633-8_37

579

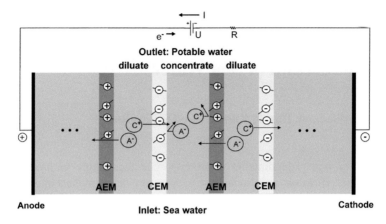

Fig. 1 Simplified structure of a desalination stack in two dimensions. A periodic arrangement of anion-exchange-membrane (AEM), diluate channel, cation-exchange-membrane (CEM) and concentrate channel is embedded between anode and cathode. After removing the ions, drinkable water can be extracted at the end of the diluate channel

With the increased computing power available on systems like *Hornet*, it now becomes possible to assess interaction of all the effects numerically. In this project we investigated the overall process of electrodialysis for sea water desalination including the driving electric field, the flow through the channel and the interaction of species within the flow and the exchange of ions with the membranes.

A schematic illustration of an electrodialysis stack is shown in Fig. 1. It consists of alternatingly arranged anion (AEM) and cation exchange membranes (CEM). Anodes and cathodes are used at the ends of the stack to apply an electrical field. This electric field separates the anions and cations from the sea water by driving them in opposite directions. AEM and CEM are selective permeable membranes, they allow either anions or cations to pass through respectively. Thus, two kinds of channels are observed in the stack. One, where the ions are driven towards membranes that allow their passing, is called the diluate channel, as ions are removed from it. And the other one, where ions are driven towards membranes that block their passing is called the concentrate channel, as ions are collected by it. At the outflow of the stack, desalinated water is extracted from the diluate channels, this water is fed again into the stack several times until the desired concentration level is obtained.

Ion exchange membranes (AEM and CEM) in the stack are separated by a complex structure called spacer. This spacer structure is shown in Fig. 2. It acts as a mechanical stabilizer and also induces turbulence. The effective transport of ions through the membranes and also the total energy consumption of the stack is mainly influenced by the geometry of this spacer.

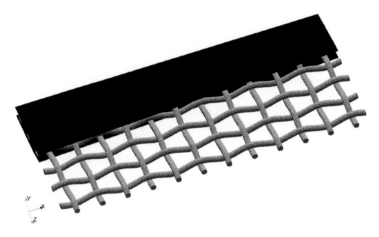

Fig. 2 Structure of a spacer used to mechanically stabilize the fluid channels of the stack

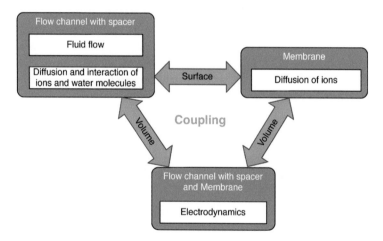

Fig. 3 Multi-physical heterogeneous system required in the simulation of the water desalination process

An overview of the physical subsystems and its interactions are shown in Fig. 3. The following discussion is mainly concerned with simulations of the hydrodynamics and mass transport in the flow channel. Results will also be shown for flow overlayed by a fixed electric field, and using a simple membrane model that serves as a boundary condition to the species.

To study the physical phenomena in the spacer filled channel, a suitable numerical method is required. We chose a Lattice Boltzmann Method (LBM) for the numerical simulations, due to its advantage of easily integrating complex geometries and its high performance on large scale HPC systems [1]. For the simulations of mass transport in aqueous ionic mixtures we use a multi-species extension of the

Lattice Boltzmann method, which is described in [2]. The multi-species LBM model recovers the Maxwell-Stefan equations for diffusive fluxes and the incompressible Navier-Stokes equations for the transport of ions and mixture in the macroscopic regime. This solver is embedded in our highly scalable parallel framework *APES* (Adaptable Poly-Engineering Simulator) [3], which is based on an octree data structure for the mesh [4].

In this work, we focus on the performance of our implementation of the numerical schemes on the *Cray XC40* system *Hornet*. The paper is organized as follows: In Sect. 2 we briefly describe the governing equations and numerical method. Section 3 is concerned with performance and scalability of single-fluid LBM and multi-species LBM schemes. In Sect. 4 the simulation results of single electrodialysis unit comprising dilute and concentrate channel with membranes are presented. Finally, we summarize this work and highlight planned future work in Sect. 5.

2 Governing Equations and Numerical Method

In this section, we briefly review the governing equations of mass transport for an aqueous ionic mixture. For a detailed discussion we refer the reader to [2]. The mass transport is governed by a mass conservation equation for the species concentrations (**w** denotes the number density averaged velocity for the mixture)

$$\partial_t n_k + \nabla \cdot n_k \mathbf{w} = -\nabla \cdot \mathbf{J}_k, \tag{1}$$

which is closed by the Maxwell-Stefan closure relation for the diffusive fluxes $\mathbf{J}_k$ (in terms of the Maxwell-Stefan diffusivities $D_{k,l}$ and number density fraction χ_k)

$$\nabla \chi_k = \frac{1}{n} \sum_{l \neq k} \frac{1}{D_{k,l}} (\chi_k \mathbf{J}_l - \chi_l \mathbf{J}_k), \tag{2}$$

where n and n_k are total number density and species k number density respectively. The momentum balance for the mixture in the low Mach limit is given by the incompressible Navier-Stokes equation with external forcing term $\mathbf{F}$

$$\partial_t \mathbf{v} + \nabla \cdot (\mathbf{v} \otimes \mathbf{v} + pI) = \nu \nabla^2 \mathbf{v} + \mathbf{F}. \tag{3}$$

These equations are solved numerically by a multi-species lattice Boltzmann method [2]. Besides the generalizations of the lattice Boltzmann method for aqueous mixtures, the algorithm has a simple stream-collide structure for each species:

$$f_i^k(\mathbf{x} + \mathbf{e_i}\Delta t, t + \Delta t) = f_i^k(\mathbf{x}, t) - \omega^k(f_i^k(\mathbf{x}, t) - f_i^{eq,*,k}(\mathbf{x}, t)) \tag{4}$$

This makes the method presented in [2] a perfect candidate for massively parallel implementations. The verification of this method with various boundary conditions are presented in [5].

3 Performance and Scalability

This section describes the performance of *Musubi* [6] on the *Cray XC40* system *Hornet* at *HLRS*. *Hornet* provides 3944 compute nodes with two *Intel Haswell* processors each. With 12 cores per processor this results in 24 cores per node. For our performance analysis, up to 2048 compute nodes or 49,152 cores are used. In this analysis only *MPI* parallelism is considered. The code is compiled with Intel 15 compiler as Cray compiler has some run time issues with our code. The issues related to Cray compiler are reported to Cray support team at HLRS.

An internode performance is measured for single-fluid LBM and multi-species LBM with three species. The scaling analysis are performed on fully periodic cubic simulation test case which results is exact problem size and communication surface on each core. A refinement of a cube by one level always results in increase in problem size by a factor of 8 due to doubling element count in each direction. This is in some sense a worst case scenario for the MPI communication, since there are no obstacles anywhere, that could reduce the communication surfaces. In our analysis, problem sizes from 512 elements up to 68 billion elements are covered.

Our solver framework ensures nearly perfect balancing due to space-filling curve used for element ordering. Two settings, the classical single-fluid flow and the multi-species flow consisting of water, sodium and chloride, are investigated. In both models a *D3Q19* layout with a *BGK*-like collision operator is used. Streaming and collision steps are solved at each time step for both models. However, the multi-species LBM model used in this work requires the solution of an additional cell local linear equation system, increasing the required number of floating point operations per lattice update. The single-fluid solver requires only around 160 floating point operations per cell (= lattice) update, while a simulation with 3 species requires 850 operations per update. This is a factor of 5 more in the number of operations, while the memory consumption grows only by a factor of 3 to account for all species. The operation count increases more than linearly with the number of species, as the size of the equation system to be solved increases with the number of species (hence the total operation count is proportional to the square of the number of species). Also, the amount of data to be exchanged via *MPI* increases linearly with the number of species. Expectations therefore are an increased sustained performance in the multi-species simulation, while scalability might be slightly decreased.

584   K. Masilamani et al.

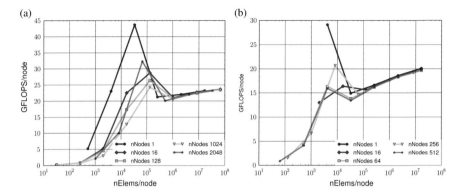

Fig. 4 Internode performance of the single-fluid (**a**) and multi-species (**b**) LBM with periodic cube

In this section, the performance of our solver is measured in terms of Giga gloating point operations per second (GFLOPS). To represent the scaling behavior of our code, the performance is presented per execution unit i.e. GFLOPS/node to get a clearer impression of the performance independent of the number of used execution units. An ideal parallel execution is expected to just replicate the serial behavior on each execution unit. However, the execution performance is influenced by cache usage, non-computational implementation overheads, vector lengths, communication times and so on. The performance maps in Fig. 4 shows the achieved performance per node over the problem size per node for various total number of nodes for single-fluid LBM (Fig. 4a) and multi-species LBM (Fig. 4b). The performance map combines both, weak scaling and strong scaling. Weak scaling can be measured by the vertical comparison of points between different lines for the node count i.e. fixing the number of elements per node. The closer the points are located to each other the better the weak scaling. Strong scaling on the other hand is not as easily seen in the performance map, but can be derived by moving to the left when increasing the number of nodes. This reduces the number of elements per process with increasing node counts, as required by strong scaling with a fixed overall problem size.

Weak scaling for a problem size of 16,777,216 elements per node is shown in Fig. 5a. As can be seen, the weak scaling is almost perfect for single fluid as well as three species, with only a small drop in the parallel efficiency for larger counts of compute nodes.

Strong scaling for a problem size of 16,777,216 elements and 134,217,728 elements are shown in Fig. 5b. Similar to the weak scaling, the simulation with three species shows comparable scalability to the single fluid simulation, though both drop more significantly in this case. The performance increase in single-fluid simulation at 256 nodes and 2048 nodes for problem size of 16,777,216 and 134,217,728 respectively due to cache effects as the problem size is getting small

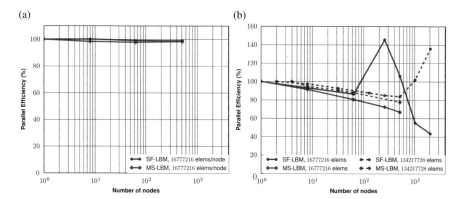

Fig. 5 Weak (**a**) and strong (**b**) scaling performance of the single-fluid and multi-species simulation with periodic cube

enough to fit into cache i.e. 65,536 elements per node. This cache region is not seen for multi-species simulation since it requires more nodes than the single-fluid to fit the same problem size.

3.1 Dynamic Load Balancing

The performance of our code with the spacer structure on *Hermit* system is presented in [7]. In multi component flow simulation with spacer structure, the computational load imbalance arises from fluid element which requires boundary treatment like inlet, outlet and membrane. This imbalances were resolved by load balancing algorithm. In Musubi, the space filling curve partition algorithm (SPartA) [8] is deployed to handle the computational cost associated with boundary treatment. For load balancing, the weights are provided for each element. In our application weights per element is computed from computational time spend on compute kernel and boundary treatment. SPartA uses this weight to compute optimal split position such that weights are equally distributed among the processes. To ensure optimal load balancing, this algorithm is applied dynamically.

Figure 6 shows the run time of compute kernel and set boundary routines for 1000 time steps with total problem size of five million elements with spacer structure distributed on 32 cores. Due to initial distribution, almost perfect computational load is obtained for compute kernel but boundary elements causes load imbalances which can be seen in the Fig. 6a. After load balancing, Fig. 6b the element are redistributed and perfect balancing is obtained. The element distribution before and after load balancing is shown in Fig. 7.

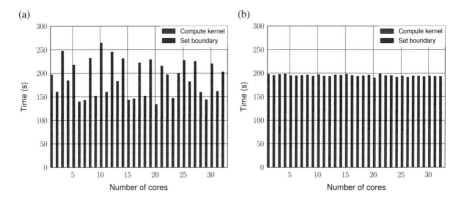

Fig. 6 Time spend on compute kernel (*blue*) and set boundary (*red*) routine for 1000 time steps with total problem size of five million elements distributed on 32 cores. (**a**) Before load balance. (**b**) After load balance

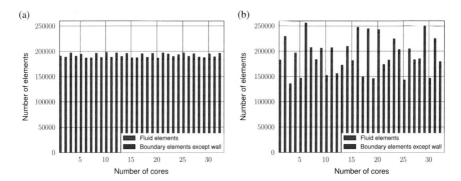

Fig. 7 Element distribution before and after load balance. *Red color* represents fluid elements and *blue color* represents the boundary elements without wall boundaries. (**a**) Before load balance. (**b**) After load balance

4 Multi-Species Simulation

In this section, the results of targeted simulation setup i.e. single electrodialysis unit as shown in Fig. 8 is presented. The verification of single-fluid LBM with spacer structure against experiment was published in [7]. In [7], the results of multi-species LBM for the transport of ionic species (*Na-Cl*) in H_2O mixture through a single spacer element was presented.

Let us consider two simple flow channels with the same *Na-Cl* and H_2O mixture with membranes in between the channels. We impose a non-zero sodium-chloride concentration profile at the inlet of the channel and advect it downstream by imposing a mixture pressure drop along the channel. Besides the advection in downstream direction, concentration gradients lead to diffusion according to the Maxwell-Stefan relation in cross-stream-direction. Additionally we impose a fixed

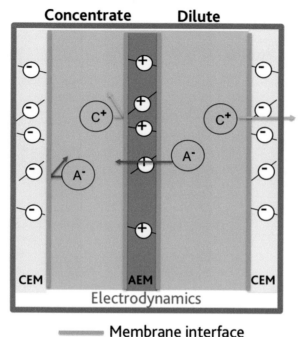

Fig. 8 Single electrodialysis unit with one dilute and concentrate channel with ion exchange membranes

electric field perpendicular to the main flow direction, which drives sodium and chloride ions into opposite directions (according to their specific charges) towards the membranes as in Fig. 9. Between two channels, the transport properties of anion exchange membrane is specified i.e. it absorbs the negatively charged anions and reflect the positively charged cations. At the outer side of both channels, the transport properties of cation exchange membranes are specified. Due to this arrangement, the ions gets depleted from right channel and increases in the left channel resulting in concentrate and dilute channel respectively.

The concentration profile of Na and Cl on the dilute and concentration channel after several time steps are shown in Fig. 9. Due to applied electric field in positive y-direction, the Na and Cl ion are driven in positive and negative y-direction respectively. The anion exchange membrane on the middle of two channels absorbs the Na from right channel and reflects Cl from left channel. Similarly, the cation exchange membrane on the left channel reflects Na and on the right channel absorbs Cl. This results in concentrate channel on the left and dilute channel on the right as in Fig. 9.

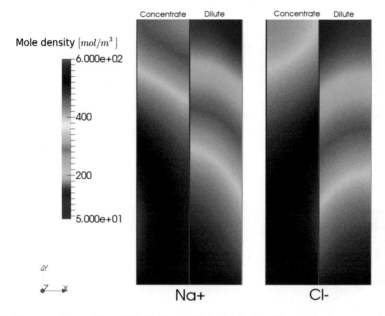

Fig. 9 Concentration profiles of *Na* (*left*) and *Cl* (*right*) in the dilute and concentrate channel coupled with ion exchange membranes

Now, it is straight forward to apply the same setting to simulate with spacer structure in dilute and concentrate channel. Thus, we can deploy Apes framework to simulate electrodialysis process. The results with same simulation settings with spacer structure are shown in Fig. 10. The upper and lower channel corresponds to concentrate and dilute channel respectively.

5   Summary and Outlook

This work is concerned with the simulations for an electrodialysis process to desalinate water. We briefly describe the governing equations and physical effects to consider. The main focus in the presented work is the flow simulation in the channels, which contain a complex structure and thus a non-trivial flow field. Specifically the strong and weak scaling performance is presented for up to 2048 compute nodes or 49,152 MPI processes. This scaling analysis was done for single fluid simulations as well as for simulations with three species for periodic domain. The scaling analysis shows that both numerical schemes scale fine on *Hornet* and large production runs where done on large process counts.

(a)   (b)

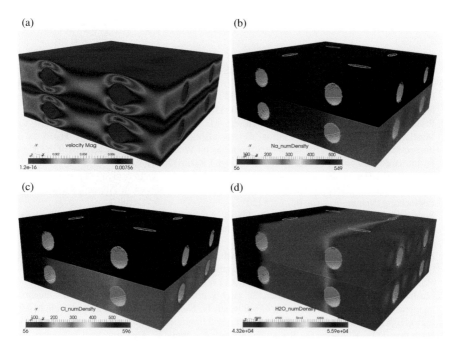

(c)   (d)

Fig. 10 (**a**) Velocity magnitude of mixture and concentration profiles of Na (**b**), Cl (**c**) and H_2O (**d**) in the dilute and concentrate channel with spacer structure at simulation of 2 s

The multi-species LBM is deployed to simulate transport of ionic species (sodium-chloride) in an aqueous solvent through a spacer structure under a fixed external electrical field.

In our ongoing work we focus on extending the simulations of fluid flow with overlayed electrical fields from fixed external field to a full interaction of ions from electrodynamics. To obtain stable coupled simulations results, this requires better boundary conditions with higher accuracy. Especially for liquid electrolytes, which are the major research topic in this project, further boundary conditions especially at membrane interface are under investigation. Overall, we will use the results of our research to provide an engineering prediction model, which covers all relevant design parameters of the electrodialysis process.

Acknowledgements This work was funded by the German Federal Ministry of Education and Research (Bundesministerium für Bildung und Forschung, BMBF) in the framework of the HPC software initiative in the project HISEEM. We thank *HLRS* for the granted compute time on the *Hornet* system and are grateful for their and *Cray*'s kind support.

References

1. Bernsdorf, J.: Lattice-Boltzmann simulation of reactive flow through a Porous media catalyst. In: Recent Progress in CFD, Japan, No.08-07 JSAE Symposium, pp. 19–23 (2007)
2. Zudrop, J., Roller, S., Pietro, A.: Lattice Boltzmann scheme for electrolytes by an extended Maxwell-Stefan approach. Phys. Rev. E **89**, 053310 (2014). doi:10.1103/PhysRevE.89.053310. http://link.aps.org/doi/10.1103/PhysRevE.89.053310
3. Roller, S., Bernsdorf, J., Klimach, H., Hasert, M., Harlacher, D., Cakircali, M., Zimny, S., Masilamani, K., Didinger, L., Zudrop, J.: An adaptable simulation framework based on a linearized octree. In: High Performance Computing on Vector Systems: Proceedings of the High Performance Computing Center Stuttgart. Springer, Berlin (2011)
4. Klimach, H., Hasert, M., Zudrop, J., Roller, S.: Distributed Octree mesh infrastructure for flow simulations. European Congress on Computational Methods in Applied Sciences and Engineering, Vienna, Austria, pp. 1–15 (2012)
5. Zudrop, J., Masilamani, K., Roller, S., Pietro, A.: A robust lattice Boltzmann method for parallel simulations of multicomponent flows in complex geometries. Submitted to Journal of Computers and Fluids and its under review
6. Hasert, M., Masilamani, K., Zimny, S., Klimach, H., Qi, J., Bernsdorf, J., Roller, S.: Complex fluid simulations with the parallel tree-based Lattice Boltzmann solver Musubi. J. Comput. Sci. **5**(5), 784–794 (2013). doi:10.1016/j.jocs.2013.11.001
7. Masilamani, K., Klimach, H., Roller, S.: Highly efficient integrated simulation of electro-membrane processes for desalination of sea water. In: Nagel, W.E., Kroner, D.B., Resch, M.M. (eds.) High Performance Computing in Science and Engineering. Springer, Switzerland (2014)
8. Harlacher, D.F., Klimach, H., Roller, S., Siebert, C., Wolf, F.: Dynamic load balancing for unstructured meshes on space-filling curves. In: IEEE 26th International Symposium on Parallel and Distributed Processing, Workshops and PhD Forum (IPDPSW), pp. 1661–1669 (2012)

Part VI
Transport and Climate

Christoph Kottmeier

Climate, being defined as the statistics of successive weather events for a period of 30 years and beyond, cannot be predicted for specific locations and times, since the coupling between processes on different scales interacting in an unpredictable way limits predictability. The relevant processes range from microturbulence and cloud microphysics to large hemispheric circulation systems and involve land-surface interactions with the atmosphere. Climate also evolves in time, since it responds to varying boundary conditions (land use), both external (changing solar radiation flux, land use) and internal (atmospheric composition) factors.

The slowly varying components of the climate systems, i.e. the oceans, ice sheets and soil characteristics cause a memory effect for the atmospheric changes. Limited area climate models are applied in the projects HRCM at KIT (IMK-TRO) with CCLM, LUCCI (IMK-IFU) with WRF and WRFCLIM at the University of Hohenheim to get higher resolution in regions of interest. Their computational and data storage requirements are quite similar to those of global models. The CPU time requirements also increase due to ensemble modeling, which becomes more and more a standard approach. The HPC requirements for simulations of natural systems like the atmosphere and the oceans in general are still increasing, since the model grid resolution is globally much too coarse to cover all energy-containing scales of motion in the atmosphere, in particular the mesoscale (10–1000 km) and the convective scale (100 m–10 km). It is aimed that approximations, such as parameterizations, can be avoided in the future and that net effects of small scales are calculated at grid resolution. The real bottleneck in regional climate modelling is storage place, however, since numerous interim model output is needed to monitor and analyse predictions. The three projects presenting their results here aim at different regions and fields of application. HRCM addresses mid Europe,

C. Kottmeier (✉)
Institut für Meteorologie und Klimaforschung, Karlsruher Institut für Technologie (KIT),
Wolfgang-Gaede-Straße 1, 76131 Karlsruhe, Germany
e-mail: christoph.kottmeier@kit.edu

in particular urban scales, as well as decadal prediction in Europe and Africa. WRFCLIM studies the sensitivity of climate change to land-surface interaction in Southwest Germany. LUCCi, finally is focused on the regional climate and impacts of advanced land use/land cover data in a river basin in Vietnam.

By contributing to the CORDEX initiative of regional climate change modeling for Europe and Africa, HCRM and WRFCLIM also provided input to the fifth IPCC assessment report on climate change. Regional climate change projections additionally are used to enable estimates of climate change consequences in various economic and social sectors. This development means that compartment crossing modelling will become more important in the future.

Application of the Regional Climate Model CCLM for Studies on Urban Climate Change in Stuttgart and Decadal Climate Prediction in Europe and Africa

H.-J. Panitz, G. Schädler, M. Breil, S. Mieruch, H. Feldmann, K. Sedlmeier, N. Laube, and M. Uhlig

Abstract To study various aspects of the climate on the regional scale, IMK-TRO employs the regional climate model (RCM) COSMO-CLM on the High Performance Computing Systems (HPC) CRAY XE6 and CRAY XE40, operated by the HLRS. The research focus is on decadal predictability, extremes and high resolution experiments within individual research projects (MiKlip, KLIMOPASS and KLIWA). Also within MiKlip the effects of soil and vegetation processes on decadal climate predictions are investigated using a via OASIS3-MCT coupled system of COSMO-CLM and the Soil-Vegetation-Atmosphere-Transfer model (SVAT) VEG3D. Ensembles are built with different techniques to quantify the uncertainty of climate projections and predictions and to assess the quality of the models. KLIMOPASS includes projections of the future climate, considering periods up to the mid of the twenty-first century. High resolution (2.8 km) experiments are performed for the State of Baden–Württemberg to study extremes and their potential added value. Altogether simulations for Germany, Europe, and Africa are performed with varying temporal and spatial resolutions from 50 to 2.8 km. The required Wall-Clock-Time (WCT) reach from 100 to 650 node-hours per simulated year.

1 Introduction

Within the evolving climate, the demand for reliable climate predictions and projections is higher than ever. The working group "Regional Climate and Water Cycle" of the Institute of Meteorology and Climate Research—Troposphere Research (IMK-TRO) at the Karlsruhe Institute of Technology (KIT) (www.imk-tro.kit.edu) uses the CRAY XE6 and the CRAY XC40 at the HLRS high performance computing

H.-J. Panitz (✉) • G. Schädler • M. Breil • S. Mieruch • H. Feldmann • K. Sedlmeier, N. Laube • M. Uhlig
Institut für Meteorologie und Klimaforschung Forschungsbereich Troposphäre (IMK-TRO),
Karlsruher Institut für Technologie (KIT), Karlsruhe, Germany
e-mail: hans-juergen.panitz@kit.edu

© Springer International Publishing Switzerland 2016
W.E. Nagel et al. (eds.), *High Performance Computing in Science and Engineering '15*, DOI 10.1007/978-3-319-24633-8_38

facilities to investigate past, present and future climate. With the climate version of the COSMO model (CCLM) of the German Weather Service (DWD), simulations are performed on a regional scale for Europe/Germany and Africa to assess the changes and variability in climate variables of interest.

The work is embedded in various national and international research programs and projects: e.g. MiKlip, KLIWA and KLIMOPASS.

2   The CCLM Model

The regional climate model (RCM) CCLM is the climate version of the operational weather forecast model COSMO (Consortium for Small-scale Modeling) of DWD. It is a three-dimensional, non-hydrostatic, fully compressible model. The model solves prognostic equations for wind, pressure, air temperature, different phases of atmospheric water, soil temperature, and soil water content.

Further details on COSMO and its application as a RCM can be found in [1, 2] and on the web-page of the COSMO consortium (www.cosmo-model.org).

3   Regional Climate Simulations Using the HLRS Facilities

3.1   Combined Effects of Urban Planning and Climate Change on the Climate of the Stuttgart Metropolitan Area (KLIMOPASS)

A growing concern of current urban planning is the possibly negative impact of climate change on urban climate, and whether adaptation or mitigation measures are able to counteract this influence. In a master thesis finished early this year [3], the combined effects of urban planning scenarios together with climate change were analysed for the Stuttgart Metropolitan Area (SMA). SMA was chosen for its unique situation in an orographic basin (so called "Kessellage") which favors poor air quality and increased thermal load. Based on publicly available data, two opposite future urban planning scenarios have been developed. The best case scenario presumes an increase of vegetated surfaces and less energy consumption in urban areas, the worst case scenario assumes an enlargement and denser development of urban and industrial areas. It is assumed that these two scenarios span roughly the possible impacts of urban development on future climate (Fig. 1 vertical). An additional projection simulation with the current land use distribution was done in order to separate the respective effects of climate change and land use change. The time periods are 1971–1975 for the control and 2021–2025 for the projections. To quantify the impact of climate change alone, one simulation for the control period and one for the projection period with the same (reference)

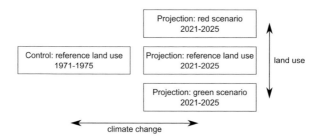

Fig. 1 Simulation concept to separate climate change and land use change

land use have been conducted as well (Fig. 1 horizontal). In view of the current emissions of greenhouse gases and climate policies, representative concentration pathways (RCP) 8.5 [4] seems to be the appropriate emission scenario. This scenario assumes continuously high greenhouse gas emissions. Originally, it was planned to cover the time periods from 1971–2000 and 2021–2050. But during the simulations we encountered problems within the used model system, whose corrections caused operational delays. Furthermore, we also encountered problems with the performance and data input of our jobs on both HPC platforms, "Hermit" and "Hornet", (details for "Hornet" are described in Sect. 4). These latter problems led to unpredictable crashes of our jobs and delayed the study further. However, these platform specific problems have been reported to the CRAY staff at HLRS who is working on solutions and who also provided workarounds. For the analysis, high-resolution climate simulations with a horizontal grid spacing of 0.025° (approx. 2.8 km) were performed using the RCM CCLM, coupled with the Soil-Vegetation-Atmosphere-Transfer (SVAT) model VEG3D developed at the IMK-TRO [5]. To better represent the different types of urban land use, specific additional land use classes, based on the "local climate zones" (LCZ) [6], have been implemented in VEG3D. Furthermore, a parameterization of anthropogenic heat input has been incorporated in the SVAT.

In the city center of Stuttgart, the simulations showed a clear increase of summer temperature by 0.87 K due to climate change. The amount of additional changes caused by the two scenarios is smaller, with an increase by 0.15 K for the worst case scenario and a decrease by 0.06 K for the best case scenario. Small, but marked differences caused by climate change were also found for global radiation, specific humidity and total cloud cover. However, these changes are considerably smaller than the climate change signal for the two land use scenarios in the city center. When looking at the entire SMA on the other hand, at some grid points that were subject to changes in land use, differences to the reference simulation are in the order of magnitude of the climate change signal. Overall however, this preliminary study shows that the influence of climate change on the climate of the SMA is larger than that of the two land use scenarios considered here.

3.2   The MiKlip Program

MiKlip (Mittelfristige Klimaprognosen; www.fona-miklip.de) is a research program
dedicated to the development of a model system which is able to predict the climate
development up to a decade ahead. Decadal predictions are a new field of research
with pioneering papers published less than a decade ago [7–9]. The predictability on
interannual to decadal timescales arises from the initialization of slow components
of the earth system-like the ocean, and partly land surface processes or sea ice.
Whereas global decadal predictions contributed already to the simulation suite
for the last IPCC report in 2013 [10], the regionalization of these predictions is
a completely novel research field. The aim is to provide climate informations to
potential users beyond the weather or seasonal forecasts services. Within MiKlip
Module C, the feasibility of the regional downscaling of decadal predictions is tested
for two main regions: Europe (projects DecReg and REGIO_PREDICT, among
others) and Africa (project DEPARTURE). Within MiKlip several generations of
decadal hindcasts (retrospective forecasts) for the period 1961–2015 were generated
with the global model MPI-ESM-LR using improved initialization procedures.
The first generations were named baseline0 (b0) and baseline1 (b1) [11]. On the
global scale it could be shown that the skill of the initialized decadal predictions
outperforms the skill of uninitialized climate projections [12]. The challenge for
the regionalization is to prove an added value compared to the global predictions,
from which most of the predictability is inherited from processes acting outside
the domain of the regional model. During a first phase of MiKlip the first world-
wide regional prediction ensemble was generated [13] with a grid resolution of
0.22° (ca. 25 km) and 5 starting dates every 10 years (1961,1971,1981,1991,2001)
using the global b0 ensemble as boundary forcing. The ensemble was produced by
a consortium at the German Climate Computing Center (DKRZ), the DWD and
HLRS. The main task was to establish the feasibility and basic skill characteristics.
Mieruch et al. [13] showed that compared to the global hindcasts generally the skill
was maintained for mean temperature and precipitation with improvement of the
ensemble spread, which is an important measure for the reliability of forecasts.
Predictable signals for the year-to-year variability can be expected to a lesser degree
than for multi-year averages depending on the forecast length (lead time; e.g. mean
year 2–5, cf. [14]). During the second phase the ensemble generation was largely
extended by switching to the b1 global forcing (see Sect. 3.2.1). The experiments
from the first phase were repeated to compare the regional performance of the
ensemble generations. In addition a larger ensemble with annual starting dates of
the hindcast between 1960 and 2011 was generated with a coarser grid resolution
(0.44°, ca. 50 km). The research topics for this second phase are: comparison of
the skill of the b0 and b1 ensemble, creating more robust skill estimates using the
larger sample size of the coarser ensemble and to establish recommendations for
an operational decadal prediction system, for instance with respect to the model
resolution.

Such an operational prediction system requires a reliable computing infrastructure and a large workspace for the data management. A program chain covering the access to the global forcing data, the pre-processing, the actual climate simulation, the post-processing and the delivery of the standardized output data to a data base for the analysis has been established. Actual forecasts are a time critical effort. For an efficient throughput a pipelining for the generation of several parallel ensemble members is necessary.

3.2.1 Decadal Regional Predictability (DecReg)

DecReg aims at the analysis and characterization of decadal regional predictions. In the years 2012 and 2013 we generated a part of a ten members regional decadal prediction ensemble (1961–2010) using the regional climate model CCLM at the CRAY XE6. This b0 ensemble has been analyzed for the standard variables temperature and precipitation [13].

In the second project phase the CRAY XE6 has been used to create a part of the successor ensemble b1. The difference between the two baselines is, as mentioned before, the global driving data. Whereas for b0 only the ocean was initialized in the global MPI-ESM-LR model, the b1 version was additionally initialized with an assimilated atmospheric state. Assuming the functioning of the initialization, b1 should be superior to b0. However, regarding the complex nature of the problem, it is a-priori not clear, if b1 is in fact the superior model version over Europe. For the global MPI-ESM-LR model it has been shown that the b1 ensemble is superior to the b0 ensemble mainly near the equator [11]. Thus, one main task in 2014 was the comparison of these two model versions for both, the MPI-ESM-LR and CCLM, regarding the predictive skill over the European continent.

We analyzed

- large regions, namely the eight regions, 1 = British Isles, 2 = Iberian Peninsula, 3 = France, 4 = Central Europe, 5 = Scandinavia, 6 = Alps, 7 = Mediterranean, 8 = Eastern Europe
- summer and winter temperature and precipitation
- detrended data, i.e. we removed the long-term trend using a simple linear regression
- the first 5 years after the initialization.

Figure 2 shows exemplarily summer precipitation time series for region 6, i.e. the Alps. The left panel shows the CCLM b0 and the right panel the b1 results. E-OBS [15] observational data is shown as a black line, whereas the ensemble mean of the model is plotted as a blue line. The light blue shading represents the ensemble spread, i.e. the standard deviation of the ensemble. Following the described data pre-processing, the essential part of the analysis is to measure the predictive skill of the hindcasts. To quantify the predictive skill we developed a new skill score, taking account of the ensemble mean hindcast skill and the ensemble consistency, simultaneously. The model yields two informations, the skill of the ensemble mean

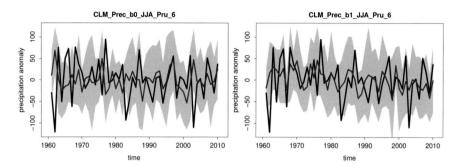

Fig. 2 Summer precipitation anomalies for the region 6, the Alps. *Left*: CCLM baseline0, *right*: CCLM baseline1. *Black*: E-OBS, *blue*: CCLM, *light blue*: CCLM ensemble spread

using the E-OBS observations as a reference, and the model consistency, i.e. how many ensemble members show a similar outcome, which gives an indication regarding the certainty of this result. The used skill score is based on Bayes theorem (e.g. [16]). Here, we don't want to go into to much technical details and concentrate on the meaning of the score. We use the most simple variation of the score, which yields the *probability of correctly forecasting a point in time being above or below the long-term median.* Thus, being a probability and being a binary information it is clear that any skill below a probability of 0.5 is useless, because it would be not superior to guessing. From this point of view, one has a good "null"-model in mind, i.e. guessing.

Figure 3 shows an example analysis of the skill. We analyzed only the first 5 years after the initialization in each decade. The ordinate depicts the skill score, which is perfect if it is unity and not better than guessing at 0.5. The abscissa represents the eight regions defined above. The continuous black line describes the average skill, which can be obtained by using a shuffled E-OBS data set, in the sense of a bootstrap surrogate test as a forecast, which turns out to be not better than guessing. However, due to the finite length of the data and the inherent noise, there is a considerable scatter of the skill obtained by the surrogates, which is plotted as a light blue shading (standard deviation). That means, anything within this light blue tube is likely to be produced by chance. The top left panel of Fig. 3 shows the results for CCLM b0 summer precipitation for lead years 1–5. The top right panel shows the results for CCLM b1 and the two bottom panels represent the results from the driving model MPI-ESM-LR. A real improvement from b0 to b1 cannot be observed.

Concluding, we can say that decadal predictions, which are still a quite novel field, show up to now not the predictive skill to be useful in the sense of economical, political or societal decision making. However, it is important to note that this conclusion is made by us for the European region, depending on the used ensembles, the observational reference, and the data pre-processing as well as the used skill score. Nevertheless, we have seen that there exists the potential to be useful, because some points in time are predictable and others not. Hence, there is much room for improvement. On the one hand, the models must be improved and the ensembles

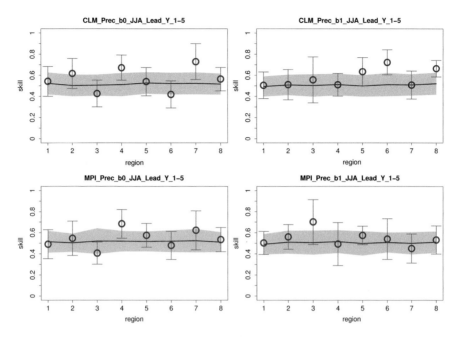

Fig. 3 Skill of the climate models, i.e. the probability of correctly forecasting a point in time being above or below the long-term median for the eight regions. The *black line* represents the skill of a surrogate data set and the *bluish shading* depicts the respective scatter

enlarged for a better signal-to-noise ratio. On the other hand, the processes behind the predictability must be better understood. That means, there might be a physical reason, why the model is able to forecast some points in time and others not. Thus, it is of utmost importance to understand the physical processes, which causes the predictability, and to explore the model's capabilities to represent the physical mechanisms responsible for the evolution of climate variables on timescales of years to a decade.

Beside the analyses of the baseline versions, DecReg evaluates the role of the soil as a slow component of the climate system by analyzing simulations with different SVATs in different configurations and initializations. Therefore, several sensitivity studies are performed to enlarge the ensemble further. For b0 and b1 over Germany and Europe (with 7 and 25 km resolution), selected realizations are repeated with the SVAT VEG3D instead of the standard SVAT TERRA-ML [17] implemented in CCLM (CCLM_TERRA-ML). A short summary of the difference between TERRA-ML and VEG3D is given in Sect. 3.2.2. The coupling of CCLM and VEG3D (CCLM_VEG3D) was done via the OASIS3-MCT coupler.

Earlier studies with the coupled system and its setup are described in [18]. For DecReg the high resolution simulations are of particular interest. First results of the coupled system over an extended area over Germany with 7 km resolution are presented. The simulation was done on the CRAY XE6 for the decade 2001–2010

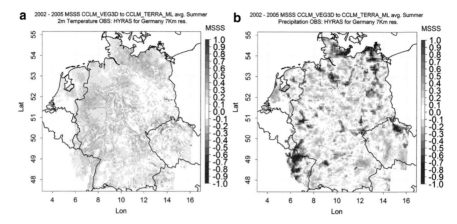

Fig. 4 MSESS of the summer mean 2 m temperature (*left*) and monthly precipitation sums of the summer months (*right*) for CCLM_VEG3D to CCLM_TERRA-ML with HYRAS observations over Germany 7 km for the years 2002–2005. (**a**) MSSS 2002–2005 T_2M. (**b**) MSSS 2002–2005 TOT_PREC

with b0 as forcing. The 2 m temperature and precipitation are compared to the HYRAS data [19] only for the years 2001–2006 due to the temporal restrictions of the observations. One main issue of the CCLM simulations within MiKlip is a cold wet bias which occurs during summer time. The CCLM_VEG3D clearly reduces the bias of the 2 m temperature, especially for summer, by 0.3 K. Unfortunately, the wet bias still exists and is slightly enhanced in summer (in the annual average CCLM_VEG3D and CCLM_TERRA-ML are nearly similar).

The MSESS (as defined in Sect. 3.2.2, Eq. (1)) indicate in reddish colors the higher agreement of one model with the observations compared to another. In Fig. 4a, the summer temperature MSESS for the period 2002–2005 (years 2–5) is shown. The MSESS is reduced compared to the lead year 2001 (not shown), but the advantage of using CCLM_VEG3D instead of CCLM_TERRA-ML is still quite obvious. For the precipitation, the Fig. 4b displays no clear signal, the MSESS is nearly 0 in the average. Beside the ongoing evaluation of the 7 km simulation another set of 25 km simulations over Europe is performed on the CRAY XE6 for the decade 2001–2010 with b1 global forcing.

3.2.2 Decadal Climate Predictability in West Africa (DEPARTURE)

The main goal of DEPARTURE is to investigate the decadal climate predictability of the West African Monsoon system (WAM). The potential decadal predictability in the WAM region arises from slowly varying processes in the climate system like the evolution of the sea surface temperature (SST), the changes in land cover and soil characteristics, as well as the changes of greenhouse gas and aerosol concentrations.

One boundary condition influencing the decadal variability of the West African Monsoon is the interaction between the soil, the vegetation and the atmosphere: the characteristics of soil and vegetation have an influence on the radiation balance and affect the exchange of latent and sensible heat between the surface and the atmosphere. In a RCM this interactions are described in a SVAT. Results of recent studies indicate that TERRA-ML is not able to represent the soil-vegetation-atmosphere interaction sufficiently well (e.g. [20]). Thus, to improve the description of this process, TERRA-ML was replaced by a more sophisticated SVAT VEG3D in the CCLM. Compared to TERRA-ML, this new SVAT includes an explicit vegetation layer. Within this vegetation layer, temperature and humidity values are calculated, determining the latent and sensible heat fluxes between the surface and the atmosphere. The hydraulic conductivity in the soil is described with the parameterization of van Genuchten [21]. The thermal conductivity is a function of the soil water content (approach described in [22]).

In order to study the influence of the interactions between the soil and the vegetation on the decadal climate predictability of the African monsoon, ERA-Interim [23] driven CCLM simulations, one using TERRA-ML, the other one using VEG3D, were performed for the decade 2001–2010. Their results were compared among each other and with observed climate data [24]. Figure 5a, b show the RMSE based Mean Square Error Skill Score (MSESS) (Eq. (1)) for the monthly mean 2 m temperatures and the yearly sum of total precipitation. The MSESS is defined as

(a)   (b)

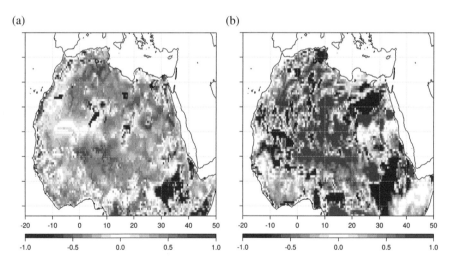

Fig. 5 *MSESS* of VEG3D compared to TERRA-ML for the monthly mean 2 m temperature (*left*) and the yearly sum of total precipitation (*right*) for the decade 2001–2010 calculated for the land grid points. (**a**) MSESS 2 m temperature. (**b**) MSESS precipitation

follows:

$$MSESS = 1 - \frac{\Sigma(VEG3D - Obs)^2}{\Sigma(TERRA - Obs)^2} \tag{1}$$

with:

VEG3D: Simulation results of CCLM combined with VEG3D
TERRA: Simulation results of CCLM combined with TERRA-ML
Obs: Observed climate data [24]

If the VEG3D simulations achieve a higher agreement with observations the
MSESS becomes positive, indicated in red colours, if TERRA-ML matches better
the *MSESS* becomes negative (blue colours). The results show that the use of
VEG3D improves the simulation results for the 2 m temperature nearly all over West
and North Africa. Because of the explicit vegetation layer, considered in VEG3D,
the surface does not heat up and cool out too strongly during warm seasons and
cooler seasons, respectively. This behavior results in a damped and more realistic
annual cycle of temperature, matching with the observations much better [18]. Due
to this more realistic lower boundary condition, the use of VEG3D improves the
simulation results for the total precipitation in most parts of West and North Africa
as well. Thus, the results of this study illustrates the large influence of the soil-
vegetation-atmosphere interaction on the decadal variability of the West African
Monsoon system.

4 Computational Requirements

The finer resolution of regional models as well as the need for ensembles and
scenarios require a large amount of computer resources both in computing time
and data storage space which depend on various factors. During the period from
April 2014 until March 2015 a total amount of about 256,238 node-hours were
needed for all simulations. About 80 % of this total amount are apportioned to
CRAY XE6 (Hermit), about 20 % to CRAY XC40 (Hornet). The distribution of
total computational time to the main research programs are shown in Table 1.

The storage capacity amounts to 2–3 Tbyte per 10 years simulation time
including model input, model output, and post-processed data.

Table 1 Computational requirements

Research program	Node h
MiKlip	194,080
KLIMOPASS	54,639
KLIWA	7,519

5 Implementation of CCLM on CRAY XC40 "HORNET"

From the purely technical point of view the migration from the CRAY XE6, "HERMIT", system to the new system CRAY XC40, "Hornet", went without problems for the user. It was a big advantage that the user's environment (Home-directory and workspace/filesystem) as well as the batch system "Torque" did not change. It was only necessary to re-compile the CCLM model, changes of compiler options were not needed. The next step was to study the impact of the change of the computing system on the results of the CCLM model. For this purpose CORDEX Africa evaluation simulations [25] have been repeated on "Hornet" for the period 1989 till 1996, using the identical forcing data, model version and model configuration as on "Hermit". Fortunately, from a climatological point of view, the differences between the results were negligible. For the area averaged mean annual sum of precipitation and the area averaged annual means of sea level pressure and 2 m temperature, we got differences less than 1 mm, of about 0.01 hPA, and of about 0.002 K, respectively.

An objective comparison of the model's performance on both CRAY systems was not possible. The reason for this is a large Wall-Clock-Time (WCT) variability, which we experienced on both systems, "Hermit" and "Hornet", even if model runs are repeated under identical conditions with respect to the forcing data and the setup of the simulations. An example of this WCT variation on "Hornet" is shown in Fig. 6.

It illustrates the results of an exercise where we investigated the scalability of CCLM on "Hornet". For a simulation of a duration of 1 month, only the number of nodes had been varied between one and 30 in order to investigate the scalability of CCLM on "Hornet" and to find an optimum computational configuration for future climate simulations. The 6 and 20 nodes cases had been performed twice, the 30 node case even four times. These cases demonstrate the variability of the WCT even under identical computational and model conditions. For the 30 nodes case a

Fig. 6 Scalability of CCLM on "Hornet" showing the Wall-Clock-Time (WCT) in min as a function of "Hornet" nodes for a typical 1 month simulation using CCLM. Note that the abscissa is not linear. The different markers for 6, 20, and 30 nodes cases represent different runs under identical model conditions using the same number of nodes

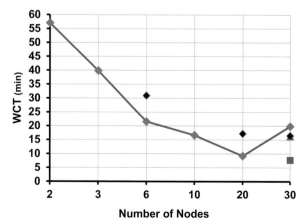

factor of about three occurred between the fastest and slowest run (about 7 min WCT and about 20 min WCT, respectively). Generally, our long-term climate simulations, which have a typical length of 30 years, are treated as jobchains of 30 yearly jobs being performed consecutively. "Up-scaling"the results from the scalability tests, we have the choice between about 1.5 h and about 4 h as WCT for each of the yearly jobs when using 30 nodes. And, actually, the runtimes of such climate simulations confirmed the experiences from model scaling. Thus, advices like "Users should always specify a realistic value for the walltime" (https://wickie.hlrs.de/platforms/index.php/CRAY_XC40_Batch_System_Layout_and_Limits) are hard to follow, at least for applications like regional climate modelling. Since the scalabiltiy tests had been performed under the identical model configuration, and thus also identical numerical conditions, and since regional climate modelling is always accompanied by a rather large amount of input and output (I/O), we concluded that the WCT variability has to be attributed to the performance of the network between CRAY XC40 and the Lustre filesystem. Of course, these experiences had been reported to and discussed with the CRAY staff at HLRS. Basically, they confirmed our conclusion but attributed the WCT variability to the restricted number of interconnects between the filesystem and the computing system. The consequence is that, if there is "high traffic" on these interconnects due to other applications, then the performance of I/O intensive runs diminish considerably. In the meantime, the number of interconnects has been increased, but the WCT of RCM applications did not really become more stable.

6   Future Work

The HLRS high performance computing facilities have been essential to our group's research and will continue to be so in the future. A series of demanding projects, in terms of computational as well as storage requirements, is ahead of us. The second phase of the MiKlip project, which will probably start in autumn 2015, aims at providing an operational decadal prediction system, including regional decadal predictions. This will involve (a) improving the long-term memory components of the model system, specifically the soil-land surface component, (b) analyse and optimize the prediction ensemble in terms of ensemble size and selection of ensemble members in order to provide a prediction ensemble, which is well balanced between optimal skill and reliability on the one hand, and efficient generation procedures on the other hand, and (c) performing and analyzing extended regional hindcasts. All of this will require considerable amounts of computation time.

Another topic is the implementation of a water isotope module in our regional climate model with the ultimate goal of performing long-term paleoclimate simulations for comparison with isotope proxy data. In the first stage, we will implement an isotope module in CCLM and perform sensitivity and validation simulations, using currently available observations. Due to the additional computational load caused by the isotope module, a substantial amount of computation time will be necessary. We also plan to prepare the first very long-term (order of centuries) regional climate simulations, which will involve some technical adjustments. The third topic is regional climate simulations with very high resolution, in the order of 3 km and less. It turned out that such runs can provide better results, e.g. for temperature, precipitation, humidity and radiation, which is important for hydrology (simulation of small river catchments) and urban areas.

The projects described here require careful planning ahead, but presently it is very difficult for us to plan our simulations due to the still quite unpredictable performance of the Hornet and its environment (see Sect. 5).

References

1. Baldauf, M., Seifert, A., Förstner, J., Majewski, D., Raschendorfer, M., Reinhardt, T.: Operational convective-scale numerical weather prediction with the COSMO model: description and sensitivities. Mon. Weather Rev. **139**(12), 3887–3905 (2011)
2. Rockel, B., Will, A., Hense, A.: The regional climate model COSMO-CLM (CCLM). Meteorol. Z. **17**(4), 347–348 (2008)
3. Imhof, H.: Combined effects of urban planning and climate change on the climate of the Stuttgart metropolitan area. Master's thesis, Karlsruhe Institute of Technology (2015)
4. van Vuuren, D.P., Edmonds, J., Kainuma, M., Riahi, K., Thomson, A., Hibbard, K., Hurtt, G.C., Kram, T., Krey, V., Lamarque, J.-F., Masui, T., Meinshausen, M., Nakicenovic, N., Smith, S.J., Rose, S.K.: The representative concentration pathways: an overview. Clim. Chang. **109**(1–2), 5–31 (2011)
5. Schädler, G.: Numerische Simulationen zur Wechselwirkung zwischen Landoberfläche und atmosphärischer Grenzschicht. Ph.D. thesis, Karlsruhe Institute of Technology (1990)
6. Stewart, I.D., Oke, T.R.: Local climate zones for urban temperature studies. Bull. Am. Meteorol. Soc. **93**(12), 1879–1900 (2012)
7. Smith, D.M., Cusack, S., Colman, A.W., Folland, C.K., Harris, G.R., Murphy, J.M.: Improved surface temperature prediction for the coming decade from a global climate model. Science **317**(5839), 796–799 (2007)
8. Keenlyside, N.S., Latif, M., Jungclaus, J., Kornblueh, L., Roeckner, E.: Advancing decadal-scale climate prediction in the North Atlantic sector. Nature **453**, 84–88 (2008)
9. Pohlmann, H., Jungclaus, J.H., Koehl, A., Stammer, D., Marotzke, J.: Initializing decadal climate predictions with the GECCO oceanic synthesis: effects on the North Atlantic. J. Clim. **22**(14), 3926–3938 (2009)
10. Stocker, T.F., Qin, D., Plattner, G.-K., Tignor, M., Allen, J., Boschung, S.K., Nauels, A., Xia, Y., Bex, V., Midgley, P.M.: Climate change 2013: the physical science basis. Contribution of working group I to the fifth assessment report of the intergovernmental panel on climate change. Technical report, Cambridge University Press, Cambridge/New York (2013)
11. Pohlmann, H., Müller, W.A., Kulkarni, K., Kameswarrao, M., Matei, D., Vamborg, F.S.E., Kadow, C., Illing, S., Marotzke, J.: Improved forecast skill in the tropics in the new MiKlip decadal climate predictions. Geophys. Res. Lett. **40**(21), 5798–5802 (2013)
12. Meehl, G.A., Goddard, L., Boer, G., Burgman, R., Branstator, G., Cassou, C., Corti, S., Danabasoglu, G., Doblas-Reyes, F., Hawkins, E., Karspeck, A., Kimoto, M., Kumar, A., Matei, D., Mignot, J., Msadek, R., Navarra, A., Pohlmann, H., Rienecker, M., Rosati, T., Schneider, E., Smith, D., Sutton, R., Teng, H., van Oldenborgh, G.J., Vecchi, G., Yeager, S.: Decadal climate prediction: an update from the trenches. Bull. Am. Meteorol. Soc. **95**(2), 243–267 (2014)

13. Mieruch, S., Feldmann, H., Schädler, G., Lenz, C.-J., Kothe, S., Kottmeier, C.: The regional MiKlip decadal forecast ensemble for Europe: the added value of downscaling. Geosci. Model Dev. **7**(6), 2983–2999 (2014)

14. Goddard, L., Kumar, A., Solomon, A., Smith, D., Boer, G., Gonzalez, P., Kharin, V., Merryfield, W., Deser, C., Mason, S.J., Kirtman, B.P., Msadek, R., Sutton, R., Hawkins, E., Fricker, T., Hegerl, G., Ferro, C.A.T., Stephenson, D.B., Meehl, G.A., Stockdale, T., Burgman, R., Greene, A.M., Kushnir, Y., Newman, M., Carton, J., Fukumori, I., Delworth, T.: A verification framework for interannual-to-decadal predictions experiments. Clim. Dyn. **40**(1–2), 245–272 (2013)

15. Haylock, M.R., Hofstra, N., Tank, A.M.G.K., Klok, E.J., Jones, P.D., New, M.: A European daily high-resolution gridded dataset of surface temperature and precipitation. J. Geophys. Res. **113**, D20119 (2008)

16. Sivia, D., Skilling, J.: Data Analysis: A Bayesian Tutorial. Oxford University Press, Oxford (2006)

17. Schrodin, E., Heise, E.: A new multi-layer soil model. COSMO Newsl. **2**, 149–151 (2002)

18. Panitz, H.-J., Schädler, G., Breil, M., Mieruch, S., Feldmann, H., Sedlmeier, K., Laube, N., Uhlig, M.: High resolution climate modeling with the CCLM regional for Europe and Africa. In: High Performance Computing in Science and Engineering '14, pp. 561–574. Springer, New York (2014)

19. Rauthe, M., Steiner, H., Riediger, U., Mazurkiewicz, A., Gratzki, A.: A central European precipitation climatology? Part I: Generation and validation of a high-resolution gridded daily data set (HYRAS). Meteorol. Z. **22**(3), 235–256 (2013)

20. Kohler, M., Schädler, G., Gantner, L., Kalthoff, N., Königer, F., Kottmeier, C.: Validation of two SVAT models for different periods during the west African monsoon. Meteorol. Z. **21**(5), 509–524 (2012)

21. van Genuchten, M.Th.: A closed-form equation for predicting the hydraulic conductivity of unsaturated soils. Soil Sci. Soc. Am. J. **44**(5), 892–898 (1980)

22. Johansen, O.: Thermal conductivity of soils. Ph.D. thesis, University of Trondheim (1975). (CRREL Draft Translation, 1977), ADA044002

23. Dee, D.P., Uppala, S.M., Simmons, A.J., Berrisford, P., Poli, P., Kobayashi, S., Andrae, U., Balmaseda, M.A., Balsamo, G., Bauer, P., et al.: The ERA-interim reanalysis: configuration and performance of the data assimilation system. Q. J. R. Meteorol. Soc. **137**(656), 553–597 (2011)

24. Willmott, C.J., Matsuura, K., Legates, D.R.: Global air temperature and precipitation: regridded monthly and annual climatologies (version 2.01) (2004)

25. Panitz, H.-J., Fosser, G., Sasse, R., Sehlinger, A., Feldmann, H., Schädler, G.: Modelling near future regional climate change for Germany and Africa. In: High Performance Computing in Science and Engineering '12, pp. 375–390. Springer, Berlin/Heidelberg/New York (2013)

High-Resolution WRF Model Simulations of Critical Land Surface-Atmosphere Interactions Within Arid and Temperate Climates (WRFCLIM)

Josipa Milovac, Oliver-Lloyd Branch, Hans-Stefan Bauer, Thomas Schwitalla, Kirsten Warrach-Sagi, and Volker Wulfmeyer

Abstract The interaction of the Earth's surface with the atmosphere is a key component that needs to be well represented in atmospheric models to increase the accuracy of numerical weather prediction and climate projections. The Weather Research and Forecasting (WRF) model was applied to investigate the model sensitivity to its physics parametrizations and the impact of land use changes in temperate continental and arid regions. Results show that the model is more sensitive to the land surface model than to the boundary layer parametrization. Furthermore, the impact of land surface processes is not constrained to the lower boundary layer. It extends up to the boundary layer top, influencing the strength and the location of convection and precipitation. The study underscores the demand for research on parametrization schemes for model applications at high resolution based on data sets from novel observing systems of the atmosphere.

In a land use change impact study the influence of deforestation on the weather in the temperate climate of Germany has been investigated. The results show a decrease in temperature, an increase in humidity, and overall a decrease in precipitation.

Large-scale plantations in coastal arid regions can contribute to mitigation methods of anthropogenic climate change, through storage of atmospheric CO_2. Results of WRF simulations indicate that rainfall events can be triggered by large arid plantations, but the strength and probability of such events is governed by regional climate. This indicates that impacts can be controlled via intelligent placement of such schemes. Thus, desert plantations have the potential not only to mitigate climate change but also to increase precipitation in arid regions and reverse desertification.

J. Milovac • O.-L. Branch • H.-S. Bauer (✉) • T. Schwitalla • K. Warrach-Sagi • V. Wulfmeyer
Institute of Physics and Meteorology, University of Hohenheim, Garbenstrasse 30, 70599 Stuttgart, Germany
e-mail: josipa.milovac@uni-hohenheim.de; hans-stefan.bauer@uni-hohenheim.de

© Springer International Publishing Switzerland 2016   607
W.E. Nagel et al. (eds.), *High Performance Computing in Science and Engineering '15*, DOI 10.1007/978-3-319-24633-8_39

1 Introduction and Motivation

Anthropogenic greenhouse gases such as CO_2 will bring about significant changes in the Earth's climate, with substantial social, economic and environmental consequences. There will be a multitude of impacts on regional energy and water cycles; and of particular concern are changes in the intensity and frequency of extreme events such as droughts or extreme precipitation. The impacts will differ between regions, therefore end users like federal agencies, hydrologists, and farmers require high resolution climate information.

The resolution of most global climate simulations is more than 100 km and therefore too coarse for almost all end users and decision makers. Downscaling of global projections, applying high resolution climate models is essential for achieving results at regional scales. An improved simulation of the spatial distribution of temperature and precipitation including wet-day frequencies, dry periods and extremes can be expected.

WRFCLIM aims to investigate and to improve the performance of regional climate simulations in Europe with the Weather Research and Forecast (WRF) model [1]. The model is operated from 12 km down to the convection permitting scale of 3 km, for advancing process understanding [2–5]. Special attention is paid to soil-vegetation-atmosphere (SVA) feedback processes, which are essential for cloud formation and precipitation. In spite of the overall success in simulation, systematic errors remain in low and high mountain ranges, which need to be corrected to produce acceptable simulations for most applications.

Within this context the WRFCLIM objectives are to

- provide high resolution climatological data to the scientific community
- support the quality assessment and interpretation of the regional climate projections for Europe through the contribution to the climate model ensemble for Europe (EURO-CORDEX[1]) for the next IPCC (International Panel of Climatic Change[2]) report.

The main objectives of the convection permitting simulations of WRFCLIM are as follows:

- Replace the convection parametrization by a dynamical simulation of the convection chain to better resolve the processes of the specific location and actual weather situation physically.
- Gain an improved spatial distribution and diurnal cycle of precipitation through a better SVA feedback simulation and a more realistic representation of orography to support the interpretation of the 12 km climate simulations for local applications e.g. in hydrological and agricultural management.

---

[1] http://www.euro-cordex.net.

[2] http://www.ipcc.ch.

The strength of SVA feedback varies across seasons and regions, and it will likely increase in a changing climate [6, 7]. Therefore, it is essential to investigate the whole model chain across all components of the SVA system. By this effort, SVA feedback under a changing climate can be understood and weather and climate models can be improved. This is a pre-requisite for regional climate models (RCMs) to reach the predictive skill in reproducing the current climate variability including extreme events [8, 9] and to project these to the future. A model system that can be applied seamlessly on horizontal and temporal scales is proposed as a very promising tool to be able to project climate variability [10]. WRF offers the capability and infrastructure for the development of such a seamless modelling approach. An objective of the DFG funded Research Unit 1695 "Agricultural Landscapes under Climate Change—Processes and Feedbacks on a Regional Scale"[3] is the development and verification of a seamless atmosphere-land surface model system based on WRF including an advanced representation of SVA feedback processes with emphasis on the water and energy cycling between croplands, the planetary boundary layer (PBL) and the free atmosphere. The WRF model is validated with high-resolution case studies to improve the process understanding over a wide range of temporal scales.

One of the major human factors influencing the climate is the land use/cover change. By changing land surface characteristics, the energy and water exchange between the land surface and the atmosphere is affected. The quantification of this impact has provoked much research due to its potential for carbon sequestration. Within the WRFCLIM project two simulations were performed in order to analyse the influence of the land use/cover change on weather and climate, with special emphasis on precipitation. The experiments were conducted for two different climate environments, in moderate and dry hot climate conditions.

Large-scale, high-resolution climate simulations like WRFCLIM are an example of applications which need to be executed on the largest and fastest available computing systems like the high-performance computing systems at HLRS. For the 12-km resolution, the model domain covers 13 million grid cells whose physics and dynamics are simulated with a 1 min time step. Increasing resolution towards convection permitting simulations requires even more resources. Other computer clusters have neither the required cores-h to finish such climate simulations in a reasonable time nor the needed storage capacity.

In the following chapter, a technical description of the simulations within WRFCLIM at HLRS which were done between April 2014 and April 2015 and a summary of the current analysis of their results are reported.

---

[3]https://klimawandel.uni-hohenheim.de/.

2 WRF Simulations at HLRS

This section reports the technical details in of the recent simulations on the CRAY XE6 in Sect. 2.1 and summarizes their results in Sect. 2.2.

2.1 Technical Description

During 2014 WRF was applied with a focus on SVA feedback in Germany and in Israel/Oman. WRF offers the choice of various physics parametrizations to describe sub-grid processes like e.g. cloud formation, radiation and turbulent transport. In WRFCLIM it is usually applied with the land surface model NOAH [11] and NOAH-MP [12], the Morrison two-moment microphysics scheme [13], the Yonsei University (YSU) planetary boundary layer parametrization [14], the Kain-Fritsch-Eta convection scheme [15] in case of grid cells larger than 4 km. The radiation is parametrized with the RRTMG shortwave and longwave radiation schemes [16].

The sensitivity studies on parametrization schemes (Sect. 2.1.1) and on the land use in a dry, hot climate (Sect. 2.1.3) of WRFCLIM in 2013/2014 were extended to gain further insights into process understanding. The model settings and the required system resources were similar as introduced in [17]. Furthermore, two simulations were performed for studying an impact of the land use/cover change on the model representation of the surface variables, precipitation and the feedback processes between the land surface and the atmosphere.

The resources used on CRAY XE6 are summarized in Table 1 for May 2014 to 15 April 2015. More details are available in the following sub-sections.

2.1.1 Sensitivity to Parametrization Schemes

The WRF model offers multiple physics parametrization schemes that can be combined in many different ways. The schemes differ in complexity and approaches, ranging from simple and efficient to more sophisticated and computationally more costly, as well as from newly developed schemes to well-tested ones. They are responsible for representing the sub-grid-scale processes that cannot be resolved by the model itself. Such processes include e.g. the radiative and turbulent transports that occur on significantly smaller scales than the model resolution. The foremost issue in weather and climate modeling efficiently is to determine the most appropriate model configuration, which is spatially and temporarily dependent. Therefore, sensitivity studies of parametrization schemes are an unavoidable part of improving the model performance.

Table 1 Technical details of the WRF simulations performed from May 2014 to April 2015

Simulation, Chapter	No. of CPUs	$\Delta x, \Delta y$ [km]	Simulation Period	No. of grid cells	Δt [s]	No. of simulations	Data storage	CPUh
BL and LSM experiments, Sect. 2.1.1	320	6 km, 2 km	21.08.–16.09.2009	$194 \times 217 \times 89$, $271 \times 271 \times 89$	12	1	1.7 TB	51,000
Impact/ISRAEL, Sect. 2.1.3	224	2 km	15.06.–31.08.2014	$200 \times 200 \times 89$	12	4	2.4 TB	79,000
LU experiments, Sect. 2.1.2	320	6 km, 2 km	01.04.–23.06.2013	$194 \times 217 \times 89$, $271 \times 271 \times 89$	12	2	9.5 TB	181,000

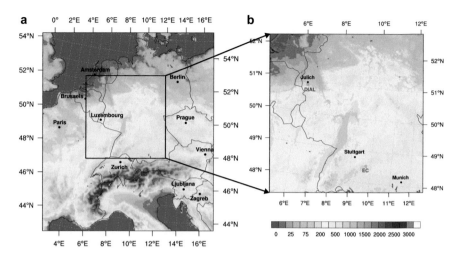

Fig. 1 Topographic representation of the parent domain (**a**) at the $6 \times 6 \, \text{km}^2$ horizontal resolution ($1164 \times 1302 \, \text{km}^2$), and the child domain (**b**) at the $2 \times 2 \, \text{km}^2$ horizontal resolution ($542 \times 542 \, \text{km}^2$), with the measurement sites used for evaluation added in *red*

An extensive analysis of the WRF model sensitivity to the four PBL schemes[4] and the two LSMs[5] was performed on the CRAY XE6 at HLRS. The horizontal resolution of the parent and child domains were 6 and 2 km (Fig. 1), with 89 vertical levels. Time step for integration was 36 s in the parent and 12 s in the child domain. With 320 processors, a total of 51,342 CPU hours were used for the simulation and 1.7 TB of the storage space was occupied (see Table 1).

Simulations for a dry day and additionally for a convective day case study were conducted on the CRAY XE6 at HLRS. The dry case simulation results were compared with the sophisticated measurements performed with the Differential Absorption LIDAR (Light Detection and Ranging, DIAL) and the available radiosonde (RS) measurements. The measurements were done within the TransRegio (TR) 32 FLUXPAT campaign in autumn 2009. In the convective case, simulations were used for process understanding and studying the model discrepancies in representing convection initiation, cloud formation and precipitation.

2.1.2 Sensitivity to Land Use/Cover Change

The ability of the WRF model to simulate the influence of a land use change which includes a conversion of forests to croplands/natural vegetation mosaic on weather

---

[4]ACM2 (The asymmetric convective model, version 2; [18]), MYJ (Mellor-Yamada-Janjic; [19]), MYNN (Mellor-Yamada-Nakanishi-Niino level 2.5; [20]) and YSU [14].

[5]NOAH [11] and NOAH-MP [12].

over the 54 day period was studied. Two simulations were carried out with WRF version 3.6.1 on the CRAY XE6 at the HLRS. The domains for the simulations were exactly the same as in the sensitivity study of WRF to parametrization schemes (Fig. 1). Based on the results of the aforementioned sensitivity study, the WRF model in both cases was configured with the most promising MYNN PBL scheme and NOAH-MP LSM.

Due to continuous increase of the model resolution, a correct representation of the land surface in meteorological models is becoming increasingly important. LSMs rely on parameters such as e.g. albedo, emissivity or roughness length, which are based on the land use maps [21]. WRF originally uses 1 km MODerate resolution Imaging Spectroradiometer (MODIS) land use data set retrieved in 2001. A more recent land use data set with increased resolution was applied in these two simulations. The original Coordinated Information on the European Environment (CORINE) land use data at the 100 m horizontal resolution from 2006 was used in the Control run, while in the Impact run all the forests categories were replaced with the croplands/natural vegetation mosaic. Due to this conversion, about 29.5 % or almost 1/3 of the land use/cover within the child domain was changed (Fig. 2). This was the only difference between the two simulations.

The simulations were run in the "prognostic mode" as described in [17], with the simulation period being 84 days, starting with 1 April 2013 and ending on 23 June 2013. The model was forced with the European Centre for Medium-Range Weather Forecasts (ECMWF) operational analysis data, available every 6 hours. The first month was considered as the spin-up run and therefore excluded from the subsequent analysis. Simulations were run on 320 processors and the whole procedure took about 181,830 CPU hours to be completed. The model output occupied $\approx$ 9.5 TB storage space (see Table 1).

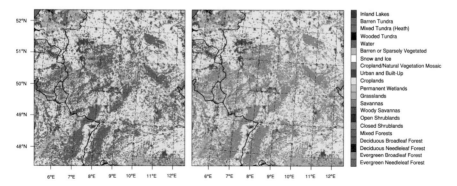

Fig. 2 Representation of the land use/cover change for the child domain at the 2×2 km^2 horizontal resolution (542×542 km^2). The *left panel* shows the original land use/cover from the CORINE land use/cover data set used for the control run and the *right panel* demonstrates the changed land use/cover (forests $\Rightarrow$ cropland/natural vegetation mosaic) that was used in the impact run

2.1.3   The Impact of Irrigated Biomass Plantations on Mesoscale Climate in Arid Regions

The introduction of large plantations ($>100 \times 100$ km) into arid regions has provoked much research due to their potential for carbon sequestration, agricultural development and weather modification. However, it is important to assess if the installation of large plantations and the resulting changes in land surface characteristics, have an impact on the climate.

This investigation was achieved via simulations within a dynamical downscaling experiment, using the coupled WRF-NOAH atmospheric/land surface models. First of all, the model was calibrated for arid conditions and verified with plantation and desert observations in Israel to assess the model output in terms of land surface exchanges. The model on the whole performed very well with only small biases, so we can have confidence that our simulations of plantations are realistic.

Next, the simulations of very large plantations were carried out over arid regions in Oman and Israel region (see Fig. 3) over 38 days during the 2012 summer season (JJA). The domain size was 200 by 200 grid cells, the model integration time was 12 s and the number of processors used was 224. The storage space required was 2.4 TB. Two model case studies were considered—with plantation and without plantation (Impact and Control).

Results from the two regions and scenarios were compared in terms of precipitation, surface temperatures, wind speeds and so on, as well as the prevailing climatic conditions which tend to constrain convection and rainfall. The model runs were then extended to 90 days to allow for robust statistical correlations between daily rainfall events and background climate conditions such as wind speed or radiation.

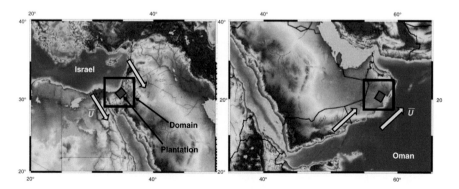

Fig. 3 Maps of plantation simulations in Israel and Oman showing domain and prevailing wind direction

2.2 *Results*

Some of the most interesting results from the aforementioned simulations performed on the CRAY XE6 at the HLRS are represented in the following three sub-chapters.

2.2.1 Sensitivity to Parametrization Schemes

The sensitivity of the WRF model to 4 PBL (the non-local ACM2 and YSU, and the local MYJ and MYNN) schemes and 2 LSMs (NOAH and NOAH-MP) has been studied extensively in cloudless and convective weather conditions. The two case studies showed a significant sensitivity of the model to its initial configuration.

The dry case study analysis was performed for 8 September 2009. The humidity profiles simulated with the model were compared with the high resolution DIAL measurements [17] and with the radiosonde (RS) measurements performed at 16 UTC at the same site as the DIAL. On Fig. 4 the RS temperature measurements are compared with the six simulations in a form of the skew T-logp diagram to investigate the convective PBL (CBL). This type of diagram allows for the calculation of the lifting condensation level (LCL), which can be defined as the level at which a raising air-parcel becomes saturated, and under favourable conditions it coincides with the anticipated cloud base. The observed LCL of 787 hPa is overestimated by all the simulations. The LCL heights are lower in cases of NOAH

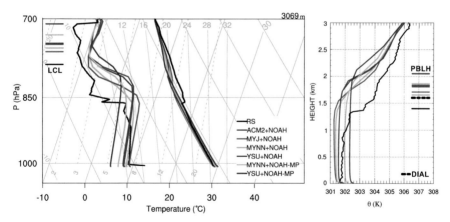

Fig. 4 Skew T-logp diagram of 16 UTC 8 September 2009 sounding (*left panel*), showing vertical profiles of the temperature (right profiles) and dew point temperature (left profiles) as simulated with the six WRF runs along with the RS measurements (the RS launched at 16 UTC). Lines perpendicular to the left vertical axis of the diagram are the LCL values in hPa. On the *right panel* the potential temperature profiles valid at 16 UTC as simulated with the six model simulations and measured with the RS. The perpendicular lines are the PBL heights as simulated and estimated from the RS and, additionally, from the DIAL measurements. Line colours depicted in the legend located in the lower right corner of the *left panel*

with the local schemes (MYNN: 756 hPa, MYJ: 763 hPa) than with the non-local schemes (YSU: 747 hPa, ACM2: 749 hPa). When applying NOAH-MP, the height of the LCL is significantly increased, by 37 hPa for YSU and 25 hPa for MYNN.

Dew point temperature profiles show that the non-local PBL schemes and NOAH-MP simulate lower amounts of moisture within the CBL and consequently a higher PBL than the local PBL schemes and NOAH. The CBL is slightly colder with the local schemes, as well as with NOAH than with the non-local PBL schemes and NOAH-MP (Fig. 4, right panel). Furthermore, the results indicate that NOAH-MP increases the PBL height for both the local MYNN and non-local YSU schemes, but in all six cases the PBL height is more than 300 m too high comparing to the RS measurements. Within the CBL the model is more sensitive to the LSM choice, while the characteristics of the inversion layer are more sensitive to the PBL schemes.

A day chosen for the convective case study was 25 August 2009. A frontal system was passing over Germany during the day, causing high atmospheric instability. At certain places thunderstorms events were followed by heavy amount of convective precipitation. This study showed that the model simulation of the convective processes such as e.g. convective initiation and cloud formation are strongly dependent on the model configuration. Figure 5 depicts the total accumulated precipitation during 24 h period starting on 6 UTC 25 August 2009. The location, the pattern

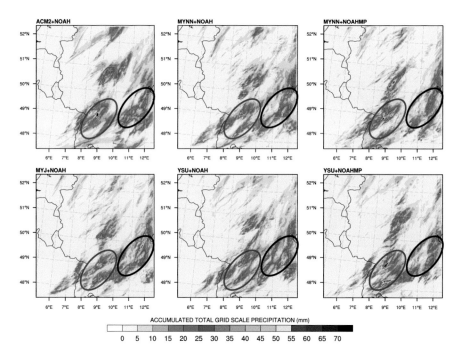

Fig. 5 Total grid scale precipitation accumulated during a period from 6 UTC 25 August until 6 UTC 26 August 2009 as simulated with different combinations of the PBL schemes and the LSMs. The scheme combination is displayed in the *upper left corner* of the corresponding panel. The first precipitation pattern is marked by a *red* ellipse and the second is by a *black* one

and the amount of the precipitation is highly sensitive to the model configuration. Among the PBL schemes, the most intensive precipitation event is accounted with the ACM2 scheme. The first precipitation pattern is occurring around midday and the convection processes are not strong with any other PBL scheme but with ACM2. However, NOAH-MP intensify the precipitation for the first pattern significantly for both MYNN and YSU. This indicates that for this specific case, the processes at the land surface have a higher impact on the convective processes than the processes within the CBL. The second precipitation pattern was happening later in the afternoon. In this case, NOAH-MP decreases the amount of precipitation occurrence by more than 30 mm. The convective case study showed that the model is more sensitive to LSMs than to the PBL schemes in representing the convective processes and the precipitation.

This sensitivity study highlights that the processes at the land surface strongly affect not only the lower PBL, but also the interfacial layer. They impact on convective processes and the entrainment of air from the free atmosphere to the PBL which controls the PBL evolution. In models, the parametrizations of the entrainment fluxes are calibrated based mostly on large-eddy simulations (e.g. in the YSU PBL scheme). Such parametrizations need extensive evaluation and validation using realistic measurements. For that purpose, extensive observations of flux profiles are a prerequisite.

2.2.2 Sensitivity to Land Use/Cover Change

The impact of the land use/cover change, from forests to croplands and natural vegetation mosaic, on weather in western and southwestern Germany has been analysed using the WRF model simulations. The model with the original CORINE land use data (the Control run) have been compared with the simulation using the modified land use (the Impact run) in terms of temperature, humidity and precipitation.

Figure 6 shows the difference in potential temperature and the mixing ratio at the first model level between the Control and the Impact run averaged over the 54 day period (from 1 May until 23 June 2013) for the child domain. The results indicate a decrease in temperature and an increase in humidity in the Impact run. The maximum averaged differences are 0.4 K and 0.32 g kg^{-1}, for the potential temperature and the mixing ratio, respectively. These preliminary results indicate that the influence of the land use/cover change is significant even during a relatively short time period.

The impact of the deforestation on the amount of precipitation is depicted on Fig. 7. The figure shows the difference in the accumulated daily precipitation averaged over the 54 period (upper panel) and the differences spatially averaged over the domain (lower panel).

The time averaged difference shows mostly a decrease of the precipitation amount for up to 2 mm in the regions directly affected by the land use/cover change. The regions affected the most are those in which the evergreen needleless

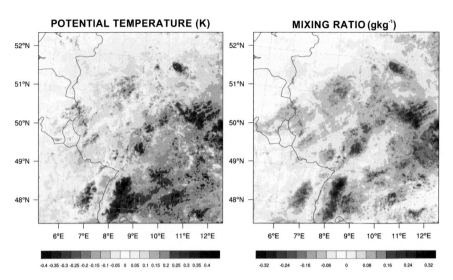

Fig. 6 The difference in potential temperature (*the left panel*) and in mixing ratio (*the right panel*) between the Control and the Impact run averaged over the 54 day period (from 1 May until 23 June 2013)

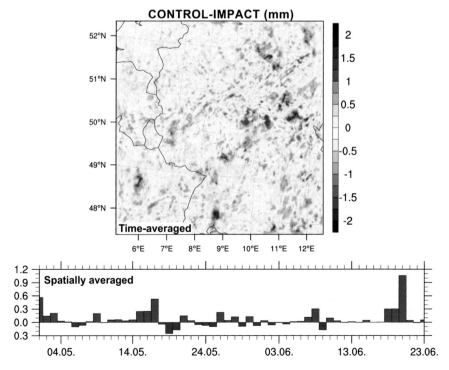

Fig. 7 The precipitation difference in mm between the Control run and the Impact run: *upper panel* shows the values averaged over the 54 day period and the *bottom panel* shows the daily values spatially averaged over the whole child domain

forest conversion took place. Furthermore, the spatially averaged difference in the precipitation amount between the Control and the Impact run (lower panel of Fig. 7) shows that on most of the days within the 54 day period the amount of the precipitation was decreased due to the land use/cover change.

Although these preliminary results have to be considered with caution because the 54 day period used for averaging might be too short for drawing any final concluding remarks on the impact of this kind of the land use/cover change, this first study encourages enhanced studies on land use change. Deforestation impacts the Earth's radiation balance through the release of the stored CO_2 to the atmosphere which causes warming, as well as through the increase in surface albedo which results in cooling effects. The land use/cover change applied here caused the albedo to increase, which in return resulted in the cooling effect. But this land use/cover conversion will cause substantial changes in amount of the CO_2 emitted in the atmosphere, which supposed to induce the warming effects on a longer time scale. Therefore, further analyses are necessary to obtain detailed insights into the impact of this kind of land use/cover change on e.g. temperature, humidity and precipitation.

2.2.3 The Impact of Irrigated Biomass Plantations on Mesoscale Climate in Arid Regions

To assess the impact of adding plantations, the total precipitation increase from adding the plantation is shown for the 38 day period (Fig. 8). It is apparent that a large increase in rainfall occurs in Oman because of the plantation, which demonstrates a clear possibility for weather modification. However, very little impact occurs in Israel.

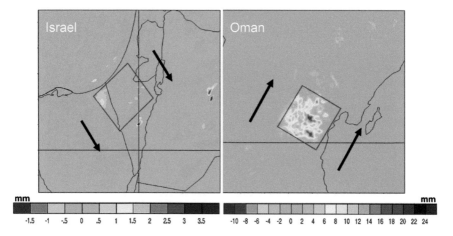

Fig. 8 The increase in precipitation in Israel and Oman through the addition of the plantation within the desert. This was calculated by subtracting the total control precipitation from impact

The mechanism underlying the increase in precipitation is a surface heating effect over the canopy and an increase in roughness which leads to an increase in turbulence and also to convergence and ascent of moist air. The reason for the lack of impact in Israel is that the air is much drier above 1000 m altitude than in Oman, which tends to suppress the upward motion of surface air parcels and the development of clouds. Israel is therefore too stable in summer for impacts.

When the background conditions were correlated with wet and dry days over 90 days it was clear that the occurrence of precipitation is strongly dependent on the amount of radiation at the land surface. This is evidenced by statistical two-sample T-tests (Fig. 9) where a clear connection between the strength of radiation and wet/dry days is evident. This is as expected because greater radiation leads to greater heating and convergence.

Furthermore it became evident that the strength of air convergence (which leads to ascent and cloud development) is dependent on wind speed as shown in statistical correlations (see Fig. 10). The stronger the wind speed, the less convergence occurs and therefore a reduced probability of rainfall follows.

In summary we can say that an increase in rainfall would occur if biomass plantations were cultivated within arid regions, but that certain climate conditions are necessary for impacts—sufficient humidity, high surface radiation, moderate wind speeds, and an unstable temperature profile. Therefore a regional assessment of these factors could help to identify ideal locations for plantations to increase local rainfall in arid regions. An initial global analysis of arid regions shows that the arid south-western United States and Mexico, particularly Baja California is the most promising location for strong impacts.

The model validation results were published in the Hydrology and Earth System Sciences journal [4]. The weather modification results formed the basis for the Doctoral thesis "The Impact of Irrigated Biomass Plantations on Mesoscale Climate in Coastal Arid Regions" [22] at the University of Hohenheim, which was awarded Summa Cum Laude in November 2014.

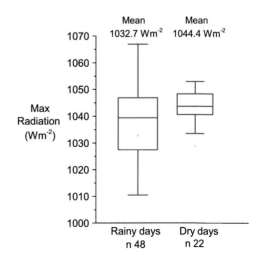

Fig. 9 Statistical analysis of the occurrence of wet and dry days over the Oman plantation, with respect to the amount of maximum daily surface radiation. This was carried out with a two-sample T-test and a confidence level of 95 % ($p = 0.05$)

Fig. 10 Correlation between
wind speed and air
convergence strength. The
plot shows that higher wind
speeds tend to weaken
convergence and therefore the
probability of cloud
development and rainfall
impacts

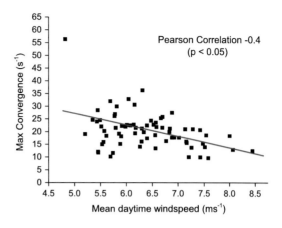

3   Outlook

The climate projections within the ReKliEs-De project are currently in the preparation phase on the new CRAY XC40 system. An ensemble consisting of five global climate projections from the CMIP5 Project (Coupled Model Intercomparison Project Phase 5) on 200 km resolution will be conducted with the WRF model version 3.6.1 for 1960 until 2100 within the BMBF-funded ReKliEs-De project on regional climate change and the DFG funded Research Unit 1695 on regional climate change in agricultural landscapes. The simulations will be downscaled to 45 km resolution, and further to a 12 km grid over Europe. One simulation will be further downscaled to 3 km resolution until 2030. Vertical resolution of the simulations is set to 50 levels. The first future climate simulation is going to start in May 2015.

Acknowledgements This work is part of the Project RU 1695 funded by DFG and supported by a grant from the Ministry of Science, Research and Arts of Baden-Württemberg (AZ Zu 33-721.3-2). The future simulations in preparation are part of the ReKLies-De project funded by the BMBF (Federal Ministry for Education and Research). DIAL measurements were performed within the Project Transregio 32, also funded by DFG and provided by Andreas Behrendt and Florian Späth (University of Hohenheim). Model simulations were carried out at HLRS on the CRAY XE6 within WRFCLIM and we thank the staff for their support.

References

1. Skamarock, W.C., Klemp, J.B., Dudhia, J., Gill, D., Barker, D.M., Duda, M.G., Huang, X.Y., Wang, W., Powers, J.G.: A description of the advanced research WRF version 3. NCAR Technical Note TN-475+STR, NCAR, Boulder (2008)
2. Warrach-Sagi, K., Schwitalla, T., Bauer, H.S., Wulfmeyer, V.: A regional climate model simulation for EURO-CORDEX with the WRF model. In: Sustained Simulation Performance 2013, pp. 147–157. Springer International Publishing, Berlin (2013)

3. Greve, P., Warrach-Sagi, K., Wulfmeyer, V.: J. Appl. Meteorol. Climatol. **52**, 2312 (2013)
4. Branch, O., Warrach-Sagi, K., Wulfmeyer, V., Cohen, S.: Hydrol. Earth Syst. Sci. **18**, 1761 (2014)
5. Bauer, H.S., Schwitalla, T., Wulfmeyer, V., Bakhshaii, A., Ehret, U., Neuper, M., Caumont, O.: Tellus A, **67**, 25047 (2015). doi:10.3402/tellusa.v67.25047
6. Dirmeyer, P.A., Cash, B.A., Kinter, J.L., Jung, S.T., Marx, L., Towers, P., Wedi, N., Adams, J.M., Altshuler, E.L., Huang, b., Jin, E.K., Manganello, J.: J. Hydrometeor. **13**, 981–995 (2012)
7. Taylor, C.M., de Jeu, R.A.M., Guichard, F., Harris, P.P., Dorigo, W.A.: Nature **489**, 423 (2012)
8. Lenderink, G., van Meijgaard, E.: Nat. Geosci. **1**, 511–515 (2008)
9. Zolina, O., Simmer, C., Belyaev, K., Gulev, S.K., Koltermann, P.: J. Clim. **26**, 2022 (2013)
10. Palmer, T.N., Doblas-Reyes, F.J., Weisheimer, A., Rodwell, M.J.: Bull. Am. Meteorol. Soc. **59**, 459 (2008). http://dx.doi.org/10.1175/BAMS-89-4-459
11. Chen, F., Dudhia, J.: Mon. Weather Rev. **129**, 569 (2001)
12. Niu, G.Y., Yang, Z.L., Mitchell, K.E., Chen, F., Ek, M.B., Barlage, M., Kumar, A., Manning, K., Niyogi, D., Rosero, E., Tewari, M., Xia, Y.: J. Geophys. Res. **116**, D12109 (2011)
13. Morrison, H., Thompson, G., Tatarskii, V.: Mon. Weather Rev. **137**, 991 (2009)
14. Hong, S.Y., Noh, Y., Dudhia, J.: Mon. Weather Rev. **134**, 2318 (2006)
15. Kain, J.S.: J. Appl. Meteorol. **43**, 170 (2004)
16. Iacono, M.J., Delamere, J.S., Mlawer, E.J., Shephard, M.W., Clough, S.A., Collins, W.D.: J. Geophys. Res. **113**, D13103 (2008). doi:10.1029/2008JD009944
17. Warrach-Sagi, K., Bauer, H.S., Schwitalla, T., Milovac, J., Branch, O., Wulfmeyer, V.: High-resolution climate predictions and short-range forecasts to improve the process understanding and the representation of land-surface interactions in the WRF model in Southwest Germany (WRFCLIM). In: High Performance Computing in Science and Engineering '14, pp. 575–592. Springer International Publishing, Berlin (2014)
18. Pleim, J.: J. Appl. Meteorol. Climatol. **46**, 1383 (2007)
19. Mellor, G., Yamada, T.: J. Atmos. Sci. **31**, 1791 (1974)
20. Nakanishi, M., Niino, H.: J. Meteorol. Soc. Jpn. **87**, 895 (2009)
21. Pineda, N., Jorba, O., Jorge, J., Baldasano, J.M.: Int. J. Remote Sens. **25**, 129 (2004)
22. Branch, O.L.: The impact of irrigated biomass plantations on mesoscale climate in coastal arid regions. Doctoral thesis, Institute of Physics and Meteorology, University of Hohenheim, 138 p. (2014)

Do We Have to Update the Land-Use/Land-Cover Data in RCM Simulations? A Case Study for the Vu Gia-Thu Bon River Basin of Central Vietnam

Ngoc Bich Phuong Nguyen, Patrick Laux, Johannes Cullmann, and Harald Kunstmann

Abstract The accuracy of land-use/land-cover (LULC) data plays a critical role for improving regional climate simulations. In this study, we implement three observed (2001, 2005, and 2010) and one projected high-resolution LULC data for the Vu Gia-Thu Bon (VGTB) basin of Central Vietnam into the Weather Research & Forecasting (WRF) model. As an option in WRF, the Noah land-surface model is used to model the land's state variables. Simulated hydro-meteorological variables such as precipitation and temperature from the updated and WRF-default LULC runs are compared with observations in the VGTB basin. The results give evidence that the updated LULC data lead to an improved representation of annual precipitation and average temperature for the region. Even though there are still some inconsistencies in representing the maximum and minimum temperatures as well as precipitation in the dry and wet season we can conclude that using the updated LULC data improves the regional climate simulations.

1 Introduction

Since the interaction between land surface and the atmosphere is characterized by the partitioning of net radiation between the two turbulent energy fluxes (sensible and latent heat), as well as the aerodynamic roughness of the land surface, the effects of LULC on climate depend mainly on the land cover types. Furthermore,

N.B.P. Nguyen (✉) • J. Cullmann
IHP/HWRP Secretariat, Federal Institute of Hydrology, Am Mainzer Tor 1, 56058 Koblenz, Germany
e-mail: nguyen@bafg.de; cullmann@bafg.de

P. Laux • H. Kunstmann
Karlsruhe Institute of Technology (KIT), Institute for Meteorology and Climate Research, Atmospheric Environmental Research (IMK-IFU), Kreuzeckbahnstrasse 19, 82467 Garmisch-Partenkirchen, Germany
e-mail: patrick.laux@kit.edu; harald.kunstmann@kit.edu

© Springer International Publishing Switzerland 2016
W.E. Nagel et al. (eds.), *High Performance Computing in Science and Engineering '15*, DOI 10.1007/978-3-319-24633-8_40

[4] showed the physical linkage between the surface and the atmosphere, and demonstrated how even slight changes in surface conditions can have a pronounced effect on weather and climate. He said, that "...the albedo, and the fractional partitioning of atmospheric turbulent heat flux into sensible and latent fluxes are shown to be particularly important in directly affecting local and regional weather and climate". As a consequence, the accuracy of representing LULC data in a key issue for regional climate models.

In this study, we implemented updated (observed and projected) LULC data into the WRF model and analysed the differences for the VGTB basin in Central Vietnam to evaluate the importance of the representation of LULC in regional climate simulations. For the evaluation, the probability of the minimum, maximum and average surface temperature as well as precipitation during the dry and wet season is investigated. The regional climate model WRF is driven by ERA-interim reanalysis data for the period 2001–2012 and the ECHAM5 global climate model, driven by the A1B emission scenario for the period 2001–2020.

2 Data and Methods

2.1 Land-Use/Land-Cover Data

In this section, we describe both the updated LULC data and the WRF-default LULC data. The WRF-default LULC data is obtained by the U.S. Geological Survey Land Use/Land Cover System (USGS-GLCC) and created by using 1 km Advanced Very High Resolution Radiometer (AVHRR) satellite images spanning the period April 1992 through March 1993 by means of an unsupervised classification technique [1]. Based on this, over 82 % area of the VGTB basin is covered by cropland, while forest accounts for only 11.2 %. A minority of the urban land class is found in this region accounting for about 0.1 % of the total LULC (Table 1). Several studies [5, 7] have mentioned that the LULC data was not updated and not accurate for all regions and some land cover types.

The updated LULC data were obtained by preprocessing and classifying remote sensing data, which has been achieved in the project *Land Use and Climate Change interactions in Central Vietnam* (LUCCi). The classification has been performed for the years 2001, 2005, 2010 and a scenario for the year 2020 (hereinafter referred to as LULC-2001, LULC-2005, LULC-2010, and LULC-2020). The data include Spot

Table 1 The fraction of land use classes as represented in the WRF-default LULC and updated LULC maps within the VGTB basin

	WRF-default LULC (%)	Updated LULC (%)
Cropland	82	25–28
Forest	11.2	49–52
Grassland	4.1	17–18
Urban	0.07	1.3–1.5

5 data in a 2.5 m horizontal resolution covering approximately 90 % of the VGTB basin, Landsat ETM+ data (30 m horizontal resolution) covering the entire basin, as well as Road Network vector data, FIPI forest map vector data, and River network vector data from 2001, 2005 and 2010. After various preprocessing steps such as geometric corrections, cloud and shadow removal, radiometric corrections, and gap-filling the land-use map is created based on a supervised classification algorithms. The classification has been checked using ground trusting activities in the VGTB basin, and the resulting map has been refined based on auxiliary information from local stakeholders. After that, the accuracy of the LULC data has been assessed, showing an overall accuracy of about 82 % of correctly classified pixels according to the calculated confusion matrix. These LULC maps have a spatial resolution of 30×30 m and consists of six classes following the IPCC nomenclature, i.e. forest, cropland, grassland, water, settlements, and one additional class representing all other land-use classes. More information about the entire procedure is given in [6].

To integrate the updated LULC maps of the VGTB basin into the WRF model, the LULC data are spatially aggregated to a 1 km resolution fitting better to the resolution of the regional climate simulations of 5 km. For the aggregation of land-use classes, the dominant classes within the coarser 1 km grid are selected. These 1 km aggregated LULC maps (hereinafter referred to as LUCCi LULC maps) exhibit in the following proportions of the classes in Table 1. The area of cropland in the LUCCi LULC data were much smaller than in the WRF-default LULC data, which accounted for only 25–28 % of the total area. The areas of forest and grassland increase remarkably at around 50 % and at around 18 %, respectively; while the urban area accounts for only around 1.5 %. Considerable differences between the WRF-default LULC and the LUCCi LULC data can be seen for this region (Fig. 1 and Table 1).

2.2 Climate Data

The surface observation data in this study include monthly temperature and precipitation values at surface during the period 2001–2012, obtained from the observation stations in the VGTB basin. The observation data were used to investigate the climate trend and evaluate the simulations derived from modelling. In the VGTB basin, the number of available hydro-meteorological stations is limited [9]. There is only three meteorological stations and twenty-two hydrological stations available. Due to missing values and inconsistent time series, only seventeen rain gauge stations and three meteorological stations have been selected for this study. Figure 2 shows the locations of the stations. The meteorological stations are highlighted in red, the rain gauges in white.

To obtain high-resolution climate information for this region, we downscaled ERA-interim and ECHAM5 (A1B) data using the regional climate model WRF. ERA-interim is a global atmospheric reanalysis data and ECHAM5 (A1B) is the data from an atmospheric general circulation model, driven by the A1B emissions

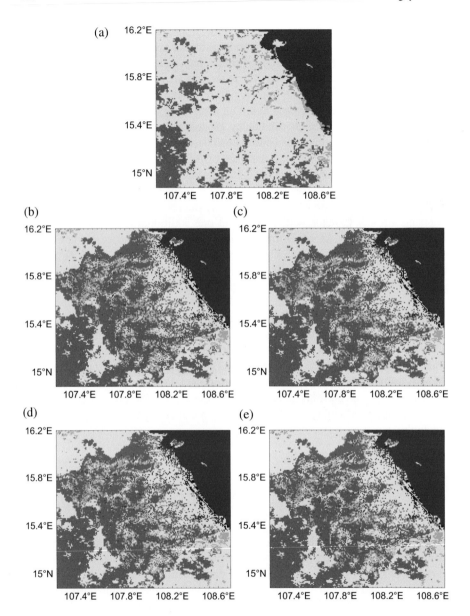

Fig. 1 Land-use/land-cover maps of the VGTB basin. (**a**) WRF-default LULC, (**b**) updated LULC-2001, (**c**) updated LULC-2005, (**d**) updated LULC-2010, (**e**) updated LULC-2020

scenario. The WRF model used one mother domain and two nested domains with a resolution of 45, 15 and 5 km, respectively. To capture the most crucial large-scale circulations, the main domain covered Southern Asia from 5°S to 27°N latitude and 90°E to 130°E longitude (see Fig. 3). The first nested domain covered South-

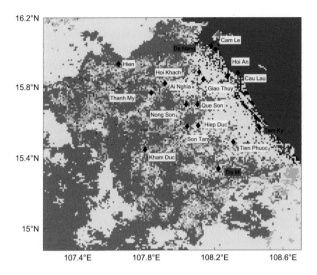

Fig. 2 Hydro-meteorological observation network in the VGTB river basin. The meteorological stations are highlighted in *red*, the rain gauges in *white*

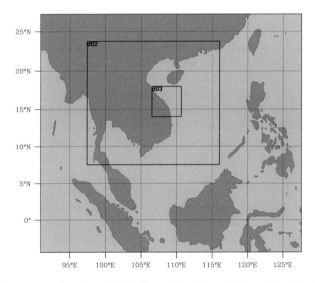

Fig. 3 Domain setup used for the regional climate simulations. Horizontally, 90 × 80 grid points with a resolution of 45 km (D01), 136 × 124 grid points with a resolution of 15 km (D02), and 88 × 82 grid points with a resolution of 5 km (D03) are used

Eastern Asia and parts of the Eastern Sea and the second nested domain focused on the VGTB basin. We used the cumulus parametrization scheme from Betts-Miller-Janjic, the microphysics from the WSM 3-class simple ice scheme, the boundary layer from the YSU scheme and the land surface scheme from the Unified Noah land-surface model [2, 3]. In this land-surface model, based on the atmospheric information from the surface layer scheme, radiative forcing from the radiation scheme, and precipitation forcing from the microphysics and convective schemes, together with internal information on the land's state variables, the heat and moisture fluxes were calculated for all land points. The soil is stratified into 4 layers, for which soil temperature and moisture is modelled. The layer thicknesses are 10, 30, 60 and 100 cm from the top down. For the modelling, it also includes evapotranspiration, soil drainage, and run-off, taking into account different vegetation categories, monthly vegetation fraction, and soil texture. The scheme provides sensible and latent heat fluxes to the WRF planetary boundary-layer (PBL) scheme. In addition, monthly variations in vegetation fraction and albedo are interpolated and stored in the lower boundary file [8]. The details of the Noah LSM are described in Chen and Dudhia [1].

3   Results

The updated LULC maps shows significant differences in the study area compared to the WRF-default LULC map. Accordingly, we expect big differences in the regional climate simulations due to the updated LULC compared to the WRF-default LUC data. In this study, we have considered different observed LULC data, i.e. LUCCi LULC-2001, LULC-2005, LULC-2010 and LULC-2020, however, the climate model between the LUCCi LULC maps differ only marginally. The distinctions are 1–3 % for forest and cropland, 1 % for grass and only 0.2 % for urban area. Therefore, we only show results based on the LUCCi LULC-2001 and the WRF-default LULC data.

Figure 4 shows the differences between the WRF-default LULC and the LUCCi LULC-2001 for the albedo (a), the roughness length (b), the latent heat flux (c), and the sensible heat flux (d). Albedo of the LUCCi LULC-2001 is higher than that of the WRF-default LULC in the urban area along the coastline and in the Western parts of the VGTB basin where the roughness length is decreased considerably. The increase of albedo in the urban area is around 0.15 and in the Western of the VGTB basin is around 0.1. In the lowland area near the coastline and the Southern of the VGTB basin, there is a slight decrease of the albedo by around 0.05. The area where the roughness length increases 0.2 m due to the change from cropland to forest shows no change in albedo. Figure 4c reveals that the use of the LUCCi LULC-2001 leads to a slight decrease of the latent heat flux over the basin. The

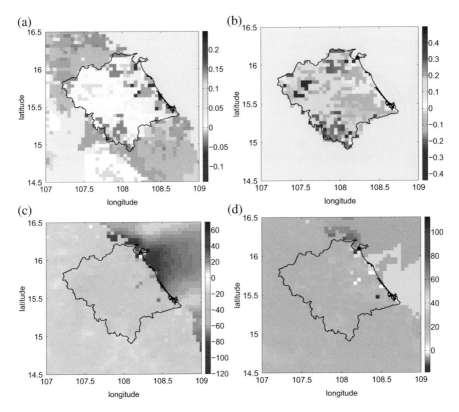

Fig. 4 Mean difference of albedo, roughness length, and latent and sensible heat flux for the year 2001. (**a**) Albedo, (**b**) Roughness length, (**c**) Latent heat flux, (**d**) Sensible heat flux

albedo of the urban area increases around 20 W/m². In contrast, the sensible heat flux increases by around 30 W/m² over the basin (Fig. 4d). The highest increase in the sensible heat flux can be observed over the urban area.

In order to find out if the LUCCi LULC data can improve the numerical simulations, the observed data were compared with both the LUCCi LULC-2001 and the WRF-default LULC data. The selected hydro-meteorological stations were used to assess the total annual precipitation and the surface temperature (average, maximum and minimum) during the period 2001–2012. In order to obtain the results of the model at a station, the average values of the surrounding 9 grid-points next to the station is calculated.

Figure 5 shows the accumulated probability of precipitation during 2001–2012 in the VGTB basin. The data were derived from the observed data (black), the WRF-default LULC (blue) and the LUCCi LULC-2001 (red). The comparison was conducted for the dry season (Feb-May)(a), the rainy season (Sep-Dec)(b) and for the annual precipitation (c).

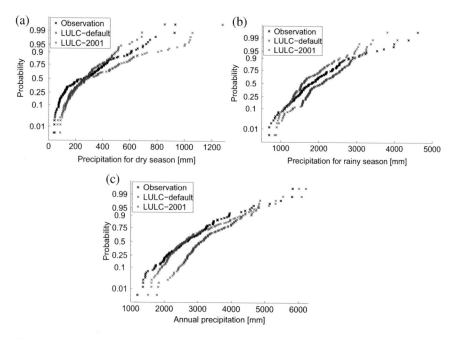

Fig. 5 Probability of total precipitation. (**a**) Precipitation for the dry season, (**b**) precipitation for rainy season, and (**c**) annual precipitation

In general, the precipitation of the LUCCi LULC-2001 exhibits a good agreement with the observation in annual precipitation, overestimates in the dry season and underestimates in the rainy season. The precipitation of the WRF-default LULC is always higher compared to the observations, affecting both the rainy season and the annual precipitation. In the dry season, 90 % probability level of the observed precipitation is below 500 mm. The WRF-default LULC exhibits similar magnitudes compared to the observed data, while the precipitation of the LUCCi LULC-2001 is much higher accounting for above 800 mm at the 90 % probability level. The results reveal that the LUCCi LULC-2001 leads to higher precipitation amounts than expected. Therefore, the updated LULC data does not show any improvement compared to the WRF-default LULC during the dry season. In the rainy season, 90 % of the observed precipitation is below 2500 mm. The precipitation of the LUCCi LULC-2001 is lower and the precipitation of the WRF-default LULC is higher than the observation. For annual precipitation, the LUCCi LULC-2001 shows similar values with the observation amounting around 3300 mm at the 90 % probability level. At the same level, the precipitation amount of the WRF-default LULC is increased by around 500 mm. The LUCCi LULC-2001 is

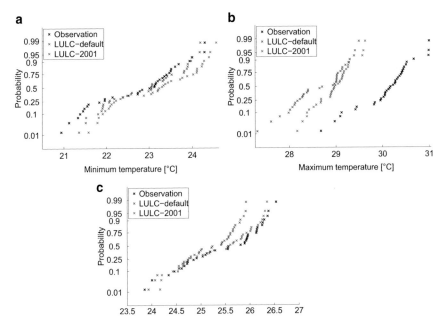

Fig. 6 Probability plots of surface temperature. (**a**) minimum temperature, (**b**) maximum temperature, and (**c**) average temperature

generally in a better agreement with the observations than the WRF-default LULC. Our findings demonstrate the need to use the updated LULC data for simulating precipitation in Central Vietnam.

Figure 6 presents the probability of the minimum (a), maximum (b) and average (c) surface temperature at three meteorological stations during the period 2001–2012. The same applied to Fig. 5, the results from the WRF-default LULC are presented in blue, from the LUCCi LULC-2001 presented in red, and the observed data are presented in black. The minimum temperatures of the WRF-default LULC are consistent with the observed data, while the ones of the LUCCi LULC-2001 often appear overestimated. Maximum surface temperature in both the LUCCi LULC-2001 and the WRF-default LULC are underestimated if compared with the observations. Nevertheless, the average surface temperature of the LUCCi LULC-2001 is in good agreement with the observed data.

Figure 7 shows the mean annual surface temperature and the annual precipitation in the VGTB basin using the WRF-default LULC (blue-dash line) and the LUCCi LULC-2001 (red line) from 2001 to 2020 using ECHAM5 (A1B) as lateral boundary conditions. As shown in Fig. 7a, the surface temperatures of the LUCCi LULC-2001 are often increased compared to the WRF-default LULC. The maximum difference can be observed for 2012. Then, the WRF-default LULC

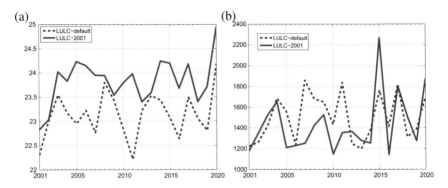

Fig. 7 Interannual mean surface temperature (**a**) and precipitation (**b**) for the VGTB basin

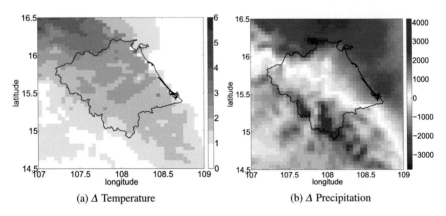

(a) Δ Temperature   (b) Δ Precipitation

Fig. 8 Mean annual differences between the LUCCi LULC-2001 and the WRF-default LULC (LUCCi LULC-2001 minus WRF-default LULC) for surface temperature (**a**) and precipitation (**b**) during the period 2011–2020 using ECHAM5 (A1B)

shows the lowest surface temperature value at the same time. The results indicate a remarkable increase in the surface temperature over this region if the LULC map is updated. For the first half of the analyzed period, precipitation of the LUCCi LULC-2001 is reduced compared to the result derived from the WRF-default LULC, whereas for the period from 2015 to 2020, there is no remarkable difference between the WRF-default LULC and the LUCCi LULC-2001 in precipitation. According to the Fig. 8, the mean surface temperature and the annual precipitation are strongly affected by LULC data. With the use of the updated LULC data, the temperature increases over the basin from 1 to 3°C. The lowland is drier with a decline of precipitation in 1000–2000 mm, whereas the precipitation considerable increases in the highland area, specially in the Southern area, reaching values up to 3000 mm.

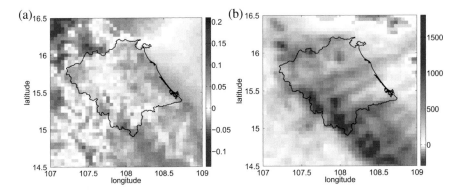

Fig. 9 Mean annual difference between 2011–2020 and 2001–2011 (2001–2020 minus 2001–2011) of surface temperature (**a**) and precipitation amount (**b**)

Figure 9 shows the mean differences of surface temperature and annual precipitation for the periods 2011–2020 and 2001–2010. The results show a slight increase of the surface temperature for the near future in the order of 0.2°C. Precipitation is expected to increase considerably in the highlands and gradually reduced in the lowlands. The largest increase of precipitation is expected in the Northern parts of the highlands. There, the amount of precipitation is expected to be increased up to 1500 mm.

Although there are still some inconsistencies in the results for precipitation and surface temperature, we believe that the updated LULC improved the overall quality of the regional climate simulations for the VGTB basin of Central Vietnam. Particularly for regions, where the LULC has changed tremendously during the past decades and relatively old LULC maps are used in the land surface models embedded in the RCMs, we suggest the usage of updated LULC information to model the current climate. For regional climate change projections, however, this remains a challenging task and would require the usage of LULC change projections accordingly.

4 CPU Usage and Storage Capacities for This Study

Since the beginning of using the ForHLR system in Stuttgart in January 2015, we have simulated the regional climate in the VGTB basin with the current LULC data in the WRF model (WRF-default LULC) and the LUCCi LULC data obtained from LUCCi project. The number of simulated months is presented in Table 2 and the benchmark of the WRF simulating two days at ForHLR is illustrated in (Fig. 10).

Table 2 Number of simulated months using the regional climate model WRF

Experiments	Perturbed runs	Months for perturbed runs	Long-term transient runs	Months for long-term runs	Total
LULC-default	5	15	1	240	315
LULC-2001	5	15	1	240	315
LULC-2010	5	15	1	240	315
LULC-2020	5	15	1	240	315
					1260

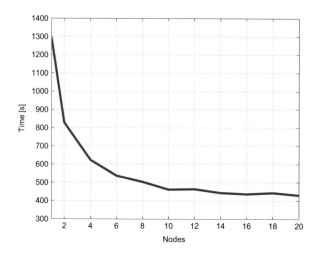

Fig. 10 Benchmark of the WRF simulating the 2 days at ForHLR

The simulations were performed using 200 CPUs for each month. The computing time for the three domains for one month was 4 h which results in $4 \times 200 = 800$ CPUh per month. This means that for computing for 1260 months $800 \times 1260 = 1{,}008{,}000$ CPUh were consumed.

Acknowledgements This research is funded by the Federal Ministry of Education and Research (research project: Land Use and Climate Change Interactions in Central Vietnam (LUCCi), reference number 01LL0908C). The provision of CPU and storage capacities at Karlruhe Institute of Technology (KIT), Steinbuch Centre for Computing (SCC) and Karlruhe Institute of Technology (KIT), Institute of Meteorology and Climate Research (IMK-IFU) is highly acknowledged.

References

1. Chen, F., Dudhia, J.: Coupling an advanced land surface-hydrology model with the penn state-ncar mm5 modeling system. part i: Model implementation and sensitivity. *Mon. Weather Rev.* **129**(4), 569–585 (2001)

2. Laux, P., Lorenz, C., Thuc, T., Ribbe, L., Kunstmann, H. et al.: Setting up regional climate simulations for southeast Asia. In: High Performance Computing in Science and Engineering Ś12, pp. 391–406. Springer, Berlin/Heidelberg (2013)

3. Laux, P., Thuc, T., Kunstmann, H. et al.: High resolution climate change information for the lower mekong river basin of southeast Asia. In: High Performance Computing in Science and Engineering Ś13, pp. 543–551. Springer International Publishing (2013)

4. Pielke, R.A., Avissar, R.: Influence of landscape structure on local and regional climate. Landsc. Ecol. **4**(2–3), 133–155 (1990)

5. Pitman, A.: The evolution of, and revolution in, land surface schemes designed for climate models. Int. J. Climatol. **23**(5), 479–510 (2003)

6. Schultz, M., Avitabile, V.: 2010 land use/ land cover map of the vu gia-thu bon (vgtb) river basin, central Vietnam - version 2.0. Technical report, 2012

7. Sertel, E., Robock, A., Ormeci, C.: Impacts of land cover data quality on regional climate simulations. Int. J. Climatol. **30**(13), 1942–1953 (2010)

8. Skamarock, W.C., Klemp, J.B., Dudhia, J., Gill, D.O., Barker, D.M., Duda, M.G., Huang, X.-Y., Wang, W., Powers, J.G.: A description of the advanced research wrf version 3. NCAR Technical Note, 2008

9. Souvignet, M., Laux, P., Freer, J., Cloke, H., Thinh, D.Q., Thuc, T., Cullmann, J., Nauditt, A., Flügel, W.-A., Kunstmann, H. et al.: Recent climatic trends and linkages to river discharge in central vietnam. Hydrol. Process. **28**(4), 1587–1601 (2014)

Part VII
Miscellaneous Topics

Wolfgang Schröder

The research areas which have been tackled in the previous chapters are related to solid state physics, fluid mechanics, aerodynamics, flows with chemical reactions, thermodynamics, structural mechanics, and so forth. The following contributions even widen the field where numerical simulations represent useful tools to gain novel results and as such to improve the scientific knowledge in the various areas. The papers evidence the close relationship between mathematics and computer science and the development of novel scientific models. Compact mathematical descriptions are solved by highly sophisticated and efficient algorithms on a high-end machine. This interdisciplinary collaboration between several scientific fields defines the extremely intricate numerical challenges and as such drives the progress in fundamental and applied research. The subsequent articles represent an excerpt of the numerous projects being linked with HLRS. They show that the outcome of the computations is more than just some quantitative results. The findings confirm fundamental physical models and help to derive new theoretical approaches.

The G-CSG, the Institute of Radiology, and the Institute of Bio-Statistics and Mathematical Modeling of the Goethe University of Frankfurt performed advanced simulations within biophysical applications which require advanced algorithms and implementations on massively parallel high performance computers. The experimental basics, the modeling and simulation details, and the biophysical meaning of the estimation of the diffusion constant of a major player in the replication of the genetic information of the Hepatitis C virus (HCV), namely the NS5a viralprotein, are discussed. NS5a movement is restricted to the surface of the Endoplasmatic Reticulum (ER, a medusa-hair like important cell compartment). Therefore, the dynamics of NS5a is described by surface PDEs (sPDE) which mimic experimental FRAP time series. The sPDEs were performed with UG, a software framework for the numerical solution of problems that are described by systems of PDEs based

W. Schröder (✉)
Institute of Aerodynamics, RWTH Aachen University, Wüllnerstr. 5a, 52062 Aachen, Germany
e-mail: office@aia.rwth-aachen.de

on finite volume methods and multigrid solvers. A substantial amount of single PDE evaluations were performed on the HLRS Hermit and Hornet computers for various experimental data sets and for various geometric setups in the context of the parameter estimations. To this end, the consumption of a vast number of core hours made it possible to derive valid final values for the diffusion constant of NS5a which could enter spatio-temporal resolved models of HCV replication dynamics at a cellular level. These methods could give fruitful stimulus to the young field of spatio-temporal resolved computational virology. Interest from the experimental side for further similar simulations could give rise for extended need for core hours on HLRS HPC computers.

The Institute of Materials and Processes of the Karlsruhe University of Applied Sciences in Technology and Economy conducted large-scale phase-field simulations. In material science, simulations are a common tool for the understanding of the underlying behavior of different classes of materials. Due to the growing complexity, the simulation domains and the computational effort steadily increase. Various applications of the phase-field method ranging from the solidification of ternary eutectics and pure ice systems to the interaction of multiple liquid phases on fibres are presented. These topics share the need of a large number of cores to investigate the decisive physical effects in adequate time. The findings consist of an overview for this wide range of applications and the scaling behavior of the used software frameworks.

Large scale simulations of planetary interiors were performed by the Institute of Planetary Research of the German Aerospace Center and the Department of Planetary Geodesy of the Technical University of Berlin. Due to the massive increase of computational power over the past decades numerical models of planetary interiors have become one of the principal tools to investigate the thermo-chemical evolution of terrestrial bodies. Large scale numerical simulations are commonly used to analyze the interior heat transport, surface tectonics, and chemical differentiation of planetary bodies across the Solar System. In the study of DLR and TU Berlin large scale focused simulations using the mantle convection code Gaia in spherical and Cartesian geometry are presented. The results were obtained on the HLRS system Hornet running on up to 54×10^3 computational cores. The strong scaling results show an optimal speedup for a grid with 55 million computational points corresponding to 275 million unknowns.

The Institute of Aerodynamics and Gas Dynamics (IAG) together with the Institute of Space Systems (IRS) of the Stuttgart University have investigated the feasibility of fully three-dimensional coupled PIC-DSMC simulations for the expansion of a plasma in vacuum. Laser ablation is used to remove material from a solid surface with a laser beam. The numerical study of this process has been directed towards direct laser-solid interactions, tackled by molecular dynamics simulations which have been conducted in the past. The project report "Coupled PIC-DSMC Simulations of a Laser-driven Plasma Expansion" from Copplestone, Binder, Mirza, Nizenkov, Ortwein, Pfeiffer, Fasoulas, and Munz describes the results obtained with an MPI-based high-order Particle-in-Cell scheme coupled with a Direct Simulation Monte Carlo method to handle the complex phenomena, which usually are simulated using disjoint techniques. The focus of this work was directed towards the PIC solver and its parallel performance on HPC systems, the influence of deposition methods and load balancing issues. The numerical settings according to preliminary DSMC simulations show a spatial under-resolution for the PIC solver. For this reason, future work will be directed towards setups with lower densities, through higher numbers of simulation particles and lower values for the MPF as well as higher temperatures as reported in the literature.

On Estimation of a Viral Protein Diffusion Constant on the Curved Intracellular ER Surface

M.M. Knodel, A. Nägel, S. Reiter, M. Rupp, A. Vogel, M. Lampe, P. Targett-Adams, E. Herrmann, and G. Wittum

Abstract Advanced simulations within biophysical applications ask for advanced algorithms and implementations which are running efficiently on massively parallel high performance computers. The software framework UG fulfills these preconditions. Therefore, we present insight into the experimental basics, the modelling and simulation details, and the biophysical meaning of the estimation of the diffusion constant of a major player in the replication of the genetic information of the Hepatitis C virus (HCV), namely the NS5A viral protein. NS5A movement is restricted to the surface of the Endoplasmatic Reticulum (ER, a medusa-hair like important cell compartment). Hence, the dynamics of NS5A are described by surface PDEs (sPDE) which mimic experimental FRAP (fluorescence recovery after photobleaching) time series data. The sPDE computations were performed with UG upon large unstructured grids representing realistic reconstructed ER surfaces. We explain the context of the parameter estimations which asked for a substantial amount of single sPDE evaluations which we performed on the HLRS

The authors "E. Herrmann and G. Wittum" contributed equally.

M.M. Knodel
Goethe-Center for Scientific Computing, Frankfurt University, Frankfurt, Germany

Institute of Radiology, University Medical Center Erlangen, Erlangen, Germany
e-mail: mknodel@gcsc.uni-frankfurt.de

A. Nägel • S. Reiter • M. Rupp • A. Vogel • M. Lampe (✉) • G. Wittum
Goethe-Center for Scientific Computing, Frankfurt University, Frankfurt, Germany
e-mail: anaegel@gcsc.uni-frankfurt.de; sreiter@gcsc.uni-frankfurt.de;
rupp@gcsc.uni-frankfurt.de; vogel@gcsc.uni-frankfurt.de; lampe@gcsc.uni-frankfurt.de;
wittum@gcsc.uni-frankfurt.de

P. Targett-Adams
Medivir AB, Department of Biology, Huddinge, Sweden
e-mail: paul.targett-adams@medivir.com

E. Herrmann
Department of Medicine, Goethe University Frankfurt, Frankfurt, Germany
e-mail: herrmann@med.uni-frankfurt.de

© Springer International Publishing Switzerland 2016
W.E. Nagel et al. (eds.), *High Performance Computing in Science and Engineering '15*, DOI 10.1007/978-3-319-24633-8_41

Stuttgart Hermit and Hornet supercomputers for various experimental data sets and for various geometric setups. This enabled us to derive valid final values for the diffusion constant of NS5A on the ER surface. The estimated diffusion constant values are intended to enter spatio-temporal resolved models of HCV replication dynamics at a cellular level.

1 Introduction

We present the computational context of a parameter estimation study in the field of spatio-temporal resolved computational virology [17] which we performed on the HLRS Hermit and Hornet supercomputers. Namely, we describe the preconditions, the partial differential equation (PDE) model, the evaluation methods and the meaning of the derived biophysical meaningful results [17] of the estimation of the diffusion constant of the so-called NS5A protein [1, 22] of the Hepatitis C virus (HCV) [19] replication. In particular, we describe the reasons why these studies consumed a substantial amount of core hours on the HLRS supercomputers.

HCV is the main reason for liver transplantations and causes enormous economic costs for the health insurances which even enhanced caused by recently developed efficient, but expensive new medicine, so called direct antiviral agents (DAA) [18].

NS5A is a major player of the replication of the viral RNA (vRNA) of HCV which anchors to the surface of the Endoplasmatic Reticulum (ER) [7, 25, 27], a medusa-hair like organised intracellular tubular bilipid membrane structure corresponding to various metabolic purposes. As in experiment, our surface PDE (sPDE) simulations restrict the NS5A movement to the ER surface.

Our computations are mimicking so-called experimental FRAP (fluorescence recovery after photobleaching) time series [11, 23] data which describe the dynamics of the NS5A protein. We derived the diffusion constant for a substantial amount of experimental time series cross-combined with various realistic reconstructed ER geometries based on experimental fluorescence z-stacks. The realistic reconstructed ER surfaces based on experimental data [27] were deblurred with Huygens SVI [13, 26, 30] and reconstructed using NeuRA2.3 [12]. The FRAP experiments are modelled with sPDEs which are evaluated with UG [9, 24, 29], a software framework for the numerical solution of problems that are described by systems of partial differential equations based on finite volume methods and massively parallel multigrid solvers. The derivation of the biophysical meaningful diffusion constant of the viral protein asked for a substantial amount of computation time. This reasons on the one hand side on the application of the Gauss-Newton algorithm which causes for each single setup a high number of single sPDE evaluations upon unstructured grids which bear a high number of nodes already at base level (about 10^6). The necessary investigation of the numerical stability of the sPDE computations (submitted by us recently [15]) itself and their application within the parameter adjustment can be considered to cause nearly a multiplicative factor for the core hours. In particular, the combination of the experimental setups (20 time series from [11], 10 geometric setups based on [27]) augments the needed core hours as well. The biophysical meaningful results stimulate the young field of spatio-temporal resolved research within computational virology.

The NS5A diffusion constant parameter estimation already by now allowed for meaningful biophysical results which we submitted recently [17]. The parameter estimation of the dynamics of the NS5A protein is a major element of complete spatio-temporal resolved models of the HCV vRNA replication cycle. Therefore, the main computation time (in the order of a few million core hours) so far was spent on the parameter estimation. Hence, this part will be the major part of our upcoming descriptions.

Nevertheless, we will also give brief insight into some qualitative parameter studies which we did already on our way to develop a complete model of the vRNA cycle [16]. These studies are up to now more qualitative, i.e., they still have to be adapted to the experimental values [14].

2 Experimental Environment and Mathematical Model

When a virus hijacks a body own host cell, its aim is to defraud and abuse the metabolism for creating new virus particles which again will attack other cells. Plus stranded RNA viruses [21] like polio, Hepatitis C (HCV) [19], Dengue or Yellow fever virus use comparably simple mechanisms to replicate themselves: After entering a target host cell, they uncoat their genetic information, the so-called viral RNA (vRNA). The vRNA mimics to be body own messenger RNA so that the ribosomes translate viral polyproteins. These polyproteins will then cleave into nonstructural and structural proteins, NSP and SP. The NSPs create an environment which replicates the vRNA. This environment has characteristic dynamic spatial properties. The SPs will build hulls around the replicated vRNA to form new virus particles which subsequently infect other cells.

As consequence for the host, such procedures cause different degrees of cell stress according to the aggressivity of the virus. While the dangerous polio virus is restricted at the moment to few poor countries, the life-threatening hemorrhagic fever causing Dengue and Yellow fever viruses may enter US and Europe within the next years caused by climatic changes. In comparison, HCV is less aggressive, however, it causes severe liver damages within a time range of about 10–20 years [19]. Therefore, HCV is the main reason for liver transplantations. Nevertheless, HCV is also one of the best investigated targets of virus research, resulting in concise qualitative knowledge, e.g., by quantitative experimental values [14].

HCV replication inside single host cells harbours strong spatial implications and dependencies [19, 25]: The NSPs anchor to the ER surface directly after their cleavage. Their movement is therefore restricted to this curved 2D manifold embedded within the complete 3D space. In part, they form vesicular structures, called membranous web(s), which grow on the ER surface. The membranous webs act as vRNA replication complexes. Within these replication complexes, the vRNA gets copied, leaves the complexes, translates new viral proteins at other ribosomes at the ER surface which again replicate the vRNA. Therefore, during this so-called vRNA replication cycle, the cell gets randomly and heterogeneously filled with (unconnected and in part moving) web regions [32], viral proteins and vRNA (which in part gets caught by SPs to form new virus particles).

A major non-structural protein of HCV is the so-called NS5A viral protein. It is responsible for a bunch of processes related to HCV replication [1, 22, 25, 31].

Computational virology so far was restricted to ODE compartment models like [2, 5, 6, 8] to our best knowledge even though, e.g., recently developed efficient NS5A blocking DAAs show strong implications onto the spatial distribution of NS5A [10, 20, 28].

Indeed, few is known quantitatively about spatial movement properties of any ingredient of HCV replication. A detailed quantitative spatio-temporal resolved, mathematical-biophysical understanding of the underlying processes and dynamics of and between the ingredients of virus replication at a cellular level is still beyond present knowledge by now.

However, there exist since years data which could help to improve this topic: Data are as well available for the geometric setup, namely fluorescence z-stacks with stained ER surfaces [27] and as well, data are available describing the dynamics of the movement of NS5A [11] based on so-called experimental FRAP time series. The stained z-stacks based on markers of the ER surface can be used to build geometric setups for simulations related to NS5A movement. At the beginning of a FRAP experiment [23], the ingredient (NS5A) fluoresces equally all over the cell. However, afterwards, a strong laser deletes the fluorescence in a special region. This region is called the FRAP ROI (region of interest). The observation of the re-filling with the ingredient from the unbleached region around allows for the determination of the mobility of the ingredient. Assuming the ingredient to have diffusive behaviour, the comparison of the experimental intensity to the integrated luminosity respectively concentration of corresponding simulations allows for the estimation of the diffusion constant of the ingredient. Since the measurement process itself is based on a soft laser beam in each time point of measurement, also this influence has to be taken into account within the model when adjusting the diffusion constant of the simulations for matching the experimental values.

Therefore, our aim was to combine the already available data for the geometric setup and the FRAP time series data describing the dynamics of NS5A movement in order to derive a diffusion constant of NS5A on the ER surface. The diffusion constant that recapitulates the best the experimental values of the dynamics of the integral of the luminosity in the FRAP ROI can be assumed to be the diffusion constant of NS5A. The assumption of a pure diffusive process is a first approach, since the knowledge about viral protein dynamics is strongly limited. The optimal parameter was derived using the Gauss-Newton procedure [3].

Since the context of the calculations is the development of a complete spatio-temporal resolved model of the vRNA viral replication cycle of HCV, we also give short insight into the complete model which is under construction and the corresponding simulations [16]. This model also incorporates the replication centres of the vRNA (the membranous "web"s) which grow on top of the ER surface.

2.1 Technical Steps

Our approach to model the HCV vRNA cycle and to estimate the NS5A diffusion constant splits into the following base stones:

- Reconstruction of the ER geometries based on z-stacks from [27]. Based on these data, we reconstructed about 5 ER surfaces [15] using NeuRA2.3 [12]. We assigned to each one of the 5 basic geometric setups 2 exemplary FRAP ROIs. Thus, about 10 geometric basic setups were available.
- Simulation of the NS5A dynamics on the ER surface mimicking FRAP time series [11]. We were able to fit our simulation data to about 10 time series for two different cell types, i.e. all in all, we had about 20 time series for parameter optimisations which were realised by the following steps:
  - Establishing a numerical stable framework of NS5A dynamics on the ER surface within the UG4 software, in particular checks of the refinement stability within the multigrid based calculations [15]. Hence, for testing the numerical stability, different levels of refinement were checked for single sPDE calculations and for a defined subset of sPDE calculations.
  - Using the Gauss-Newton procedure to derive the best diffusion constant for each combination of the 20 time series with each one of the 10 basic geometric setups. Hence, about 200 basic setups asked for parameter estimation.
  - The calculations were done considering various different biophysical scenarios. Therefore, also a reaction term was introduced into the diffusion sPDE to take into account different model scenarios of the experimental reality. For details we refer to [17]. This simulations which served "on the fly" for numerical stability tests caused additional global multiplicative factors.
  - The single results were averaged to get final parameter values for the NS5A diffusion constant for different cell types and different biophysical scenarios.

2.2 Computational Domain and sPDE

The geometric basics of our sPDE calculations are five realistic reconstructed ER geometries. In Fig. 1 we show screenshots of one of these ER geometries. ([17] depicts screenshots of all reconstructed ER geometries.) The geometries harbour about 10^6 nodes at base level [15].

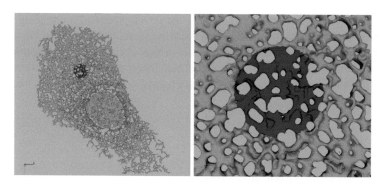

Fig. 1 Geometric basics: One of the reconstructed ER geometries. FRAP ROI subdomain marked in *red*. At the *right hand side*: Magnification. Extracted from [15]

The basic sPDE of the parameter estimations which was evaluated at a very high number at the big ER geometries referring to the FRAP experiments reads:

$$\frac{d}{dt}c_{ns5a}(t, \mathbf{x}) = D_{ns5a}\Delta_{(T)}c_{ns5a}(t, \mathbf{x}) \tag{1}$$

where $\Delta_{(T)}$ represents the Laplace-Beltrami-Operator which is the projection of the Laplace operator to the tangential space of the two-dimensional hypersurface which is embedded into the complete 3D space. $c_{ns5a}(t, \mathbf{x})$ is the NS5A concentration and D_{ns5a} the diffusion constant which we estimated.

The consideration of the intensity reduction caused by the measurement process asked for the introduction of an additional reaction term leading to additional computational resources consumption. We had to perform various test with and without this reaction term. The derivation of exact formulae and numbers in this context is extensively discussed in [17].

We performed comparison calculations for the parameter optimisation on a 2D continuum planar geometry which has at base level simply 37 vertices [15]. Also here, the FRAP ROI has circular structure. The 2D plane calculations were performed to test for the influence of the curved surface manifold on the estimated parameter results.

3 Numerical Simulation

3.1 Basics of the Simulation Environment

We reconstructed about 5 geometries of complete ERs. The surfaces of the ERs harbour about 500,000 up to 1,500,000 nodes. This means that we have—without refinement—about 10^6 degrees of freedom (DoFs) [15]. The meshes are triangular meshes. We performed calculations with one and two refinements, and we also checked with 3 refinements for special cases.

The 2D geometry for comparison calculations (only few nodes at base level) had to be refined up to 8 (two Millions of DoFs) resp. 9 levels [15].

In Table 1, we see the number of the degrees of freedom (DoF) at different levels. The table also shows the number of the triangles (elements/faces) of the geometries. For more details, c.f. [15]. The ER geometries consist out of the FRAP ROI $\mathcal{F}$ and the unbleached region $\mathcal{U}$, i.e., $\mathcal{E} = \mathcal{U} \cup \mathcal{F}$.

Table 1 DoF number and face number of the realistic reconstructed geometries of the Endoplasmatic Reticulum and the 2D planar geometry at level L, extracted from [15]

Geo	L	DoF $\mathcal{E}$	DoF $\mathcal{U}$	DoF $\mathcal{F}$	faces
$\mathcal{E}_1$	0	815,111	794,128	20,983	1,636,803
	1	3,270,924	3,187,566	83,358	6,547,212
	2	13,092,959	12,761,035	331,924	26,188,848
	3	52,378,665	51,054,345	1,324,320	104,755,392
$\mathcal{E}_2$	0	1,212,622	1,174,204	38,418	2,430,181
	1	4,861,079	4,708,431	152,648	9,720,724
	2	19,448,536	18,840,730	607,806	38,882,896
	3	77,785,622	75,360,708	2,424,914	155,531,584
$\mathcal{E}_3$	0	170,209	140,022	30,187	340,108
	1	680,786	560,740	120,046	1,360,432
	2	2,722,264	2,243,664	478,600	5,441,728
	3	10,886,516	8,975,464	1,911,052	21,766,912
$\mathcal{E}_4$	0	601,706	591,336	10,370	1,208,661
	1	2,414,802	2,373,711	41,091	4,834,644
	2	9,666,977	9,503,511	163,466	19,338,576
	3	38,675,259	38,023,311	651,948	77,354,304
$\mathcal{E}_5$	0	728,636	699,338	29,298	1,463,597
	1	2,924,907	2,808,345	116,562	5,854,388
	2	11,708,240	11,243,894	464,346	23,417,552
	3	46,838,070	44,985,132	1,852,938	93,670,208
2D	0	37	24	13	36
	1	133	96	37	144
	2	505	384	121	576
	3	1,969	1,536	433	2,304
	4	7,777	6,144	1,633	9,216
	5	30,913	24,576	6,337	36,864
	6	123,265	98,304	24,961	147,456
	7	492,289	393,216	99,073	589,824
	8	1,967,617	1,572,864	394,753	2,359,296
	9	7,867,393	6,291,456	1,575,937	9,437,184

In the standard case we performed our diffusion equation computations for the parameter optimisation [17] for all geometric setups with a time step width of about 0.1 for a time range of about 120 s. (In all cases, further flexible adaptive refinement was enabled if the Newton solver would not have converged.)

In the case of the complete but so far qualitative model [16], some parameter studies already have been done. This is due to our aim of making the models output more realistic. In the future, our model system will be extended to the big and complete ER geometries. The geometry which we use for the (so far not so much time consuming) heuristic 3D surface model is based on about 6000 vertices, we refine one and two times for basic calculations and more often for test calculations. In this case, we are calculating up to 10 different concentrations which are coupled with nonlinear diffusion-reaction equations. Hence we have about 60,000 DoFs at base level also in this case.

3.2   Discretisation and Solver Properties

The basics of the standard setup, the discretisation of the sPDEs and the solving with the parallel multigrid solvers, all done within UG4, was organised as follows [15]:

- Standard setup: onefold spatial refinement, DoFs: 2–5 Millions
- Time discretisation: Implicit Euler
- Spatial discretisation: vertex-centered Finite Volumes
- Nonlinear Newton Solver
- Linear Solver of Newton Solver: BiCGStab
- Preconditioner of Linear Solver: GMG (geometric multigrid)
- Pre- and post-smoother of GMG: SGS (symmetric Gauss-Seidel)
- Base Solver of GMG: BiCGStab
- Preconditioner Base Solver: ILU
- sPDEs calculated with excellent UG4 (scalability, usability)

3.3   Overview of the Basic Fit Setups

Let us summarise all calculations we performed for estimating the diffusion constant of NS5A on the ER surface [17]:

Roughly speaking we have *200 basic parameter sets* for fixed level refinement to be evaluated as product of ERs, FRAP ROIs and experimental time series. For each setup, the Gauss-Newton procedure has to call various times the basic sPDE calculations. Further global multiplicative factors arise from different biophysical interpretations of the experiments, in particular referring to the intensity reduction caused by the measurement process (details c.f. [17]).

3.4   Single Work Steps/Contributions and Combination

In the forthcoming, we estimate the core hours we applied for the parameter optimisation. It is a very rough estimation, but we depict in principle all steps we had to do and which lead to the amount of core hours used.

Each of the following steps will contribute with a special coefficient. At the end, a formula and a table will be given for the sake of combining all the forthcoming numbers.

- Basic sPDE calculations: The average run times for each one of the 5 realistic reconstructed ER geometries for a single run on a single geometry for simulating about 120 s of experiment (with a basic time step size of 0.1 s and for one spatial refinement) are about 70 Core hours.[1]
  70 Core hours per single run at refinement level one ($r = 1$, $\delta_t = 0.1$):
  $$T_b|_{\delta_t=0.1}^{r=1} = 70\,h$$
- Spatially refined sPDE calculations: The meshes are triangular surface meshes. We performed calculations with one and two refinements. (We checked also with 3 refinements for special cases.) The calculation times for spatial refinement level 2 are of about a factor 4 enhanced in comparison to level one for the basic sPDE calculations, because triangular grid refinement causes about a factor 4 in DoF number enhancement, i.e., enhances the needed core time analogously.
  280 Core hours per single run at refinement level two ($r = 2$, $\delta_t = 0.1$):
  $$T_b|_{\delta_t=0.1}^{r=2} = 280\,h$$
- As stated above, we reconstructed 5 geometries of complete ERs [27].
  5 reconstructed geometries.
  $$G = 5$$
- Into these geometries, we have to put the region which is bleached (FRAP ROI) in comparison to the rest of the ER—where afterwards the viral protein diffuses inside. (The latter process is measured experimentally to be compared to our simulation data.) We constructed for each geometry two different FRAP regions [17].
  2 different localisations of FRAP ROIs at level 1, only one at level 2 and higher
  $$F_{ref=1} = 2, F_{ref=2} = 1$$
- Our experimental partners preserved 20 experimental time series [11] which describe the re-filling of a region which was made "empty" of (stained) NS5A by a strong laser beam based on the FRAP method [23]. We have to fit our model equations to each one the experimental results of these 20 time series.
  20 experimental time series.
  $$t_s = 20$$
- Different models for interpretation of experimental basics referring to the measurement process, c.f. [17]:

---
[1]some geometries need more, some less, caused by the different node number of the reconstructed geometries.

– Factor 4 for the case of level 1 refinement (standard setup time and space refinement).
– Factor 1 for level 2 and higher

$M_1 = 4, M_2 = 1$

• *Gauss-Newton-Procedure:*
The Gauss-Newton procedure asks usually for about 5–10 Newton steps, each Newton step asks (for a single parameter) for a base calculation and two derivative calculations as well as 4 line search calculations.

– In average, we have to do about 7 Newton steps for the case of one parameter to be estimated.
$N = 7$
7 Newton steps
– One parameter to be estimated asks for about 7 single evaluations per Newton step: 1 Base, 2 Derivatives for one parameter estimation, 4 line searches
$C_{para=1} = 7$
7 calculations one parameter evaluations Base, Deriv, Meriv, 4 Line search (*para* = parameter number)

We have to do 7 Newton steps of which each one asks for about 7 calculations for one parameter. This causes about 49 PDE calculations per ER with fixed FRAP region and per time series. These values are rough averages.
49 Gauss Newton single runs for one parameter to be estimated

• While we performed extended calculations for our standard setups time step size $\delta_t = 0.1$s and spatial refinement level 1 and also for level 2, and we also had to do test calculations for smaller time steps and spatial refinement 3. This is due to the fact that even though we could only see small deviations in the region of few per mill when refining the time step size more, we have to do test parameter optimisations also for more fine time step size. Similar aspects hold true for higher spatial refinements. As well, we performed evaluations for refinement level 3 and normal time step size, however for only one FRAP ROI, i.e., for only one geometric setup.

$t_s|_{\delta_t=0.01} = 1, F|_{ref=1,\delta_t=0.01} = 2$
$t_s|_{refs=3} = 1, F|_{ref=3,\delta_t=0.1} = 1$

The evaluation times for the basic sPDE is about 110 Core hours for refinement level 1, time step size 0.01. For the case of refinement level 3 (time step size 0.1 s), we used about 1120 h.
$T_b|_{\delta_t=0.01}^{r=1} = 110\,h$
$T_b|_{\delta_t=0.1}^{r=3} = 1120\,h$

We summarise the multiplication procedure in Table 2 which takes into account all tasks denoted.

Table 2 Multiplicative factors of Core hour consummation of parameter estimation calculations for complete results production

Refinements	δ_t	N	C_p	T_b	t_s	g	F	M	Sum
1	0.1	7	7	70	20	5	2	4	$2.74400 \cdot 10^6$
2	0.1	7	7	280	20	5	1	1	$1.37200 \cdot 10^6$
3	0.1	1	1	1120	1	5	1	1	$0.05600 \cdot 10^6$
1	0.01	1	1	110	1	5	1	1	$0.00550 \cdot 10^6$
1	0.001	1	1	160	1	1	1	1	$0.00016 \cdot 10^6$
									$4.12231 \cdot 10^6$

Written as formula, we get for the complete time S, with refinements r (1,2,3) and the three modes of δ (1: 0.1 s, 2: 0.01 s and 3: 0.001).

$$S = \sum_{r=1}^{3} \sum_{\delta=1}^{2} \sum_{p=1}^{2} N \cdot C|_{r,\delta} \cdot T_b|_{r,\delta} \cdot t_s|_{r,\delta} \cdot g \cdot F|_{r,\delta} \cdot M|_{r,\delta} \tag{2}$$

The easiest way to evaluate the formula is via Table 2.

Hence, about four Millions of core hours were consumed for the estimation of the diffusion constant of the NS5A protein on the ER surface [17]. About another 300.000 core hours were consumed for comparison calculations on a 2D planar continuum geometry. We omit a detailed derivation.

Further about 100.000 core hours were consumed for the qualitative complete model parameter studies.

Hence, all in all, the HCV project so far consumed about 4.5 Millions of core hours on the HLRS hermit and hornet supercomputers. Anyhow, already by now biophysical meaningful results were derived, namely the size of the diffusion constant of the NS5A protein on the ER surface [17].

Further amount will be needed in future calculations for parameter estimations referring to other ingredients of the vRNA replication cycle of HCV for making the model completely realistic, establishing a broad and reliable framework for enabling substantial input into experimental HCV research, ready for extension to other viruses.

4 Results

4.1 NS5A Diffusion Constant Estimation on the ER Surface

We give some insight into the methods of the derivation of the results concerning the parameter fit study of the NS5A diffusion constant estimation on the ER surface.

4.1.1   Refinement Studies of Single sPDE Evaluations

We compare the integrated FRAP ROI intensities

$$\mathcal{I}_{L,\delta_t}(t) = \int_{\mathcal{F}} c_{ns5a}(t, \mathbf{x}) d\mathbf{x} \qquad (3)$$

where L is the refinement level in space and δ_t the time step size. c_{na5a} is evaluated with (1) (for simplicity, we did not denote L and δ_t on the right hand side of (3)). The presented results are the relative differences of the form

$$\mathfrak{R}_{L,\bar{L},\delta_t,\bar{\delta}_t}(t) = \left| \frac{\mathcal{I}_{L,\delta_t^a}(t) - \mathcal{I}_{\bar{L},\delta_t^b}(t)}{\breve{a}\mathcal{I}_{L,\delta_t^a}(t)} \right| \qquad (4)$$

where $L, \bar{L}$ are different refinement levels in space and $\delta_t, \bar{\delta}_t$ the time step sizes.

   We present one of our refinement studies of a single sPDE computations from [15], namely for spatial refinement from 1 to 3 levels, Fig. 2. The diffusion constants used for this studies were chosen heuristically and are not explicitly estimated values [15].

   For extended details of the refinement studies in space and time, we refer to [15].

4.1.2   Simulation of a FRAP Experiment

A screenshot of a typical simulation scenario of the NS5A dynamics on the ER surface (ER geometry V) is shown in Fig. 3. Corresponding movies are attached as supplemental to [17]. Examples of the FRAP ROI integral intensity evolution of the simulations in comparison to the experimental data from [11] are shown in [17].

Fig. 2 Comparing (integrated) FRAP ROI concentrations (relative differences $\mathfrak{R}_{L,\bar{L},\delta_t,\delta_t}$ for different levels $L, \bar{L}$) under variation of spatial refinement of geometry for one of the 5 ER geometries. The time step size is $\delta_t = 0.1 s$. Extracted from [15]

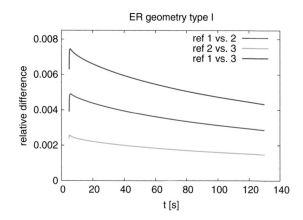

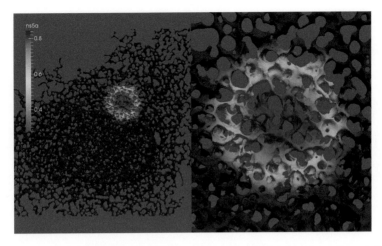

Fig. 3 Concentration of the diffusive moving viral protein restricted to the surface of the ER. Screenshot of the spatially resolved concentrations on ER geometry V

4.1.3 Averaged Results and Refinement Stability

The final value of the diffusion constant was averaged over the basic setups geometry—time series. Since we have 2 cell types with respective 10 time series [17], we had to average over respectively two times 100 basic setups.

We checked for refinement stability corresponding to the averaged values of the diffusion constant in relation to the refinement level. Whereas for the big ER geometries, the refinement from level 1 to 2 caused only a minor change already, the 2D geometry calculations asked for more refinements, caused of course by the small coarse grid.

Caused by the big experimental errors, level 1 computations were sufficient precise for the ER surface case and level 8 for the planar 2D case [17].

The entire final and averaged values concerning the estimation of the biophysically meaningful NS5A diffusion constant are presented in [17].

4.2 Qualitative Model of vRNA Cycle

In Fig. 4 we present a simulation movie screenshot of a qualitative sPDE diffusion-reaction model of some of the most important ingredients of the vRNA cycle. Figure 5 shows the time evolution of the integral of the concentration of the ingredients within the subdomain of a web region.

Details of the qualitative model will be published soon [16].

The future aim would be to incorporate realistic parameters within such a model for all ingredients which is a highly non-trivial task.

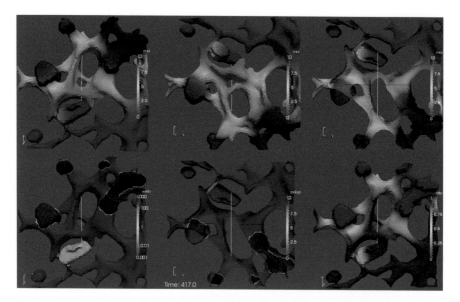

Fig. 4 Screenshot of an evaluation of the concentrations of the components of the viral RNA replication cycle in the sPDE model on the ER surface incorporating the replication regions and ribosomes

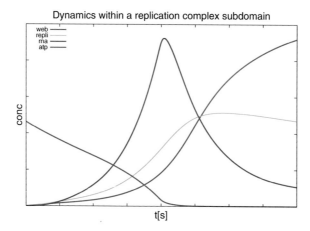

Fig. 5 Evaluation of the concentrations of the components of the viral RNA replication cycle within a replication complex subdomain

5   Discussion and Conclusions

We obtained biophysical meaningful results for the diffusion constant of the Hepatitis C NS5A protein on the Endoplasmatic Reticulum curved surface [17]. This parameter estimation builds a way to spatio-temporal resolved modelling and simulation within virology research. These results are intended to enter extended models of the vRNA replication cycle [16].

The afore described methods could be applied within future projects to other viruses which belong like HCV also to the same family of so-called plus stranded RNA viruses [21], e.g., to the Dengue virus [4].

The combination of the excellent scalable and applicable Finite Volume/multigrid framework UG4 with the HLRS supercomputers allows for such computations as presented within this study. Similar progress would not have been possible a few years ago—nor the simulation software framework nor adequate supercomputers were available.

Our modelling and simulation framework UG combined with the HLRS supercomputers could give impact onto experimental virological research in the middle run and to antiviral drug development in the long run.

Acknowledgements This work has been supported by the Goethe-Universität Frankfurt. We thank K. Xylouris (G-CSC) for very fruitful discussions on the evaluation of the simulation results, R. Dutta-Roy (Karolinska Institute, Stockholm, Sweden) for profound explanations of FRAP experimental setup and data analysis and Wouter van Beerendonk (Huygens SVI, Netherlands) for his very friendly support in Huygens usage, backgrounds, and licensing. The HLRS Stuttgart is acknowledged for the supplied computing time on the Hermit and Hornet supercomputers. John McLauchlan, Glasgow University, is acknowledged for providing the FRAP time series data [11] as experimental basis of the parameter estimation procedures.

References

1. Belda, O., Targett-Adams, P.: Small molecule inhibitors of the hepatitis c virus-encoded ns5a protein. Virus Res. **170**(1–2), 1–14 (2012). Review
2. Binder, M., Sulaimanov, N., Clausznitzer, D., Schulze, M., Hüber, C.M., Lenz, S.M., Schlöder, J.P., Trippler, M., Bartenschlager, R., Lohmann, V., Kaderali, L.: Replication vesicles are load- and choke-points in the hepatitis c virus lifecycle. PLoS Pathog. **9**(8), e1003561 (2013)
3. Björck, A.: Numerical Methods for Least Squares Problems. SIAM, Philadelphia (1996)
4. Chatel-Chaix, L., Bartenschlager, R.: Dengue virus and hepatitis c virus-induced replication and assembly compartments: The enemy inside - caught in the web. J. Virol. **88**(11), 5907–5911 (2014)
5. Dahari, H., Ribeiro, R.M., Rice, C.M., Perelson, A.S.: Mathematical modeling of subgenomic hepatitis c virus replication in huh-7 cells. J. Virol. **81**(2), 750–760 (2007)
6. Dahari, H., Sainz Jr. B., Perelson, A.S., Uprichard, S.L.: Modeling subgenomic hepatitis c virus rna kinetics during treatment with alpha interferon. J. Virol. **83**(13), 6383–6390 (2009)
7. Friedman, J.R., Voeltz, G.K.: The er in 3d: a multifunctional dynamic membrane network. Trends Cell Biol. **21**(12), 709 (2011). Review
8. Guedj, J., Rong, L., Dahari, H., Perelson, A.S.: A perspective on modelling hepatitis c virus infection. J. Viral Hepat. **17**(12), 825–833 (2010). Review
9. Heppner, I., Lampe, M., Nägel, A., Reiter, S., Rupp, M., Vogel, A., Wittum, G.: Software framework ug4: parallel multigrid on the hermit supercomputer. In: Wolfgang E. Nagel, Dietmar H. Kröner, Michael M. Resch (eds.) High Performance Computing in Science and Engineering '12. Springer, Berlin (2013)

10. Idea, I., Zhanga, L., Chena, M., Inchauspeb, G., Bahlc, C., Sasagurid, Y., Padmanabhana, R.: Characterization of the nuclear localization signal and subcellular distribution of hepatitis c virus nonstructural protein ns5a. Gene **182**(1–2), 203–211 (1996)

11. Jones, D.M., Gretton, S.N., McLauchlan, J., Targett-Adams, P.: Mobility analysis of an ns5a-gfp fusion protein in cells actively replicating hepatitis c virus subgenomic rna. J. Gen. Virol. **88**(2), 470–475 (2007)

12. Jungblut, D., Queisser, G., Wittum, G: Inertia based filtering of high resolution images using a gpu cluster. Comput. Vis. Sci. **14**, 181–186 (2011)

13. Kano, H, van der Voort, H.T.M., Schrader, M., van Kempen, G.M.P., Hell S.W.: Avalanche photodiode detection with object scanning and image restoration provides 2-4 fold resolution increase in two-photon fluorescence microscopy. BioImaging **4**, 87–197 (1996)

14. Keum, S.J., Park, S.M., Park, J.H., Jung, J.H., Shin, E.J., Jang, S.K.: The specific infectivity of hepatitis c virus changes through its life cycle. Virology **433**, 462–470 (2012)

15. Knodel, M.M., Nägel, A., Reiter, S., Rupp, M., Vogel, A., Targett-Adams, P., Herrmann, E., Wittum, G.: Multigrid analysis of spatially resolved hepatitis c virus protein properties. CVS - Proceedings of the European Multigrid Conference (2014, submitted to)

16. Knodel, M.M., Reiter, S., Rupp, M., Vogel, A., Targett-Adams, P., Herrmann, E., Wittum, G.: Mechanistic dynamics of hepatitis c virus replication in single liver cells: Simple full 3d (surface) pde models (2015, in preparation)

17. Knodel, M.M., Nägel, A., Reiter, S., Rupp, M., Vogel, A., Targett-Adams, P., McLauchlan, J., Herrmann, E., Wittum, G.: Quantitative analysis of hepatitis c ns5a viral protein dynamics on the er surface (2015, submitted)

18. Kohli, A., Shaffer, A., Sherman, A., Kottilil, S.: Treatment of hepatitis c: a systematic review. J. Am. Med. Assoc. **312**(6), 631–640 (2014)

19. Moradpour, D., Penin, F., Rice, C.M.: Replication of hepatitis c virus. Nat. Rev. Microbiol. **5**, 453–463 (2007)

20. Mottola, G., Cardinali, G., Ceccacci, A., Trozzi, C., Bartholomew, L., Torrisi, R., Pedrazzini, E., Bonatti, S., Migliaccio, G.: Hepatitis c virus nonstructural proteins are localized in a modified endoplasmic reticulum of cells expressing viral subgenomic replicons. Virology **293**(1), 31–43 (2002)

21. Paul, D., Bartenschlager, R.: Architecture and biogenesis of plus- strand rna virus replication factories. Wo. J. Virol. **2**(2), 1-000 (2013)

22. Reiss, S., Rebhan, I., Backes, P., Romero-Brey, I., Erfle, H., Matula, P., Kaderali, L., Pönisch, M., Blankenburg, H., Hiet, M.S., Longerich, T., Diehl, S., Ramirez, F., Balla, T., Rohr, K., Kaul, A., Bühler, S., Pepperkok, R., Lengauer, T., Albrecht, M., Eils, R., Schirmacher, P., Lohmann, V., Bartenschlager, R.: Recruitment and activation of a lipid kinase by hepatitis c virus ns5a is essential for integrity of the membranous replication compartment. Cell Host Microbe **9**(1), 32–45 (2011)

23. Reits, E.A.J., Neefjes, J.J.: From fixed to frap: measuring protein mobility and activity in living cells. Nat. Cell Biol. **3**, E145 - E147 (2001)

24. Reiter, S., Vogel, A., Heppner, I., Rupp, M., Wittum, G.: A massively parallel geometric multigrid solver on hierarchically distributed grids. Comput. Vis. Sci. **16**(4), 151–164 (2013)

25. Romero-Brey, I., Merz, A., Chiramel, A., Lee, J.Y., Chlanda, P., Haselman, U., Santarella-Mellwig, R., Habermann, A., Hoppe, S., Kallis, S., Walther, P., Antony, C., Krijnse-Locker, J., Bartenschlager, R.: Three-dimensional architecture and biogenesis of membrane structures associated with hepatitis c virus replication. PLoS Pathog. **8**(12), e1003056 (2012)

26. Scientific Volume Imaging B.V., Hilversum, Netherlands. Huygens compute engine. Software (2014). http://www.svi.nl/HuygensSoftware

27. Targett-Adams, P., Boulant, S., McLauchlan, J.: Visualization of double-stranded rna in cells supporting hepatitis c virus rna replication. J Virol. **82**(5), 2182–2195 (2008)

28. Targett-Adams, P., Graham, E.J., Middleton, J., Palmer, A., Shaw, S.M., Lavender, H., Brain, P., Tran, T.D., Jones, L.H., Wakenhut, F., Stammen, B., Pryde, D., Pickford, C., Westby, M.: Small molecules targeting hepatitis c virus-encoded ns5a cause subcellular redistribution of their target: insights into compound modes of action. J. Virol. **85**(13), 6353–6368 (2011)

29. Vogel, A., Reiter, S., Rupp, M., Nägel, A., Wittum, G.: Ug 4: Comput. Vis. Sci. **16**(4), 165–179 (2013)
30. van der Voort, H.T.M., Brakenhoff, G.J.: 3-d image formation in high-aperture fluorescence confocal microscopy: a numerical analysis. J. Microsc. **158**(1), 43–54 (1990)
31. Wilby, K.J., Partovi, N., Ford, J.A., Greanya, E., Yoshida, E.M.: Review of boceprevir and telaprevir for the treatment of chronic hepatitis c. Can. J. Gastroenterol. **26**(4), 205–210 (2012)
32. Wölk, B., Büchele, B., Moradpour, D., Rice, C.M.: A dynamic view of hepatitis c virus replication complexes. J. Virol. **82**(21), 10519–10531 (2008)

Application of Large-Scale Phase-Field Simulations in the Context of High-Performance Computing

Johannes Hötzer, Marcus Jainta, Marouen Ben Said, Philipp Steinmetz, Marco Berghoff, and Britta Nestler

Abstract In material science, simulations became a common tool for the understanding of the underlying behaviour of different classes of materials. Due to the growing complexity of problems at hand, the simulation domains, and therefore the computational effort is steadily increasing. We presents various application of the phase-field method; ranging from the solidification of ternary eutectics and pure ice systems to the interaction of multiple liquid phases on fibers. All these topics have in common, that they need a large number of cores to investigate the decisive physical effects in adequate time. We show an overview of the results for this wide range of applications and the scaling behaviour of the used software frameworks.

1 Introduction

In the recent years, numerical simulations has been established in engineering science and became a powerful tool beside experiments. This is driven by the decrease of computational cost and developments of high performance computing (HPC). Especially in the field of material science, simulations gain more importance allowing a new insight in microstructure evolution [19, 34]. Increasing the knowledge of microstructure properties is essential for designing new materials, that satisfy the novel and specific industrial requirements. A well established method for multiphysic simulations is the phase-field method (PFM). The main advantage of PFM is the solving of free boundary problems without interface tracking, remeshing and complicated boundary conditions at phase boundaries. Moreover, it allows the

J. Hötzer (✉) • M. Jainta • M. Ben Said • P. Steinmetz • M. Berghoff • B. Nestler
Institute of Applied Materials, Reliability of Components and Systems (IAM-ZBS), Karlsruhe

Institute of Technology (KIT), Haid-und-Neu-Str. 7, 76131 Karlsruhe, Germany

Institute of Materials and Processes, Hochschule Karlsruhe Technik und Wirtschaft, Moltkestr. 30, Karlsruhe, Germany
e-mail: johannes.hoetzer@kit.edu; marcus.jainta@kit.edu; marouen.bensaid@kit.edu; philipp.steinmetz@kit.edu; marco.berghoff@kit.edu; britta.nestler@kit.edu

© Springer International Publishing Switzerland 2016
W.E. Nagel et al. (eds.), *High Performance Computing in Science and Engineering '15*, DOI 10.1007/978-3-319-24633-8_42

consideration of multi-phase systems and coupling of several physical fields, such as temperature, concentration, velocity, elasticity and plasticity [30].

In this work, we present applications of the phase-field method for different physical problems, requiring the usage of HPC. The following simulation studies are conducted on the HLRS clusters hermit and hornet. The underlying model for these applications is described in Sect. 2. In Sect. 3 we describe the directional solidification of ternary eutectic systems. The computations allow a detailed analysis of the pattern formation during solidification of alloys. The detailed physics of this work is published in [21]. The scaling performance of large-scale 3D microstructure simulations is exemplary shown in Sect. 4. In the next section (Sect. 5), the dendritic solidification of pure ice particles is shown. Finally, Sect. 6 presents simulations of droplets on fibers. Due to the required large domains and long simulation times for all those applications, these results can only be obtained by using highly parallelized code and HPC facilities.

2   Phase-Field Method

In the phase-field method, the spatio-temporal evolution of an interface is described by a smooth field for each phase (or order parameter) $\phi(x, t)$. It adopts constant values, e.g. $\phi(x, t) = 1$ or $\phi(x, t) = 0$, in the bulk phase regions and establishes a diffuse interface in the transition region. The position of the mathematically sharp interface is located at the level set $\phi(x, t) = 1/2$.

The coupling of the dynamic evolution of the phase-fields to the transport of heat or solute across the interface is implicitly given by defining local free energy or entropy densities, which depend on the phase state ϕ as well as the temperature T and composition.

A set of N different phase-fields, denoted as $\boldsymbol{\phi} = (\phi_1(x, t), \ldots, \phi_N(x, t))^T$ is introduced, where each vector component represents the volume fraction of the occurring phase. The method uses a diffuse interface which thickness depends on a parameter $\epsilon \geq 0$.

Starting from a Lyapunov functional, the evolution equations for the phase-fields are calculated via variational derivation according to an Allen-Cahn type approach. These equations describe the minimization of the free energies respectively grand-potentials or the maximization of the entropy in the system. Depending on the problem at hand, different paths are given in the related sections.

3   Ternary Eutectic Directional Solidification

The microstructure of metals strongly influences the mechanical properties of components and is affected by the process parameters. In order to directly control them, an understanding of the underlying physical processes is crucial. In simulations, the changes of physical as well as process parameters can be easily conducted, compared with experiments. Due to the widely adjustable microstructures in ternary eutectics, they are often used to study the physical behavior during directional solidification of alloys [9, 10, 14].

In a ternary eutectic system, the melt, consisting of three chemical species, transforms into three solid, mostly intermetallic, phases at a defined concentration and temperature. In our study, we describe the evolution of four different phases starting with a grand-chemical functional

$$\Psi(\boldsymbol{\phi}, \boldsymbol{\mu}, T) = \int_{\Omega} \left(\varepsilon a(\boldsymbol{\phi}, \nabla\boldsymbol{\phi}) + \frac{1}{\varepsilon}\omega(\boldsymbol{\phi}) \right) + \psi(\boldsymbol{\phi}, \boldsymbol{\mu}, T) d\Omega \qquad (1)$$

described in detail in a recent article [21].

The solidification rate of the growing phases and the direction are controlled by an analytical temperature gradient. Starting from a randomized Voronoi filling, the microstructure solidifies in one direction and meanwhile different patterns parallel to the growth front evolve.

Ruggerio et al. [33] predict five idealized phase arrangements from geometrical considerations and Lewis et al. [23] depict them graphically. In Fig. 1 these idealized patterns are shown. The phases arrange in three lamellae (Fig. 1a), one fibrous phases and two lamellae (Fig. 1b), two fibrous phases and one lamella (Fig. 1c), two fibers embedded in a matrix (Fig. 1d) as well as three fibers (Fig. 1e).

By performing systematical parameter variations of the concentrations in the liquid and solid phases or the surface energies, similar patterns can also be reproduced in the simulations. Four of these five patterns are found in sections parallel to the growth front. These results are discussed in [20].

In real ternary eutectic alloys the occurring patterns are much more complicated, but can be attributed to this basic patterns [9, 10, 14]. Systematical variation of the base sizes parallel to the growth front has shown a strong influence of the boundaries on the occurring patterns. The characteristic quantities of the microstructure can be analyzed with principal component analysis (PCA) of two-point correlations. The quantitative analysis yields the conclusion that large domains are required to achieve physical results with a statistical reliability and illustrate the to necessity of using

(a)   (b)   (c)   (d)   (e)

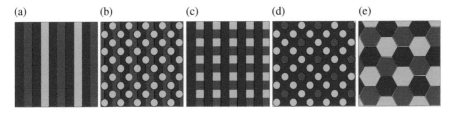

Fig. 1 Idealized patterns in ternary eutectic directional solidification parallel to the growth front, following [23]. (**a**) Three lamellae, (**b**) one fiber phases and two lamellae, (**c**) two fibers and one lamella, (**d**) two fibers embedded in a matrix, (**e**) three fibers

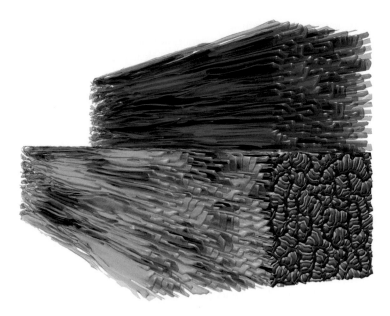

Fig. 2 Simulation result of the ternary eutectic system Ag-Al-Cu with a size of 1200×1200×6350 cells. The phases forming lamellar brick-like structures are exempted *left* and *above*

high performance computing and supercomputers to resolve representative volume elements (RVE).

In Fig. 2 a simulation result of the ternary eutectic system Ag-Al-Cu with a size of 1200 × 1200 × 6350 cells is depicted. The three solid phases grow from rear left to right front, driven by an analytical frozen temperature gradient. Different chained brick-like structures embedded in a matrix form parallel to the growth front. The simulations are in good accordance with experimental micrographs. A further comparison of representative properties of the formations is presented in [21]. The lamellae, building the brick-like structures, are exempted left and above. While growing, several split and merge events of the solid phases occur.

Massive parallel 3D simulations provide a detailed view into the three-dimensional structure while growing. Obtaining the same information from experiments requires a large technical effort, using synchrotron tomography to resolve the different phases [32].

Furthermore, simulations provide the advantage to precisely conduct parameter changes and process conditions, under well-defined conditions. Microstructure properties are determined depending on the concentrations, the velocity of temperature gradient or physical parameters, like the diffusion constants or the surface energies.

4 Scaling

To study ternary eutectic directional solidification [20, 21] highly optimized kernels and massive parallel codes are necessary to simulate realistic domain sizes. Therefore we optimized the phase-field model [7, 21] to the target hardware using vectorization and buffering techniques as well as the massive parallel WALBERLA framework [13, 17, 22].

The WALBERLA framework is written in C++ and uses domain decomposition techniques for the parallelization on current supercomputers such as the Hermit at the HLRS. The simulation domain is therefore decomposed in equal sized blocks which are distributed to the cores. The message passing interface is used for the ghost layer exchange. Also a static load-balancing based on blocks is implemented for the phase-field module, due to the more compute intensive diffuse interface, which is located at a defined region for the directional ternary eutectic solidification simulations. The computing domain is decreased using a moving window technique in the growing direction [21, 35]. To reduce the resulting output, it is possible to write the surface mesh of the phases, instead of a voxel output.

The model is solved using two sweeps, the phase-field itself and the chemical potential sweep. Each sweeps iterates over the domain and calculates the next time step. The temperature is described by an analytic function. The sweeps are optimized by using buffering techniques for staggered values such as gradients. Further the code was vectorized using compiler intrinsics to fully exploit the SSE and AVX registers of current CPUs. Figure 3 shows the time in seconds per 100 iterations for three different block sizes per core and the used optimizations on the HLRS system Hermit. We used for each run all cores in the node to ensure the environment of a production run. To make use of the AMD Turbo Core technology

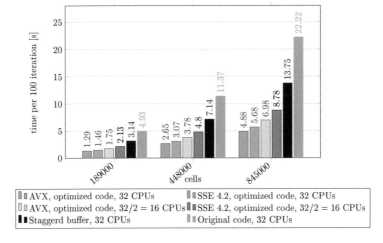

Fig. 3 Comparison of the runtime per iteration for different optimized code versions and domain sizes on the HLRS system Hermit

and to avoid sharing of the floating point unit (FPU), we disabled the half of the cores within a node and therefore doubled number of blocks to compare the time for the same total domain size. The three used block size are 15^3, 30^3 and 25^3 cells. For the runs we used four optimized versions of the code which are based on each other. The standard code without optimizations, a buffer optimized version to store gradients and staggered values which are used twice and a SSE as well as a AVX version with explicit use of intrinsics and restructured code are compared.

Using buffering techniques nearly half the required calculation time by using more memory to store the later required values. The explicit vectorization further decreases the calculation time more than twice in this case. The difference between SSE and AVX is less than expected, using all cores or only the half. This nearly similar values for SSE and AVX is the research of further investigations. One possibility might be the mixed usage of SSE and AVX intrinsics, and the not fully vectorized code which can lead to performance drawbacks. Also the shared FPU could lead to this drawback in the case of AVX, but the results using only the half of the cores show no evidence for this theory. The usage of half of the cores by exploiting the AMD Turbo Core technology does not lead to an improve of the total calculation time compared to the usage of all cores. However, the results show that the Turbo and the non shared FPU leads to an improved computation time instead of doubling it, by using only half of the cores.

Based on the performance results of one node, we studied the efficiency for weak scaling with up to 32,768 cores on the Hermit supercomputer. Therefore we compared the efficiency for different block sizes per core for the AVX and SEE vectorization. For the measurements block sizes of 15^3, 20^3 and 25^3 were used. Figure 4 shows the efficiency starting from 8 cores up to 32,768. For all runs, the

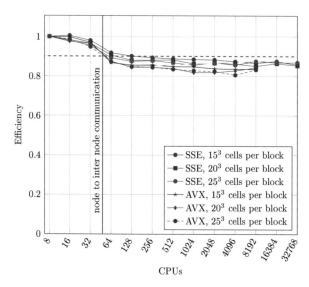

Fig. 4 Efficiency of a phase-field code for different block sizes and vectorization with up to 32,768 cores on the HLRS system Hermit

efficiency is higher than 80 %. After distributing the job over a single node of 32 cores, the efficiency drops by 9 % to 17 % due to the use of the slower CRAY Gemini interconnect compared to node internal network. Afterwards, the efficiency remains nearly constant. The efficiency of the 15^3 SEE version is the best. For all domain sizes, the efficiency is in the same range. It should be noted, that the efficiency of the runs of the SSE versions are better than for the AVX ones. Also the smaller domain size show a better efficiency in the SSE and AVX case. This can result from load imbalances due to the used optimizations, which depend on the domain sizes. This load imbalances will be the study of future work.

Using buffering and vectorization techniques we could decrease the computation time by 75 % compared to the standard code. Also the efficiency shows good results with over 80 % for up to 32,000 cores.

5 Simulation of Pure Ice

Water is a common substance in the surrounding nature. The phase transitions between the liquid and the solid form, the crystalline material ice, is the most commonly experienced one for mankind. In the scientific community, different fields have contributed to describe the behaviour of water during the freezing process at the microscale. The focus in materials science and process engineering for many applications, lie on the detailed description of the crystallisation and melting process. All applications have in common, that it is first necessary to understand the correct growth behaviour of pure ice. The freezing of ice exhibits many classical features of crystal growth and solidification from the melt like facetting, the occurrence of a roughening transition and dendritic growth. A lot of work is done in this field by M. Maruyama and co-workers. They use hydrostatic pressure to effectively control the melting temperature fast and in the whole setup simultaneously [26]. They conclude, that the equilibrium shape of pure ice depends on the pressure and exhibits prism faces (fully facetted at high or rounded at low pressure). The morphology of the growth shape can be described as flat hexagonal dendrites ranging from seaweed to needle-like growth. To capture the whole range of growth forms of ice, the properties of the water–ice interface depend on pressure p and temperature T. In a first attempt to describe the growth behaviour in the pure water-ice system with undercoolings of up to 6 K, these values are assumed to be constant.

The general phase-field model described in Sect. 2, has been applied previously to dendritic growth of pure Ni [2, 31]. In the following, the formulation is adapted towards the growth of crystal ice.

To ensure thermodynamic consistency, the model for crystal growth is coupled to temperature evolution and relies on the formulation of an entropy density functional

$$\mathscr{S}(\boldsymbol{\phi}, e) = \int_\Omega \left(s(e, \boldsymbol{\phi}) - \left(\varepsilon a(\boldsymbol{\phi}, \nabla \boldsymbol{\phi}) + \frac{1}{\varepsilon} w(\boldsymbol{\phi}) \right) \right) dx. \tag{2}$$

The growth shape of a crystal in equilibrium with its liquid state results from the co-occurrence of solid–liquid interface energy anisotropy and kinetic anisotropy. For the purpose of modelling crystalline anisotropies, that depend only on the interface normal vector, we use real-valued spherical harmonics, that are fitted to available data [18].

5.1 Modelling of the Far Field

To take the effect of a long-range temperature field into account without increasing extraordinarily the size of the simulation domain, a new boundary condition is applied. The temperature in a boundary cell at time $t + \Delta t$ is calculated assuming that the overall temperature profile decays normal to the boundary like a Gaussian function. For the left boundary in x direction this reads

$$T = T_t^l(x) = T_\infty + A \cdot \exp(a\,(x-c)^2). \tag{3}$$

T_∞ is the given temperature far away of the ice–liquid interface in the undercooled liquid. The three unknown parameters A, a and c are calculated at the time t by solving the system of Eq. (3) for the three temperature values in the three neighbouring cells closest to the boundary, which results after some forward calculations:

$$vT_0 = T_\infty + \left(\frac{T_1 - T_\infty}{T_2 - T_\infty}\right)^3 \cdot (T_3 - T_\infty).$$

5.2 Simulation: Free Solidification of Ice

The general setup is similar in both, 2D and 3D configuration of free solidification of ice into an undercooled melt. We investigate three different undercooling temperatures of the liquid; 2, 4 and 6 K below the melting temperature. To save computational time, we utilise the crystal symmetry of ice and therefore calculate 1/4 of the original domain in 2D and 1/8 in 3D. To achieve this, we use Neumann boundary conditions at the lower end of the simulation domain for both, the phase-field as well as the heat equation, and set the part of the initial nucleus at that corner. For the other boundaries we impose the extrapolation condition for the temperature and Dirichlet condition $\phi_0|_{\partial\Omega} = \phi^l$ for the phase-field.

The initial ice nucleus is set to a size of fifty times of the critical nucleus for the given undercooling. This leads to $R_{2K} = 1.238 \cdot 10^{-6}$ m, $R_{4K} = 6.143 \cdot 10^{-7}$ m and $R_{6K} = 4.065 \cdot 10^{-7}$ m. The initial condition for the temperature is given by an analytical distribution depending on the size of the initial nucleus. The resolution of the simulation domain and therefor the utilised number or cores depends on the undercooling. Each simulation domain is distributed to multiple cores by domain decomposition, with a local domain size of $25 \times 25 \times 25$ cells, leading to a maximum number of 13.824 cores.

5.3 Free Solidification of Ice in Two Dimensions

The simulation domain is a square with an edge length of $84.5\,\mu$m, the number of time steps is set to 10^8 leading to a simulated time of 0.26, 0.062 and 0.026 s for the three different temperatures. In Fig. 5, we show the growth shapes of the ice crystals for the three undercoolings. To illustrate the shape, we display contour lines for $\phi_\alpha = 0.5$ at equidistant points in time. Starting from a bulky dendrite for small undercooling, we observe a more pronounced, pointy dendrite for large undercooling. To identify the size and shape difference of the dendrites at the same point in time, we mark the contour for $t = 0.026$ s in each of the plots.

5.4 Free Solidification of Ice in Three Dimensions

The size of the simulation domain in the case of the three-dimensional setting is smaller than the 2D settings due to computational limitations. The edge length is $23.24\,\mu$m in each of the three spacial dimensions. As in the two-dimensional settings, we have different spacial resolutions and different time steps depending on the undercooling. For the three settings, the total simulation time is $t_{2K} = 10.15$ ms, $t_{4K} = 1.652$ ms and $t_{6K} = 0.330$ ms. Figure 6 shows an ice dendrite grown into undercooled water of 4 K below melting temperature, 1.652 ms after

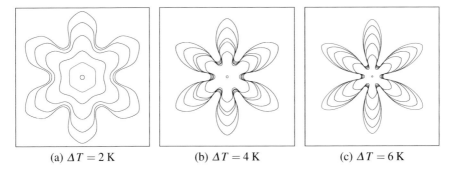

(a) $\Delta T = 2$ K   (b) $\Delta T = 4$ K   (c) $\Delta T = 6$ K

Fig. 5 *Contour lines* at different time-steps of the growing ice crystal for three different undercoolings. The *black lines* are equidistant in time with (**a**) $\Delta t = 0.065$ s, (**b**) $\Delta t = 0.015$ s and (**c**) $\Delta t = 0.005$ s. The *red contour line* shows the shape of the different dendrites for the same point in time ($t = 0.026$ s). The *boxes* represent a part of the simulation domain with edge length of 74 μm

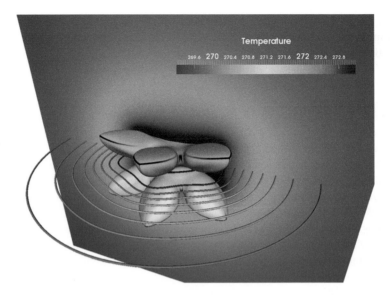

Fig. 6 Ice dendrite growing into an undercooled melt of 4 K after $t = 1.652$ ms. The solid part in shades of *blue* indicates the contour of the dendrite, *darker blue* refers to areas with high curvature at the surface to underline the edges of the dendrite shape. The *colours* in the plane represent different temperatures in the liquid, ranging from an undercooled state (*blue*) to the melting temperature (*red*). The tubes in the foreground illustrate temperature contours in the third dimension, where the temperature distance between two tubes is 0.4 K

the start of the simulation. The colour on the backplane indicates the resulting temperature distribution in the liquid. The plane has an extent of 46.5 μm in both directions after reconstruction by mirroring the simulation domain at two lower boundaries. The distinctive shape of the dendrite is underlined by blue shades, where darker colours refer higher curvature. The temperature contour lines show a nearly spherical temperature field justifying the assumption for the 3D extrapolation boundary condition.

The velocities for the steady state growth of the three dimensional dendrite tips are in good agreement with theoretical values predicted from the Langer and Müller-Krumbhaar [24].

6 Compound Droplets on Fibers

In the recent years, a wide range of studies concerned with wetting behaviors of droplets hanging on horizontal fibers is presented. Those systems reflect completely different wetting properties than sessile droplets on flat solid substrates. Theoretical, experimental and numerical investigations show, that one can identify two equilibrium shapes: the clamshell and the barrel shape [4, 5, 27–29, 37]. Whereas barrel shapes can be described analytically [4], numerical approaches are necessary for the description of clamshell shaped droplets. The authors of [5, 27] also investigate the transition between both equilibrium configurations. A droplet adopts a barrel shape when its volume is large or when the fiber radius is small. A volume decrease or a fiber radius increase leads to a clamshell configuration. This transition can also be induced using electrowetting [8, 11, 12]. In such processes, the droplet volume and the fiber are constant but the contact angle at the liquid/fiber/air interface is affected by the electrical field. The public domain software *Surface Evolver* (SE), developed by Brakke [3], has been established for numerical investigations of the equilibrium morphologies of single droplets on cylindrical fibers [6]. Recent works [15, 16, 25] demonstrated that droplets on fibers may constitute the starting point of an open digital microfluidics and that optofluidic devices can be designed by using optical fiber networks. However, the evaporation and contamination of the droplets constitute the major difficulties for such devices. This can be counterbalanced by the encapsulation of the core water droplets into an oil shell leading to the creation of compound droplets. To our knowledge so far, such wetting problems, consisting of two or more droplets with different physical properties can not be numerically solved by the SE software. Therefore, we introduce a multiphase-field model with appropriate boundary conditions [1] enabling the numerical investigation of multi-droplet systems, in particular of compound droplets.

Here, we study the wetting behavior of compound droplets on horizontal thin fibers. The considered systems consist on water or soapy water droplets encapsulated by an oil drop. Beside experimental studies, we perform numerical simulations to show that compound droplets can be formed on fibers and that they take specific equilibrium shapes. We focus on the various contact lines formed when different phases meet and we investigate their equilibrium states. We find out, that depending on the surface tensions, the triple contact lines can remain separate or merge together and form quadruple lines. The nature of the contact lines affects the wetting behavior of the compound droplets on fibers. Indeed, both experimental and numerical results show that, during the detachment process, depending on whether the contact lines are triple or quadruple, the characteristic length is the inner droplet radius or the fiber radius.

6.1 Multiphase-Field Model

The numerical simulations presented in the next section are based on a multi-phase-field model of Allen–Cahn type with a well designed boundary condition reflecting the wetting properties of compound droplets [1]. The model is extended by an energy density term reflecting the gravity effect. The free energy of the system reads

$$\mathscr{F}(\boldsymbol{\phi}) = \int_{\Omega} \left(\epsilon a(\boldsymbol{\phi}, \nabla \boldsymbol{\phi}) + \frac{1}{\epsilon} w(\boldsymbol{\phi}) + g(\boldsymbol{\phi}) + f_{gv}(\boldsymbol{\phi}) \right) d\Omega + \int_{\partial_s \Omega} f_w(\boldsymbol{\phi}) \, dS.$$

$$(4)$$

Here, Ω is the domain of consideration and the fiber is denoted by $\partial_s\Omega$. The vector components ϕ_α, $\alpha \in \{w, o, A\}$, refer to the water, oil and air phases, respectively. When considering soapy water instead of water, the subscript s is used. For more details we refer to [1, 36].

Minimizing the free energy functional (4) by use of variational calculus methods, leads to the following time dependent evolution equations for each phase α in the system in Ω

$$\epsilon(\nabla \cdot a_{,\nabla\phi_\alpha}(\phi, \nabla\phi) - a_{,\phi_\alpha}(\phi, \nabla\phi)) - \frac{1}{\epsilon}w_{,\phi_\alpha}(\phi) - g_{,\phi_\alpha}(\phi) + f_{gv,\phi_\alpha}(\phi) - \lambda_1$$
$$= \tau\epsilon\partial_t\phi_\alpha \tag{5}$$

with the natural boundary condition

$$-\epsilon a_{,\nabla\phi_\alpha}(\phi, \nabla\phi) \cdot \boldsymbol{n} - f_{w,\phi_\alpha}(\phi) - \lambda_2 = 0 \quad \text{on } \partial_S\Omega. \tag{6}$$

The time relaxation parameter τ is set to unity in all simulations. The notation $a_{,\nabla\phi_\alpha}$, $a_{,\phi_\alpha}$, $w_{,\phi_\alpha}$, $g_{,\phi_\alpha}$ and f_{w,ϕ_α} is used to indicate the partial derivatives $\partial/\partial\nabla\phi_\alpha$ and $\partial/\partial\phi_\alpha$ of the functions $a(\phi, \nabla\phi)$, $w(\phi)$, $g(\phi)$ and $f_w(\phi)$, respectively. The divergence of the vector field $a_{,\nabla\phi_\alpha}(\phi, \nabla\phi)$ is denoted by $\nabla \cdot ()$ and the time derivative $\partial\phi_\alpha(\mathbf{x}, t)/\partial t$ is written as $\partial_t\phi_\alpha$. The normal to the fiber, $\partial_s\Omega$, is denoted by $\boldsymbol{n}$. λ_1 and λ_2 are Lagrange multipliers, according to the constraint $\sum_\alpha \phi_\alpha = 1$.

The evolution equations (5) and (6) are discretized using finite differences on a regular grid with equidistant spacings in all spacial directions. For the discretization in time, the forward Euler scheme is implemented, which requires the control of the time step Δt for ensuring numerical stability.

6.2 Result and Discussion

In order to investigate the influence of the nature of the contact lines on the wetting behavior of compound droplet systems, we consider experimentally and numerically, droplets with different surface tensions, densities and volumes as shown in Fig. 7. The compound droplets are close to detachment, since the capillary forces start to dominate the capillary force. Both systems contain $0.5\,\mu l$ of water. Pure water is colored in blue and soapy water in red. We increase the volume of the oil drop gradually until the drop is ready to fall. In the experiments the observed critical volume of oil is $3.9\,\mu l$ for pure water and of $3.5\,\mu l$ for soapy water. Numerical simulations predict a critical volume of $4.2\,\mu l$ for pure water and $3.5\,\mu l$ for soapy water. We notice, that pure water can hold larger volumes of oil than soapy water. Additionally, Fig. 7 shows that the shapes of the drops are different, depending on the nature of the inner droplet. Even if the volumes are slightly different, the simulations also lead to these different shapes which agree to those

Fig. 7 Two droplets close to detachment: (*upper row, left* to *right*) simulation results of compound droplets with 0.5 μl of pure water (*blue*) and 4.4 μl of oil, and a compound droplet with 0.5 μl of soapy water (*red*) and 3.5 μl of oil on a fiber (diameter $d = 200$ μm). (*Lower row, left* to *right*) photograph of the experiment showing a compound droplet with 0.5 μl of pure water and 3.9 μl of oil, and a compound droplet with 0.5 μl of soapy water and 3.5 μl of oil on a fiber (diameter $d = 200$ μm)

observed experimentally, as it can be seen in Fig. 7. Oil is literally hanging from the water droplet. The presence of two distinct triple lines allows the oil to move independently from the water. Whereas, in the case of soapy water, since there is only two quadruple line, the system moves as a whole under the fiber. The insight in Fig. 8 magnifies this effect. The shape of the soapy water/oil compound droplet is similar to the one of a single droplet only made of oil. The few differences are due to the surface tension and the contact angle that are influenced by the presence of a quadruple contact line.

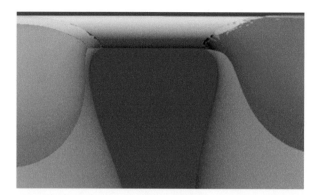

Fig. 8 Insight in the triple and quadruple junction regions. *Left*: pure water/oil compound droplet. Two distinct triple lines (water/oil/fiber and water/air/fiber) are formed. *Right*: soapy water/oil compound droplet. At the fiber, the four phases soapy water, oil, air and fiber meet and form a quadruple line

Additionally we investigate the detachment process, depending on whether the contact lines are triple or quadruple. The simulation results show excellent agreement with the experimental investigations as reported in [36]. We find that, the characteristic lengths, for the pure water/oil and the soapy water/oil compound drops, are the inner droplet radius and the fiber radius, respectively. For soapy water/oil systems, the maximal total volume of the compound drop, before it falls down due to gravity, is constant independently on the inner droplet radius. The total volume depends only on the fiber radius. Whereas, for the pure water/oil systems, the inner droplet volume influences significantly the detachment behavior, as it acts like a capillary. Those different detachment behaviors are due to the establishment of a quadruple line along the fiber and due to the existence of two distinct triple lines in the latter case. We mention, that a parameter study is conducted in order to mimic the experiments numerically, as the model acts very sensitive on some parameters. The simulations are performed on $560 \times 120 \times 140$ and $560 \times 160 \times 160$ grid cells running on 673 and 1025 CPU's for about 36 h, respectively.

References

1. Ben Said, M., Selzer, M., Nestler, B., Braun, D., Greiner, C., Garcke, H.: A phase-field approach for wetting phenomena of multiphase droplets on solid surfaces. Langmuir **30**(14), 4033–4039 (2014)
2. Berghoff, M., Nestler, B.: Scale-bridging phase-field simulations of microstructure responses on nucleation in metals and colloids. Eur. Phys. J. Spec. Top. **223**(3), 409–419 (2014)
3. Brakke, K.A.: The surface evolver. Exp. Math. **1**(2), 141–165 (1992)
4. Carroll, B.J.: The accurate measurement of contact angle, phase contact areas, drop volume, and Laplace excess pressure in drop-on-fiber systems. J. Colloid Interf. Sci. **57**(3), 488–495 (1976)

5. Carroll, B.J.: Equilibrium conformations of liquid drops on thin cylinders under forces of capillarity. A theory for the roll-up process. Langmuir **2**(2), 248–250 (1986)
6. Chou, T.H., Hong, S.J., Liang, Y.E., Tsao, H.K., Sheng, Y.J.: Equilibrium phase diagram of drop-on-fiber: coexistent states and gravity effect. Langmuir **27**(7), 3685–3692 (2011)
7. Choudhury, A.: Quantitative Phase-Field Model for Phase Transformations in Multi-component Alloys (Schriftenreihe des Instituts für Angewandte Materialien, Karlsruher Institut fuer Technologie). KIT Scientific Publishing, Karlsruhe (2013)
8. Ruiter, R.: Manipulation of Drops with Electrowetting: From Morphological Transitions to Microfluidics. Universiteit Twente (2014). doi:10.3990/1.9789036536394
9. Dennstedt, A., Ratke, L.: Microstructures of directionally solidified Al-Ag-Cu ternary eutectics. Trans. Indian Inst. Metals **65**(6), 777–782 (2012)
10. Dennstedt, A., Ratke, L., Choudhury, A., Nestler, B.: New metallographic method for estimation of ordering and lattice parameter in ternary eutectic systems. Metallogr. Microstruct. Anal. **2**(3), 140–147 (2013)
11. Dufour, R., Dibao-Dina, A., Harnois, M., Tao, X., Dufour, C., Boukherroub, R., Senez, V., Thomy, V.: Electrowetting on functional fibers. Soft Matter **9**, 492–497 (2013)
12. Eral, H.B., de Ruiter, J., de Ruiter, R., Oh, J.M., Semprebon, C., Brinkmann, M., Mugele, F.: Drops on functional fibers: from barrels to clamshells and back. Soft Matter **7**(11), 5138 (2011)
13. Feichtinger, C., Habich, J., Köstler, H., Röde, U., Aoki, T.: Performance modeling and analysis of heterogeneous lattice Boltzmann simulations on CPU-GPU clusters. Parallel Comput. **46**, 1–13 (2015)
14. Genau, A., Ratke, L.: Morphological characterization of the Al-Ag-Cu ternary eutectic. Int. J. Mater. Res. **103**(4), 469–475 (2012)
15. Gilet, T., Terwagne, D., Vandewalle, N.: Digital microfluidics on a wire. Appl. Phys. Lett. **95**(1), 10–13 (2009)
16. Gilet, T., Terwagne, D., Vandewalle, N.: Droplets sliding on fibres. Eur. Phys. J. E **31**(3), 253–262 (2010)
17. Godenschwager, C., Schornbaum, F., Bauer, M., Köstler, H., Rüde, U.: A framework for hybrid parallel flow simulations with a trillion cells in complex geometries. In: Proceedings of SC13: International Conference for High Performance Computing, Networking, Storage and Analysis, p. 35. ACM (2013)
18. Handel, R., Davidchack, R.L., Anwar, J., Brukhno, A.: Direct calculation of solid-liquid interfacial free energy for molecular systems: TIP4P ice-water interface. Phys. Rev. Lett. **100**(3), 036104-1–036104-4 (2008)
19. Hecht, U., Gránásy, L., Pusztai, T., Böttger, B., Apel, M., Witusiewicz, V., Ratke, L., De Wilde, J., Froyen, L., Camel, D., et al.: Multiphase solidification in multicomponent alloys. Mater. Sci. Eng. R. Rep. **46**(1), 1–49 (2004)
20. Hötzer, J., Jainta, M., Steinmetz, P., Dennstedt, A., Nestler, B.: Die Vielfalt der Musterbildung in Metallen. Horizonte **45** (2015)
21. Hötzer, J., Jainta, M., Steinmetz, P., Nestler, B., Dennstedt, A., Genau, A., Bauer, M., Köstler, H., Rüde, U.: Large scale phase-field simulations of directional ternary eutectic solidification. Acta Mater. **93**, 194–204 (2015)
22. Köstler, H., Feichtinger, C., Rüde, U., Aoki, T.: A geometric multigrid solver on Tsubame 2.0. In: Efficient Algorithms for Global Optimization Methods in Computer Vision, pp. 155–173. Springer, Berlin (2014)
23. Lewis, D., Allen, S., Notis, M., Scotch, A.: Determination of the eutectic structure in the Ag-Cu-Sn system. J. Electron. Mater. **31**(2), 161–167 (2002)
24. Langer, J.S., Müller-Krumbhaar, J.: Stability effects in dendritic crystal growth. J. Cryst. Growth **42**, 11–14 (1977)
25. Lismont, M., Vandewalle, N., Joris, B., Dreesen, L.: Fiber based optofluidic biosensors. Appl. Phys. Lett. **105**(13), 133701 (2014)
26. Maruyama, M.: Relation between growth and melt shapes of ice crystals. J. Cryst. Growth **318**(1), 36–39 (2011)

27. McHale, G., Newton, M.I.: Global geometry and the equilibrium shapes of liquid drops on fibers. Colloids Surf. A Physicochem. Eng. Asp. **206**(1–3), 79–86 (2002)
28. McHale, G., Käb, N., Newton, M., Rowan, S.: Wetting of a high-energy fiber surface. J. Colloid Interf. Sci. **186**(2), 453–461 (1997)
29. McHale, G., Newton, M.I., Carroll, B.J.: The shape and stability of small liquid drops on fibers. Oil Gas Sci. Technol. **56**(1), 47–54 (2001)
30. Moelans, N., Blanpain, B., Wollants, P.: An introduction to phase-field modeling of microstructure evolution. Calphad **32**(2), 268–294 (2008)
31. Nestler, B., Danilov, D., Galenko, P.: Crystal growth of pure substances: phase-field simulations in comparison with analytical and experimental results. J. Comput. Phys. **207**(1), 221–239 (2005)
32. Requena, G., Cloetens, P., Altendorfer, W., Poletti, C., Tolnai, D., Warchomicka, F., Degischer, H.P.: Sub-micrometer synchrotron tomography of multiphase metals using kirkpatrick-baez optics. Scr. Mater. **61**(7), 760–763 (2009)
33. Ruggiero, M.A., Rutter, J.W.: Origin of microstructure in the 332 K eutectic of the Bi-In-Sn system. Mater. Sci. Technol. **13**(1), 5–11 (1997)
34. Steinbach, I.: Phase-field model for microstructure evolution at the mesoscopic scale. Annu. Rev. Mater. Res. **43**(1), 89–107 (2013)
35. Vondrous, A., Selzer, M., Hötzer, J., Nestler, B.: Parallel computing for phase-field models. Int. J. High Perform. Comput. Appl. **28**(1), 61–72 (2014)
36. Weyer, F., Ben Said, M., Hötzer, J., Berghoff, M., Dreesen, L., Nestler, B., Vandewalle, N.: Compound droplets on fibers. Langmuir **31**(28), 7799–7805 (2015)
37. Wu, X.F., Dzenis, Y.A.: Droplet on a fiber: geometrical shape and contact angle. Acta Mech. **185**(3–4), 215–225 (2006)

Large Scale Numerical Simulations of Planetary Interiors

Ana-Catalina Plesa, Christian Hüttig, Maxime Maurice, Nicola Tosi, and Doris Breuer

Abstract The massive increase of computational power over the past decades has established numerical models of planetary interiors to one of the principal tools to investigate the thermo-chemical evolution of terrestrial bodies. Large scale computational models have become state of the art to investigate the interior heat transport, surface tectonics and chemical differentiation of planetary bodies across the Solar System and beyond. In the present work we present large scale numerical simulations performed using the mantle convection code Gaia in spherical and Cartesian geometry. The results have been obtained on the HLRS system Hornet running on 54×10^3 computational cores. The strong scaling results show an optimal speedup for a grid with 55 million computational points corresponding to 275 million unknowns.

1 Introduction

The slow creep of material driven by thermal and chemical buoyancy inside planetary mantles is the most important mechanism through which planetary bodies loose their heat. Both thermal and chemical convection are the most dominant dynamical processes influencing planetary differentiation and the formation of surface geological structures as well as magnetic field generation e.g., [1]. Measurements from various space missions or geochemical sample analysis of meteorites and terrestrial rocks suggest the existence of local to global scale chemical heterogeneities in the interior of planetary bodies [2, 6, 8, 25]. However such heterogeneities are difficult to model in global convection simulations in 3D spherical geometry, mainly due to the high spatial resolution required and the additional complexity which they introduce. However, over the past decades, large improvements both in processor

A.-C. Plesa (✉) • C. Hüttig • M. Maurice • D. Breuer
German Aerospace Center, Institute of Planetary Research, Berlin, Germany
e-mail: a_ples01@uni-muenster.de; christian.huettig@dlr.de; maxime.maurice@dlr.de; doris.breuer@dlr.de

N. Tosi
Department of Planetary Geodesy, Technical University Berlin, Berlin, Germany
e-mail: nicola.tosi@tu-berlin.de

© Springer International Publishing Switzerland 2016   675
W.E. Nagel et al. (eds.), *High Performance Computing in Science and Engineering '15*, DOI 10.1007/978-3-319-24633-8_43

power and amount of available memory have made possible to tackle highly complex scenarios using parameters appropriate to model the interior dynamics of terrestrial bodies of the Solar System and exoplanets. Numerical simulations modeling vigorous convection at extreme Rayleigh numbers with highly variable viscosity and including chemical heterogeneities have become state of the art to investigate the differentiation and cooling of a terrestrial planetary mantle.

In this work we present the recent developments and results using the mantle convection code Gaia in various geometries. We discuss results obtained running large scale numerical simulations on the new Cray XC40 platform at HLRS, Hornet.

2 Models and Methods

2.1 Physical Model

Thermo-chemical mantle convection is modeled by solving the conservation equations of mass, momentum, energy and composition e.g., [19]. The equations are scaled with the mantle thickness D as length scale, the thermal diffusivity κ as time scale, the temperature drop across the entire mantle ΔT as temperature scale and the contrast in chemical density $\Delta\rho$ as compositional scale. Assuming a Newtonian rheology for a Boussinesq fluid with an infinite Prandtl number, as the mantle of terrestrial bodies is considered highly viscous with negligible inertia, the non-dimensional equations of thermo-chemical convection are the following e.g. [7]:

$$\nabla \cdot \mathbf{u} = 0, \tag{1}$$

$$\nabla \cdot \left[\eta(\nabla\mathbf{u} + (\nabla\mathbf{u})^T)\right] + (RaT - Ra_C C)\mathbf{e}_r - \nabla p = 0, \tag{2}$$

$$\frac{\partial T}{\partial t} + \mathbf{u} \cdot \nabla T - \nabla^2 T - \frac{Ra_Q}{Ra} = 0, \tag{3}$$

$$\frac{\partial C}{\partial t} + \mathbf{u} \cdot \nabla C = 0, \tag{4}$$

where η is the viscosity, p the dynamic pressure, $\mathbf{u}$ the velocity, t the time, T the temperature, C the chemical component and $\mathbf{e}_r$ is the unit vector in radial direction. The variables in the above equations are scaled using the relationships to physical properties presented e.g. in [3]. The thermal and compositional Rayleigh numbers, Ra and Ra_C respectively, in Eq. (2) are defined as follows:

$$Ra = \frac{\rho g \alpha \Delta T D^3}{\kappa \eta_{ref}} \quad \text{and} \quad Ra_C = \frac{\Delta\rho g D^3}{\kappa \eta_{ref}}, \tag{5}$$

where ρ is the reference density, $\Delta\rho$ the density contrast, g the gravitational acceleration, α the thermal expansivity, ΔT the temperature contrast between inner and outer boundaries, D the thickness of the mantle, κ the thermal diffusivity, and η_{ref} the reference viscosity. The internal heat sources Rayleigh number in Eq. (3), Ra_Q, is given by:

$$Ra_Q = \frac{\rho^2 g \alpha Q_m D^5}{\kappa k \eta_{ref}},\tag{6}$$

where Q_m is the heat production rate and k is the thermal conductivity.

The viscosity is calculated according to the Arrhenius law for mixed diffusion and dislocation creep [13]. The non-dimensional formulation of the Arrhenius viscosity law for temperature, pressure and strain rate dependent viscosity is given by Roberts and Zhong [18]:

$$\eta(\varepsilon, T, p) = \left(\frac{\varepsilon}{\varepsilon_{ref}}\right)^{\frac{1-n}{n}} \exp\left(\frac{E + pV}{n(T + T_0)} - \frac{E + p_{ref}V}{n(T_{ref} + T_0)}\right),\tag{7}$$

where E and V are the activation energy and activation volume, respectively, and T_{surf} the surface temperature. The variables T_{ref}, p_{ref}, and ε_{ref} are the reference temperature, pressure and strain rate, respectively. The factor n introduces the non-linear rheology. For dislocation creep n is equal to 3.5 while for diffusion creep it is 1. Due to the strongly temperature-dependent viscosity, a stagnant lid will rapidly form on top of the convecting mantle. Additionally a pseudo-plastic approach can be used such that the stagnant lid undergoes plastic yielding if the convective stresses are high enough to overcome the imposed yield stress σ_y. An effective viscosity can thus be defined as follows:

$$\frac{1}{\eta_{eff}} = \frac{1}{\eta(\varepsilon, T, p)} + \frac{1}{\frac{\sigma_y}{2\dot{\varepsilon}}} \quad \text{with} \quad \sigma_y = \sigma_0 + z\frac{\partial\sigma}{\partial z},\tag{8}$$

where z is the depth and $\dot{\varepsilon}$ is the second invariant of the strain rate tensor e.g., [12]. The plastic yielding introduces a plate-like behavior of the lithosphere with the surface being recycled when the lithosphere undergoes plastic failure represented by a strong reduction of the near-surface viscosity, which will necessarily allow for the surface material to flow.

2.2 Technical Realization

Gaia is a fluid flow solver that uses a fixed mesh in arbitrary geometries (Fig. 1) to solve the conservation of mass, momentum and energy [Eqs. (1)–(3)]. The code is mainly employed for modeling Stokes-flow with strongly varying viscosity, as

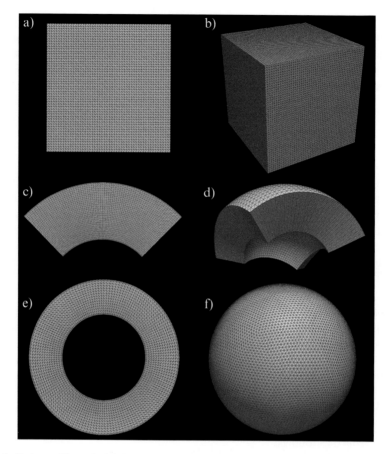

Fig. 1 Various grids used with Gaia: (**a**) 2D Cartesian box; (**b**) 3D Cartesian box; (**c**) 2D regional cylinder; (**d**) 3D regional spherical shell; (**e**) 2D cylinder; (**f**) 3D full spherical shell

appropriate for mantle convection simulations, although during the last years the equations were extended to include rotation and to support flows driven by inertia (finite Prandtl number).

Written in C++, Gaia is library-independent enabling its easy portability on all type of systems [9]. The code uses the finite-volume method to discretize the governing equations and can handle both regular and fully irregular Voronoi grids [10, 15] with various geometries as shown in Fig. 1. A fully implicit second-order scheme is applied for the temporal discretization [5]. The user can choose between various time-stepping criteria using either a flexible time-step, fulfilling the Courant-Friedrichs-Lewy condition, which can adapt according to the problem or a fixed time-step. Another time stepping mechanism, usually employed for steady-state problems, relates the next time step to the inverse of the velocity magnitude change, such that the time steps increase considerably when approaching the steady-state solution.

Gaia solves for velocity and pressure simultaneously, by including the three velocity components (u, v, w) and pressure p in one single system of equations. The code uses a selective Jacobi preconditioner, which is only applied to the velocity part of the matrix. For further details, we refer the readers to [11].

The code brings its own iterative solvers where the user can choose between BiCGS, BiCGS(l), Jacobi and TFQMR, with the BiCGS(4) method being often the best choice for typical mantle convection scenarios. Additionally, the code can employ an under-relaxation method for the velocity field taking into account a fraction of the previous velocity value to compute the current one [5]. This approach is used to speed up convergence for steady-state problems or to couple in non-linear effects.

Convection driven by active chemical fields is modeled with the particle in cell method (PIC) e.g. [16, 21, 24]. Equation 4 is modeled by solving a transport equation for tracer particles which carry various chemical species and move according to the velocity field computed on the fixed grid. The particles are then used to update various compositional fields at grid resolution by interpolating their values at cell centers. The method has the advantage over classical grid-based methods of being essentially free of numerical diffusion and provides the advection of an arbitrary large number of different compositional fields with less computational effort.

An efficient domain decomposition of a given computational mesh results into p equal volumes with each being mapped to a computational core. Minimizing the area between the sections attributed to each computational core, leads to a minimized overhead of data exchange between processors. The so-called halo-cells, or ghost-cells, are additional cells in each domain and form an overlapping zone that is used for data exchange between processors. Each domain has its border consisting of halo-cells located on the same position as the active cells of the neighboring domains. A first measure of parallelization is the ratio between halo-cells and active grid-cells as it determines the amount of communication between processors. While for 3D Cartesian box grids, due to their regular nature, an optimal domain decomposition can be easily achieved, the problem becomes highly complex when 3D spherical shells are involved. In Gaia we use the so-called Thomson-points to laterally decompose the sphere by distributing points, all assumed with equivalent potential energy, on the surface of the sphere and minimizing the global potential field energy. This method is known as the Thomson Problem [22]. The closest "Thomson" point defines the domain of every grid cell. The domain decomposition determined by this approach leads to equal volumes of the spherical regions for each domain, this being important to balance the computational effort of each processor [9]. However, this approach is limited, since the correct calculation of the Thomson locations on the unit sphere requires itself supercomputers for more than 16 points. Gaia uses publicly available pre-computed positions from the Cambridge University as input for up to $p = 4352$ to efficiently map the computational domains for 3D spherical shell grids. Using additional radial slices of the spherical shell, the number of equal volume computational domains can be enhanced. However, a radial slice of the spherical shell is expensive in terms of halo-cells and is efficient only when a large number of computational cores is used.

3   Results and Discussion

3.1   *Code Performance*

The ability of a numerical code to scale with a large number of computational cores is often difficult to achieve. Because of this and in order to reduce computational costs numerical simulations of interiors dynamics often use limited parameter ranges. During the test phase of the new HLRS system Hornet, we have tested the performance of our code using up to 54×10^3 computational cores (dual socket Intel(R) Xeon(R) CPU E5-2680 v3 @ 2.5 GHz having two sockets per node with 12 cores each). To this end we have performed numerical simulations using a 3D Cartesian box regular grid with 55 million computational points (275 million unknowns) and a 3D spherical shell fully irregular grid with 12 million computational points (60 million unknowns) and additional 240 million tracers. The strong scaling achieved with the two setups is shown in Fig. 2a. For the 3D Cartesian box grid we used up to 54×10^3 cores, while the 3D spherical shell simulation

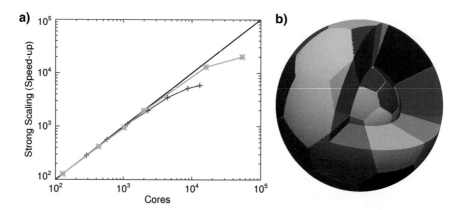

Fig. 2 (**a**) Speed-up factor when running the A4 case shown in [26] on both a regular 3D Cartesian box grid with 55 million computational points (*green line*) and a fully irregular 3D spherical shell grid with 12 million computational points (*red line*) on up to 54×10^3 computational cores; (**b**) Domain decomposition of a 3D spherical shell grid using 32 Thomson points

has been performed with up to 13×10^3 cores. The domain decomposition for the 3D spherical shell grid used 2 radial and 4352 lateral slices (4352 being the largest amount of available Thomson-points). Figure 2b shows a typical domain decomposition for the 3D spherical shell with 32 domains.

3.2 Application: Mantle Convection in Terrestrial Planets

Gaia is one of the six large scale projects selected by HLRS to run on the new XC40 platform Hornet. In the following two projects we illustrate the importance of such large scale computational systems for improving our knowledge about the physical processes occurring in the interior of terrestrial bodies.

In the first project, we investigate the heat transport efficiency at high convection vigor (i.e., Rayleigh number of 10^{11}). The setup, including a viscosity contrast of 10^6 over the entire mantle, is highly relevant for the mantle of the Earth. Although at the beginning of the simulations, mantle plumes with a broad plume head are formed, the convection planform rapidly changes to a small scale pattern with thin mantle plume stencils (Fig. 3). These fine structures justify the chosen mesh resolution of 55 million computational points.

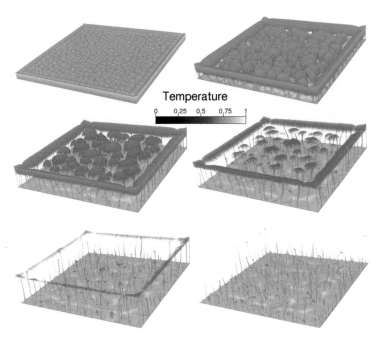

Fig. 3 Temperature evolution in a system heated from below and from within at $Ra = Ra_Q = 10^{11}$ and $\Delta\eta = 10^6$

In the second project, we investigate the overturn of an unstable density profile and is motivated by the fact that the earliest stage of planetary evolution, large amount of energy is available to melt the entire mantle producing a global magma ocean. The subsequent cooling of the mantle causes the magma ocean to freeze from the core mantle boundary to the surface due to the steeper slope of the mantle adiabat compared to the solidus (melting temperature) [20]. Assuming fractional crystallization, dense cumulates will be produced close to the surface due to iron enrichment in the residual liquid. A gravitationally unstable mantle thus forms, which is prone to overturn [4]. The style of the overturn as well as the subsequent configuration of the density profile are ultimately important for the subsequent thermo-chemical evolution of a planetary body and may have a major impact on the later surface tectonics and volcanic history. Using published density profiles for the initial unstable stratified mantle, we obtain, upon mantle overturn, a stable stratification which inhibits thermal convection and thus is difficult to reconcile with late volcanic activity (Fig. 4). Moreover, in the absence of mantle upwellings, possible mantle geochemical reservoirs which may form during the early stages of planetary evolution are difficult to be sampled [17, 23]. However, reservoir-signatures have been recorded for Mars in the so-called SNC meteorites, which have tapped the mantle reservoirs over the entire history of the planet.

Therefore, in a further step we have started to investigate the cooling and interaction of a liquid magma ocean with the proceeding solidification of the underlying mantle. To this end we have conducted preliminary simulations in 2D box geometry in order to understand the conditions for the onset of solid-state thermal convection before complete mantle solidification. This process would have important implications for the crystallization sequence, and in particular for the density gradients which in turn will necessarily affect the overturn itself and the subsequent thermo-chemical evolution. In fact, preliminary results show that a sufficiently high Rayleigh number guarantees the onset of solid-state convection prior to complete crystallization of the mantle [14]. Thus convective mixing may reduce density gradients prior to complete solidification and in such case, the density contrast at the end of the crystallization sequence may be strongly diminished (Fig. 5).

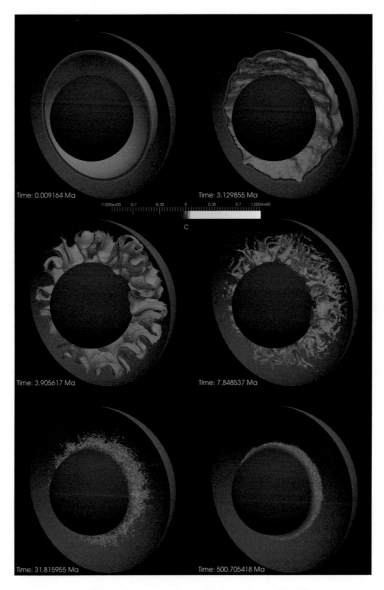

Fig. 4 Overturn and evolution of an unstable stratified density profile. The parameters used are listed in [17]

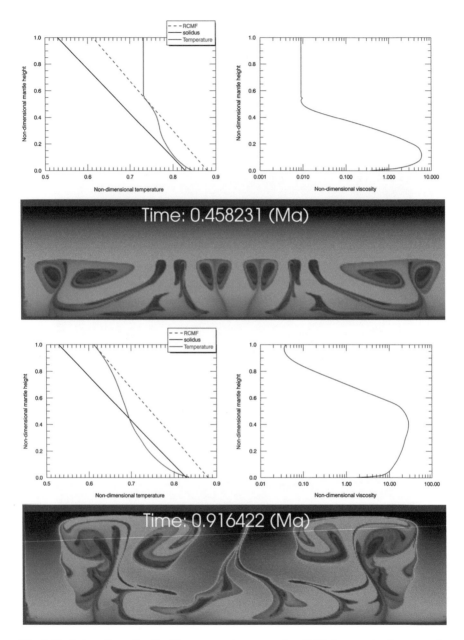

Fig. 5 Evolution of a cooling liquid magma ocean on top of a solidified convecting lower mantle

4   Conclusions and Outlook

In this work, we have shown that our mantle convection code, Gaia, can efficiently run large scale computational tasks using meshes consisting of up to 54 million points on more than 5×10^4 computational cores. Such high resolution simulations of interior dynamics of planetary bodies performed on the new Cray XC40 (Hornet) machine at the HLRS supercomputing center in Stuttgart offer the tools necessary to tackle complex scenarios for planetary evolution. Gaia has become one of the few codes world-wide able to efficiently run large scale simulations of interior dynamics and thanks to high-performance computational centers like HLRS with its massive parallel computing resources high resolution mantle convection models can be employed to better understand the physical processes active in the interior of terrestrial bodies using realistic parameter values.

In future work, we will investigate the cooling behavior of a liquid magma ocean and its interaction with the proceeding mantle solidification in 3D spherical geometry. Our goal is to quantify under which conditions solid state convection will reduce and possibly erase chemical heterogeneities before complete mantle crystallization. Even in a thin, fully crystallized part in the lower mantle an unstable, partly built, density profile may contribute to increase the Rayleigh number above the critical one and thus drive the onset of solid-state convection. Active compositional fields may thus play an important role not only for the subsequent planetary evolution after the complete crystallization of the magma ocean but also for the initial cooling stage of a liquid mantle and onset of solid-state convection.

Acknowledgements This work has been supported by the Helmholtz Association through the research alliance "Planetary Evolution and Life" and through the grant VG-NG-1017, by the Deutsche Forschungsgemeinschaft (grant TO 704/1-1), and by the Interuniversity Attraction Poles Programme initiated by the Belgian Science Policy Office through the Planet Topers alliance. Computational time has been provided by the High-Performance Computing Center Stuttgart (HLRS) through the project Mantle Thermal and Compositional Simulations (MATHECO).

References

1. Breuer, D., Moore, W.B.: Dynamics and thermal history of the terrestrial planets, the moon, and Io. Treatise Geophys. **10**, 299–348 (2007)
2. Charlier, B., Grove, T.L., Zuber, M.T.: Phase equilibria of ultramafic compositions on Mercury and the origin of the compositional dichotomy. Earth Planet. Sci. Lett. **363**, 50–60 (2013)
3. Christensen, U.: Convection with pressure- and temperature-dependent non-Newtonian rheology. Geophys. J. R. Astron. Soc. **77**, 343–384 (1984)
4. Elkins-Tanton, L.T., Parmentier, E.M., Hess, P.C.: Magma ocean fractional crystallization and cumulate overturn in terrestrial planets: implications for Mars. Meteorit. Planet. Sci. **38**, 1753–1771 (2003)
5. Ferziger, J.H., Perić, M.: Computational Methods for Fluid Dynamics, 2nd revised edn. Springer, Berlin (1999). ISBN: 3-540-65373-2

6. Foley, C.N., Wadhwa, M., Borg, L.E., Janney, P.E., Hines, R., Grove, T.L.: The early differentiation history of Mars from ^{182}W–^{142}Nd isotope systematics in the SNC meteorites. Geochim. Cosmochim. Acta **69**, 4557–4571 (2005)

7. Grasset, O., Parmentier, E.M.: Thermal convection in a volumetrically heated, infinite Prandtl number fluid with strongly temperature-dependent viscosity: implications for planetary thermal evolution. J. Geophys. Res. **103**, 18,171–18,181 (1998)

8. Hofmann, A.W.: Mantle geochemistry: the message from oceanic volcanism. Nature **385**(6613), 219–229 (1997)

9. Hüttig, C., Stemmer, K.: Finite volume discretization for dynamic viscosities on Voronoi grids. Phys. Earth Planet. Inter. (2008) doi:10.1016/j.pepi.2008.07.007

10. Hüttig, C., Stemmer, K.: The spiral grid: a new approach to discretize the sphere and its application to mantle convection. Geochem. Geophys. Geosyst. **9**, Q02018 (2008). doi:10.1029/2007GC001581

11. Hüttig, C., Tosi, N., Moore, W.B.: An improved formulation of the incompressible Navier-Stokes equations with variable viscosity. Phys. Earth Planet. Inter. **220**, 11–18 (2013). doi:10.1016/j.pepi.2013.04.002

12. Ismail-Zadeh, A. Tackley, P.J.: Computational Methods for Geodynamics. Cambridge University Press, Cambridge (2010)

13. Karato, S., Paterson, M.S., Fitz Gerald, J.D.: Rheology of synthetic olivine aggregates: influence of grain size and water. J. Geophys. Res. **91**, 8151–8176 (1986)

14. Maurice, M., Tosi, N., Plesa, A.-C., Breuer, D.: Evolution and Consequences of Magma Ocean Solidification. European Geoscience Union, EGU2015-884 (2015). Retrieved on April 18, 2015, from http://meetingorganizer.copernicus.org/EGU2015/EGU2015-884.pdf

15. Plesa, A.-C.: Mantle convection in a 2D spherical shell. In: Rückemann, C.-P., Christmann, W., Saini, S., Pankowska, M. (eds.) Proceedings of the First International Conference on Advanced Communications and Computation (INFOCOMP 2011), Barcelona, pp. 167–172, 23–29 October 2011. ISBN:978-1-61208-161-8. Retrieved on November 3, 2011, from http://www.thinkmind.org/download.php?articleid=infocomp_2011_2_10_10002

16. Plesa, A.-C., Tosi, N., Hüttig, C.: Thermo-chemical convection in planetary mantles: advection methods and magma ocean overturn simulations. In: Rückemann, C.-P. (ed.) Integrated Information and Computing Systems for Natural, Spatial, and Social Sciences. IGI Global, Hershey (2013)

17. Plesa, A.-C., Tosi, N., Breuer, D.: Can a fractionally crystallized magma ocean explain the thermo-chemical evolution of Mars? Earth Planet. Sci. Lett. **403**, 225–235 (2014). doi:10.1016/j.epsl.2014.06.034

18. Roberts, J.H., Zhong, S.: Degree-1 convection in the Martian mantle and the origin of the hemispheric dichotomy. J. Geophys. Res. E: Planets **111**, E06013 (2006)

19. Schubert, G., Turcotte, D.L., Olson, P.: Mantle Convection in the Earth and Planets. Cambridge University Press, Cambridge (2001)

20. Solomatov, V.S.: Fluid dynamics of a terrestrial magma ocean. In: Canup, R.M., Righter, K. (eds.) Origin of the Earth and Moon. University of Arizona Press, Tucson (2000)

21. Tackley, P.J., King, S.D.: Testing the tracer ratio method for modeling active compositional fields in mantle convection simulations. Geochem. Geophys. Geosyst. **4**(4), 8302 (2003). doi:10.1029/2001GC000214

22. Thomson, J.J.: On the structure of the atom: an investigation of the stability and periods of oscillation of a number of corpuscles arranged at equal intervals around the circumference of a circle; with application of the results to the theory of atomic structure. Philos. Mag. Ser. 6 **7**, 237–265 (1904)

23. Tosi, N., Plesa, A.-C., Breuer, D.: Overturn and evolution of a crystallized magma ocean: a numerical parameter study for Mars. J. Geophys. Res. Planets **118**, 1–17 (2013). doi:10.1002/jgre.20109

24. van Keken, P.E., King, S.D., Schmeling, H., Christensen, U.R., Neumeister, D., Doin, M.-P.: A comparison of methods for the modeling of thermochemical convection. J. Geophys. Res. **102**, 22477–22495 (1997)

25. Weber, R.C., Lin, P.Y., Garnero, E.J., Williams, Q., Lognonne, P.: Seismic detection of the lunar core. Science **331**, 309–312 (2011). doi:10.1126/science.1199375
26. Zhong, S., McNamara, A.K., Tan, E., Moresi, L., Gurnis, M.: A benchmark study on mantle convection in a 3-D spherical shell using CitcomS. Geochem. Geophys. Geosyst. **9**, Q10017 (2008). doi:10.1029/2008GC002048

Coupled PIC-DSMC Simulations of a Laser-Driven Plasma Expansion

S. Copplestone, T. Binder, A. Mirza, P. Nizenkov, P. Ortwein, M. Pfeiffer, S. Fasoulas, and C.-D. Munz

Abstract In the field of material processing or spacecraft propulsion, laser ablation is used to remove material from a solid surface with a laser beam. The numerical study of this process has been directed towards direct laser-solid interactions, tackled by molecular dynamics simulations which have been conducted in the past. An additional field of interest arises, when considering the interaction of a laser beam and the plasma created by former laser impacts. For this purpose, an Message Passing Interface parallelized, high-order Particle-in-Cell scheme coupled with a Direct Simulation Monte Carlo method is used to handle the complex phenomena, which usually are simulated using disjoint techniques. The complete scheme is constructed to run on three-dimensional unstructured hexahedra, where for the Particle-in-Cell solver, a highly efficient discontinuous Galerkin method is used to calculate the electromagnetic field. Simulations under realistic settings require the use of high performance computing, where the parallel performance of the coupled solver plays the most important role. This work offers insight into such an undertaking by simulating the expansion of a plasma plume in three dimensions using this coupled algorithm.

1   Introduction

Laser ablation is a method for the removal of material from a, usually metallic, surface. Two applications where this is performed are spacecraft propulsion and material processing, each with different goals. To capture such a process numerically, three-dimensional molecular dynamics simulations have been conducted in the past [18]. These, however, were aimed at direct laser-solid interactions and their time and length scales reside within the nano scale. Further interest lies within

S. Copplestone (✉) • P. Ortwein • C.-D. Munz
Institute of Aerodynamics and Gas Dynamics (IAG), University of Stuttgart, 70569 Stuttgart, Germany
e-mail: copplestone@iag.uni-stuttgart.de; munz@iag.uni-stuttgart.de

T. Binder • A. Mirza • P. Nizenkov • M. Pfeiffer • S. Fasoulas
Institute of Space Systems (IRS), University of Stuttgart, 70569 Stuttgart, Germany
e-mail: fasoulas@irs.uni-stuttgart.de

© Springer International Publishing Switzerland 2016   689
W.E. Nagel et al. (eds.), *High Performance Computing in Science and Engineering '15*, DOI 10.1007/978-3-319-24633-8_44

the interaction of successive laser pulses, which interact with the afore created plasma due to laser impacts on the surface. Simulations regarding this process have been performed in the past, but were restricted to disjoint techniques [7, 12, 16], which either utilize the Particle-in-Cell (PIC) or Direct Simulation Monte Carlo (DSMC) method. These are limited by one or more of the following drawbacks: only considering neutral particle species, neglecting electromagnetic effects or simplified chemical reaction laws, initial settings or boundary conditions (BC). Combined PIC-DSMC simulations for different applications have been reported in the literature, but were restricted to quasi-neutrality [13] or one- and two-dimensional setups [17]. In this work we will present investigations regarding the feasibility of fully three-dimensional coupled PIC-DSMC simulations for the expansion of a plasma in vacuum.

The organization of this work is as follows: after a brief introduction of the model, the PIC and DSMC solver are discussed in Sects. 2 and 3, respectively. In Sect. 4, simulation results are presented and performance considerations regarding high performance computing (HPC). Finally, Sect. 5 will be devoted to a summary and conclusions.

The core of the model is Boltzmann's equation

$$\left(\frac{\partial}{\partial t} + \mathbf{v} \cdot \nabla + \frac{1}{m^s} \mathbf{F} \cdot \nabla_{\mathbf{v}} \right) f^s(\mathbf{x}, \mathbf{v}, t) = \left. \frac{\partial f}{\partial t} \right|_{Coll}, \tag{1}$$

which describes basic particle kinetics, where $f^s(\mathbf{x}, \mathbf{v}, t)$ is the six-dimensional Particle Distribution Function (PDF) in phase-space for each species s with mass m. It describes the amount of particles per unit volume, which are found at a certain point $(\mathbf{x}, \mathbf{v})$ in phase-space and time t. The left hand side of (1), where $\mathbf{F}$ is an external force field, is solved using a deterministic Particle-in-Cell [6] method, while the right hand side, where the collision integral $\left. \frac{\partial f}{\partial t} \right|_{Coll}$ accounts for all particle collisions in the system, is solved by applying a non-deterministic Direct Simulation Monte Carlo [3] method.

The PDF is approximated by summing up a certain number of macro particles N_p and is given by

$$f^s(\mathbf{x}, \mathbf{v}, t) \approx \sum_{n=1}^{N} \omega_n \delta (\mathbf{x} - \mathbf{x}_n) \, \delta (\mathbf{v} - \mathbf{v}_n)$$

where the δ-function is applied to position and velocity space, separately, and the macro particle factor (MPF) $\omega_n = N_{phy}/N_{sim}$ is used to describe the ratio of simulated to physical particles. For each particle of each species, the change in relativistic momentum is calculated by

$$\frac{d\gamma \mathbf{v}}{dt} = \frac{\mathbf{F}_L}{m_0^s} \tag{2}$$

where $\gamma = \left(1 - \left(\frac{v}{c}\right)^2\right)^{-\frac{1}{2}}$ is the Lorentz factor (c is the speed of light), $\mathbf{F}_L$ the Lorentz force and m_0^s the particle mass at rest. The set of equations is numerically solved using the procedures given in [11, 14] and are briefly described in the following.

2   PIC Solver

The PIC solver is used to find an approximation of the left hand side of Eq. (1) and consists of two parts, namely a particle pusher that updates the particle properties in time and a field solver that evolves the electromagnetic fields in time. Initially, the particle pusher advances the particle positions in time by evaluating the external force field $\mathbf{F}$ at each particle position and solving Eq. (2) for which a standard Runge-Kutta (RK) scheme [4] is applied. In the PIC context, the external force field in Eq. (1) resembles the Lorentz force acting on charged particles and is given by

$$\mathbf{F} = \mathbf{F}_L = q^s\,(\mathbf{E} + \mathbf{v} \times \mathbf{B})\;,\tag{3}$$

where $\mathbf{E}$ and $\mathbf{B}$ are the electromagnetic fields and q^s the particle charge. These time-dependent electromagnetic fields are evolved in a second step and are calculated using Maxwell's equations, here, in purely hyperbolic form [10]

$$\frac{\partial \Phi_E}{\partial t} = \chi \left(\frac{\rho}{\varepsilon_0} - \nabla \cdot \mathbf{E}\right)\tag{4}$$

$$\frac{\partial \Phi_B}{\partial t} = -\chi c^2 \nabla \cdot \mathbf{B}\tag{5}$$

$$\frac{\partial \mathbf{E}}{\partial t} = c^2 \nabla \times \mathbf{B} - \frac{\mathbf{j}}{\varepsilon_0} - \chi c^2 \nabla \Phi_E\tag{6}$$

$$\frac{\partial \mathbf{B}}{\partial t} = -\nabla \times \mathbf{E} - \chi \nabla \Phi_B\;,\tag{7}$$

where Φ_E and Φ_B are the divergence correction terms for the electric and magnetic fields, respectively. The factor χ, which adjusts the divergence correction velocity, is tuned in order to propagate divergence errors out of the computational domain. The PHM system is solved using a Discontinuous Galerkin Spectral Element Method (DGSEM) as described in [14] and which will be presented shortly. Consider an arbitrary three-dimensional physical space Ω, where Eqs. (4)–(7) are re-written in order to satisfy a conservation form

$$\frac{\partial \mathbf{U}}{\partial t} + \nabla \cdot \mathbf{F}\,(\mathbf{U}) = \mathbf{S}\;,\tag{8}$$

where **F** represents the physical flux vector, the solution and the source term vectors are

$$\mathbf{U} = (\Phi_E, \Phi_B, \mathbf{E}, \mathbf{B})^T \,, \tag{9}$$

and

$$\mathbf{S} = \frac{1}{\varepsilon_0} (\chi\rho, 0, \mathbf{j}, \mathbf{0})^T \,, \tag{10}$$

respectively. In the DGSEM context, Eq. (8) is transformed from the physical space Ω to the reference space $\mathcal{E} \in [-1, 1]^3$ giving

$$\frac{\partial \mathbf{U}}{\partial t} + \frac{1}{J} \nabla_\xi \cdot \tilde{\mathbf{F}} = \mathbf{S} \,, \tag{11}$$

where J is the Jacobian determinant of the transformation, $\nabla_\xi\cdot$ the divergence operator with respect to the reference space and $\tilde{\mathbf{F}}$ the transformed flux vector. This equation is then multiplied by a test function ϕ and integrated over $\mathcal{E}$ leading to

$$\int_\mathcal{E} J \frac{\partial \mathbf{U}}{\partial t} \phi\, d\boldsymbol{\xi} + \underbrace{\int_{\partial\mathcal{E}} \tilde{\mathbf{F}} \cdot \mathbf{N}\, \phi\, ds}_{\text{Surface Integral}} - \underbrace{\int_\mathcal{E} \tilde{\mathbf{F}} \cdot \nabla_\xi \phi\, d\boldsymbol{\xi}}_{\text{Volume Integral}} = \int_\mathcal{E} JS\, \phi\, d\boldsymbol{\xi} \tag{12}$$

where integration by parts has been used to split the divergence integral into two parts, a surface integral and a volume integral. These two integrals are of specific importance to the parallelization model applied here, because only the surface integral is responsible for the inter-cell coupling between DG cells. In an Message Passing Interface (MPI) parallelism, the only message needed for communication is built from this surface integral, which leads to a highly efficient scheme with remarkable scaling performance in HPC contexts [1, 14]. Equation (12) is advanced in time utilizing the same RK method as for the particles.

3 DSMC Solver

The DSMC method approximates the right hand side of Eq. (1) by modelling binary particle collisions in a probabilistic and transient manner. The DSMC module applies an additional change in particle velocity after the PIC solver determines the change in velocity by Lorentz forces. The main dichotomy compared to the PIC solver is the non-deterministic calculation of changes in particle velocity utilizing random numbers in a collision process. Additionally, chemical reactions may occur in such collision events. The primordial concept of DSMC was developed by Bird [3] and is commonly applied to the simulation of rarefied and neutral gas flows.

The collision operator in (1) is given by

$$\frac{\partial f}{\partial t}\Big|_{Coll} =$$

$$\int W(\mathbf{v}_1, \mathbf{v}_2, \mathbf{v}_3, \mathbf{v}_4)\{f(\mathbf{x}, \mathbf{v}_1, t)f(\mathbf{x}, \mathbf{v}_2, t) - f(\mathbf{x}, \mathbf{v}_3, t)f(\mathbf{x}, \mathbf{v}_4, t)\}d\mathbf{v}_1 d\mathbf{v}_2 d\mathbf{v}_3 , \tag{13}$$

where W represents the probability per unit time in which two particles collide and change their velocities from $\mathbf{v}_1$ and $\mathbf{v}_2$ to $\mathbf{v}_3$ and $\mathbf{v}_4$. The DSMC method, however, does not solve this collision integral directly, but rather applies a phenomenological consideration of the collision process of simulation particles in a statistical framework. In a first step, a particle pair for the collision process is found by examining each cell and applying a nearest neighbour search in the form of an octree cell division. An alternative method is the random pairing of all particles in each cell, but with additional restrictions to the cell size. The collision probability is modelled by choosing a cross section for each particle species using microscopic considerations, e.g., quantum mechanics. As with the PIC method, macro particles are simulated instead of real particles to reduce computational effort. The collision probability of two particles, 1 and 2, is determined by methods found in [2, 3], which yields

$$P_{12} = \frac{N_{p,1}N_{p,2}}{1 + \delta_{12}}\omega_n \frac{\Delta t}{V_c S_{12}}(\sigma_{12}g_{12}) , \tag{14}$$

where δ_{12} is the Kronecker delta, V_c the cell volume, Δt the time step, σ the cross section, S_{12} the number of particle pairs of species 1 and 2 in V_c and g the relative velocity between the two particles considered. This probability is compared to a pseudo random number $R \in [0, 1)$ and if $R < P_{12}$, the collision occurs, otherwise it does not. Subsequent events such as chemical reactions or relaxation processes are computed in the same manner, but using additional probabilities.

This may change the internal energy of particles, i.e., their rotational, vibrational energy and electronic excitation. Chemical reactions are modelled via the Arrhenius law or quantum-mechanic considerations, which lead to dissociation, recombination, exchange reaction, excitation or ionization. Macroscopic properties like temperature or density are calculated by sampling particle positions and velocities over time within each cell.

4   Plasma Plume Expansion

The underlying physics of an expanding plasma plume following a laser ablation process can be very diverse, depending on numerous settings concerning the laser, the target surface and the system environment. Due to the high non-linearity of the process, many different phenomena might occur [16]. Two environmental scenarios are commonly regarded: an expansion into a rarefied medium [15, 16] or vacuum [9, 19]. Here, the focus will be directed to the latter. However, the basic behaviour of the expansion process is the following. The impacting laser creates a highly dense plasma of electrons, ions and neutral particles, which undergoes rapid expansion. The high energy electrons quickly diffuse into space due to their low mass but the heavier particles accelerate slowly leading to a charge separation and a corresponding electric field. The ions are then accelerated further within this process, due to the static electric field developing and the energy shifted from the electrons. In the simulations considered here, an existing plasma is initialized and starts to expand as soon as the simulation begins.

4.1 Computational Setup

The numerical setup is chosen similar to preliminary investigations concerning only the DSMC solver, whereas here, emphasis is appointed to the PIC solver. Figure 1 illustrates the setup, where the computational domain is constructed by cubic cells creating a cuboid domain of size $x \times y \times z = 120 \times 120 \times 210\,\mu\mathrm{m}^3$. Two species of equal number are considered in the simulation, electrons and singly charged aluminium ions, which yields an ionization degree of unity. The plasma is initialized in form of a cylinder of radius $r = 20\,\mu\mathrm{m}$ and height $h_z = 70\,\mu\mathrm{m}$ in z-direction, within which the particles are homogeneously distributed and their velocities are chosen according to a Maxwell–Boltzmann distribution. The initial conditions for the quasi-neutral plasma are given in Table 1. The bottom of the domain is represented by a perfectly electric conducting boundary condition, all other sides are open boundary conditions.

The numerical details for the two basic cases are given in Table 2. The number of degrees of freedom (DOF) for the DG solver are calculated from the number of cells in the domain and the polynomial degree via $DOF = N_{\mathrm{cells}}(p+1)^3$ and the integration points are Legendre-Gauss distributed within each cell. An approximation of the smallest wavelength to be resolved within the simulation can be derived by considering the Debye length λ_D of the initial plasma and the points per wavelength (PPW) needed by the DG scheme [5]. The calculation of the source terms for Maxwell's equations are carried out in two different ways, either using a shape function [8] or a Dirac δ-function. This yields different modifications to the test cases and significantly influences the parallel performance.

4.2 PIC-DSMC Feasibility and Preliminary Results

The main goal of the realistic setup is to clarify whether and how three-dimensional simulations of an expanding plasma plume can be conducted with a coupled PIC-DSMC algorithm. The different time steps required by the schemes, for PIC $\Delta t \approx$

Fig. 1 Schematic of the simulation setup with a cylindrical particle distribution in space. The particles (*blue*) are located near a perfectly electric conducting wall (*gray*), while the remaining boundaries are open

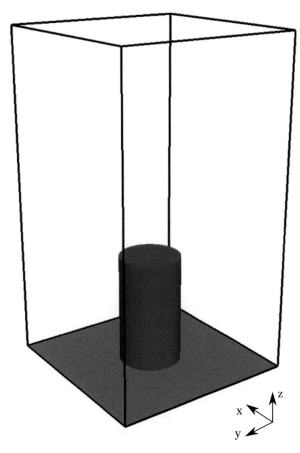

Table 1 Initial conditions

Property	Value
Particle number	$N_{e^-} = N_{Al^+} = 1 \cdot 10^5$
Electron density	$n_{e^-} = 5 \cdot 10^{24} \ 1/m^3$
Particle weight	$\omega_n \approx 4.4 \cdot 10^6$
Electron temperature	$T_{e^-} = 2 \cdot 10^4 \ K$
Ion temperature	$T_{Al^+} = 1 \cdot 10^4 \ K$
Debye length	$\lambda_D = 2.52 \cdot 10^{-9} \ m$

Table 2 Basic numerical settings

Case	Model	Cells $\cdot 10^3$	L_{cell} μm	*DOF* $\cdot 10^6$	p	Δt $\cdot 10^{-17}$ s	PPW $\cdot 10^{-3}$
1	PIC	896	1.5	307	6	1.64	11.8
2	PIC+DSMC	112	3	14	4	5.78	4.2

$\mathcal{O}(10^{-17}s)$ and for DSMC $\Delta t \approx \mathcal{O}(10^{-12}s)$, show that they operate on different time scales, which becomes important when considering the parallelism of the code. The very small time step restriction of the PIC solver is problematic when larger time scales are of interest, a problem which may be handled by increasing the mesh size at a later simulation time.

Case 2 is used to analyse the plume expansion for the coupled PIC-DSMC solver and is run up to 250 ps simulation time. As a first step, the DSMC module is only used for elastic collisions between the particle species and other effects, such as chemical reactions are neglected. As shown in Table 2, the resolution for the electromagnetic solver in terms of PPW is far and under from a standard PIC setting. Hence, the results are very sensitive to the numerical settings, e.g., time step and number of DOF and, are therefore of qualitative use. However, these simulations give a brief estimation of the costs involved to resolve a simulation and this will be considered in future work.

Figure 2 depicts the energy distribution over time, where the potential energy shows that the expansion process is electrostatically dominated. An electrostatic potential is created and drives the main ion acceleration process. The kinetic energy is coupled to this process and shows the different behaviour of the light (electrons) and heavy (ions) particles under the influence of such a field.

Figure 3 illustrates the electric field $\mathbf{E}_z$ and particle distribution at $t = 35$ ps (14 % simulation time) and, additionally, the electron and ion PDFs for the z-velocity component are plotted along the z-direction. The electric field distribution shows how the electrostatic potential starts to build up: a negatively charged wall at $z = 0$ and a positively charged plasma front at $z > 70$ μm, which is slowly moving in the positive z-direction. This is mainly due to the charge separation, which occurs when the lighter electrons diffuse into space but which the heavier ions cannot follow. Hence, the electrons are distributed within the whole space, their velocity vectors being scattered and undirected due to their high mobility. However, they are partially held back by the expanding plasma front causing them to lose energy [16]. The ions, on the other hand, rely on the electric potential as well as the density gradient to be accelerated, which occurs in a radial direction away from the plasma plume centre. Therefore, they are more densely packed within the electric potential. These

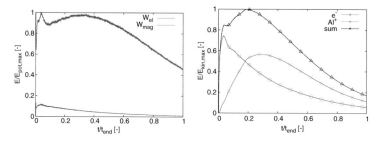

Fig. 2 Potential field energy (*left*) and particle kinetic energy (*right*) over time. Normalization values are $E_{kin,max} = 2.5 \cdot 10^{-3}$ J, $E_{pot,max} = 1.9 \cdot 10^{-4}$ J and $t_{end} = 250$ ps

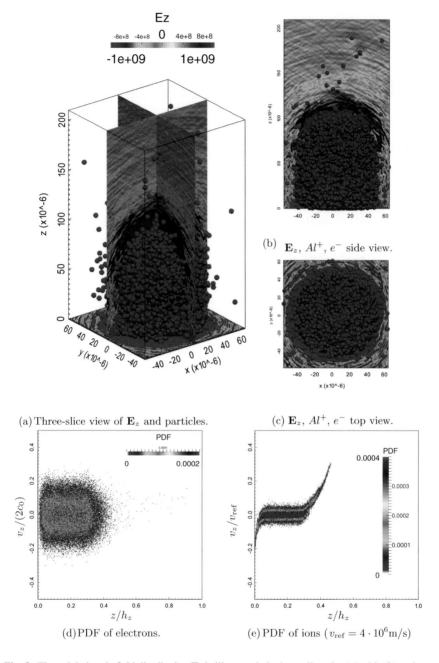

(a) Three-slice view of $\mathbf{E}_z$ and particles.

(b) $\mathbf{E}_z$, Al^+, e^- side view.

(c) $\mathbf{E}_z$, Al^+, e^- top view.

(d) PDF of electrons.

(e) PDF of ions ($v_{ref} = 4 \cdot 10^6$ m/s)

Fig. 3 The axial electric field distribution $\mathbf{E}_z$ is illustrated via three-slice plot (**a**), side (**b**) and top (**c**) view, where electrons (ions) are marked as *blue* (*red*) *spheres*. PDF distribution of electrons (*d*) and ions (*e*) for the z-velocity component in z-direction. All images are snapshots of the system conditions taken at $t = 35$ ps and only half of the total number of particles is displayed in (**a**–**c**) for purposes of lucidity

conditions are supported by the PDFs that shows the phase space of the particles for v_z as plotted against z. The electrons immediately diffuse and reach high velocities, which depends strongly on the chosen deposition method. Due to the negatively charged wall, a plasma sheath develops that keeps the electrons away from direct contact with the wall. The positive and negative peaks of the electric potential induce a bending of the ion PDF in $\pm z$-direction that mainly drives their acceleration.

4.3 Performance

The separate parallel performance of the PIC and DSMC algorithms have been shown in a preceding publication [14]. The computationally disparate costs of the PIC and DSMC methods necessitate subjecting the parallelism of the PIC solver to critical scrutiny. Therefore, emphasis is placed on the parallel performance of the pure PIC solver and the coupled PIC-DSMC solver with regard to a realistic scenario of an expanding plasma plume. Contrary to homogeneous setups with periodic BC or stationary processes, non-periodic BC and a heterogeneous spatial particle distribution distort the conditions and can lead to impaired parallel performance due to load imbalance. The relative parallel efficiency is calculated via

$$\eta = \frac{240}{N_{\text{proc}}} \frac{t_{240}}{t} , \tag{15}$$

where t is the wall time of the current run with N_{proc} processors and t_{240} the reference wall time for a simulation that is run on ten compute nodes, each consisting of 24 processors. A minimum number of 10 nodes was chosen as a basis for comparison, as they possess the minimum amount of memory needed for Test Case 1. Figure 4 illustrates how the scheme performs in the range of 10–400 nodes, each featuring 24 processors. Test Case 1 shows a steady drop in efficiency for increasing number of cores when a shape-function is used for deposition, as the amount of communication increases. This drawback is diminished by switching to a cell-local deposition as, e.g., a δ-function realizes. An efficiency of unity or larger is retained, which might indicate caching effects for higher numbers of processors, where the number of cells and particles, hence, the problem size each processor has to handle is reduced to a point where it fits into cache memory nearer to the processor core. Another effect might be due to lessened load imbalances that occur at lower numbers of processors due to the heterogeneous particle distribution.

For Test Case 2, which is only carried out with a δ-function, the DSMC module is active and a clear drop in efficiency is visible for 100 and more nodes, indicating a load imbalance build up. Even though the DSMC module resides only on a local cell using particle data, where no additional communication is needed, the additional calculations have to be carried out by the particle holding processors, therefore, increasing the imbalance. This problem is tackled by the following load balancing considerations.

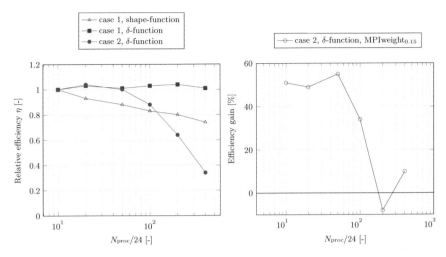

Fig. 4 Parallel efficiency (*left*) obtained from scaling investigations conducted on the Hornet system of the HLRS, where the number of processors chosen were between 240 and 9600 and the efficiency gain by was manual selection of the weighting factor (*right*)

The number of cells assigned to each processor is determined by weighting the particles within each cell and choosing a weighting factor, which has a default value of 0.02, i.e., a particle is computationally worth 2 % compared to a DGSEM element. The estimated cost for each cell is then calculated and distributed among the chosen number of processes. For 9216 CPUs, the weighting factor optimum has been found to be 0.15 for the settings chosen in Test Case 2 and the efficiency gain relative to the default value of 0.02 is shown in Fig. 4 for the processor range chosen here. The positive effect of adjusting the weighting factor is restricted to a lower number of processors and can even become negative, e.g., for the case of 4800 CPUs. This is because the particles in the test case being considered are densely packed at the beginning of the simulation and cells with many particles cannot be split further due to the limit of one cell per MPI process. This disadvantage can be overcome by increasing the total number of cells and clearly shows that an appropriate amount of cells per processor for a given number of particles must be chosen for the scheme to perform correctly. Because the optimal weighting factor depends on the number of processors, cells and simulation conditions that change over time, it is indispensable to calculate this factor dynamically and adjust the load distribution in order to achieve satisfactory balancing.

5   Summary and Conclusions

Preliminary investigations regarding the feasibility of fully three-dimensional coupled PIC-DSMC simulations of a laser-induced plasma expansion process have been conducted. The focus of this work was directed towards the PIC solver and its parallel performance on HPC systems, the influence of deposition methods and load balancing issues. The numerical settings according to preliminary DSMC simulations show a spatial under-resolution for the PIC solver. For this reason, future work will be directed towards setups with lower densities, through higher numbers of simulation particles and lower values for the MPF as well as higher temperatures as reported in the literature. Additionally, the absence of an ambient medium may lead to different outcomes and, hence, its incorporation into the simulation might be mandatory.

Acknowledgements We gratefully acknowledge the Deutsche Forschungsgemeinschaft (DFG) for funding within the projects "Kinetic Algorithms for the Maxwell-Boltzmann System and the Simulation of Magnetospheric Propulsion Systems" and "Coupled PIC-DSMC-Simulation of Laser Driven Ablative Gas Expansions". The latter being a sub project of the collaborative research center (SFB) 716 at the University of Stuttgart. Computational resources have been provided by the Bundes-Höchstleistungsrechenzentrum Stuttgart (HLRS).

References

1. Atak, M., Beck, A., Bolemann, T., Flad, D., Frank, H., Hindenlang, F., Munz, C.-D.: Discontinuous Galerkin for high performance computational fluid dynamics. In: Nagel, W.E., Kröner, D.H., Resch, M.M. (eds.) High Performance Computing in Science and Engineering '14, pp. 499–518. Springer International Publishing, New York (2015)
2. Baganoff, D., McDonald, J.D.: A collision selection rule for a particle simulation method suited to vector computers. Phys. Fluids A **2**, 1248–1259 (1990)
3. Bird, G.A.: Molecular Gas Dynamics and the Direct Simulation of Gas Flows. Oxford University Press, Oxford (1994)
4. Carpenter, M.H., Kennedy, C.A.: Fourth-order 2N-storage Runge-Kutta schemes. NASA Tech. Memo. **109112**, 1–26 (1994)
5. Gassner, G., Kopriva, D.A.: A Comparison of the dispersion and dissipation errors of Gauss and Gauss-Lobatto discontinuous Galerkin spectral element methods. SIAM J. Sci. Comput. **33**(5), 2560–2579 (2011)
6. Hockney, R.W., Eastwood, J.W.: Computer Simulation Using Particles. McGraw-Hill Inc., New York (1988)
7. Itina, T.E., Hermann, J., Delaporte, P., Sentis, M.: Laser-generated plasma plume expansion: combined continuous-microscopic modeling. Phys. Rev. E **66**, 066406 (2002)
8. Jacobs, G.B., Hesthaven, J.S.: High-order nodal discontinuous Galerkin particle-in-cell method on unstructured grids. J. Comput. Phys. **214**(1), 96–121 (2006)
9. Mora, P.: Thin-foil expansion into a vacuum. Phys. Rev. E **72**, 056401 (2005)
10. Munz, C.-D., Schneider, R., Voß, U.: A finite-volume particle-in-cell method for the numerical simulation of devices in pulsed power technology. Surv. Math. Ind. **8**, 243–257 (1999)
11. Munz, C.-D., Auweter-Kurtz, M., Fasoulas, S., Mirza, A., Ortwein, P., Pfeiffer, M., Stindl, T.: Coupled particle-in-cell and direct simulation Monte Carlo method for simulating reactive plasma flows. C. R. Mec. **342**(10–11), 662–670 (2014)
12. Nedelea, T., Urbassek, H.M.: Particle-in-cell study of charge-state segregation in expanding plasmas due to three-body recombination. J. Phys. D Appl. Phys. **37**(21), 2981 (2004)
13. Oh, D.Y.: Computational Modeling of Expanding Plasma Plumes in Space Using a PIC-DSMC Algorithm. Massachusetts Institute of Technology, Department of Aeronautics and Astronautics, Cambridge (1997)

14. Ortwein, P., Binder, T., Copplestone, S., Mirza, A., Nizenkov, P., Pfeiffer, M., Stindl, T., Fasoulas, S., Munz, C.-D.: Parallel performance of a discontinuous Galerkin spectral element method based PIC-DSMC solver. In: Nagel, W.E., Kröner, D.H., Resch, M.M. (eds.) High Performance Computing in Science and Engineering '14, pp. 671–681. Springer International Publishing, New York (2015)
15. Romagnani, L., Bulanov, S.V., Borghesi, M., Audebert, P., Gauthier, J.C., Löwenbrück, K., Mackinnon, A.J., Patel, P., Pretzler, G., Toncian, T., Willi, O.: Observation of collisionless shocks in laser-plasma experiments. Phys. Rev. Lett. 101, 025004 (2008)
16. Sarri, G., Murphy, G.C., Dieckmann, M.E., Bret, A., Quinn, K., Kourakis, I., Borghesi, M., Drury, L.O.C., Ynnerman, A.: Two-dimensional particle-in-cell simulation of the expansion of a plasma into a rarefied medium. New J. Phys. 13(7), 073023 (2011)
17. Serikov, V.V., Kawamoto, S., Nanbu, K.: Particle-in-cell plus direct simulation Monte Carlo (PIC-DSMC) approach for self-consistent plasma-gas simulations. IEEE Trans. Plasma Sci. 27(5), 1389–1398 (1999)
18. Sonntag, S., Trichet Paredes, C., Roth, J., Trebin, H.-R.: Molecular dynamics simulations of cluster distribution from femtosecond laser ablation in aluminum. Appl. Phys. A 104(2), 559–565 (2011)
19. Thaury, C., Mora, P., Héron, A., Adam, J.C.: Self-generation of megagauss magnetic fields during the expansion of a plasma. Phys. Rev. E 82, 016408 (2010)